Y0-AEU-377

SENSORS AND TRANSDUCERS/ AUTOMOTIVE ELECTRONICS SERIES

PT-68

Edited by

Ronald K. Jurgen

RECEIVED
NOV 29 2005
HL LIBRARY

Published by
Society of Automotive Engineers, Inc.
400 Commonwealth Drive
Warrendale, PA 15096-0001
U.S.A.
Phone (412) 776-4841
Fax: (412) 776-5760

Copyright © 1997 Society of Automotive Engineers, Inc.

ISBN 0-7680-0117-X
Library of Congress Catalog Card Number: 97-80548

All rights reserved. Printed in the United States of America.

Permission to photocopy for internal or personal use, or the internal or personal use of specific clients, is granted by SAE for libraries and other users registered with the Copyright Clearance Center (CCC), provided that the base fee of $5.00 per article is paid directly to CCC, 222 Rosewood Dr., Danvers, MA 01923. Special requests should be addressed to the SAE Publications Group.
ISBN 0-7680-0117-X/97 $5.00

SAE Order No. PT-68

Introduction

A Wealth of Choices

The ability to sense and measure accurately a variety of parameters under harsh automotive environments has made possible the ever increasing application of electronic controls in vehicles. Intense sensor development over the years has given rise to many different types of sensors and innovative ways to apply them in the real automotive environment.

This book, the first in a new Progress in Technology series on automotive electronics, presents a compilation of papers that offers a meaningful insight to the wealth of sensor choices and their innovative designs and applications. The papers have been purposefully grouped according to the parameter being measured. This approach—rather than a grouping by automotive applications like engine control, suspension control, etc.—was selected with the rationale that designers are primarily interested in being able to measure a specific parameter, no matter in what application that measurement will be used. (Those interested in what types of sensors are used for what applications are referred to Table 1 in the final paper in the book, "Future Sensing in Vehicle Applications," by Randy Frank.)

There are two additional groupings: one on multiple parameter sensors and the other on so-called smart sensors that include signal-conditioning and other functions bundled with the basic sensor.

The variety of choices for sensing parameters is evident in all parameter categories. For example, linear and angle position sensors include types based on magnetoresistance, the Hall effect, ultrasonics, and infrared, capacitive, and optical techniques. Pressure sensors include capacitive, piezoresistive, thin-film, and fiberoptic types. Innovative ways in which sensors are used are amply documented. Here are a few examples:

- Paper 971103 describes a tire monitoring system in which pressure and temperature sensors are embedded in each tire of a vehicle. Data from the sensors are transmitted with antennas, one mounted on each wheel rim and the other rigidly mounted in the corner assembly. The corner antennas are powered remotely by resonating them at a low frequency.

- Paper 950348 describes a unique system for detection of side-impact collisions. A pressure sensor and associated circuits for signal-conditioning and decision-making are located inside the door cavity of a vehicle to monitor a dynamic pressure change. An intruding object or vehicle causes a rapid volume reduction of the cavity inside the door and leads to an adiabatic increase in pressure within the cavity. The increased pressure causes air to leak and the air mass inside the door decreases.

- Paper 962103 describes a knock detection system that uses an in-cylinder optical probe integrated directly into the engine spark plug to monitor the radiation from the burned gas in the combustion process. A second probe is built into a steel body and installed near the end gas region of the combustion chamber. It measures the radiant emission from the end gas in which knock originates. Knock can be detected by the spark plug optical probe as a high-frequency ripple.

- Paper 970856 details how the influence of variable air-fuel ratio inside a spark-ignition engine is examined by the use of an ionization sensor. The measured ion currents are used for predicting the local air-fuel ratio in the vicinity of the spark plug.

Randy Frank, in his paper "Future Sensing in Vehicle Applications," describes emerging sensing technologies using advanced materials, radar, infrared, chemical, and next-generation magnetic sensors as well as new sensor applications including night vision, obstacle detection/collision avoidance, unauthorized vehicle entry, and passenger position/size detection in smart air bag systems.

The papers in this book should provide a valuable reference to anyone interested in learning about sensors and how they are applied innovatively in vehicles.

* * * * *

This book and the entire automotive electronics series is dedicated to my friend Larry Givens, a former editor of SAE's monthly publication, Automotive Engineering.

Ronald K. Jurgen, Editor

Table of Contents

Introduction

Pressure Sensors

Linear and Angle Position Sensors

Flow Sensors

Temperature Sensors

Gas Sensors

Speed and Acceleration Sensors

Engine Knock Sensors

Torque Sensors

Air-Fuel Ratio Sensors

Fuel Level Sensors

Force Sensors

Various Parameter Sensors

Multiparameter Sensors

Smart Sensors

A Look to the Future

PRESSURE SENSORS

970608

A Monolithic Integrated Solution for MAP Applications

Rajan Verma, Ira Baskett, Mahesh Shah, Theresa Maudie,
Dragan Mladenovic, and K. Sooriakumar
Motorola

Copyright 1997 Society of Automotive Engineers, Inc.

ABSTRACT

A monolithic sensing solution for manifold absolute pressure (MAP)' is presented. This work includes examination of design, fabrication, temperature compensation, packaging and electromagnetic compatibility (EMC) testing of the fully integrated monolithic sensor. The circuit uses integrated bipolar electronics and conventional IC processing. The amplification circuit consists of three op-amps, seven laser trimmable resistors, and other active and passive components. Also discussed is a summary of an automotive application MAP sensor general specification, test methods, assembly, packaging, reliability and media testing for a single chip solution.

INTRODUCTION

Ever increasing requirements for better fuel economy, safety, and comfort in automobiles has put demand on the sensor industries to develop a high quality, more reliable, and lower cost sensor for use in high volume manufacturing. The stringent requirement of the Corporate Average Fuel Economy (CAFE) regulations makes it necessary for sensors to be incorporated into automotive electronics. There are several sensors used in today's automobile to fulfill the above needs [1,2,3], and among them are many silicon based sensors. One of these is a MAP application. The first silicon based MAP sensor was incorporated in automobiles in the 1980s [4].

Two of the technologies developed in early 80s for MAP application were capacitive and piezoresistive (PRT) pressure sensors. The capacitive pressure sensor known as SCAP (silicon capacitive absolute pressure sensor) was incorporated in Ford Motor Company vehicles, and the piezoresistive sensor was incorporated by number of other automobile manufacturers. However, these technologies utilized sensing elements with no signal conditioning. In some cases the sensor included a resistor network for temperature compensation but did not incorporate signal conditioning on the same chip.

An integrated sensor offers a cost effective solution. A die size offers an opportunity to reduce the package size, especially important where expensive materials are used for housing sensors [5]. In addition, an integrated sensor is less susceptible to outside interference where wire interconnects between the transducer and control circuitry introduce the coupling of EMI into the system. An integrated sensor also improves yield and reliability by having fewer number of connections where failure can occur [6]. The connections are usually exposed to harsh media environments and therefore more susceptible to corrosion and other potential failures [6].

To date most silicon based automotive pressure sensors do not have integrated circuitry on the same chip. The availability of reliable low cost integrated technology coupled with silicon micromachining has increased the number of potential applications for fully integrated pressure sensors. In recent years, development of new technologies for better media and environmental protection, along with better sensor fabrication techniques has resulted in a robust piezoresistive pressure sensors.

MAIN SECTION

PRINCIPLES OF OPERATION - MAP sensors measure the vacuum in the intake manifold. When the engine goes through an intake cycle, a given cylinder receives the fuel-air charge from the intake manifold. The pressure measurement from the intake manifold is provided to the engine control unit (ECU), which then calculates the MAF (mass air flow) rate from the pressure measurement using the following equation [7]

$$\text{Mass Air Flow} = \frac{n \cdot \text{MAP} \cdot \text{displacement} \cdot \text{RPM}}{T_{\text{Charge}}} \qquad \text{(Eqn. 1)}$$

Table 1. General MAP requirements.

Transfer Function	Vout = Vs (P • K_1 - K_2) ± Error, K_1 & K_2 are constants
Low pressure requirement (kPaA)	15
High pressure requirement (kPaA)	250
Ratiometricity	1% ± 0.5% for 1% Vs change
Power supply (V)	Typically 5.0 ± regulated
Response time (ms)	≤15
Sink (mA)	.08 to 1
Source (mA)	.20 to 5
Thermal cycle - unpowered	200 to 700 cycles, -40/125°C, 60 min/cycle
Pressure/Temp Cycle	200 to 3000 cycles, -40/125°C, 0.5 to 1.5 hr/cycle
Hot Storage (Powered)	100 to 1000 hrs, 125°C
Hot Storage (Unpowered)	500 to 1000 hrs, 125°C
Cold Storage (Unpowered)	96 to 1000 hrs, -40°C
Humidity	96 to 1000 hrs, 60 to 85°C, 85 to 90% RH, with or without bias
Drop	1 to 5 drops of 1 meter
Mechanical Shock	5 to 100 g pulses of 10 msec
EMC/EMI (susceptibility)	50 to 200 V/M, 1 to 1000 MHz

where n is an empirically determined factor, usually about 0.6, MAP is the intake manifold absolute pressure, displacement is the volume of the cylinders multiplied by the number of cylinders, RPM is the engine rotation in revolutions per minute, and T_{charge} is the temperature of the air/fuel charge. The data is used to adjust the vehicle's injector pulse width, thereby insuring optimum engine stoichiometry and preventing a lean burn.

The MAP sensor general specification and testing requirements varies for the different automakers. The specifications usually depend on their algorithm, technology, and system requirements. A typical MAP sensor specification and testing requirements is shown in Table 1. The typical pressure range for the MAP sensor is 105 kPa full scale. However, in the case of turbo charged engines, the pressure range is typically 250 kPaA full scale.

MONOLITHIC DESIGN - Figure 1 shows the top view of an integrated pressure sensor. The monolithic sensor contains op-amps, and passive components including SiCr resistors for laser trimming. This is an analog device which uses bipolar integrated circuit technology. The single-chip MAP sensor uses a single series temperature compensation of span resistor which provides a varying common mode voltage for use in temperature compensation of offset trimming. A total of three op amps are used in the sensor design. The first two form an instrumentation amplifier to isolate the transducer output from the resistor network. The third is the output buffer with level shifting divider R_8 and R_9, and zero pressure offset pedestal set by R_{10} and R_{11} (see Figure 2). TC of offset is corrected by shifting the negative side of the transducer differential output with temperature. Zero pressure offset is trimmed at the divider (R_8 and R_9) and pull-up resistor to Vcc (R_7), which allows minor adjustments independent of gain.

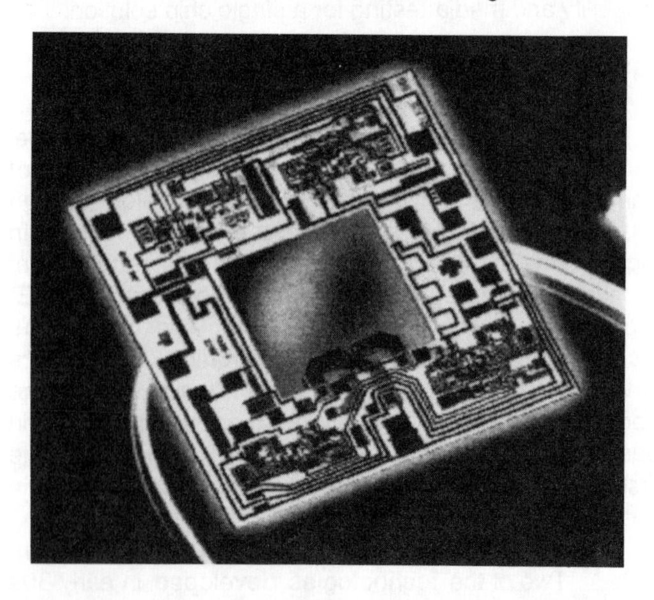

Figure 1. Top view of an integrated pressure sensor.

The sensing transducer design is a single piezoresistive element. It consists of a diaphragm and a piezoresistive element located near the edge of diaphragm at a 45% angle [8]. The diaphragm size is about 1000 microns and thickness is about 20 microns. The die size is approximately 3 mm^2..

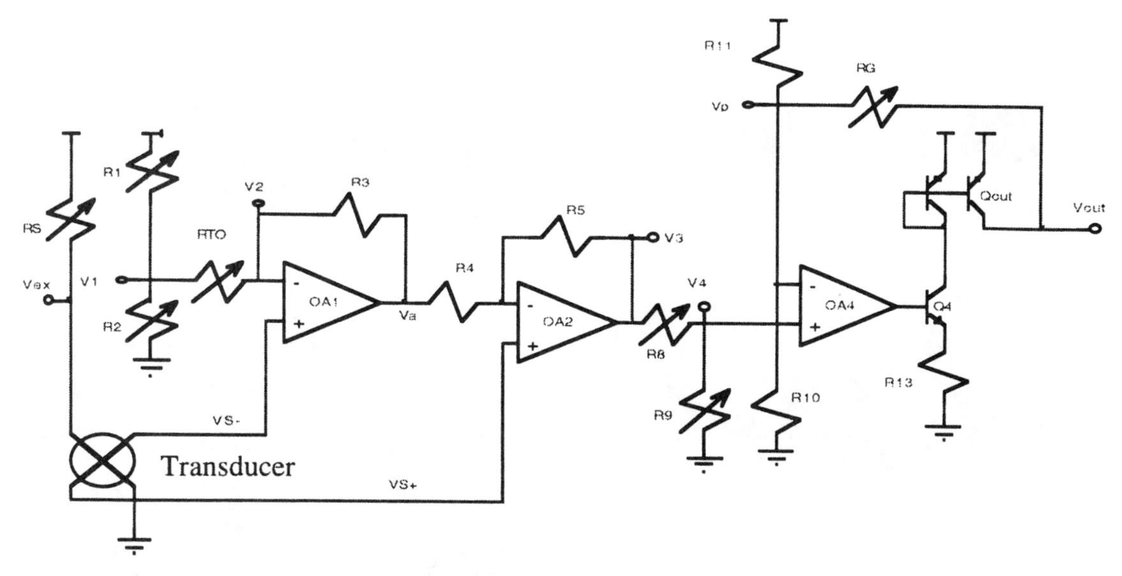

Figure 2. Signal condition for transducer

SOURCE/SINK - The first single-chip MAP sensor was designed to interface directly with the A/D inputs of a microprocessor. The output signal encompasses the upper rail voltage minus a saturation voltage for the PNP output driver and ground plus a similar voltage to allow for a reasonable output leakage current across the load resistor for maximum resolution by the microprocessor A/D. The sensor output is ratiometric when µP Vref-hi supplies the positive supply voltage and Vref-lo is ground. The high input impedance of the microprocessor combined with the source only output of the sensor requires a filter of 51kΩ in parallel with 50pf to insure cancellation of high frequency noise.

Since the first single-chip MAP sensor was designed to drive only the high impedance input of a microprocessor, additional current requirements (for EMI suppression, corrosion prevention, or implementation of a logic function) require additional source current / sink current drive capability. This is easily accomplished by adding an op amp buffer at the sensor output. The change in current drain is minimal. Accuracy is not affected since the only error is the op amp input offset voltage of only a few millivolts over temperature. Any load, accuracy, and current drain can be accommodated by the choice of a suitable op amp for the buffer.

LASER TRIMMING - All system level resistors are SiCr which has a thermal coefficient of resistance (TCR) of near zero. The TCR is important for the series span compensation resistor (Rs) since span decreases with temperature. Since the transducer input resistance increases with temperature, the series resistor will cause the voltage across the transducer to increase. This increase in voltage across the transducer counteracts a loss in sensitivity. The room temperature voltage must be set to a value which will cause the excitation voltage (Vex) to increase at a proper rate.

Next, the pressure to the device is set to the minimum level, R1 or R2 is trimmed to set V1=V2, and the offset is adjusted to 0.2V by adjusting the divider R8 or R9. The divider value is set to approximately cancel the transducer common mode voltage. With pressure applied, Rg is trimmed to set the desired sensitivity. The input network of OA4 allows gain to be adjusted without changing the previously trimmed offset.

The device must then be heated to trim TC of offset. At elevated temperature, V2 is now above V1, and current will flow into RTO. RTO is trimmed at the minimum pressure and elevated temperature to achieve the same offset voltage as set at room temperature, 0.2 V in this case.

PROCESS - The MAP sensor consists of bipolar integrated circuit and a micromachined sensing element on a single monolithic chip. Both bipolar processing and sensor fabrication are well established technologies. However, marrying these technologies presents some challenges, since the sensor fabrication requires non-conventional IC processing such as a deep etching of the silicon to form a thin diaphragm (see Figure 3).

A typical sequence of fabrication steps illustrating the technique is shown in Figure 3 for a generic integrated piezoresistive sensor (IPS). The X-ducer™ and bipolar devices were fabricated using conventional IC diffusion processing. Following the diffusion, SiCr thin films and interconnect metalization is deposited and patterned. Aluminum is used as the interconnect metalization. Once the device fabrication is

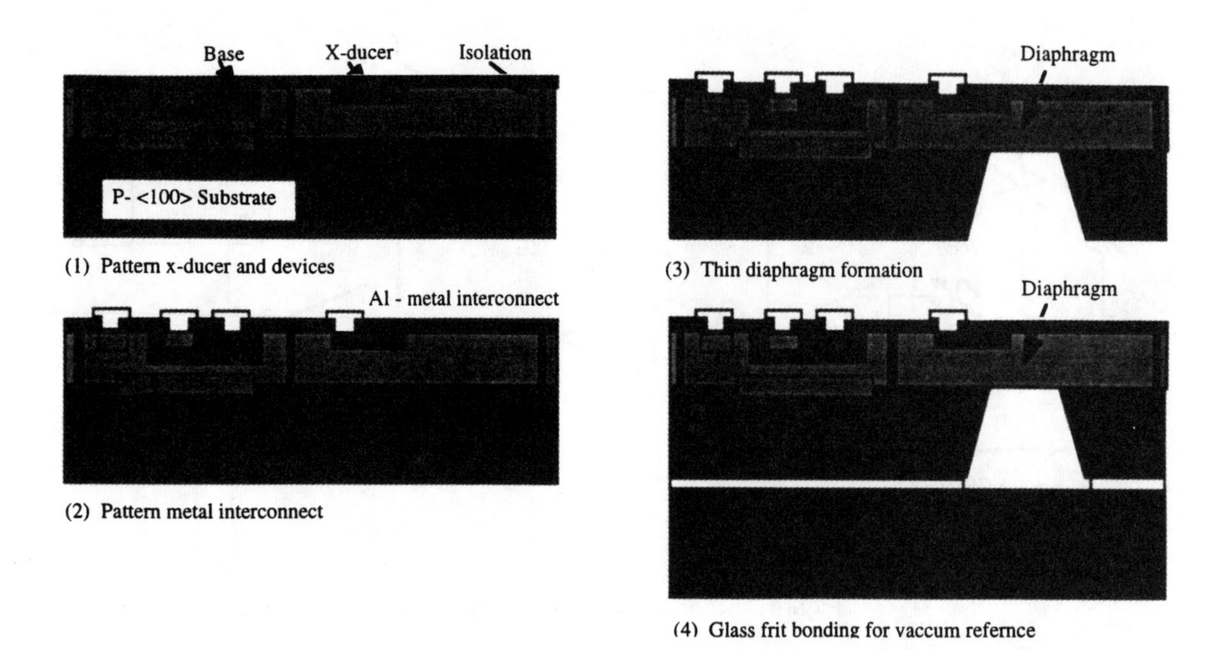

Figure 3. Process sequence for generic integrated absolute pressure sensor

completed, a deep anisotropic cavity is etched into the silicon from the back side of the wafer to form the diaphragm. The active wafer is then frit bonded to a constraint in an evacuated chamber, forming an absolute reference cavity below the diaphragm.

EMC PERFORMANCE - Electromagnetic compatibility is a major issue for automotive applications. The specifications and bench test set up for EMI testing is not standardized across the industry. The test method discussed here is most severe and adequate enough to meet most of the requirements, i.e., device capable of lower susceptibility or greater resistance to the present electromagnetic (EM) signals.

Figure 4 describes the test setup. More detail of the setup, testing conditions and test results are described elsewhere [9]. The requirement for testing integrated sensor involves a 200 V/m radiated immunity test from 10 kHz to 1 Ghz. The test can be extended to higher frequency to evaluate influence of the 1 - 18 GHz signals.

Tested pressure sensors met the requirement and did not exceed the AC ripple or DC offset susceptibility criteria over the 200 MHz to 1GHz frequency range. Measurements were repeated several times over time periods as long as several days with the same test setups, and results show very similar values.

PACKAGING - One of the great challenge in semiconductor sensor manufacturing is the packaging. Unlike IC packaging, the pressure sensor can be directly exposed to harsh media. Thus, it requires more than the concepts that evolved out of conventional IC packaging technology. The suitable material and mounting techniques to provide include: (a) mechanical support, (b) electrical interface, (c) environmental protection, and (d) media interface (see Figure 5). Typical requirements for sensor packaging are discussed below.

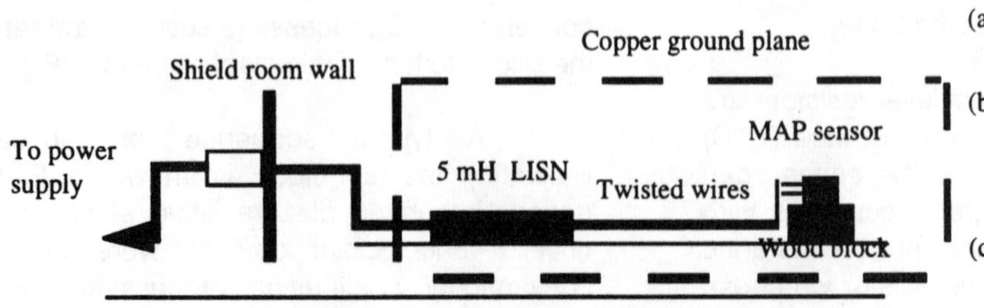

Figure 4. EMC test setup

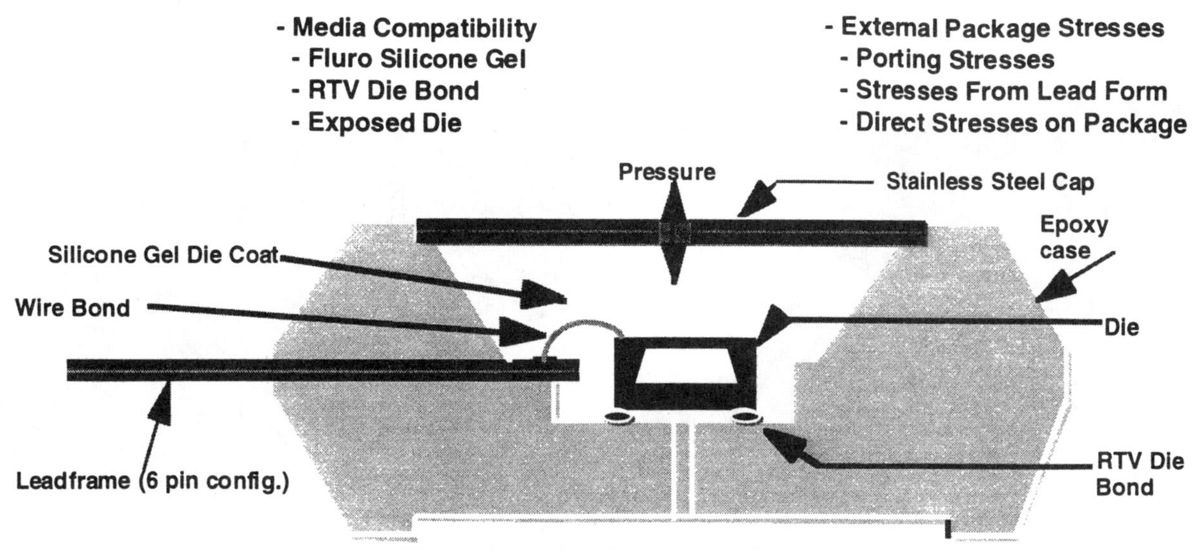

Figure 5. Basic chip carrier package.

The number of leads in sensor packaging varies according to the product and its application. The IPS package described here is designed with eight pins. The actual number of pins used by the customer is only three. The additional pins are used for laser trimming and are not connected in the application. The leads are designed with width of 1.27 mm (50 mil) and 2.54 mm (100 mil) spacing. For Surface Mount and other packages (Piston fit), the leads are formed to create a gull-wing shape or can be formed for a through hole solder joint. The piston fit package is designed to accept an O-ring to create either a radial pressure seal or a surface seal using a soft material such as silicone. Several versions of the piston fit package are shown in Figure 6. The size, spacing and the shape of the leads follow a standard practice, thus no special requirements for pad layout or via hole is required during PWB layout. The solder bond pad sizes and solder paste application will be same as other semiconductor components.

Conventional semiconductor components are typically shipped with leads that are solder dipped or tin plated after overmolding. This is not a major issue since the packages are not open to the atmosphere. For

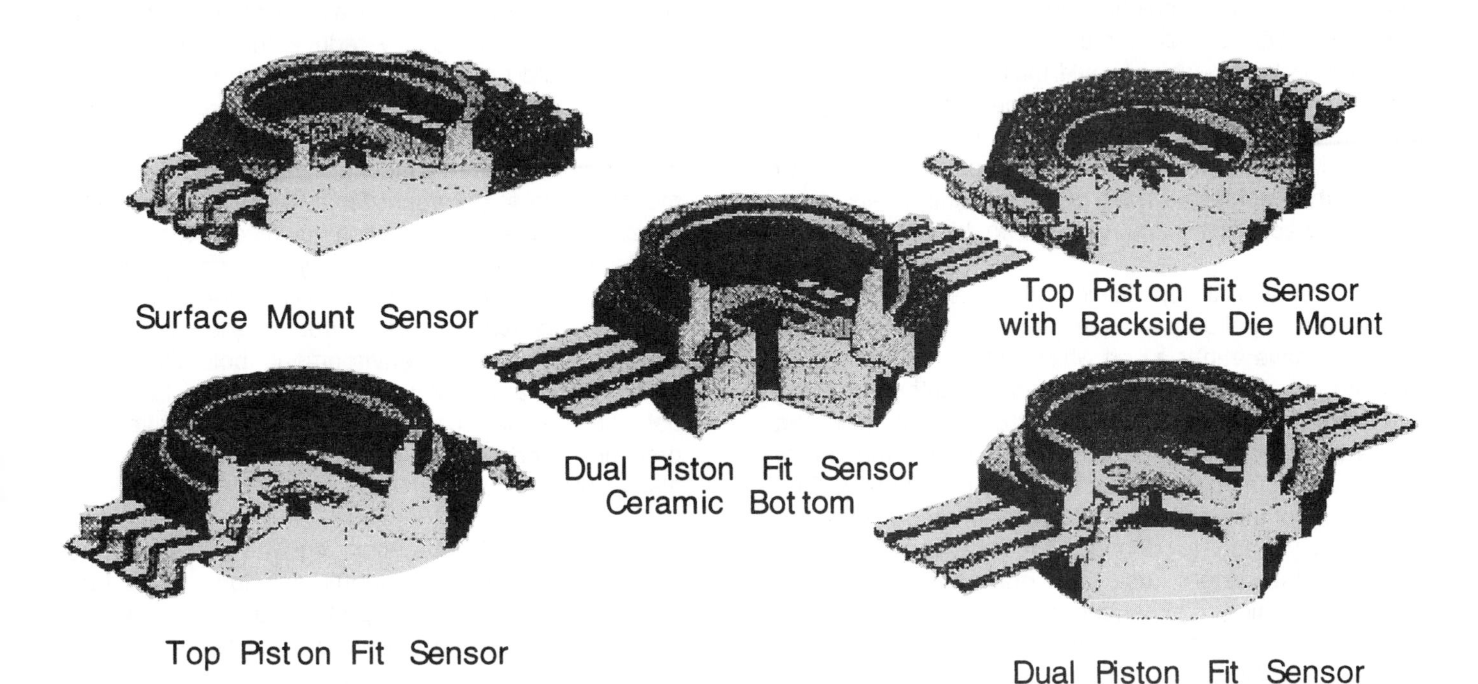

Figure 6. Surface mount and piston fit packages. The top and dual piston fit packages are designed to accept an o-ring for the pressure interconnect.

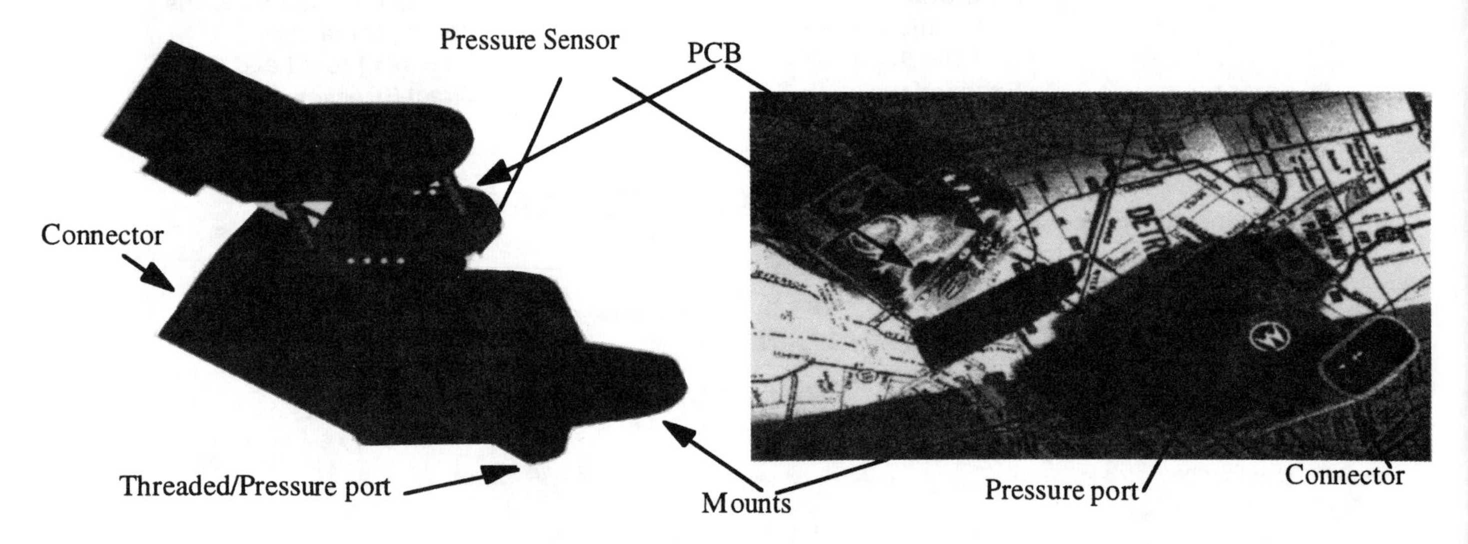

Figure 7. Conceptual housing options for user implementation.

pressure sensor packages, there is always an opening for pressure interface. The lead configuration makes it difficult to perform solder dipping or tin plating. Without such a treatment, the underlying Ni layer may not pass the solderability requirement. An acceptable solution to the solderability issue is to provide a flash of Au on the solderable portion of the leads, which would protect the underlying Ni. In surface mount assembly with Sn-Pb solder, the presence of Au is known to form Au-Sn intermetallic. An excessive amount of Au is likely to cause embrittlment of the solder joint, which will result in lower fatigue life. A solution adopted by Motorola and many other sensor manufacturers is to use a flash of Au on the lead frame, which would maintain solderability of the lead at the same time introduce fairly insignificant amount of Au in the Pb-Sn solder. In a typical solder joint, this amount of Au will result in approximately 1% of Au in the solder; which is significantly less than commonly acceptable 5% Au in the solder joint.

For a MAP application, the sensor needs to sense the manifold vacuum pressure. Therefore, it needs to have either a port which will be connected to vacuum hose, or it will be mounted in another housing with a port. These housings are either directly mounted to the manifold or externally mounted under the hood. The direct mount configuration is shown in Figure 7. The advantage of this technique is that the method does not require any hose to connect to the sensor thereby reducing the system cost. The draw back is that the sensor housing experiences significantly higher temperatures. This higher temperature may result in different housing materials and could influence the electrical specification.

The MAP sensor can also mounted direct on the ECU, eliminating the need for a wire harness and external connector. This will result in a lower system cost and less source and sink current requirements since the external connector is eliminated. For this configuration, the pressure hose needs to be extended from the manifold.

RELIABILITY AND MEDIA TESTING - To ensure accurate testing, knowledge of the application, lifetime requirements, and what constitutes a failure is crucial. A physics-of-failure approach can significantly reduce the development cycle time and produce a higher quality product [10]. The focus of the physics-of-failure approach includes an understanding of the application, lifetime expectation, failure mechanism(s), and lifetime models. The requirement for a typical MAP or BAP pressure sensor application involves testing to temperature extremes, thermal cycle, humidity, media exposure, vibration, shock, cyclic pressure, and overpressure testing [11]. Through reliability testing and knowledge of the environment, potential failure mechanisms are uncovered. A complete listing of potential failure mechanisms that may affect a pressure sensor device has been presented elsewhere [12].

The MAP application requires the sensor to survive in a fuel or aqueous solution. The fuel exposure typically is performed at elevated temperature to a wide variety of fuel types. A test matrix of several fuels based on ASTM guidelines that includes various additives such as methanol, water, or acids and a test procedure have been discussed elsewhere [13]. Acid testing either performed independent to the fuel or following exposure has proven to be an affective test scheme for product development. The nitric and sulfuric acid tests are a concern due to NO_x + water = HNO_3 and SO_x + water = H_2SO_4. A proposed test scheme for product development involves a variety of material types and environments. Not all materials from the same family will respond the same to the testing. Actual media test results will be published elsewhere.

A key aspect to reliability and media testing involves the determination of what constitutes a failure. The definition of an electrical failure can range from catastrophic, to exceeding a predetermined limit, to just a small shift. The traditional pre to post electrical characterization before and after the test interval can be enhanced by *in situ* monitoring. In situ monitoring may expose a problem with the sensor device during testing that may go undetected once the media or another environmental factor is removed. For example, swelling of polymeric materials when exposed to certain media or environment, may result in shift in the device output. Such a failure mechanism can only be detected by insitu monitoring. Response variables during environmental testing can include: electrical, visual, analytical, or a physical characteristic such as swelling or weight change. A typical definition of failure for the MAP and BAP application is to both be within the error budget after the exposure. In addition, the output voltage shift from the initial value needs to be within a predetermined value.

CONCLUSION

A monolithic integrated pressure sensor has been developed for MAP applications. The integrated pressure sensor is small in size and provides better performance than its predecessors. The sensor can be directly mounted on the ECU PWB or underhood. It shows great resistance to harsh media thereby making it well suited for automotive applications. Additionally, the monolithic integrated pressure sensor shows very good immunity to surrounding electromagnetic fields since the need for an interconnect is eliminated.

ACKNOWLEDGMENT

The authors would like to thank John Trice, Randy Frank, Wendy Chan, Demetre Kondylis, Dan Wallace, Kelvin Blair, Aristide Tintikakis, Wayne Chavez, and Gary Beaudin for their encouragement and valuable inputs.

X-Ducer is a trademark of Motorola.

REFERENCES

[1] Peter Kleinschmidt and Frank Schmidt, "How Many Sensors Does a Car Need?," Sensors and Actuators A, 31 (1992) 35-45.

[2] J. Binder, "New Generation of Automotive Sensors to Fulfill the Requirements of Fuel Economy and Emission Control," Sensors and Actuators A, 31 (1992) 60-67.

[3] M H Westbrook, "Future Developments in Automotive Sensors and Their Systems," J. Phys. E: Sci. Instrum. 22 (1989) 693-699.

[4] William G. Wolber and Kensall D. Wise, "Sensor Development in the Microcomputer Age," IEEE Transaction on Electron Device, vol. ED-26, NO. 12, December 1979.

[5] J. M. Giachino, "Smart Sensor for Automotive Applications," Proceedings of the 16th International Symposium on Automotive Technology and Automation, 1984.

[6] J. M. Giachino, "The Challenge of Automotive Sensors," The 1984 IEEE Solid-State Sensor Conference.

[7] Meyer, E.W., Jr.; Graham, K.A,; Kenyon, B.F.; Kissel, W.R. "Outstanding Convergence 80.

[8] Ira Baskett, Randy Frank, and Eric Ramsland, "The Design of a Monolithic, Signal Conditioned Pressure Sensor," IEEE 1991 Custom Integrated Circuits Conference.

[9] Dragan Mladenovic, Rajan Verma, and Randy Frank, "EMI Considerations for Automotive Sensors," SAE Sensors & Actuators 1997, To be published.

[10] Jimmy H. Hu, "Methodology for Developing Reliability Validation Tests for Automotive Electronics", Proc. EEP, Vol. 10-2, (1995), pp. 627-633.

[11] "Guide to Manifold Pressure Transducer Representative Test Method", SAE J1346, (1981).

[12] T. Maudie, D. J. Monk, D. Zehrbach, and D. Stanerson, "Sensor Media Compatibility: Issues and Answers", Sensors Expo, Anaheim, CA (1996).

[13] T. Maudie, "Testing Requirements and Reliability Issues Encountered with Micromachined Structures", Proceedings of the Second International Symposium on Microstructures and Microfabricated Systems, Chicago, IL, ECS (1995), 223.

ELSEVIER

JSAE Review 17 (1996) 281–286

Study on the measurement of oil-film pressure of engine main bearing by thin-film sensor – The influence of bearing deformation on pressure sensor output under engine operation

Yuji Mihara [a], Makoto Kajiwara [b], Takayuki Fukamatsu [b], Tsuneo Someya [a]

[a] *Engine Research Laboratory, Musashi Institute of Technology, 1-28-1 Tamazutsumi, Setagaya-ku, Tokyo, 158 Japan*
[b] *Graduate school, Musashi Institute of Technology, 1-28-1 Tamazutsumi, Setagaya-ku, Tokyo, 158 Japan*

Received 29 January 1996

Abstract

In a previous paper, oil-film pressure in an engine main bearing was measured by a using thin-film Manganin sensor. This sensor had adequate sensitivity to strain and temperature in terms of measurement principle. In this study, more accurate measurement of oil-film pressure was allowed by the formation of the newly developed thin-film strain sensor made of Ni–Cr–Al at the same location as that of the Manganin pressure sensor, and the correction of measurement errors caused by strain. The characteristics of bearing metal surface strain varied by engine operating conditions are also shown in this paper.

1. Introduction

Deformation of bearing and the pressure dependency of lubricating oil viscosity have been considered in recent years for the analysis of conditions of lubrication for sliding bearings, and measurements of oil-film pressure distribution, using a bearing test rig, have been conducted for the verification of results of analysis. However, the measurement of oil-film pressure on the engine main bearing has been considered difficult due to the change in sliding surface profile and the decrease of shaft and engine block rigidity that would be caused by the installation of pressure sensors. The authors et al. succeeded in the measurement of oil-film pressure in actual engine operation by forming a thin-film pressure sensor of 3 to 4 μm thickness directly onto the main bearing sliding surface by means of sputtering, using the Physical Vapor Deposition (PVD) method [1]. The strain of main bearing and the temperature amplitude of the main bearing sliding surface in actual engine operation were measured, and the validity of the measured values of the oil film pressure were verified, as reported in the previous paper [1], as this sensor had adequate sensitivity to strain and temperature in terms of the principle of measurement. As a result, it was found that measurement errors caused by bearing strain were greater than those caused by the temperature amplitude. In this study, a thin-film strain sensor that would not be affected easily by the oil-film pressure and temperature was developed and formed on the main bearing sliding surface at the same location as that of the Manganin pressure sensor. The pressure correction factor was determined from the measured value of strain detected by the strain sensor, in order to allow more accurate determination of the measurement error caused by the bearing deformation. The measured values of the main bearing oil-film pressure in actual engine operation and the correction method are described, together with a comparison of the amount of bearing metal surface strain varied by the engine operating conditions.

2. Thin-film sensors

2.1. Principle of measurement and error factor

The principle of measurement for the thin-film pressure sensor and the thin-film strain sensor is as follows. Quite small change in sensor resistance caused by the change in pressure, strain or temperature is to be measured as the change in voltage by means of a Wheatstone bridge circuit.

0389-4304/96/$15.00 © 1996 Society of Automotive Engineers of Japan, Inc. and Elsevier Science B.V. All rights reserved
PII S0389-4304(96)00022-7

JSAE9631641

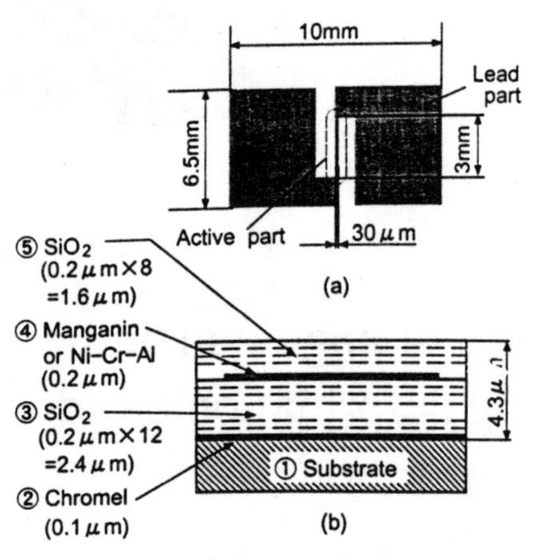

Fig. 1. Structure of thin film sensor.

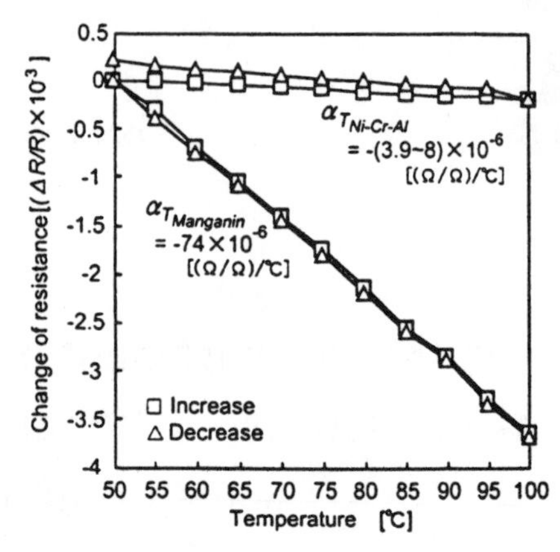

Fig. 3. Temperature characteristics of thin-film sensors.

The output of such a sensor can be expressed by the following equation:

$$\Delta R/R = (\Delta R/R)_P + (\Delta R/R)_\epsilon + (\Delta R/R)_T \quad (1)$$

where $\Delta R/R$ = rate of change in resistance, P = pressure, ϵ = strain, T = temperature.

In case of the pressure sensor, the sensitivity to the strain in the right-hand second term and the sensitivity to the temperature in the third term must be minimized as much as possible. For the strain sensor, on the other hand, the sensitivity to the pressure in the right-hand first term and the sensitivity to the temperature in the third term must be minimized as much as possible.

2.2. *Thin-film sensor structure and shape*

2.2.1. *Thin-film pressure sensors*

The structure of the pressure sensor film prepared in this study is shown in Fig. 1(b). Two kinds of substrate materials were used [①], namely, the stainless steel (SUS304) used as an equal stress beam shown in Fig. 4, and the aluminum alloy bearing material (with steel back metal) used in the test engine. In case of the aluminum metal, the Chromel (90 Ni, 10 Cr) [②] is first coated by 0.1 μm thickness as the middle layer, aiming at increasing the strength in adhering the insulation film [③] and the substrate material [①]. Next, SiO_2 [③] is coated by 0.2 μm each for 12 times to form a total film thickness of 2.4 μm or so as the insulation film. The multiple film coating is done because the insulation film will be destroyed if the film is formed continuously by more than 0.2 μm, due to the temperature rise of the bearing in sputtering. The amount of film formation per coating varies according to the size and material of the bearing. Next, the Manganin alloy [④] (82 Cu, 15 Mn, 3 Mn Wt%) is coated by 0.2 μm. Then SiO_2 is coated by 0.2 μm each for 8 times with a total thickness of about 1.6 μm [⑤] over a Manganin film [④] to prevent destruction of the sensor element that may be caused by the crank shaft contact or the metallic powder containd in the lubricating oil. The reason for the formation of multiple protection layers is the same as that for the multiple insulation layers. The shape of the sensor is as shown in Fig. 1(a) with a length of 3 mm, width 30 μm and thickness 0.2 μm. The lead part is so designed that its area can be maximized as much as possible against the pressure sensing area so that the resistance change caused by the pressure can be minimized. The method of forming the thin-film sensor on the stainless beam is to set the film thickness per coating in the range of 0.7 to 0.8 μm for both the insulation and protection layers. Other film formation methods are the same as those on the aluminum bearing.

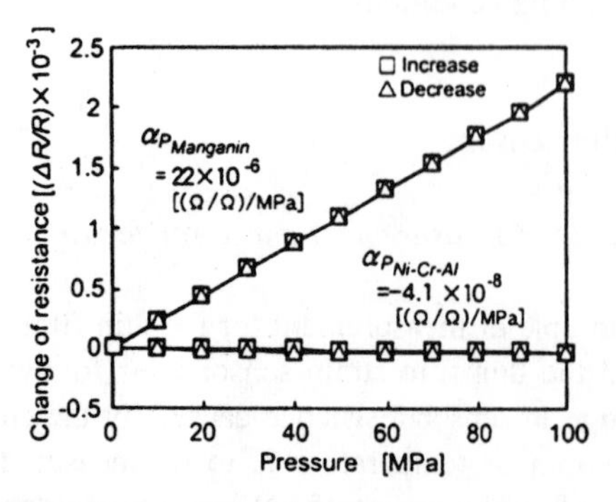

Fig. 2. Pressure characteristics of thin-film sensors.

2.2.2. *Thin-film strain sensor*

The sensor is made by a sputtering target alloy with the composition rates of 70.5 Ni, 18.5 Cr and 11.0 Al (Wt %). The shape of the sensor and the film structure are the same as those of the pressure sensor. It is found by checking on properties of various materials [2,3] used in this study that this alloy has characteristics of extremely low sensitivity to both the pressure and temperature (Fig. 2, Fig. 3). Hence, it has become possible to measure the strain alone while ignoring the effects of the oil-film pressure and temperature amplitude.

2.3. Output characteristics of pressure sensor and strain sensor

2.3.1. Pressure characteristics

Figure 2 shows the result of pressure calibration using a pressure vessel. The test pressure was set in the range of 0 to 100 MPa. The pressure sensitivity of the Manganin sensor was

$$\alpha_{P\ \mathrm{Manganin}} = 22.0 \times 10^{-6} (\Omega/\Omega)/\mathrm{MPa}.$$

The nonlinearity and the hysteresis were 1% F.S. or smaller. In the case of Ni–Cr–Al, the pressure sensitivity was

$$\alpha_{P\ \mathrm{Ni-Cr-Al}} = -4.1 \times 10^{-8} (\Omega/\Omega)/\mathrm{MPa},$$

and the nonlinearity and the hysteresis were 3% F.S. or smaller, showing that the pressure sensitivity is quite low.

2.3.2. Temperature characteristics

Figure 3 shows the result of temperature calibration, with the test temperature range of 50 to 100°C. The temperature sensitivity of the Manganin sensor was

$$\alpha_{T\ \mathrm{Manganin}} = -74 \times 10^{-6} (\Omega/\Omega)/°\mathrm{C},$$

and the nonlinearity and the hysteresis were 3% F.S. or smaller. In the case of Ni–Cr–Al, however, the nonlinearity and the hysteresis were greater than those of Manganin, while the temperature sensitivity was smaller as

$$\alpha_{T\ \mathrm{Ni-Cr-Al}} = -(3.9 \text{ to } 8.0) \times 10^{-6} (\Omega/\Omega)/°\mathrm{C}.$$

2.3.3. Strain sensitivity

A thin-film sensor and a strain gage (120 Ω with the gage factor $K_{\mathrm{Strain\ Gage}} = 2$) were installed on the equal stress beam shown in Fig. 4 [4], and the gage factor (K_x) of each thin-film was calculated by the following Eq. (2) according to the measured results (Fig. 5) of the strain gage's output (ϵ) and the thin-film sensor where the load (W) was applied to the tip of the beam.

$$K_x = (\epsilon_x/\epsilon) \cdot K_{\mathrm{Strain\ Gage}} \quad (2)$$

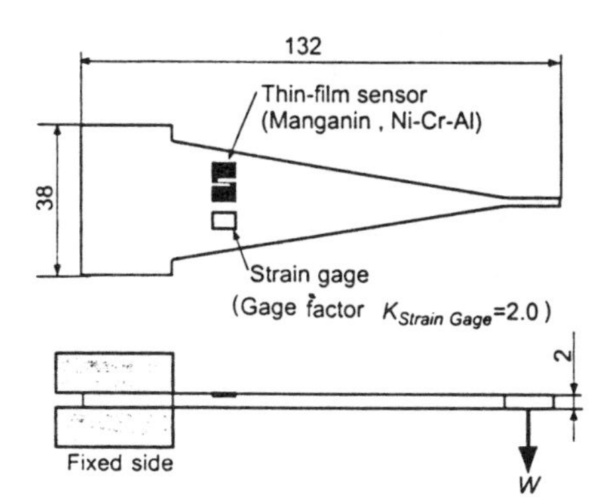

Fig. 4. Testing device of gage factor.

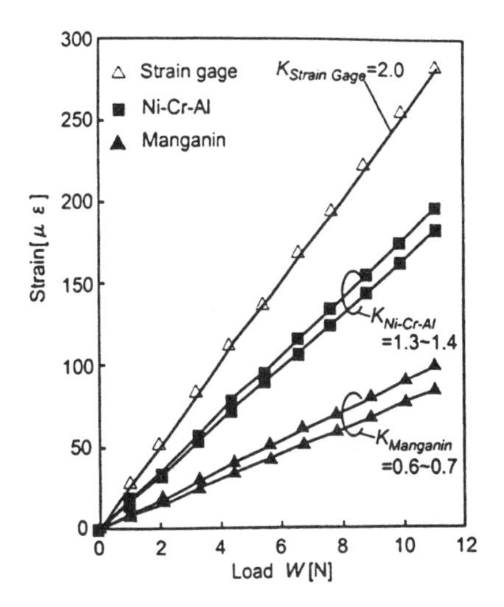

Fig. 5. Comparison of gage factor.

The strain output (ϵ_x) differs among the individual thin-film sensors as shown in Fig. 5. The results show that $K_{\mathrm{Manganin}} = 0.6$ to 0.7 for Manganin, while $K_{\mathrm{Ni-Cr-Al}} = 1.3$ to 1.4 for Ni–Cr–Al.

2.4. Effect of bearing deformation on oil-film pressure measured value

2.4.1. Pressure correction factor calculation method

The pressure correction factor is calculated by the following Eq. (3), according to the main bearing strain output measured in engine operation, with the Ni–Cr–Al layer formed on the engine main bearing.

$$P_\epsilon = \frac{(K_{\mathrm{Manganin}}/K_{\mathrm{Ni-Cr-Al}}) \cdot (\Delta R/R)_{\mathrm{Ni-Cr-Al}}}{\alpha_{P\ \mathrm{Manganin}}} \quad (3)$$

where P_ϵ = pressure correction factor, $(\Delta R/R)_{\mathrm{Ni-Cr-Al}}$ = measured amount of strain (converted into the rate of resistance change)

Using this equation for correction, the oil-film pressure correction factor against the error caused by the effect of strain shown in ③ of Fig. 10 is calculated.

2.5. Effect of temperature on measured value of oil-film pressure

2.5.1. Effect of temperature amplitude

The pressure-temperature coefficient is approximately −3.4 (MPa/°C), according to results of the sensitivity of Manganin to the pressure and temperature sensitivity measured in 2.2. However, the temperature amplitude per cycle in engine operation becomes about 0 (K) when a thin-film thermocouple is formed on the bearing sliding surface at the same location as that of the thin-film pressure sensor. Therefore, it is judged that a pressure measurement error does not occur due to temperature amplitude.

Table 1
Test engine specification

Water-cooled in-line 4 gasoline engine	
Bore × stroke (mm)	80 × 79
Stroke volume (cm^3)	1588
Journal diameter (mm)	51.920
Bearing width (mm)	17
Diametral clearance (mm)	0.022–0.045
Bearing material	aluminum alloy

2.5.2. *Zero-line of oil-film pressure measured values*

The zero-line moves as the engine oil temperature rises due to the temperature sensitivity of the Manganin sensor as shown in Fig. 3. Therefore, the minimum value per cycle was set as the zero-line (0 MPa) in the experiments.

2.6. *Zero-line of strain measured values*

The travel of zero-line caused by the rise of the engine oil temperature is small, as the temperature sensitivity of the strain sensor is low, as shown in Fig. 3. Therefore, the sensor output with the engine stopped is set as the zero-line.

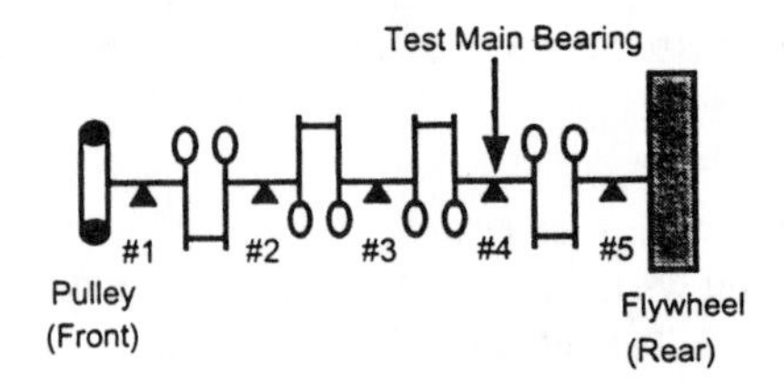

Fig. 6. Position of test main bearing.

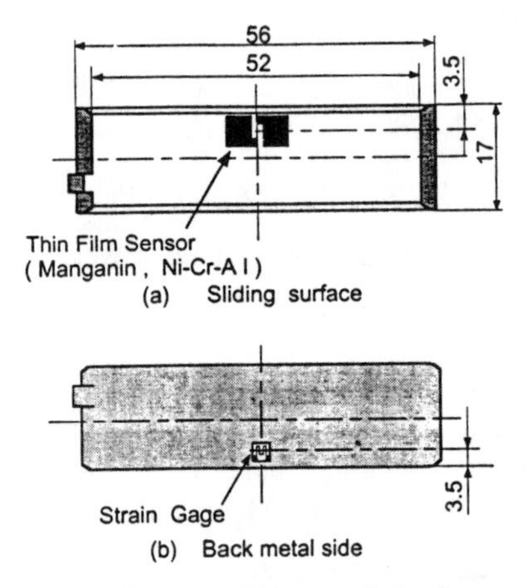

Fig. 7. Sensor position on main bearing.

3. Experiments

3.1. *Test engine and test equipment*

The test engine used in the experiments was a four-cylinder gasoline engine (Table 1), and the experiments were conducted on the fourth main bearing (called #4) counted from the front (Fig. 6). Both the thin-film pressure sensor and the thin-film strain sensor were formed vertically in the downward direction, with their longitudinal centers of sensor units installed at a location 3.5 mm from the engine front side bearing end (Figs. 7 and 8). A strain gage was adhered on the back metal of the bearing immediately beneath the thin-film strain sensor, and the output of the two strain sensors compared. The bearing load working on #4 was measured by adhering a strain gage to the cap bolt [5] (Fig. 8).

3.2. *Measured results and the corrections*

3.2.1. *Strain working on bearing*

Figure 9 shows the experimental results at the engine speed of 3000 rpm with the half load, and the corrected results of oil-film pressure measured values. The amount of strain on the bearing sliding surface detected by the thin-film strain sensor was convered by the following Eq. (4) for the calculation of the true value of amount of strain,

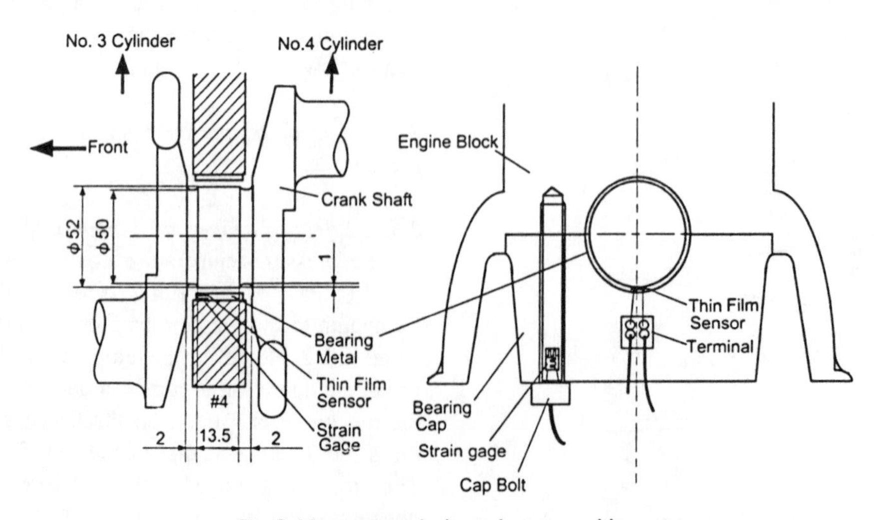

Fig. 8. Measurement device and sensor position.

considering the sensitivity of the sensor to the strain ($K_{Ni-Cr-Al}$ = 1.3 to 1.4), and the fact that the gage factor set for the gage amplifier used in the measurements was 2.0.

$$\epsilon_{REAL} = \left(K_{Gage\ amp.} / K_{Ni-Cr-Al} \right) \cdot \epsilon_{Measurement} \tag{4}$$

where ϵ_{REAL} = converted value of strain output, $\epsilon_{Measurement}$ = strain value measured by Ni–Cr–Al, $K_{Gage\ amp.}$ = gage factor set for strain amplifier ($K_{Gage\ amp.}$ = 2.0).

Figure 9 [①] represents the strain output from the strain gage adhered to the back metal of bearing, while [②] represents the strain output from the thin-film strain gage, converted by Eq. (4). The absolute maximum value of each of the above is approximately 70 μ ε, but their signs are opposite. These outputs do not agree completely, though the third and fourth cylinder pressures rise, and the amount of strain increases significantly in the region of − 160° CA (Crank Angle) to 120° CA for both, where the bearing load [⑥] increases. Therefore, the relationship between the bearing metal sliding surface and the back metal is not so simple as that of stretched side versus shrunk side. The deformation of metal alone can also occur as verified.

3.2.2. Value of pressure corrected against strain

Figure 9 [③] represents the pressure values corrected by substituting the measured results of the thin-film strain gage into Eq. (3). The strain of the bearing is almost always negative as shown in Fig. 9 [②]. Hence the output of measured oil-film pressure is smaller than the actual value. Therefore, Fig. 9 [③] is added to the measured value as the correction value of pressure.

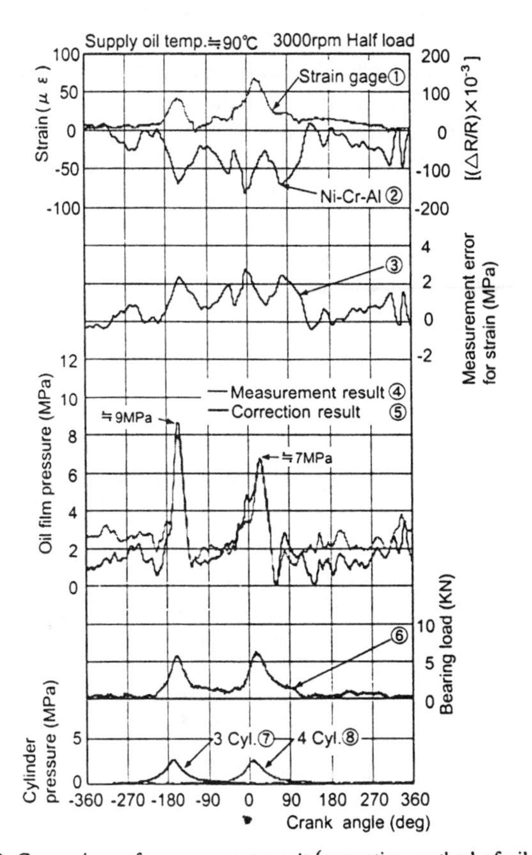

Fig. 9. Comparison of measurement result (correction method of oil film pressure).

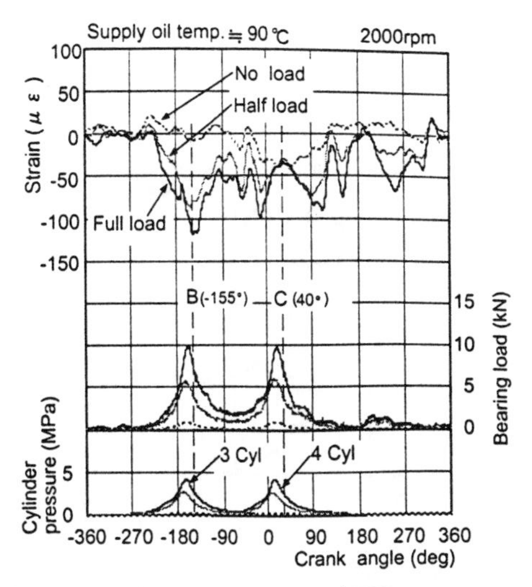

Fig. 10. Comparison of metal surface strain (2000 rpm constant, effect of load condition).

3.2.3. Measured values and corrected values of oil-film pressure

Figure 9 [④ grey line] represents the measured values of oil-film pressure, while [⑤ black line] represents the corrected values of oil-film pressure obtained by adding the correction value [③] to the values in [④]. The zero line for [⑤] is set by the method described in 2.5.2. As a result, the oil-film pressure values in [⑤] became greater due to the increase of #4 main bearing load [⑥] caused by the increase of cylinder pressure [⑦ and ⑧]. It was also found that an oil-film pressure of about 9 MPa occurred

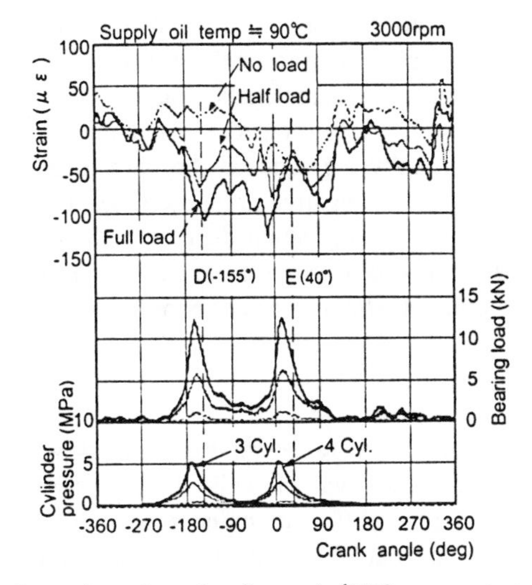

Fig. 11. Comparison of metal surface strain (3000 rpm constant, effect of load condition).

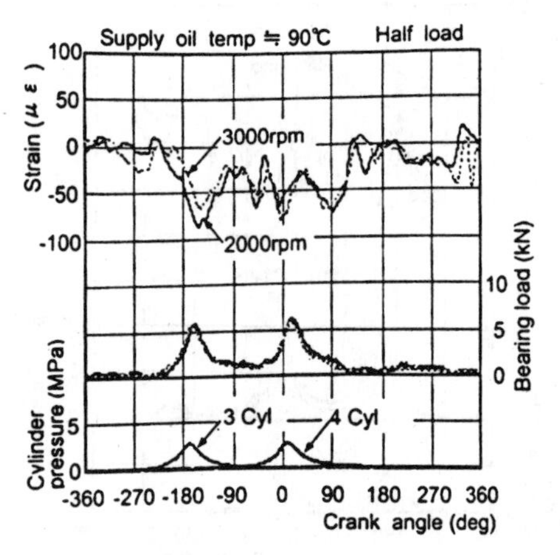

Fig. 12. Comparison of metal surface strain (load constant, effect of engine speed).

where the third cylinder was working, and approximately 7 MPa where the fourth cylinder was working.

3.3. Comparison of amount of strain on bearing sliding surface under different operating conditions

3.3.1. Effect of load

Figure 10 shows the comparison in amount of strain where the engine load is varied at the engine speed of 2000 rpm. Under the no load condition, a strain of about $-40\ \mu\ \epsilon$ is found at the crank angle C (40° CA) where the fourth cylinder combustion pressure is working, but hardly any strain is found at other crank angles. As the load increases, negative strain output increases according to the increase in bearing load caused by the increased third cylinder pressure. The value of such strain becomes greatest at the crank angle B (−155° CA), which is approximately $-120\ \mu\ \epsilon$ with the full load. On the other hand, the bearing load increases at a crank angle (approx. 40° CA) where the fourth cylinder pressure increases, but the strain starts to decrease sharply thereafter and the amount of strain becomes practically the same regardless of the load, which is a characteristic feature. The tendency of amount of strain according to the change of load at 3000 rpm is basically the same as the case of 2000 rpm (Fig. 11) – that is, the negative strain increases at the angle D (−155° CA) while the strain decreases sharply at the angle E (40° CA) according to the increase in bearing load, and the amount of strain becomes similar regardless of load.

3.3.2. Effect of engine speed

Figure 12 shows the results of experiments where the engine speed alone was varied, while the cylinder pressure was kept constant at different engine speeds. The experimental results up to 3000 rpm show no significant change in strain output by the change in engine speed. This is presumably due to the fact that the experiments were conducted on #4 main bearing, having no significant change in bearing load caused by the intertia force.

4. Concluding remarks

(1) Thin-film sensors were formed on the sliding surface of the engine main bearing, by which the measurements of oil-film pressure and metal strain were made.

(2) It was found that the thin-film element formed with the sputtering target having the composition rates of 70.5 Ni, 18.5 Cr and 11.0 Al (Wt %) is effective as the sensor element having low sensitivity to pressure and temperature, while having adequate sensitivity to strain.

(3) As a result of determining the strain sensitivity of each sensor material by means of the beam, it was found that the strain sensitivity is 0.6 to 0.7 for Manganin, and 1.3 to 1.4 for Ni–Cr–Al.

(4) The thin-film Manganin pressure sensor is affected by the strain of the bearing sliding surface as it is sensitive to strain. More accurate measurement of oil-film pressure was allowed owing to the formation of the thin-film strain sensor at the same location as that of the pressure sensor, and the correction of measurement errors causd by strain.

(5) By measuring the strain on the bearing metal surface by means of the thin-film strain gage, it was found that some negative strain occurs at the crank angle at which mainly the cylinder pressure increases. It was found, however, that the strain decreases sharply where the fourth cylinder is working, though the negative pressure increases further due to the increase in main bearing load caused by the increased engine load where the combustion pressure in the third cylinder is working.

References

[1] Mihara, Y. et al., Development of Measuring Method for Oil Film Pressure of Engine Main Bearing by Thin-Film Sensor, JSAE Review, Vol. 16, No. 2 (1995).

[2] Kaneoya, R., Studies of a High Accuracy Ni–Cr Thin-Film Resistor (in Japanese), IEICE Trans., Vol. 52-C, No. 11 (1969).

[3] Nakamura, M., Study on the Measurement of Instantaneous Temperature and Development Thin-Film Pressure sensor (in Japanese), Master's thesis of Musashi.I.T. (1995).

[4] Shinohara, J. et al., Application of Film Technology for Sensors (in Japanese with English summary), Ishikawajima-Harima Engineering Review, Vol. 29, No. 5 (1989).

[5] Mihara, Y. et al., A Study of Main Bearing Load and Deformation in a Multi-cylinder Internal Combustion Engine (in Japanese with English summary), Proc. JSAE, No. 943 (1994).

971103

An Inductively Coupled Method for Remote Tire Pressure Sensing

Robert C. Mortensen, Larry D. Ridge, and Randy J. Hilgart
SSI Technologies, Inc.

Mark B. Monson and Robb A. Peebles
Locus Inc.

ABSTRACT

This paper describes a method for measuring and transmitting tire pressure and temperature using solid state temperature and pressure sensors embedded in each tire. Data is transmitted with closely coupled field antennas, one mounted to the wheel rim and the other rigidly mounted in the corner assembly. The sensors are remotely powered by resonating the antenna in the corner assembly at a low frequency. Digitized sensor data is transmitted back to the antenna in the corner assembly.

This approach is advantageous for several reasons. The use of low frequency field antennas eliminates the crosstalk with other tires, systems, or vehicles. The vehicle battery powers the system rather than a battery located in the wheel. The system is capable of providing continuous measurement with a high update rate. Finally, the system electronics can be reduced to custom integrated circuits, resulting in components which are small, light weight, and cost competitive.

INTRODUCTION

A vehicle's driving safety characteristics and fuel economy are partially a function of proper vehicle tire inflation. The obvious importance of proper tire inflation has created a need for continuous, real time information regarding tire pressure and temperature. Several companies have developed systems using sensors and radio frequency (RF) communication techniques. These systems are prone to unwanted cross talk with other vehicle electronic systems. The RF based systems are powered remotely by a battery and subsequently have a reduced update rate in an effort to conserve battery life. Another system utilizes the wheel speed sensors of the Anti-Lock Brake System (ABS). This system, while low cost in nature, does not have satisfactory pressure sensing resolution and is not compatible with runflat or low profile tires.

A solution to the deficiencies of using RF systems or ABS based algorithms is a system which is based on low frequency field antennas which inductively couple power between an antenna in the corner assembly and a transponder/sensor located within each vehicle tire. The objective of this paper is to explore the limitations of the current state of the art and describe an approach in which real time, continuous data can be provided in an economical manner.

TECHNOLOGY BACKGROUND

Currently, there are several commercial approaches for determining tire pressure and temperature, all with performance limitations.

One approach utilizes the ABS wheel speed sensors to compare the effective rolling radius of an individual tire to that of the other three tires. Complex algorithms are applied to discount natural wheel speed differentials which are encountered while turning corners and performing other normal driving maneuvers. There are numerous limitation of this technology which may limit its penetration in the automotive marketplace. These limitations include:

- To date, system engineers have been unable to meet the goal of 13.8 kPa resolution for measuring tire pressure. Current state of the art provides 55.2 kPa resolution.
- The system does not provide tire identification. It merely indicates the presence of a deflated tire. The driver must manually check each tire for proper inflation to determine which tire is in question.
- System response is poor. Given the complicated nature of the system algorithms, numerous self checks are required to ensure that the system does not misdiagnose a tire fault. This results in a slowing of system response to the driver. If a tire rapidly deflates, the system may not prompt the driver in a timely fashion.
- The system is not compatible with runflat tires or low profile tires.
- The system does not provide pressure and temperature data.
- The car must travel a speed greater than 4.8 km/hr to function.

Another technical approach involves RF based communication. The RF based systems utilize a switch or sensor to determine in-tire pressure levels. The switch or sensor can be mounted to the rim or within the tire valve stem. A signal is transmitted from the tire to the on board RF receiver which redirects the signal to the driver information display module. Unlike the ABS based system, this system has the advantage of providing actual pressure measurements. Much like the ABS based system, there are numerous performance problems which are inherent to RF technology.

- The potential for unwanted cross talk with other wheels, systems, and vehicles.
  The power source for RF communication is a battery packaged within the tire. This is problematic for several reasons:
  - The battery has a limited life and may become a service issue. Some RF system suppliers recommend replacing the batteries at the same time as tire changes.
  - To extend the battery life, performance requirements have been relaxed. The system may not operate at speeds below 5 km/hr and the update rate has been reduced as low as 1 reading per minute. The systems may not provide an output unless the tire goes below a predetermined threshold.
  - Costly lithium batteries are required to reach a 10 year life.
- Wheel identification is provided through the use of unique identification codes at each wheel, but if wheels are rotated the system must be manually reset, resulting in another service procedure.
- The package containing the pressure sensor and RF electronics is large and heavy, primarily due to the weight of the battery. The system may require a unique counterbalancing method at each wheel.

The inductive coupling method for determining tire pressure and temperature described herein avoids the performance problems with existing technologies and does so in a cost competitive manner.

THEORY OF OPERATION

The purpose of the Remote Tire Pressure Sensing system is to provide information regarding the pressure and temperature within a vehicle tire. A drawing illustrating the configuration of the system is shown in Figure 1.

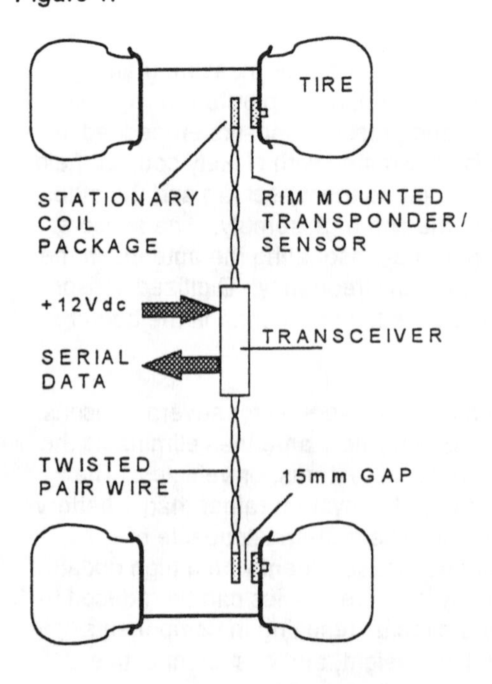

Figure 1

RTPS System Configuration

The system consists of 3 sub-systems: the transceiver circuit, a low frequency field antenna located in each corner assembly, and a transponder circuit located on each rim. The system operates using low frequency field antennas located in each corner assembly and each tire. The antenna located in the corner assembly is referred to as the transceiver antenna and the antenna located in the tire is referred to as the transponder. Communication and power transfer occur simultaneously as the transponder and transceiver antennas pass with each wheel rotation.

The transceiver circuit and antenna simultaneously provide power, and receive data from a remotely located transponder. The transceiver (capable of controlling all 4 vehicle tires) creates a 180 kHz drive frequency which provides power to each transponder. The return signal from the transponder is a Bi-Phase Shift Keyed (BPSK) 90 kHz carrier. As the data packet is received from the transponder it is checked for errors, formatted, and provided to the vehicle ECU. The system can provide either an analog or digital signal. If the data packet contains errors, it is discarded and the next data packet is used. The loss of one or even several of these packets does not significantly affect system performance. The system transfers numerous packets (30/second at low vehicle speeds) each wheel revolution.

The transponder measures the pressure and temperature of the gas within a vehicle tire and transmits the data to the transceiver. The schematic of the transponder circuit is shown in Figure 2.

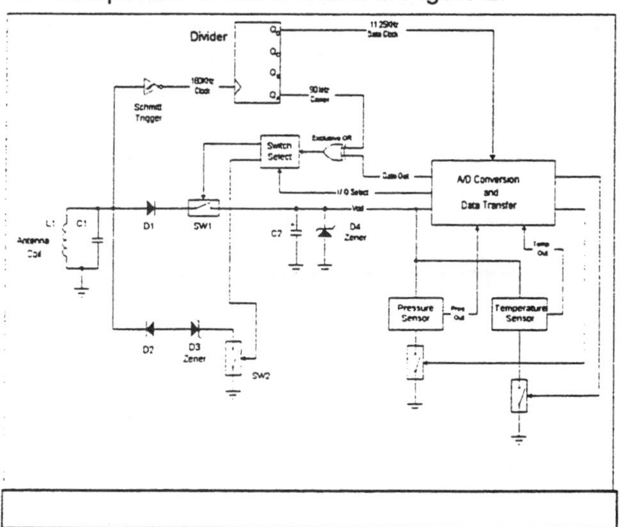

Figure 2
Transponder Schematic

The transponder circuit consists of four main sections: the power supply, sensors, microprocessor, and data modulation.

The power supply is created by coupling energy through antenna coil L1. The antenna is resonated with C1 at the drive frequency f_{drive} (180kHz). The coil voltage peaks are rectified through diode D1, charging the supply capacitor C2. Vdd is switched on when C2 reaches a predetermined voltage value (V_{on}), and is limited to V_{zener}.

Pressure and temperature sensor readings are made only when the supply voltage is above V_{on}. Since the voltage can be anywhere between V_{on} and V_{zener} when acquiring a reading, both sensor and the analog to digital converter (ADC) are ratiometric. To conserve energy, each sensor is powered with a pulse long enough for its output to settle and for the ADC to acquire a reading.

Currently, a microprocessor controls the transponder. The microprocessor includes analog to digital capability and an externally clocked serial port. The basic operation of the processor is to:

1. Turn on sensor switches to power the pressure and temperature sensors.
2. Measure the pressure and temperature using the A/D converter.
3. Digitize the analog data.
4. Copy the data to the serial port.

The return signal from the transponder is a Bi-Phase Shift Keyed (BPSK) 90 kHz carrier. The serial communication interface (SCI) is set up for 9 bit data transmission. By transmitting 2 consecutive 9 bit data words, the SCI can simulate an 18 bit data word to the transceiver. Each 18 bit data word provides information regarding bit error detection and either pressure or temperature sensor bits. The SCI alternates pressure and temperature readings.

Some of the key features of this system are

- ***Low Frequency Magnetic Field Communication...***The magnetic field falls off within inches of the stationary antenna, eliminating the possibility of unwanted crosstalk with other wheels, systems, or vehicles.
- ***Automatic Wheel Identification...***The stationary coil provides the means for wheel identification so that tires can be rotated without loss in the ability to identify tire location.
- ***High Update Rate...***The RTPS system provides continuous information to the vehicle computer. The system provides pressure and temperature output every revolution at low speeds.
- ***No Secondary Power Source...***The power source is the car battery which delivers power through the stationary coil antenna. This is superior to the RF based systems which use costly lithium batteries located in each wheel.
- ***Digital Based System...***The control circuit facilitates system diagnostics for analyzing and pinpointing system faults. Each pressure reading undergoes error detection to ensure that a false signal is not sent to the vehicle computer. The digital approach also facilitates programming changes to accommodate customer preference.

- ***Continuous Operation...***Many systems provide information only when pressure falls below a predetermined threshold. RTPS provides data every 1-4 wheel revolutions. This will facilitate an active display for the instrument panel.
- ***Active Wheel Speed Output...***RTPS can provide wheel speed information to be used for self test features of other systems such as the speedometer or ABS wheel speed sensors.
- ***Low Mass & Size...***The mass of the production intent transponder in each wheel will be less than 25 grams, below the threshold which requires special counter balancing measures.

RESULTS

PRESSURE MEASUREMENT-The graphs in Figures 3 and 4 illustrate to system performance over the full scale pressure ranges. As can be seen from the graphs the system is capable of very linear, accurate measurement of pressure. Graphs of the temperature function would show similar accuracy. Subsequent analysis of the electronics has proven that the error induced by the electronics is negligible. Therefore, system performance will be a function of the performance of the pressure and temperature devices.

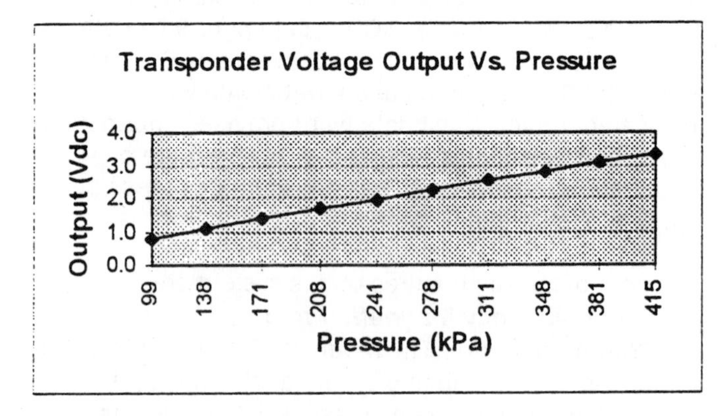

Figure 3
Transponder Output Voltage vs. Pressure

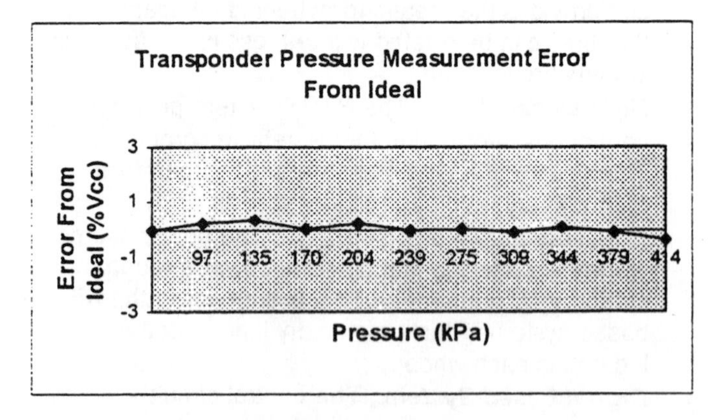

Figure 4
TRANSPONDER ERROR FROM IDEAL

SUPPLY CURRENT VS. GAP-One key aspect of the system is the ability to generate sufficient power to drive the transponder circuit. Power transfer capability was evaluated using a variety of mechanical constructions, varying drive voltages, and varying gaps between the respective field antennas. Figure 5 illustrates how the current supplied to the transponder varies with mechanical gap between field antennas. These results indicate a robust ability to transfer power. Irrespective of the gap distance, the current supplied to the transponder is well above the current consumption of the transponder circuit. The current consumption to supply ratio will be even more favorable when the transponder electronics are reduced to an integrated circuit.

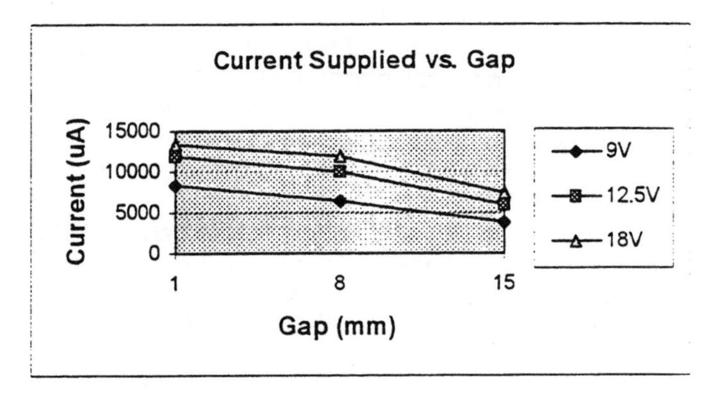

Figure 5
Current Supply vs. Antenna Gap

CONCLUSION

This paper describes an alternative approach to remotely measure tire temperature and pressure. The application of inductive coupling to facilitate tire pressure and temperature measurement addresses the weaknesses found in alternative technologies, such as RF based systems and ABS based algorithms. Issues such as unwanted field communication, battery life limitations and reduced system performance are addressed with the system described in this paper.

Future activities will emphasize commercializing the core technology. The circuits used for the transceiver electronics and transponder will be reduced to custom integrated circuits. The component package size for the antenna and transponder will be miniaturized for acceptance in most automotive applications.

Further, this technology has application beyond that of tire pressure and temperature sensing. In essence, RTPS is a technology for remotely sensing the physical properties of any structure which is spinning. For instance strain gage applied to a rotating shaft could remotely monitor stress strain characteristics of the shaft. An accelerometer could be applied the a rotating shaft to determine such common problems as dull cutting tools or worn bearings. The combination of solid state technology with remote measurement offers numerous opportunities in the automotive and industrial markets.

REFERENCES

[1] "Run-Flat Tires will Jack Up Tire-Pressure Monitor Sales", The Hansen Report on Automotive Electronics, vol.7, No. 5, June 1994, p. 1, 6

[2] "Sumitomo develops deflation warning system", Tire Business, July 1991, p. 16

[3] Hiroaki Nishimura, Masaaki Katsumata, "Development of Digital Tire Pressure Display Device", SAE paper 851237 (1985)

[4] Patrice Gay and Denis Bugnot, "Real Time Tire Pressure Electronic Monitoring System with Dashboard Possibilities", SAE paper 890106, pp. 241-250

962200

Touch Mode Capacitive Pressure Sensors for Automotive Applications

Wen H. Ko and Qiang Wang
Case Western Reserve Univ.

ABSTRACT

The principle, simulation, design, characteristics and application of touch mode capacitive pressure sensors for embedded monitoring of tire pressure are presented. In touch mode operation, the diaphragm of the Capacitive pressure sensor is touching the substrate structure. The advantages of this mode of operation are: near linear output, large over-range pressure and robust structure that make it capable to withstand harsh industrial field environment. When properly packaged, the device can be used to measure fluid flow, force, acceleration, and displacement, etc. in automotive applications.

INTRODUCTION

Capacitive pressure sensors are known to have no turn-on temperature drift, high sensitivity, robust structure and less sensitive to environment effects. However its output is nonlinear with respect to input changes and the sensitivity in the near linear region is not high enough to ignore many stray capacitance effects. In the normal mode of operation, the diaphragm is kept at a distance away from the substrate as shown in Fig. 1-a. If the sensor is deigned to operate in the pressure range where the diaphragm is allowed to contact the substrate with a thin layer of insulator (t_m), as shown in Fig. 1-b, then the device is a touch mode capacitive sensor. The touch mode device was developed to withstand harsh industrial environment, and with one or two orders of magnitude higher sensitivity than the normal mode near linear operation, so that some of the stray capacity effects can be neglected. This paper presents the principle, computer simulation, design, characteristics and application of touch mode capacitive pressure sensors. The sensors can be embedded in the tire or packaged as conventional sensors.

PRINCIPLE

The basic element of a capacitive pressure sensor is an equivalent parallel plate capacitor with clamped edges, where the diaphragm would deform responding to a differential pressure applied to two sides of the diaphragm, as shown in Fig. 1. The capacitance, neglecting the fringe effect, is

$$C = \varepsilon \frac{A}{d} \tag{1}$$

where ε is the permittivity of the media between the two plates; A is the area of the electrode plate; d is the gap space between the two plates. The upper plate of the capacitor, known as the diaphragm, deforms when a differential pressure between the external environment and the inside chamber is applied. The general equation relating the deflection of a rectangular diaphragm, without residual stress, in normal operation region can be expressed as:

$$\begin{aligned} &D\left(\frac{\partial^4 w}{\partial x^4} + 2\frac{\partial^4 w}{\partial x^2 \partial y^2} + \frac{\partial^4 w}{\partial y^4}\right) \\ &= P + N_x \frac{\partial^2 w}{\partial x^2} N_y \frac{\partial^2 w}{\partial y^2} + 2N_{xy}\frac{\partial^2 w}{\partial x \partial y} - x\frac{\partial w}{\partial x} - y\frac{\partial w}{\partial y} \end{aligned} \tag{2}$$

where D is flexural rigidity, $D=E/[12(1-\nu^2)]$; P is the differential pressure; w is deflection at point (x,y); Nx, Ny and Nxy are direct and shear stresses in a plane parallel to the plate surface; E is Young's modulus; and ν is Poisson's ratio.

In the normal operation mode of a capacitive sensor, the diaphragm does not contact the substrate electrode. The output capacitance is nonlinear due to its inverse relationship with the gap ($d_0 - w$), (d_0 is initial gap), which is a function of pressure P, as given in equation (2). This nonlinearity becomes significant for large deflection, [$(w_o/h) > 0.3$, w_0 is center deflection], and large sensing capacitance regions. Many efforts have been made to reduce the nonlinear characteristics of capacitive sensors either by modifying the structure of sensors or by using special non-linear converter circuits. In the touch mode capacitive sensor, as shown in Fig. 1-b, the major component of the sensor capacitance is that of the touched area where the effective gap is the thickness of the thin insulator layer, t_m, on the substrate electrode. In this touch mode operation region, the capacitance varies with pressure nearly linearly and the sensitivity (*dC/dP*) is much larger than that in the near linear region of a normal mode device. A typical C-P characteristic of a capacitive pressure sensor covering normal and touch mode regions is shown in Fig. 2. It has four regions, i.e. normal, transition, linear and saturation regions. The touch mode capacitive pressure sensors (TMCPS) operate in the region III-linear region. They were developed to meet the manufacturing and operational conditions of industrial applications. Its cross-section is shown in Fig. 1-b. After the diaphragm touches the substrate and as the pressure increases, the sensor capacitance is mainly determined by the capacitance of the touched area instead of the capacitance in the untouched "normal operation area". In region III, Fig. 2, the change of touched area is almost proportional to pressure, thus the C-P characteristics is nearly linear. The support of the substrate to the diaphragm after touch, enables the devices to have very large over-load protection. In summary, the advantages of TMCPS are: nearly linear C-P characteristics, large overload protection, high sensitivity and simple robust structure that can withstand industrial handling and environment.

COMPUTER SIMULATION — FINITE ELEMENT MODELING

It is important to have a good understanding of the deflection, stress and strain of the diaphragm in order to design capacitive pressure sensors properly. The output capacitance of a capacitive pressure sensor can be calculated by integration over the deformed diaphragm with a series of equivalent parallel plate elementary capacitors where the effective gap is the difference between the zero pressure gap and the local deflection. In the normal mode, the deflection can be calculated from equation (2) by numerical approximation analysis [1, 2]. In the touch mode device, however, equation (2) is no longer valid. Hence the finite element modeling (FEM) of diaphragms is used to determine the touch mode deflection of the diaphragm. A uniform mesh of 24 x18 elements, with 8 nodes for each element, in a quarter of the diaphragm is used as shown in Fig. 1-c. By geometrically nonlinear computing using ABAQUS, the deflection, stress and capacitance versus pressure at each node can be calculated for a set of sensor parameters *a, b, h,* and *d*, as shown in Fig. 1, and material properties *E* and *ν*. GAP element in ABAQUS is used to model the touch mode operation. Post data analyses based on information obtained on deflection at each node is developed to find the capacitance-pressure behaviors to predict the performance of simulated capacitive sensors [3,4].

From the FEM simulation, the performance of the capacitive sensor can be estimated as the device parameters are varied. The dimensionless deflection at the center point of diaphragms is shown in Fig. 3, where the normalized deflection ratio (w_o/h) is plotted against the normalized pressure (Pa^4/Eh^4) for two (b/a) ratios. The curves in Fig. 3 can be used to determine the touch point pressure (P_t) of known diaphragms by letting the center deflection equal to the initial gap (d_o-t_m). For a set of (*a,b*) and diaphragm material, the touch point can be selected by (*h*) and (*d*) using curves in Fig. 3. After touch, the stress in the touched area is tensile and the stress in the untouched area has three sub-regions as shown in Fig. 4. Fig. 5 shows the deflection along y-axis as the pressure is increased in equal steps. The deflection along x-axis has similar behavior. Beyond the touch point, the shape of the untouched portion of the diaphragm remains nearly the same. However, the touched area *A* equals to ($a + bP - cP^3$) where a,b,c are constants [3]. *A* increases nearly linearly at first, then approaches saturation for larger pressures. This explains the C-P characteristics in region III and IV in Fig. 2.

The average sensitivity (*dC/dP*) is defined as the slope of the straight line obtained by linear curve fitting over the working pressure range. The nonlinearity is defined as

$$\frac{|\Delta C_+| + |\Delta C_-|}{2(C_{max} - C_{min})} 100\% \qquad (3)$$

where ΔC_+ and ΔC_- are the maximum and minimum deviation from a straight line characteristics, respectively; C_{max} and C_{min} are the maximum and minimum capacitance in the operation range, respectively.

The sensitivity is proportional to ($1/t_m$). The sensitivity is also inversely proportional to the thickness of the diaphragm for a fixed touch point as shown in Fig. 6. Fig. 7 shows (*dC/dP*) sensitivity and nonlinearity versus touch point pressure of a typical device.

These simulation results can be used to predict the trend of the sensor's performance when the device parameters are varied. The understanding and quantitative curves can be used as a design tool to arrive at a specific desired performance. The results of the diaphragm deflection and stress in both touched and untouched regions are useful for micro-actuators incorporating diaphragms.

FABRICATION AND PACKAGE

For molded in-package pressure monitoring used in industries, such as tires, the sensors have to survive high temperature, high pressure manufacturing processes, and to operate in harsh environment of repeated shock, vibration, stress, and temperature cycling for tens of years. It is desired to have high sensitivity and good linearity to simplify the reading circuitry. The TMCPS was developed to meet these requirements. A rectangular diaphragm was selected to reduce the effect of the maximum stress near the corners on the sensor performance. With the help of finite element modeling, the parameters of TMCPS can be designed to meet device performance specifications and fabrication constrains. A flow chart indicating major fabrication process steps for the sensor chips is given in table 1. An isolation layer, t_m, is deposited over the substrate electrode to provide isolation in touched range and to aid the anodic bonding of silicon to the glass substrate to form a hermetically sealed cavity for absolute pressure measurement. This layer plays an important role in the TMCPS. Sputtered #7740 Pyrex glass is used as the insulating layer. The substrate material could be glass or silicon. A special process was developed to fabricate hermetic electrical feed-throughs to connect the electrodes inside the hermetically sealed cavity to external bonding pads. With carefully selected thickness of the isolation layer, #7740 glass substrate with thin sputtered glass can be anodically bonded to the silicon to achieve hermetical sealing [5].

The tire sensor chip is to be packaged on printed circuit board or ceramic substrate together with other components of the system. Multiple layer package process was developed to protect the sensor from shock and interfering stress, and yet to transmit the pressure within the desired accuracy over the life of many decades. The resulting package is shown in Fig. 8. It was molded into rubber material at 300°C, 400% over-pressure for several hours and then evaluated in the field with no degradation in performance [5,6].

EXPERIMENTAL RESULTS

Hundreds of TMCPS, with device parameters and material properties given in Table 2, have been fabricated in Electronics Design Center, CWRU, for laboratory and field tests in the last two years. A typical P-V characteristic is shown in Fig. 9. The CP-10 CMOS C-V converter circuit [7] is used in the sensor evaluation to convert capacitance into voltage. The measured non-linearity, as defined in equation (3), is 0.6% in the rang of 80-110 PSI.

The hysteresis was found in some sensors. The hysteresis of sensors is defined as the difference between the outputs when the pressure is cycled upward and downward, which can be expressed as follows:

$$H = \frac{\max|V_F - V_B|}{V_{max} - V_{min}} 100\% \qquad (4)$$

where V_F and V_B are sensor's outputs when the forward and backward pressure are applied, respectively.

The hysteresis of TMCPS is mainly due to the quality of the isolation layer. Since the diaphragm presses against this layer during the operation, non-elastic deformation of the sputtered glass layer results in the sensor hysteresis. However, after a few high pressure cycles (400% over-pressure), the glass layer is stabilized hence the hysteresis is reduced significantly. Two hysteresis curves of a typical sensor is shown in Fig. 10. After a few 400PSI pressure cycles, the hysteresis can hardly be observed. The sensors have been successfully molded in the industrial packages and are undergoing simulated life tests in proving grounds. Fig. 11 shows a TMCPS' P-V characteristic before and after molding process, which consists of 300°C and 400% of full scale over-pressure for one to two hours [6].

CONCLUSION

The FEM and computer simulation, design, fabrication, evaluation and field testing of touch mode capacitive pressure sensors in industrial applications are reported. The absolute pressure measuring device is demonstrated to have the advantages of good stability, low power consumption, robust structure, large overload ability and high (*dC/dP*) sensitivity. The basic device design can be used to measure pressure from 10^{-4} to 10^{3} PSI full scale with only changes of several process parameters. It can withstand manufacturing package temperature up to 300°C for several hours after packaging, and may have 200% to 200,000% full scale over pressure protection. This new mode of operation may supplement the piezoresistive and normal mode capacitive pressure sensor in industrial applications where mechanical and electrical stability are important.

The initial step in developing a CAD simulation program for capacitive pressure sensor has been made. The modeling of touch mode capacitive sensors (neglecting the build-in stresses of the diaphragm) was developed to obtain a set of data base for a family of sensors over a range of device parameters. From the data accumulated, the effects of device parameters on the device performance can be estimated. These results would be useful to capacitive sensor designers in estimating the performance of device before fabrication and also may serve as a guide for adjustment of parameters to arrive at a desired performance in mass production.

With proper package, the fluid flow, force, acceleration and displacement can be converted into pressure. Therefore the device can be used to monitor measure flow, force, acceleration and displacement in automotive applications. The design and simulation of the diaphragm would be applicable to diaphragms for other sensors, actuators, and microsystems. The understanding on the deflection and stress of a diaphragm under various loads, with small or large deflection, will be useful for diaphragm design to accurately predict the behavior of the element under normal or touch modes operation.

Reference

1. S.P. Timosheko and S. Woinowsky-Krieger, Theory of Plates and Shells, 2nd Ed. (McGraw Hill, New York, 1970).
2. M.D. Giovanni, Flat and Corrugated Diaphragm Design Handbook, (Mercel Dekker, New York, 1982) p. 178-206.
3. Xiang X. Huang, "Modeling of Large Deflection and Touch Mode Capacitive Sensors", MS Thesis, EEAP Dept., Case Western Reserve Univ., Cleveland, Ohio, USA, (1994).
4. Xiaoyi Ding, "Mechanical Properties of Silicon Films and Capacitive Microsensors:, Ph.D. Thesis, CWRU, May 1990
5. Yang Wang, "Industrial Capacitive Pressure Sensor", MS Thesis, EEAP Dept., Case Western Reserve Univ., Cleveland, Ohio, USA, (1994).
6. Wen H. Ko, "Capacitive absolute pressure sensor for industrial applications", Patent No. 5,528,452, June, 1996
7. W.H.Ko and G.J.Yeh, "An integrated interface circuit for capacitive sensors", *Microsystem Technology*, (Springer International) Vol.1, No.1, Oct. 1994, pp.42-47. Patent No. 4,820,971, April, 1989

Table 1. TMCPS Process Flow Chart

Step	Name	Description
Silicon Process		
1	Oxidation	Generate masking layers
2	Photolithography I	Define Gap area
3	RIE Silicon etch	Form gap
4	Boron diffusion	Define the thickness of diaphragm
5	Photolithography II	Define the etch back area
Substrate (glass) Process		
6	Electrode I	Form substrate electrode by sputtering Cr/Pt/Cr
7	Isolation	Sputter 7740 glass over the substrate electrode
8	Electrode II	Form contact pad for the diaphragm
9	Pad	open windows for wire bonding pads
Final Process		
10	Anodic bonding	Bond silicon wafer to the substrate wafer
11	Silicon etch	Form diaphragm
12	Dicing	Dicing sensor chip

Table 2. Touch Mode Capacitive Pressure Sensor Parameters

Parameter	Symbol	Value	Unit
Diaphragm Length	a	1500	μm
Diaphragm Width	b	457	μm
Diaphragm Thickness	h	5.0-8.0	μm
Gap	g	5.0-8.0	μm
Insulator Layer Thickness	t_m	0.01-3.0	μm
Young's Modulus of Si (P^+)	E	1.3E+11	Pa
Poison's Ratio	ν	0.3	

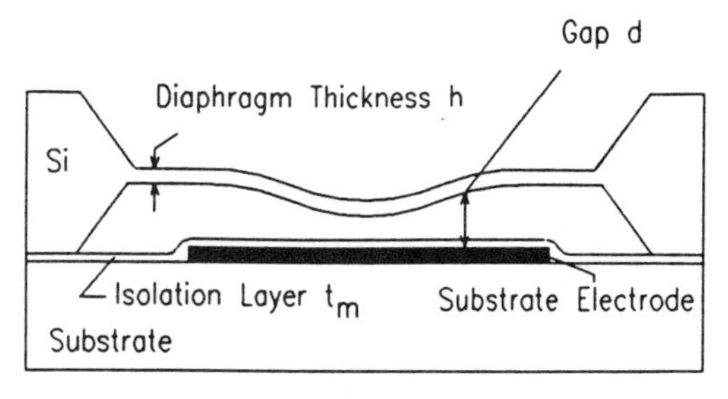

(a) Normal mode of operation

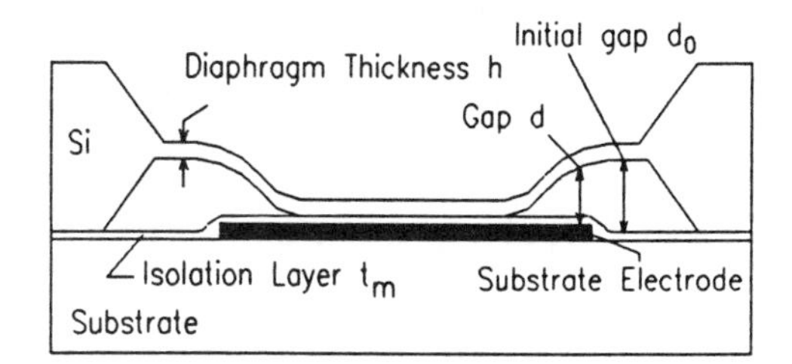

(b) Touch mode of operation

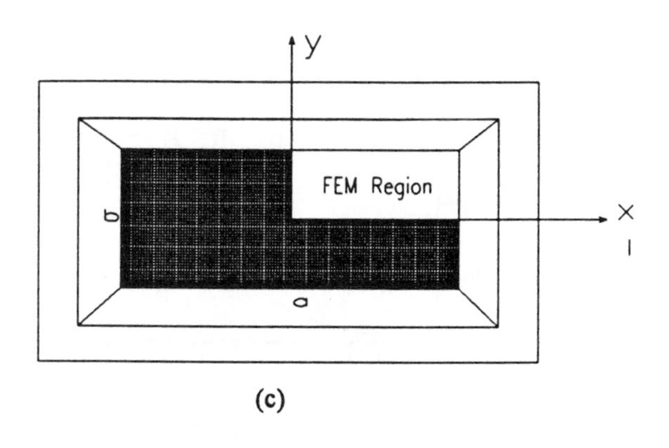

(c)

Figure 1 .Typical capacitive pressure sensor, structure a,b - dimension of the diaphragm; d - gap, h - thickness;

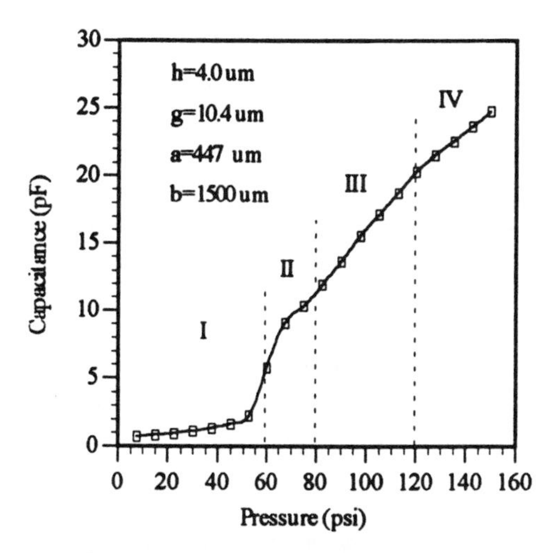

Figure 2. Typical C-P Characteristics of a TMCPS
I normal region ; II transition region;
III linear region; IV Saturation region.

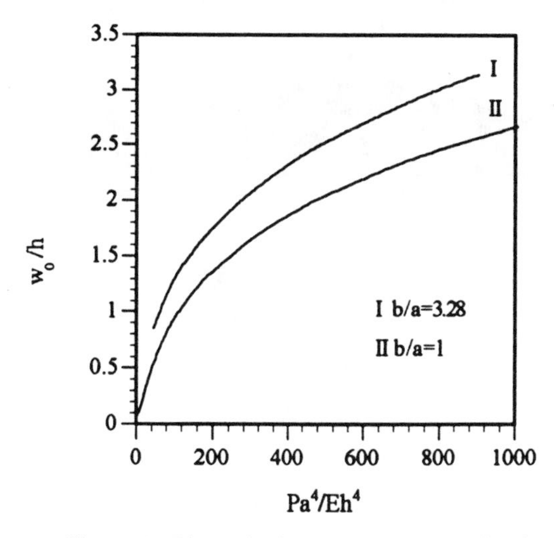

Figure 3. Dimensionless Center Point Deflection

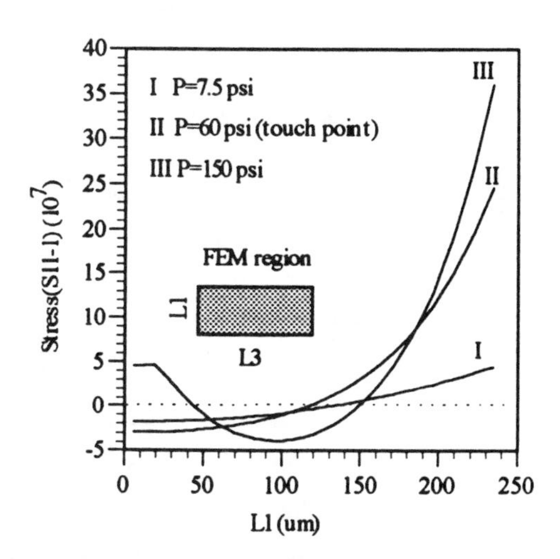

Figure 4. Stress versus coordinate L1 Touch point P_t=60 psi
a=457 μm b=1500 μm h=6.0 μm g=7.95 μm

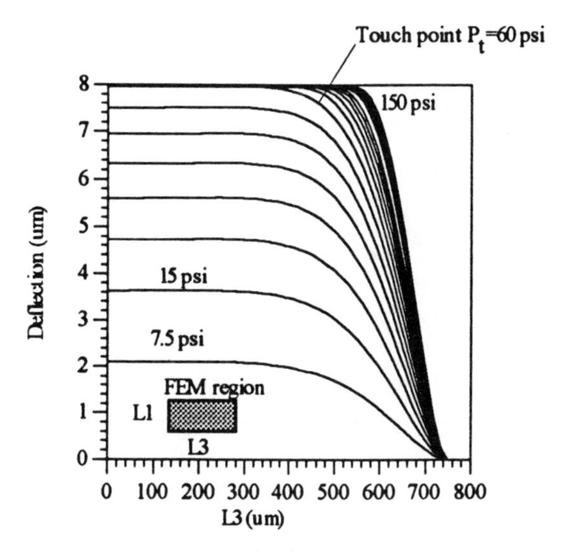

Figure 5. Deflection with different pressures along coordinate L3
a=447 μm b=1500 μm gap=7.95 μm h=6.0 μm

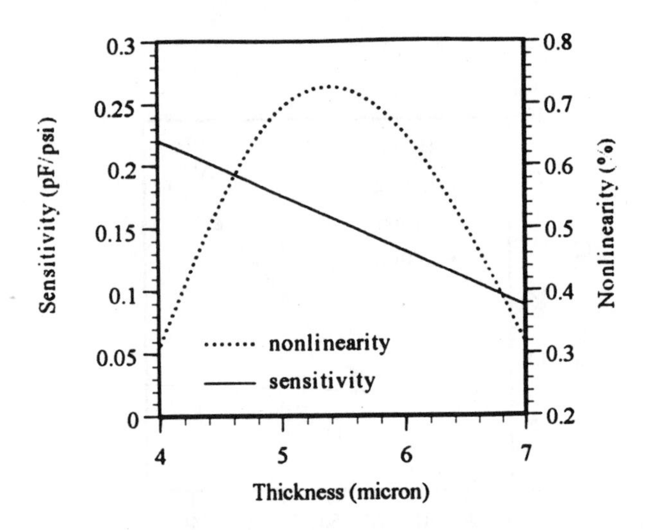

Figure 6. Sensitivity and nonlinearity versus diaphragm thickness

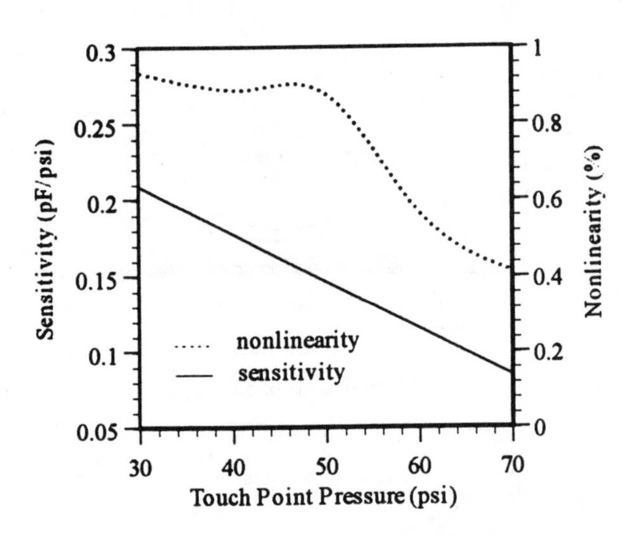

Figure 7. Sensitivity and nonlinearity versus touch point P_t

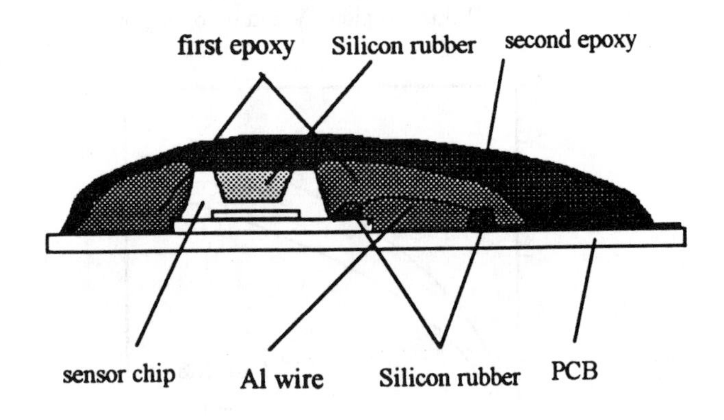

Figure 8. Package of TMCPS for the industry application

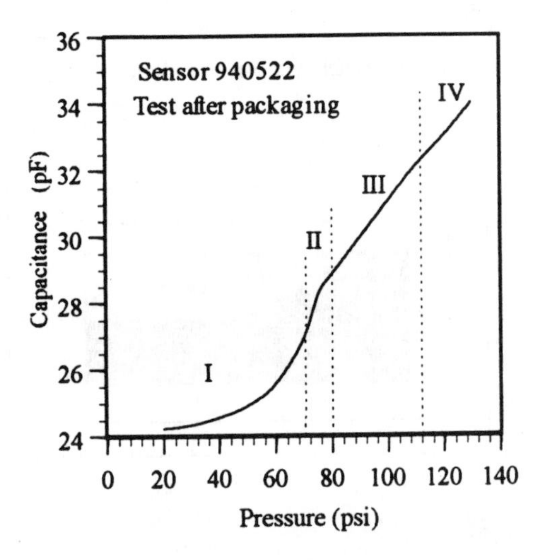

Figure 9. Typical C-P characteristics of a TMCPS
I normal region II transition region
III Linear region IV saturation region

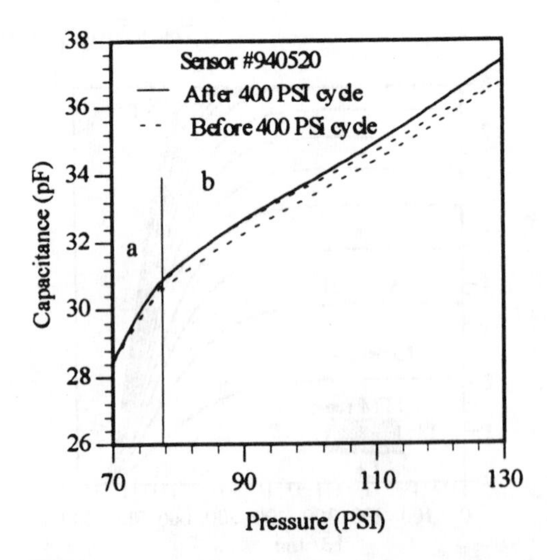

Figure 10 Hysteresis in the touch mode operation, region (a) - normal mode; region (b) - touch mode

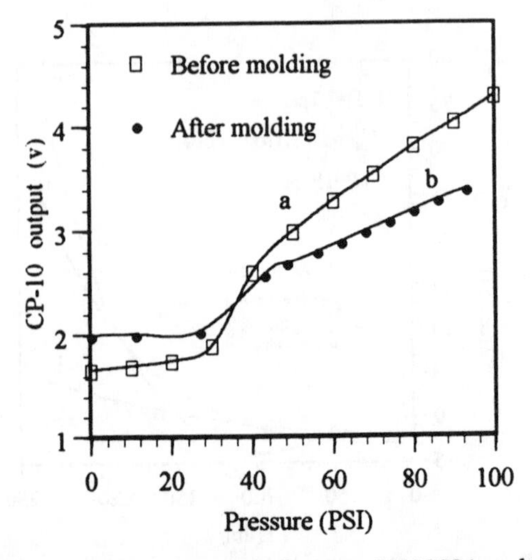

Figure 11. Field test results of sensor #9210084 molded in an industrial package. Curve (b) includes some pressure attenuation due to overall package.

952084

Multi-Channel and Portable Fiber Optic Combustion Pressure Sensor System

Thomas Poorman, Sergiey Kalashnikov, and Marek T. Wlodarczyk
OPTRAND INC.

Adam Daire, Wayne Goeke, Richard Kropp, and Pradip Kamat
Keithley Instruments, Inc.

ABSTRACT

An eight-channel fiber-optic combustion pressure system is described intended for continuous monitoring and control applications in diesel- and natural gas-engines. The portable control/monitoring unit of the system offers capabilities of real-time data acquisition and triggering from a shaft position or TDC sensors. Several processing functions are offered including calculations of peak pressure (PP), Indicative Mean Pressure (IMEP), and location of peak pressure (LPP). The system allows for 50 kHz, burst mode transfer of multi-sensor data to a PC. OPTRAND's commercially available combustion sensors for 1000 and 3000 psi maximum pressure ranges can interface to the unit with optical patch cables available up to 100 meters in length. The system offers 0.1Hz to 15 kHz frequency response, 1% accuracy at constant temperature, and maximum uncooled sensor housing temperature of 300°C. Performance data are presented from seven-month endurance tests conducted at customers' engine cites indicating excellent sensor durability and fatigue-free operation.

INTRODUCTION

In-cylinder pressure transducers are currently being considered for advanced engine control and monitoring systems used in natural gas and diesel-fuel burning engines. Since cylinder pressure is the fundamental thermodynamic variable, it can be used to determine a variety of engine parameters for closed-loop controls. For example, cylinder pressure sensors are commonly used to determine apparent rate of heat release and indicated mean effective pressure (IMEP) [1]. Also cylinder pressure history is used to determine the best air/fuel ratio in closed-loop controls, thereby significantly increasing fuel efficiency and reducing emission levels of polluting gases [2]. In diesel engines in-cylinder pressure sensors are best suited to adjust an engine's operating state on a cylinder-to-cylinder basis based on the start-of-combustion information.

There are two driving forces behind increased interest in continuous cylinder pressure measurement in large-bore engines. The first one has to do with the Clean Air Act Amendments (CAAA) of 1990 which will force most of U.S. operators of Stationary Reciprocating Internal Combustion Engines (SRICE's) to monitor and report their emissions on a continuos basis. A key aspect of the CAAA is that some emission sources will need to be equipped with Enhanced Monitoring/Compliance Certification (EM/CC) capability. EM/CC requires that a mechanism, procedure, or system to be implemented that assures that unit is complying with an emission limit or standard on a continuous basis. While cylinder pressure information does not provide a direct information on emissions, maintaining pressures in all cylinders at nominal levels is a prerequisite for low emission levels [3]. In addition, cylinder pressure is the best predictor for Parametric Emission Monitoring Systems (PEMS) which are used to predict emission levels of combustion emission gasses such nitrogen oxides (NOx), volatile organic compounds (VOCs), and carbon monoxide (CO) without a need of using direct monitoring techniques [4].

The second benefit of continuous pressure monitoring in combustion engines is in the area of engine diagnostics and health monitoring. In-cylinder pressure sensors provide a direct and deterministic misfire detection, while indirect techniques are limited by their inability to distinguish misfire from factors such as incorrect spark-timing and rough driving conditions. Similarly, in-cylinder pressure sensing is best suited to detect high frequency knock signals without being complicated by factors such as cylinder-to-cylinder variability, shock, vibrations, and signal phase-delays, plaguing externally mounted sensors. In the area of engine health diagnostics, pressure trend analysis can extend maintenance schedules and provide warning before developing engine problems.

EXISTING SENSORS FOR CYLINDER PRESSURE MONITORING

Conventional electronic pressure transducers, such as piezoresistive or piezoelectric, are not suited for continuous high temperature pressure measurements encountered in combustion engines. Fundamentally, strain or capacitance gages used in pressure transducers exhibit large unrepeatable and unpredictable changes in gauge output at temperatures typically greater than 125°C or 250°C, respectively. These changes are caused by such effects as alloy segregation, phase changes, selective oxidation, and diffusion, and ultimately lead to premature failure of the gauge or lead wires. While water cooling has been used to increase temperature ranges of piezo sensors, uncooled sensors are preferred for industrial applications. Fiber-optic sensors, on the other hand, are potentially very well suited for high temperature applications. Due to the resistance of fused silica to extreme temperatures, fiber optic sensors can potentially operate at temperatures up to 800°C. Electrical passiveness makes them immune to EMI and ground-loop problems. Due to very low transmission losses from optical fibers, sensor interface devices can be located away from high temperature areas, as far as hundreds of meters.

Several fiber optic sensors have been described in the literature potentially applicable to high-temperature pressure measurements [5]-[6]. However, the majority of these sensors have been designed for specialized military applications and their prices are too high for large scale commercial applications. Among fiber-optic sensors suitable for low-cost commercial applications, the simple intensity-modulated sensor we reported in the past [7] offers the most promise. It utilizes an optical fiber in front of a flexing diaphragm for optical reflection measurement of pressure-induced deflections. By employing this sensing principle coupled with a hermetically sealed sensor structure to eliminate diaphragm oxidation under high temperatures, we recently demonstrated [7] that such a sensor can operate under prolonged exposure to high temperatures, and that the sensor can readily detect misfire or knocking in an automotive engine. Our second-generation sensor system [8], presently available commercially, has been already used in engine research applications and large-bore engine monitoring and control.

FIBER OPTIC PRESSURE SENSOR SYSTEM

The fiber optic pressure sensor system reported here is dedicated to control and monitoring of multi-cylinder engines. It consists of fiber optic pressure sensors connected to a pressure monitoring unit directly or via optical patch cables. The pressure monitoring unit communicates to the engine control system or PEMS in a number of customer dependent protocols and formats. The unit is capable of connecting to up to eight fiber optic pressure sensors. Each of the sensors is mounted directly or via a spark plug into the cylinder and transmits the pressure profile to the unit. The monitoring unit multiplexes one cylinder at a time, collects pressure data at a specified sampling rate and transmits analyzed data to the control system or PEMS.

The monitoring unit, called the MultiPSI 8000, provides capabilities of sensor interfacing, calibration, health monitoring, as well as data acquisition and processing. It can be connected to an engine shaft encoder. The encoder senses a signal pulse (TTL levels) when it detects the Top Dead Center (TDC) as well as a pulse for the holes that are a degree apart for engine speed measurement. When the TDC pulse is received, the data collection for each cylinder starts and data is collected for the specified period. The sampling rate is dependent on the mode of data collection that is selected by the user. The data collected is also stamped with the encoder signal from the holes that are a degree apart. Basic specifications of the MultiPSI 8000 monitoring unit are summarized in Table 1 below.

Measurement Performance:	
Pressure Range:	0 - 1000, 0-3000 psi
Overpressure:	Twice pressure range
Resolution:	12 bits.
Frequency Response:	0.1 Hz to 15 kHz.
Linearity & hysteresis:	± 1 % full scale
Temp. Coefficient:	0.03%/°C (typical)
Measurement Speed:	
Measurement Rate:	Real time: 2000 readings/sec. Burst mode: 50k readings/sec
Measurement Trigger:	Manually from front panel. External over RS232 port, based on encoder TDC signal.
Analog:	0.1 Hz to 15 kHz
Engine Specifications:	
Speed:	to 10000 rpm.
Angular Resolution:	1 degree.
Encoder Output:	L levels 0 to 5 V.
Power:	
Input Power:	12-24 VDC or 110 VAC
Analog Output:	0 - 5 V or 4 - 20 mA
Communications Link:	RS232C.

Table 1. Sensor system specifications.

SENSOR DESIGN

The present sensor operates on the principle of optical monitoring of diaphragm deflection exposed to combustion pressure [7]. Based on the use of a single optical fiber and a metal diaphragm, the sensor head design is robust and low cost. As schematically shown in Fig. 1, it consists of three basic elements: 1) diaphragm, 2) sensor housing, and 3) fiber/ferrule subassembly.

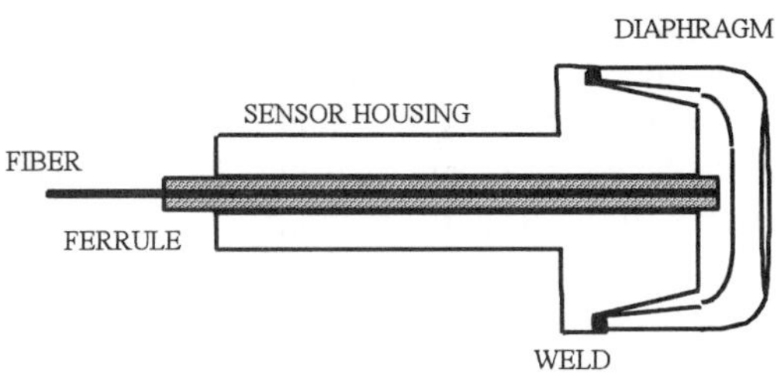

Figure 1. Sensor head construction.

The response of the sensor to pressure depends on two physical phenomena. The first one is displacement of diaphragm due to pressure. As explained in detail previously [7], the dependence of diaphragm deflection on pressure is essentially quadratic with deflection decreasing with increasing pressure. For relatively thick diaphragms required for combustion pressures, deflections are typically very small resulting in highly linear relation between pressure and deflection.

The second phenomenon that affects the present fiber optic sensor response is the dependence of optical signal change on diaphragm-to-fiber separation. It is in general non-linear and depends on fiber size and numerical aperture as well as separation between fiber and diaphragm at zero pressure [7] Depending on the choice of these parameters a highly linear relation between signal change and fiber deflection can be obtained.

From the point of view of sensor reliability the diaphragm is the most critical element of a sensor. It must maintain excellent mechanical properties at extreme operating conditions and function repeatably over as many as tens of thousands of hours. Its reflectivity must also remain nearly unchanged over the sensor life time. Compared to the previously used flat disk construction [7], the present sensor uses a more durable and rugged hat-shape diaphragm with varying thickness across its diameter - as schematically shown in Fig. 1. A high strength alloy (Inconel) has been used as a diaphragm material. This proprietary diaphragm design has been selected so it can withstand hundreds of millions of deflections without yielding or mechanical creep. Other benefits of the present construction include excellent linearity of the pressure response and reduced sensitivity to direct flame effects.

OPERATION OF MultiPSI 8000 MONITORING UNIT

The MultiPSI 8000 monitoring unit consists basically of five components: (1) a mother board, (2) up to 8 channel boards, (3) alphanumeric display, (4) power supply, and (5) an enclosure box with fiber optic and electrical connectors. Figures 2 and 3 show schematic diagrams of the mother board and the channel board.

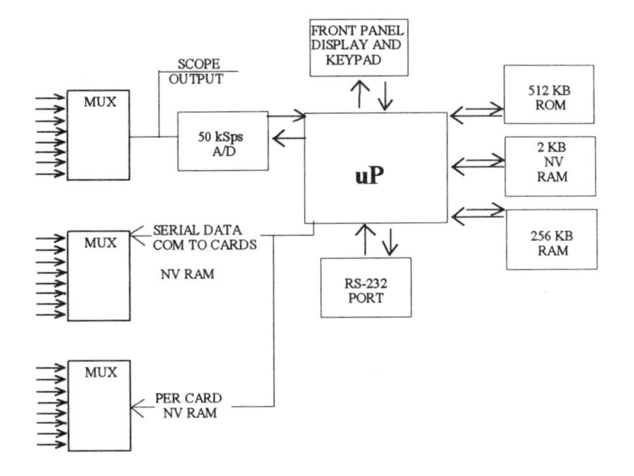

Figure 2. Schematic diagram of MultiPSI mother board

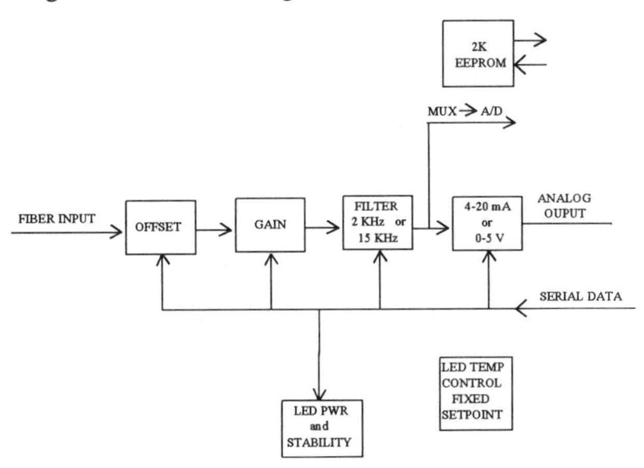

Figure 3. Schematic diagram of MultiPSI channel board

The design of the motherboard is based on Keithly Instruments Digital Voltmeter 2001. The motherboard comes with eight expansion slots for up to eight channel boards - each for a single fiber optic pressure sensor. The mother board provides four basis functions: (1) sensor calibration and diagnostics, (2) data acquisition and sensor output, (3) display control, (4) front panel key pad interfacing, and (4) signal processing, as described in more details in Section 5 below.

The pressure monitoring unit connects to a maximum of eight pressure sensors. The unit can also accept up to four TTL level inputs, two of which are assigned for the flywheel encoder signal. The connection to the control system or a PEMS is via an RS232C port. The unit also has an analog pressure output of either 0 to 5 volts or 4 to 20 mA. It has the capability of storing up to one hundred thousand data points.

The pressure monitoring unit operates in three modes: free running, real time, and burst mode. The user can calibrate the unit for specific sensors, configure the trigger mode, and configure the cylinder scan rate as well as the order of the scan. These modes are described below.

Free Running Mode: In this mode the meter measures the pressure and display it at about five readings per second. These are calibrated readings based on the fact that the meter is calibrated to the sensor. This is the default mode.

Real Time Mode: In this mode the module measures the pressure, calibrates the reading, and outputs the reading as a 0 to 5 volt or 4 to 20 mA signal. This mode is started by pressing the "Real Time" key on the front panel or sending an equivalent command over the RS232C port. The user has the capability of determining how fast data should be sampled by the unit. The typical sampling rate is between 2,000 and 5,000 readings per second.

Burst Mode: In this mode the meter measures up to 50,000 readings per second and stores the readings in memory. A maximum of 50,000 readings can be stored in memory. The burst mode can be started either from the front panel by pressing the "Store" key or over the RS232C port. The burst mode reading rate can be configured by pressing the "Configure" key and then the "Store" key. Data collection and storage are initiated by a TDC signal. After burst mode has been initiated for a given cylinder, readings can be recalled from memory using the "Recall" key. The same functionality can be accessed over the RS232C port. The data collected can be analyzed to provide average peak pressure per cylinder as well as the average angle between the TDC and the peak pressure location for a given cylinder. The averages are computed either over all 50,000 readings or over 100 cycles, whichever the user selects.

Configure Trigger: The user can configure his trigger for the different modes of operation by pressing both the "Configure" and "Trigger" keys simultaneously. The meter can be manually triggered from the front panel when the user presses the "Trigger" key or electronically over the RS232C port. The user can also configure the meter to be automatically triggered by an internal trigger. The other trigger mechanism is the signal from the TDC encoder.

Configure Channel: The user can scan eight channels for a given mode of operation. The user will press the "Configure" and "Channel" keys simultaneously and the display reads "Scan Chan=X,X,X...". The user may scan channels sequentially, as selected, or in any given order.

Calibration:

Each channel on the monitoring unit is calibrated for each sensor. The user specifies the channel number being calibrated and follows a calibration procedure. This requires first disconnecting the sensor and storing the corresponding optical signal intensity. The display then prompts the user to connect the sensor and enter calibration number uniquely assigned to each sensor. At that time the unit and given channel are calibrated to the sensor.

Diagnostic:

The unit on an ongoing basis evaluates the health of the sensor. This involves monitoring current flowing through the sensor needed for calibrated sensor performance. In addition, sensor output signals can be compared over time to their pre-established standards. Signal levels, signal rise rate, mean effective pressure, etc. can be evaluated. If the meter detects sensor problems, the front panel will display the channel effected and communicate the information over the network (RS232C port), if applicable.

Data Analysis:

The monitoring unit passes analyzed data either to the control system or PEMS. Data collection modes have been described above. Data analysis methods per cylinder are described below.

Cylinder Peak Pressure (PP): Peak pressure data is averaged over 100 cycles (PP). The coefficient of variation (COV) is also calculated.
Location of Peak Pressure (LPP): Location of peak pressure with respect to TDC in degrees is averaged over 100 cycles. The coefficient of variation (LPP COV) is also calculated.
Indicated Mean Effective Pressure (IMEP): Average pressure per cycle (IMEP) is calculated and then averaged over 100 cycles. The coefficient of variation is also calculated.
Number of Mis-fires (NM): The unit determines the percentage of mis-fires over the latest samples of "N" cycles for each cylinder.
Number of Engine Revolutions (NER): The unit computes the number of revolutions during the data collection period.

LABORATORY TEST RESULTS

The basic laboratory tests of fiber optic combustion pressure sensors include linearity and hysteresis comparisons against reference piezoelectric and piezoresistive sensors, temperature soaking and cycling, and sensitivity dependence on temperature. Fig. 5 demonstrates a comparison of the typical room temperature dynamic pressure responses, over 0 to 1000 psi pressure range, of OPTRAND's OPS 300-1000 sensor and a reference piezoelectric transducer (PZT) (Kistler Model 6051B).

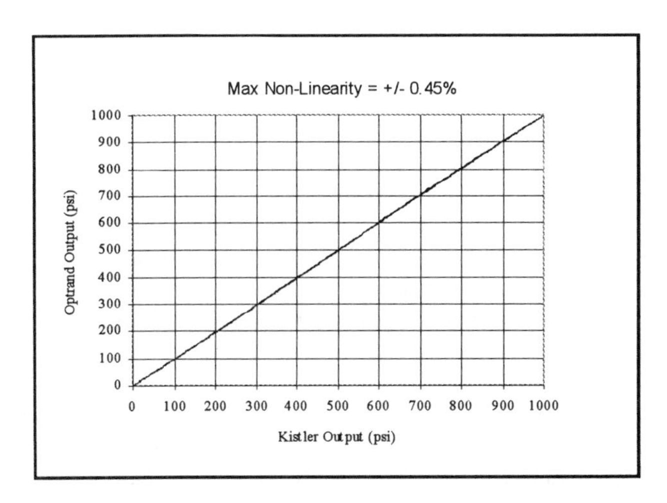

Figure 4. Dynamic pressure comparison between OPS 300 and a piezoelectric transducer.

Typical fiber optic sensor linearity is of the order of +/-0.7% over 1000 psi pressure range. Fiber optic sensors showed negligible hysteresis under room-temperature conditions. When the fiber optic sensor is connected to different interface units, a typical full scale error is of the order of +/- 1%.

Pressure detection sensitivity dependence on the sensor housing temperature was investigated under laboratory conditions. An unheated PZT (Kistler Model 6051) was used as a reference. The fiber optic and the reference sensors were installed in a common pressure chamber. Sensor sensitivity was established by heating the fiber optic sensor housing over a 250 °C temperature range and monitoring peak pressure changes with a data acquisition system. A typical value of temperature coefficient of sensitivity of OPS 300 is around +0.03%/ °C.

The final laboratory test involved a short exposure (few seconds) to the direct effect of propane torch flame. Using a rotating disk with a small opening, both our sensor and a reference piezoelectric transducer (Kistler 5061B) were exposed to a periodic thermal shock due to direct exposure to 800 °C flame. As shown in Fig. 5, both sensors show a typical error of about 0.8%.

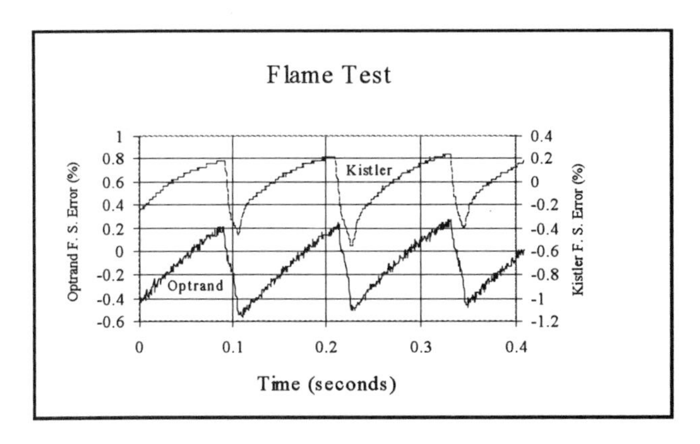

Figure 5. The effect of direct flame heating for OPS 300 and a piezoelectric transducer.

As a part of the experiment, the effect of flame quenching/shield adopter was evaluated. When OPTRAND sensor was installed behind an adopter consisting of a metal disk with a small cylindrical passage [9], the flame effect was virtually non-detectable, as shown in Fig. 6.

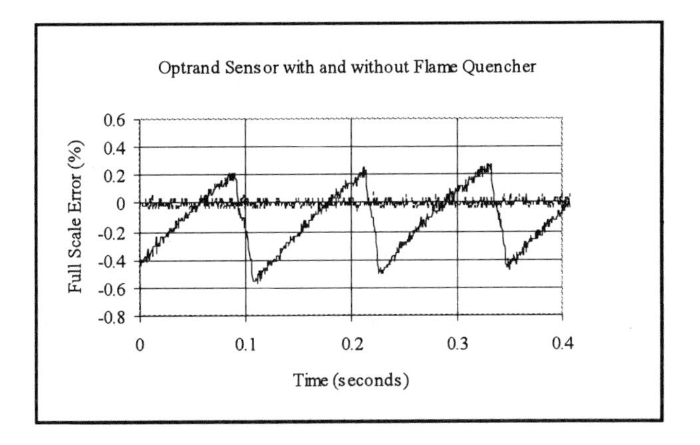

Figure 6. OPS 300 response to direct flame heating with and without a flame quenching adopter.

LONG-TERM ENGINE TEST RESULTS

We report here the results of two continuing field studies conducted on large-bore, stationary engines. One engine is used in a natural gas pipe line compressor station and the other is employed for electricity generation. In the first test OPTRAND's OPS 300-1000 sensors have been installed into cylinder heads of two stroke engines (typically running at 250 RPM and peak pressure around 400 psi) with reference PZTs (Kistler Model 6121) located in different ports on the same engine cylinders. Several sensors have been under tests with one sensor installed about nine months ago and other sensors about seven months ago. Fig. 7 demonstrates a comparison chart for a sensor after 50 Million and 90 Million pressure cycles.

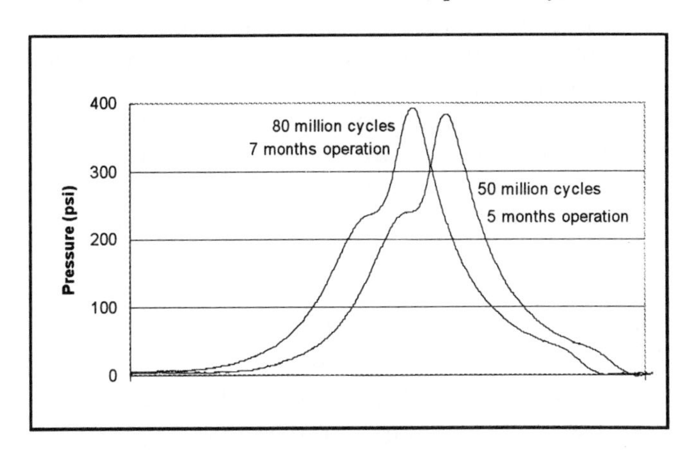

Figure 7. Pressure traces of OPS 300-1000 after 50 and 90 Million of cycles of engine operation

The second test has been performed on a four stroke engine at SouthWest Research Institute, Saint Antonio, Texas. Both sensor and diagnostic outputs are being monitored during this test. The sensor housing temperature was estimated to be around 200 °C. Unlike pipeline engines, the generator engine operates at high speed (1800 RPM) and high pressures (peak pressure around 900 PSI), subjecting sensors to higher stress levels and larger number of cycles. One of the sensors has already accumulated approximately 100 Million pressure cycles over 2,000 hours of operation, without any signs of deteriorating performance. The experiment continues aimed at demonstrating 4,000 hours of operation or about 200 Million of cycles.

SUMMARY AND CONCLUSIONS

We have described the design and performance of a multi-channel fiber optic combustion pressure system dedicated to Parametric Emission Monitoring. The system offers capability of monitoring up to eight sensors and four trigger signals at the same time. The MultiPSI 8000 portable monitoring unit allows for sensor calibration, health monitoring, information input and display, as well as digital data transfer. It can calculate on a real-time basis Peak Pressure, Indicative Mean Pressure (IMEP), and location of peak pressure (LPP). The system allows for 50 kHz, burst mode transfer of multi-sensor data to a host PC.

We have demonstrated performance data collected at two large-bore engines sites over long-term endurance tests. One test was conducted on pipeline compressor engine while the second one was done on an electricity generating engine. Both pressure output data and sensor health monitoring information were collected in one of the tests. Obtained fiber sensor accuracy is comparable to much more expensive piezoelectric reference transducers. A maximum accumulated number of hours for one of the sensors tested on a pipeline engine is around 5000 for sensor housing temperatures ranging from 160°C to 260°C. The corresponding number of pressure cycles is around 90 million cycles exceeding considerably a maximum allowable number of cycles specified by piezoelectric sensor manufacturers. The total number of hours for a generator engine is so far near 2000 hours with around 100 million cycles accumulated. Sensor health monitoring output indicate small fluctuations in sensor current during operation without any long term degrading effects present.

REFERENCES

[1] E. H. Gassenfeit and J. D. Powell, ``Algorithms for air-fuel ratio estimation using internal combustion engine cylinder pressure," SAE paper no. 890300, presented at the 1989 SAE International Congress and Exposition, Detroit, MI, February, 1989.

[2] K. Sawamoto, Y. Kawamura, T. Kita and K. Matsushita, ``Individual cylinder knock control by detecting cylinder pressure," SAE paper no. 871911, presented at the 1987 SAE International Congress and Exposition, Detroit, MI, February, 1987.

[3] "Stationary Emissions Monitoring For Pipeline Engines," W. E. Liss, M. P. Whelan, R. Lott, Gas Research Institute, May 1994.

[4] G. M. Beshouri, "An Assessment of the Effectiveness of Parametric Monitoring and Analysis for Prediction of Engine Emissions," 8th International Reciprocating Machinery Conference, Denver, Colorado, 1993.

[5] M. Lequime and C. Lecot, "Fiber optic pressure and temperature sensor for down-hole applications," Proc. SPIE, Vol. 1511, 1991.

[6] J.W. Berthold, W.L. Ghering and D. Varshneya, "Design and characterization of a high-temperature, fiber-optic pressure transducer," J. Lightwave Tech., Vol. LT-5, No. 7, 1993.

[7] G. He, A. Patania, M. Kluzner, D. Vokovich, V. Astrakhan, T. Wall, and M. Wlodarczyk, "Low-Cost spark plug-integrated fiber optic sensor for combustion pressure monitoring," SAE paper no. 930853, presented at SAE International Congress and Exposition, Detroit, MI, March, 1993.

[8]. T. Poorman, S. Kalashnikov, and M.T. Wlodarczyk, "Commercially Available Low-Cost Fiber-Optic Combustion Pressure Sensor," EUROPTO'95, Munich, June 19-23, 1995.

[9]. A. L. Randolph, "Cylinder-Pressure-Transducer Mounting Techniques to Maximize Data Accuracy," SAE paper no. 900171, presented at SAE International Congress and Exposition, Detroit, MI, March, 1990.

940634

SOI Type Pressure Sensor for High Temperature Pressure Measurement

Yuji Hase, Mikio Bessho, and Takashi Ipposhi
Mitsubishi Electric Corp.

ABSTRACT

An SOI type pressure sensor has been developed which can measure pressure at high temperature environments above 150°C.

SOI stands for Silicon On Insulator. A single-crystalline silicon layer is located on an insulating layer formed on a silicon substrate. The piezoresistors of the SOI type pressure sensor are made from the single-crystalline silicon layer which is isolated from the silicon substrate by the insulating layer. There is no leakage current from the piezoresistors. The SOI structure is made by the laser-recrystallization-method.

The properties of the SOI type pressure senor are as good as conventional semiconductor pressure sensors.

INTRODUCTION

Semiconductor pressure sensors have advantages such as small size, high accuracy, long time stability and mass-productivity. Semiconductor pressure sensors are used for many automotive applications, for example, manifold pressure measurement, exhaust gas pressure, and oil pressure. But conventional pressure sensors cannot measure pressures above 150°C. It is necessary for new automotive control systems to measure pressure in high-temperature environments and high-temperature pressure media.

Piezoresistors pressure sensors are insulated from the Si substrate by the pn junctions. At high-temperature environments above 150°C, a leakage current occurs at the pn junction, and the insulation from the Si substrate become unstable.

The SOI structure is highly suitable structure in insulating the piezoresistors from the Si substrate. Piezoresitors are completely insulated from the Si substrate by their SOI structure and above 150°C pressure measurement is stable.

Several methods of forming SOI type pressure sensors have been reported: depositing a polysilicon layer on an insulation layer, recrystallizing polysilicon with a laser or electron beam, burying the insulating layer by implanting oxygen (SIMOX), and binding two silicon wafers with an insulation layer [1-7]. As the piezoresistance coefficient of polysilicon is less than single-crystalline silicon, piezoresistors of single-crystalline silicon are desirable

We have developed an SOI type pressure sensor whose piezoresitors are formed from single-crystalline silicon using a laser recrystallization method. Our laser recrystallization method can form recrystalline silicon with as good a crystal quality as bulk single crystalline silicon. [8]

The SOI type pressure sensor can be used for exaust gas pressure measurement, combustion pressure measurement and oil pressure measurement in automatic transmissions. This paper describes the structure, the properties and the fabrication process of the SOI type pressure sensor.

STRUCTURE

Fig.1 shows the structure of the SOI type pressure sensor, Fig.1a shows the plan, and Fig.1b the cross section. The piezoresistors are located at the edges of the diaphragm. Four piezoresistors are connected in a full bridge circuit. As shown Fig.1b,the sensor chip is mounted on a pedestal. The pressure medium is led to the underside of the sensor chip through a pressure feed pipe.

Fig.2 shows the cross section of the SOI type pressure sensor. The piezoresistors are isolated from the silicon substrate by a SiO_2 layer. The piezoresistors are located at edges of the diaphragm. The sensor chip is 2.5mm square and the diameter of the diaphragm is 1.5 mm. By changing the thickness of the diaphragm, several pressure ranges can be measured.

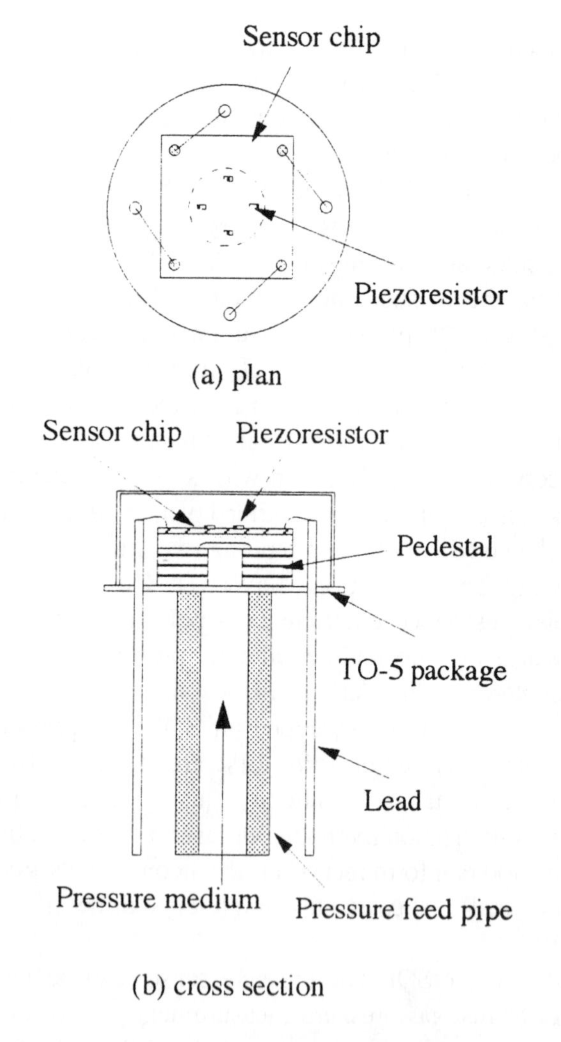

Fig.1 Structure of SOI type pressure sensor

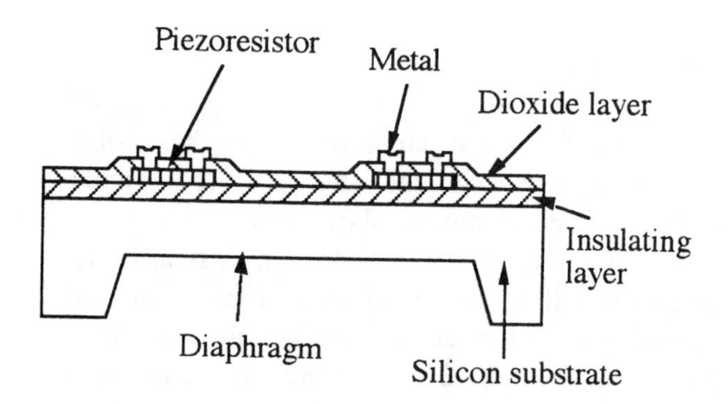

Fig.2 Cross section of sensor chip

FABRICATION

The following fabrication process is used to make the SOI type pressure sensor. Fig.3 shows the fabrication process flow of the SOI type pressure sensor.[8]

1. A dioxide layer is deposited on (100) silicon substrate to form the insulating layer.

2. Seeds which connect to the Si substrate are formed in the dioxide layer.

3. Polysilicon is deposited on the dioxide layer. Polysilicon and Si substrate are connected through the seeds.

4. A cw argon laser beam scans the polysilicon layer to make recrystalline silicon.

5. Piezoresistors are patterned from the recrystalline silicon layer.

6. Piezoresistors are made by implantation of elements such as Boron.

7. A dioxide layer is deposited above the surface of the sensor chip, then a metal layer is deposited and photolithograpied to connect the piezoresistors.

8. The diaphragm is fabricated by etching the underside of the silicon substrate.

Apart from the laser recrystallization process, most of the processes are standard semiconductor processes, so the SOI type pressure sensor can be produced as cheaply as conventional semiconductor pressure sensors.

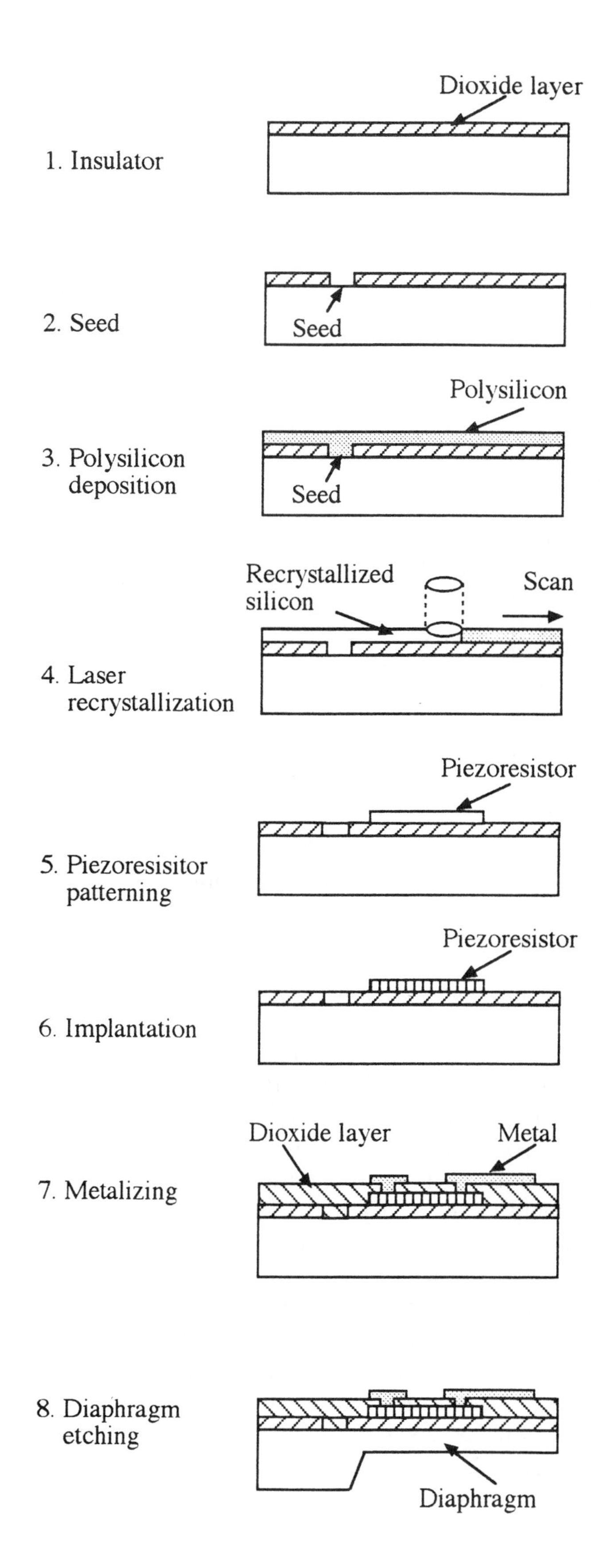

Fig3. Fabrication process flow of SOI type pressure sensor

TEST and RESULTS

(1) Sensitivity

Fig.4 shows the pressure sensitivity of the SOI type pressure sensor. The non-linearity was 0.4 % F.S. Conventional semiconductor pressure sensors (pn type) have a linearity of 0.1 % F.S., and the SOI type has a linearity as good as the pn type.

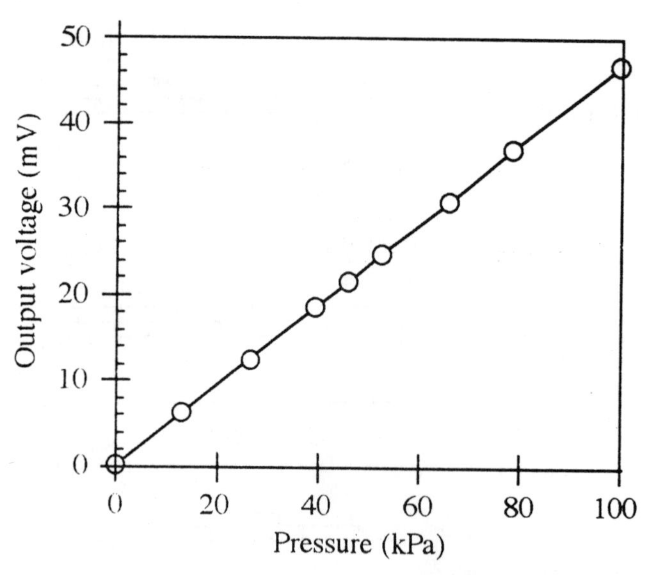

Fig.4 Pressure vs Output

(2) Temperature properties

Fig.5 shows the temperature properties. Fig.5a shows the offset voltage drifts from the offset voltage at 25°C. The offset voltage of the SOI type didn't change above 150°C, but the offset voltage of the pn type is unstable.

Fig.5b shows the sensitivity dependance on temperature. The y axis indicates the rate of sensitivity shift at each temperature to base sensitivity, which is sensitivity at 25°C. Upto 150°C both the pn type and the SOI type have a negative sensitivity dependance on temperature. Above 150°C the SOI type retains similar sensitivity properties

Fig.5c shows the reliability of the offset voltage at 200°C. After 1000 hours, the offset voltage drift was 2mV.

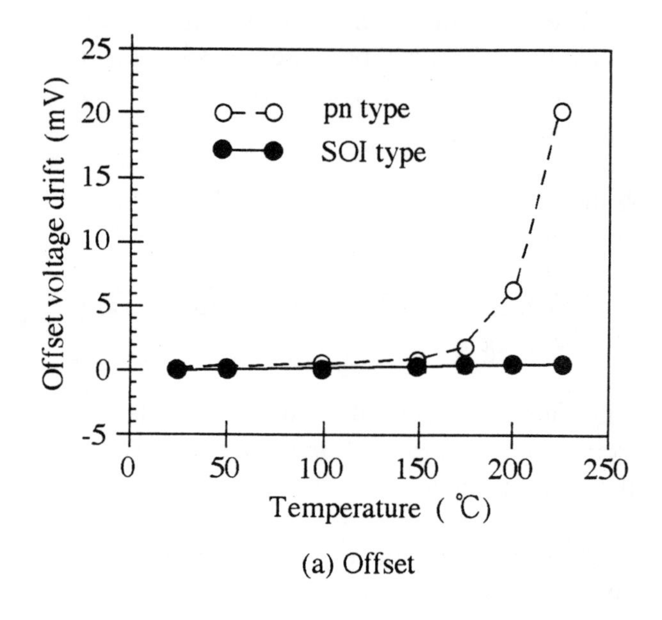

(a) Offset

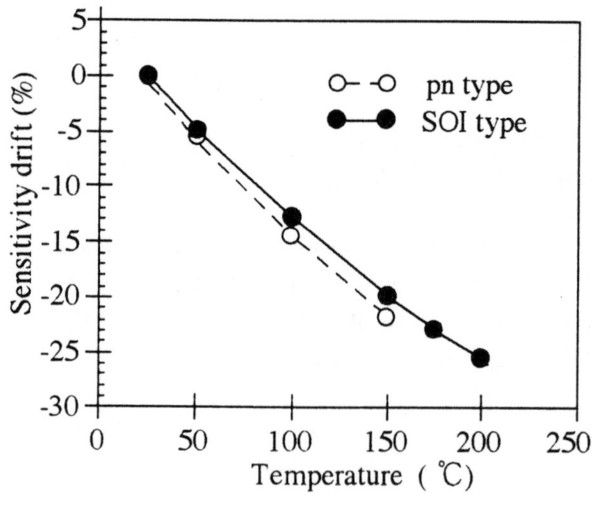

(b) Sensitivity

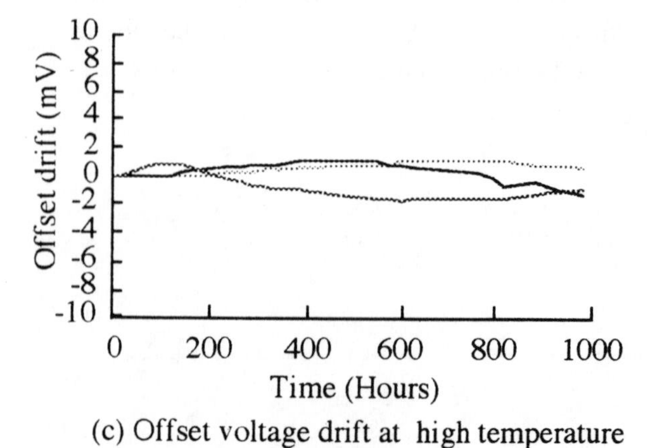

(c) Offset voltage drift at high temperature

Fig.5 Temperature properties

DISCUSSION

The "seeds" which connect the polysilicon to the silicon substrate make a single crystalline silicon in the laser recystallization method. In laser recrystallization several fabrication processes are added to the usual semiconductor process, but this method has a cost advantage compared with other SOI wafers which have a high initial cost.

The sensitivity dependance on temperature of the SOI type is equal to the pn type. It is possible that the same temperature compensation of the pn type applies to the SOI type, because the sensitivity dependance on temperature of the SOI type shows the same trend above 150°C.

CONCLUSION

We have developed a SOI type pressure sensor for high temperature pressure measurement. The SOI type pressure sensor can measure pressure above 150°C. The SOI type pressure sensor can be use for new automotive control systems.

Because the SOI type pressure sensor piezoresistors are isolated from the pressure medium, it is possible to measure the pressure of electrically conductive media (e.g. water, oil).

We can mass-produce SOI type pressure sensors with high sensitivity and high accuracy at a low cost because of using laser recrystallization method to fabricate the SOI structure.

REFERENCE

(1) T. Fukazawa, M. Mizukoshi, A. Asai and K. Hara: High-Temperature Semiconductor Pressure Sensor for Automobile, SAE No.860473

(2) J. Suski, V. Mosser and J. Goss: Polysilicon SOI Pressure Sensor, Sensor and Actuators, 17.(1989), 405-414

(3) G. S. Chung, S. Kawahito, M. Ashiki, M. Ishida and T. Nakamura: Novel High-performance Pressure Sensor Using Double SOI Structure, Transducers '91 Digest of technical papers, (1991), 676-681

(4) P. J. French, H. Muro, T. Shinohara, H. Nojiri and H. Kaneko: SOI Pressure Sensor, Sensors and Actuators, A35, (1992) 17-22

(5) K. Petersen, J. Brown, T. Vermeulen, J.Mallon, J.R. and J.Bryzek: Ultra-Stable, High-Temperature Pressure Sensors Using Silicon Fusion Bonding; Sensors and Actuators, A21-A23(1990), 96-101

(6) E. I. Givargizov, A. B. Limanov, G. D. Prijakin and V. I. Vaganov: Silicon-on-Insulator (SOI) Structure for Pressure Sensor, Sensors and Actuators, A28, (1990), 96-101

(7) B. Diern, R. Truche, S. Viollet-Bosson and G. Delapierre: 'SIMOX' (Separation by Ion Implantation of Oxygen): a Technology for High-temperature Silicon Sensors, Sensors and Actuators, A21-A23, (1990), 1003-1006

(8) K. Sugahara, S. Kusunoki, Y. Inoue, T. Nishimura and Y. Akasaka: Orientation Control of the Silicon Film on Insulator by Laser Recrystallization, Jounal of Applied Physics Vol.62 No.10, (1987), 4178-4181

950348

Physically Different Sensor Concepts for Reliable Detection of Side-Impact Collisions

Alfons Härtl, Gerhard Mader, Lorenz Pfau, and Bert Wolfram
Siemens Automotive

ABSTRACT

This paper describes new concepts to detect side impact collisions. Based on the specific system requirements for side impact detection, two physically different concepts will be described and compared to each other.

Acceleration sensing principles, applied in today's single point sensing systems, were adapted to cope with the unique requirements for side collision detection.

A more advanced and completely new concept is based on the sensing of the pressure change within the cavity of the impacted door.

Based on these sensing principles, different system configurations will be illustrated. The performance of both sensing principles will be compared on the basis of available crash and misuse test conditions.

In conclusion, it can be stated that the aforementioned sensing principles support the rigid firing requirements of a timely airbag deployment.
However, the selection of the system configuration and the physical sensing principle has to account for the individual deformation behavior of the vehicle's side structure.

INTRODUCTION

The effort within the automotive industry to improve the safety of passenger cars has been successful. Air Bags have been widely introduced to protect the occupant in frontal accidents, which will help to reduce injury numbers in head-on collisions. However, the field of supplemental restraint systems will remain challenging.

Even so, side collisions only account for 20% of all accidents, they show proportionally higher occurrence of accidents with severe and fatal injuries of approximately 50%. Air Bag systems will now be applied to protect passengers in side collisions. The first (mechanical) side air bag system has recently entered the market.

The "sensor" within the system represents a key component, which has to cope with requirements even tougher than those known for frontal impact sensors: Not only a robust distinction between fire and no fire/misuse events has to be possible, but in the case of a severe side collision, the fire decision is required within 5 ms or less.

In cooperation with different car manufactures, Siemens has developed, analyzed and proven out different sensor concepts, which have shown in several crash tests, that they absolutely satisfy the above described requirements.

This paper describes two physically different approaches.

THE PRESSURE CONCEPT

A satellite (pressure sensor plus circuitry for signal conditioning and decision making) is located inside the door cavity to monitor a dynamic pressure change.

An intruding object causes a deformation of the door and thus reduces the volume of the cavity inside the door. The rapid volume reduction leads to an adiabatic increase of the pressure within the cavity of the door. Due to the increased pressure, air starts to flow through leakage and the air mass inside the door decreases. An interesting effect observed is that the pressure amplitude (especially in the first several milliseconds) is quite insensitive to a change of the leakage size. This behavior has been proven experimentally, as well as, with a theoretical model.

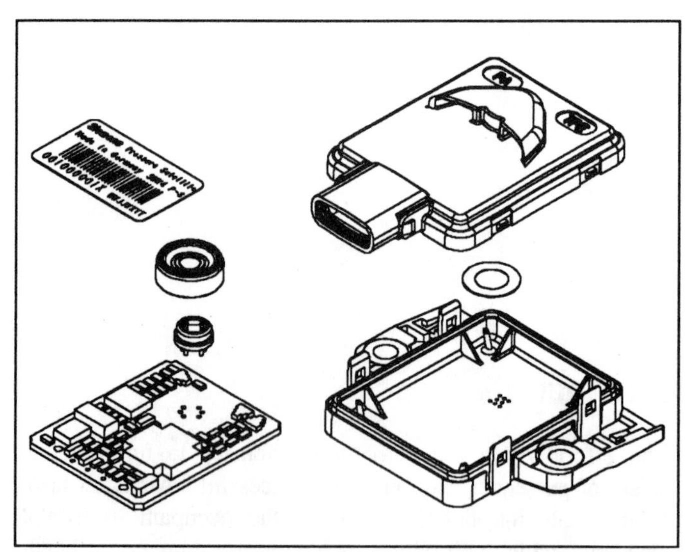

Figure 1: Pressure satellite

The pressure sensor used is a silicon micro machined component, which senses absolute pressure. The device is made up of two silicon layers with an evacuated cavity between them, to provide a pressure reference. Changes of external pressure will lead to a deflection of the silicon membrane (2nd silicon layer). Piezo resistors, implemented into the membrane, will change their value and therefore, provide information about the external pressure.

Amplification and filtering will generate the input for the decision making circuitry, where the actual pressure signal will be compared against defined firing conditions (e.g. pressure threshold and simple integral of p(t)). The measured signal is a relative value of $\Delta p/p_o$, with p_o representing the actual atmospheric pressure and Δp the dynamic pressure change. Typical pressure signals (Δp) for severe side-impacts fall in the range of 20...200 millibar (0,3...3,0 psi) with p_o=1000 mbar.
Upon recognition of a firing condition, the bag will be deployed. A remark to note is, that the comparable noise level to those pressure levels is far beyond a starting jet airplane.

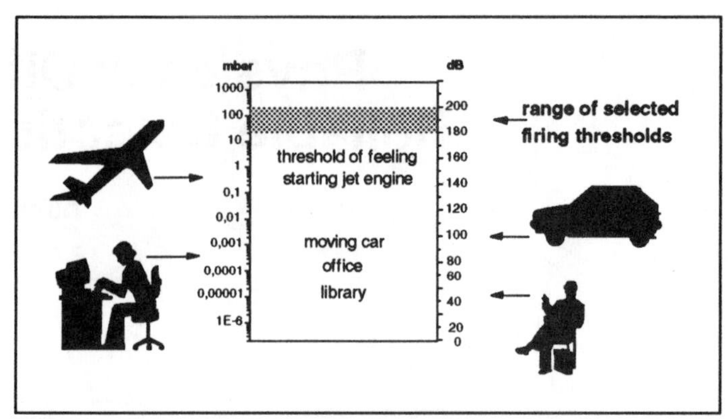

Figure 2: Pressure range, firing threshold

Another interesting fact is the possibility of a self test with this type of sensor. Due to the fact that the atmospheric pressure is contained within the output signal, high end and low end limits for the output signal can be defined and failures of the sensor (e.g. a leak of the cavity) can be detected and communicated to a warning indicator.

EXPERIMENTAL RESULTS FOR PRESSURE SENSING

During the development and prove out-phase, Siemens has worked with many different OEM's. Our pressure sensors were used in crash test on more than 30 different platforms.

Many different crash conditions have been tested (e.g. FMVSS214, EEVC, car-to-car crashes at various angles, pole impacts and truck impacts etc.). In addition to fire and nofire tests, several misuse and abuse conditions were tested (e.g. bicycle impact, foot kicks, hammer blow, frontal crash, door opening into rigid objects, door slamming and sound tests, etc.).

Figure 3 shows different fire conditions and their pressure responses. For this specific door, a fire threshold of Δp= 34mbar (p_o=1000 mbar). As shown, the signals exceeded the threshold in all cases before 5 ms.

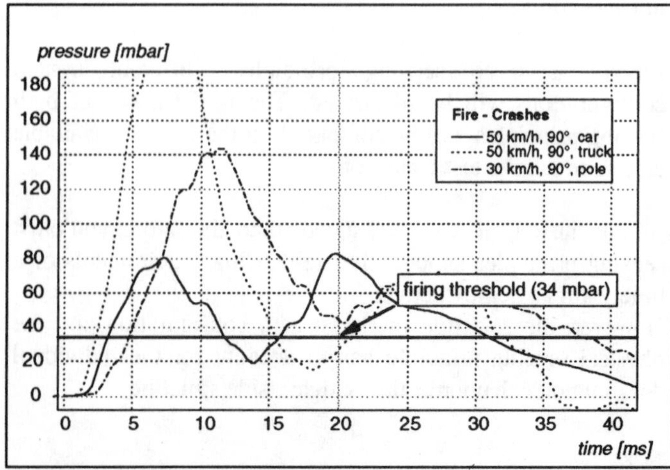

Figure 3: Example for pressure signals - fire cases

The pressure signals show all, a typical behavior for fire conditions, where a steep increase of pressure leads to a fairly high peak value before the signal is reduced, due to leakage later in the event.

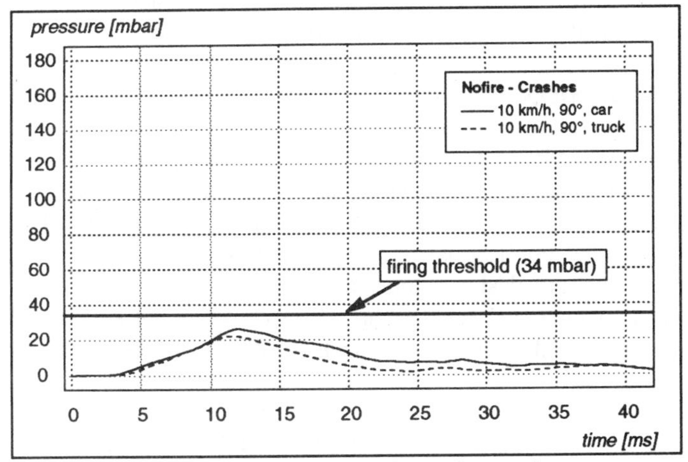

Figure 4a: Example for pressure signals - nofire-crashes

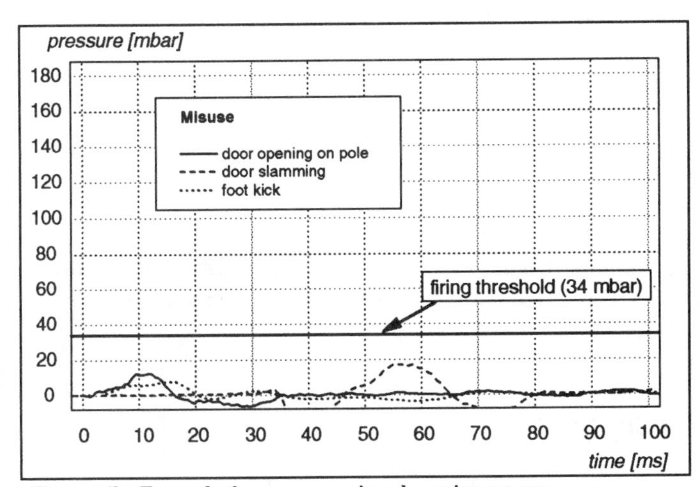

Figure 4b: Example for pressure signals - misuse tests

Figure 4 explains various misuse and nofire conditions. These events describe the low end of the fire threshold.

The comparisons of both figures 3 & 4 indicate, that a fire/nofire decision, containing a considerable safety margin, could be attempted easily with a pure threshold. High sophisticated algorithms are not necessary to distinguish these crash conditions.

More complicated crash situations might require flexible criteria for the fire decision. This can be easily implemented by the use of a simple crash algorithm.

THE ACCELERATION CONCEPT

Acceleration Sensing for side collision detection uses the well known technology of single point sensing systems. An accelerometer is the heart of a g-satellite, which is located either inside the door or close to the door structure (e.g. B-pillar, cross car beam, etc.). The signal provided by the state-of-the-art accelerometer is fed into the microcontroller. An algorithm based on physical meaningful criteria can distinguish between fire and nofire situations. In case of a fire condition, the bag will be ignited.

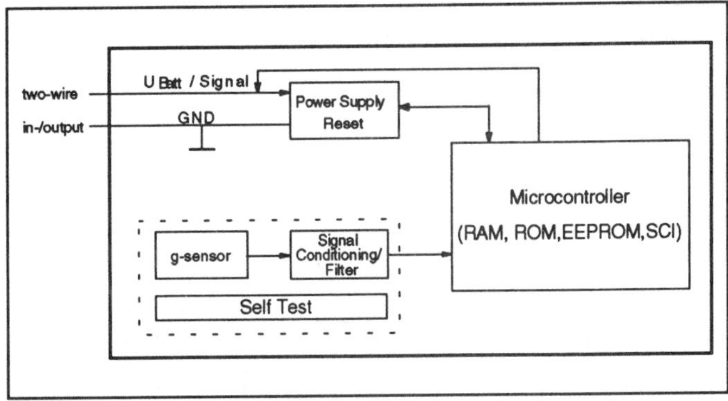

Figure 5: Block diagram of g-satellite

Depending on the location of the satellite, the g-range of the accelerometer necessary for a reliable detection of side collisions will vary. Inside the door, much higher values can be seen. The accelerometer should, therefore, provide at least a range of ± 200 g. In some cases, depending on the mounting position and door structure, a 500 g device has proven to be the right choice.

EXPERIMENTAL RESULTS FOR G-SENSING

It is common behavior to collect acceleration data from various locations during a side crash in order to determine the best feasible location. In close cooperation with car manufacturers all over the world, Siemens was able to establish a large crash data bank for side collisions. Based on the available crash data, algorithms have been developed and g-sensor satellites have participated in several hundred crash tests as well as thousands of misuse tests.

As a result of extensive testing and simulation, it can be said that positioning of the sensing device (g-sensor) is extremely critical as it relates to the crash performance.

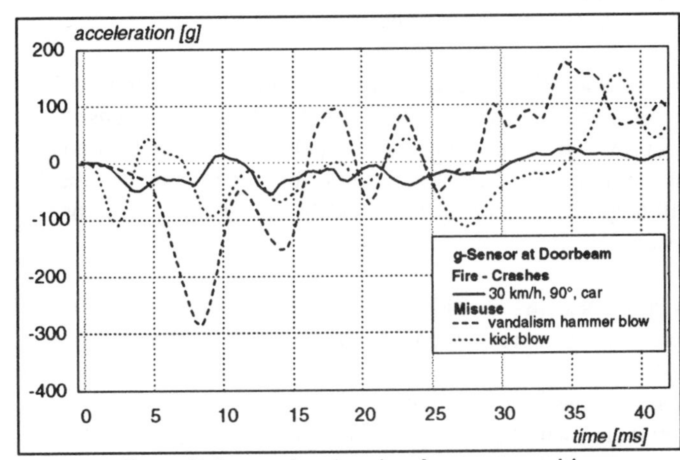

Figure 6: Example for acceleration signals - sensor position: doorbeam

Figure 6 shows fire and nofire/misuse conditions for a g-sensor located inside a door for one specific platform. Here it becomes very difficult to distinguish the abuse conditions from some fire events within the required time. A sensor located inside the door can see considerable g-forces, especially if the door is hit directly at the sensors location.

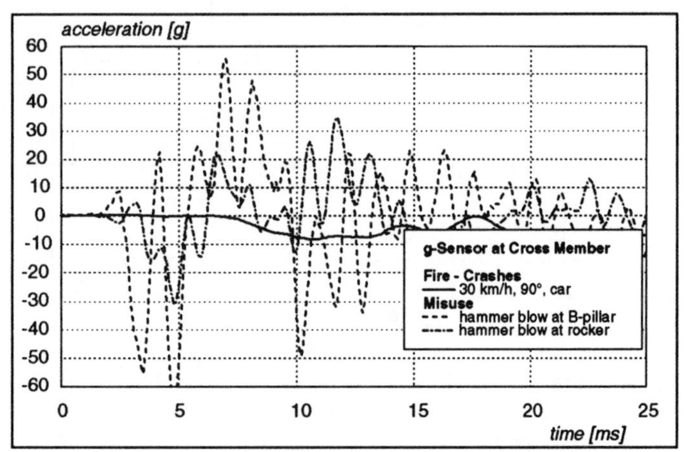

Figure 7: Example for acceleration signals - sensor position: cross member

Figure 7 reflects a scenario where the sensor is located outside the door at a cross member. The g-forces seen in the different events show much lower levels compared to the door location. A clear distinction between fire and nofire/misuse events is possible with a mainly velocity based algorithm.
If the car structure is stiff enough, the signals from an additional lateral g-sensor inside the central control unit can be combined with the satellite signals for the discrimination of fire crashes.

COMPARISON

The goal of a safety device has been clearly defined: Best crash performance has to be targeted in order to protect the passengers in all possible events.

The most feasible location for a specific platform has to be determined by testing. Packaging limitations have to be taken into account many times over.

In case a location outside the door is able to satisfy the crash requirements, the acceleration based technology (g-sensing) has proven it's reliability in single point sensing applications. Further improvements of algorithms for faster firing decisions will help this technology to be widely used within side airbag systems.

In case a sensing location inside the door turns out to be the best solution, a pressure based sensing system will provide the best fit.
Siemens has shown that this totally new approach is convincing with fast firing decisions and simple firing criteria compared to highly complex algorithms. Fire and nofire events can reliably be distinguished and packaging inside the door is easily facilitated, due to the fact, that the pressure signal is homogeneous within the door cavity.

Possible concerns of this new technology (e.g. influence of leakage, rust proofing sprayed on sensor) have been investigated and proven by testing to be insignificant.

CONCLUSION

Both systems have their advantages if they are applied in the right manner. It is fascinating to see that sensing devices available today (g-sensor) can be adapted and advanced toward new applications. It is even more fascinating to investigate and design a total new sensing system (pressure sensor), which will fill the gap where conventional systems (sensing inside the door) cannot provide satisfactory performance.

Both sensor concepts are available today and work on today's vehicle chassis without major structural modifications. By applying them to our cars, we will further improve the safety of the vehicles on our roads.

ACKNOWLEDGEMENT

The authors would like to express their gratitude to the OEM's who tested our side-impact sensors in crash tests.

REFERENCES:

[1] A. Härtl, G. Mader, L. Pfau, R. Muhr
"New Sensor Concept for Reliable Detection of Side-Impact Collisions", The 14th International Technical Conference on Enhanced Safety of Vehicles (ESV) 94-S6-0-14, (May 1994)

[2] J. Franz, U. Kippelt, "Numerical and Experimental Simulation of Different Loadcases of Side Impacts", The 14th Int. Conference on Enhanced Safety of Vehicles Proceedings 94 S6 W27, (1994)

[3] Rudolf Muhr, "Dynamische Drucksimulation bei Kompressionsversuchen, ein dynamisches System", diploma thesis, University of Regensburg, Germany, Institute of Theoretical Physics, (1994)

LINEAR AND ANGLE POSITION SENSORS

970601

High Accuracy Semiconductive Magnetoresistive Rotational Position Sensor

Yasushi Ishiai, Noriyuki Jitousho, and Tetsuhiro Korechika
Matsushita Electronic Components Co., Ltd.

Joe LeGare and Sumitake Yoshida
Panasonic Industrial Co.

Copyright 1997 Society of Automotive Engineers, Inc.

Recently there is demand for rotation sensors capable of high-accuracy detection and very low-speed detection of rotation at high temperature for automobile use. To meet this requirement, a rotation sensor using an InSb thin-film magneto-resistors with good thermal stability has been developed.

This sensor transduces magnetic flux change due to gear rotation to resistance change. It is composed of InSb thin-film magneto-resistors fabricated by a newly developed process and signal shaping circuits where resistor signals are converted to digital signals using no amplifier. Accordingly, the signals are independent of the measured frequencies, making possible very low speed (0 to 20 Hz) detection. The sensor stably operates in the temperature range from -40 to 150 degree C for thousands hours. There is no need for a shielded harness due to the digital output signal. It can be used for high temperature environments like in the engine compartments for rotation detection of ignition systems, electronically controlled AT systems, antiskid braking systems, etc.

INTRODUCTION

Trends, such as regulations for automobiles, like CAFE, OBD II, and LEV, along with energy and safety consciousness, and the need high-accuracy in navigation systems, etc., have increased demand for improved rotation sensors. These sensors need the following characteristics: (1) capability of very-low-speed detection at 0 - 20 Hz and (2) capability of high-accuracy detection at high temperature. In order to meet these requirements, we have been studying the "semiconductor magneto-resistor" that enables high-sensitivity sensor output and thus greatly-reduced peripheral circuits.

This semiconductor magneto-resistor (=semiconductor MR) was already reported in 1950's[1], especially in the period when the deposition process of InSb thin-film , as III-V family semiconductor thin-film, first became the subject of research[2]. The semiconductor MR, whose magnetic sensitivity is in proportion to electron mobility of semiconductor, limits its material to InSb and InAs. Therefore, it is an important point regarding how the crystalline semiconductor thin-film material is formed. Traditionally, slices of a bulk single-crystal wafer or those formed by the growth of the high-quality InSb thin-film on Mica substrate are transferred onto another substrate [3]. In either construction, because of the use of an adhesive layer, the problems, mainly caused by the difference in expansion coefficients, made it inadequate for the high-temperature specifications as in the electronic package applications. Therefore, we have developed the InSb thin-film MR based on the following basic technologies:

(1) The technology for the growth of InSb thin-films with excellent crystallinity directly on the Si substrate.

(2) The electrode material & process technology for ensuring reliable connections over a wide range of temperature.

(3) The heatproof packaging technology.

These technologies as well as the characteristics of a complete rotation sensor for automotive use are described below.

INSB THIN-FILM MAGNETO-RESISTORS

The basic structure of InSb thin-film magneto-resistors is shown in Fig. 1. It is basically constructed, after the growth of InSb thin-film on the substrate, by depositing a number of ladder-structure electrodes called, "shorting bar electrodes". In this construction, whose principle of operation is shown in Fig.2 (a), a phenomenon is used that the current path is elongated by the Lorentz

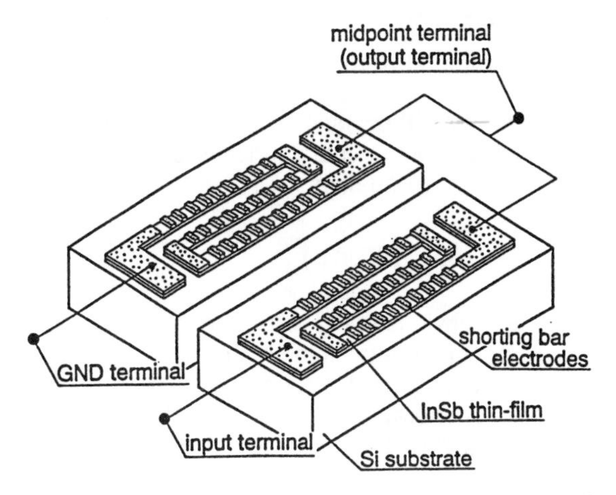

Fig.1 Basic structure of InSb thin-film magneto-resistors

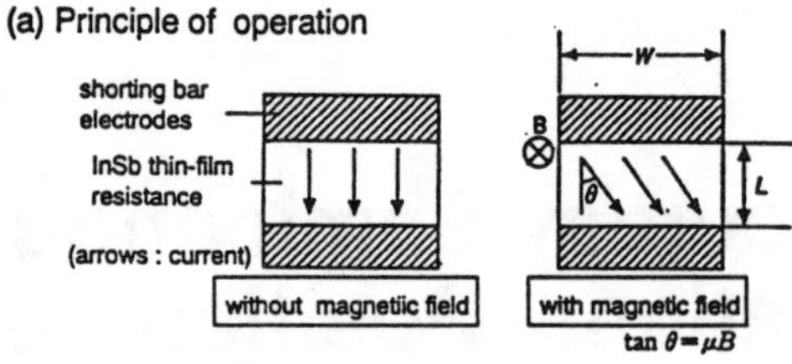

(b) Resistance change under magnetic field

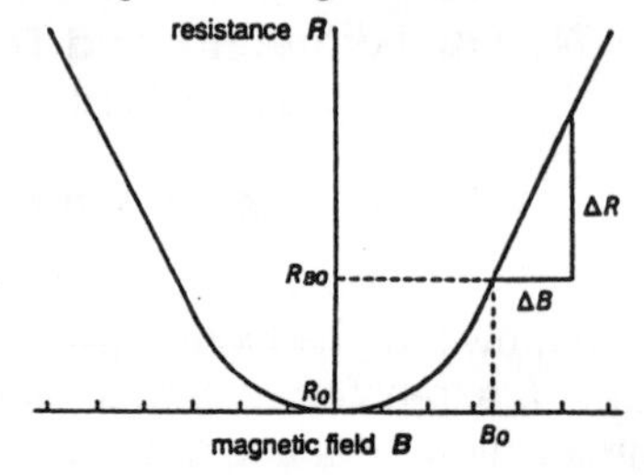

Fig.2 Principle and basic characteristics of InSb thin-film magneto-resistors
(a) Principle of operation
(b) Resistance change under magnetic field

force applied to the carrier when the magnetic field is applied in the transverse direction relative to the film surface of the InSb thin-film between the shorting bar electrodes. Here, as shown in Fig. 2 (b), the resistance change against the magnetic field differs between the low magnetic field region and the high magnetic field region. With adequate accuracy it can be expressed by the following equations:

(low-magnetic-field characteristics)

$$R_B / R_0 = \frac{\rho_B}{\rho_0}(1+g(\mu B)^2) \tag{1}$$

(high-magnetic-field characteristics)

$$R_B / R_0 = \frac{\rho_B}{\rho_0}(G+\frac{W}{L}\mu B) \tag{2}$$

where, R_B: resistance with magnetic field, R_0: resistance without magnetic field, μ: electron mobility of the semiconductor, B: applied magnetic field, g and G: functions of element's shape (L/W) respectively, L: element length, W: element width, ρ_B / ρ_0: magneto resistive effect (nearly equal to 1).

Generally, the change over point from square-law characteristics to linear characteristics is given by the following empirical equation:

$$\mu B = 0.65 \quad \text{(SI units)} \tag{3}$$

As is obvious in Fig. 2 (b), the resistance change of this element is greater in the high-magnetic-field region, and accordingly it is necessary, in practical use, to set the point of operation in the linear (high field) region by placing a magnet on the backside of the element.

Furthermore, as is obvious in Equation 2, in this semiconductor magneto-resistor, the magnetic sensitivity or the magnitude of resistance change is in proportion to the element's shape (*W/L* ratio) and the electron mobility of semiconductor used. Since the magnitude of electron mobility, in particular, is a key factor of the element's characteristics, InSb, which has the highest value of μ among semiconductors, is used. In addition, as regards the merits of thin-films, the following points are indicated: (1) capability of high impedance, (2) improvement in heat resistance ,and (3) great reduction in the material (wafer) cost.

We have developed the technology for the growth of InSb thin-film with excellent crystallinity and thus realized the InSb thin-film magneto-resistors with high sensitivity and high heat resistance. The fabrication process is described below.

FABRICATION PROCESS OF INSB THIN-FILM MAGNETO-RESISTORS - The fabrication process of InSb thin-film magneto-resistors is shown in Fig. 3. As for the fabrication technology for the InSb thin-film with high electron mobility or excellent crystallinity, studies have been reported by several groups. Studies about heteroepitaxy on the InSb on GaAs substrate are the most popular, and as for growth technology, MBE [4,5] , MOCVD [6,7], and sputtering [8] have been reported. In addition, examples of growth on CdTe [9] and sapphire [10] have also been reported. As a fabrication technology of InSb thin-film in practice, the method that, after the deposition on the Mica substrate, transfers it onto another substrate via an adhesive layer is being used in manufacturing hall elements [11]. Among them, taking into account the substrate cost, heat resistance, and the priority of proper orientation of the surface of the InSb thin-film, we have decided to use the Si(111) substrate.

Prior to the growth of the InSb thin-film, the surface washing treatment of Si substrate is an important point. Traditionally, in heteroepitaxy of InSb thin-film on Si substrate, a method that obtains an atomically clean surface by heating the substrate at high temperature above 900°C in a vacuum chamber and sublimating SiO existing on the surface is used [12]. In this method, the high-temperature and the ultra-high vacuum that are required, results in poor mass-productivity. Since the optimum temperature in the growth of the InSb thin-film is relatively low (430°C), we adopted the hydrogen termination treatment where the Si surface structure can be maintained sufficiently stable up to this temperature.

In the fabrication process, we use the 4-inch Si(111) substrate (resistivity ≧ 1 kΩ·cm). After treatment of the substrate surface by RCA washing, it is immersed in dilute HF solution. The hydrogen termination treatment is applied removing the oxidized layer on the substrate surface and terminating the unoccupied valence on the Si surface with hydrogen atom. After that, the substrate is rinsed with ultra pure water and introduced into the vacuum chamber.

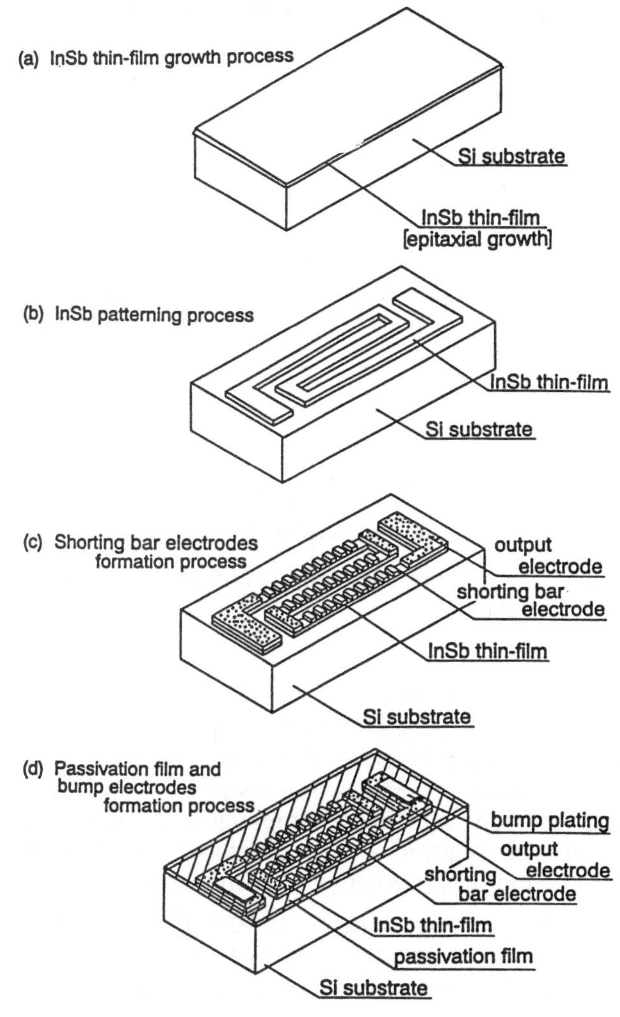

Fig.3 Fabrication process of InSb thin-film magneto-resistor

Then, the growth process of the InSb thin-film starts. In this process, since the lattice mismatch between InSb and Si is as great as about 19%, a simple one-step growth technology will result in a polycrystalline thin-film, the two-step growth technology is adopted where in (1) after the preliminary growth layer is deposited at low temperature, (2) the main layer is grown at normal growth temperature. In addition, in this growth technology, the vacuum evaporation method (three temperature method) is used where In and Sb are separately evaporated by

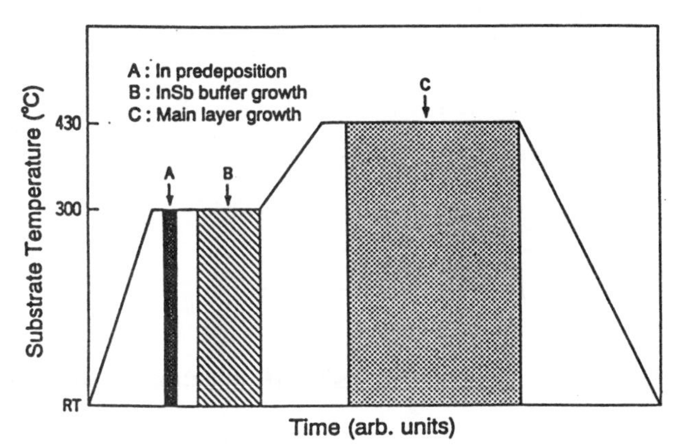

Fig.4 Typical substrate temperature profile for growth of InSb film on Si(111)

resistance heating. At the first stage of study, the typical substrate temperature profile for growth was a simple two-temperature type with low temperature and high temperature, as shown in Fig. 4.

(1) Deposition of the preliminary growth layer - The hydrogen-terminated Si(111) substrate described above is introduced into the vacuum chamber and, after lowering the vacuum pressure below 5×10^{-5} Pa, the substrate temperature is raised to 300°C. Then, as shown in Fig. 4 A, about 0.5 nm thick of In film is predeposited under the pressure of less than 1×10^{-4} Pa at the deposition rate of about 0.1 nm/s. Here, if the substrate temperature is too low, judging from the observation of the surface morphology of the InSb thin-film after the following main growth, a discontinuous mass crystal will be formed. Therefore, it is necessary to maintain the appropriate temperature. If the predeposited In film is too thin, it will not play a sufficient role as the core formation material, and if it is too thick, flocculation will be caused. Therefore, it is necessary to control the appropriate thickness.

After pre deposition of In, maintaining the substrate temperature at 300°C, In and Sb are co-evaporated as shown in Fig. 4 B. Here, with the atomic ratio Sb/In = 3 and the In deposition rate of approximately 0.1 nm/s, the approximately 25 nm-thick InSb thin-film is deposited. The change of reflection high-energy electron diffraction (RHEED) pattern of a series of low-temperature preliminary growth layers from the early stage of growth is as follows:

As In is fed from the early stage of growth, the pattern intensity of the substrate Si(111) gradually decreases and, on the coevaporation of In and Sb, the intensity of SI(111) rapidly decreases and finally disappears. On the other hand, InSb(111) gradually emerges and finally the complete orientation pattern of InSb(111) is observed.

Here, the epitaxial growth of InSb(111)//Si(111), InSb<110>//Si<110> is confirmed. After that, as shown in Fig.4 B → C, the substrate temperature is elevated to 430°C for the main growth.

(2) Deposition of the main growth layer - After maintaining the substrate temperature at 430°C, as shown in Fig. 4 C, an approximately 3 μm-thick InSb thin-film is deposited with the atomic ratio Sb/In = approximately 2.5 and the In deposition rate of approximately 0.3 nm/s (the InSb growth rate 0.75 nm/s).

The RHEED pattern and SEM image after completion of the main growth are shown in Fig. 5 (a)(b). Fig 5 clearly shows that, in the film after the two-step growth with the two-temperature profile with low temperature and high temperature as shown in Fig. 4, although strong and clear streaks are observed, the boundary exists between the domains of epitaxial growth. That is, although the epitaxial growth is achieved satisfying InSb(111)//Si(111), InSb<110>//Si(110), domains with two direction exist: one is the exactly same direction as Si and the other is the direction rotated by 180deg around the Si(111) plane. In measurement of electron mobility (μ) of the film in this state using the van der Pauw technique, μ was found to be $\mu = 3.63 \text{m}^2/\text{V}\cdot\text{s}$, which is not fully satisfactory.

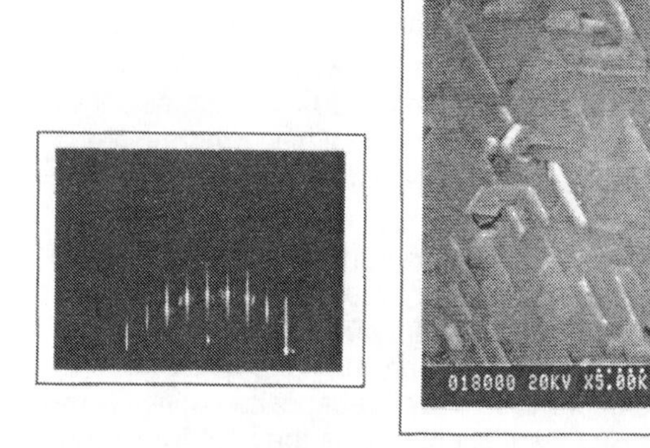

(a) RHEED pattern (Si [110] azimuth) (b) SEM image

Fig.5 Structure of InSb film after typical growth process

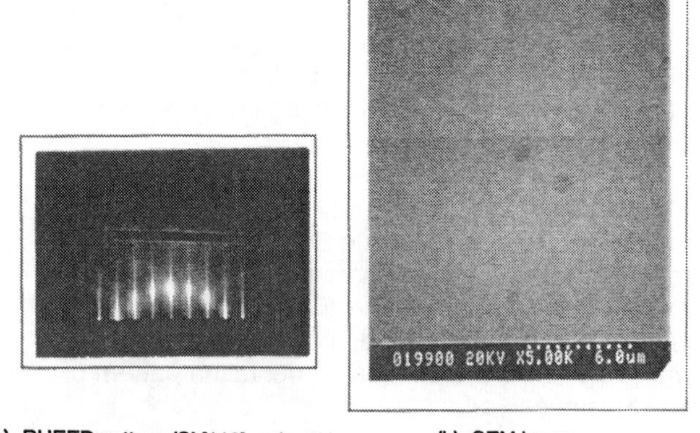

(a) RHEED pattern (Si [110] azimuth) (b) SEM image

Fig.6 Structure of InSb film after optimized growth process

In this regard, we particularly focused on the temperature profile during the predeposition to clarify the relation between this profile and the crystallinity of the growing film and thereby optimized the conditions of the main growth. Thus, as shown in Fig.6 (a) (RHEED pattern) and (b) (SEM image), we realized the epitaxial growth of InSb thin-film on the Si substrate where clear streaks of InSb(111) with Kikuchi lines are observed and the domain boundary shown in Fig. 5 disappears to form the nearly single domain with excellent surface smoothness.

For the completed InSb thin-film, μ was measured and the value nearly equal to the theoretical value[13] was obtained (maximum value $\mu = 5.99 \text{m}^2/\text{V}\cdot\text{s}$). In addition, this growth process of the InSb thin-film is well reproducible and enables the consistent fabrication of thin-films with excellent crystallinity and high electron mobility.

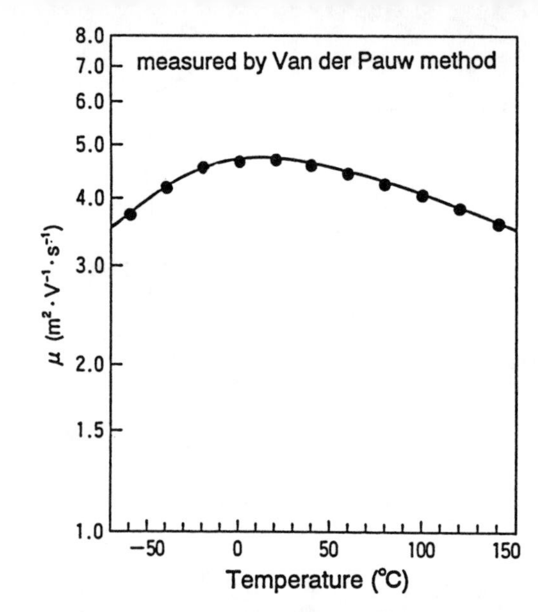

Fig.7 Temperature dependence of mobility of typical InSb thin-film

Fig. 7 shows an example of the temperature dependence of electron mobility of typical InSb thin-films (thickness: 3 μm) fabricated by this process. Electron mobility has its maximum value near room temperature and decreases in both high temperature and low temperature regions. The temperature dependence of electron mobility, with carrier concentration at room temperature of approximately $1\times10^{22} \text{m}^{-3}$, is shown by the following theoretical equations[14,15], where electron mobility under scattering by ionized impurity is μ_I, electron mobility under scattering by lattice is μ_L, electron mobility under scattering by dislocation is μ_D, resultant electron mobility under these scattering mechanisms is μ_H, electron mobility considering surface scattering on the thin-film is μ_{TH}:

$$\mu_I = 1.10T^{0.73} \tag{4}$$

$$\mu_L = 1.09\times105T^{-1.68} \tag{5}$$

$$\mu_D = 8.165\times10^{11}\frac{T}{N_D} \tag{6}$$

$$\mu_H = \left(\frac{1}{\mu_I}+\frac{1}{\mu_L}+\frac{1}{\mu_D}\right)^{-1} \tag{7}$$

$$\mu_{TH} = \frac{\mu_H}{1+\dfrac{2.06\times10^{-3}\cdot\mu_H T^{0.5}}{d}} \tag{8}$$

N_D: dislocation density, d: thickness of the InSb thin-film.

Based on Equations 4-8, the theoretical values of the temperature dependence of electron mobility are obtained and, by comparison between the theoretical values and our measurement values, the dislocation density is considered to be approximately $10^{13}m^{-2}$. This is because dislocations generated by interfacial mismatch with the Si substrate still remain in the growing thin-film. In addition, it is found that scattering by dislocations in the low-temperature region and scattering by the lattice in the high-temperature region are dominant factors of electron mobility respectively. The differences in electron mobility due to thin-film quality is great mainly in the low-temperature region, and the better quality of thin-film results in smaller decreases in the low-temperature region. In the high-temperature region, since scattering by the lattice is dominant and nearly uniform, the differences are not very great.

After completion of growth of the InSb thin-film, through the photo lithograph process shown in Fig. 3 (b), the InSb thin-film is processed in the longitudinal direction of the element by the wet process, then the formation process of shorting bar electrodes with the ladder-structure shown in Fig. 3 (c) starts. After the uniform coating of electrode material, the desired ladder shape of electrodes is obtained by the photo lithograph process. In this process, since a number of ladder-structure electrodes are formed on the InSb thin-film, the important points are as follows:

(1) to minimize contact resistance between the electrodes and the InSb thin-film,

(2) to minimize stress by the electrode material on the InSb thin-film, and

(3) to prevent erosion of InSb thin-film in processing the electrodes (wet process).

We formed the ohmic contact with low contact resistance on the InSb thin-film and, from the viewpoint of chemical selectivity (on InSb) during the wet process, adopted the Cu/Cr laminated electrode. As for the above-mentioned point (1), in addition to HCl solution treatment used in the removal treatment of the InSb oxidized layer prior to the electrode formation, $(NH_4)_2S_x$ (ammonium polysulfide) treatment was introduced to keep the surface of the InSb thin-film in an unoxidized condition and to realize low contact resistance and, as for (2), for the evaporation formation, total stress was reduced mainly by the electrode thickness control and, as for (3), the process was stabilized by an appropriate selection of treatment solution.

After the formation of shorting bar electrodes, as shown in Fig. 3 (d), a passivation film electrodes was formed over the electrodes. For the passivation film, low-expansion polyimide ($\alpha: 5 \times 10^{-6}$/deg) was used.By using the low-expansion material, as previously described, we minimized the stress on the InSb thin-film and suppressed the change of the forbidden band width shown in the following equation. In particular, expansion coefficient of the upper layer material is great to increase the forbidden band width when tensile stress is applied and to increase effective mass of electron and to deteriorate the magnetic characteristics of the element.

$$\frac{dE_g}{dP} = 1.58 \times 10^{-10} eV \cdot m^2 / N$$

(Left side: change of forbidden band width per unit stress, stress dependence: tensile (+), compressive (-))

Stress applied to the InSb thin-film in burning this polyimide is at most approximately (tensile stress) and, in the wide range of temperature, has no adverse effect on the element's characteristics. After spin coating of this polyimide material, plating windows are made through the conventional photo lithograph process and it is burnt in vacuum. By the polyimide passivation film thus formed, under specifications for electronic packaging reliability, sufficient reliability is assured. Then, by electrolytic plating, Au/Ni is plated to complete the element.

Fig. 8 shows the characteristics of resistance change with magnetic field of the completed element (μ:4.5m^2/V·s, room temperature value). Under 0.4 T, the resistance is three times as high as the resistance without a magnetic field.

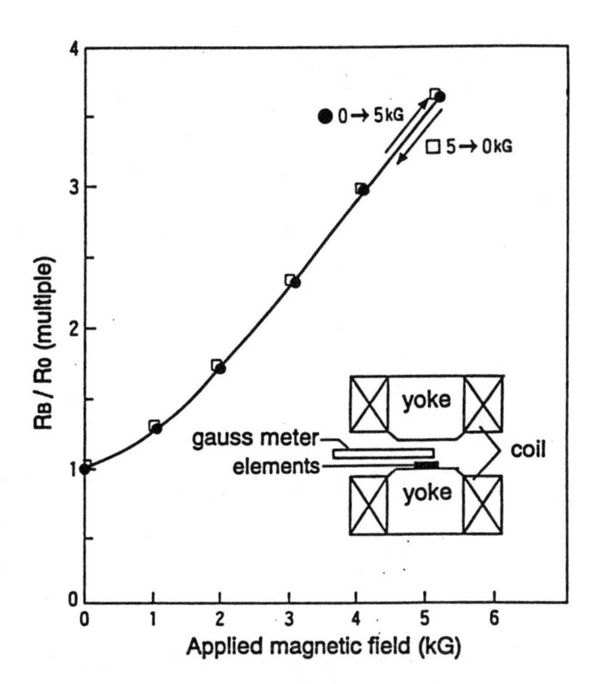

Fig.8 Characteristics of resistance change under magnetic field of InSb thin-film magneto-resistors

Fig. 9 shows the relation between the output voltage and the air gap (between the detection gear and the element surface) at room temperature in the differential element. A differential wheatstone bridge configuration is used to compensate for the large temperature dependence of the InSb element's impedance characteristic. It is understood that output voltage of SMR is sufficiently high under the usual condition (air gap is about 1.5mm).

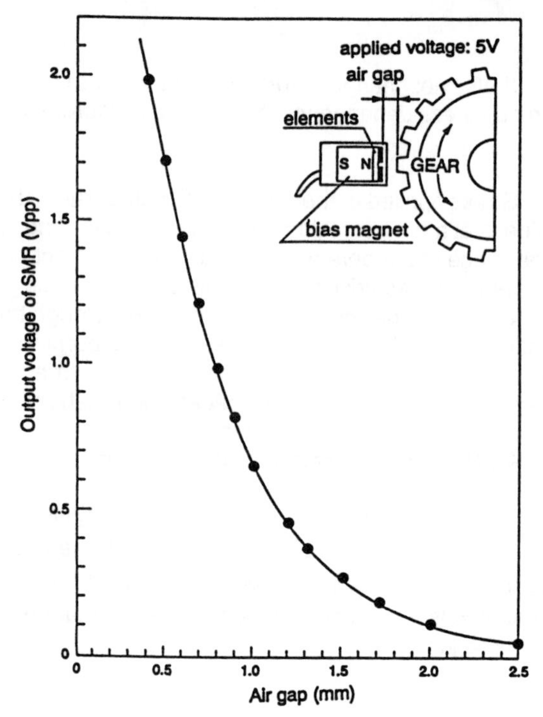

Fig.9 Characteristics of output voltage vs. air gap (at 5V applied)

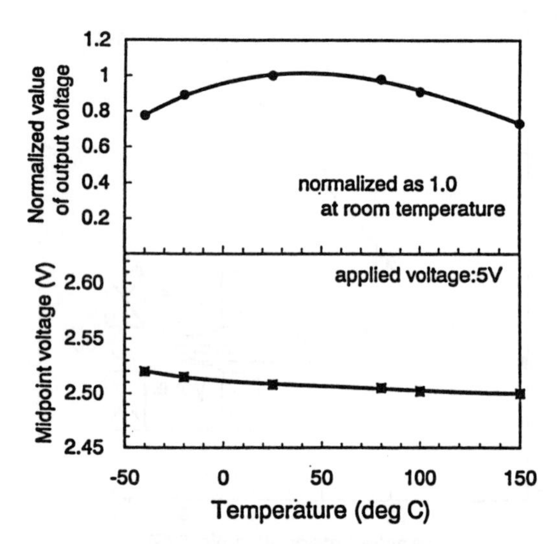

Fig.10 Characteristics of dependence of output voltage (normalized at 1 at 25 degree C value) on temperature

In addition, Fig. 10 shows the typical characteristics of temperature dependence of output voltage (normalized at room temperature value) and midpoint voltage in the differential element. Output voltage has a similar temperature dependence to that of electron mobility and has its maximum value near room temperature. It also tends to decrease in both the low-temperature and high-temperature regions. Midpoint voltage tends to slightly diverge in the low-temperature region and converge in the high-temperature region. This is caused by heterogeneity of the element including differences in the thin-film quality between the two elements. This needs to be further reduced in the future.

HEAT-PROOF PACKAGING TECHNOLOGY - When the InSb thin-film magneto-resistors are used as a rotation sensor for automobiles, the packaging technology is important factor.

The construction of inner lead bonding parts is shown in Fig. 11. In packaging, important points are as follows:

(1) to assure the stable connection in the temperature range of -40 ~150℃ as required for automotive use, and

(2) to reduce the package thickness (including mold) of the upper surface of the element (in order to allow the distance between the element and the detection gear to be as small as possible).

Usually, in sensor elements for automobile use, wire-bonding→transfer products such as hole IC are popularly used. However, we require the upper surface package thickness to be less than 200 μm as described in (2) and transfer products do not necessarily satisfy this requirement. In this regard, we adopted the TAB packaging that enables a thinner package.

As shown in Fig.11, since the InSb thin-film magneto-resistors form the detection part with two-chip differential type, two device holes are made to keep the two chips at a certain distance (to fit the detection gear pitch) and, as for the ILB (Inner Lead Bonding) connection, a Au-Sn eutectic is formed between Sn plating on the Cu foil lead of TAB and Au electrodes on the element connection parts. After that, a resin coating is applied in order to protect the element surface and the ILB parts. This completes the detection element.

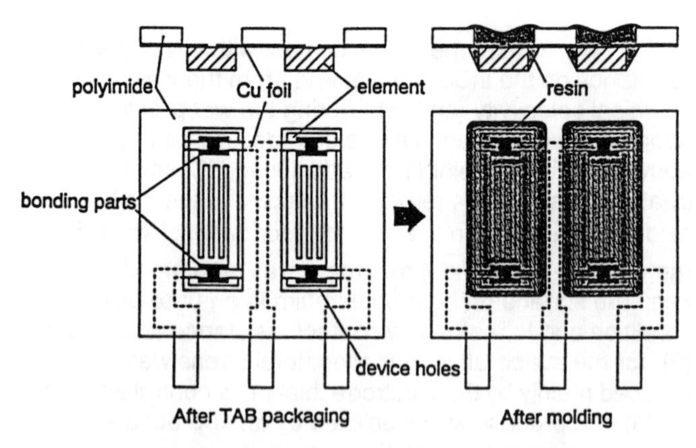

Fig.11 Construction of inner lead bonding parts to InSb thin-film MR

Since TAB packaging has not been applied to automotive area so far, durability verification was needed. In order to assure the stability of the connections, as previously described in (1), the design assurance of the structure particularly in terms of its consistency was an important point. Regarding the design assurance, we performed theoretical analyses in terms of thermal factor (thermal stress), vibration factor (resonance resistance), and connection factor (lead disconnection). Fig. 12 shows the evaluation models for analyzing thermal stress. In this figure, (a) is the face structure model of TAB and (b) is the cross-sectional structure model. Using this model, the stress distribution was obtained. According to the analytical results, the maximum stress occurs on the Cu foil in the OLB (Outer Lead Bonding) parts indicated with the dotted line in this figure, but breakage will not occur because it is below the breaking strength at the use temperature limit of 150°C. At this point, in terms of equivalent safety ratio at 150°C, as shown in Table 1, a safety ratio greater than 1.5 will be assured by the forming height more than 0.1 mm. As for the vibration factor, a resonance frequency more than 20 kHz shall be assured in design and, as for the connection factor, the connection conditions shall be defined to avoid breakage of the lead. Thus, the suitability of the design was theoretically evaluated and it was found that it is sufficiently reliable in specifications for automobile use. This was confirmed by reliability tests.

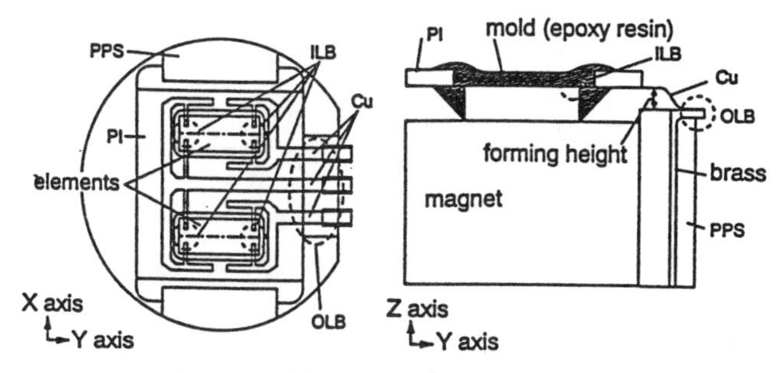

Fig.12 Model for analyzing thermal stress to TAB structure

Table 1 Safety ratio thermal stress analysis for TAB structure

(a) TAB face structure model		
forming height	safety ratio by thermal stress analysis	
(mm)	X-axis direction	Y-axis direction
0	2.50	2.50
0.1	2.50	2.50
0.5	2.50	2.50
(b) TAB cross-section structure model		
forming height	safety ratio by thermal stress analysis	
(mm)	Y-axis direction	Z-axis direction
0	1.09	1.85
0.1	1.60	2.50
0.5	2.01	2.50

ROTATION SENSOR WITH INSB THIN-FILM MAGNETO-RESISTORS

CONSTRUCTION OF ROTATION SENSOR - The construction of rotation sensor with the above-mentioned InSb thin-film MR as the detection part is shown in Fig. 13.

The InSb thin-film magneto-resistors are placed on the Sm-Co bias magnet. The peripheral circuit parts are constructed on the flexible printed circuit (FPC), and the OLB connection is made by welding at the junction terminal connecting the element and the peripheral circuit. After that, by ultrasonic bonding, the resin cap is integrated with the functional parts to complete the sensor

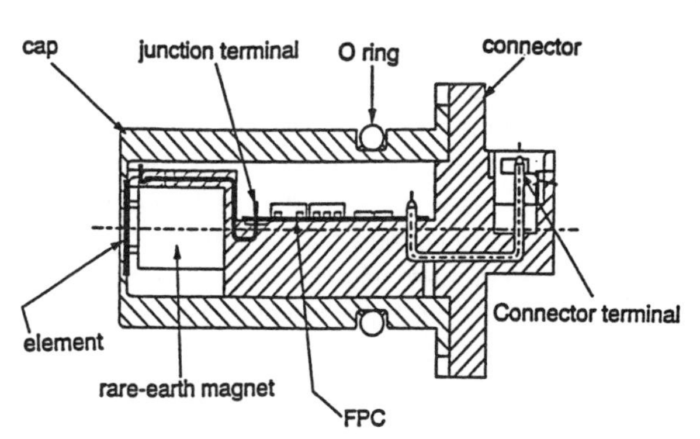

Fig.13 Construction of rotation sensor with InSb thin-film MR

unit. In this rotation sensor, construction of the peripheral driving circuit is shown in Fig. 14. Since output sensitivity of the InSb thin-film MR is high, the waveform shaping circuit (for converting into a rectangular wave) is constructed without an amplifier, using a simple circuit consisting only of a single comparator. For this peripheral circuit, each component is chosen to sufficiently meet the required heat resistance, pressure resistance, etc.

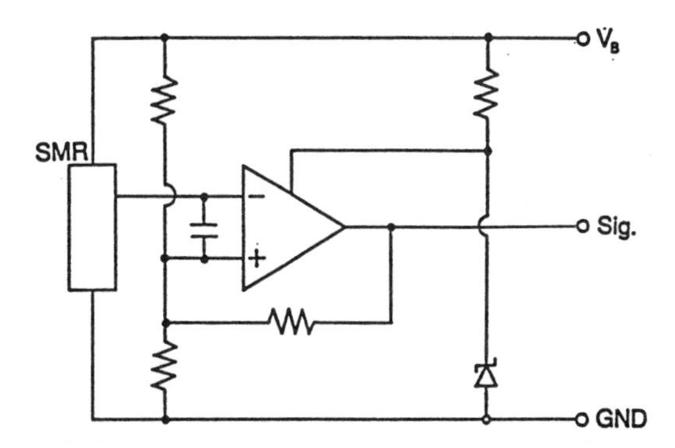

Fig.14 Construction of peripheral driving circuit

Since an amplifier is not used, the following important points must be considered.

(1) to optimize the magnetic circuit part,

(2) to reduce the gap between the element surface and the unit surface within the sensor unit,

and in order to assure reliability such as heat resistance,

(3) to assure structural reliability.

As for (1), optimization of the pitch between the two elements and the shape of magnet and clarification of optimum operating magnetic field of the element (resistance change per unit magnetic flux is greatest in the range of 0.3~0.4 T) were performed to increase the difference in resistance between the two parts of the differential element as much as possible. As for (2) and (3), as previously described about TAB, it was designed to assure theoretically sufficient reliability.

ROTATION SENSOR CHRACTERISTICS - Fig. 15 shows the output signal of the InSb thin-film MR and the sensor output signal after signal shaping at temperatures of -40, 25, and 150°C. In addition, Fig. 16 shows the resistance change of the element after the heat cycle test between -40°C and 150°C under 5V drive. Then Fig. 17 shows the resistance change of the element after the high-temperature voltage bias test at 150°C, 5V. The InSb thin-film magneto-resistors have sufficient temperature stability (and achieve stable operation) for use as a rotation sensor.

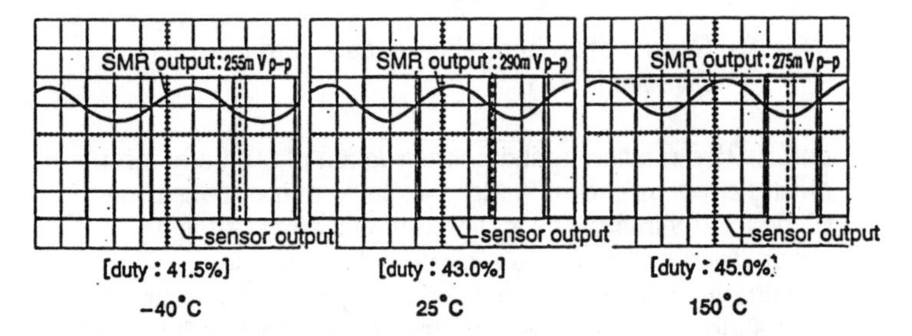

Fig.15 Output signal of InSb MR and sensor output signal after signal shaping

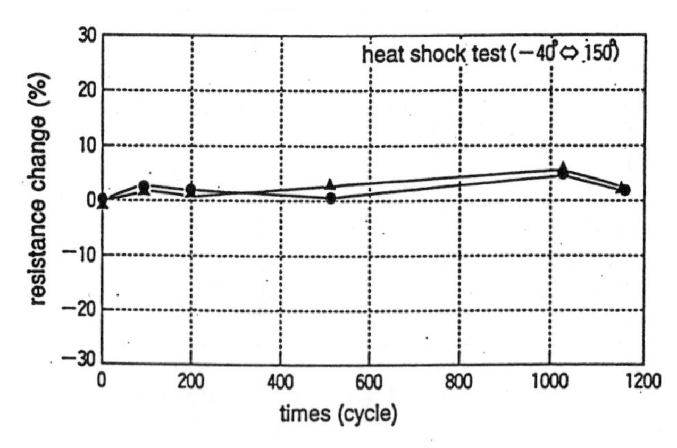

Fig.16 Resistance change of InSb MR after heat shock test

Table 2 shows the reliability test results as a rotation sensor for automobile use. Table 3 shows the basic specifications of this InSb thin-film MR type rotation sensor. Considering the absence of fatigue deterioration, it clearly has sufficient performance for use as a rotation sensor for automotive applications.

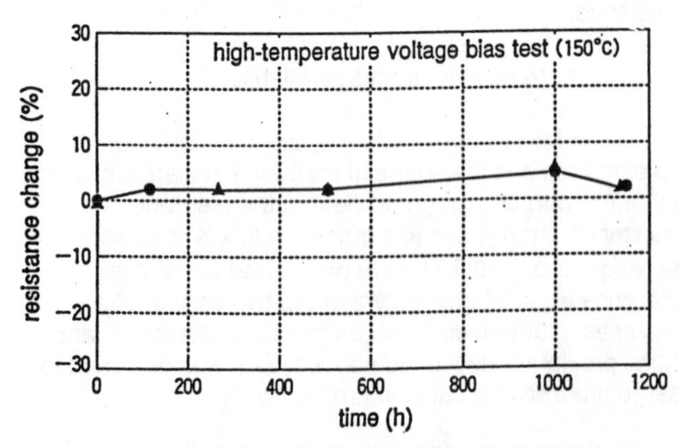

Fig.17 Resistance change of InSb MR after high-temperature voltage bias test

Table 2 Reliability test results of rotation sensor

test items	conditions	results
high temperature	150°C, 500h	1000h OK
low temperature	-40°C, 500h	500h OK
high temperature and high humidity	85°C, 90~95%, 500h	500h OK
heat shock	-40°C, 150°C, 1h each, 1000cycle	2400cycle OK
high temperature voltage bias	150°C, 5V, 1000h	1000h OK
electrostatic charge	±1kv, 10 times for each terminal	OK
surge durability	70V, 200ms	OK
EMI test (strong electric field)	2~200 MHz, 100 V/m	OK

Table 3 Specification of InSb thin-film MR type rotation sensor

[Basic Specifications]

operating temperature	-40~150°C
operating voltage	5±0.25V
consumption current	10 mA or less
output voltage	High:4.8 V, Low:0.8 V
output current	10 mA or less
duty	50±20%
air gap	1.5 mm or less
detection frequency	15kHz or less
insulation resistance	500M ohm or more

CONCLUSION

We have developed the technology for the epitaxial growth of InSb thin-films with high electron mobility directly on Si substrates and , using this technology, have realized InSb thin-film magneto-resistors with high heat resistance and high sensitivity and incorporated these into a rotation sensor with sufficient reliability for automobile use. In the future, we plan to establish the technology for mass production for automotive applications such as electronic control AT system, engine ignition system, and ABS.

ACKNOWLEDGMENTS

Finally, we greatly appreciate the staff of Ion Engineering Center K.K., whose facilities were used in developing this InSb thin-film MR rotation sensor, particularly in the basic study on InSb/Si epitaxial growth technology.

REFERENCES

[1] W.Thomson Phil. Trans. 146, p.736 (1956).

[2] H.Lippmann and F.Kuhrt Z.Naturforsch 13a, p.462 (1958).

[3] Toshiaki Fukunaka, et al. : The application and properties of thin InSb films for highly sensitive magneto-resistance elements. Toyo-tsushinnki Technical Report. No.40 (1987).

[4] J.L.Davis and P.E.Thompson : Molecular beam epitaxy growth of InSb films on GaAs. Appl. Phys. Lett. 54, p.2235 (1989).

[5] J.R. Soderstrom, et al. : Molecular beam epitxy growth and characterization of InSb layers on GaAs substrates. Semicond. Sci. Technol. 7, p.337 (1992).

[6] C. Besikci, et al. : Anomalous Hall effect in InSb layers grown by metalorganic chemical vapor deposition on GaAs substrates. J. Appl. Phys. 73, p.5009 (1993).

[7] D.L. Partin, et al. : Growth of high mobility InSb by metalorganic chemical vapor deposition. J. Electronic Ater. 23, p.75 (1994).

[8] T.S. Rao, et al.:Effect of substrate temperature on the growth rate and surface morphology of heteroepitaxial indium antimonide layers growth on (100) GaAs by metalorganic magnetron sputtering. J. Appl. Phys. 65, p.585 (1989).

[9] J.C. Chen, et al. : Low-temperature heteroepitaxial growth of InSb on CdTe by metalorganic chemical vapor deposition. Appl. Phys. Lett. 53, p.773 (1988).

[10] Toshinori Takagi, et al. : Preperation of InSb and GaAs thin films by ionized-cluster beam deposition. Vacuum. 22, No.7 p.15 (1979).

[11] Ichiro Shibasaki Nikkakyo-Geppou (in Japanese). May, p.12 (1988)

[12] J.I. Chyi, et al. : Molecular beam epitxy growth and characterization of InSb on Si. Appl. Phys. Lett. 54, No11, p.1016 (1989).

[13] A.Many, et al. : Semiconductor surfaces. (North Holland, Amsterdom), p.307 (1965)

[14] P.K. Chiang and S.M. Bedair : Growth of InSb and InAs1-xSbx by OM-CVD. J.Electrochem. Soc.131, p.2422 (1986).

[15] D.L. Dexter and F.Seitz : Effect of dislocations on mobilities in semiconductors. Phys.Rev. 86, p.964 (1952).

[16] Tetsuo Kawasaki, et al. : Epitaxial growth of InSb on hydrogen-terminated Si (111). Extended Abstract. J. Appl. Phys. 29a-ZR-1 (1993).

970382

Steering Wheel Angle Sensor for Vehicle Dynamics Control Systems

Jürgen Gruber
Robert Bosch GmbH

Copyright 1997 Society of Automotive Engineers, Inc.

ABSTRACT

Since 1995, Bosch has produced the Vehicle Dynamics Control Systems for passenger cars. In these systems the **Steering Wheel Angle Sensors** are used as the reference for other sensors, and for the whole system. The main requirements for these sensors are a high level of inherent safety and the absolute measuring of the steering wheel position, within the whole steering range, immediately after power on. The other requirements are as usual for automotive applications mass producibility, low cost and a high resistance against environmental influences. This paper describes a new steering wheel angle sensor that is under development, in comparison to the sensor in production.

INTRODUCTION

The first VDC-systems which have been in the market since 1995 use different steering wheel angle sensors.

To provide improvements over the existing products, Bosch developed its own steering wheel angle sensor, that is able to meet the demands of vehicle dynamics control systems.

This paper describes two types of sensors, where one of them, the steering wheel angle sensor SAS1, has been in production since 1996, the other, the SAS3, which is based on a different measuring principle, is under investigation and will go into production in 1999.

Both sensors transmit their message via „Controller Area Network“ CAN, a common and safe bus system, so that it is possible to use the information also for other systems, such as vehicle navigation and suspension control.

The center position of the steering wheel is stored as an offset in an EEPROM after the sensor has been mounted into the vehicle, or after new adjustment of the steering system has been made in a workshop. The measured angle is corrected by the offset.

DESCRIPTION OF THE SENSORS

STEERING WHEEL ANGLE SENSOR SAS1: The Bosch steering wheel angle sensor SAS1 uses a digital measuring principle. It has two code systems which are based on magnets and associated Hall elements, switched by rotating metal discs.

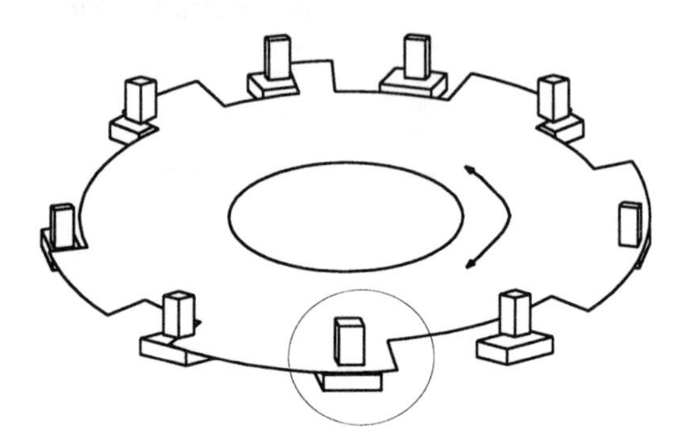

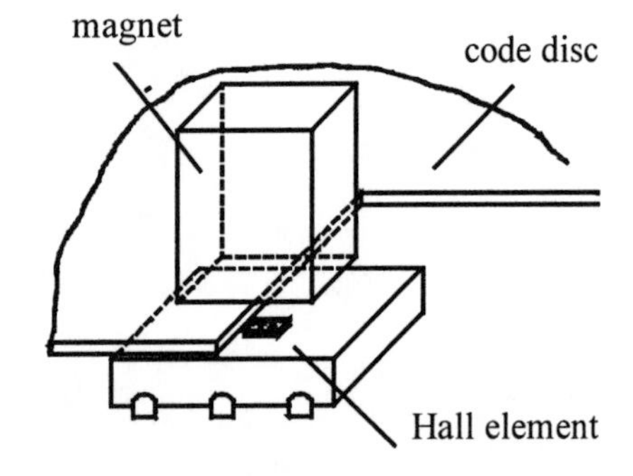

Figure 1: Measuring principle of SAS1

One of the code systems, the finecode system, is to detect the angular position of the steering wheel within one revolution, the other system, the coarsecode system, enables the sensor to know in which of the possible revolutions the steering wheel is positioned. The two code systems are transmitted 4:1 by gearwheels, whereby the steering wheel angle sensor can measure the absolute angle within a range of 1,440 degrees. This is sufficient for common passenger cars.

The values of both code systems are given to a µ-processor that does all plausibility checks and which transmits the angle values of the steering wheel via CAN interface.

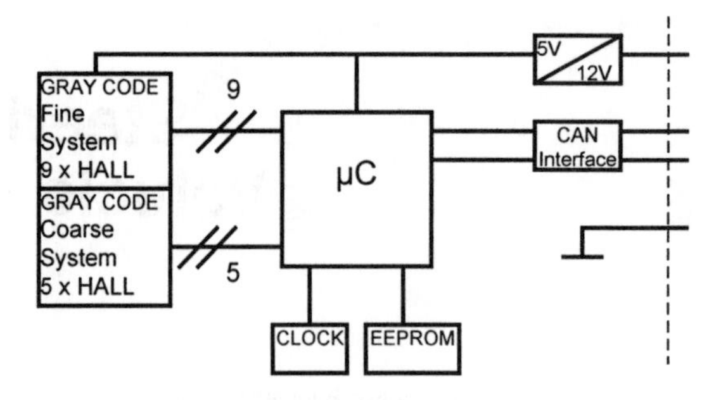

Figure 2: Block circuitry of SAS1

The finecode system is directly driven by the steering column and detects any movement of the steering wheel.

It is based on a GRAY code using 9 magnet Hall elements. The GRAY code is generated in such a way that from one step to the next exactly one bit is changed. This concerns in like manner the transition from 357.5° to 0°, where also only one bit is changed.

The coarsecode system is driven by a transmission of 4:1 of the steering column. This means that the coarsecode disc rotates once when the steering wheel rotates four times. The coarsecode is needed immediately for the absolute position of the steering system after the engine is started. The coarsecode system uses a much simpler code of 5 magnet Hall elements which are switched on after one another, and again switched off in the same order. Therefore there are ten different permissible coarsecode combinations.

STEERING WHEEL ANGLE SENSOR SAS3: The steering wheel angle sensor SAS3 is based on anisotropic magnetoresistive measuring elements AMR, which have an electrical resistance corresponding to the direction of an outside magnetic field. The measuring principle is analog.

To eliminate influences of temperature, the measuring elements consist of a couple of AMR elements, which are turned 45° to each other. The output voltages of the AMR elements are processed by an evaluating circuitry that is located in the same housing with the AMR elements.

The sensor uses two magnets which are driven by tooth wheels and which have different ratios to the steering wheel.

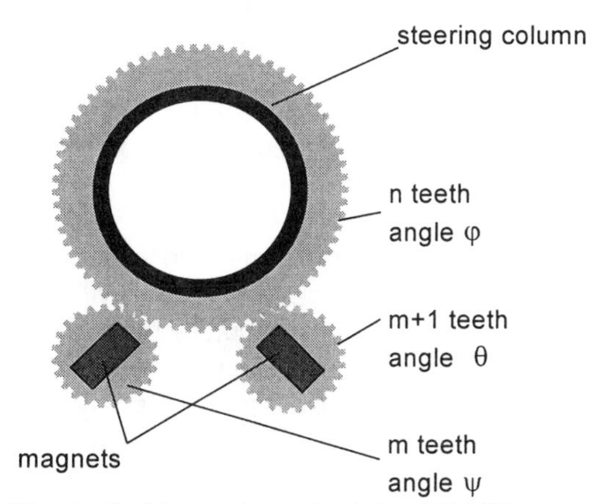

Figure 3: Measuring principle of SAS3

The respective direction of the magnetic flux is measured by the AMR elements.

By computing the angle values of these two measuring elements, one can calculate the absolute position of the steering wheel within the whole steering range.

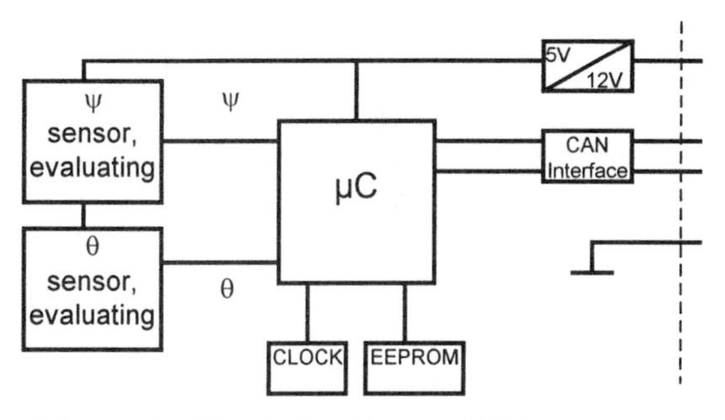

Figure 4: Block circuitry of SAS3

The calculation is described as follows:

With
- n number of teeth of steering column
- m number of teeth of measuring wheel 1
- $m+1$ number of teeth of measuring wheel 2
- φ angle of steering column
- α, ψ angle of measuring wheel 1
- β, θ angle of measuring wheel 2

Since the AMR elements are able only to measure in a range of 180°, the angles of the measuring wheels are:

$\alpha = \psi + i \cdot 180°$, $\beta = \theta + j \cdot 180°$ with i, j as integer values.

The steering wheel angle is:

$$\varphi = \frac{m}{n} \cdot \left(\psi + i \cdot 180°\right) \qquad [1]$$

and $$\varphi = \frac{m+1}{n} \cdot \left(\theta + j \cdot 180°\right) \qquad [2]$$

[1] = [2]

$$\frac{m}{n} \cdot \left(\psi + i \cdot 180°\right) = \frac{m+1}{n} \cdot \left(\theta + j \cdot 180°\right)$$

$$\Rightarrow \quad m \cdot i - (m+1) \cdot j = \frac{(m+1) \cdot \theta - m \cdot \psi}{180°} = k \qquad [3]$$

where k has to be an integer value because m, i and j are all integers.

In reality k is no integer value, for the angles ψ and θ are inaccurate due to measuring failures. By setting k to the next integer value the total failure of the calculated steering angle can be reduced considerably. The quality of the measured values ψ and θ can be judged by the difference of k from the next integer value. If this difference exceeds a predetermined value, the sensor can recognize a failure and send an error message to the system control unit.

with [3] $\Rightarrow k = m \cdot i - (m+1) \cdot j$ [4]

The only relevant solution of the equation [4] is

$$i = j = -k \qquad [5]$$

The steering wheel angle can be calculated as the average value of the equations [1] and [2], by which the total angle failure of the sensor is reduced additionally:

$$\varphi = \frac{m \cdot \psi + (m+1) \cdot \theta - (2m+1) \cdot k \cdot 180°}{2n}$$

This calculated angle is transmitted to the system control unit after adding the stored calibration offset value of the center position.

REQUIREMENTS OF THE VDC SYSTEM AND REALIZATION BY THE STEERING WHEEL ANGLE SENSORS

ABSOLUTE MEASURING OF THE STEERING WHEEL ANGLE: The VDC system needs the absolute value of the steering wheel angle immediately after power on without any movement of the steering wheel or the car.

SAS1: The steering wheel angle sensor SAS1 uses the information of the finecode system which is a combination of 9 bit, and determines the associated angle by using a list of values.

SAS3: In the SAS3 the steering wheel angle is calculated by the μ-processor from the two angles ψ and θ of the measuring elements.

MEASURING RANGE IS EQUIVALENT TO THE STEERING RANGE OF THE VEHICLE: Only by measuring the absolute steering wheel angle within the whole steering range, it is possible to have full system support directly after power on.

SAS1: The steering wheel angle sensor SAS1 needs an additional code system, the coarsecode system which is transmitted 4:1 of the finecode system

SAS3: The SAS3 does not need any further effort to calculate the absolute steering wheel angle within the whole steering range. The measuring range of the SAS3 is determined by the ratios of the toothed wheels, and works within four revolutions of the steering wheel.

INHERENT SAFETY OF THE STEERING WHEEL ANGLE SENSOR: None of the systems with access to the steering wheel angle sensor includes plausibility checks at standstill of the car. Plausibility checks while driving need a certain time during which systems must be out of operation. Therefore positive information of the steering wheel angle sensor is relevant. In the VDC system, the steering wheel angle sensor is used as a reference for other sensors and for the whole system, because it indicates the driver's desire which cannot be checked by a system.

SAS1: The SAS1 uses a 9 bit code which provides 512 possible combinations, from which 144 bit combinations are permitted and which have to occur in a predetermined sequence. In case a combination or a sequence is not permitted a failure message is sent to the system control unit and the VDC is switched off. If a coarsecode failure occurs the sensor remains in function while the ignition is on, and calculates the number of rotations of the steering wheel by means of software. An error however is stored in the EEPROM, which means that the system will not be available after the next power on.

SAS3: The SAS3 uses the geometrical and mathematical interrelationships of the two measured angles. To each position of one of the two magnet wheels permitted positions of the other magnet wheel are set correspondingly. All other values lead either to a correction by the algorithm or to a failure message, depending on the deviation of the measured value from the predetermined value.

RESOLUTION: The VDC system requires a resolution of 2.5° of the steering wheel angle, but there are other systems like car navigation or suspension control which need a higher resolution.

SAS1: The SAS1 is designed especially for use in the VDC system and has therefore a resolution of 2.5°, which led to the mentioned number of 144 bit combinations for 360°.

SAS3: Due to the analog measuring principle of the SAS3 the resolution is better than 1°. With this high resolution it is possible to calculate the angular velocity of the steering wheel even for very slow rotation by differentiation of the angle values. Therefore this sensor can be used in various other systems.

ACCURACY: The VDC system requires an accuracy of 2.5° of the steering wheel angle sensor.

SAS1: In the SAS1 the accuracy depends on the mechanical clearances and on the precise positioning of the Hall elements. An accuracy of 2.5° is realized.

SAS3: The accuracy of the SAS3 is only influenced by mechanics. The position of the sensor elements is not important, because there is a large area of homogeneous magnetic field direction in the center of the magnets. Failures which do not exceed a predetermined value are corrected by the algorithm, thus increasing the accuracy.

COSTS: For automotive applications, it is important to offer high performance devices at low costs.

SAS1: The SAS1 needs 14 Hall elements and precision manufacturing. Therefore the potential for costs reduction is limited.

SAS3: The SAS3 achieves costs reduction, because it needs only two measuring elements and magnets, and because the precision of positioning the measuring elements is not important.

TORQUE TO ROTATE: The additional torque to the steering column impedes the reset of the wheels and has to be reduced to a minimum.

SAS1: The metal code disc located between the magnets and the Hall elements is attracted by the magnets. This causes a certain friction between the disc and the sensor housing and results in a torque of approximately 0.08 Nm maximum.

SAS3: The only metal parts of the SAS3 are the magnets, which are integrated in the two measuring gear wheels. Therefore no magnetic force can generate any friction.

INTERFACE: For the VDC system it is important that the angle is transmitted via a monitored and safe bus system. The information of the two steering wheel angle sensors SAS1 and SAS3 is sent to the system control unit via CAN, a common bus system with separate monitoring. In order to calibrate the sensors after mounting them into the vehicle, CAN interface is used.

SAS1: The only interface of the SAS1 is the CAN interface.

SAS3: The SAS3 has additional customer specific interfaces using the steering angle in other systems.

MOUNTING IN THE CAR: The steering wheel angle sensor is mounted in the car between the steering wheel and the steering gear. Relevant are easy installation and removal for service.

SAS1: The SAS1 needs a closed ring design, because the 9 Hall elements are positioned on the circumference of the sensor. Therefore, it must be mounted axially to the steering column.

SAS3: For the SAS3 there is the possibility to use an open shape of the sensor. Thus it can be positioned radially to the column. Only a sun gear which may belong to the steering column or to the roll connector, is needed. It is planned to design a version of SAS3 with the packaging of SAS1 for mounting the sensor on the steering column to replace the SAS1 in current projects, if there is a costs or performance advantage.

CONCLUSION

Both Bosch steering wheel angle sensors comply with the requirements of vehicle dynamics control systems: inherent safety, measuring of the absolute angle and measuring range of the whole steering range. Both sensors represent a very high safety level.

The SAS3 has advantages in performance, higher resolution and accuracy, possible output of angular velocity, customer specific interface, reduced torque to rotate, higher flexibility of installation in the vehicle and low costs. Since the development of this sensor is still in process, additional customers' requirements for other systems can be discussed. The Bosch steering wheel angle sensor SAS3 is planned to be used in numerous projects with VDC and other systems.

952085

Techniques for Distance Measurement

Miles Upton
Cambridge Consultants Ltd.

ABSTRACT

This paper presents a comparison of the available electronic techniques for distance measurement suitable for the off-highway environment. The main techniques described are:

- ultrasonic
- passive infrared
- laser radar
- FMCW radar
- impulse radar
- capacitive
- vision

The techniques are compared across a number of key parameters, including sensitivity, measurement range and cost. A brief description of some of the disadvantages is also included. Finally the paper discusses the use of sensor fusion techniques to overcome the inherent disadvantages of particular techniques.

INTRODUCTION

The need for accurate distance measurement is increasing in the off-highway vehicle market, with applications as diverse as height sensing for combine harvester headers [1] and obstacle detection for long extension booms. Collision warning techniques are being rapidly developed in the on-highway environment [2], and these can be applied to off-highway vehicles in applications such as forestry (tree avoidance) and building sites. Research by the UK Health and Safety Executive has shown that vehicle collisions are a major cause of death and injury in quarries, including reversing accidents, dump trucks travelling forwards whilst tipping and loaders hitting overhead power lines [3]. In all these cases, the addition of a collision warning sensor would have drastically reduced the number of accidents.

For each application area, the specific requirements vary and it is usually necessary to tailor the sensing technique to the application. This paper presents an overview of the currently available electronic distance measurement techniques. The aim is to give a comparison of different sensing technologies across a number of key parameters so that the reader can choose the most appropriate technique for a particular application. Note that this paper considers mainly the problem of non-cooperative distance sensing - by non-cooperative we mean that the object to be sensed has not been modified to make it responsive to a particular sensing technique. The exception to this is vision systems, which need to be cooperative (or have significant prior knowledge) to give any distance measurement.

The main techniques considered are ultrasonic, passive infrared, laser radar, frequency modulated-continuous wave (FMCW) radar, impulse radar, capacitive and video-based vision systems. In each case the key parameters compared are:

- sensing range - the maximum range over which the technique can be used
- resolution - the relative change in distance that can measured
- directionality - the width of the beam over which the sensor is sensitive, usually defined in terms of the angle at which the power is 3dB down on the peak.
- response time - how quickly the sensor can respond to a change in the distance
- cost - these figures are given as an indication only and are based on high-volume (100,000+) manufacture. Cost of any particular sensor will largely depend upon the precise specification required, including any environmental hardening and signal processing.

- size - again this is an indication only. Precise size will be implementation dependent and will vary with the constraints of a particular application.
- environmental immunity - the off-highway environment is particularly severe, so this heading describes any particular weaknesses of the various techniques.

For each sensing technique a brief description of the sensing technique is included and any specific advantages and disadvantages are also described.

It is often impossible to obtain the ideal sensor for a particular application and it may be more appropriate to use more than one type of sensing technique to achieve a desired performance requirement. This approach is called sensor fusion. This paper concludes with a discussion of sensor fusion and gives some examples of where it could be used to advantage.

ULTRASONICS

Ultrasonic distance transducers work by measuring the time-of-flight of a short burst of sound energy. Typically, separate transmitters and receivers are used and the transmitter is excited with a burst of high frequency (40 - 80kHz) signal. The distance to an object can be evaluated simply by measuring the time interval between transmitting a pulse and receiving a reflection.

Table 1 gives the main features of ultrasonic sensors. The main advantage of ultrasonic sensors are their relatively low cost and small size. The major disadvantage is that certain targets give a very poor reflection and this can lead to them being undetected. For example, lightly packed soil and grass are very difficult to detect. Another disadvantage is the variation with temperature of the propagation time in air for sound waves. This can be significant (0.17 %/°C at 20°C) and should be compensated for in any sensor intended for use over a wide temperature range.

PASSIVE INFRARED

Passive infrared (PIR) sensors measure the thermal energy emitted by objects in the vicinity of the sensor. Due to the passive nature of the sensor it is not possible to determine the distance to an object very precisely. The main role for these sensors is in warning systems, such as detecting a person in the path of a reversing vehicle. They are already used extensively as intruder detectors in security systems.

Table 2 gives the main parameters for PIR sensors. As with ultrasonic sensors, the main advantages are low cost and small size. A major disadvantage is the slow response time which limits their use in collision warning systems where the vehicle operator would often not be given enough prior warning to avert a collision.

LASER RADAR

There are two types of laser radar:

- pulsed, which uses short, high power pulses of infrared light and the distance to an object is determined from the time of flight of the pulse
- continuous wave (CW), where the light is amplitude modulated with a sine wave of ~100MHz. The distance to the object can be deduced from the phase difference between the outgoing and reflected light.

The pulsed technique is the most popular, because the signal processing is easier and it is possible to achieve longer ranges than with the CW technique. The main application of laser radar is in long range directional distance measurement as the light beam is usually focused. A fuller description of laser radar can be found in [4].

Table 3 gives the main parameters for pulsed laser radar. The technique offers long range, high directionality and fast response time, but it is costly and it is prone to external environmental effects such as poor visibility and mud on the sensor. The range of this technique in also limited in many applications by the need to keep the laser power below eye-safety levels.

FMCW RADAR

FMCW radar uses a high frequency electromagnetic carrier (typically microwave frequencies or higher) which is frequency modulated, usually with a sawtooth type signal. This is transmitted and the reflected signal is compared with the transmitted signal to give a frequency difference which is proportional to the distance to the object [5]. The Doppler shift on the returned signal can also be used to determine the relative speed of the other object. Current systems being developed for on-highway use work at 24GHz, 77GHz and 94GHz. This technology is widely applicable and it is particularly useful in environments where visibility is poor.

Table 4 gives the main features of FMCW radar. A major disadvantage of this technique is the relatively high cost of the electronics at microwave and millimetre wave frequencies. Apart from the ability to 'see' through mud and spray, another advantage of this technique is that it can used for both narrow and wide beamwidths, allowing the beamwidth to be modified to suit a particular application. The disadvantage of wide beamwidths is a loss in range.

Alternatively, a relatively narrow, long-range beam can be electronically scanned to give the benefits of longer range and wide beamwidth, but at the expense of complexity and additional cost.

IMPULSE RADAR

Impulse radar is similar to FMCW radar in that an electromagnetic signal is radiated, but instead of using a continuous wave, the radiated signal consists of very short pulses. The receiver samples the returned signal at defined intervals after the pulse is transmitted to determine whether an object is present.

One of the major advantages of impulse radar over FMCW radar is that similar resolution can be achieved using much lower frequency electronics with a consequent saving in cost. It has similar advantages over optical techniques in terms of immunity to environmental conditions (fog, mud etc).

Table 5 shows the major parameters of impulse radar-based systems. A disadvantage of this technique compared with FMCW radar is a reduction in maximum range (50m compared with 200m), which occurs because it is not possible to have a wideband antenna which is also narrow beamwidth. It is also susceptible to external electromagnetic interference due to the broadband receiver required.

CAPACITIVE

Capacitive sensors use electrodes excited at low frequencies (typically 5kHz) and objects are detected by variations in the capacitance between the electrodes when the objects are in close proximity (distances less than 2m). A capacitive sensor is sensitive to most objects that contain some significant dielectric material. One of the major advantages of this technique is that it will work down to very short ranges.

Table 6 gives the main parameters of capacitive systems. Despite their limited range, they are very low cost and the sensor head can often be integrated into existing plastic work to reduce manufacturing costs. One of the major applications is in slow-speed collision warning during manoeuvring operations. They are also robust to external environmental effects such as mud, spray etc. A disadvantage is that capacitive techniques cannot be used to measure absolute range, because the variation in the capacitance will depend upon the dielectric properties of the object being sensed.

VISION SYSTEMS

Vision systems for distance measurement are based upon using a video camera to view an object as it moves. The position of an object can then be determined from the movement across or up and down the field of view. Normally, visual clues (eg white crosses) are needed on the object to give the image processing software something to track. Actual distance measurement depends upon detailed prior knowledge of the geometry of the object to be tracked. Further details of vision systems used in the on-highway environment can be found in [6].

Table 7 gives the main parameters for vision systems. Typically, these systems could be used for docking-type operations, but the high cost and high sensitivity to external environmental effects make their use unlikely in most vehicle applications. One of the major problems is the significant processing power needed to process the image.

COMPARISON OF TECHNIQUES

The techniques described above can be broadly classified into three categories:

1. optical - passive infra-red, laser radar and vision
2. electromagnetic - FMCW radar, impulse radar and capacitive
3. acoustic - ultrasonics

Each technique will have preferred applications where it is better suited. For example, FMCW radar is probably the best technique for long-range distance measurement, despite its relatively high cost. Capacitive sensors are ideally suited to robust, short-range applications.

Optical techniques all suffer similar disadvantages in that they are sensitive to external environmental effects (rain, mud etc) which limits their usefulness in many applications. Of the optical techniques, both passive infra-red and vision cannot give a direct measurement of distance to an object. Therefore, for general applicability laser radar is the most useful of these techniques, despite its relatively high cost.

The electromagnetic techniques all have the same advantage over optical techniques in that they are generally immune to the external environment. There is a significant cost difference between the most expensive technique (FMCW radar) and the cheapest (capacitive), and this is reflected in the relative range of the techniques, with FMCW radar being long range (100m +) and capacitive being short range (< 2m). FMCW radar can be used at short and medium ranges with wider beamwidths, so this is the most flexible of all the techniques discussed.

Ultrasonic techniques can give high resolution for a relatively low cost, but they suffer the disadvantage of being sensitive to external variations in temperature. However, in applications where only short-term relative measurements of distance are required, ultrasonics is often well-suited.

SENSOR FUSION

In certain applications, one sensing technique alone is inadequate to achieve the technical performance required. In this case it is appropriate to consider using more than one type of sensor to cover the flaws in a single sensor approach. The signals from both sensors can then be processed to give a composite result which is more accurate or reliable. The disadvantage of this approach is the additional cost of a second sensor, and the increase in the cost and complexity required for signal processing.

A typical requirement for a bulldozer is to warn of potential collisions with people whilst ignoring collisions with other objects. A passive infra-red sensor can be used to differentiate between human and inanimate objects, but it is incapable of determining distance. Combining this with an impulse radar to detect the distance to the person would make a more intelligent and reliable sensor which could determine the likely accident potential. It would also avoid false alarms caused by distant, but large thermal radiators such as the sun.

A major application of sensor fusion is in situations where a sensor must detect a wide variety of objects of differing properties. Ultrasonic techniques are insensitive to acoustically 'soft' materials (eg cloth) and to objects at an oblique angle to the sensor. Electromagnetic techniques are insensitive to materials with very low dielectric constants (eg a dry plastic pipe). In these examples, the addition of a laser radar sensor would overcome the disadvantages of the other two techniques.

CONCLUSIONS

This paper has presented an overview of various techniques which can be used for non-cooperative distance measurement. The techniques have been compared across a variety of parameters including cost, size and range. The aim has been to give the reader the opportunity to assess the relative merits for a particular application. Where appropriate the disadvantages of each technique have been described.

To overcome disadvantages in a particular technique, the use of sensor fusion has been described. This allows multiple sensing techniques to be used at the expense of additional cost and signal processing complexity.

REFERENCES

[1] Romes, R, "Electrohydraulic Header Control", *Off-Highway Engineering*, SAE, Supplement to Volume 102, no. 9 of *Automotive Engineering*, September 1994.

[2] Kawai, Mitsuo, "Collision Avoidance Technologies", SAE Convergence '94, 94C038, Sept. 1994.

[3] Quarry Fact File, "Ten Years of Death", UK Health and Safety Executive, May 1993.

[4] Yangisawa, T, et al. "Development of a Laser Radar System for Automobiles", SAE International Congress and Expo, 920745, Feb 1992.

[5] Woll, Jerry, "Radar Based Vehicle Collision Warning System", SAE Convergence '94, 94C036, Sept. 1994.

[6] Saneyoshi, Keiji , et al, "3-D Image Processing for Recognition of Road Geometry and Obstacles", *Journal of Automotive Engineers of Japan,* Volume 46, no. 4, 1992.

Table 1 - Parameters for Ultrasonic Sensors

Sensing range	10m maximum
Resolution	10mm resolution achievable
Directionality	30° beamwidth minimum
Response time	Limited by velocity of sound, eg 60msec for object at 10m
Cost	$15
Size	30mm diameter for transducer
Environmental immunity	Good in poor visibility, but dependent upon variations in sound propagation with temperature

Table 2 - Parameters for Passive Infra-Red Sensors

Sensing range	10m maximum
Resolution	Poor, usually based upon a threshold of object detection
Directionality	Typically 90° beamwidth, can be improved with lenses
Response time	Slow, typically 1 second
Cost	< $10
Size	Depends upon lenses, typically 20mm square
Environmental immunity	Poor, but better than other optical systems in poor visibility

Table 3 - Parameters for Laser Radar Sensors

Sensing range	100m maximum, depending on laser power and eye safety issues, 0.5m minimum due to pulse length
Resolution	1mm minimum
Directionality	1° beamwidth achievable
Response time	Fast, typically 10msec
Cost	$50+
Size	50mm diameter by 100mm long
Environmental immunity	Poor, susceptible to bad visibility and dirt. Good in areas of high electromagnetic interference.

Table 4 - Parameters for FMCW Radar Sensors

Sensing range	150m+, depending upon beamwidth
Resolution	10mm
Directionality	Minimum 2°, wide beams possible
Response time	Fast, typically 1msec
Cost	$200+
Size	250mm by 150mm by 100mm typically
Environmental immunity	Good, unaffected by poor visibility

Table 5 - Parameters for Impulse Radar Sensors

Sensing range	50m maximum, 0.5m minimum
Resolution	10mm
Directionality	25° minimum beamwidth minimum, but wider if necessary
Response time	Fast, typically 1msec
Cost	$100+
Size	250m by 100mm by 100mm typically
Environmental immunity	Good, unaffected by poor visibility and spray

Table 6 - Parameters for Capacitive Sensors

Sensing range	2m maximum
Resolution	1cm
Directionality	Poor, typically 90° or higher beamwidth
Response time	Fast, typically 1msec
Cost	$1
Size	Small, depends on area to be sensed
Environmental immunity	Good in all conditions, but can be affected by salt spray

Table 7 - Parameters for Vision Sensors

Sensing range	100m+
Resolution	Poor without visual clues
Directionality	Good, depends on lenses used
Response time	Medium, typically 100ms+ due to processing time
Cost	$200+
Size	Camera typically 40mm diameter by 100mm long
Environmental immunity	Poor, susceptible to fog, spray and mud

960508

Ultrasonic Sensing of Head Position for Head Restraint Automatic Adjustment

Andrew J. Massara
Lear Seating Corp.

Copyright 1996 Society of Automotive Engineers, Inc.

ABSTRACT

In order to provide a safer seating environment, an automatically adjusting head restraint system (AHRS) was developed. The system will serve as a passive head restraint in the event of a rearward impact. Using ultrasonic transducer technology, the AHRS, a microprocessor controlled four way power head restraint mechanism, locates the top of occupant's head and adjusts the head restraint to an appropriate location. This paper demonstrates the effectiveness of employing ultrasonic sensor technology to automatically position a four way power head restraint system. The application and the considerations related to its implementation are discussed.

INTRODUCTION

Adjusting the head restraint on a seat is an ergonomically difficult task but a necessary one for safety and comfort. Most consumers are unaware that misalignment of the head restraint mechanism increases the risk of spinal injury due to excessive flexure of the vertebral column about the neck during a rearward impact. Insurance claims show that over sixty percent of all permanent impairments from automobile accidents occur in the neck and skull area.* Eighty-five percent of all neck injuries seen clinically are a result of rear end collisions.** In the endeavor to enhance safety of the seating environment, an Automatic Head Restraint System (AHRS) was developed to serve as a passive restraint device. The AHRS locates the occupant's head and automatically adjusts the head restraint to the correct position for safety and comfort.

* Brian O'Neill, B.Sc.; William Haddon, Jr. MD.; Albert B. Kelly; and Wayne W. Sorenson, Ph.D. "Frequency of Neck Injury Claims in Relation to the Presence of Head Restraints" : Automobile Head Restraints, A.J.P.H. March 1972.

** "X-ray Study of the Human Neck Motion Due to Head Inertia Loading" SAE Paper #942208

Automatic adjustment of the head restraint should significantly reduce operator error or misuse of head restraints and should reduce neck injuries.

This paper discusses the components of an AHRS, sensor technologies investigated to provide feedback to the adjustment mechanism, related considerations of their implementation, control logic development, and the effects of body position and hair styles on the feedback system.

TYPICAL AHRS COMPONENTS

An AHRS is comprised of four basic systems, a four way power head restraint mechanism, an ultrasonic sensor(s), control logic and power distribution circuitry. The existing prototype system consists of the following components:

- mechanical height adjustment mechanism
- linear, pneumatic fore/aft adjuster with the support pneumatics *
- an ultrasonic sensor and the support electronics (microprocessor based)
- relay power distribution circuit

*AHRS technology is not restricted to linear actuating head restraints. Linear actuation was selected for the ease of implementing an AHRS with this geometry. Other geometry can be accommodated.

The first iteration of the AHRS employs an off the shelf instrument grade transducer from Massa Products Corporation for proof of concept. Both the sensor and the electronics were supplied by Massa. The transducer emits a 75 kHz pulse in a 20 degree conical beam, capable of locating objects between 127 to 457 mm from its face. Tests revealed two inadequacies of this sensor. The first is the beam angle is too narrow, unable to detect occupants when seated as little as 64 mm out of position **(see Fig. 1).** The second being its range characteristics. The sensor is "blind" to objects closer than 127 mm to its face, forcing the sensor placement far to the back of the head restraint, making its packaging that more difficult. The sensor is "blind" to objects inside of 127 mm from its face due to the amount of settling time the transducer requires after transmitting a pulse before it can receive a reflected signal, this phenomena is called ring time.

A second iteration prototype AHRS will employ two transducers strategically placed to accommodate reasonably out of position occupants. The frequency selected for the transducer, (100 kHz), reduces the attenuation of the reflected wave due to hair to a minimum. The dual transducer array will provide a much wider sensing area, accommodating occupants ± 90 mm out of position, and will provide for a more precise means of locating the top of the head. Sensor arrangement and control logic is still under development for optimization of system repeatability and reliability.

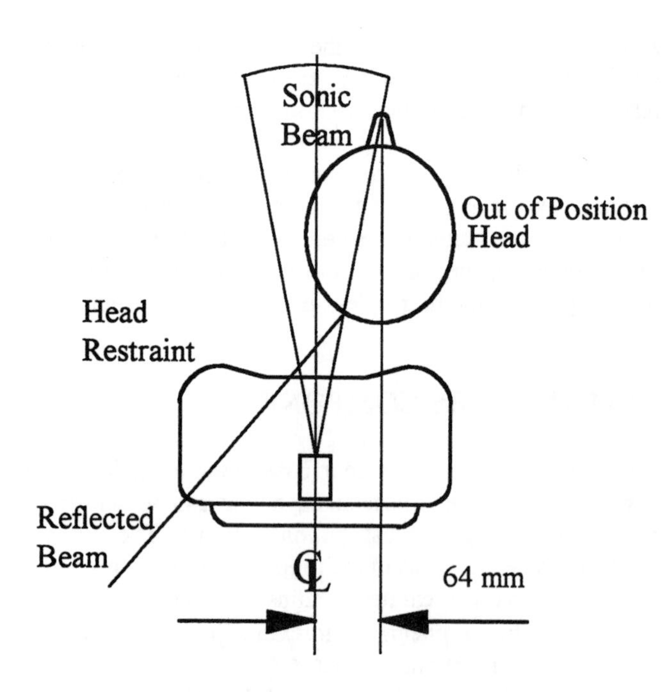

Top View of Head Rest and Head Form
Fig. 1

SENSOR TECHNOLOGIES INVESTIGATED AND THEIR CONSIDERATIONS

Only three commercially available sensing technologies could be considered for the AHRS due to the nature of the application, infra-red, laser, and ultrasonic.

INFRA-RED - The infra red sensing method operates on a beam principle comprised of a transmitter and receiver requiring opposing collinear positioning. Infra-red beam sensors are commonly used in industrial applications to detect presence. This type of device provides a digital output, the state change triggered upon disturbance of the beam. AHRS head location using this technology would require use of at least two sensors to define the height and fore/aft envelope with transmitters and receivers mounted facing each other on opposite sides of the head restraint in a collinear arrangement. The geometry of this packaging results in an ergonomically unfriendly head restraint **(see fig. 2).** False head locations would also result from the use this configuration from the system's inability to distinguish between beam disruptions resulting from hair or the scull.

LASER - Laser proximity sensing would operate in a similar fashion to ultrasonic, projecting a beam from behind the head towards the skull and detecting distance and presence by reflection. Laser technology is cost prohibitive and also limited by its inability to penetrate hair. False readings of the surface of thick hair as scalp would result.

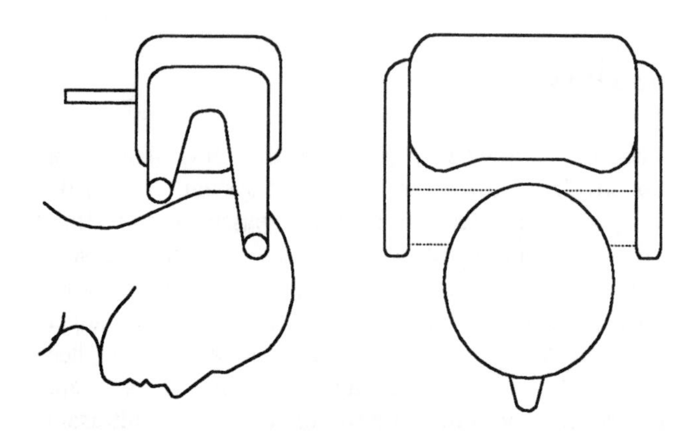

Side and Top View of Infra Red and Laser Based Sensing Arrays
Fig. 2

ULTRASONIC - Ultrasonic sensing operates on the principle of TDR, time domain reflectometry. When a pulse of sound is emitted from a transducer, it propagates away from the source at the speed of sound of the medium it is passing through. If an object is in the beam angle of the sensor, and it provides an adequate surface(s) to reflect a strong enough signal for the transducer to detect, the time of flight between transmission and detection provides a distance measurement (assuming the speed of sound in the carrier medium is constant.) This magnitude of time is what the control system interprets as a distance measurement. Ultrasonic sensors proved to be the only technology which was cost effective and capable of penetrating hair to give a positive skull location a majority of the time. Ultrasonic transducers also offer design flexibility, and the ability to be produced in a wide variety of shapes, sizes, operating frequencies, and sensing range capabilities.

CONTROL LOGIC DEVELOPMENT

Before control logic can be developed, reset parameters, triggers and the sequence of events must be established. For example, an AHRS's may require the following trigger conditions to be "made" before adjusting:

- occupant sensor made
- vehicle placed in drive
- 15 seconds elapsed after transmission enters drive begin adjustment sequence

The reset position for the AHRS is the lowest and most retracted position. The AHRS returns to this position after the vehicle is turned off to aid ingress and egress and also to assure the greatest driver visibility over the passenger seat when unoccupied. The sequence of events for the AHRS is:

- Raise head restraint to its highest position and begin sensing
- If the head is detected, adjust head restraint forward until preset distance is reached or end of travel is attained; end sequence
- If nothing is detected, lower the mechanism while sensing until head is detected
- adjust head restraint forward until preset is reached or end of travel is attained; end sequence
- If nothing is detected over entire height adjustment, end at reset position

If nothing is detected throughout the height adjusting sequence, the occupant is assumed to be too short for detection and therefore unable to be accommodated within the seat's targeted anthropometric constraints. If the sensor detects presence at its highest position, the occupant is assumed to be of the tallest percentile and the accommodates for fore/aft adjustment. Manual override of the AHRS should be provided regardless of the sensing configuration or control scheme to account for faulty adjustment resulting from occupant motion during the adjustment sequence or for comfort reasons.

The method of sensing and number of sensors incorporated into the system also affects the control logic and the flexibility of the system. For example, if ultrasonic sensors were mounted in the head liner and head rest, and position sensors built into the tracks, recliner and head rest mechanism, one could readily determine the location of the head and adjust the head restraint to the correct position directly. This method of sensing, however, would be costly and complicated. After considering all the alternative configurations, it was decided that the most simple and cost effective means of head location is to use one or two ultrasonic sensors mounted at the top of the head restraint and no position sensors.

EFFECTS OF BODY POSITION

Signal detection is limited by the geometry of the surface which the transducer is trying to detect. Similar to light reflecting off a mirror, the only geometry which permits the sound pulse to be reflected back to its source is a surface perpendicular to the source. Because the head is spherical, part of the head can be in the path of the signal, and still go undetected **(see Fig. 3)**. Orientation of the transducer and design of its beam angle, therefore, requires careful consideration to obtain reliable results. **Figure 4** portrays the intended second iteration prototype configuration. Observe that the angle at which the sensors are oriented in **figure 3** (side view) orchestrates the height adjustment of the mechanism. The dual sensor arrangement verses the single sensor arrangement provides a wider envelope to accommodate out of position occupants up to $\pm$ 90 mm off the center of the seat.

EFFECTS OF HAIR ON THE SOUND PULSE

A test was performed to determine the effects of hair on the signal using a 100 kHz ultrasonic transducer and its support electronics, a metal scale, a rigid fixture, an oscilloscope and a group of subjects with various hair styles. The test consisted of sitting a subject in front of the sensor which was rigidly secured to a scale mounted in the fixture. The transducer was mounted 152 mm from the end of the scale facing the subject. The subjects were placed directly in front of the sensor, placing their scalp against the scale's end. Transducer output was read on the oscilloscope, and digitally stored. Four general hair style groups were identified and are represented in table 1 with their respective errors for detection of the scalp.

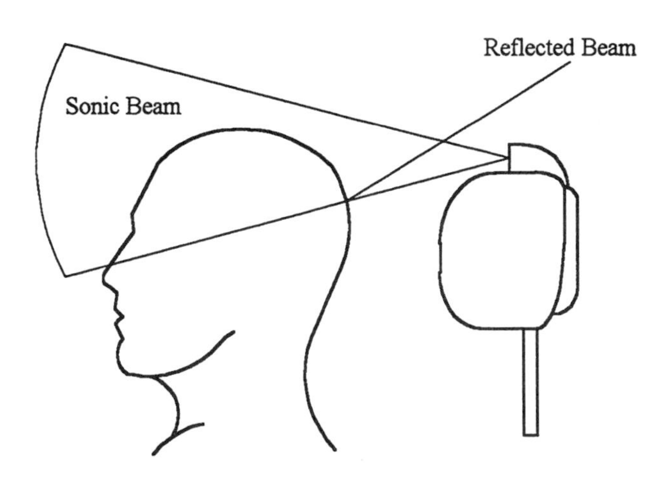

Fig. 3

	Hair Style	% Error	mm Equivalent
1	short flat hair	10%	15
2	curl intensive hair	16%	24.5
3	long straight hair	10%	15
4	thin/balding	6%	9

Table 1

The best results were obtained from group 4. Group 4 had the least amount of hair and provided the highest quality reflective surface. Thick, curl intensive hair can act as a false target providing enough of a reflected signal from its apparent surface to be interpreted as the scalp. Short flat and long straight hair styles also can create an apparent surface but reside closer to the scalp, and therefore are associated with lower errors.

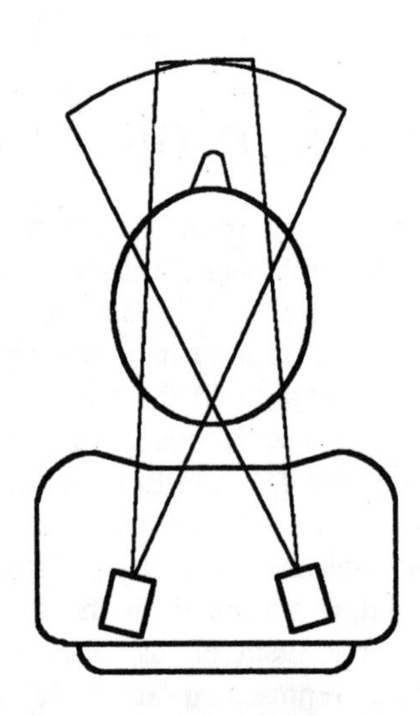

Top View of Head Restraint and Head Form with Dual Sensor Array
Fig. 4

CONCLUSION

Clinical studies and other data demonstrate the high rate of occurrence of spinal injuries resulting from rearward impacts. A properly adjusted head restraint minimizes injuries. An automatically adjusting head restraint offers a solution to the ergonomic difficulties associated with head restraint adjustment and enhances the safety of the seating environment for occupants who do not choose to adjust their head restraints. Ultrasonic transducer technology provides the most cost and performance effective means of locating an occupant's head for automatic head restraint adjustment.

ACKNOWLEDGEMENTS

I would like to acknowledge Massa Products Corporation for their contributions to the design and development of the AHRS.

940631

High Precision Angular Position Sensor

Shigeru Kato and Jun-ichi Nakaho
Tokai Rika Co., Ltd.

ABSTRACT

The authors have developed an angular position sensor which outputs an analog voltage in accordance with the angular position of rotation. The sensor detects the absolute angular positions of rotation and offers high precision along with high durability. The sensor detects the angular positions by means of optical detection of the relative distance between a spiral slit and a circular slit, which both have a coaxial center and are transparent. The nonlinearity of this sensor is less than ±0.5% and the detection range of the angular position is more than 355°. The operating temperature range is -30℃ to +80 ℃, with a thermal drift within 1° at the center of the detection range.

INTRODUCTION

Steering sensors which detect angular positions of the rotating steering wheel are used for suspension control, traction control, power steering control, and 4-wheel steering control[1,2]*. These steering sensors are of two types: pulse output sensors and analog output sensors. The pulse output sensor gives an output of counted pulses in proportion to rotation. The analog output sensor gives an output of analog voltage in proportion to rotation. A typical example of pulse output sensors is the optical rotary encoder, which is composed of a rotating disc, light sources, and photodiodes[3]. A typical analog sensor is a potentiometer composed of a carbon-film resistor and a rotating wiper[3]. The rotary encoder mainly detects relative angular positions and requires processing by a CPU in order to detect the absolute angular positions. It, therefore, cannot detect the absolute angular positions until it detects the neutral pulse which indicates the neutral position of the steering wheel after the vehicle has started off. Furthermore, erroneous detection of the absolute angular positions may occur due to noise pulses. Consequently, use in systems which require only relative angular positions is desirable. The potentiometer can detect absolute angular positions by a change in resistance, but there are problems with accuracy and durability due to wear of the carbon film by the wiper.

To resolve these problems, the authors have developed a new angular position sensor which offers high durability and precise detection of absolute angular positions.

* Numbers in parentheses designate references at end of paper.

PRINCIPLE OF ANGULAR POSITION DETECTION

Theoretically, the absolute angular positions of rotation may be detected through optical detection of a spiral slit whose distance from the center varies linearly in accordance with rotation[4,5]. If the center of the slit coincides with the center of rotation, this method would present good linearity because the changes in the position of the spiral slit would be proportional to the angle of rotation. In actuality, however, it is extremely difficult to make the two centers coincide precisely and this method is therefore not practical.

When the center of the slit does not coincide with the center of rotation, the slit position, which changes by rotation, exhibits undulating characteristics. To correct these undulations, the authors have added a circular slit with a center identical to the spiral slit. By detecting the relative distance of the slits from the center of rotation, the eccentric error of the spiral slit is canceled by the circular slit and good linearity is obtained.

Figure 1 is a schematic diagram of the sensor developed on the basis of the concept just described. The main components are a rotating disc, two light-emitting diodes (LEDs), two position-sensitive devices (PSDs), and an electronic circuit. The disc is opaque except for the areas of the spiral and circular slits, which are optically transparent and have the same center. Each LED emits light to one of the slits. The PSDs are oriented to detect the positions of light spots that vary in the radial direction of the disc, with each PSD detecting the position of the light transmitted via the corresponding slit. The electronic circuit processes the signals from the PSDs, outputs voltage corresponding to the angle of rotation, and controls the drive currents for the LEDs.

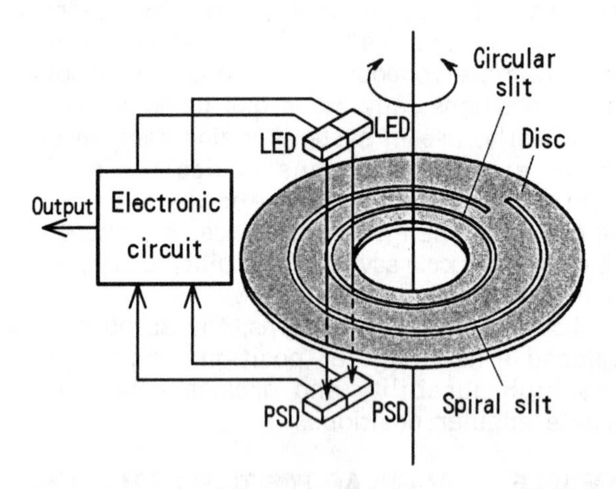

Figure 1. Schematic Diagram of the Angular Position Sensor

The shape of the photo-sensitive area of the PSD illustrated in Figure 2 is a narrow rectangle. The three-layer structure of the PSD consists of a p-type silicon layer, an i-type (intrinsic) silicon layer, and an n-type silicon layer[6]. This is what is known as a pin photodiode and it outputs current from both ends of the p-type layer according to incident light. The PSD is longitudinally sensitive and enables detection of the position of the center of the incident light by means of the two currents in that direction.

The position x of a light spot is calculated by

$$x = \frac{L}{2} \cdot \frac{I_1 - I_2}{I_1 + I_2}, \qquad (1)$$

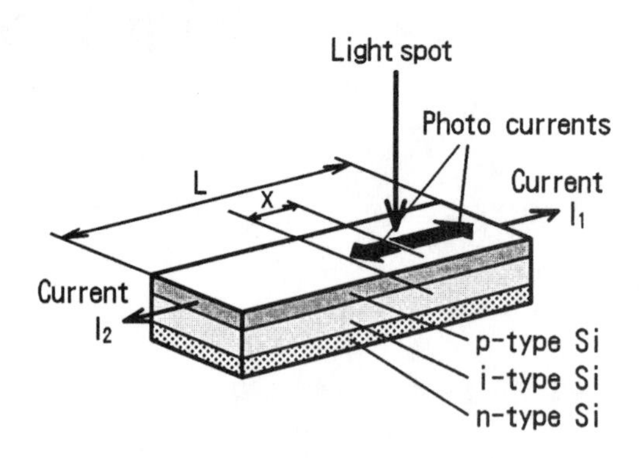

Figure 2. Structure of a Position-Sensitive Device (PSD)

where

L is the detection length of the PSD,
I_1 is the output current at one end of the PSD,
I_2 is the output current at the other end of the PSD.

The total current $I_1 + I_2$ is proportional to the light intensity. The PSD detects the position of the light spot regardless of the light intensity. Its detection error is a very small value of about 15μm.

The sensor detects the angular positions in the following manner. One of the two PSDs optically detects the changing positions of the spiral slit by the light of the corresponding LED. The other PSD similarly detects the changing positions of the circular slit. When the center of rotation corresponds with the center of the slits, the position signal of the circular slit detected by the PSD does not change, because the distance between the center of rotation and this slit is constant regardless of rotation, whereas the position signal of the spiral slit detected by the PSD changes linearly in accordance with rotation. In this case, the angular positions of rotation can be detected precisely only by the spiral slit. In actuality, however, it is impossible to detect the angular positions of rotation with precision because it is difficult to make the center of the slit coincide with the center of rotation. For this reason, the position signal of the circular slit is used to correct the eccentricity of the center of rotation and the spiral slit. Therefore, the signal of the circular slit cancels the error of the spiral slit through calculation of the difference in the positions of the slits. In this way, the precise angular positions of rotation can be detected easily because there is no need to align the centers accurately.

THEORETICAL CONSIDERATIONS

What follows is a theoretical verification of the principle just described.

As illustrated in Figure 3, the curve of the spiral slit in polar coordinates (r, θ) is given by the following equation,

$$r = r_0 + q \cdot \theta, \tag{2}$$

where

$q = p/(2\pi)$,
p is the pitch of the spiral slit,
r_0 is the radius at $\theta = 0$.

The curve of the circular slit is given by

$$r = r_c, \tag{3}$$

where r_c is the radius of the circular slit.

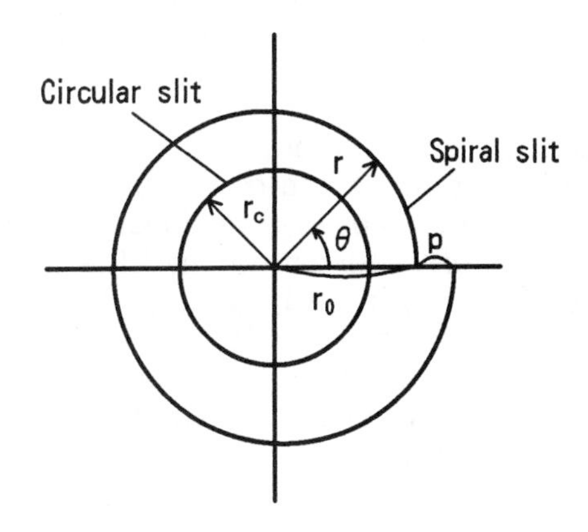

Figure 3. Curves of a Spiral and Circular Slit

As shown in Figure 4, when the center O' of the slits is apart from the center O of rotation by the polar coordinates (δ, α), the distance R_s of the spiral slit from the center of rotation is given by

$$R_s = \delta \cdot \cos(\theta - \alpha) + \sqrt{r_p^2 - \delta^2 \cdot \sin^2(\theta - \alpha)} \tag{4}$$

with

$$r_p \fallingdotseq \frac{r_s + \sqrt{r_s^2 + 4 \cdot q \cdot \delta \cdot \sin(\theta - \alpha)}}{2} \tag{5}$$

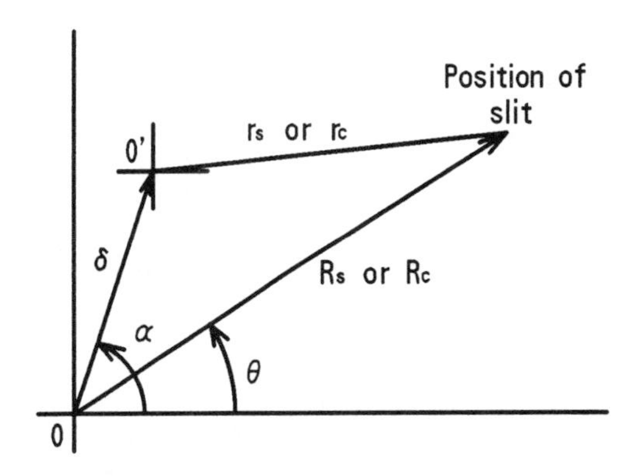

O : Center of rotation
O' : Center of each slit

Figure 4. Relation of the Distance of the Slit and the Eccentricity

and

$$r_s = r_0 + q \cdot \theta, \tag{6}$$

if $r_c \gg \delta$.

In the case of the circular slit, the distance R_c of the slit from the center of rotation is given by

$$R_c = \delta \cdot \cos(\theta - \alpha) + \sqrt{r_c^2 - \delta^2 \cdot \sin^2(\theta - \alpha)}. \tag{7}$$

The relative distance d of the two slits is obtained by

$$d = R_s - R_c \fallingdotseq (r_s - r_c) + \frac{q \cdot \delta \cdot \sin(\theta - \alpha)}{r_s}. \tag{8}$$

The first term of EQ (8) represents the relative distance of the two slits and the second term represents the error due to the eccentricity of their centers.

Figure 5 shows the changes in the positions of the spiral slit and the circular slit as well as in their relative distance, with calculations using the following conditions: r_0=30.75mm, p=2.5mm, r_c=27mm, δ=0.5mm, and α=270°. The positions of the spiral and circular slits due to rotation show undulations arising from the influence of the centers of rotation and the slits on eccentricity, but the differential distance of the spiral and circular slits shows good linearity.

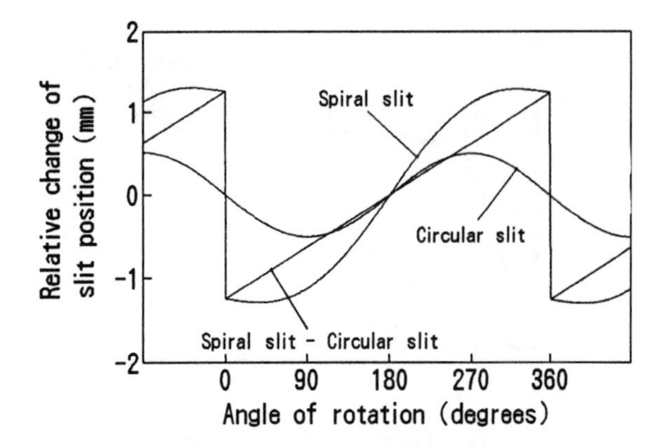

Figure 5. Relative Changes of the Positions of Slits versus Angle of Rotation (Theoretical)

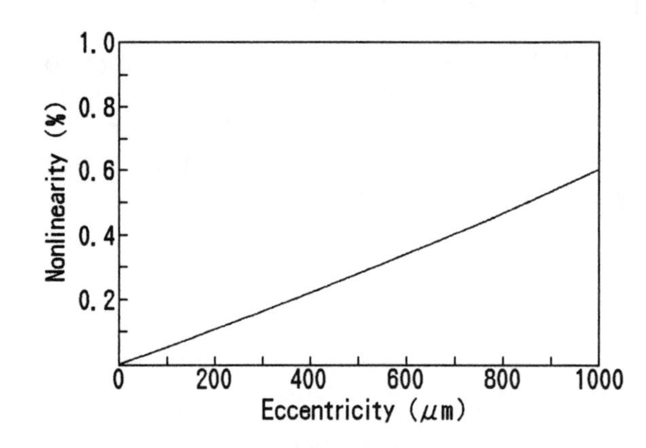

Figure 7. Nonlinearity of the Sensor versus Eccentricity (Theoretical)

Figure 6 shows the relation of the error of the differential distance of the spiral and circular slits to the eccentric distance δ between the centers of rotation and the slits. This error is estimated by $p \cdot \delta/(2\pi r_0)$ and becomes $1.29\times10^{-2}\times\delta$ [mm] under the conditions described earlier. This result means that the influence of the eccentricity δ is reduced to about 1/100. When δ =0.5mm, for example, the error is a very small value of about 6.5μm.

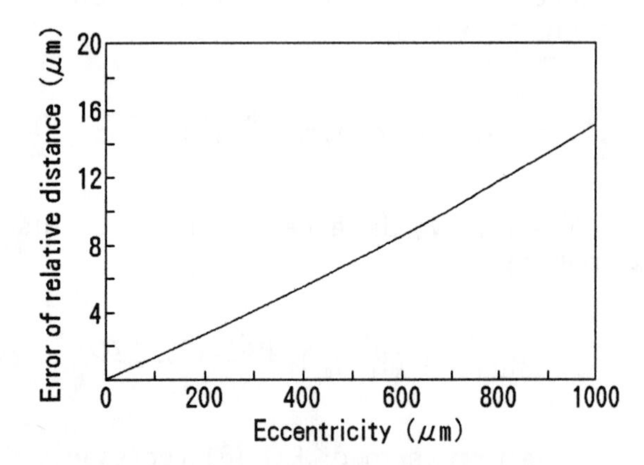

Figure 6. Error of the Relative Distance of Two Slits versus Eccentricity (Theoretical)

Figure 7 shows the relation of the nonlinearity of the sensor to the eccentric distance δ. This nonlinearity is estimated by $100\cdot\delta/(2\pi r_0)$, and becomes $5.18\times10^{-1}\times\delta$ [%]. For example, when δ =0.5mm, the nonlinearity is a very small value of about 0.26%. In fact, it is eminently possible to reduce the eccentricity to less than 0.5mm, therefore providing theoretical verification that this method has high precision.

ELECTRONIC CIRCUIT

In general, the signals output from a PSD are processed by EQ (1). This equation requires a division operation, but an analog amplifier for division is both expensive and imprecise. For this reason, a circuit which does not require a dividing amplifier has been developed. If the total current I_1+I_2 output from a PSD is always constant, the division in EQ (1) is meaningless. Because the total current I_1+I_2 is proportional to the light intensity incident on the PSD, the authors have employed a feedback circuit which controls the drive current of the LED to maintain the total current I_1+I_2 at a constant value. Figure 8 is the block diagram of the circuit employing this method. The circuit can be made using inexpensive electronic components such as general ICs, transistors, and the like.

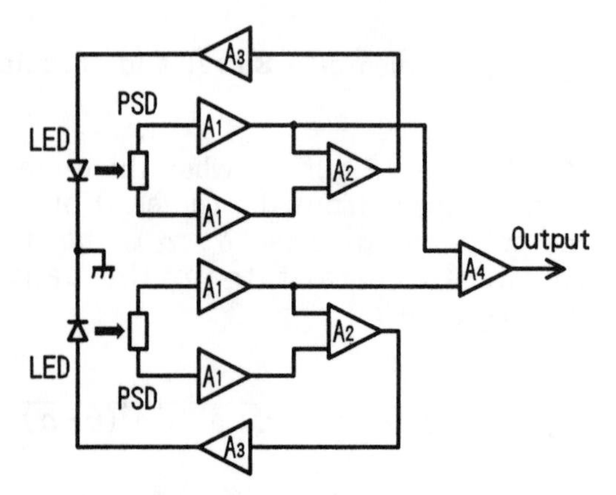

A_1 :Current-voltage converters
A_2 :Summing amplifiers
A_3 :LED drivers
A_4 :Differential amplifier

Figure 8. Block Diagram of the Circuit

The circuit consists of four current-voltage converters, a differential amplifier, two feedback circuits, and a power source (not illustrated). The current-voltage converters convert output currents from the PSDs to voltages. The differential amplifier calculates the difference between the output signals of the two PSDs. The feedback circuits control the drive currents for the LEDs to keep the total currents of the PSDs constant, respectively. The power source supplies stable voltages to the circuits.

This circuit offers high precision and low temperature dependence, and is low in cost.

EXPERIMENTAL RESULTS

Figure 9 shows the external view of a sensor developed for use as a steering sensor.

Figure 10 is a plane view of the disc. The outer diameter of the disc is ϕ 70mm, the inner diameter (for passing through the steering shaft) is ϕ 40mm, the minimum radius r_0 of the spiral slit is 30.75mm, the pitch p of the spiral slit is 2.5mm, and the radius r_c of the circular slit is 27mm. The detection length for each PSD is 3.5mm.

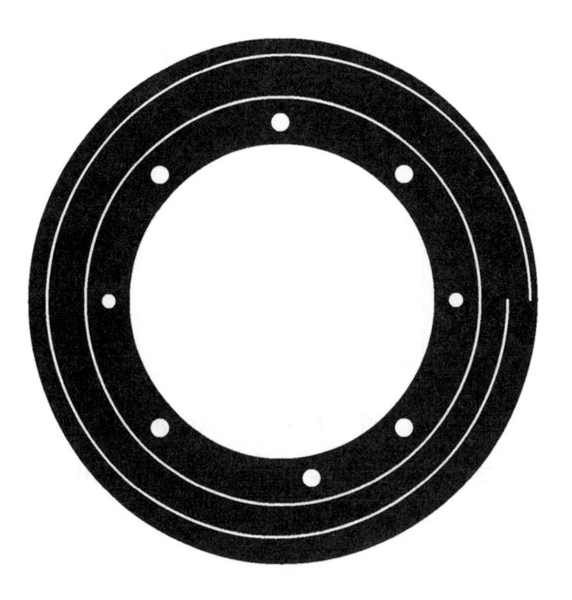

Figure 10. Plane View of the Disc

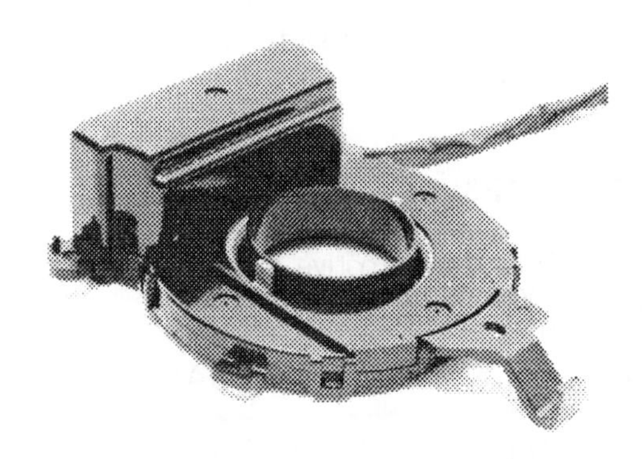

Figure 9. External View of the Developed Sensor

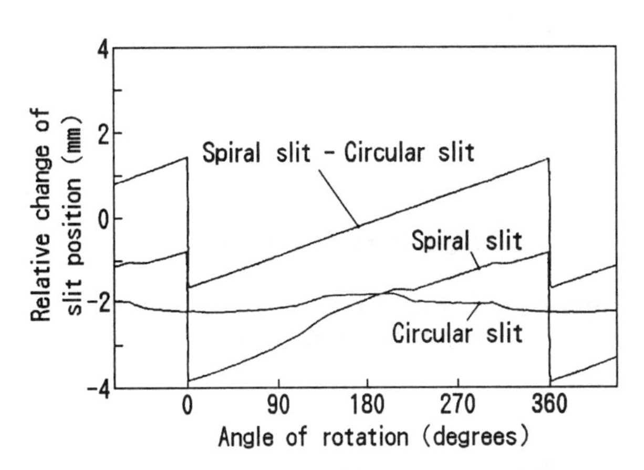

Figure 11. Relative Changes of the Positions of Slits versus Angle of Rotation (Experimental)

Figure 11 shows the signals of the position changes of both slits and of the difference between the slits. The signals of both slits show undulations and noise, but the signal for the difference of the slits exhibits good linearity. The undulations result in the eccentricity of the centers, and the noise causes the clearance of the shaft and bearing. It is thus experimentally proven that this principle cancels out the influence of the eccentricity and the clearance, thereby obtaining good linearity.

Figure 12 shows the characteristics of output voltage versus rotation angle and Figure 13 shows the error. This sample is adjusted for an output of 5V per turn. The nonlinearity is within ± 0.5%, error is within ±2°, and the detection range is more than 355°.

Figure 14 shows thermal drift for temperature changes of -30 ℃ to +80℃. The thermal drift at the angular position of 180° is within 1° and at all angular positions is within 4°. The performance values of this sensor are summarized in Table 1.

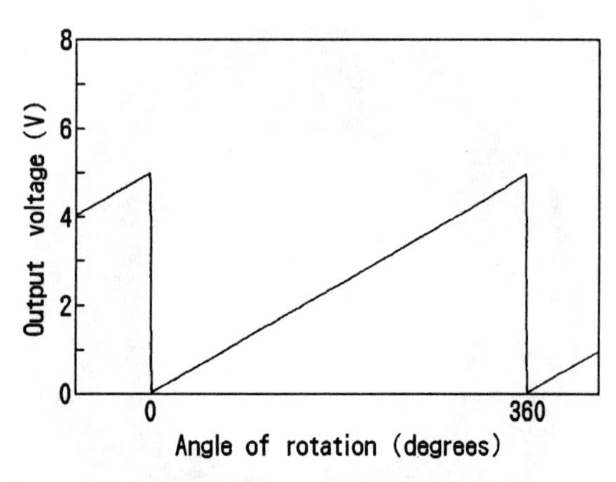

Figure 12. Output Voltage versus Angle of Rotation (Experimental)

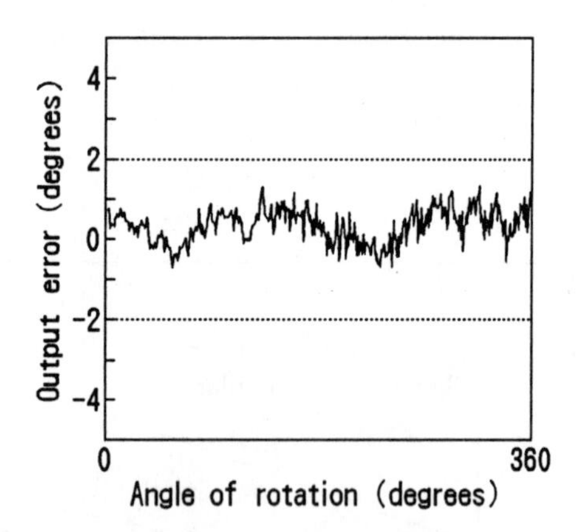

Figure 13. Output Error versus Angle of Rotation (Experimental)

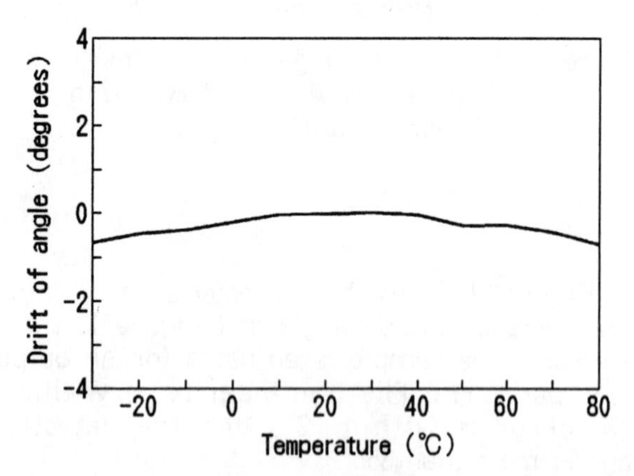

Figure 14. Drift of Angle versus Temperature (Experimental)

Table 1. Performance Values of the Sensor

Sensitivity		5/360 V/°
Error at 25℃	Nonlinearity	Max. ±0.5%
	Error of angle	Max. ±2°
Range of detection		Min. 355°
Thermal drift in operating temperature range		Max. 1° (at 180°)
Operating temperature range		-30℃ to +80℃
Storage temperature range		-40℃ to +90℃

CONCLUSIONS

The developed sensor outputs a linear analog voltage proportionate to the angular position in accordance with rotation. It offers excellent linearity, a wide detection range, a wide operating temperature range, and low thermal drift. In consideration of the performance figures obtained, it is relatively low in cost. It can, therefore, replace currently used steering sensors and is also applicable as a rotary sensor.

The method that has been developed can detect not only rotary positions but also linear positions and can be used for various types of position sensors.

REFERENCES

1. R.F.Wells, "Automotive Steering Sensors", SAE Paper 900493, 1990.
2. R.F.Wells, "Non-Contacting Sensors for Automotive Applications", SAE Paper 880407, 1988.
3. J.R.Carstens, Electrical Sensors and Transducers, Prentice Hall, Englewood Cliffs, N.J., 1992.
4. Jpn.Laying-Open Pat. 60-236020, 1985.
5. T.Arai and Y.Matsui, "Trial Manufacture of the Angle Position Detector Utilizing the PSD and Its Application to the Two Dimentional Motion with a Constant Speed", Proceedings of the 4th Sensor Symposium, Tsukuba, Japan, 1984, pp.255-257.
6. Specification sheets on PSDs, Hamamatsu Photonics K.K., Hamamatsu, Japan.

910270

A Hall Effect Rotary Position Sensor

Robert Bicking, George Wu, Joe Murdock, Don Hoy, and Rusty Johnson
MicroSwitch
Freeport, IL

ABSTRACT

Rotary position sensors (RPS) currently are applied widely in engine management systems for throttle position sensing and are being considered for other applications such as drive-by-wire. The potentiometer RPS relies on contact between a resistive element and a wiper and thus has an inherent wear mechanism.

This paper describes a noncontacting RPS which is essentially a drop-in replacement for the present device. A linear output Hall Effect IC is key to achieving the required functionality, but it must be combined with a magnetic actuation scheme which provides a very linear and stable magnetic field as a function of input angle and packaging which provides long life and control of mechanical tolerances. Each of these design elements will be discussed.

INTRODUCTION

Rotary position sensors (RPS) are widely used in automobiles with electronic engine control systems to provide analog, DC voltage proportional to the position of the throttle plate over the range 0° to 90°. The basic approach is straightforward, a linear resistance element with a wiper attached to the shaft. Design improvements have made this device very reliable (1). Nevertheless, the quest for continuous improvement and longer life has led to a search for a simple, reliable, noncontacting RPS.

One approach to constructing a noncontacting RPS is described herein. It consists of a linear Hall effect IC which is actuated by a shaped magnet mounted on the rotor. Because of its simplicity, it potentially can be cost-competitive with the existing approach.

MAGNETIC ACTUATION

It is necessary to design a magnetic actuation scheme capable of generating a magnetic field (B) which is linear as a function of angular rotation (θ) over at least a 90° range and further, doesn't drop below the value B_{90} between 90° and 120°. This is shown in Figure 1. As shown, this requires operating the sensor with bi-polar magnetic fields, so that B = O occurs at 45°.

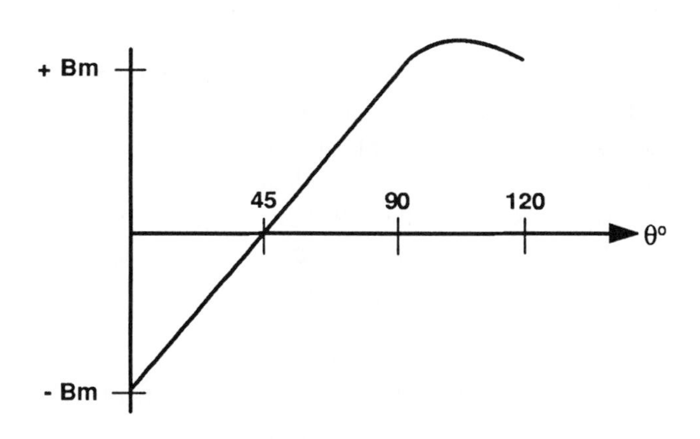

Fig. 1 Magnetic Field vs Input Angle

Permanent magnet materials have cost proportional to volume so that it is desirable to use as small a magnet as practical. Considerations in material selection include field strength (B), stability, temperature coefficient of B and the ability to shape the material.

The approach used herein is a quasi-elliptical shaped molded magnet attached to the input rotor. The Hall IC is located in proximity to it as shown in Figure 2. The spacing between the magnet and the Hall IC is a tradeoff between minimizing positional sensitivity (which requires a relatively large gap) and maximum field strength (for which the gap should be minimized). With the particular magnet material used (Neodymium), the measured sensitivities near the worst case 0° and 90° angles (referring to Figure 2) are x = 0.6°/mil, y = 0.3°/mil and, z = 0.2°/mil. It should be noted that y and z represent misalignments and are nominally zero.

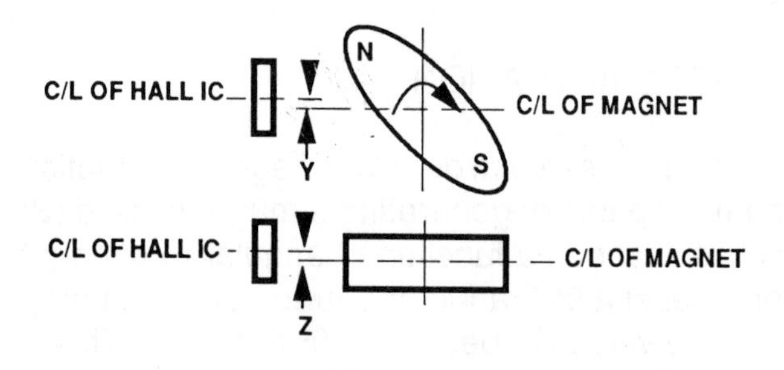

Fig. 2 Magnet - Hall IC Relationship

LINEAR HALL EFFECT SENSOR

The linear Hall effect sensor used incorporates both thin film on-wafer trim and thick film application specific trim to achieve high sensitivity, low null and sensitivity temperature coefficients and unit to unit interchangeability. It is capable of operation from a 5V supply over the -40° to 150°C range (2). A specification summary is given below.

Linear Hall IC Specifications
Operating Temperature Range (°C): -55 to 150
Supply Voltage (VDC): 4.5 to 12
Vo (o gauss)(VDC): 2.500 +/- 0.025
Sensitivity (MV/gauss): 5.0 (trimmed in RPS)
Temperature Coefficient (%/°C): +/-0.03 (null & sensitivity)

ERROR SOURCES

MECHANICAL TOLERANCES & WEAR - Mechanical tolerances of the rotor and housing (which constrains the rotor) can show up as an output shift if the shaft moves or wobbles. The worst case error will occur near the 0° and 90° shaft angles. The return spring in the RPS will force the rotor to housing gap to be zero consistently so that no error will result. Wear of the bearing surfaces will show as a calibration shift. Selection of appropriate materials can minimize both initial tolerances and the effects of wear.

STABILITY - This is primarily a function of the magnet chosen since the Hall IC exhibits stability comparable to other silicon ICs. Magnets are available with excellent stability on the order of 0.02 - 0.04%/10 years.

LINEARITY - By using shaped magnets, linearity of better than 1% may be achieved so that linearity is not an issue. Because this is a noncontacting device, wear is not expected to degrade the linearity or cause erratic or discontinuous output.

TEMPERATURE EFFECTS - One of the potential problem areas in applying a Hall RPS is that the closed throttle output must be very stable. Otherwise, the electronic engine control system may interpret an output shift during warm-up as a driver-initiated throttle command. Unfortunately, the maximum potential RPS temperature error occurs at the 0° (closed throttle) and 90° (wide open throttle) positions, since the effects of both the magnet field strength temperature coefficient (TC) and Hall IC span TC will be greatest.

The Hall TC has both a predictable nominal value and an unpredictable part to part variation so that complete cancellation of the magnet TC by the Hall TC cannot be assumed. A measure and compensate approach could be used to minimize temperature errors but this would add cost. The approach selected simply trades deadband around the 0° and 90° throttle positions for excellent temperature stability. The Hall IC exhibits excellent temperature stability when driven into saturation (only about 10mV shift over the entire temperature range). Therefore, the scale factor is adjusted such that the IC remains in saturation at θ = 0° and 90° over the entire temperature range. This results in a variable deadband and slope as a function of temperature as shown in Figure 3.

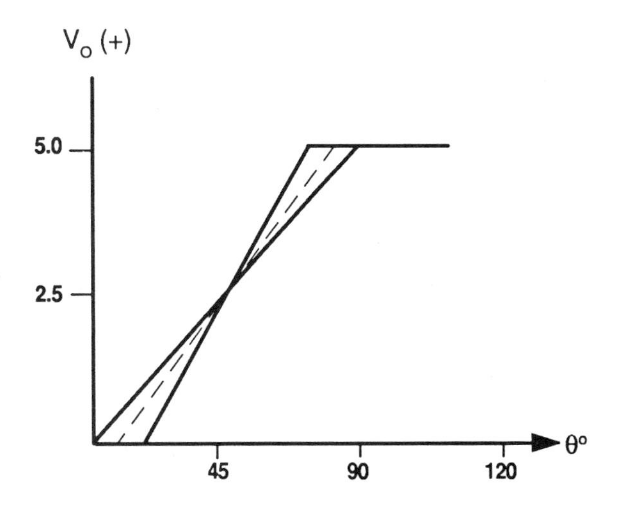

Fig. 3 Input Angle vs Output Voltage

CONCLUSIONS

A Hall effect RPS is described which combines a commercially available sensor with a shaped magnet to produce a high performance noncontacting device. It has been tested for over 10 million cycles from 0° to 80° to 0° of input angle with no noticeable performance degradation. Closed throttle output stability is achieved by calibration which trades several degrees of deadband for keeping the Hall sensor in saturation over the entire temperature range. The RPS is shown both before and after assembly in Figure 4.

REFERENCES

1. Riley, R. E., "High Performance Resistive Position Sensors", SAE Paper #890302.
2. MICRO SWITCH Data Sheet SS94B1, "Linear Hall Sensor", 1990.

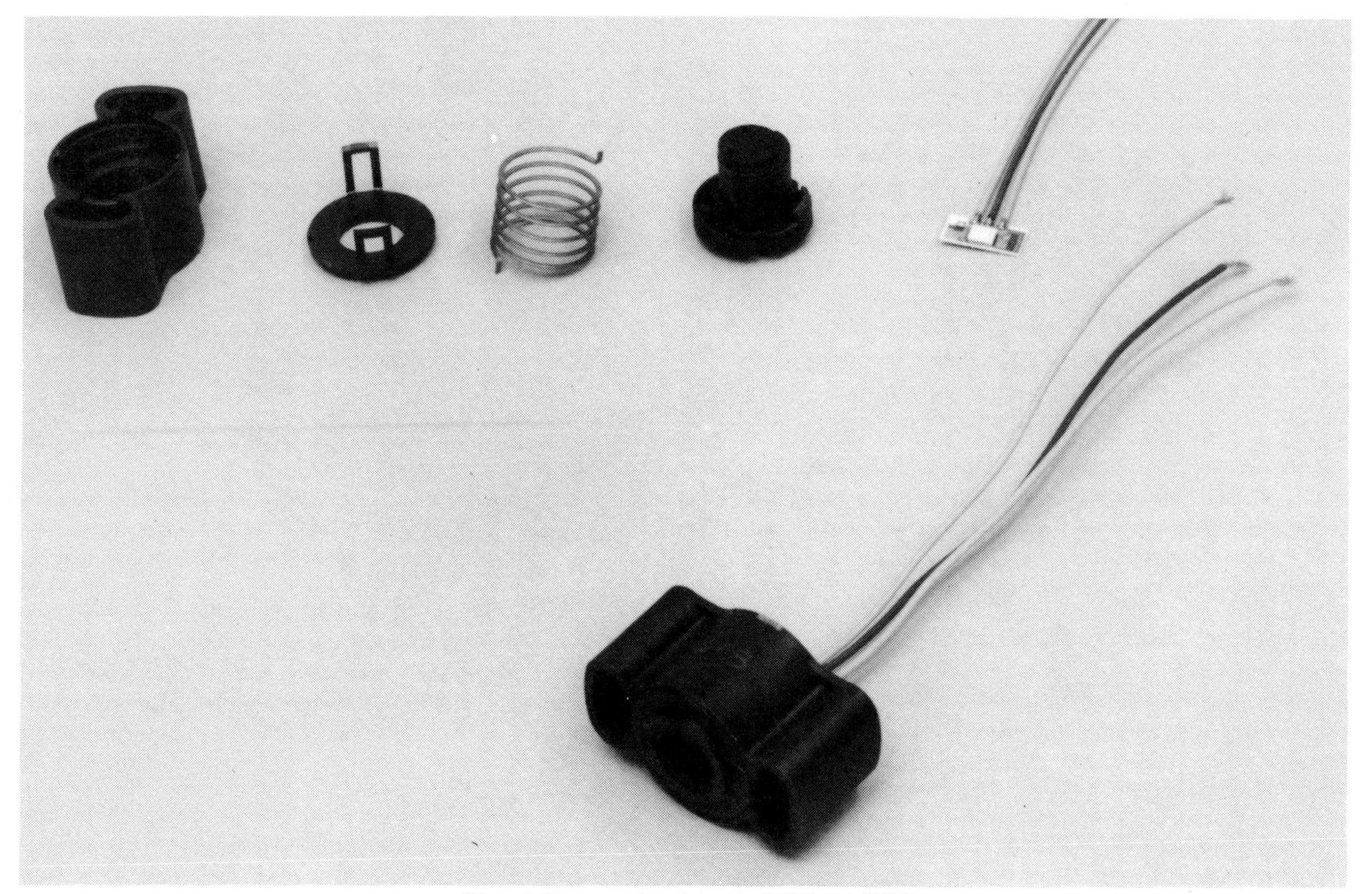

Fig. 4 Noncontacting Hall Effect RPS

870470

MRE Rotation Sensor: High-Accuracy, High-Sensitivity Magnetic Sensor for Automotive Use

Yoshimi Yoshino, Kenichi Ao, and Mitsuharu Kato
IC Engineering Dept.
Nippondenso Co., Ltd.
Shuji Mizutani
Electronics Engineering Dept.
Nippondenso Co., Ltd.

ABSTRACT

A high-accuracy automotive rotation sensor with high-sensitivity has been developed using a ferro-magnetic resistance element, called MRE, which we have recently developed. This sensor composed of MRE and a multi-pole magnet, is a compact,environmental-resistive,low rpm detectable device with high hybrid circuit density including an electric signal processing circuit.

The new sensor is especially designed for generating a double-frequency output to ensure a duty ratio for precise detection of rotation.

The sensor is now used as a vehicle speed sensor capable of measuring from low rpm to high rpm, mounted directly onto the transmission.

The sensor is not only considered to be a riable alternative to the conventional rotation sensor, but will be used in an expanding number of applications.

OVER THE YEARS, VARIOUS ROTATION SENSORS have been developed and applied widely in the field. Most of those sensors are optical or electro-magnetic induction type. However, the former has a disadvantage of environmental resistivity, and the latter incapable of measurements of low rpm. To improve the above drawback, though studies of sensors with of a hall-effect element and a magneto-resistance element using semiconductor materials are being parallelly continued steadily, but those sensors do yet serve as a complete improvement in point of temperature characteristic and sensitivity, so that it is difficult to present a compact, multi-pulse type rotation sensor.

The situation initiated the development of a new rotation sensor meeting all the requirements of the field.

COMPARISON OF ROTATION MEASUREMENT METHOD

Table 1 shows a comparative table of some major rotation measuring systems, obtained by comparative analysis of several characteristic functions. As a consequence, we have concluded that the MRE method is best suit to rotation measurements.

Firstly, we will explain MRE in comparison with semiconductive magnetic sensors.

Table 1. ComParison chart of various rotation sensing systems.

	Sensing - type adoption criteria				
Type	Multi-pulse adaptability ※	Low rpm detection	Environmental resistivity	Operating temperature range	Cost
Reed switch	poor	good	good	good	good
Photo-coupler	good	excellent	poor	poor	fair
Variable reluatance	fair	poor	excellent	good	fair
Capacitance	good	good	poor	fair	poor
Hall device	fair	excellent	good	fair	fair
M R E	excellent	excellent	excellent	good	fair

※20 Pulses per revolution

CHARACTERISTICS AND FEATURES OF MRE

MRE,compared with well-known hall-effect elements and semiconductive magneto-resistance elements,has a variety of technical and characteristic features to mention. The hall elements and semiconductive magneto-resistance elements detect magnetic field by the phenomenon that directions of electron is changed by Lorentz-force. On the other hand,MRE applies in measuring magnetic field the anisotropic magnetoresistive effect that scattering probability of electrons changes by magnetic field(1-6). As the result,MRE causes resistance change in accordance with the angle between the directions of current and magnetic field. Typical ferro-magnetic materials, when disposed in a magnetic field, have their own maximum resistance value when the current and magnetic field direction are parallel to each other, and their minimum value when they are perpendicular to each other(2,7), as shown in Fig.1 .

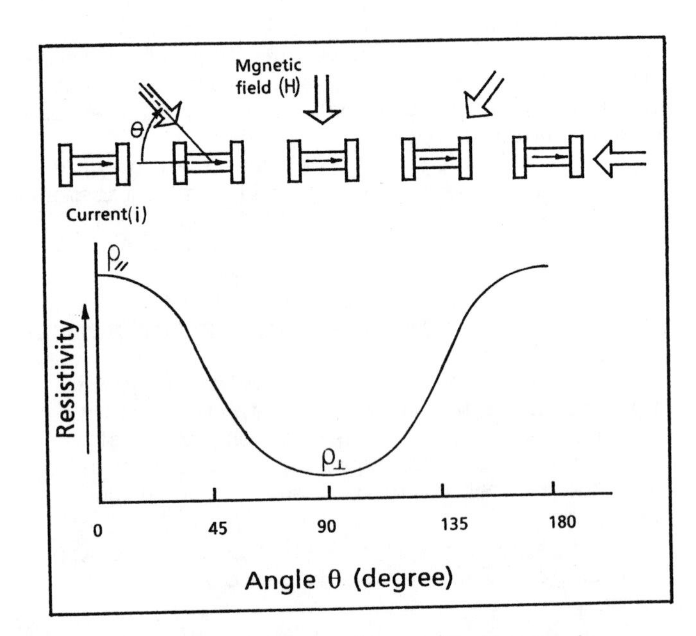

Fig . 1 MRE's Resistivity change by the angle between the current and the magnetic field

This anisotropic resistance value $\rho(\theta)$ can be obtained by the following equation:

$$\rho(\theta)= \rho_{\perp}\sin^2\theta + \rho_{//}\cos^2\theta \qquad (1)$$

wherein θ is the angle between the directions of current and magnetic field, and $\rho_{//}$ and $\rho_{\perp}$ are the resistivities parallel to and a right angle to a current flow,respectively, in magnetic field.

The above equation is well known as Viogt-Thomson fomula(1).

Fig.2 shows change in resistivity as the magnitude of a magnetic field changes, where ρ_0 is the resistivity with no magnetic field applied. From the figure, it becomes evident that the resistivity becomes saturated and constant as the magnetic field increases.

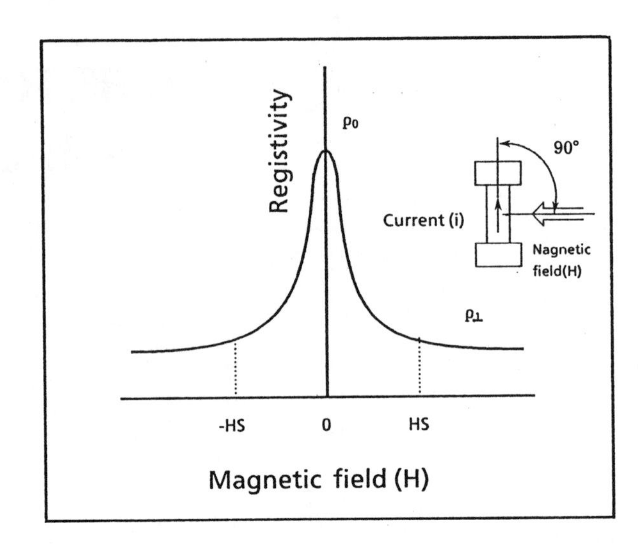

Fig . 2 MRE's Resistivity change by the magnitude of the magnetic field applied

The magnetic field at which this phenomenon occures is called "Saturation Magnetic Field(H_S)" and can be described as follows:

$$H_S = k \times t/w + H_K \qquad (2)$$

where k is constant, t is the ferro-magnetic film thickness, w is the pattern width, and H_K is the anisotropic magnetic field. The above equation indicates that H_S is proportional to the film thickness and inversely proportion to the pattern width of the resistors. The anisotropic magnetic field is an inherent value of materials and in generally speaking, N_i-F_e alloy's value is small than that of N_i-C_o alloy.

Fig.3 illustrates the relationship between the output and the magnetic flux density of MRE and the current semiconductor magnetic sensors, and Fig.4 shows temperature characteristics of those sensors. As is evident from Fig.3, the output of hall elements linearly increases as the magnetic flux density increases and the magneto-resistance element's output increases in proportional to the square of the magnetic flux density.

The MRE output rapidly increases with the increase in the magnetic flux density when the magnitude of magnetic field is relatively weak, and its output becomes unchanged any more since the magnetic field reaches H_S. The MRE output is nearly 10 times other sensors outputs in the magnetic flux density of 5 mT which is relatively weak, and then MRE is able to generate a steady output with H_S or more. This unique characteristic- allows for a simple design of a magnetic circuits.

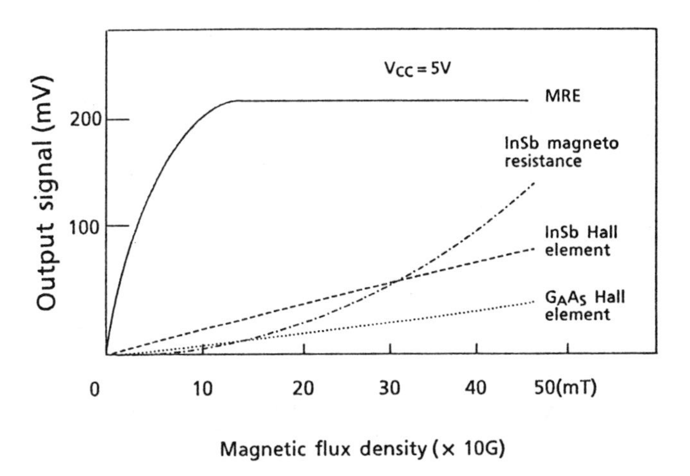

Fig. 3 Unique characteristic of MRE

In addition, as shown in Fig.4, MRE generates the linear decreasing output as the device temperature increases at a low temperature coefficiency rate of -3.5×10^{-3}/°c, relatively low compared with that of the hall element, compensation techniques then are very simplified.

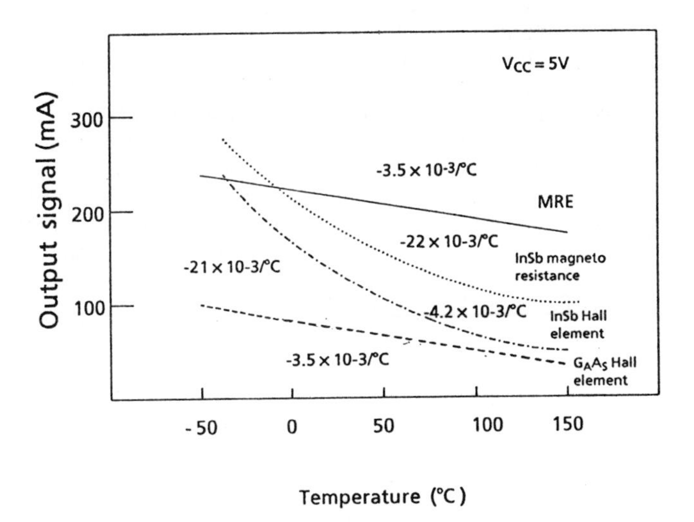

Fig. 4 Temperature characteristics of MRE and other sensors

Further more, MRE, not being semiconductor material, is operable even at a high temperature in theory. As explained above, MRE is the most appropriate sensor meeting the demand of severe automotive operating environment.

The developed MRE chip's characteristics are tabulated in Table 2. As shown by the Table, resistivity change is about 4.7 %, and no variance of resistivity observed.

Table 2. Characteristics of the developed MRE chip

Item	Conditions	Values	Unit
Sheet resistivity	B = 0mT T = 25°C	1.65	Ω/□
Resistivity change	B = 20mT T = 25°C	4.7	%
Temperature coefficiency rate of resistivity	B = 0mT 20mT	3.1×10^{-3}	/°C
Out put temperature coefficiency rate	B = 20mT	constant voltage -3.5×10^{-3} constant current -0.4×10^{-3}	/°C

The resistivities linearly increase as temperature increase with a temperature coefficiency rate of 3.1×10^{-3}/°c . On the other hand, the output linearly decreases with the rise of temperature with -3.5×10^{-3}/°c in constant voltage driving. However, in constant current driving the output temperature coefficiency rate is -0.4×10^{-3}/°c ,being about one-tenth of the coefficiency rate of constant voltage driving method.

Then, features of MRE are summariszed as follows:

MRE has;

(1) unequalled sensitivity even in a weak magnetic field.

(2) unparalleled high output in a weak magnetic field.

(3) magnetic field saturation characteristics.

(4) magnetic directivity.

(5) a wide operating temperature range.

ROTATION MEASUREMENT METHOD

Prior to the development of a rotation sensor consisting of MRE and a multi-pole ring magnet, the magnetic directivity of MRE (the relationship between magnetic field direction and resistivity change) was investigated and the experimental results are shown in Fig.5 . We chosed No.2 for sensing magnetic field, the main reasons are the following four(4):

No.2, compared with No.1 method,

(1) allows for a short distance between MRE and the magnet.

(2) assures an uniform magnetic field applied over MRE.

and No.2, compared with No.3 method,

(3) produces unparalleled high output.

(4) has satisfactory position accuracy.

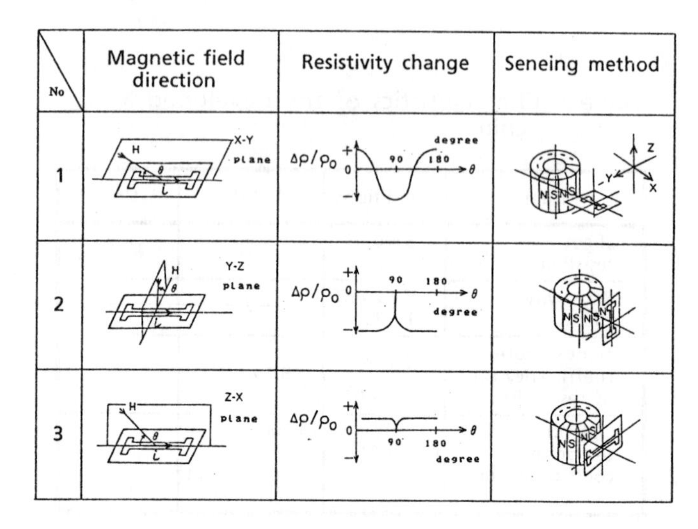

Fig . 5 Magnetic directivity of MRE

Fig.6 shows the method measuring rotation by No.2 sensing systems. The multi-pole magnet with the rotation shaft sets up against MRE.

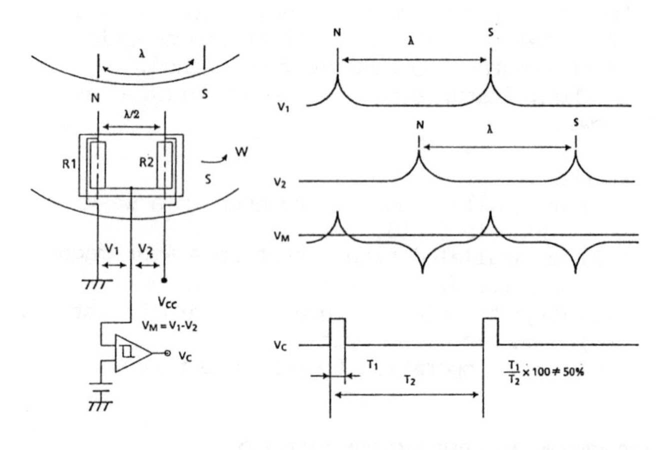

Fig . 6 Output duty of No.2 sensing system

Resistive elements are arranged in parallel with a $\lambda/2$ interval, where λ is magnet poles pitch, and constitute the half-bridge circuit.

When the " N " or " S " pole of the magnet is in position facing Resistor R1, the resistance value of R1 is large, and otherwise, its value is small. If the " N " or " S " pole comes to face Resistor R2 by rotating in direction of " W ", Resistor R2 in turns takes large value and Resister R1 takes small value. The repetition of this action with rotation of magnet generates the output signal " V_M". However, this method has too strong the output directivity, this caused output pulses with a duty ratio far less than 50% after wave-shaping processing (V_c).

To improve the duty ratio for an exact rotational detection, the system should produce the double frequency output from the foundamental frequency, then the double frequency output is to be devided exactly in half, so that the resulting duty ratio is 50%, as shown in Fig.7 .

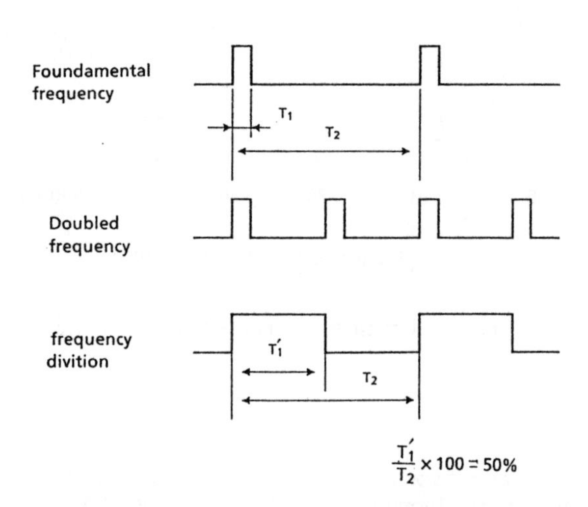

Fig . 7 Improvement of duty ratio

METHOD OF DOUBLE FREQUENCY OUTPUT

One of means for generating the double frequency output is to use a pair of MREs arranged such that both output signals are combined with each other. However, this method requires the MRE sensor to be accurately disposed one-fourth the pitch out of position with the magnet poles pitch, and the resulting signal processing circuitry will become a very complexed one. To meet the requirement, we have developed a system which produces double frequency output with only one MRE chip. We will explain the function below, with reference to Fig.8(a) through Fig.8(c).

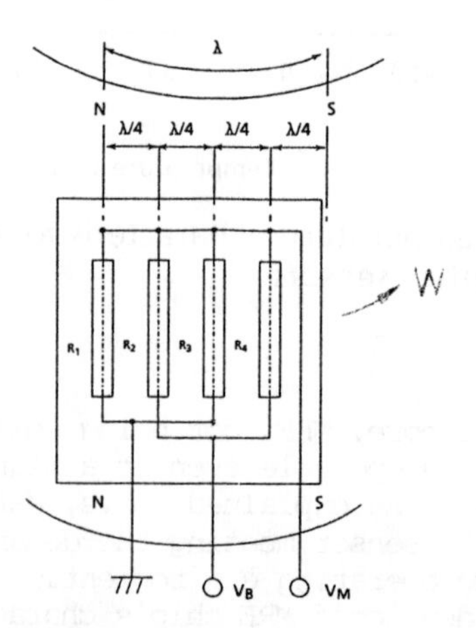

Fig . 8 (a) Ferro-magnetic resistors arrangement (half - bridge configuration)

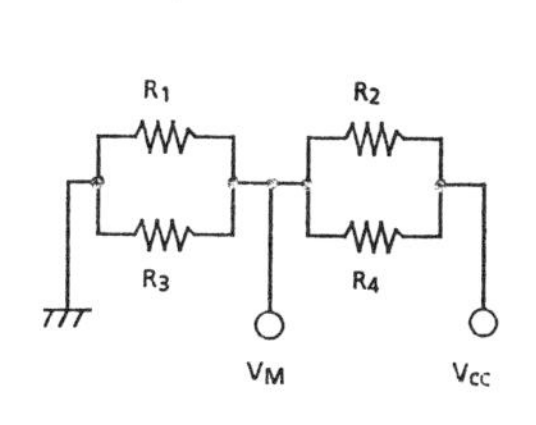

Fig. 8 (b) Equivalent circuit

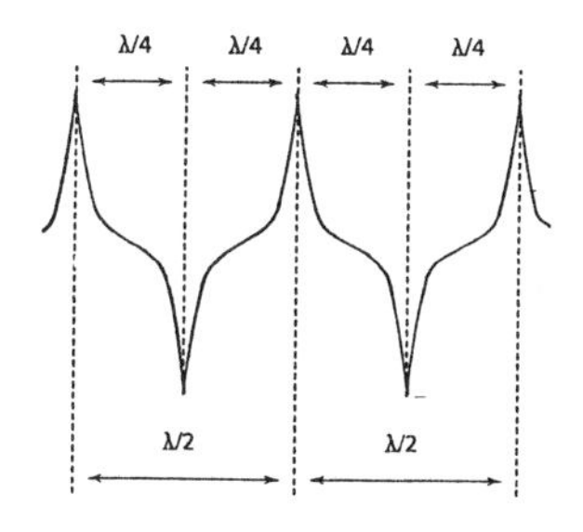

Fig. 8 (c) Double frequency obtained

Resistive elements R_1 through R_4, made of ferro-magnetic thin film,are arranged in parallel each other with an λ/4 interval, as shown in Fig. 8(a), where λ, as stated above, is the distance between the adjacent different poles. Fig.8(b) shows the equivalent half-bridge circuit. In this patterning of the resistive element, as so for explained above, each resistor is not affected by the magneto-resistance of the remaining resistor because their magnetic directivity is very strong, thus gaining the high bridge output.

When the " N " pole of the magnet is in position facing Resistor R_1, as mentioned above, the resistance value of R_1 is large and Resistor R_2, R_3 and R_4 take small values. If the " N " pole comes to face Resistor R_2 by rotating in direction " W ", Resistor R_2 inturn takes large value than those of Resistors R_1,$R_3$and R_4. Repitition of this action will thus generate the double frequency output signal as shown in Fig.8(c).

Fig.9 shows a photograph of a MRE chip for the rotation sensor.

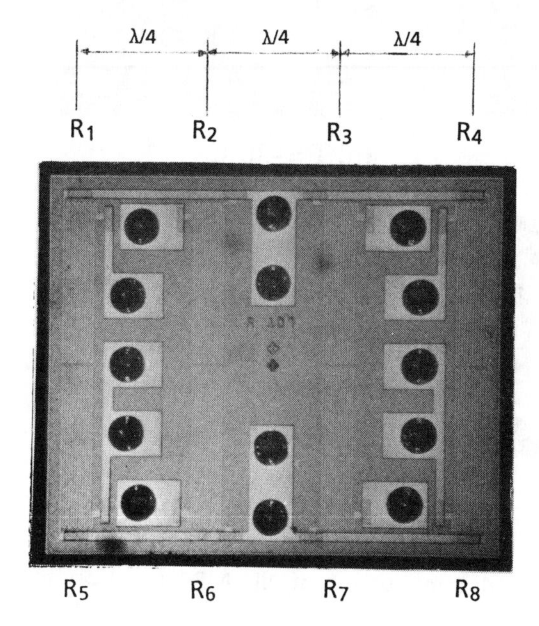

Fig. 9 MRE chip

This MRE chip is not only covered with a special protect film to meet the automotive sensor requirements, but are face-down bonding type called " flip-chip " for easy access to hybrid IC for mounting on it. As clearly seen from photograph, the MRE sensing die constitutes a parallely-connected, full-bridge circuit, thus making the sensor output large and the mutual point drift minimize because temperature coeffiency rate of Resistors $R_1$to R_8 is equall each other.

Fig.10 illustrates the output waveforms of the MRE chip, and appeares double frequency output signal.

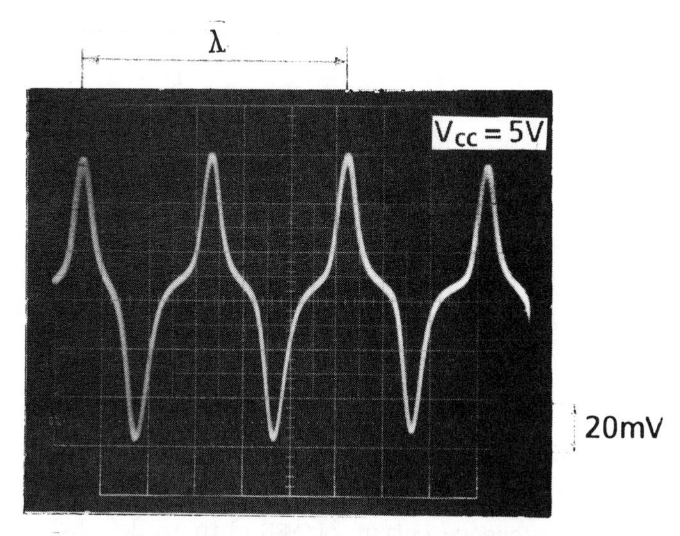

Fig. 10 The output waveforms of MRE chip

Fig.10 and Fig.11 show hybrid micro-circuit and the configuration of the MRE rotation sensor, respectively.

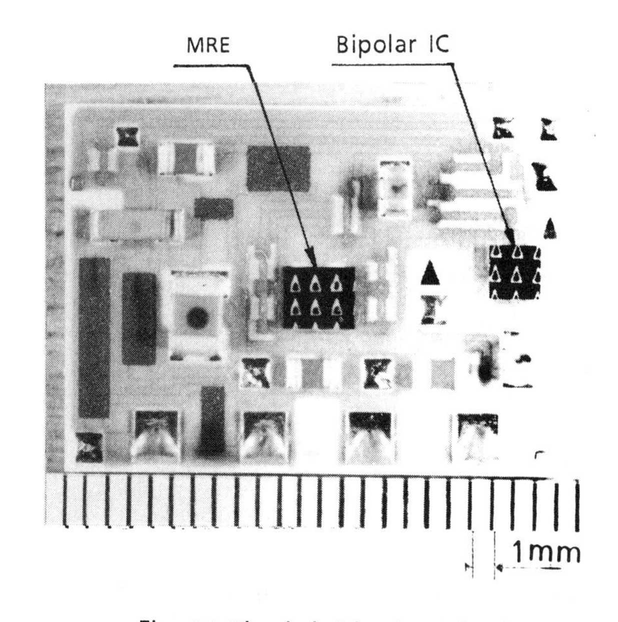

Fig. 11 The hybrid micro-circuit

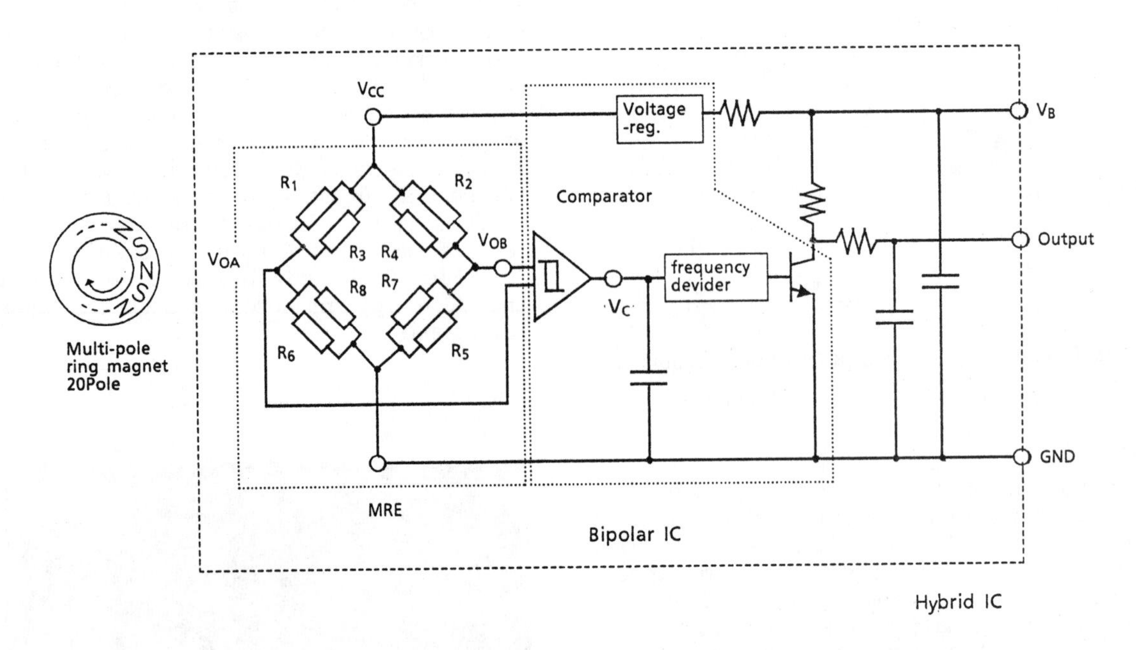

Fig. 12 The circuit configuration of developed MRE rotation senser

This hybrid micro circuit consists of one MRE chip, a monolithic IC, several thick film resisters, and chip condensers. By it, The double frequency output of MRE chip is devided and is converted to the rectangular wave form.

Fig.13 is the structure of the MRE rotation sensor. The sensor is composed of the rotary multi-pole magnet (10 " N " poles and 10 " S " poles) having diameter of 22mm, the hybrid IC consisting of one MRE chip and a signal processing circuit, and a aluminium die-cast housing.

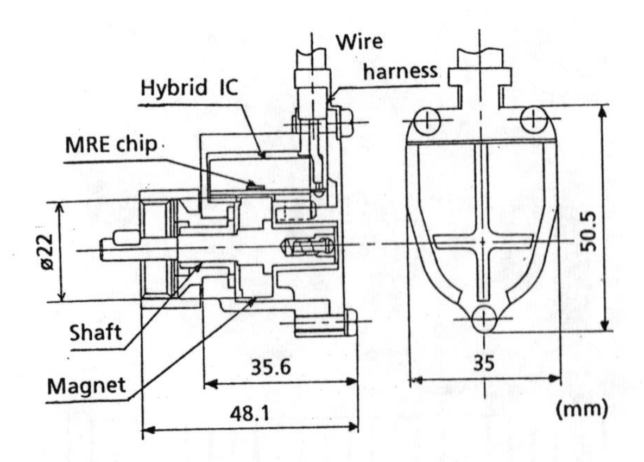

Fig. 13 Cross-section of the developed MRE rotation sensor for automotive use

Fig.14 shows the output waveforms in -30°c, and 120°c. What is evident from the waveforms, the output duty ratio reveals unchanged from -30°c to 120°c.

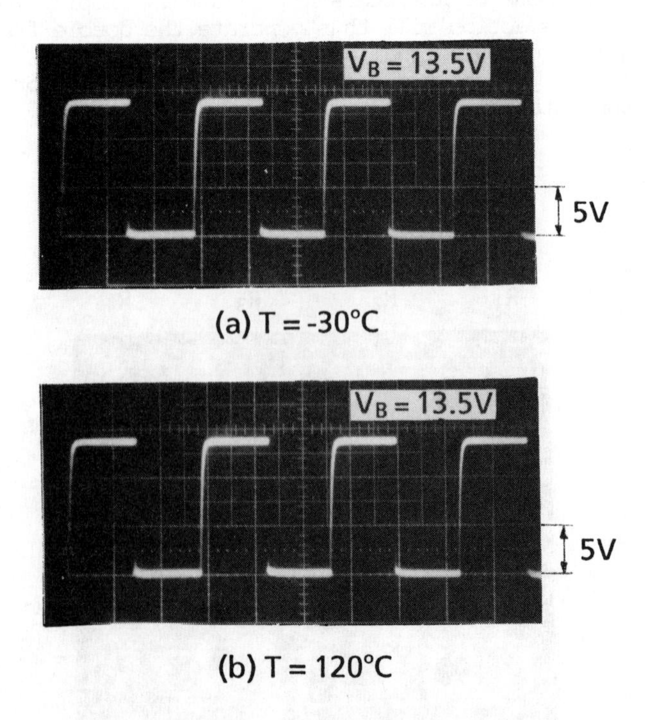

Fig. 14 The output of ND MRE rotation sensor

The sensor specifications are shown in Table 3. The sensor has already reveals its outstanding quality through a variety of reliability-tests, and has been used as the vehicle speed sensors, mounted directly onto the transmission of " SOARER " producted in Toyota Moter CO.,Ltd., and so on. This sensor has eliminated the need for a rotary cable, which has helped significantly in the freeing up of space at the rear of the meter section and in the elimination of cable noise, and produced the high-responsibility and high-accuracy of the speed meter(8) .

Table.3 Specification of the ND MRE rotation sensor.

Item	Spec.
The number of pulses	20 pulses per rotation
Output type	Open-collector duty ratio;50 ± 10%
Operating voltage	8 to 16 volt.dc.
Measurable transmission rpm	5~3,000 rpm
Operating temperature	-30~120°C
Storage temperature	-40~130°C
Vibration resistivity	196m/S2 (20 G)

CONCLUSION

The new MRE-type High-Accuracy, High-sensitivity magnetic rotation sensor, consisting of MRE sensing chip and a rotary multi-pole magnet offers product/system designers of automotive electronics a compact, environmental-resistive magnetic vehicle speed sensing system that will permit easy acces for the transmission of a vehicle. The " precise detection from low rpm to high rpm " capability often permits full range of applications, not limitted to speed sensors, such as detections of engine rpm, cranking angle and steering wheel position, and rotation sensors for electrically controlled transmission, in the field.

The versatility of the sensor also permits its applications in reserch or design in various industrial field.

ACKNOWLEDGEMENT

The autor is thankful to all those involved in the successful development of the MRE rotation sensor. A spesial ackowledgement is extended to people of Toyota Moter CO.,Ltd. for their contributions.

REFERENCE

(1) W. Thomson ; proc. Roy. Soc. London, 8 , 546 (1857).

(2) H. C. Van Elst ; Physica, 25 , 708 (1959).

(3) J. Smit ; Physica, XVI(6) , 612 (1951).

(4) F. G. West ; Nature, 188 , 129 (1960).

(5) T. Kasuya ; Prog. Theor. Physics, 16 , 58 (1956).

(6) J. Kondo ; Prog. Theor. Physics, 27(4) , 772 (1962).

(7) T. R. Mcguire , R. I. Potter ; IEEE Trans. Mag. MAG 11(4) , 1018 (1975).

(8) S. Mizutani , T. Ohtake ; SAE Paper 860410 (1986).

FLOW SENSORS

950433

Breakthrough in Reverse Flow Detection - A New Mass Air Flow Meter Using Micro Silicon Technology

Uwe Konzelmann, Hans Hecht, and Manfred Lembke
Robert Bosch GmbH

ABSTRACT

A new mass air flow meter, using a sensor element based on micro silicon technology, is presented. The sensor is able to detect the amount and direction of air flow. It is located in an aerodynamic bypass, which acts as a non-linear filter. This enables the mass air flow meter to correctly determine the air mass aspirated by a four cylinder engine, even in the case where, due to strong oscillations of the flow, timewise backflow occurs near the measuring position. Results of laboratory and engine tests will be presented.

INTRODUCTION

Thermal mass air flow meters are widely used as the primary load sensor for fuel injection systems. Conventional air flow meters, where the sensitive platinum resistors are located on a glass or ceramic substrate, offer a good relation between function and cost (for example the Bosch hot film HFM2 (1)). But it is not possible to get a correct air flow signal under every operating condition of an engine.

During the engine start, for instance, such a mass air flow meter often needs several seconds until correct air mass information is available. Also in transient conditions the sensor is unable to follow fast changing flow rates.

A dynamically fast sensor, having a small thermal mass (for example the Bosch hot wire (2)), does not have these specific problems. Still, under operating conditions where strong oscillations lead to temporal reverse flow near the sensor location, large errors in the indicated air mass occur. Therefore information of the direction of the flow is necessary to overcome large errors.

To avoid all the above mentioned problems, a new mass air flow meter was developed, which should combine the following features:

* robustness of a conventional hot film design,
* dynamic speed of a hot wire and
* reverse flow detection.

The new design is based on micro silicon technology, which enables production of a thermally fast sensor on a thin membrane and a layout of the sensor which is able to detect the direction of the flow.

MEASURING PRINCIPLE

The sensor is located in a measuring duct. It is mounted flush to the wall on a sheet metal, which is oriented parallel to the flow. A boundary layer develops on the sheet metal, and changes its thickness with the flow rate. Depending on the boundary layer thickness heat transfer rates change at the wall.

A section through the sheet metal and sensor is shown in Fig. 1. The sensitive part is a thin membrane produced by anisotropic etching on the sensor chip.

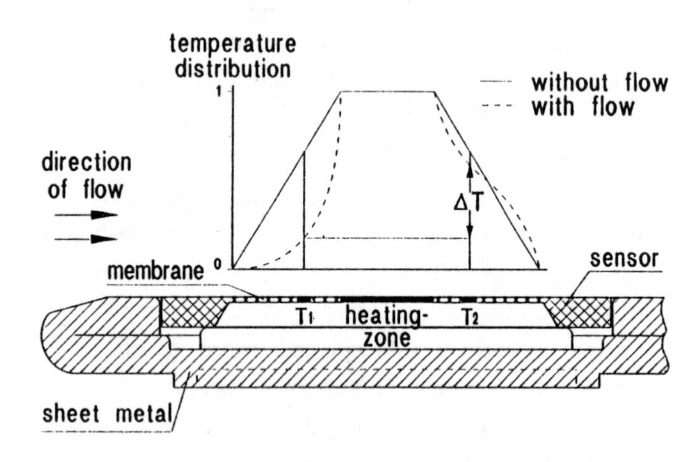

Figure 1: Measuring principle

In the middle of the membrane is a heating zone, where the temperature is held to a constant value, depending on the temperature of the ambient air. Without air flow the temperature shows approximately a linear slope with ambient temperature at the rim of the membrane. The membrane is cooled upstream of the heating zone, depending on the air flow rate. Downstream of the heating zone the temperature distribution is less changed. Two platinum resistors, which act as temperature sensors, are located upstream and downstream of the heating zone. With the help of these sensors the temperature difference ΔT on the membrane can be detected. It gives a measure of the mass air flow rate and for the direction of the air flow, as can be seen in Fig. 2.

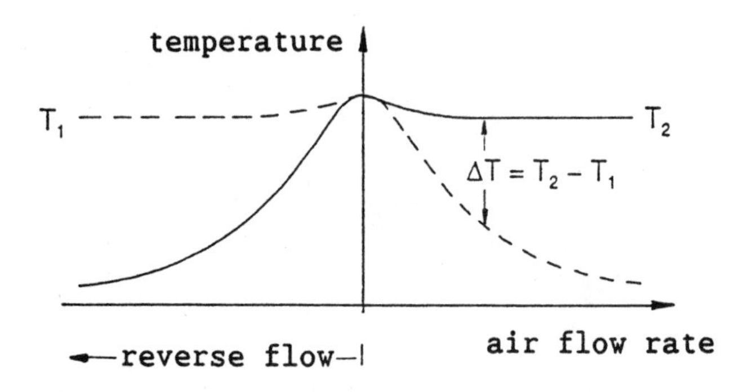

Figure 2: Temperature difference ΔT.

ELECTRIC CIRCUITRY

The electric circuit (Fig. 3) has to perform two tasks:

* control of the temperature within the heating zone, and

* the analysis of the temperature difference on the membrane.

CONTROL OF HEATING ZONE - This part of the electric circuit is similar to the Bosch hot film HFM2 circuit (1). A heater resistor R_H is controlled by a sensor resistor R_{HF}. The temperature of the heater is controlled by the sensor for the ambient air R_{LF}, which is located on the sensor chip.

ANALYSIS OF TEMPERATURE DIFFERENCE - A potentiometer P_1 is used to adjust the output voltage at zero air flow rate. The output signal of the temperature sensors R_{AB} and R_{AU} and the output signal of potentiometer P_1 are inputs for a differential amplifier. For final trimming of the sensor, the amplifier can be adjusted electrically. Finally the output signal of this amplifier is transformed to an output relative to a reference voltage U_{Ref} (ratiometric principle).

ELECTRICAL INTERFACE - The small size of the sensor element and the measuring principle lead to a current draw in all cases of less than 50mA. This allows use of only one ground wire.

With the ratiometric principle, the interface error between air flow meter and ECU is reduced.

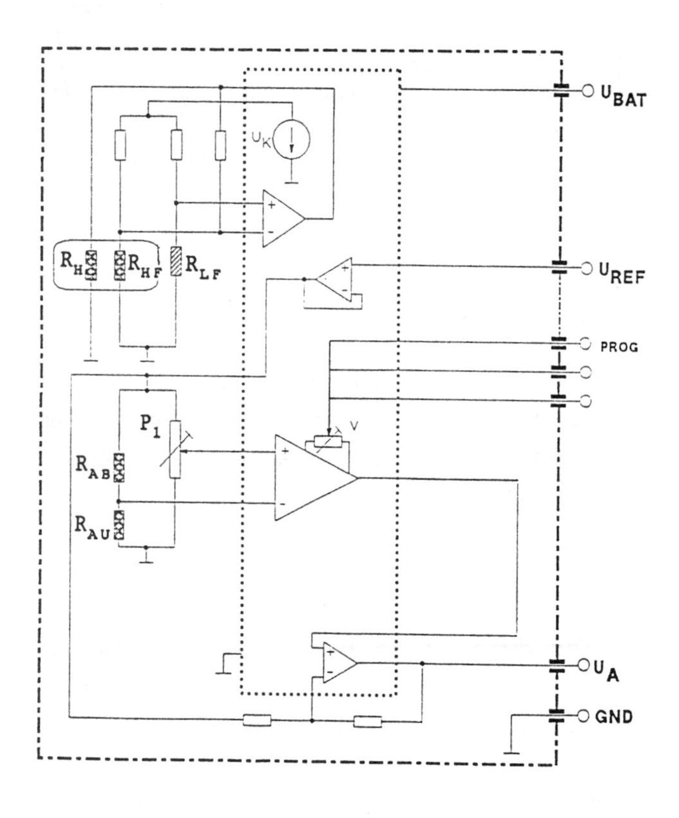

Figure 3: Electric circuit.

CHARACTERISTIC CURVE

The shape of the characteristic curve depends on the measuring principle. Offset and amplitude are given by the electric circuit (see Fig. 4).

The offset value at zero flow is normally set to 1V. Voltages above the offset value belong to flow rates in the forward direction. Reverse flow is given by values below the offset.

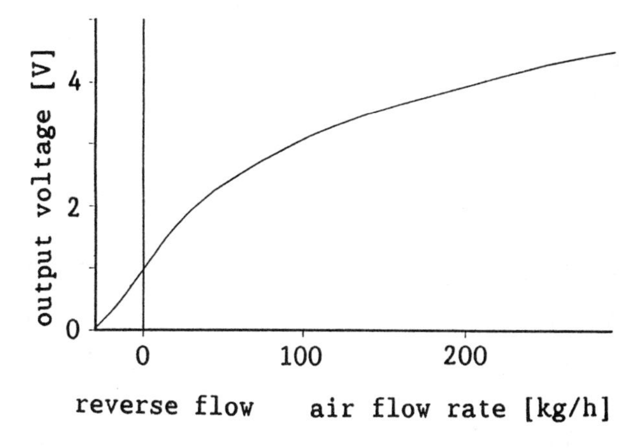

Figure 4: Characteristic curve.

DESIGN OF AIR FLOW METER

Fig.5 shows the completed air flow meter with measuring tube, wire grid as flow straightener and the plug-in sensor. Due to the plug-in principle it is of course possible to insert the plug-in sensor in different measuring tubes or in an air cleaner housing.

The compact design of the plug-in sensor leads to a small pressure drop of the new design. Fig. 6 gives an example for a inner tube diameter of 62mm, which normally allows to measure a nominal flow of 480kg/h.

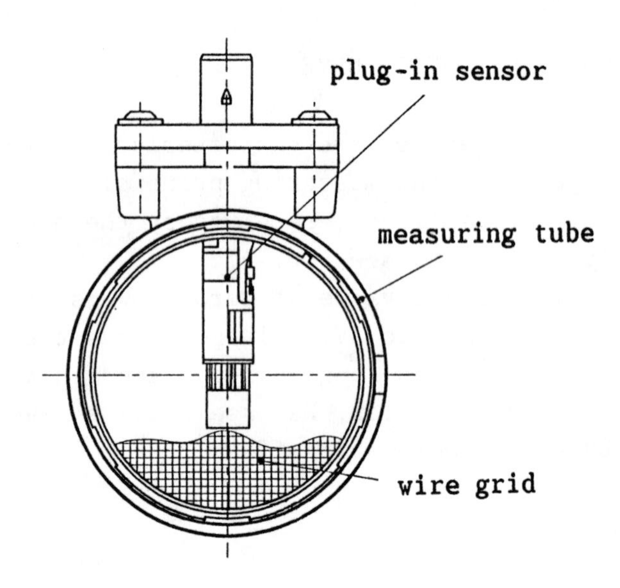

Figure 5: Plug-in sensor inserted in tube.

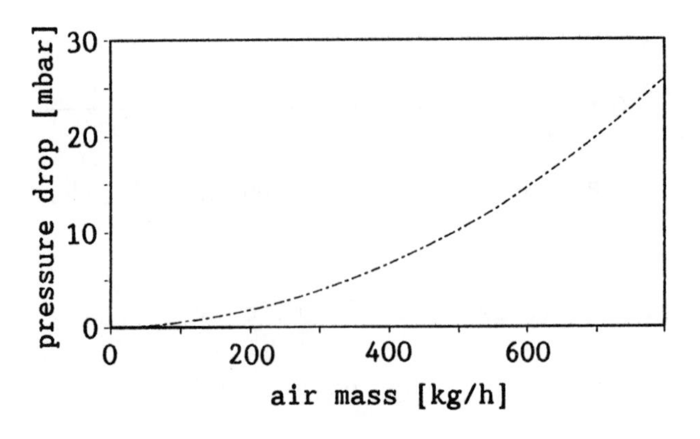

Figure 6: Pressure drop.

A cross section of the plug-in sensor is shown in Fig. 7. The mechanical parts consist of housing, sheet metal and two covers. Hybrid plate and sensor are glued onto the sheet metal. Electrical connections between connector, hybrid plate and sensor are performed by wire bonding. The sensor element is located in a bypass channel. The function of this channel is explained in the following section.

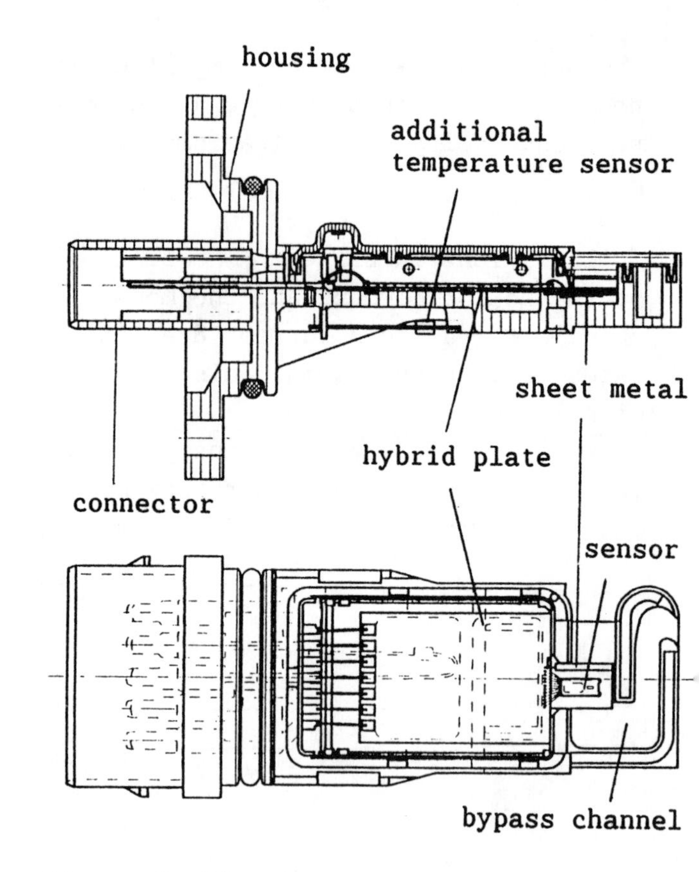

Figure 7: Plug-in sensor.

An additional temperature sensor is integrated in the air flow meter and makes the temperature of the air flow available for the fuel injection system. The cost of an extra temperature sensor can be saved.

LABORATORY TESTS

The dynamic behaviour of the new hot film is improved compared to a conventional sensor. The time constant of an air flow step from 10kg/h to 310kg/h is reduced to τ_{63}=7ms for the new design, which is about one third of a conventional air flow meter.

Turn on time is drastically reduced from several seconds for a conventional sensor to 10ms (error less than 5% of the air flow rate; flow rate 10kg/h; see Fig. 8).

When the sensor is turned on without air flow and, after a few seconds, an air flow of 10kg/h is applied, the turn on time increases. To test this behaviour it is necessary to produce an air flow step from 0kg/h to 10kg/h. With our test bench we are able to produce this step within 400ms, and this is the value we measure for the sensor, although it may be faster. Anyway, turn on times shorter than 500ms do not cause any difficulty for engine starts.

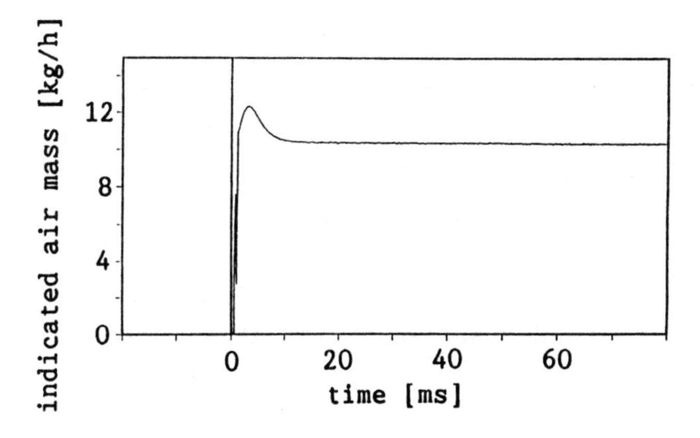

Figure 8: Turn on behaviour.

Extensive investigations were performed with pulsating air flow including reverse flow. The pulsations were generated with a loudspeaker. When the flow bench was excited with its resonance frequency, strong pulsations with back flow could be produced. The test set up is shown in Fig. 9.

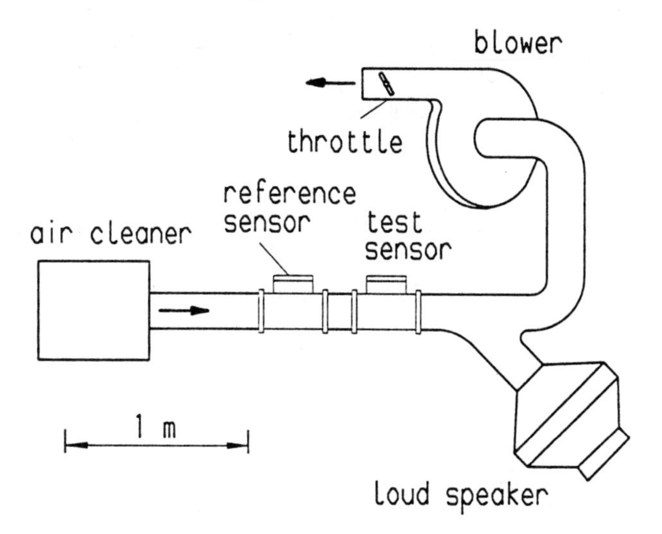

Figure 9: Test set up for pulsation investigation.

Tests of the sensor element alone showed that there was still an error of about -20% in the mean value of the indicated air flow rate, although the reverse flow was detected correctly. The reason is, that the time constant of the sensor element is still too slow, to be error free in the pulsating flow case.

To compensate for this error a bypass was designed, which should be small enough to be integrated within the plug-in sensor concept. The development was supported by flow calculations and large scale modelling of the aerodynamically important parts. Inlet and outlet of the bypass are in one plane normal to the measuring tube axis. This leads to reduced pulsation amplitudes within the bypass, because the oscillating pressure of the pulsation affects simultaneously the beginning and the end of the air column in the bypass.

As a consequence the bypass acts as a nonlinear filter function, which leads to a reduction of the pulsation amplitude near the sensor element, and a correction of the mean value of the flow passing near the sensor element.

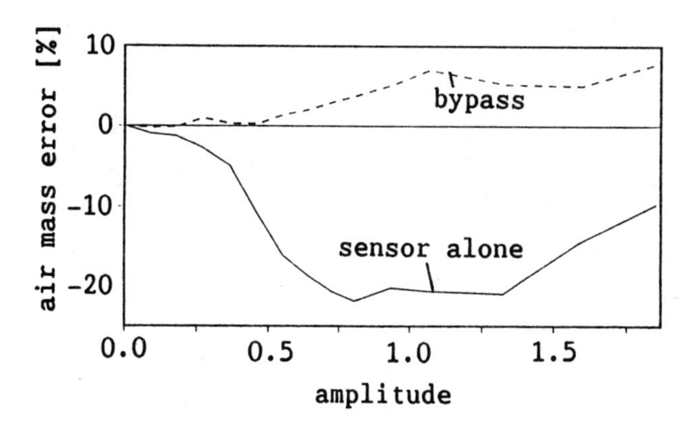

Figure 10: Indicated air mass error vs. pulsation amplitude; air flow rate 60kg/h, frequency 58Hz.

The error of the indicated air mass with and without bypass is shown in Fig. 10. The mean air flow rate is 60kg/h and the pulsation frequency is 58Hz (which is similar to a four cylinder engine at about 1800 rpm). The definition of the pulsation amplitude is shown in Fig. 11. For the amplitudes which were investigated, the air flow error with bypass remains smaller than 10%.

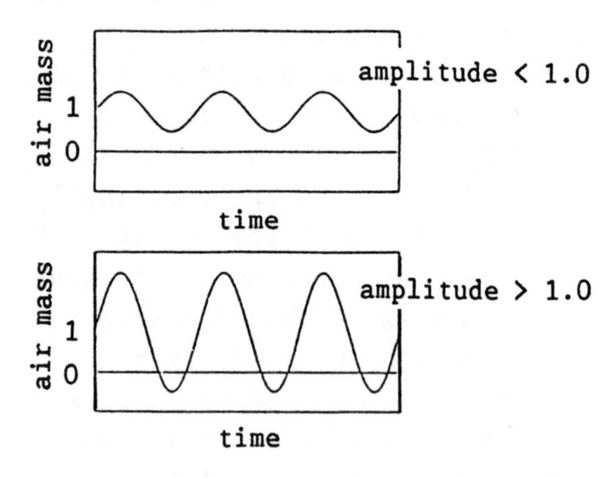

Figure 11: Definition of the pulsation amplitude.

ENGINE TESTS

Engine tests were performed with various three, four and six cylinder engines. Results for a four cylinder engine are presented. Fig. 12 shows the indicated air mass of the new hot film sensor compared with the Bosch hot wire. The hot wire is unable to detect the direction of the flow. The new hot film is fast enough to follow the time instants of reverse flow. The indicated amplitude of the new hot film is smaller, than with the hot wire. This is due to the damping effect of the bypass.

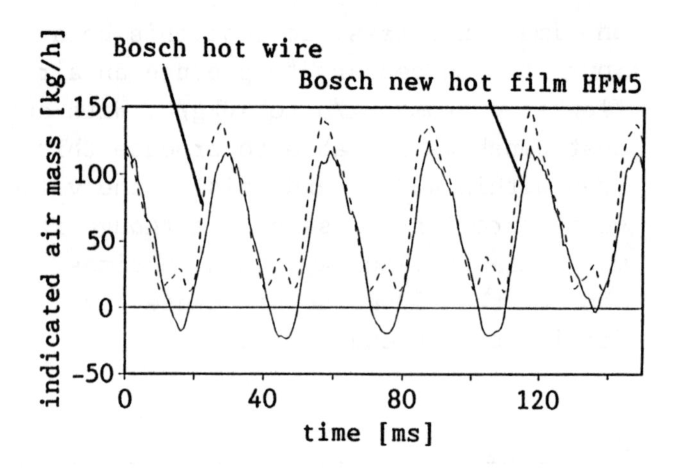

Figure 12: Pulsation investigation with a four cylinder engine, WOT, engine speed 1000 rpm.

The air mass error for the Bosch hot wire and the new design at full load is shown in Figure 13. The Bosch hot wire monitors small errors, except in the region of strong reverse flow near 1000 rpm. The new hot film HFM5 indicates small errors for all engine speeds.

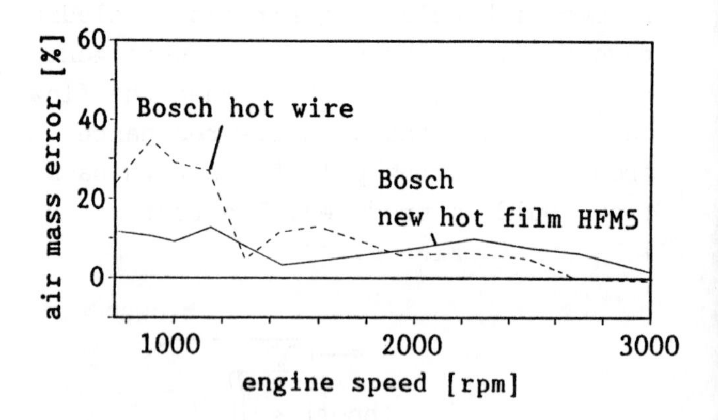

Figure 13: Air mass error against engine speed at a four cylinder engine.

INSTALLATION

The small sensor element is able to detect the turbulence in the oncoming flow, due to its thermal speed. Any flow separation upstream of the sensor increases the turbulence level and will cause a greater noise level in the signal. Pressure loss and flow instability caused by flow separations should be avoided for an optimized installation. With the help of the sensor it is easily possible to check the air flow quality and to evaluate modifications of the air duct.

When the following guidelines are respected, installation problems should not occur (see Fig. 14):

* cross sectional area may not be increased in downstream direction,

* straight air flow in front of sensor required ($L_1 > 0.5d$),

* inlet funnel reaching into the air cleaner ($L_E < 0.25 L_L$).

If, due to restricted space, flow separation cannot be avoided an additional flow straightener upstream of the sensor reduces the turbulence in the flow and, as a consequence, the noise of the signal to uncritical values.

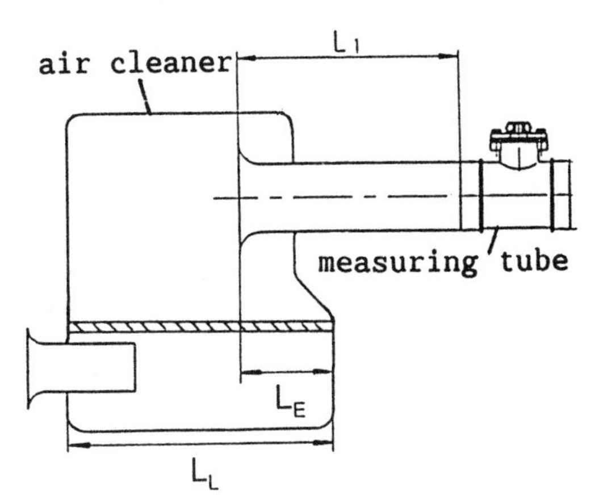

Figure 14: Installation example.

APPLICATION ASPECTS

With the new Bosch hot film sensor it is possible to use the sensor signal right after the engine start. Start correction functions can be avoided.

The air mass flow error under operating conditions with reverse flow is drastically reduced. Special software for this case is not necessary and application efforts and time can be reduced. Fig. 13 shows, that even under wide open throttle conditions only small errors occur. As a consequence, it is possible to avoid pulsation mapping during the application process.

SUMMARY

The small design of the new hot film mass air flow meter allows easy installation, with the benefit of a small pressure drop.

It combines the robustness and low cost of a conventional hot film with the speed of a hot wire sensor.

Further cost saving is possible with the integrated additional temperature sensor.

The feature of reverse flow detection offers better quality in air mass sensing and drastically reduced application efforts for three and four cylinder engines.

REFERENCES

(1) R. Sauer: Hot wire mass meter; SAE 800468

(2) R. Sauer: Hot-film mass air meter - a low-cost approach to intake air measurement; SAE 880560

940625

Hot Wire Mass Gas Flow Sensing Device

Masatoshi Sugiura
Hitachi Farmington Hills Technical Center, Inc.

Isao Okazaki
Hitachi Automotive Products (USA), Inc.

George Saikalis
Hitachi America, Ltd.

ABSTRACT

The key issue in gas metering of alternative fuel vehicle is to obtain low emission and accurate air-fuel ratio. A hot wire mass air flow sensor can directly monitor the air flow by using thermal transfer amount in an unit time to keep the hot wire at a certain temperature. Surveys were conducted regarding this method and it was verified that this method enables to monitor mass gas flow in applications such as Compressed Natural Gas(CNG) and propane(C_3H_8). The gas passage body for this sensor, which consists of a 10mm diameter bypass and a main pass has been surveyed and developed. This electrical sensing device for CNG has been completed and its performance was verified with a CNG flow test stand and a CNG engine. We have found that this thermal transfer monitoring method is not affected by a pressure change.

PRINCIPLE OF OPERATION

The electrical circuit described in Fig.1, makes an equilibrium state. When the temperature of the hot wire change, this circuit tries to increase or decrease the electrical current through the hot wire so that this circuit remains in equilibrium. Fig.2 is a simplified model of this electrical circuit. When a hot wire is surrounded in high flow, higher electrical current is required to keep the hot wire at certain temperature. Vice versa, when low flow surrounds the hot wire, lower electrical current is needed to keep the hot wire at certain temperature. In Fig.2, the power(W) dissipated to maintain a hot wire at certain temperature is described as

$$W = Ih^2 \times Rh$$

When we assume Rh constant, the power dissipated is monitored by Ih. In Fig.2, V_2 is equal to Ih multiplied by R_1. In this manner, the amount of V_2 has a relationship with how much mass gas is passing through the hot wire.

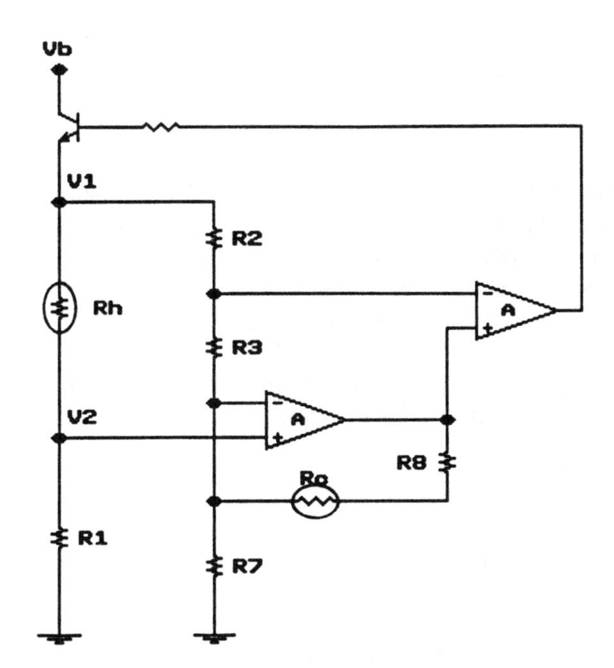

Fig.1 Circuit Diagram

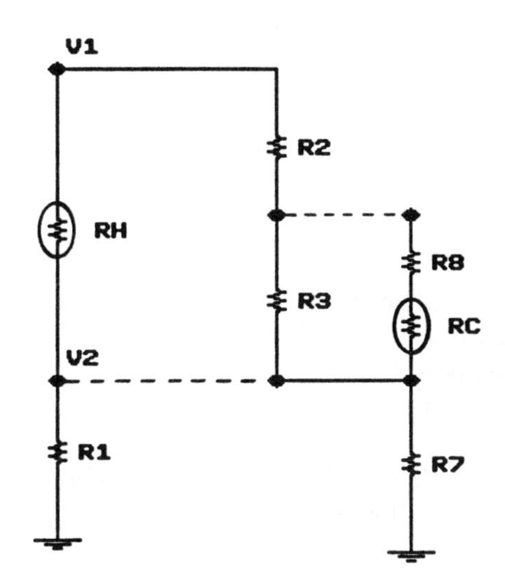

Fig.2 Simplified Circuit Diagram

MASS FLOW MEASUREMENT

When gas flows perpendicularly to the hot wire element with the velocity u, the heat transfer is given by experimental Mc Adam's equation as:

$$N_u = 2.4 + 1.6 \cdot Re^{0.35} \quad \text{------(1)}$$

$$N_u = \frac{h \cdot D}{\lambda}, \quad Re = \frac{u \cdot D}{\nu}$$

Transform of equation(1) is as follows:

$$h = 2.4 \cdot \frac{\lambda}{D} + 1.6 \cdot (\sqrt{\rho \cdot u} \sqrt{\frac{D}{\rho \cdot \nu}})^{0.35} \cdot \frac{\nu}{D}$$

$$h = \frac{\lambda}{D}(2.4 + 1.6(\frac{D \cdot u}{\nu})^{0.35}) \quad \text{------(2)}$$

where
Nu: Nusselt number
Re: Reynolds number
ρ : Specific gravity
h : thermal transfer coefficient of hot wire
λ : thermal transfer coefficient of flow
D: Hot Wire diameter
ν : Kinematic viscosity
u : velocity of flow

Therefore, the power(W) dissipated in thermal equilibrium is expressed as follows

$$W = Ih^2 \cdot Rh$$

$$= h \cdot (Th - Ta) \cdot S$$

$$= (C_1 + C_2 \cdot u^{0.35})(Th - Ta)S$$

therefore:

$$Ih^2 = \frac{\lambda}{D}(C_1 + C_2 \cdot u^{0.35})(\frac{Th - Ta}{Rh})S \quad \text{-----(3)}$$

$$(C_1 = 2.4, \quad C_2 = 1.6(\frac{D}{\nu})^{0.35})$$

where
Rh : resistance of hot wire
Th : temperature of hot wire
Ta : Ambient temperature
S : Surface area of hot wire

In equation(3), suppose (Th-Ta)/Rh and S are constant numbers, we obtain the following equations:

$$Ih = C_3\sqrt{C_1 + C_2 \cdot u^{0.35}}$$

$$Ih = C_4\sqrt{1 + \frac{C_2}{C_1} \cdot u^{0.35}} \quad \text{------(4)}$$

C_2/C_1 and C_4 are described as follows:

$$\frac{C_2}{C_1} = \frac{1.6}{2.4}(\frac{D}{\nu})^{0.35}$$

$$C_4 = \sqrt{C_1 \cdot \frac{\lambda}{D} \cdot \frac{Th - Ta}{Rh} \cdot S}$$

RESEARCH FOR GAS FLOW MEASUREMENT

Equation(3) indicates that with Kinematic Viscosity, Specific Gravity and Thermal conductivity of any particular gas, the mass flow calculation can be obtained.

In the following steps, we survey air, CNG and propane and monitor the difference in electrical current through the hot wire.

In air

$$\frac{C_2}{C_1} = 1.97$$

$$C_4 = 52.7 \times 10^{-3}$$

which gives:

$$Ih = 52.7\sqrt{1 + 1.97 \cdot u^{0.35}} \quad \text{------(5)}$$

In CNG

$$\frac{C_2}{C_1} = 1.93$$

$$C_4 = 66.0 \times 10^{-3}$$

which gives:

$$Ih = 66.0\sqrt{1 + 1.93 \cdot u^{0.35}} \quad \text{------(6)}$$

In this manner, we obtain the relation of Ih and u in Fig.3.
Fig.4 is the highlight of the relationship of Ih and mass with a model of 10mm diameter mass air flow sensor.

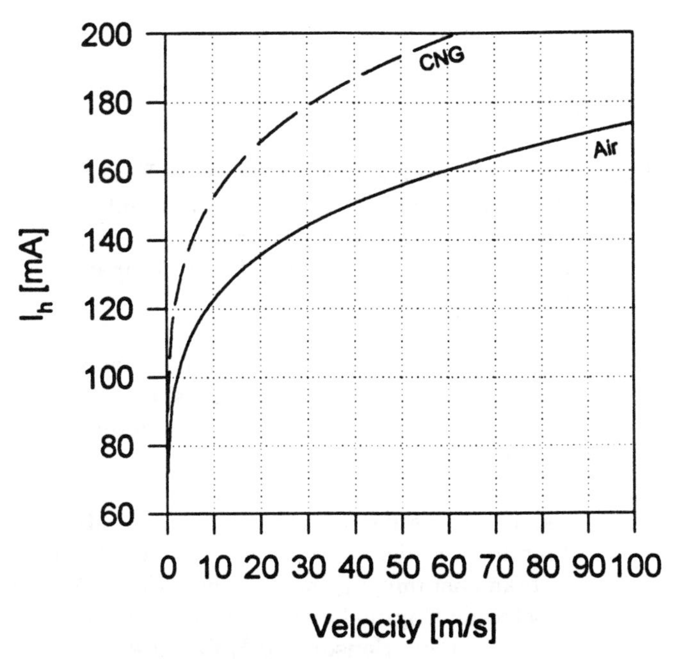

Fig.3 Relationship between velocity and Ih

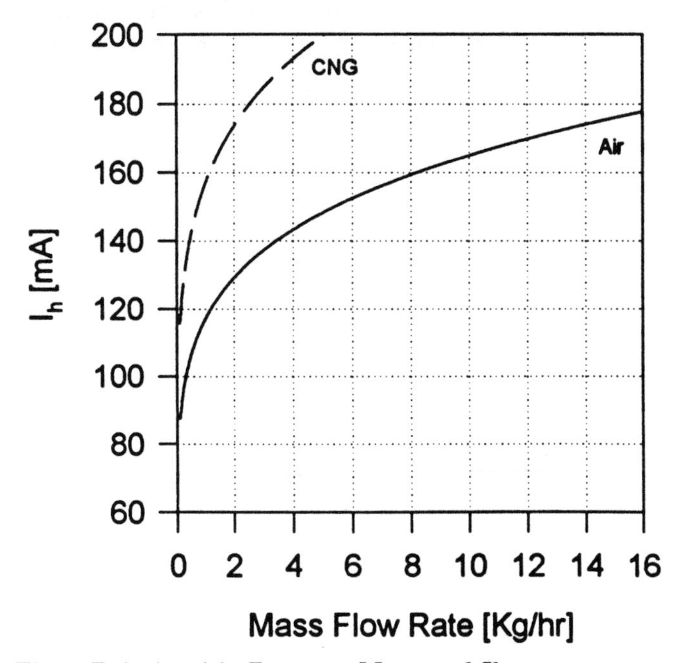

Fig.4 Relationship Between Mass and Ih

REQUIRED FLOW BAND

When gas flow is monitored with this device, the power dissipation in the hot wire must be kept in the same range as the mass air flow sensor.
Samples desired to measure from 2[kg/h] to 36[kg/h] were developped for CNG taking into consideration a flow band and thermal dissipation.
The conversion for mass flow calculation between CNG and air in thermal transfer is given as follows: 36[kg/h] of CNG is equivalent to 143[kg/h] of air, 2[kg/h] of CNG is equivalent to 8[kg/h] of air.

When a 10mm diameter mass air flow sensor is used as an experimental device, it is possible to measure up to 15[kg/h] of air.
This means in order to obtain 143[kg/h] of air, 228[kg/h] of air must pass besides this 10mm diameter mass air flow sensor.

MEASURED PERFORMANCE

In this manner, we have obtained a mass air flow sensor which is capable to flow from 8[kg/h] of air to 143[kg/h] of air and conducted flow test under air and gas.
Fig.5 indicates that the curve is following the theoretical model and based on this test, the hot wire air flow sensing device is capable for sensing mass gas flow.
Fig.6 indicates different main bore diameters and their impact on the monitoring range of mass gas flow.

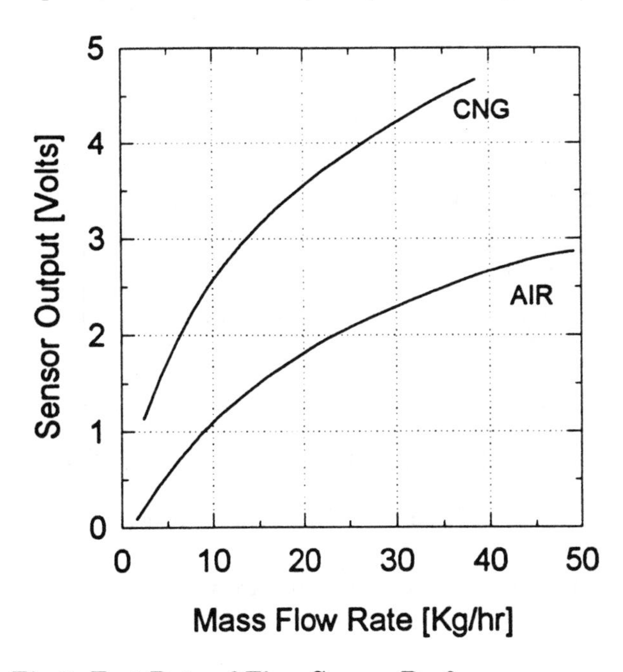

Fig.5 Test Data of Flow Sensor Performance

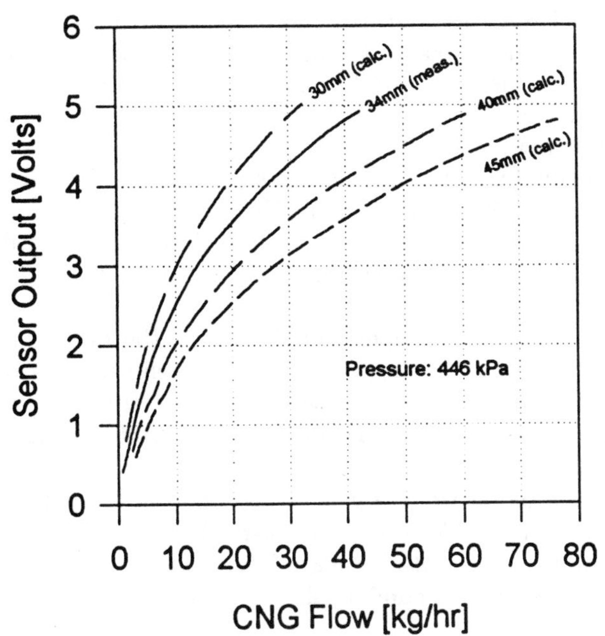

Fig.6 Performance on variety of bore

PRESSURE EFFECT

When pressure changed α times, the velocity under the same mass flow rate changes α^{-1} times. The thermal transfer under an unit time is determined how many molecules conduct thermal transfer with the hot wire and this molecule number will not change under same mass flow since however the pressure changes, the velocity is determined so that pressure multiplies with velocity stays constant($\alpha \times \alpha^{-1} = 1$).
Fig.7 indicates the relation CNG Mass Flow v.s. Output Voltage when pressure changes.

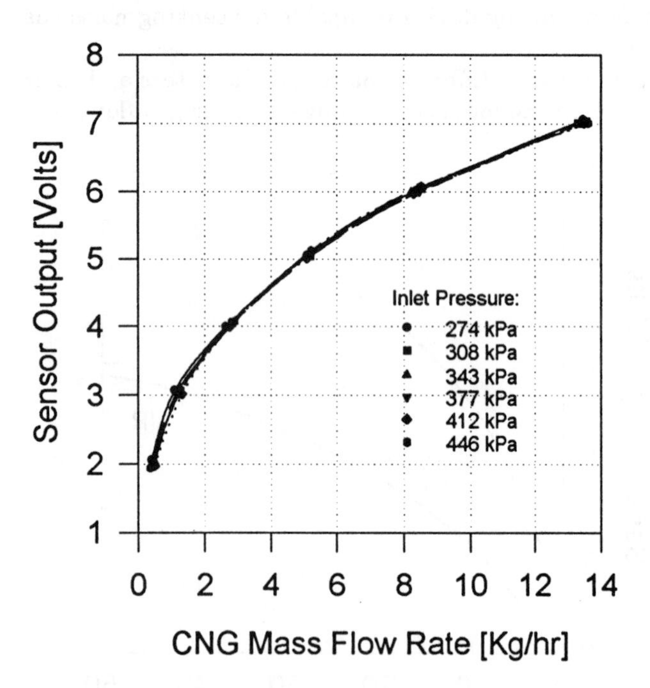

Fig.7 Pressure effect on performance

CONCLUSION

Research was done on the use of a hot wire mass air flow sensing device as a mass gas flow sensing device based on its principle of thermal dissipation. As a result, we concluded this device can be used for mass gas flow monitoring device.
To verify this, development sample was made. This sample was tested in air and CNG.
Furthermore, the pressure effect test was conducted and very low effect were observed.

As the principle of operation indicates, the electrical current through the hot wire to keep its temperature at a certain amount over a cold wire is a function of kinematic viscosity, specific gravity and thermal conductivity.
When temperature of a gas changes, power dissipation changes and its effects on the electrical current amount. As for a countermeasure for this, electrical compensation was conducted.

The survey to be conducted is the effect on performance due to chemical composition change on gaseous and also chemical durability test.

REFERENCE

[1] Donald L. Katz, Riki Kobayashi, Fred H. Poetmann, John A. Vary, Jack R. Elenbaas,and Charles F. Weinaug, "Handbook of Natural Gas Engineering", McGrawhill Book Company, Inc.,1959

[2] Sadayasu Ueno, Kanemasa Sato, Yoshishige Ohyama, and Hisamitsu Yamanaka, "Anti-Dirt Property of Hot Wire Air Flow Meter", SAE Paper #831018

[3] Takao Sasayama, Takeshi Hirayama, Matsuo Amano, Shinichi Sakamoto, Matsuyuki Miki, Yutaka Nishimura, and Sadayasu Ueno,"A New Electronic Engine Control System Using a Hot Wire Air Flow Sensor", SAE Paper #820323

[4] Corwin Snyder, Southwest Research Institute/ Gas Research Institurte "Gas Flow Test Data", 1992

970533

Study of a Sensor for Fuel Injection Quantity

Takao Iwasaki, Hayato Maehara, Oliver Berberig, Kay Nottmeyer and Takashi Kobayashi
ZEXEL Corp.

Copyright 1997 Society of Automotive Engineers, Inc.

ABSTRACT

Due to the present demand for further improved emissions and performance of diesel engines, there is a growing need to improve the control of fuel injection quantity and timing, as well as spray properties. We have developed a Micro Turbine Sensor that can measure transient injection rate and timing using micro machining technology. This sensor realizes volumetric flow measurement using a tangential turbine as the sensing element which has an outside diameter of 1mm , and which is located next to the inlet connector of the injection nozzle. The measured results are compared with a Bosch type injection rate meter. Since the tendency of measured injection rate shows fair agreement with results of the reference system, this sensor has potential as a fuel flow meter which is able to measure the injection rate and timing directly and continuously during engine operation.

INTRODUCTION

Concerning further improvement of emissions and performance of diesel engines, it is necessary to improve the control of fuel injection quantity and timing, as well as spray properties. In present conventional fuel injection systems[1] the control of injection quantity and timing depends on the mechanical setting of the fuel injection pump. Methods to measure directly have not been adopted yet. Consequently, this procedure which depends on dimensional accuracy and wear of the construction parts is not always precise. There are several injection meters in which the injection quantity is measured after injection[2]~[4]. To achieve further improvement, future systems will have to apply a new device which is able to monitor the actual parameters continuously.

The fuel flow in the injection pipe encounters severe conditions since the pressure varies rapidly within micro-seconds, reaching a maximum pressure of 100MPa. Therefore following requirements have to be satisfied for a sensor of fuel injection quantity.

(1)High response and resolution.
(2)Pressure and shock wave resistance.
(3)Minimization of the sensing element.
(4)Insensibility for changes in viscosity.
(5)No influence on injection characteristics.

We have developed a Micro Turbine Sensor that can measure transient injection rate and timing using micro machining technology. This sensor realizes volumetric flow measurement using a tangential turbine as the sensing element, which has an outside diameter of 1mm, and is located next to the inlet connector of the injection nozzle.

The purpose of this study is to examine the feasibility of a MTS application in diesel fuel injection systems and to improve the measurement accuracy such as the response and the resolution. There are several parameters which influence the response and resolution of the MTS. This paper gives attention to the effect of installation height of the turbine center above the injection pipe wall and to the influence of geometry of the turbine holder on performance of the MTS. Experimental evaluation in steady flow conditions have been carried out using a real-size turbine. To investigate the performance of the actual turbine, flow visualization using a scale-model(40:1) has been conducted. Furthermore, the response of the MTS has been examined under realistic conditions by installation between an injection pump and nozzle.

OUTLINE OF THE MICRO TURBINE SENSOR

Fig 1. shows the concept of the injection system applying a MTS. Fuel in the injection pipe close to the pump delivery valve flows backward after the end of

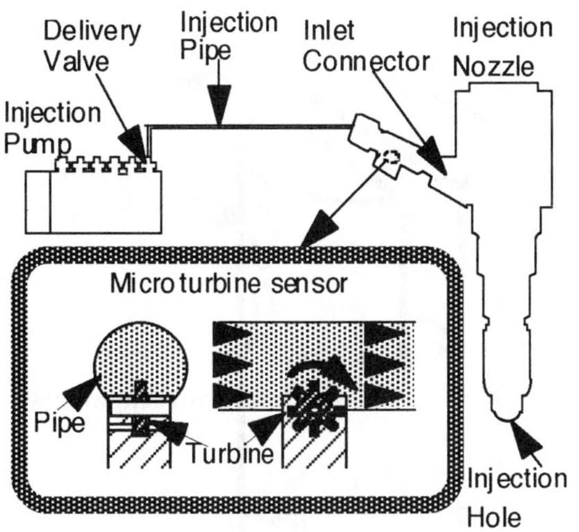

Fig.1 Concept of the injection system

injection as a consequence of delivery valve downward movement, thus relaxing the remaining pressure in the injection line. Hence, for accurate injection quantity a flow meter should be installed in the vicinity of the injection nozzle. The MTS is positioned in front of the inlet connector, as shown in Fig. 1. The center of the micro-turbine, which has 8 vanes, lines up with the pipe wall, thus making use of the Pelton turbine principle.

Details of the turbine geometry are shown in Fig. 2. Photo. 1 is a SEM image of the micro-turbine. Wire electro-discharge machining has been used for micro-turbine fabrication. The purposes of turbine miniaturization are to enable installation in an injection pipe having 2mm inside diameter, to improve its flow sensitivity as much as possible by reducing the moment of inertia and to avoid influences on the spray characteristics.

Fig. 3 shows the turbine holder that carries the micro-turbine. The turbine holder tip, 1.1mm in diameter, contains a slit to install the turbine, and drilled holes, 0.24mm and 0.15mm in inside diameter. The 0.24mm hole holds the turbine shaft, while the 0.15mm hole serves as a viewing window for detection of the turbine revolution.

A schematic arrangement of the sensing system for fuel injection quantity measurement is shown in Fig. 4. The fuel flow causes a turbine rotation which is detected by a photo-sensor using an optical fiber. In contrast with capacitive or magnetic readout principles, this mechanism does not influence measurement since no forces are involved. The detector generates square wave pulses triggered by a light beam that passes through the viewing window and which is intermittently interrupted by passing vanes of the rotating turbine. The turbine revolution speed is calculated then by counting number of pulses and dividing them by the number of vanes.

ANALYTICAL STUDY OF THE MICRO TURBINE OPERATION

In order to achieve some insight in geometry and flow parameters that might influence the MTS operation, an analysis has been conducted considering forces and torque acting upon the turbine. However, to be able to apply simple analytical equations some simplifications and assumptions have been made which are listed subsequently:

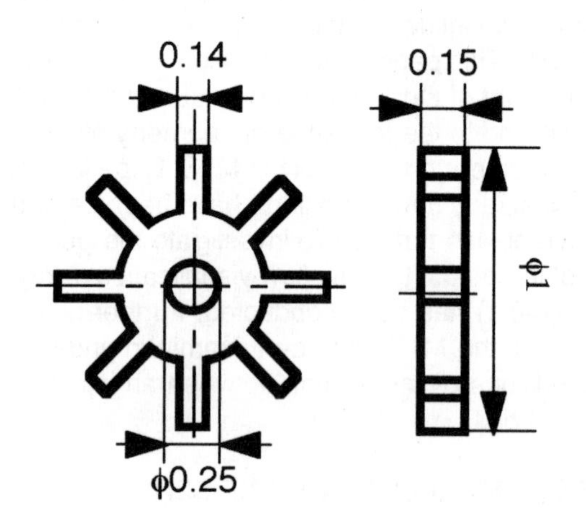

Fig.2 Micro turbine

Photo. 1 SEM image of the micro turbine

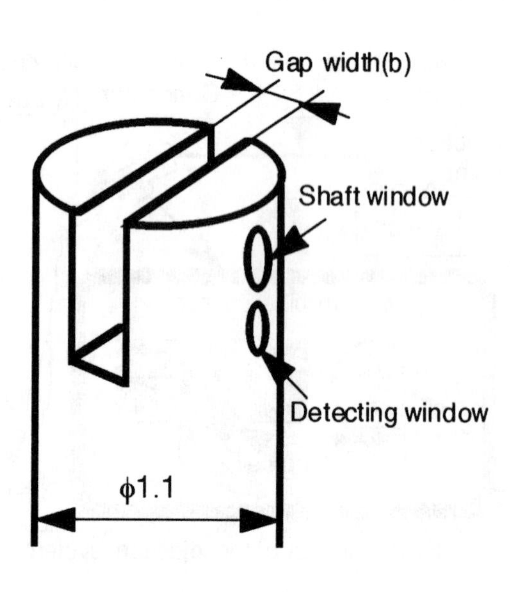

Fig.3 Turbine holder

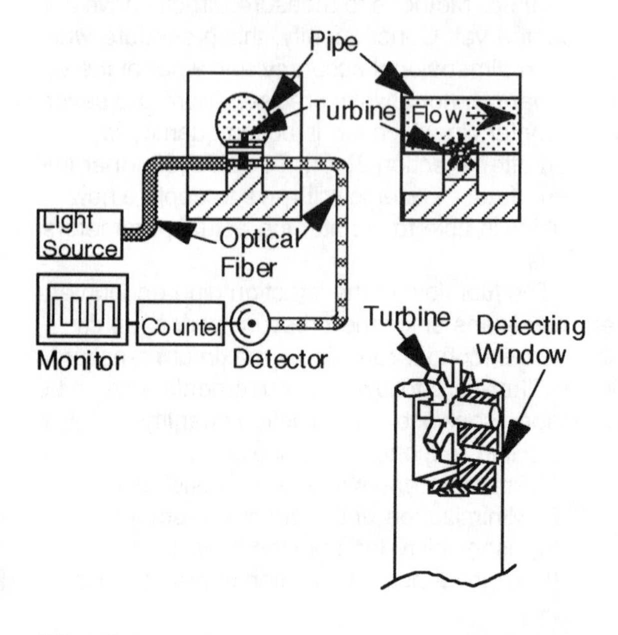

Fig.4 Schematic arrangement of sensing system

(1) The flow velocity profile is perfectly rectangular

(2) One turbine wing is fully exposed to the flow, one turbine wing is fully exposed to the resting fluid in the holder slit, all other wings are not driven or hindered by the fluid

(3) Fluid friction is considered only as shear forces in gaps, i.e. between turbine sides and adjacent holder walls, and between turbine hole and bearing pin

In Fig. 5 these simplified conditions are illustrated, which can be mathematically described by an equilibrium of involved torque:

$$M_{flow} = M_{rest} + M_{fluidfrict} + M_{inertia} \quad (1)$$

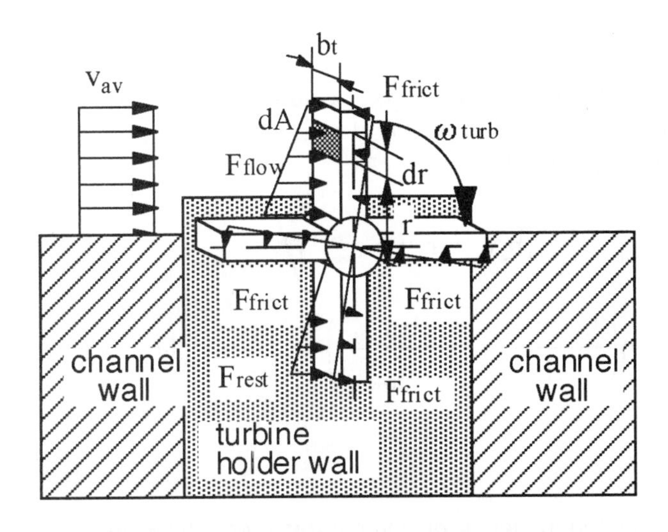

Fig.5 Torque and forces acting upon the micro turbine

Driving torque resulting from pipe flow (M_{flow})- The largest amount of forces acting upon the exposed upper wing result from a differences in velocity between the fluid and the wing, which is a function of vane radius.

$$M_{flow} = F_{flow} \cdot r$$

with

$$F_{flow} = c_d \int_{r_{root}}^{r_{tip}} p_{dyn} \cdot dA$$

$$p_{dyn} = \frac{\rho}{2}\left(v_{av} - \omega_{turb} \cdot r\right)^2$$

$$dA = b_t \cdot dr$$

where

c_d : coefficient of drag
ρ : fluid density
r : turbine radius
b_t : turbine width

Consequently, the driving torque resulting from pipe flow equals

$$M_{flow} = \frac{\rho}{2} c_d b_t \left[\frac{v_{av}^2}{2}\left(r_{tip}^2 - r_{root}^2\right) - \frac{2}{3} v_{av}\omega_{turb}\left(r_{tip}^3 - r_{root}^3\right) + \frac{\omega_{turb}^2}{4}\left(r_{tip}^4 - r_{root}^4\right)\right] \quad (2)$$

The maximum driving torque results from a turbine standing still as it occurs at the beginning of the injection stroke. If we assume the turbine still at rest at realistic average flow velocity of 23m/s, with c_d=1.15, ρ =840kg/m^3, r_{tip}=0.5mm, and b= 0.15mm the start-up turbine torque $M_{flow,max}$ becomes 3.6×10^{-6}Nm.

Retarding torque from fluid in the slit at standstill (M_{rest})- Two thirds of the turbine vanes rotate in a slit filled with the fluid. Since the slit is covered with the turbine, it is assumed that the fluid circulation within is not determined by the channel flow. Hence we assume the slit fluid to be at standstill for calculation of the dynamic pressure, while flow losses due to vortices are neglected since they can not be considered analytically.

$$M_{rest} = F_{rest} \cdot r$$

with

$$p_{dyn} = \frac{\rho}{2}\left(\omega_{turb} \cdot r\right)^2$$

$$dA = b_t \cdot dr$$

Consequently, the retarding torque from fluid in the slit at standstill is

$$M_{rest} = c_d \frac{\rho}{2} b_t \frac{\omega_{turb}^2}{4}\left(r_{tip}^4 - r_{root}^4\right) \quad (3)$$

Retarding torque from shear forces ($M_{fluidfrict}$)- For simplicity of calculation, the turbine is considered to be a full disk. However, since only the lower turbine's half is exposed to decelerating shear forces, only the area of half a disk is taken into account, equivalent to the area of one disk side.

$$M_{fluidfrict} = F_{frict,side} \cdot r_{side} + F_{frict,hole} \cdot r_{hole}$$

with

$$F_{frict,side} = \int_{r_{hole}}^{r_{tip}} \tau_{side} dA$$

and

$$\tau_{side} = \eta \cdot \frac{\omega_{turb} \cdot r}{h_{side}}$$

$$dA = 2 \cdot \pi \cdot r_{side} \cdot dr$$

and with

$$F_{frict,hl} = \tau_{hl} \cdot A_{hl}$$

and

$$\tau_{hl} = \eta \cdot \frac{\omega_{turb} \cdot r_{hl}}{h_{bear}}$$

$$A_{hl} = 2 \cdot \pi \cdot r_{hl} \cdot b_t$$

where

τ : shear stress of the fluid
η : dynamic viscosity of the fluid
h_{side} : clearance between turbine side and holder wall
h_{bear} : clearance between turbine hole and bearing shaft
A_{hl} : area of the turbine hole surface

Therefore, the retarding torque from shear forces becomes

$$M_{fluidfrict} = \pi \cdot \eta \cdot \omega_{turb} \cdot \left[\frac{\left(r_{tip}^{\ 4} - r_{hole}^{\ 4}\right)}{2h_{side}} + \frac{2 \cdot b_t \cdot r_{turb}^{\ 3}}{h_{bear}} \right] \quad (4)$$

The maximum retarding torque results from a turbine rotating at maximum possible speed, which means same velocity of the liquid and the turbine vane tips. Using the parameters of driving torque calculation and η=3.4×10^{-3}kg/(ms), h_{bear}=5μm, h_{side}=25μm, equation (3) yields a decelerating torque $M_{rest,max}$ of 2.2×10^{-6} Nm and equation (4) gives a decelerating torque $M_{fluid,max}$ of 1.0×10^{-6}Nm.

Retarding torque from turbine inertia ($M_{inertia}$)-In case of flow velocity changes, a turbine speed alteration is obstructed by inertia forces. For simplicity of calculation and to be on the save side, the turbine is again considered to be a full disk.

$$M_{inertia} = J_z \cdot \dot{\omega}_{turb} = J_z \cdot \frac{\dot{v}_{tip}}{r_{tip}}$$

with

$$J_z = \frac{m \cdot r_{tip}^{\ 2}}{2}, \qquad m = \rho_{turb} \cdot b_t \cdot \pi \cdot r_{tip}^{\ 2}$$

where

J_z : moment of inertia around the turbine hole axis
$\dot{\omega}$: angular acceleration of the turbine
m : turbine mass
ρ : turbine material density

Hence, the retarding torque from turbine inertia is

$$M_{inertia} = \frac{\rho_{turb} \cdot b_t \cdot \pi \cdot r_{tip}^3}{2} \cdot \dot{v}_{tip} \quad (5)$$

A maximum of inertial torque is reached for the steepest gradient of injection rate . For a realistic gradient of 20mm^3/deg as an example the average flow acceleration $\dot{v}_{av}$ becomes 22.9m/s^2. Assuming $\dot{v}_{av} = \dot{v}_{tip}$ and using the parameters of driving torque calculation and ρ_{turb}=7930kg/m^3, J_z yields 0.117×10^{-12}kgm^2, resulting in a retarding torque $M_{inertia,max}$ of 5.35×10^{-9}Nm. Since this worst case value is still three orders of magnitude below the other torques, it can be neglected.

Hence, equation (1) becomes

$$M_{flow} = M_{rest} + M_{fliudfrict}$$

and can be rearranged to calculate the desired value, the turbine rotational speed ω_{turb}, as a function of the other parameters.

$$\omega_{turb} = \frac{\rho}{4} c_d b_t v_{av}^{\ 2}\left(r_{tip}^{\ 2} - r_{root}^{\ 2}\right) \Big/ \left[\pi\eta\left(\frac{r_{tip}^{\ 4} - r_{hole}^{\ 4}}{2h_{side}} + \frac{2b_t r_{turb}^{\ 3}}{h_{bear}} \right) + c_d \frac{\rho}{3} b_t v_{av}\left(r_{tip}^{\ 3} - r_{root}^{\ 3}\right) \right] \quad (6)$$

Now, the influence of single parameters on the turbine speed can be examined performing variation calculations. For example the distance h_{side} between the turbine side and the holder walls is included in the denominator of the fluid friction term. This implies that the larger h_{side}, the smaller the retarding torque from fluid friction and the faster the speed of the turbine. This statement is experimentally investigated below.

EXPERIMENTAL RESULTS FOR STEADY CONDITIONS

EXPERIMENTAL SETUP - Fig. 6 shows the block diagram of the experimental setup to investigate the turbine response under steady flow conditions. The fuel in the constant-temperature tank is fed to the surgetank by a pump. The fuel flow smoothed in the surge tank is supplied to the MTS and the reference sensor through the injection pipe having 2mm inside diameter. The fuel is returned to in the constant-temperature tank then. Output data from the MTS and the reference sensor is modulated by the counter, digitized by an A/D converter and accumulated in the computer with a sampling frequency of 100kHz. The reference sensor has a specified 1% linearity in the range of 0.3 to 9 l/min. The counter can process a frequency range up to 100kHz. As a substitute for fuel, light oil has been used. Temperature fluctuations of the constant-temperature tank have been within ±1 K for the set value.

RESULTS AND DISCUSSION - A comparison of the wave form between the MTS and the reference sensor for 303K in fuel temperature and 0.6*l*/min in fuel flow rate is shown in Fig. 7. The MTS has a high resolution compared to the reference sensor.

Fig. 8 shows the turbine rotational speed to fuel flow rate at different temperatures. In the region above 0.9l/min an approximately linear relation is found, however the turbine speed slightly depends on temperature levels due to the temperature dependence of fuel viscosity.

As mentioned in the preceding discussion, since the MTS measurement resolution strongly

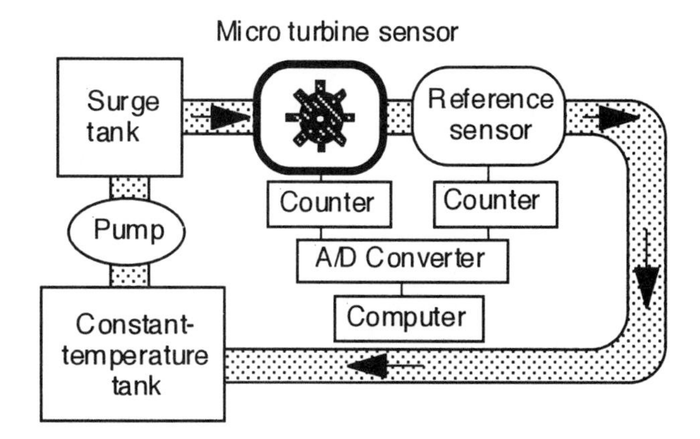

Fig.6 Block diagram for experimental apparatus

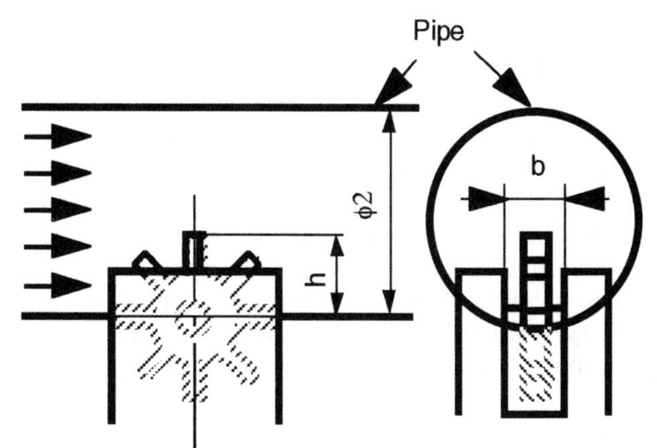

Fig.9 Geometric parameters

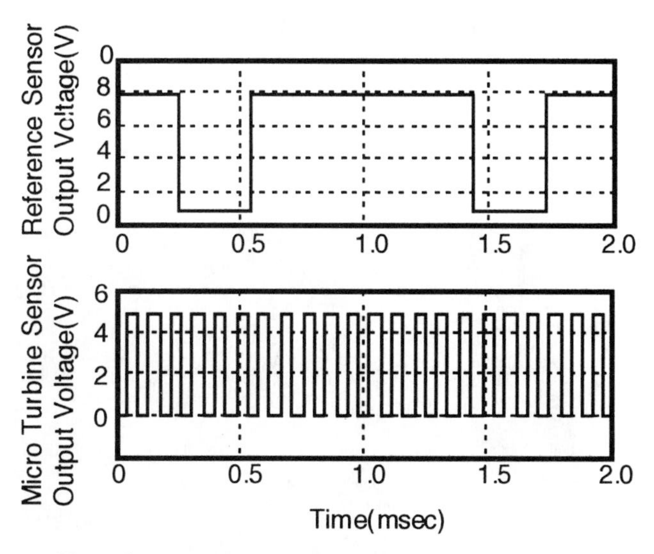

Fig.7 Comparison of wave form between MTS and the reference sensor

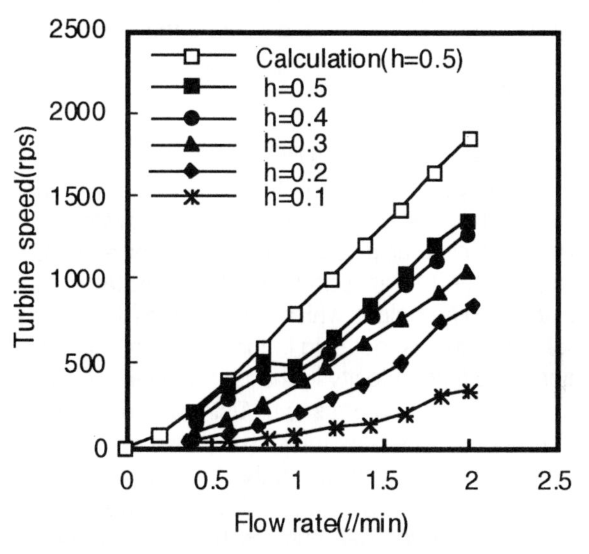

Fig.10 Comparison of experimental and analytical results(b=0.2mm)

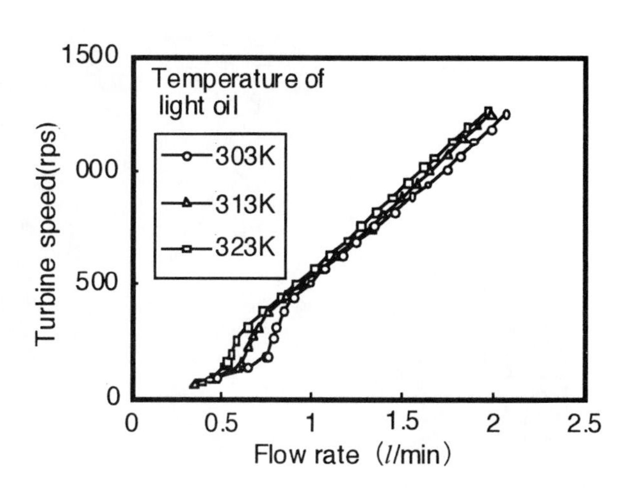

Fig.8 Flow rate and turbine speed

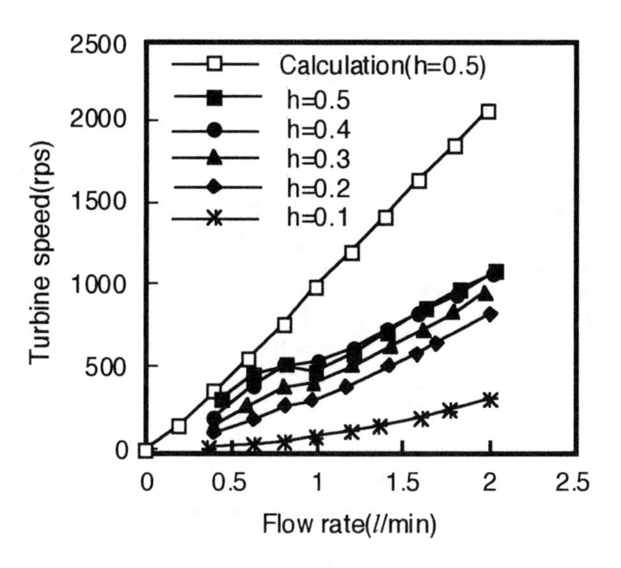

Fig.11 Comparison of experimental and analytical results(b=0.35mm)

depends on turbine speed as function of fuel flow in the injection pipe, it is necessary to improve the response of the MTS. There are several parameters influencing the response of the turbine rotation. In this paper we investigate two geometric parameters : the turbine height h above the pipe wall and the slit width b of the turbine holder as shown in Fig. 9

Fig. 10 and Fig. 11 show the effect of h and b on the turbine speed. Like in Fig. 8, horizontal and vertical axes show flow rate and turbine speed respectively. In each figure the calculation result of equation (6), in which h equals 0.5mm, is plotted.

Both experiments indicate that the turbine speed becomes faster as the height of the vane tip is increased. A comparison of the turbine speed between two holder-slit widths with 2l/min in flow rate and h=0.5mm shows that the turbine in a narrow slit(b=0.2mm) rotates 27% faster than the turbine built in a wide one(b=0.35mm). This result is in contrast to the calculations which predict a turbine speed increase of 13% in case of wide holder slit as compared to the narrow one.

In order to clarify the difference between experimental and calculation results, flow visualization using a large-scale model has been conducted.

FLOW VISUALIZATION

FLOW VISUALIZATION SYSTEM - The flow visualization system used for the present study is shown in Fig. 12. The test section of a circulating water flow channel is a circular pipe 80mm in diameter and 480mm in length made of transparent acryl. The enlarged turbine and turbine holder which is forty times larger than actual scale size is installed in the center of the test section. The flow velocity in the test section can be controlled from 0 to 0.6m/sec by using an inverter motor. A propeller type flow meter has been used to measure the velocity profile in the test section. Flow visualization has been conducted by hydrogen bubble technique. The hydrogen bubbles have been generated continuously from a cathode wire, 50 μ m in diameter, with an applied voltage of 140V. The path lines of bubbles have been observed by a camera, illuminated with a laser light sheet.

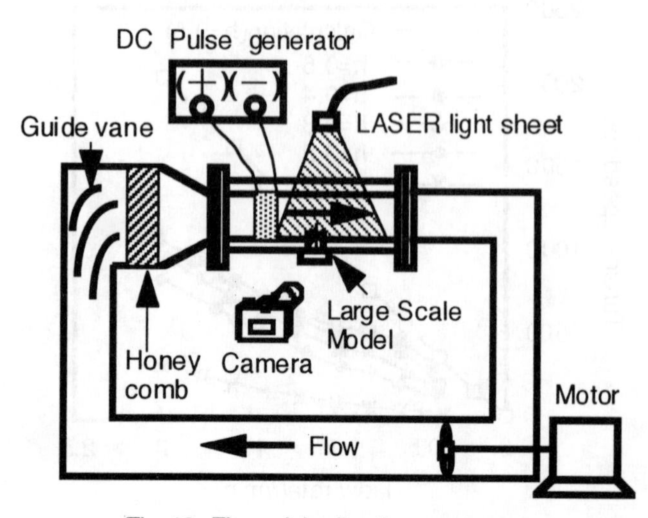

Fig. 12 Flow visiualization system

Flow visualization has been executed with the turbine fixed to the shaft. The center velocity of the test section is 0.1m/sec, resulting in a Reynolds number based on the pipe diameter of 8000. This value corresponds to an actual flow rate of 2l/min. The water temperature is 293K.

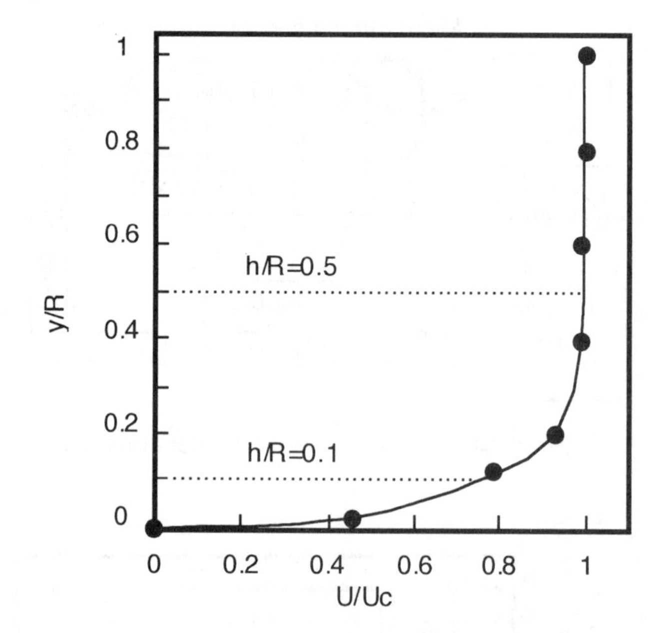

Fig. 13 Velocity profile

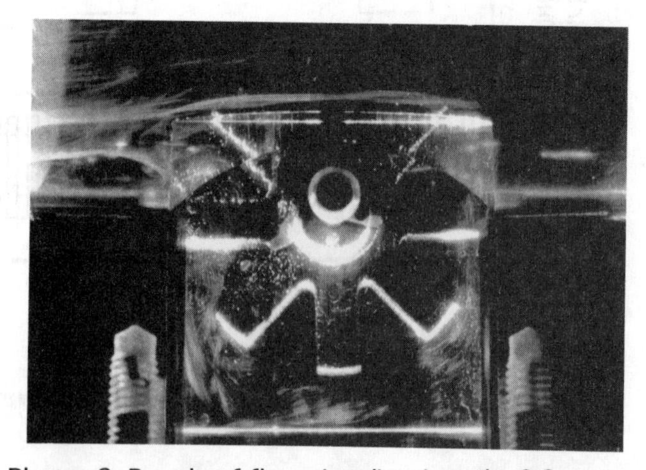

Photo. 2 Result of flow visualization (b=0.2mm)

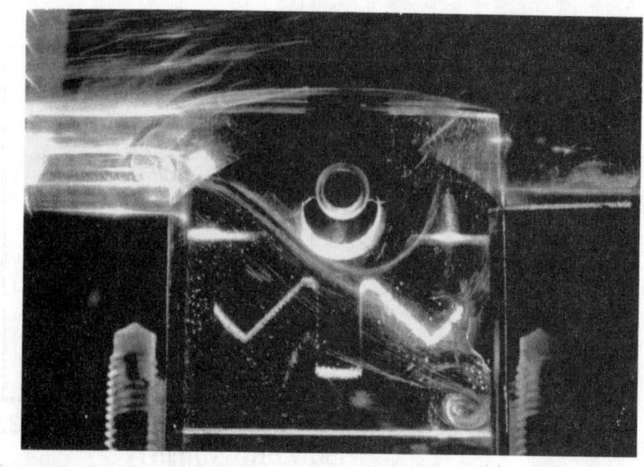

Photo. 3 Result of flow visualization (b=0.35mm)

RESULTS AND DISCUSSION - Fig. 13 shows the velocity profile inside the test section. The vertical and horizontal axis is the non-dimensional distance y/R and the non-dimensional velocity U/U_c respectively, where R represents the inside radius and U_c is the center velocity. The velocity at h/R =0.5, corresponding with the vane height of 0.5mm in the actual model, is approximately the same as the center velocity.

Results of flow visualization are shown in Photo. 2 and Photo. 3, corresponding to the actual holder slit width of 0.2mm and 0.35mm respectively.

Hydrogen bubbles are visible as white dots and lines. The main flow is heading towards the right side in these photographs. In photo. 2 it can be observed that the path lines of the hydrogen bubbles pass through the lower periphery of the turbine . While, in case of Photo. 3 the path lines appear on the side-plane of the turbine. This fact indicates that the flow magnitude in the holder slit and the flow pattern depends on the slit width.

Equation (6) assumes that the vane of the turbine pointing downwards into the holder is fully exposed to the fluid being at a standstill. However, flow visualization results show that the vane of the turbine's lower part in the holder slit is subjected to a flow force that obstructs the turbine rotation. Due to the fact that in case of wide slit width the turbine's lower part is stronger exposed to a counteracting flow, as shown in Photo. 3, the flow forces obstructing the turbine rotation are considered to be larger than in case of the narrow slit width.Hence, the differences between the experimental flow pattern in the holder slit and the analytical assumptions are one explanation for the discrepancy between calculation and experimental results.

EXPERIMENTS FOR UNSTEADY FLOW CONDITIONS

TEST CONDITIONS - Based on the results of steady flow condition experiments and flow visualization, the fuel injection rate from a diesel pump have been measured by using an actual MTS which has a vane tip height h=0.5mm and a slit width b=0.2mm. Table 1 lists the test conditions.

RESULTS - Fig. 14 shows the output signal of the MTS for the duration of one injection. It is found that the pulse width varies depending on the injection rate. The wave form of injection rate is obtained by conversion of the square wave frequency of the MTS. The result is compared then to the output from a Bosch type injection rate meter as shown in Fig. 15. The shape of

Table 1 Test condition

Pump type	Jerk type
Pump speed	750(rpm)
Injection quantity	60(mm^3/stroke)
Fuel temp.	303K

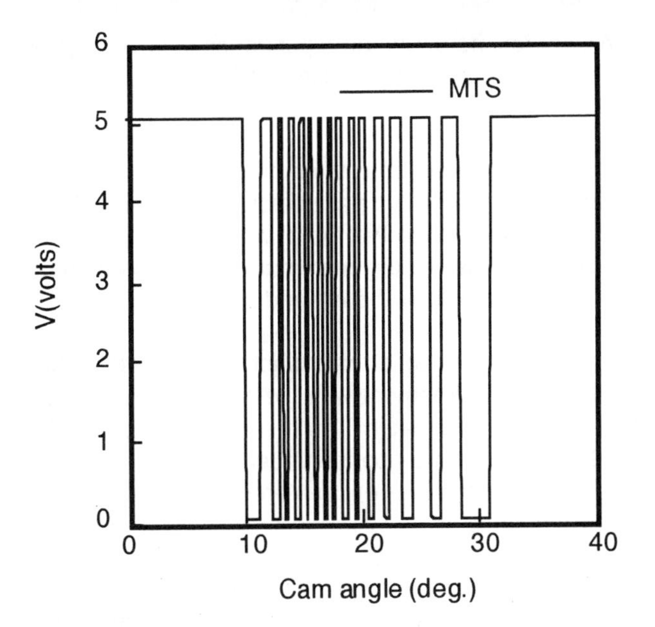

Fig.14 Out put wave form from MTS

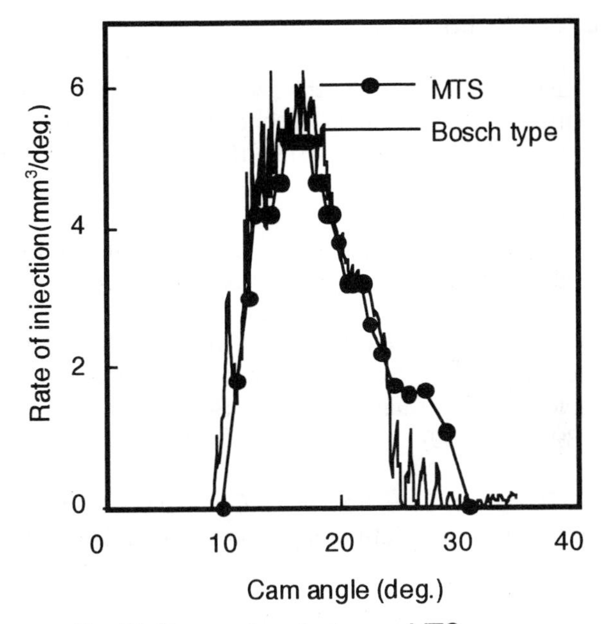

Fig.15 Comparison between MTS and Bosch type injection rate meter

injection rate function from the MTS agrees fairly with the result of the Bosch type injection rate meter. It is found that the MTS displays high response under unsteady conditions. Furthermore, the measurement resolution of the MTS for injection rate is $4mm^3$/pulse.

SUMMARY AND CONCLUSIONS

A new instrumentation method to measure the fuel injection quantity of diesel injection systems has been presented. The system's novel part is a Micro Turbine Sensor composed mainly of a micro-turbine 1mm in diameter and its holder, both extending into the injection line. The feasibility of the MTS has been examined analytically and experimentally. The following conclusions are obtained:

1. An approximately linear dependence of turbine speed on flow rate has been measured in the region above 0.9 l/min.

2. The measurement resolution of the present sensor is influenced by geometric parameters like distance from the inside tube wall to the top of vane h and slit width of the turbine holder b. The maximum turbine speed has been obtained with h=0.5mm and b=0.2mm.

3. The shape of injection rate function obtained by the MTS is in fair agreement with the results of the Bosch type injection rate meter. This results indicates that the MTS displays high response under transient and high pressure conditions as encountered inside diesel injection systems.

The next stage of investigation will be the application of the MTS inside a diesel engine to check its durability and to examine the influence of engine vibration on the MTS operation.

REFERENCES

1.H. Ishiwata, X. Li, H. Yoshikawa, N. Kitahara, "Recent Progress in Rate Shaping Technology for Diesel In-Line Pumps", SAE paper No.940194(1994)

2. W. Bosch., "The Fuel Rate Indicator: A New Measuring Instrument for Display of the Characteristics of individual Injection", SAE paper No.660749(1966).

3. A. Takamura, S. Fukushima, Y. Omori., T. Kamimoto., "Development of a New Measurement Tool for Fuel Injection Rate in Diesel Engines", SAE paper No.890317(1989)

4. G. R. Bower, D. E. Foster., "A Comparison of the Bosch and Zeuch Rate of Injection Meters", SAE paper No. 910724(1991).

TEMPERATURE SENSORS

970176

New Thermal Infrared Sensor Techniques for Vehicle Blind Spot Detection

John W. Patchell and R. Steven Hackney
A.L.I.R.T. Advanced Technology Products

Copyright 1997 Society of Automotive Engineers, Inc.

ABSTRACT

Vehicle Blind Spot Detection (BSD) represents a logical first step towards a comprehensive Collision Avoidance strategy as envisioned in plans for the Intelligent Transportation System. BSD systems have been developed based on a number of technologies, but cost and performance issues have prevented widespread commercial success, particularly for light vehicles.

This paper presents a new Vehicle Blind Spot Detector. The device is based on a recently developed, novel thermal infrared sensor technology.

A description of the operation of the sensor is provided, together with a brief comparison with existing technologies. Experimental data are also presented, showing results of operation of the sensor.

INTRODUCTION

Collision Avoidance (CA) systems will form an integral part of the Intelligent Transportation System initiative [1]. Stand-alone Blind Spot Detection represents a logical first step in the development of fully integrated CA [2].

Considerable testing has been done on related technologies [2,3], along with evaluation of prototype and commercial systems [4]. While some systems are available for heavy trucks and other large vehicles, BSD systems are not yet common and are not commercially available for light vehicles (passenger cars and small trucks).

Typically, systems with highly reliable performance are unacceptable because of total installed system cost. Conversely, most inexpensive systems cannot provide required performance across all ranges of environmental conditions and target types.

This paper presents a novel approach to infrared sensing technology with application to Blind Spot Detection. This technology appears to be relatively low cost, and holds the potential to be implemented with high operating performance.

OPERATION OF THE NEW SENSOR

The ALIRT BSD uses long wavelength (7 to 14 μm) infrared technology which senses emitted thermal energy of a vehicle. This is in contrast to reflected energy sensed by active sensors (radar, ultrasonic or short wavelength infrared).

The pyro-electric detectors used in the ALIRT BSD respond only to a change in temperature. If the temperature of the object in the sensing field is constant, the output of the detector is zero. Essentially, the detector is a differentiator that gives an output proportional to the rate of temperature change within the field of view: no temperature change equals no output. Thus, a BSD based on a single simple infrared detector will give zero output (no detection signal) for a vehicle in the host vehicle's blind spot if there is no relative motion, even though both vehicles may be moving. Conversely, a change in the temperature of the target region, for example an overpass shadow, may trigger a "false alarm."

This limitation is overcome by employing fields of view (FoV's) of two types and alternately comparing them using proprietary techniques (international patents pending). One type of FoV is aimed in the conventional manner at the blind spot while the second is aimed so that only a representative sampling of the road surface is visible -- for example immediately behind or below the host vehicle.

If no vehicle is in the blind spot, then both types of fields of view see the road surface. The difference in temperature between the two FoV's is consequently small and the output of the detector is small.

However, when a vehicle is present in one FoV and not in the other, then the output is large. When a vehicle enters the blind spot there is a positive differential between the detection and reference zones of the sensor. This differential is used to trigger an output signal (for example a light or buzzer) to warn the driver of the lane change hazard.

In the case of a change in the road temperature, as generated by the shadow of an overpass, for example, both the reference zone and the detection zone will change temperature at approximately the same time. Therefore the differential will remain zero.

One possible FoV arrangement is shown in Figure 1. The unit is shown mounted on the left side rear view mirror. The road-only element views the road directly behind the car while three lens elements are used to cover the driver's blind spot. In the initial prototype, only one of the three illustrated Detection FoV's has been implemented.

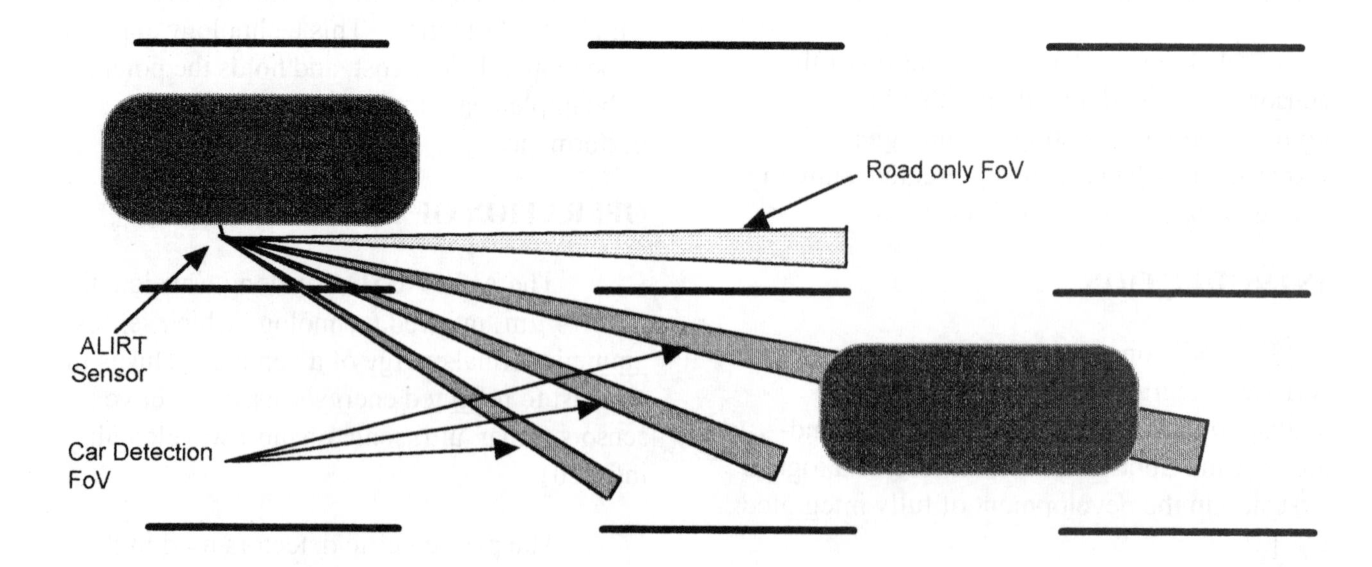

Figure 1 - Arrangement of Fields of View

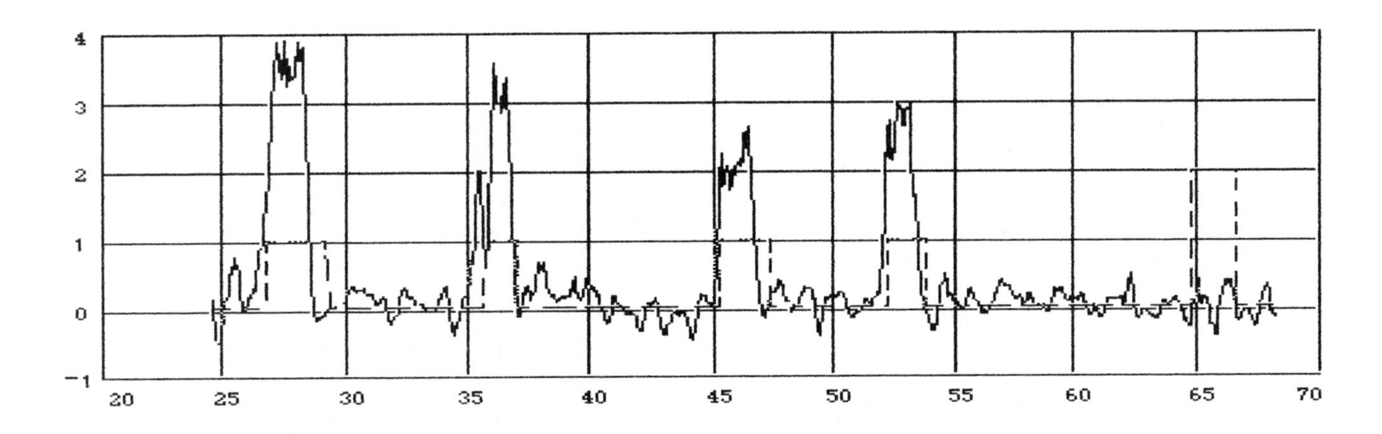

Time: 3:40 PM
Weather: Sunny, 25C
Comment: The first four events are cars. The final event (at about 65 seconds) is a bridge shadow.

Figure 2 - Detection of Cars

EXPERIMENTAL RESULTS

TEST RESULTS ACHIEVED Early testing has indicated that portions of operating vehicles are reliably warmer than the environment by at least 1 to 2C. This has been found to be true of the engine compartment, radiator and exhaust system: testing has also shown that tires will achieve approximately 2C of heating in approximately 1 km of driving in city condition. Therefore, a moving vehicle will emit significantly greater infrared than its surroundings.

Considerable testing has been conducted using a prototype device. The following graphs are typical of the results obtained: similar data have been collected across a range of

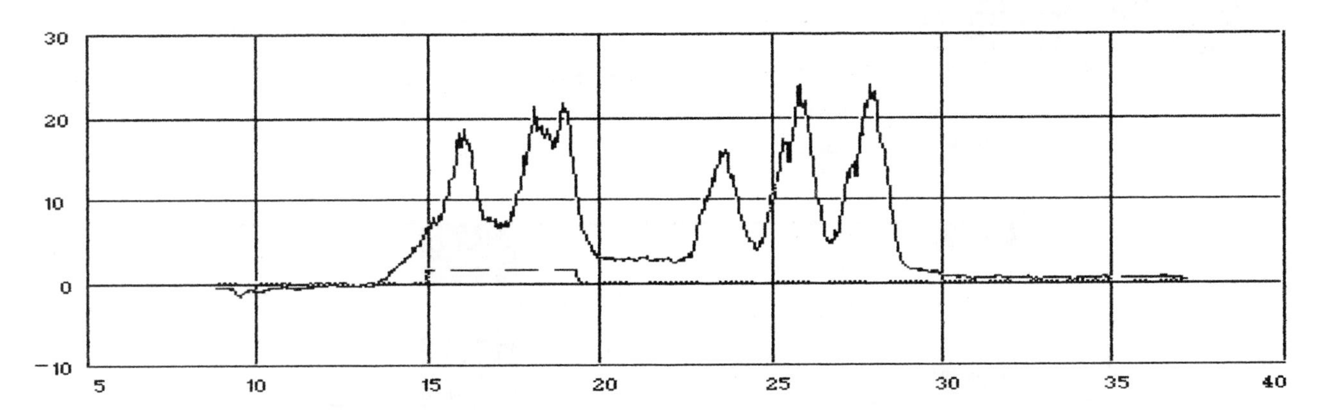

Time: 10:00 AM
Weather: 17C, Partly sunny
Comment: Transport truck signal. The first peak at 16 sec is front wheels, 18-19 sec is dual axles on rear of cab and 23, 26, and 28 sec the three axles on the trailer.

temperatures, weather conditions and times of day.

In figure 2 and 3 , the solid line is the BSD output. The broken line is a manually entered "event" channel used as a marker to record the occurrence of a vehicle in the blind spot or other significant occurrence. The horizontal axis is time in seconds.

The above data were collected using a sensor configured with a single detection field of view and a single reference field of view. The detection zone is approximately 300 mm diameter at the road surface.

Consequently, under some circumstances this configuration of the device will not detect large vehicles since the tires of the vehicle actually straddle the detection zone. This can clearly be seen in the case of the trailer of the transport truck. During a portion of the time that the trailer is resident in the blind spot, no detection is registered.

A second implementation of the device has been constructed with revised reference and detection zones . As indicated in figure 4, this version of the BSD features two distinct detection zones. This has largely eliminated the "straddle" problem found with the single-zone implementation of the original prototype.

Figure 5 shows the detection of a passenger bus using this version of the detector. With the current threshold alarm trigger of approximately 0.6C, the vehicle is continuously detected during the entire time it is in the blind spot of the host vehicle.

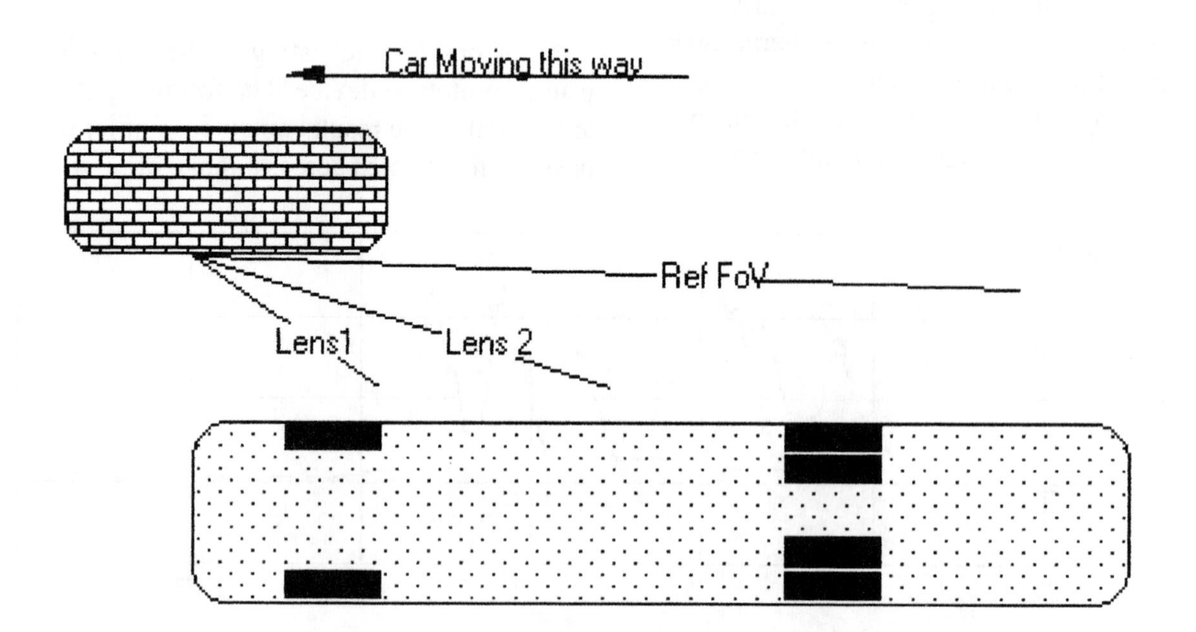

Figure 4 - Configuration of Multiple Detection-Zone BSD

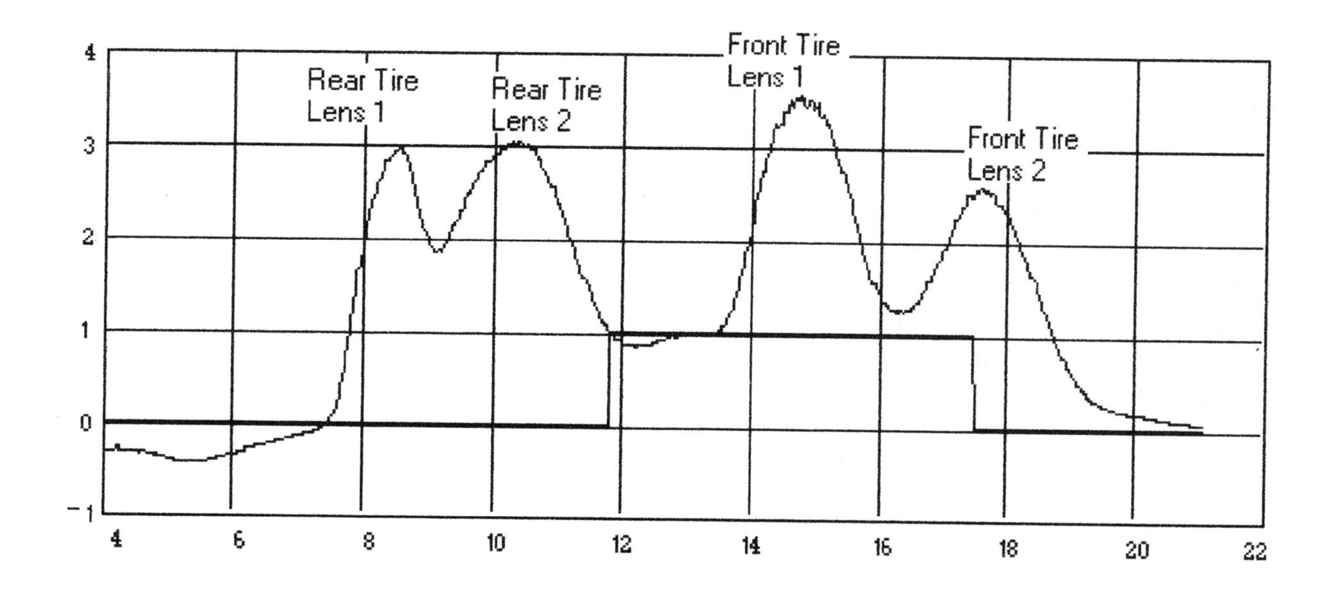

Figure 5 - Detection of a Bus Using Two Detection-Zone BSD

As can be seen from these figures the ALIRT BSD is very effective in detecting vehicles in real road conditions.

COMPARISON WITH EXISTING TECHNOLOGIES

A number of technologies have been employed in attempting to solve the BSD problem, including:

- Ultrasonic
- Active Infrared
- Laser
- Radar

These technologies are all "active" -- that is they all make use of a sensor that detects a signal which is transmitted from the host vehicle and reflected off of the target vehicle.

By contrast, the ALIRT BSD technology is passive. This strategy is expected to offer the following advantages:

System Cost: Because a passive system is inherently less complex than an active system, advantages are expected in the areas of system cost, installation cost and general reliability. Total sensor cost is estimated to be less than twenty dollars with volume manufacture.

Reliability: Near infrared has been generally found to be environmentally stable. Testing has shown the sensor to be effective in conditions of rain, winter and summer temperatures, snow and daylight or darkness. The system detects operating vehicles independently of colour, size or shape. It does not respond to parked vehicles or other non emitting roadside objects. IR transmissions are relatively unaffected by airborne moisture. Fog will affect the sensor operation less than it will affect the driver's vision.

CONCLUSIONS

A vehicle Blind Spot Detection system has been developed based on a new passive infrared technology. Testing to date indicates that the detector will be able to provide a high level of performance at relatively low cost.

REFERENCES

[1] M. Upton, "Techniques for Distance Measurement," Society of Automotive Engineers Technical Paper No. 952085, 1995

[2] S.J. Mraz, "On the Road to Smarter Cars," Machine Design, September 14, 1995

[3] E. N. Mazzae, W. R. Garrott and M. A. Flick, " Human Factors Evaluation of Existing Side Collision Avoidance System Driver Interfaces" Society of Automotive Engineers Technical Paper No. 952659, 1995

[4] "Side and Rear Object Detection Systems," Truck Engineering, November 1995.

960333

Exhaust Gas Temperature Sensor for OBD-II Catalyst Monitoring

Nobuhide Kato, Nobukazu Ikoma, and Satoshi Nishikawa
NGK Insulators, Ltd.

Copyright 1996 Society of Automotive Engineers, Inc.

Abstract

This paper describes a newly-developed, high-performance RTD,(Resistive Temperature detector), which meets OBD-II monitoring requirements. The OBD-II catalyst monitoring requirements are high temperature durability, high accuracy, and narrow piece-to-piece variation.

Catalyst monitoring methods have been reviewed and studied by checking the catalyst exotherm(1)(2).

The preliminary test results of catalyst monitoring are also described herein.

Introduction

The RTD,(Resistive Temperature detector), using positive temperature coefficient of resistance of platinum, is known as a temperature sensor having high accuracy.(3) This sensor is generally manufactured by the following process; (1) Prepare sintered Al_2O_3 substrate, (2) Print Pt resistor(paste) pattern with adjusting circuit on the substrate, (3) Fire the Pt resistor, (4) adjust the resistance by cutting the adjusting circuit (trim resistor), (5) Print glass paste covering the Pt resistor, (6) Fire the glass paste. The high temperature durability of this type of temperature sensor is not sufficient for automotive use, due to the relatively low melting point of the glass coating layer, as compared to the maximum temperature of engine exhaust gas.

In order to improve high temperature durability, the Pt resistor should be covered with a more durable material, such as Al_2O_3. Preferably, the Pt resistor should be printed on a Al_2O_3 green sheet and sandwiched with another green sheet, then co-fired.

However, in this embedded structure of Pt resistor, the resistance of the Pt resistor can not be adjusted even with adjusting circuit (trim circuit), because of the Pt resistor circuit already embedded into the Al_2O_3 sheets. Consequently, this type of sensor results in wide range piece-to-piece variation.

The measurement of gas temperature difference between the upstream and downstream of the catalyst is one of the methods for catalyst monitoring. This method requires long periods of steady state driving, i. e. : 20 to 30 minutes at 60 km/Hr, in order to obtain a saturated catalyst temperature. Therefore, it is difficult to monitor the catalyst performance with this temperature difference method, because such steady driving condition can not be obtained in actual driving conditions.

Design concept of proposed temperature sensor

Fig.1 illustrates the perspective view of proposed temperature sensor and cross-section of sensing portion.

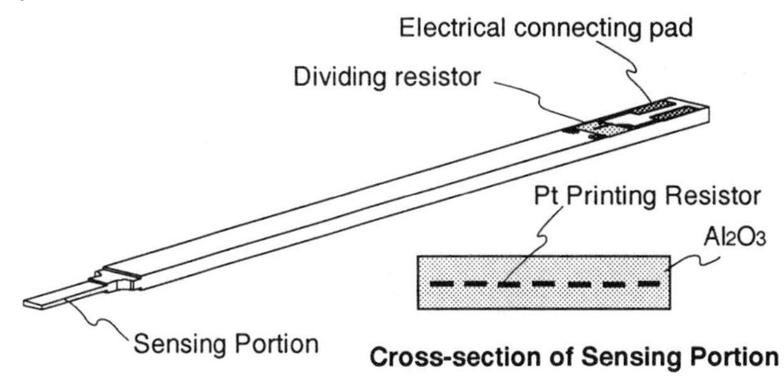

Fig. 1 Perspective view of proposed temperature sensor element

In order to obtain high temperature durability, the Pt resistor is completely embedded into the Al2O3 substrate. The Pt and Al_2O_3 cermet paste is printed on a 0.2 mm thick Al_2O_3 green sheet of 99.99% purity, and laminated with another Al_2O_3 green sheet at the sensing portion. The body portion consists of six green sheets to withstand the mechanical stress during the packaging processes. All six green sheets are laminated and co-fired at 1600℃. After sintering, the dividing resistor, (DuPont QS874), is printed on the surface of the sensor element at the opposite end of sensing portion, and then fired at 850℃.

The sensor packaging is designed so that the temperature of dividing resistor does not exceed 400℃ even under high temperature conditions.

Fig 2. illustrates the electrical circuit of the proposed sensor.

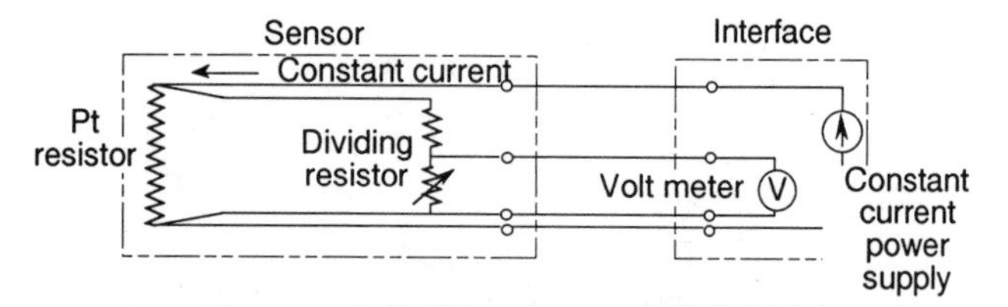

Fig. 2 Electrical circuit of proposed temperature sensor

Fundamental characteristics of proposed temperature sensor

Fig. 3 shows the TCR (Temperature Coefficient of Resistance) of the proposed temperature sensor. In Fig. 3, the output voltage is adjusted to 100 mV at 0 ℃. Each measuring point is the average value of 36 sample pcs at the melting points of Sn (231.9℃) , Zn (419.5℃) and Al (660.3℃).

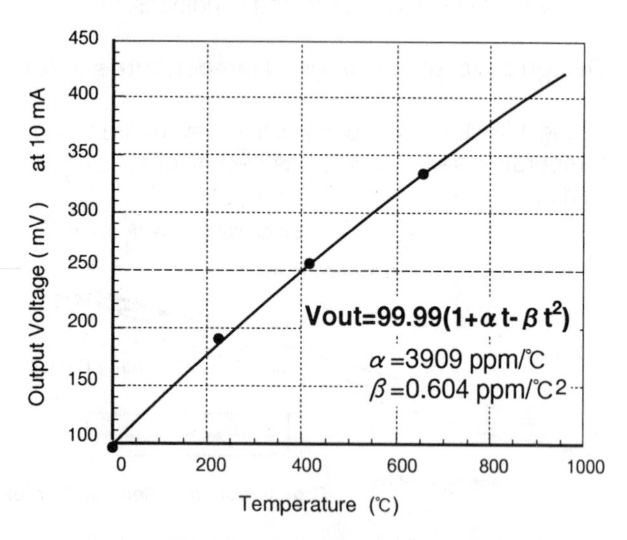

Fig. 3 TCR of proposed temperature sensor

The α and β of temperature coefficient are 3909 ppm/℃ and 0.604 ppm/℃^{2} respectively. These numbers are almost the same as that of pure Pt, 3908 ppm/℃ of α and 0.580 ppm/℃^{2} of β.

At the Al melting point, the piece-to-piece variation was determined using 36 sample piece data. The result is shown in Fig. 4.

The piece-to piece variation is approximately ±0.5% for ±5σ. This very narrow piece-to-piece variation is necessary for catalyst monitoring in which small amounts of temperature differences must be detected.

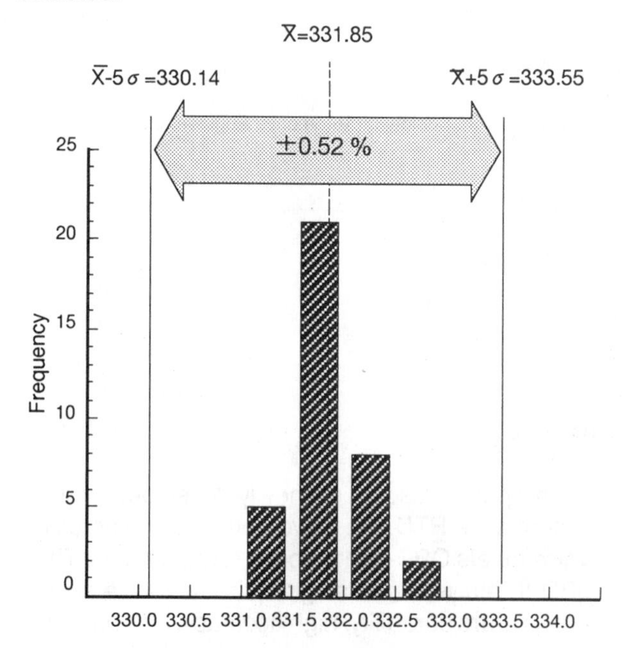

Fig. 4 Piece-to-piece variation of proposed temperature sensor

Fig.5 and 6 illustrates the response property on a burner stand, in which the average gas velocity is approximately 1.58 m/sec . The average gas velocity is calculated from inlet gas quantity (60 l/min), inner diameter of the burner stand (57 mm) and gas temperature (950℃). The measurement of absolute gas velocity is not exact, especially under high temperature conditions. The response properties of

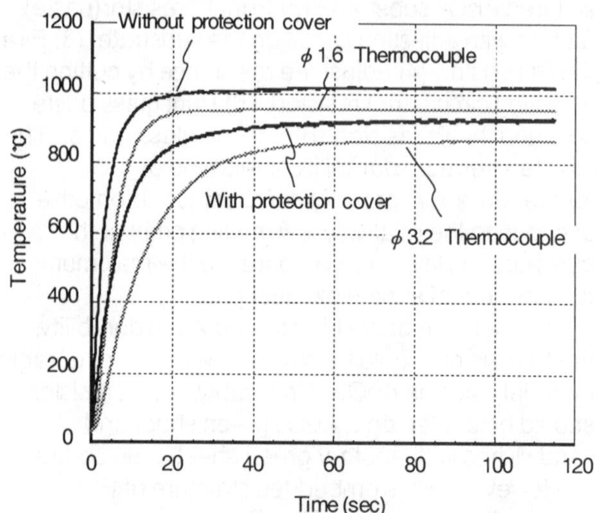

Fig. 5 Response property of proposed temperature sensor

the proposed temperature sensor, shielded and unshielded, as compared to the sheathed thermocouples are shown in the Fig. 5 as a reference.

The time constant shown in Fig.6 is calculated by using a single exponential decay.

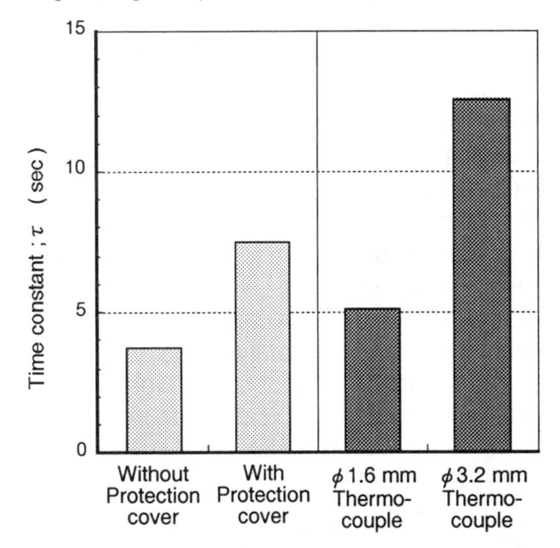

Fig. 6 Comparison of time constant

Fig.7 shows the durability test result at 1050℃ on the burner stand. In Fig. 7, the gray dots show each of the measuring points, and the filled blackened circles show their average value. A small amount of output change for each of the measuring points is thought to be measuring error. After 800 Hrs test, the output is still stable due to the Pt embedded structure and high co-fire temperature of 1600℃. This robust characteristic against high temperature soak is necessary for the monitoring of close-coupled catalyst and light-off catalyst which will be subjected to high temperature.

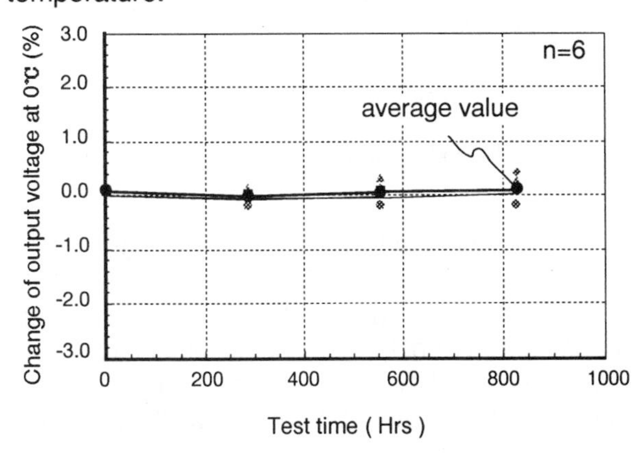

Fig. 7 Durability at 1050℃ of gas temperature

Fig.8 illustrates the appearance of proposed temperature sensor. The packaging structure is similar to oxygen sensor design, with the exception of the metal housing and protection cover.

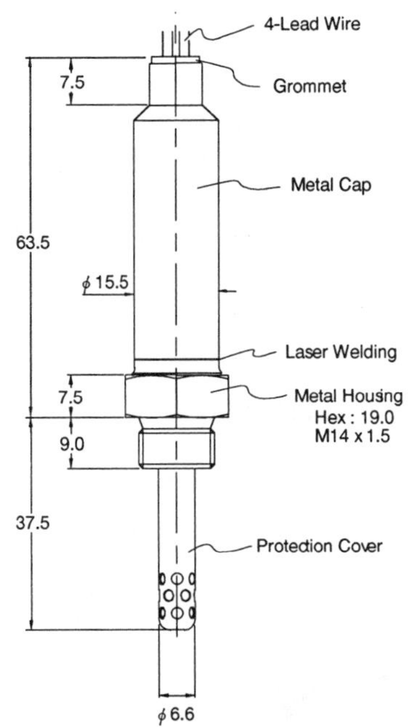

Fig. 8 Appearance of proposed temperature sensor

Concept of catalyst monitoring method

The most important point in the study of catalyst monitoring methods is that the monitoring method is theoretically independent of various driving modes.

The HC conversion efficiency of fresh and aged catalyst are typically expressed as shown in Fig. 9. If the conversion efficiency can be calculated during vehicle driving, it may be possible to judge whether the catalyst is functioning properly or not.

The conversion efficiency can be rewritten in the catalyst exothermic energy per unit quantity of gas flow and unit time. Therefore, this exothermic generated energy due to the oxidation of combustible gas, Eg (cal/sec·l), represents the conversion efficiency of the catalyst. If this Eg (cal/sec·l) is integrated during a predetermined range of catalyst temperature, for example from 100℃ to 300℃,

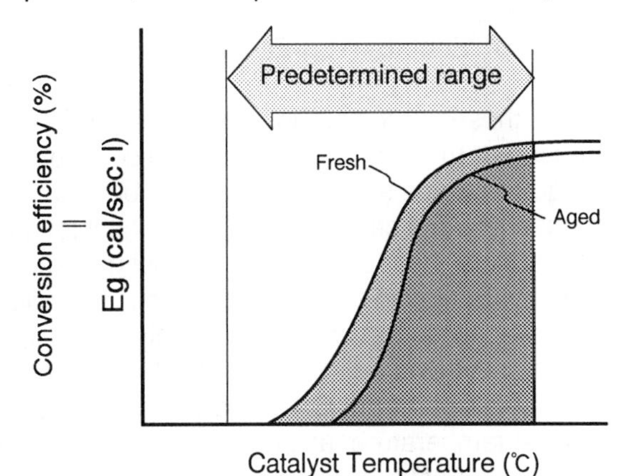

Fig. 9 Typical conversion efficiency of fresh and aged catalyst

the integrated Eg must represent the catalyst conversion efficiency and correlate to the HC emission on the FTP driving mode.

Furthermore, the integrated Eg must be independent of driving conditions because Eg itself as shown in the Fig.9 is independent of driving modes.

How to obtain the Eg

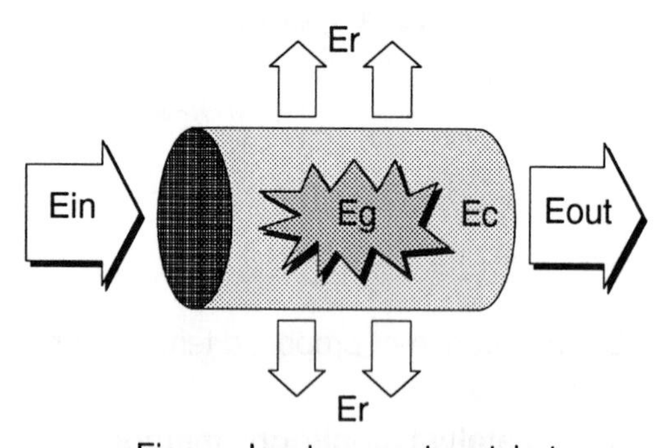

Fig. 10 Energy balance of catalyst

Fig.10 illustrates the energy balance of the catalyst. Energy balance of the catalyst is written as the following equation(1), where Ein is input energy (heat energy of inlet exhaust gas) to catalyst, Eout is output energy (heat energy of outlet exhaust gas), Eg is generated energy due to the oxidation of combustible gas at catalyst, and Ec is energy taken away to catalyst to be heated up. Er is radiation energy from catalyst.

$$(1) \quad Eout=Ein+Eg-Ec-Er$$

If these energies, Ein, Eout, Ec, and Er, can be measured, Eg must be obtained and consequently integrated Eg as shown in Fig. 9 can be obtained. In equation(1), Ein can be calculate by the inlet gas temperature, heat capacity of exhaust gas and gas quantity. Eout also can be calculated in a similar way. Ec is so-called heat capacity of catalyst, and it can be obtained in advance. Since Er is a function of catalyst temperature, Er can be calculated from the catalyst temperature, as per the relationship obtained in advance between catalyst temperature and Er.

Verification of the monitoring concept

In order to verify the concept mentioned above, the relationship between the integrated Eg and THC emission on FTP driving mode was investigated.

[Preparation / Test condition]

Fig.11 shows the exhaust system for the verification test. A 2.0L I-4 engine, 600 cc light-off(L/O) catalyst and 1700 cc main catalyst were chosen for the test. The volume of L/O catalyst, 600 cc, was selected so that the test system typifies LEV.

The L/O catalyst was installed 500 mm from the exhaust port, and the main catalyst was installed 400 mm from the end of L/O catalyst. The proposed temperature sensor-1 and sensor-2 were installed in front of and behind the L/O catalyst to monitor the L/O catalyst.

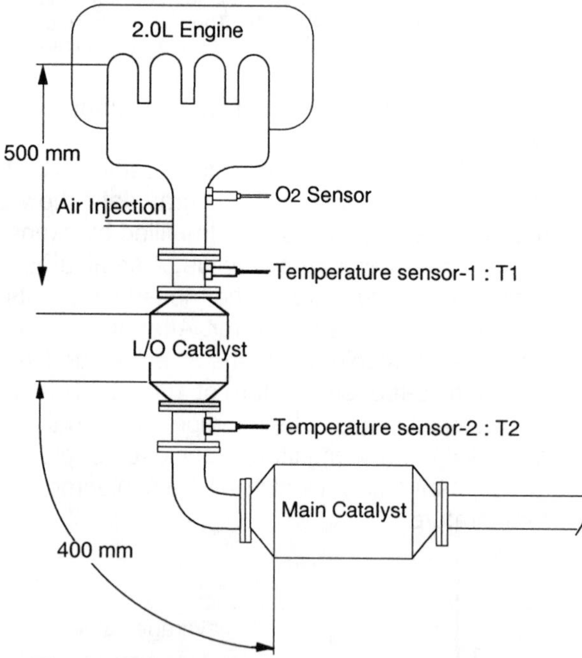

Fig. 11 Exhaust system for the verification test

[Measurement of Ec and Er in advance]

Prior to the FTP driving test, Ec of catalyst heat capacity and Er of heat radiation from the catalyst were measured. Since the Ec and Er are independent of catalyst deterioration, they can be obtained in advance.

When the vehicle runs under steady state condition and the catalyst temperature saturates, the exhaust gas heat energy is not consumed to heat up the catalyst. This means that Ec in equation(1) is zero when vehicle runs under steady state condition. At that time, if the catalyst has no loading of precious metal onto the catalyst, the generated energy due to oxidation of combustible gas, Eg, is close to zero. Therefore, when the vehicle with non-loaded catalyst runs under steady state condition, Ec and Er in equation(1) are almost zero. The equation(1) then can be rewritten under such condition to the following equation.

(2) Er=Ein-Eout

In order to obtain the Er of L/O catalyst, the vehicle without loading onto the L/O catalyst ran under 20Km/Hr, 40Km/Hr, 60Km/Hr and 80Km/Hr for 30 min. respectively. The relationship between the calculated Er and the L/O catalyst temperature represented by T2 (gas temperature behind L/O catalyst) was shown in Fig.12

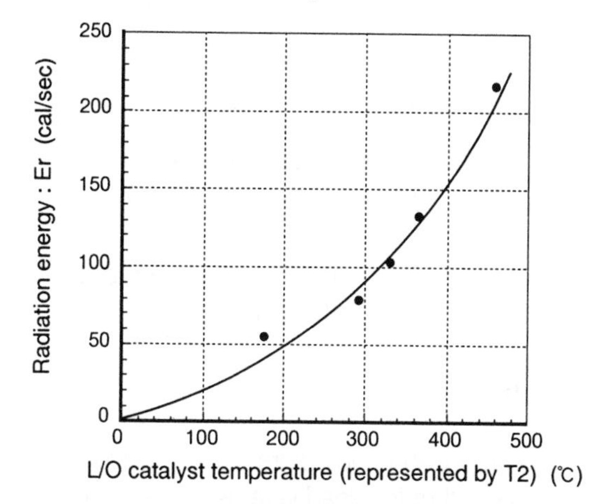

Fig. 12 Relationship between radiation energy, Er, and L/O catalyst temperature

In equation(2), Ein and Er were calculated by the following equation.

(3) Ein(cal/sec) =Q·Hs·T1

(4) Eout(cal/sec)=Q·Hs·T2

Q : gas quantity (l/sec)

Hs: Specific heat of gas (cal/l·℃)

(Specific heat of air, 0306 cal/l·℃, was used.)

T1: Inlet gas temperature (℃)

T2: Outlet gas temperature (℃)

Therefore, the equation(2) was rewritten as follows.

(5) Er(cal/sec) =Q·Hs·(T1-T2)

Next, in order to obtain Ec, the vehicle without loading onto the L/O catalyst ran under FTP driving condition, and all data was acquired every one second after engine start. In this case, only Eg is zero in equation(1). Since the equation(1) can be rewritten as follows and Er has already been known, Ec can be obtained.

(6) Ec =Ein-Eout-Er

The equation(6) is further rewritten as follows.

(7) Σ Ec = Σ (Ein-Eout-Er)

As per equation(7), Σ (Ein-Eout-Er) was calculated from room temperature up to 400℃ of L/O catalyst temperature. Fig.13 shows the result of the relationship between Σ Ec and L/O catalyst temperature. The average slope in Fig.13 shows Ec, the heat capacity of L/O catalyst.

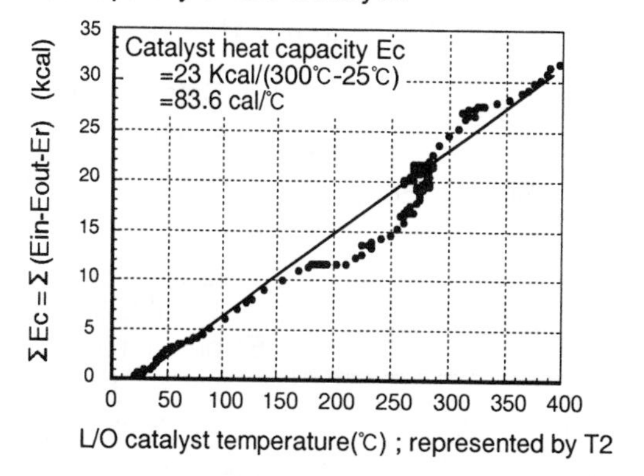

Fig. 13 Relationship between Σ Ec and L/O catalyst temperature.

Ec and Er in the equation(1) were obtained by the procedure described above.

[Measurement/Calculation of integrated Eg]

In order to obtain four level of THC emission under FTP test procedure, four catalyst systems were tested as shown in the Fig. 11. These four systems were prepared by aging with another 2.0L I-4 engine for 0, 100, 200 and 300 Hrs respectively. The aging was carried out on a combination of L/O catalyst and main catalyst. In the aging, the main catalyst was set at 15 cm downstream from the end of L/O catalyst, and the gas temperature was 750℃ just in front of the L/O catalyst.

Table 1 shows the aging condition of the catalyst system. The aging temperature of the main catalyst was higher than that of L/O catalyst, because of exothermic energy of L/O catalyst.

catalyst system	L/O catalyst	Main catalyst
A	Fresh	Fresh
B	750℃ for 100 Hrs	805℃ for 100 Hrs
C	〃 200 Hrs	〃 200 Hrs
D	〃 300 Hrs	〃 300 Hrs

* With fuel cut for 5 sec. and without fuel cut for 60 sec. were periodically repeated.

Table 1. Aging condition of catalyst system

In this measurement, 120 cc/min. of secondary air was injected for 100 seconds after the engine was started to simulate the LEV which will equip a secondary injection system to heighten catalyst activity.

Eg, generated energy, was calculated every one second after the engine start as per equation(1) in which Ec and Er were already known as mentioned above. The calculated Eg was transferred to integrated Eg with the procedure shown in Fig. 14, in order to translate to similar data in Fig. 9. Such measurement and calculation was generated for each catalyst system, A, B, C and D.

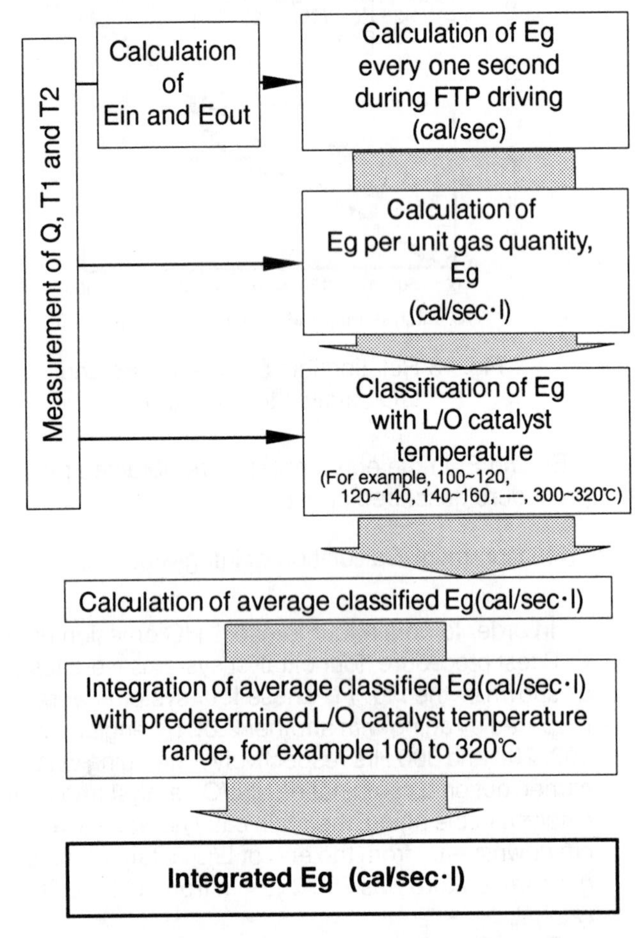

Fig. 14 Calculation process of integrated Eg

[Correlation of integrated Eg to HC emission]

The THC emission on the FTP driving mode for each catalyst system is shown in Table 2.

catalyst system	THC emission (g/mile)
A	0.058
B	0.063
C	0.070
D	0.098

Table 2. THC emission of each catalyst system

The relationship between THC emission and integrated Eg is shown in Fig. 15. Increasing integrated Eg , decreases THC , and the correlation coefficient R^2 is very high, 0.96. In this case, the integration was carried out from 100 ℃ to 320 ℃ of T2.

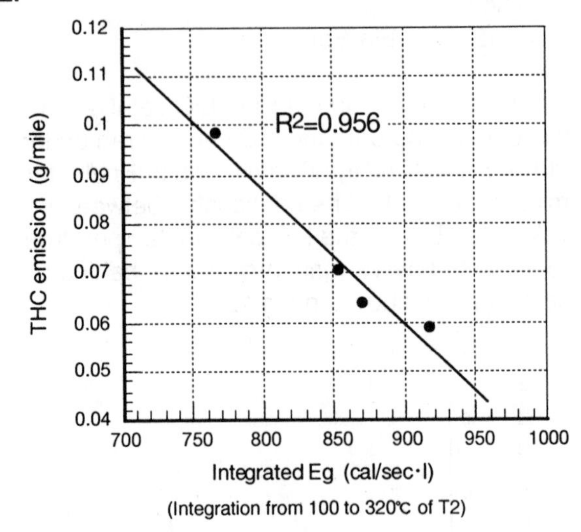

Fig.15 Correlation between integrated Eg and THC emission

Conclusion

1. The proposed temperature sensor, being composed of four-lead wire design, embedded Pt resistor into 99.99% Al_2O_3, high sintering temperature of 1600℃, and dividing resistor concept, provides high accuracy and high temperature durability which are required for catalyst monitoring.

2. Although the verification test was performed only on FTP driving mode, It can be said that Monitoring/calculating integrated Eg(cal/sec·l) of L/O catalyst during predetermined temperature range is one method to judge the catalyst deterioration. (Additional verification tests will be performed in other driving modes.)

3. Although the correlation coefficient would be decreased by including other driving modes, better correlation can be expected by using the followings;

   (1) Temperature distribution study in the L/O catalyst, investigation of the relationship between temperature distribution and T1 and T2, will aid in better assumptions of catalyst temperature.

(2) The signal of MAF(Mass Air Flow) sensor must provide more reliable information of gas quantity. (In this experiment, the gas quantity was calculated from intake manifold pressure and engine revolution.)

(3) Faster data acquisition cycle will result in more reliable information. In the experiment, one second data acquisition cycle was not enough in the acceleration portion.

References

(1) Joseph R. Theis, " Catalyst Converter Diagnosis Using the Catalyst Exotherm ", SAE Paper #942058, 1994

(2) Panagiotis D. Sparis et al, " Three-Way Catalyst Assessment Via Inlet-Outlet Temperature Measurements: A Preliminary Report ", SAE Paper #942055, 1994

(3) G.S.Iles, " Platinum Film Temperature Sensor", SAE Paper #750225, 1975

960336

Exhaust Gas High Temperature Sensor for LEV/ULEV and OBD Systems

Nobuharu Katsuki, Takashi Tamai, and Masahiro Saito
Matsushita Electronic Components Co., Ltd.

Joe LeGare and Sumitake Yoshida
Panasonic Industrial Co.

Copyright 1996 Society of Automotive Engineers, Inc.

Abstract

The purpose of this paper is to outline some of the approaches to provide an exhaust gas high temperature sensor with wide temperature detection range and fast responsiveness.

Conventional exhaust gas temperature sensors were designed only to detect an overheating catalyst, so they were unable of detecting temperatures below 600°C. Their slow responsiveness prevented them from detecting rapid catalyst temperature changes.

The development of a new thermistor material enabled the sensor to measure a wide temperature range of 300°C to 1000°C. This new sensor provides fast response time (τ = 8.7seconds.) as well as durability capability to 1000°C

Applications for this sensor include in catalyst pre-heating and OBD-II systems.

Introduction

The first thermistor material for an exhaust gas temperature sensing was developed in Japan in 1975 and has been in mass production since the early 1980's to detect the catalyst converters' overheating. Because of these slow sensor's responsiveness from their large thermal mass of the sensing portion, it was not possible to detect abnormal catalyst temperature conditions quickly. Temperature detection within a 20 second period after engine starting was not possible due to these past sensors' limitation. In the conventional application, such a quick response was not required either.

Many new emission control systems are under development to meet low emission vehicles (LEV/ULEV) and the OBD-II regulations in the US, especially for California. Due to increasing worldwide concern for the environment, more people are concerned about the greenhouse effect and destruction of the ozone layer. This trend has required the automotive industry to reduce vehicle exhaust emissions. California has introduced NMOG (non-methane organic gases) regulation for automotive emission in 1994 since NMOG is a main cause of photochemical smog. California will progressively implement tighter low emission vehicles regulations like TLEV, LEV, ULEV, and ZEV toward the year 2000. OBD regulations will be introduced simultaneously requiring the installation of a self-diagnostic function to detect malfunctions in emission control related systems to assure reductions over the vehicle's lifetime.

To meet some these requirements, heated catalyst systems are under study to heat the catalyst immediately after the engine starting so that the emission control system starts in a short period. Many catalyst heating systems are being investigated such as installation of a catalyst right under the exhaust manifold or adding an active catalyst heating like an electric or fuel heater. Alternatives for more accurate catalyst efficiency monitoring method are being evaluated to detect catalyst deterioration based upon the exothermic characteristics that result. However, since the conventional exhaust gas temperature sensors were not able to pick up such a quick temperature changes due to their large heat mass, a new exhaust gas temperature sensor with faster response was required.

To meet this market demand, thermistor-type exhaust gas high temperature sensors have been developed to help meet the OBD-II requirement for these new emission control systems. Its quick responsiveness enabled the sensor to be used in applications that required fast temperature feedback. Reducing the size of the sensing portion with 3.3 mm x 2 mm oval shape and improvement of heat transfer structure to provide fast responsiveness, allowed the sensor to detect rapid temperature changes of an object such as a catalyst. Careful material evaluation and unique structure design allow high heat and vibration durability at temperatures near 1000°C in the automotive exhaust system.

This paper describes about the newly developed exhaust gas high temperature sensor with a wide temperature range and fast responsiveness to meet these requirements.

New Material for High Temperature Thermistor

A thermistor is generally a sintered object of compound oxide. It is made through the following process; mixing two to four types of transition metal oxides such as Mn, Ni, Cr, Fe, and Co and molding into a fixed shape. This is followed by sintering at 1200°C to 1500°C temperature. Material composition of a thermistor differs based upon the temperature detection range. Usually Mn-Ni-Cr-Fe material of spinel structure is used for general purpose thermistor for operation under 300°C. For higher operating temperature ranges, a higher melting point and higher thermal stability material must be selected.

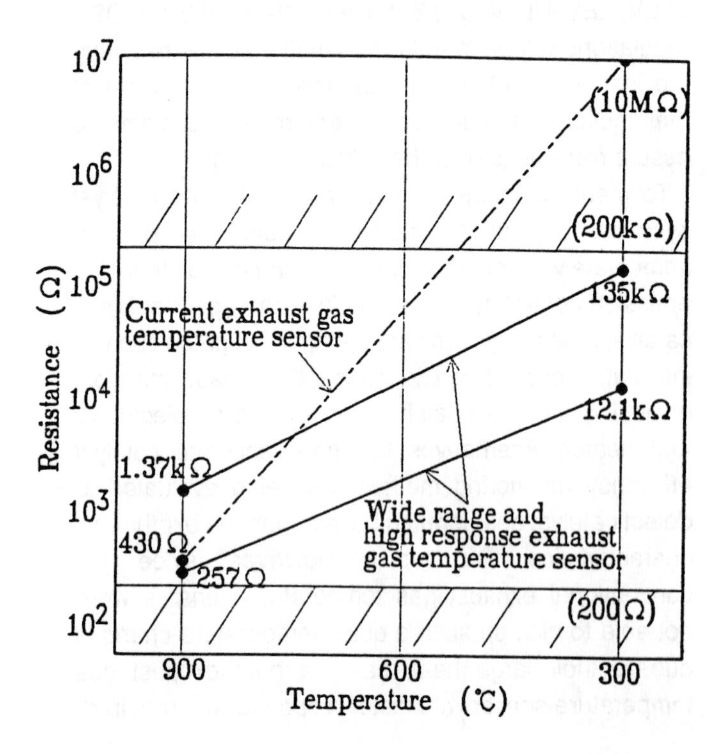

Fig.1 Resistance - Temperature Characteristics

Previous exhaust temperature thermistors used an $Mg(Al, Cr, Fe)_2O_4$ material with a spinel structure. However, it can be used for detection of only one narrow temperature band near 900°C and cannot detect temperatures under 600°C, as shown in Fig. 1. The new exhaust temperature sensor introduced here for the OBD of the LEV/ULEV systems allows temperature detection over a wide range, from 300°C to 1000°C, by using a newly developed thermistor material. This new material also allowed improved high temperature durability at 1000°C. This high temperature thermistor material has the following characteristics:

(1) Temperature detection over a wide temperature range is possible by decreasing the B constant (B_{600} = 4000K) of the material characteristics of the thermistor. As shown in Fig. 1, the resistance at 300°C is less than 200k ohms and the resistance at 900°C is more than 100 ohms.

(2) Now it is possible to make a high temperature sensor with good reduction resistance and excellent high temperature durability. This is mainly due to the successful development of an $(Al, Cr, Fe)_2O_3$ type material that has corundum structure with one positive ion site crystallographically.

Fig. 2 shows the variation of resistance in a reducing atmosphere (96% N_2 + 4% H_2 gases). This figure verifies that the corundum type material of this high temperature thermistor has much improved reduction resistance compared with the former spinel type material.

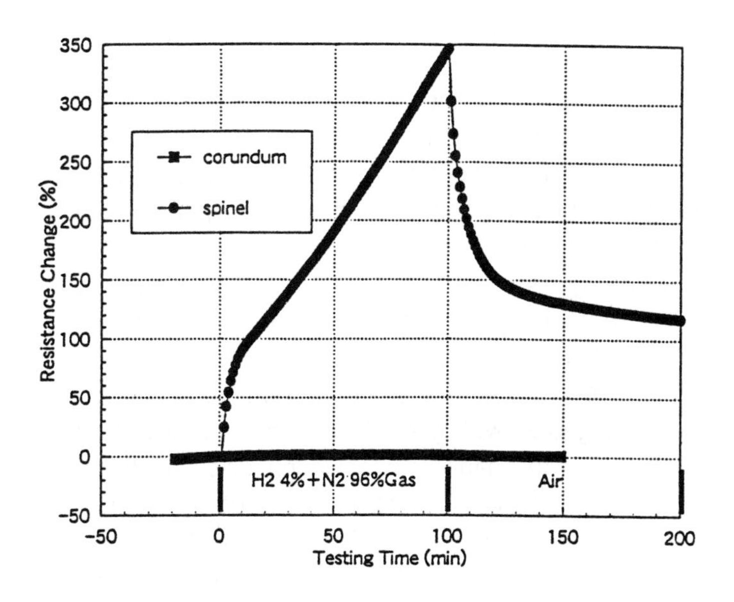

Fig.2 Resistance Drift in a Reducing Atmosphere

Sensor Structure

Fig. 3 shows dimensions and structure of the exhaust gas high temperature sensor. A high temperature thermistor device is made from a molded corundum material that is formed to an oval shape. Two platinum pipes are inserted as electrodes prior to the thermistor being sintered at a temperature near 1700°C.

The inserted platinum pipes in the sintered thermistor are connected by laser welding to the tip of a high temperature cable which is made with two core wires using heat resistant metal. A cap made of heat-resisting metallic material covers the sensing thermistor portion to protect it from the exhaust gas environment. Since the sensing portion is exposed to temperatures as high as 1000°C, it required careful design from a mechanical toughness as well as oxidation resistance stands point. We also use aging not only for the piece parts but also assembled parts to stabilize the resistance of the thermistor device.

The shape of the temperature sensing portion is an oval shape with dimensions of 3.3 mm x 2 mm. This shape improves heat transfer to the thermistor device and enables the rapid response required for the sensor.

The nut is designed with a heat resistant stainless steel and it is connected to the cable. A free spinning nut is also available by applying a flange on the cable.

Connection of cable wires and the wire harness simultaneously by crimping the edge of the collar surrounding the rubber bushing provides the wire cable with toughness against tension and also provides a waterproof seal.

Performance

(1) Resistance-temperature characteristics

The composition ratio of thermistor materials and the shape of the thermistor device mainly determine the resistance value of the sensor. When the B value becomes large, temperature resolution of the sensor becomes high. However, this causes the detecting temperature range to become small, since the B value

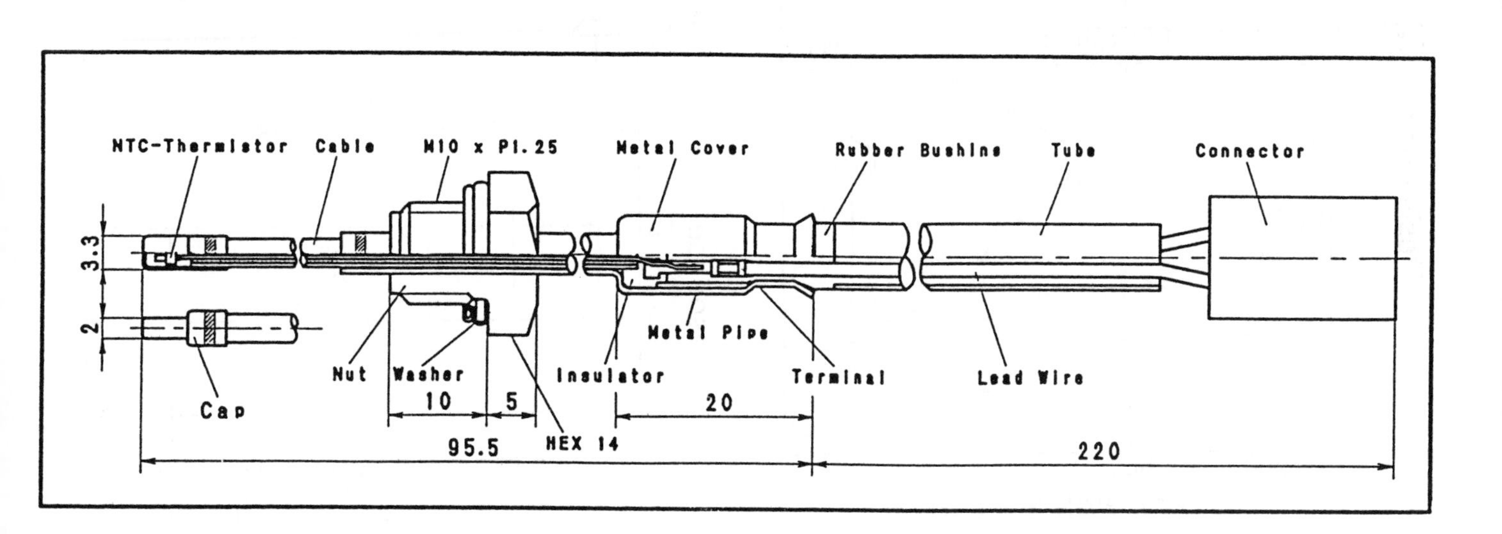

Fig. 3 Dimensions and Structure of the Sensor

indicates resistance change ratio per temperature change. It is expressed as

$$B=(\ln R2 - \ln R1)/(1/T2 - 1/T1).$$

The newly developed exhaust gas high temperature sensor contains a high temperature thermistor device with a resistance of less than 200k ohms at 300°C and greater than 200 ohms at 900°C. This allows it to be used for a wide temperature detection range from 300°C to 900°C to meet its applications. Fig. 4 shows a typical resistance-temperature characteristics.

There are two standard types of characteristics of the resistance; the high resistance type (ETS09HR) and the low resistance type (ETS09LR). The low resistance type, in particular, can detect temperatures as low as 120°C. Both types have +/-12.5°C accuracy over the whole temperature range from 300°C to 900°C.

(2) Responsiveness

The temperature sensing portion of the sensor was designed to obtain the highest responsiveness while simultaneously maintaining high temperature durability capability up to 1000°C. To obtain high responsiveness, it is essential to make the thermal capacity of the sensing portion as small as possible while making the heat conduction of the sensor from the external atmosphere to the thermistor device as big as possible.

Table 1 shows the comparison of the temperature sensing portion design of a sensor and its thermal response characteristics.

In the first step of the development, the size of the entire sensor was reduced by leaving the structure of the conventional temperature sensing portion the same. Changing the outer diameter of the sensing portion from ø6.1 to ø3.3 reduced the thermal time constant τ significantly to 10.8seconds from the previous 20seconds when measured with a flow velocity of 250 m/min. (4.1m/sec.) at 600°C.

In the second step, the shape of the sensing portion was changed to oval by eliminating the excess portion of the device that had little effect to the resistance value of the high temperature thermistor. Then a thin heat resistant cap of 3.3 mm x 2 mm oval shape was introduced to improve the heat conduction from surrounding atmosphere to the element. These design changes reduced the thermal time constant τ to 8.7seconds in order to meet our original target of less than 10seconds.

The thermal time constant τ varies considerably depending on the measurement temperature conditions and flow velocity. Fig. 5 shows the thermal response characteristics of the newly developed high exhaust gas temperature sensor with 600°C gas flow velocity ranging from 80 m/min. (1.3m/sec.) to 280 m/min. (4.1m/sec.)

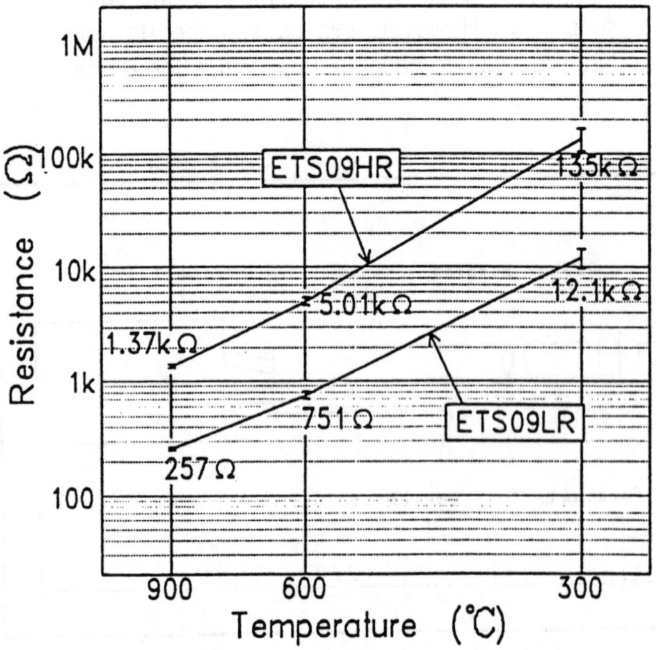

Fig. 4 Resistance - Temperature Characteristics

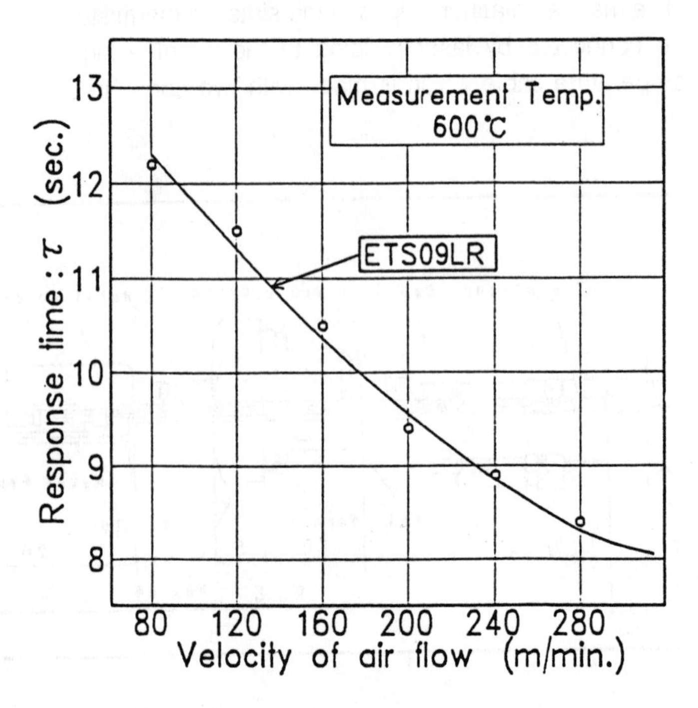

Fig. 5 Thermal Response Characteristics

Table 1 Temperature Sensing Portion Design of the Sensor

	Current Type		STEP 1		STEP 2
Dimension of Element	2, φ3.45	⇨	1, φ2.25	⇨	1, 2.25, 2
Dimension of Sensor	φ6.1		φ3.3		3.3, 2
Response Time	20 sec		10.8 sec		8.7 sec

(3) High temperature durability

Fig. 6 shows the results of high temperature durability tests. The stable corundum thermistor shows little resistance changes after a 1000hours of 900°C heat soak test. The resistance drift value is equivalent to a temperature of 6°C on average and less than 10°C including variations.

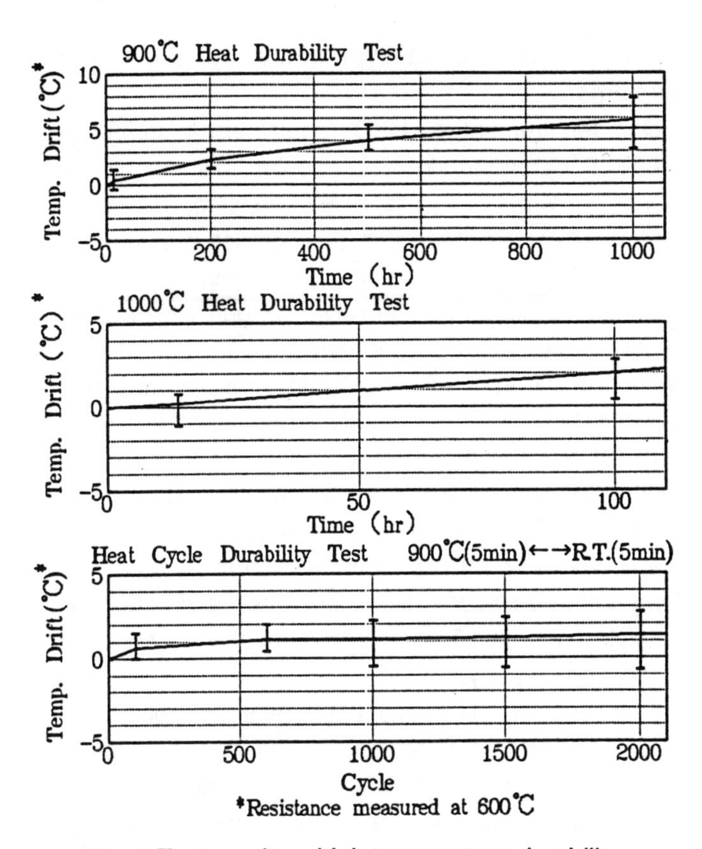

Fig. 6 Test result on high temperature durability

In the 1000°C heat soak test, which is the highest application temperature, the resistance drift after 100hours value is less than 5°C including variation.

This data indicates that there is little deterioration of the thermistor's resistance characteristics even when the sensor encounters abnormally high temperature stress.

After 2500 cycles of a heat shock test between conditions of 900°C for 5 minutes and room temperature for 5 minutes, the resistance value drift is less than +/-5°C and there is no abnormal appearance damage on the temperature sensing portion. It proves the sensor's excellent durability against rapid thermal stress caused by a large temperature changes.

(4) Specifications

As described above, this newly developed exhaust gas high temperature sensor for OBD can detect the exhaust gas temperature of an automobile over a wide temperature range with rapid responsiveness and high precision. It also has a good durability against complex environment of heat shock and vibration, thanks to a reliable design of the sensing portion. Table 2 shows the key specifications of the exhaust gas high temperature sensor.

Table 2 Key specification of Exhaust gas high temperature sensor

• Resistance		
	Temp.	Resistance
ETS09HR	300 °C	135kΩ ± 23% (±12.5 °C)
	600 °C	5.01kΩ ± 8% (±12.5 °C)
	900 °C	1.37kΩ ± 3.5% (±12.5 °C)
ETS09LR	300 °C	12.1kΩ ± 20% (±12.5 °C)
	600 °C	751 Ω ± 6.8% (±12.5 °C)
	900 °C	257 Ω ± 2.9% (±12.5 °C)
• Max. Operation Temp.		1000°C, 100Hrs.
• Response Time		8.7seconds (T=600 °C, v=250m/min)
• Durability Performance		
Heat Resistance 1		1000 °C, 100Hrs.
Heat Resistance 2		900°C,1000Hrs
Heat Shock		R.T.(5min)↔900°C(5min) 2000cycles
Vibration		40G, 250Hz, 30Hrs.
Temperature Accuracy Change After Durability		: ±12.5 °C(300°C ~900°C)

New Applications

(1) Application to catalyst heating systems

As previously described, emission control is required immediately after starting the engine in order to meet the low emission vehicle regulations. To make this possible, there are a number of system ideas to warm up the catalyst by additional heating methods such as an electric heater, or some form of a fuel heater that operates by a re-igniting uncombusted HC after mixing it with injected secondary air in a chamber near the catalyst. The exhaust gas high temperature sensor is used in these catalyst heating systems for OBD by detecting the ignition of the burner. This is to assure normal heating operation or to measure the catalyst around a minimum working temperature of about 350°C.

In order to detect proper combustion in fuel heated catalyst systems, high responsiveness of the sensor is required. Fig. 7 shows the thermal response characteristics of the low resistance type exhaust gas high temperature sensor (ETS09LR) when it is placed in the burner flame. As shown in the figure, it takes approximately 1.5seconds for the sensor to reach the resistance value of less than 1M ohms which is the

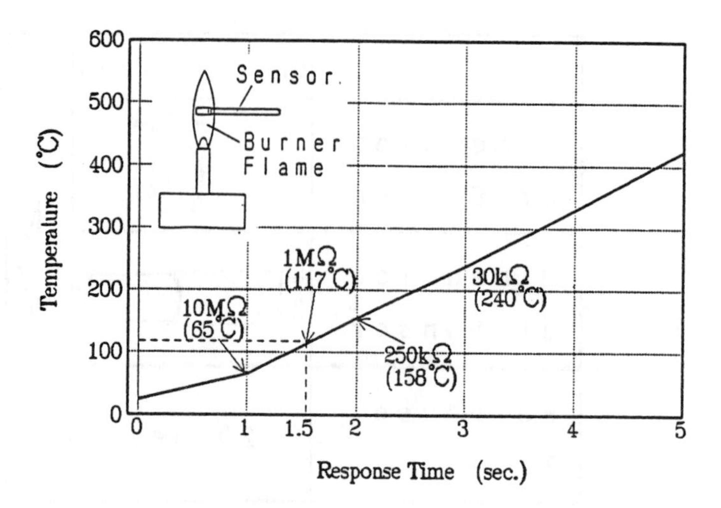

Fig. 7 Thermal response characteristics of the sensor placed in the burner flame

maximum readable resistance value. Detection of abnormal combustion requires 1000°C heat durability, and application of temperature detection of the catalyst needs wide temperature measurement ability between 300°C and 900°C. The newly developed sensor meets all of these application's requirements.

(2) Application to catalyst efficiency monitoring systems

One requirement of OBD regulations is to detect catalyst deterioration. The tighter the regulation becomes with LEV and ULEV systems, the more accurate detection of catalyst deterioration is required in the catalyst efficiency monitoring systems.

Under the current system, two oxygen sensors are mounted before and after the catalytic converter to detect deterioration of the catalyst. However, it is said that this system is not accurate enough to meet severe requirement of ULEV regulation. To improve accuracy of the catalyst efficiency monitoring, one idea under evaluation is to detect the deterioration of the catalyst using the temperature difference measured by two temperature sensors installed before and after the catalytic converter. The catalyst generates heat when it is working properly.

When an exhaust gas high temperature sensor is used in this catalyst efficiency monitoring system, it must detect the temperature difference very accurately. In order to do so, two matched sensors mounted before and after the catalyst converter are needed. Since the sensor's characteristics are initially matched, it is possible to minimize the difference of the resistance value between the two sensors even after durability. Fig. 8 shows drift of the temperature value after a durability test of a matched sensors.

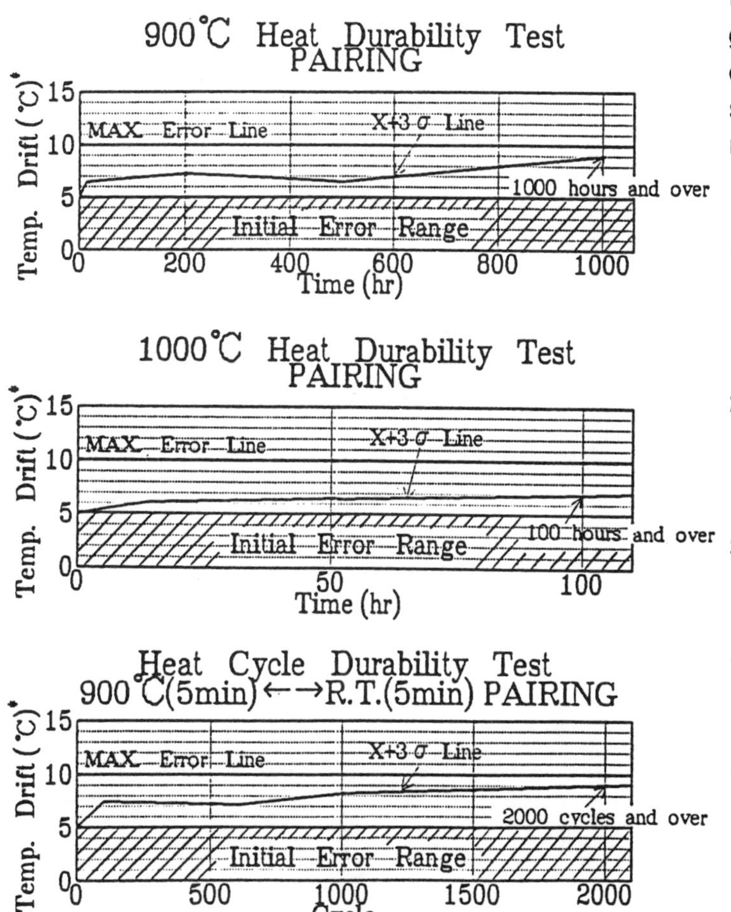

Fig. 8 Result of high temperature durability test (Paired test)

Conclusion

We have developed a thermistor type exhaust gas high temperature sensor to meet the requirements of new emission control systems presently under development for low emission vehicles (LEV/ULEV) and OBD regulations. Thanks to the newly developed high temperature thermistor of corundum system material, the exhaust gas high temperature sensor can measure a wide temperature range from 300°C to 1000°C. Also, the reduced size sensing portion along with an improvement of heat transition into the thermistor enabled the sensor to have high responsiveness of its thermal time constant as low as $\tau = 8.7$seconds. Based upon its design, it provides the high reliability needed for an automotive component like the exhaust gas temperature sensors currently being used in Japanese automotive market for single point catalyst over temperature detection. More applications of the exhaust gas high temperature sensor other than the heated catalyst system and the catalyst efficiency monitoring system are expected because of its wide temperature range up to 1000°C and its rapid responsiveness.

References

1. J. LeGare, T. Tamai,"High Temperature Measurements for On-Board Diagnostics of LEV/ULEV Systems" SAE Paper 942054, October 1994.
2. B. Pfaltzgraf, E. Otto, et al, "The System Development of Electrically Heated Catalyst (EHC) for the LEV and EU-III Legislation" SAE Paper 951072, February 1995.
3. J. R. Theis, "Catalytic Converter Diagnosis Using the Catalyst Exotherm" SAE Paper 942058, October 1994.
4. S. Yoshimoto, et al, "Environmental Conservation in the Automobile Industries" Automotive Engineering, Vol. 48, No.7 1994.
5. Nakahara, Takami, "Automobile Combustion Control Sensors" International Combustion Engine, Vol. 30, No.8 1991.
6. "Heated Catalytic Converter" Automotive Engineering, September 1994.
7. P. Langen, et al, "Heated Catalyst Converter Competing Technologies to Meet LEV Emission Standards" SAE Paper 940470, March 1994.
8. D. Eade, R. Hurley, et al, "Fast Light-off of Underbody Catalyst Using Exhaust Gas Ignition (EGI)" SAE Paper 952417, October 1995.
9. J. Hepburn, A. Meitzler, "Calculating the Rate of Exothermic Energy Release for Catalytic Converter Efficiency Monitoring" SAE Paper 952423, October 1995.
10. O.Polat, et al, "Novel Emission Technologies with Emphasis on Catalyst Cold Start Improvement Status Report on VW-Pierburg Burner / Catalyst Systems" SAE Paper 940474, February 1994.

940635

An Improved Temperature Sensing System for Diesel Engine Applications

Eric B. Andrews
Cummins Electronics Co.

Dipchand V. Nishar
Cummins Engine Co.

ABSTRACT

Thermistors are used for temperature measurement in the automotive industry because of their low cost, reliability and sensitivity to small changes in temperature. The thermistor resistance is a nonlinear function of the measured temperature. Normally, a bias resistance is used to linearize the thermistor characteristics. Unfortunately, this leads to reduced resolution at the low and high end of the temperature range. In this paper, a novel design scheme is proposed in which comparator-based logic is used to select one of four bias resistors. The resistor selected depends upon the magnitude of the temperature being measured. This approach provides improved resolution (and consequently improves the overall measurement accuracy) while requiring lesser memory and execution time than existing schemes. The proposed approach also improves the ability to diagnose open and short circuits as well as sensor faults.

INTRODUCTION

This paper describes the design of a novel temperature measurement scheme for diesel engine applications. Thermistors are used for temperature measurement in vehicles because of their low cost, reliability and sensitivity to small temperature changes [1]*. As emissions legislation becomes more and more stringent, measurement accuracy becomes paramount.

A typical thermistor resistance versus temperature characteristic is shown in Figure 1. Typically, a bias resistor is used to improve the linearity of the characteristic. A simplified schematic of one of these techniques is shown in Figure 2.

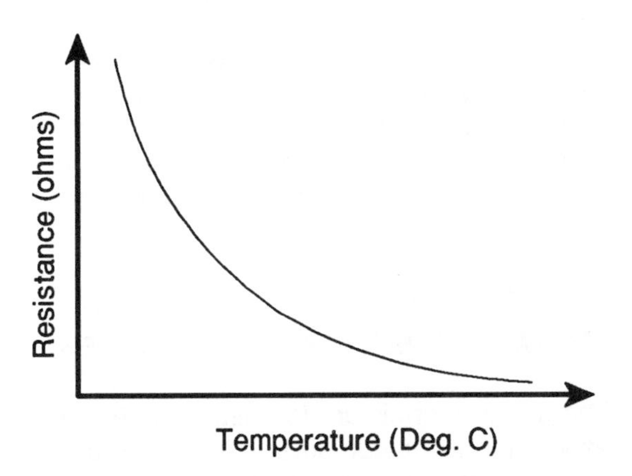

FIGURE 1. Thermistor Transfer Function

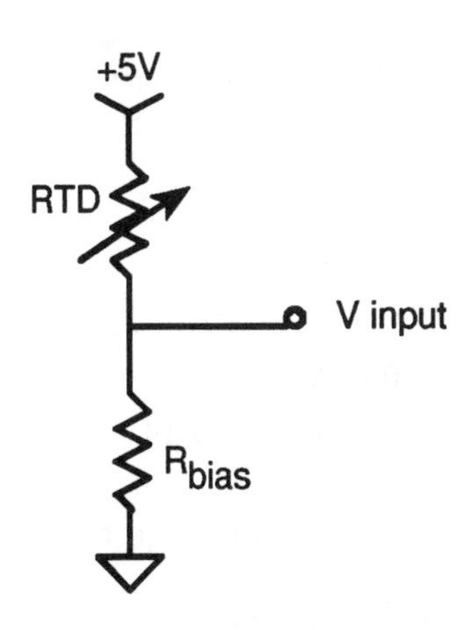

FIGURE 2. Simplified Schematic

The voltage drop across the bias resistor is used as an input to an analog-to-digital converter (ADC). The linearized input transfer function is shown in Figure 3.

* A number in brackets designates a reference at the end of the paper.

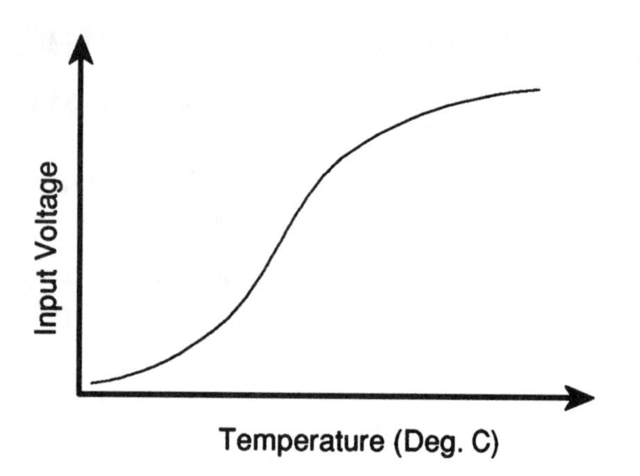

FIGURE 3. Input Transfer Function (*S* curve)

The choice of the bias resistance will improve the resolution and consequently, the overall accuracy of the temperature measurement over a limited temperature range. The output of the ADC is then converted into temperature by interpolating the ADC counts on a piece-wise linear curve fit table. This temperature measurement is used by the electronic controller resident in the vehicle to accomplish engine and emissions control functions.

TYPICAL PIECE–WISE LINEAR CURVE FIT TEMPERATURE MEASUREMENT SCHEMES

The use of thermistors for measuring temperature is widespread in the automotive industry. The resistance of a thermistor changes with a change in its ambient temperature. Since the relationship between the ambient temperature and the thermistor resistance is nonlinear (a decaying exponential), a transfer function of the temperature sensor cannot be characterized by a straight line equation [2]. The thermistor resistance can be approximated by the following equation:

$$R = \alpha\, e^{-\beta T}$$

where,
R: Thermistor resistance (Ohms)
T: Temperature (°K)
α,β: Empirical constants

The values of α and β change depending upon the temperature and are obtained from the manufacturer's literature [3]. The thermistor resistance is used in a voltage divider circuit, within the electronic control module, to translate the resistance into a voltage signal. A pull-up or bias resistor is used in the voltage divider. A typical voltage divider circuit is shown in Figure 2.

The bias resistor can be used as a design parameter to obtain some desirable features in the *S* curve— one of them being the ability to obtain a greater voltage vs. temperature resolution in certain temperature ranges. For applications such as coolant temperature, oil temperature and intake air temperature measurement, accuracy over a wide range is necessary. Selecting a single bias resistor optimizes the resolution over one temperature region at the expense of poorer resolution elsewhere. Since only one bias resistor is used in most designs, the resolution of the measurement is severely degraded at the extreme temperatures [4]. This can be seen by the small gradient (at the extremes) of the thermistor characteristic in Figure 3. The steeper the slope, the finer the measurement resolution. The shallower the slope, the coarser the resolution.

Because of the necessity to measure temperature over a wide range (typically from – 40 to +150 °C), the single resistor scheme has very little design margin for detecting open and short circuits in the system. Another drawback of the single resistor approach is that this design requires a relatively large number of straight line segments for the microprocessor to accurately approximate the *S* curve over the full range of the sensor. Figure 4 gives an overview of a typical measurement system.

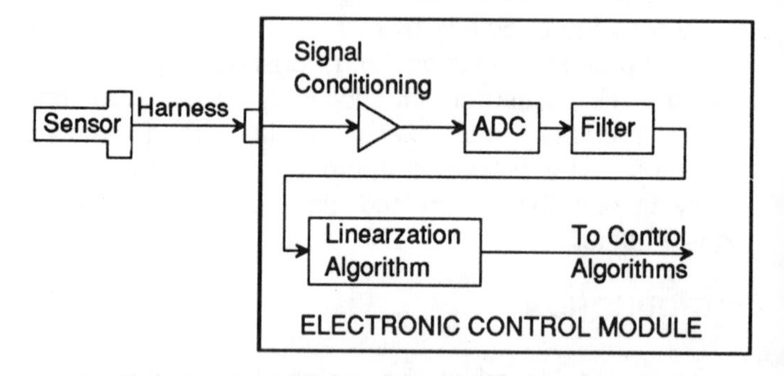

FIGURE 4. Measurement System Overview

Many techniques are available to process the sensor information. Several of these techniques employ multiple bias resistances, switched in parallel and in series, to improve the temperature measurement accuracy [5,6,7]. However, the design approach described in this paper gives the designer more flexibility to optimize measurement accuracy over a much wider temperature range.

PROPOSED DESIGN: FUNCTIONAL DESCRIPTION

The proposed design uses four bias resistors instead of one. The choice of bias resistors to measure the temperature depends upon the state of several comparators. The schematic diagram of the measurement system is shown in Figure 5. The demultiplexer (Demux) is used to isolate each bias resistor. The multiplexer (Mux) is used to input the correct voltage into the ADC.

The following paragraphs describe the function of the logic circuit. The comparator logic and state table are described in Tables 1 and 2 respectively while Figure 6 describes the logic schematic diagram for selecting the correct curve. Figure 7 shows the result of the new bias resistor switching and how it improves the resolution of the measurement.

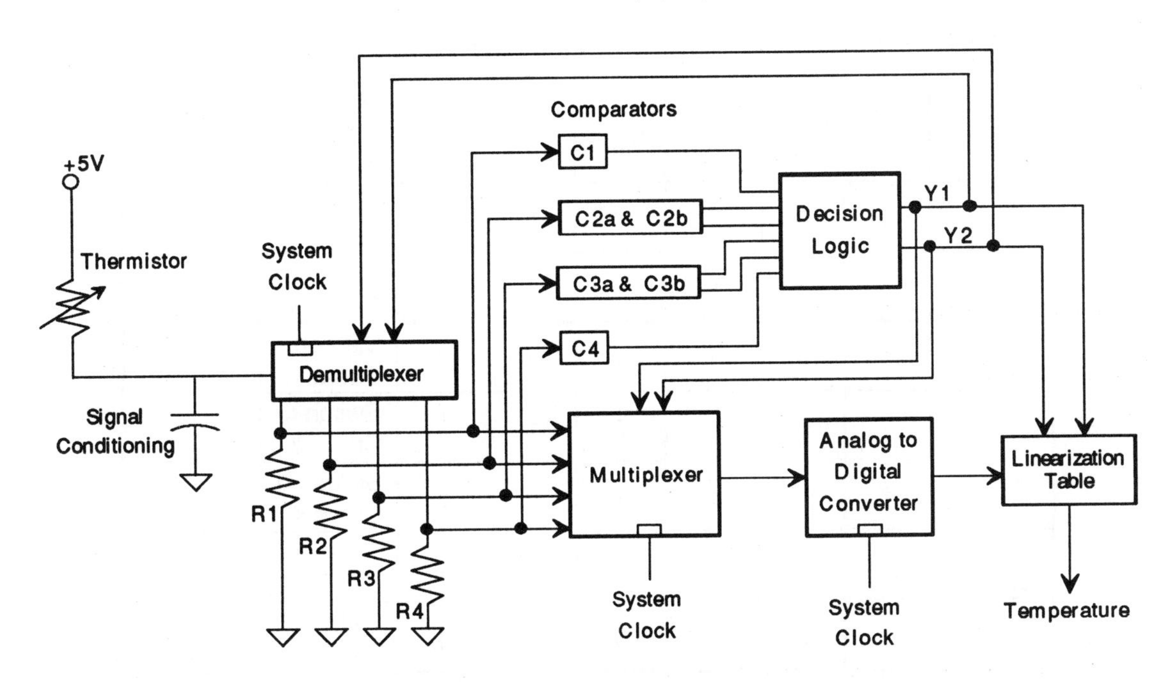

FIGURE 5. Schematic Diagram

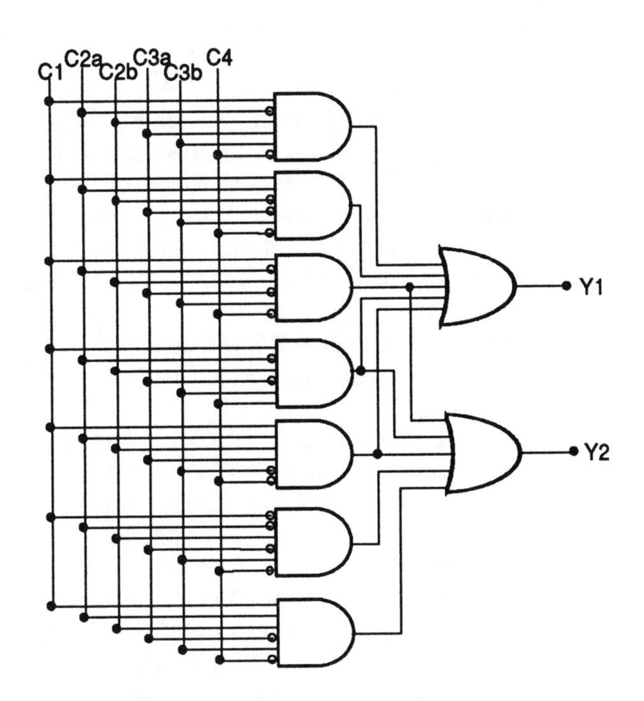

FIGURE 6. Decision Logic for Determining Control Inputs

TABLE 1 Comparator Logic

Comparator	Comparison Test	Comparator Output Y1	Y2
C1	Vin <V1max	1=True	0=False
C2a	Vin >V2min	1=True	0=False
C2b	Vin <V2max	1=True	0=False
C3a	Vin >V3min	1=True	0=False
C3b	Vin <V3max	1=True	0=False
C4	Vin >V4min	1=True	0=False

TABLE 2 State Table

Control Inputs	Comparator States						Action	Logic Output	
	C1	C2a	C2b	C3a	C3b	C4		Y1	Y2
00 Curve 1	1	0	1	0	1	0	Stay on Curve 1	0	0
	0	0	1	0	1	0	Switch to Curve 2	0	1
01 Curve 2	1	1	1	0	1	0	Stay on Curve 2	0	1
	1	0	1	0	1	0	Switch to Curve 1	0	0
	1	1	0	0	1	0	Switch to Curve 3	1	0
10 Curve 3	1	0	1	1	1	0	Stay on Curve 3	1	0
	1	0	1	0	1	0	Switch to Curve 2	0	1
	1	0	1	1	0	0	Switch to Curve 4	1	1
11 Curve 4	1	0	1	0	1	1	Stay on Curve 4	1	1
	1	0	1	0	1	0	Switch to Curve 3	1	0

Note: The undefined states do not occur during normal operation.

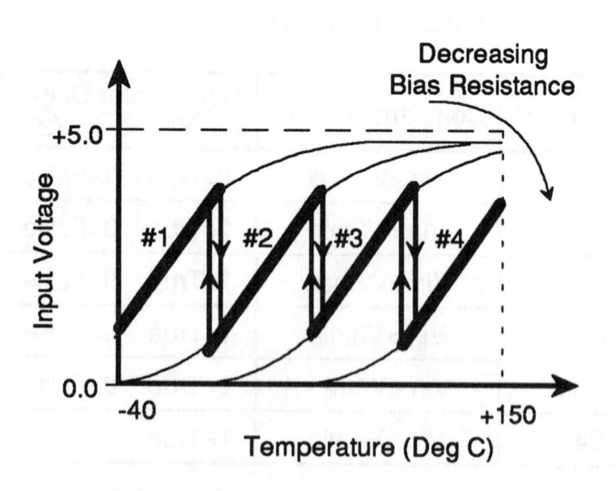

FIGURE 7. Switching Between Different Curves

SYSTEM OPERATION

At powerup, the Mux and Demux are at state 00, which corresponds to bias resistor #1 (Curve 1). Comparator #1 (C1) determines if the voltage (temperature) is greater than the operating range of Curve 1. If the voltage is less than the switching voltage for Curve 1, the output of C1 will be at logic level 1 and the temperature measurement will be accomplished using Curve 1. If the voltage is greater than the operating range of Curve 1, the output of C1 will be logic 0, and the Mux and Demux are commanded to Curve 2.

Curve 2 has two comparators. These comparators will determine whether to switch to the upper curve (3), or to the lower curve (1), or to stay on Curve 2. Comparator C2b determines if the voltage is less than the upper operating range of Curve 2. Comparator C2a tests if the voltage is above the lower operating range of Curve 2. Table 2 shows the state of the comparators needed to stay on Curve 2. If their state is 11 (i.e. both comparators are true), Curve 2 is used. If their state is

01, then the Mux and Demux will switch to Curve 1. Likewise, if the comparator states are 10, the Mux and Demux will switch to Curve 3. A hysteresis region for overlap is provided for stable transitions between curves. Figure 7 shows the hysteresis overlap designed for stable transition. The logic of the comparators was designed so that the state 00 will not exist. If that unlikely event does occur, the control would continue to operate on Curve 2 until one of the other correct states is observed.

Curve 3 has an arrangement similar to Curve 2. The two comparators C3a and C3b are set up to be similar to C2a and C2b.

Curve 4 has only one comparator, which determines if the voltage is greater than the minimum operating range of Curve 4. If the comparator state is 1, then Curve 4 is used. If the comparator state is 0, then the Mux and Demux select Curve 3.

The decision logic determines which curve is to be used. The output of the decision logic is used in three places: the Mux, the Demux and the linearization table. The Mux selects which voltage is to be used in the ADC. The Demux selects which bias resistor is to be used. Once the signal is passed through the ADC, the digital ADC counts are transformed into an equivalent temperature. The output of the decision logic is used in the linearization routine to select the slope and intercept for the straight line equation which is then used to compute the temperature.

RESULTS

Both the single bias resistor scheme and the proposed design linearize the temperature characteristic of the thermistor. Figure 8 depicts the resolution (ADC counts vs. degrees centigrade) for both the schemes. A bigger value of counts/degrees centigrade implies better resolution since a 1 °C change in temperature would result in a larger change in ADC counts. Conversely, a change of one ADC count will imply a smaller change in temperature. Thus, the temperature measurement accuracy will be less sensitive to ADC quantization errors. In this example the single resistor scheme uses a thermistor with a nominal resistance of 10k ohms (at 25 °C) and a 5k ohm bias resistor. The figure below shows that the measurement resolution is maximized at 35 °C. The resolution at the temperature extremes is much worse. For instance, the resolution is less than one count per degree centigrade at temperatures greater than 100 °C. Figure 8 also shows the resolution of the four resistor design. In this example, the resistors chosen were 50k ohms 5k ohms, 2k ohms and 400 ohms. It can be seen that the resolution at the temperature extremes has greatly improved without degrading the resolution in the other temperature regions. From the graph, it is evident that the proposed design has marked improvement, especially at the temperature extremes.

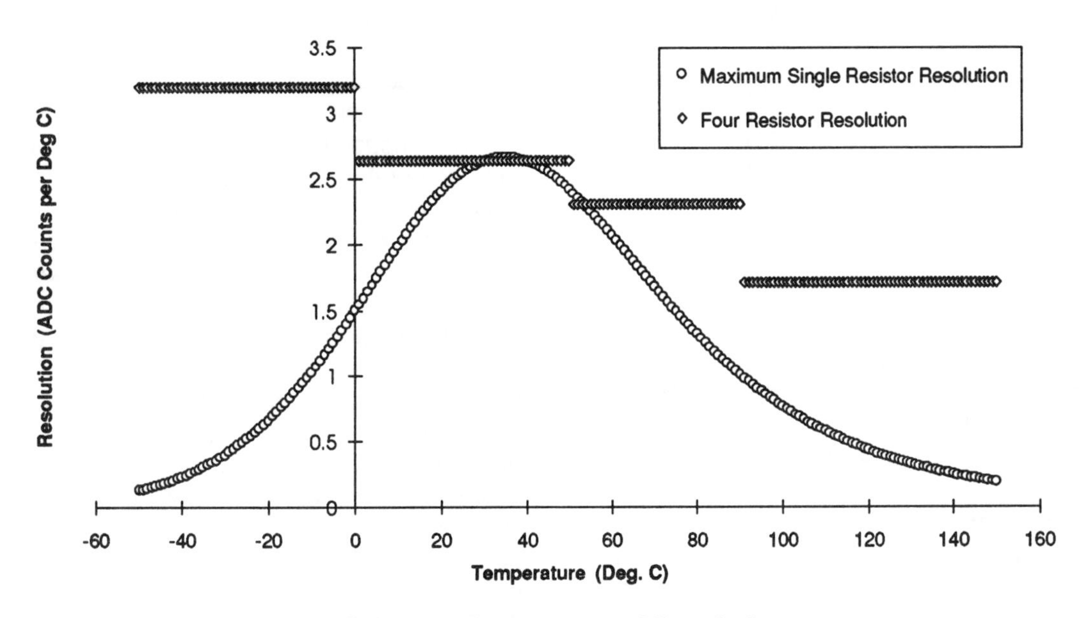

FIGURE 8. Comparison of Resolution

Figure 9 shows the normalized curve fit error associated with using straight line segments to approximate the *S* curve. The figure also shows the curve fit error associated with using several bias resistors. The results indicate that the curve fit error drops dramatically as the number of bias resistors increases. The decrease in error is less dramatic for the case of adding more line segments to approximate the characteristics of the single resistor approach. In fact, approximately 17 line segments are needed to reduce the curve fit error to the magnitude of the 4 resistor approach.

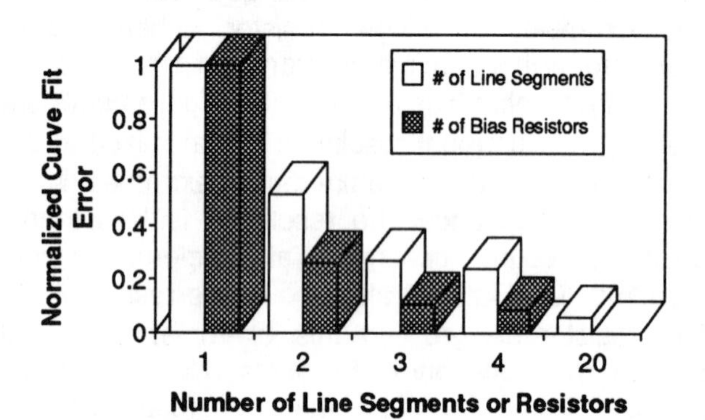

FIGURE 9. Curve Fit Error Comparison

Because of the dynamic range of the temperature measurement, the single bias resistor approach makes the detection of open and short circuits difficult. At very cold temperatures, the resistance of the thermistor is extremely high and the resulting input voltage is very low. Consequently, there is very little design margin to detect open circuit conditions as seen in Figure 10.

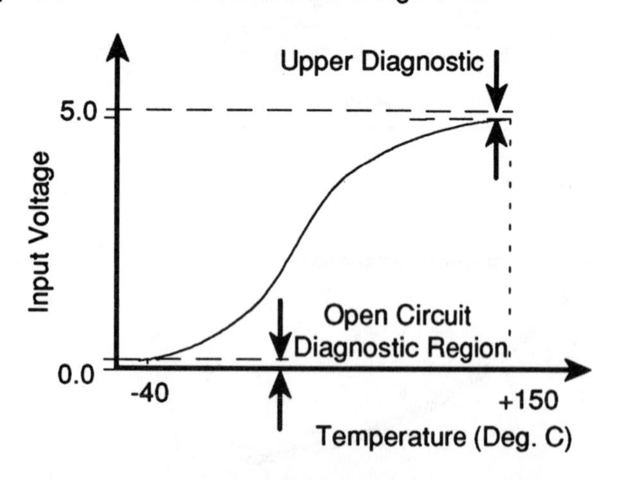

FIGURE 10. Limited Diagnostic Region

By using four different bias resistors, the ability to detect open and short circuit conditions is enhanced as seen in Figure 11.

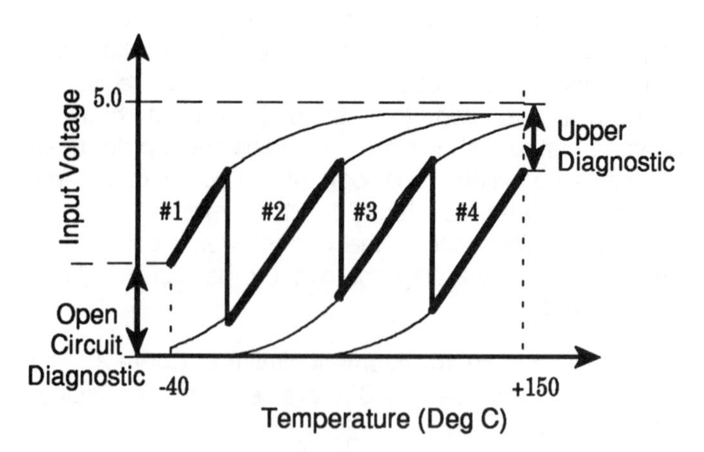

FIGURE 11. Improved Diagnostic Region

The new approach also decreases the microprocessor time required to determine the temperature. The piece-wise linear curve fit approach has to first test which table element is to be used and then do a linear interpolation to find the correct temperature. The interpolation requires four additions and two multiplications. The new design computes the temperature using a straight line equation. The processor determines the temperature by multiplying the ADC counts by a constant (slope of the line) and adding an offset (the intercept of the line).

CONCLUSIONS

Thermistors will continue to be used in the automotive industry because of their low cost, reliability and measurement sensitivity. As emissions legislation becomes more stringent, measurement accuracy becomes paramount. There are several advantages of this design over prevalent piece-wise linear curve fit approaches. First, the resolution and consequently, the overall temperature measurement accuracy are greatly improved at the temperature extremes. The improved resolution, especially at cold temperatures, will have additional emissions benefits since injection timing advance is controlled as a function of engine temperature. Therefore, accurately measuring engine temperature will improve the cold start emissions. Second, the memory requirements are reduced because the system needs only two linearization coefficients per temperature curve (slope and intercept). This is because the bias resistors have been selected to exploit the linear region of each characteristic curve. Third, by using several curves, the system can better diagnose sensor failures such as open and short circuit conditions. Fourth, since several curves are used, the system has improved linearity throughout the range of operation. Fifth, the new approach requires less processor computation to obtain the final temperature measurement. Thus, the proposed temperature measurement scheme demonstrates measurable advantages over the more prevalent single bias resistor approach.

REFERENCES

1. Rich Wegner, "Simplifying Temperature Monitoring and Control in Automotive Systems," SAE paper 922129, 1992.

2. William G Wolber, "A Worldwide Overview of Automotive Engine Control Sensor Technology," SAE paper 780207, 1978.

3. Harry N. Norton, *Handbook of Transducers*, Englewood Cliffs, NJ: Prentice Hall, 1989, pp. 386 - 388.

4. Peter J. Sacchetti and Donald R. Phillips, "A Ratiometric Temperature Sensor for High Temperature Applications," SAE Paper 800024, 1980.

5. Toshiro Hara and Masahiko Sayama, "Vehicle Driving Condition Detecting Apparatus," U.S. Patent 5,140,302, August 18, 1992.

6. Gerd Wallenfang, "Temperature Measuring Device for Recording Large Changes in Temperature," U.S. Patent 4,699,520, October 13, 1987.

7. William H. Newman and Ralph G. Burgess, "Temperature Measurements Using Thermistor Elements," U.S. Patent 5,116,136, May 26, 1992.

GAS SENSORS

970858

Performance of Thick Film NOx Sensor on Diesel and Gasoline Engines

Nobuhide Kato, Yasuhiko Hamada, and Hiroshi Kurachi
NGK Insulators, Ltd.

Copyright 1997 Society of Automotive Engineers, Inc.

ABSTRACT

This paper describes a thick film ZrO_2 NOx sensor feasible for diesel and gasoline engine applications, and introduces modification items from the previous concept design.(1)

The modification items comprise simplifying the sensing element design to reduce output terminals for package design and applying temperature control to the sensing element in order to minimize sensor performance dependency on gas temperature.

The NOx sensor indicates a stable linear signal in proportion to NOx concentration in a wide range of temperature, A/F and NOx concentration as a practical condition on both gasoline and diesel engines. The NOx sensor shows a good response in hundred msec. and a sharp signal following NOx generation in a transient state as well.

Besides, another type of a NOx sensor is proposed for low NOx measurement in a practical use, by an electromotive force(EMF) voltage instead of a pumping current.

INTRODUCTION

Automotive manufacturers face strong demands to improve an automotive emission quality in their future applications. The demands mainly consists of reducing harmful exhaust emissions(CO, HC, NOx, etc.) due to tighter emission regulations across the world, reducing CO_2 in terms of an environmental matter "greenhouse effect", and improving fuel economy to meet customers' needs.

Concerning the CO_2 reduction and the improvement of fuel economy, there is recently a trend to introduce lean-burn gasoline and diesel engines into the market. In such case of a lean-burn combustion, however, there is a task to develop a NOx catalyst capable of converting NOx emission efficiently in a lean condition as well. Therefore, there are many activities to develop the NOx catalyst and/or its control system, even though the lean-burn gasoline vehicles equipped with NOx catalysts have already been introduced into the market.(2)--(5) If there is any device capable of directly detecting NOx in an exhaust gas stream on board, it would be expected to achieve more efficient NOx conversion by the NOx catalyst or its system optimization.

In addition, regarding the reduction of the harmful emissions, having on board a measuring device for one of the harmful emission components would improve an emission control in an engine management or OBD system.

NGK Insulators, Ltd. has developed a NOx sensor for an on-board application, and presented its concept design during the 1996 SAE Congress. This paper describes a design modification to the NOx sensor from the previous concept design for practical use on gasoline and diesel engine applications, and discusses its sensor characteristics.

SENSING ELEMENT DESIGN OF NOx SENSOR

Our concept design of the NOx sensor mainly comprises of two internal cavities and three oxygen pumping cells in its sensing element made of an oxygen ion conductive ZrO_2.

Our measuring concept consists of:

(1)lowering an oxygen concentration of a measuring gas to a predetermined level in the first internal cavity, in which NOx does not decompose, and

(2)further lowering the oxygen concentration of the measuring gas to a predetermined level in the second internal cavity, in which NOx decomposes on a measuring electrode and the oxygen generated is detected as a sensor signal.

Figure 1 shows a cross-sectional view of the NOx sensor element. Each part in the sensing element functions as follows;

(1)First internal cavity:

The first internal cavity connects a measuring gas stream through the first diffusion path under a predetermined diffusion resistance. There are an oxygen pumping cell and an oxygen sensing cell inside the first internal cavity.

* *First oxygen pumping cell (Oxygen pumping cell):*
  The first oxygen pumping cell consists of a pair of first pumping (+) and (-) electrodes on ZrO_2-1 layer in order to lower the oxygen concentration to the predetermined level. The first pumping electrode (+) is platinum, and the (-) electrode is a platinum/gold alloy to reduce NOx reduction catalytic activity.
* *Oxygen sensing cell:*
  The oxygen sensing cell consists of the first pumping(-) electrode in the first internal cavity and a reference electrode in an air duct in order to monitor the oxygen concentration in the first internal cavity by generated electromotive force and feedback to the first oxygen pumping cell. The predetermined level of the electromotive force is set at 300mV corresponding to about 1ppm of oxygen concentration.

(2)Second internal cavity:

The second internal cavity connects the first internal cavity through the second diffusion path under a predetermined diffusion resistance. There are two different oxygen pumping cells and an oxygen sensing cell inside the second internal cavity.

* *Second oxygen pumping cell(Oxygen pumping cell):*
  The second oxygen pumping cell consists of the second pumping(-) electrode in the second internal cavity and the first pumping(+) electrode on ZrO_2-1 layer, in order to further lower the oxygen concentration to a predetermined level. The second pumping(-) electrode is also made of the platinum/gold alloy.
* *Oxygen sensing cell:*
  The oxygen sensing cell consists of the second pumping(-) electrode and the reference electrode in the air duct to monitor the oxygen concentration in the second internal cavity by generated electromotive force and feedback to the second oxygen pumping cell. The predetermined level of the electromotive force is set at 400mV corresponding to about 0.01ppm of oxygen concentration.
* *NOx sensing cell (Oxygen pumping cell):*
  The NOx sensing cell consists of a measuring electrode in the second internal cavity and the reference electrode in the air duct. The measuring electrode is rhodium and has NOx reduction catalytic activity. Therefore, NOx decomposes on the measuring electrode and the oxygen generated is detected as an oxygen pumping current in the NOx sensing cell. The sensor signal is in proportion to NOx concentration in the measuring gas.

In addition to the above-mentioned sensing functions, there is a heater embedded in the sensing element. The heater(-) has a terminal common with the first pumping electrode(-) to assemble the sensing element with six output terminals.

Therefore, a feature of the sensing element is the addition of the second oxygen pumping cell to stabilize the oxygen concentration in the second

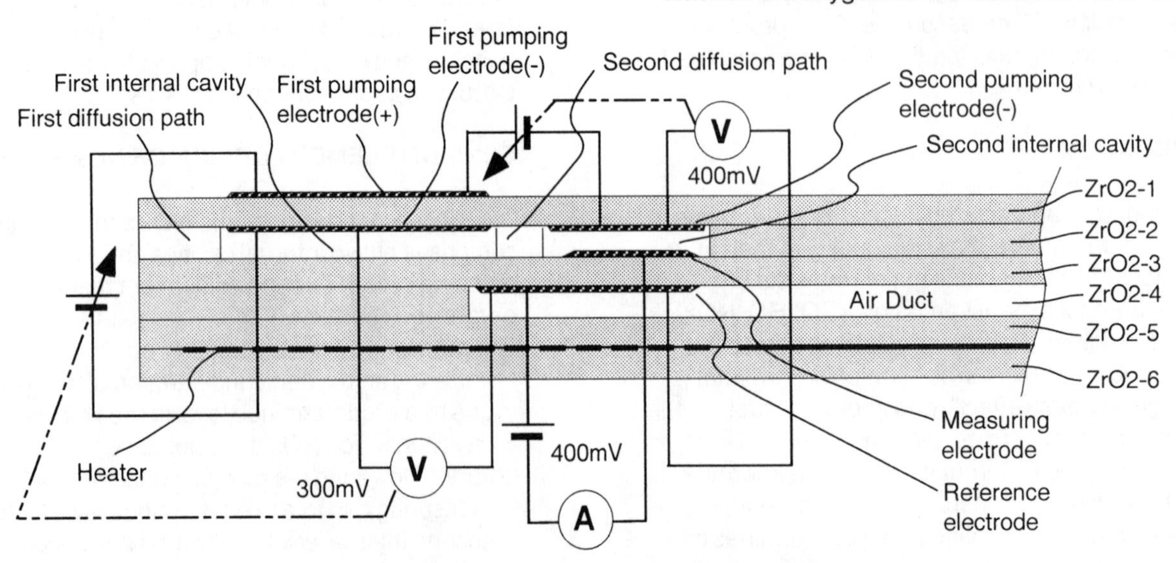

Figure 1; Cross-sectional view of NOx sensing element

internal cavity and a reduction of output terminals down to six for higher reliability and lower cost of packaging design, in comparison with the previous concept design.

Figures 2 (a) and (b) show the basic output characteristics measured on a model gas apparatus; NO sensitivity and an interference effect against various gas components.

The NO sensitivity is measured in 0.3% of O_2 and 3% of H_2O with N_2 base gas, while changing NO concentration upto 2000ppm. The gas temperature is 25℃ in this measurement. The sensor signal, a pumping current, is in proportion to NO concentration as shown in figure 2 (a).

The gas interference effect is measured in 150ppm of NO, 0.3% of O_2 and 3% of H_2O with N_2 base gas, while changing the designated interference gas concentration. Based upon the measuring results shown in figure 2 (b), there is a negligibly small interference effect against CO, C_3H_8 as a representative of HC, CO_2, SO_2. A small interference of the sensor signal is observed within the range of 0 to 20% of oxygen. The change of the sensor signal is only 0.1 μA, which is equivalent to 22ppm or +/- 11ppm of NOx as a measuring error.

DESIGN CONCEPT OF HEATER CONTROL

The NOx sensor signal is a limiting current(Ip) in the NOx sensing cell located after the gas diffusion paths.

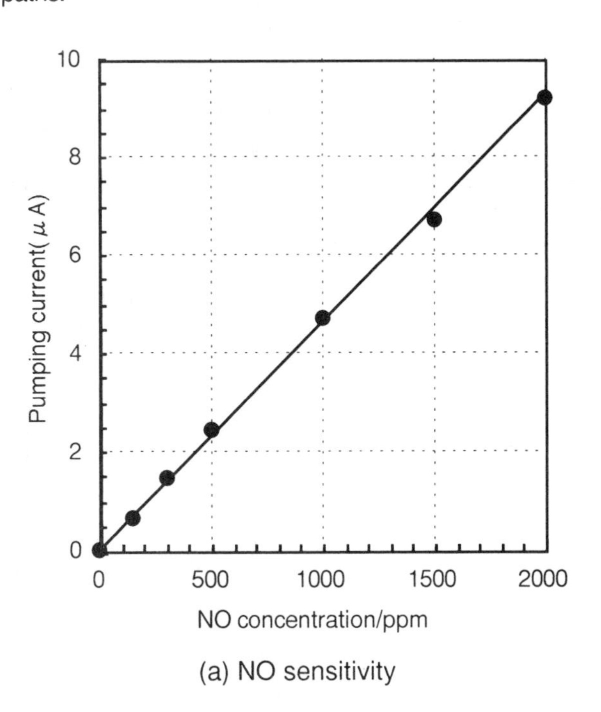

(a) NO sensitivity

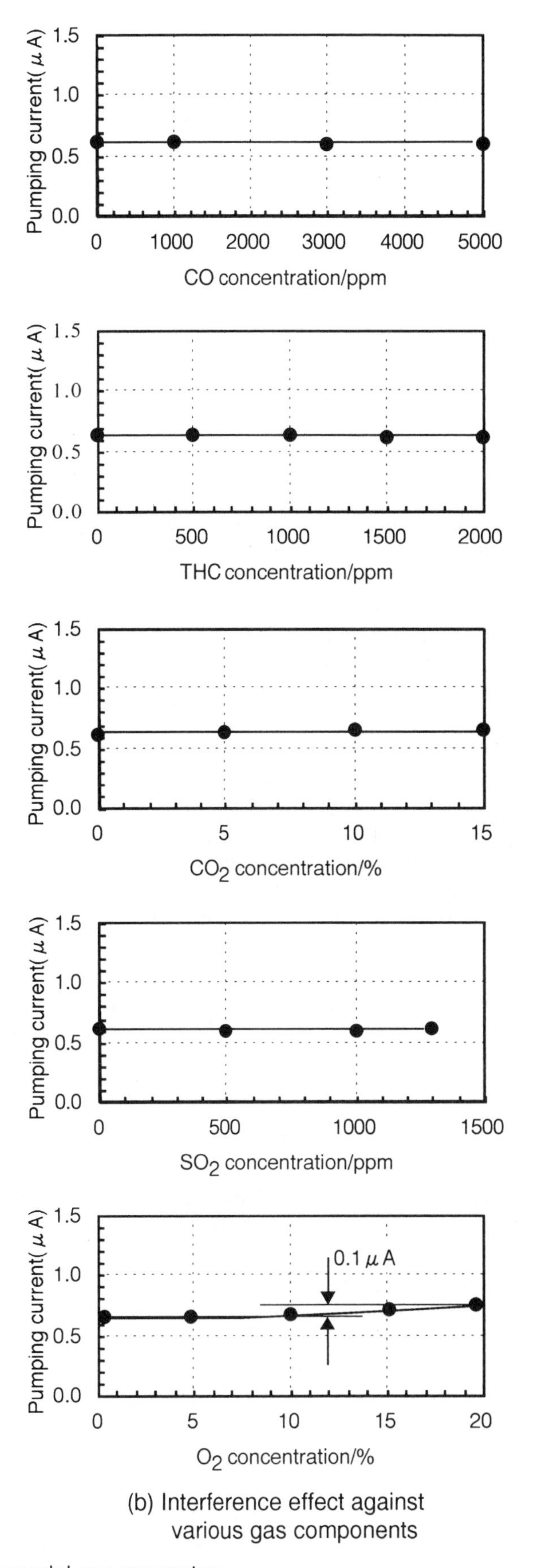

(b) Interference effect against various gas components

Figure 2; Basic output characteristics on model gas apparatus

The limiting current is expressed as:

$$Ip = 4FD/RT \times S/L \times (Poe - Pod) \quad \text{...........(1)}$$

where D = NOx diffusion coefficient,
S = effective cross section of the diffusion path,
L = effective length of the diffusion path,
Poe = partial pressure of NOx in the measuring gas,
Ped = partial pressure of NOx in the second internal cavity,
R = gas constant,
F = Faraday constant,
T = absolute temperature.

The equation(1) indicates that the NOx sensor signal depends on the NOx diffusion coefficient(D) as well as temperature. The diffusion coefficient depends on how the gas diffuses through the diffusion path.

In the case of the NOx sensor, the structure of the first and second diffusion paths is more than a mean free path, so that NOx gas diffuses as a molecule diffusion through two diffusion paths in the sensing element. This means that the NOx sensor signal(Ip) depends on a temperature by 0.75th power, according to the equation(1). Therefore, it is necessary for the NOx sensor to have some heater control in order to minimize a tip temperature fluctuation and the temperature dependency of the sensor signal, when the NOx sensor is used at various temperature conditions in an automotive exhaust gas stream. It is a common method in an exhaust gas oxygen sensor for automobiles that a heater is controlled by monitoring an impedance between its electrodes. The NOx sensor has such a heater control by monitoring the impedance of the second oxygen pumping cell.

EVALUATION RESULTS ON ENGINES

(1) Temperature dependency of sensor signal on gasoline engine

In order to verify the effect of the heater control, output characteristics of the NOx sensor were measured in several temperature conditions on a 2.0L gasoline engine. The NOx sensor was installed downstream of a catalyst, at which a gas temperature changed from 340 to 720℃. NOx concentration was adjusted by changing the EGR ratio at stoichiometric point(A/F=14.7).

The test results are shown in figures 3 (a) and (b) where the NOx sensor signal is converted to NOx concentration based upon a correlation between the sensor signal and NOx concentration at 430℃ of gas temperature condition. In this measurement, the sensor signal of all the plots indicates a good linearity in proportion to NOx concentration with a high correlation coefficient of 0.995 as shown in figure 3(a). Therefore, it is supposed that the NOx sensor with the heater control has minimum temperature dependency in its signal.

In addition, there is also a good linearity of the signal with a high correlation coefficient more than 0.999 in each of four different gas temperature conditions. However, there is a tendency of a small signal shift-up with the gas temperature increase as shown in figure 3 (b). Therefore, it is expected that the NOx sensor could show even higher accuracy with an additional temperature compensation for its signal.

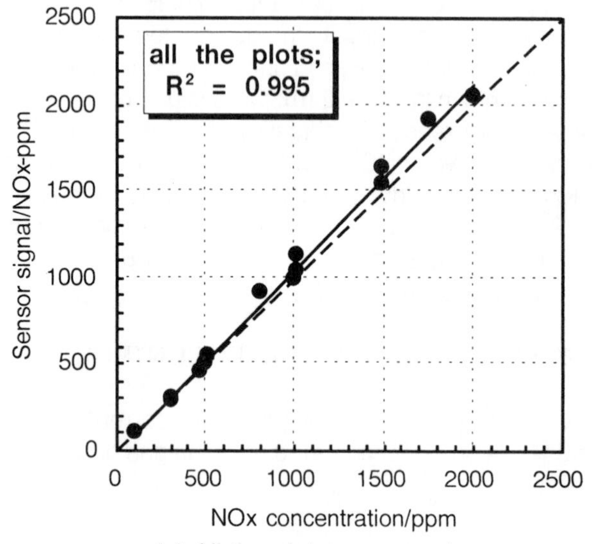

(a) All the plots measured

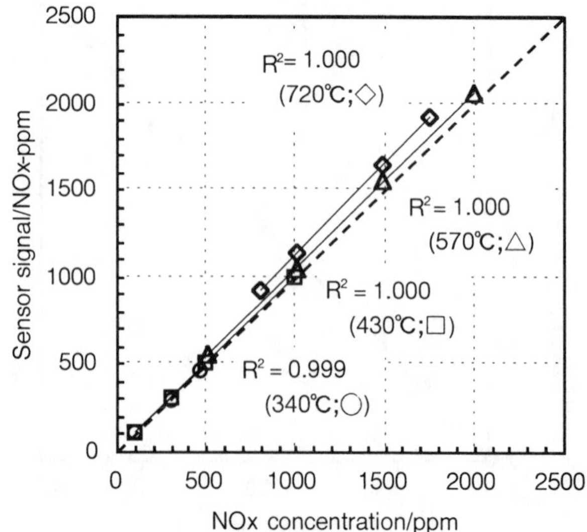

(b) Sensor signals at each gas temperature

Figure 3; Temperature dependency of NOx sensor signal on gasoline engine

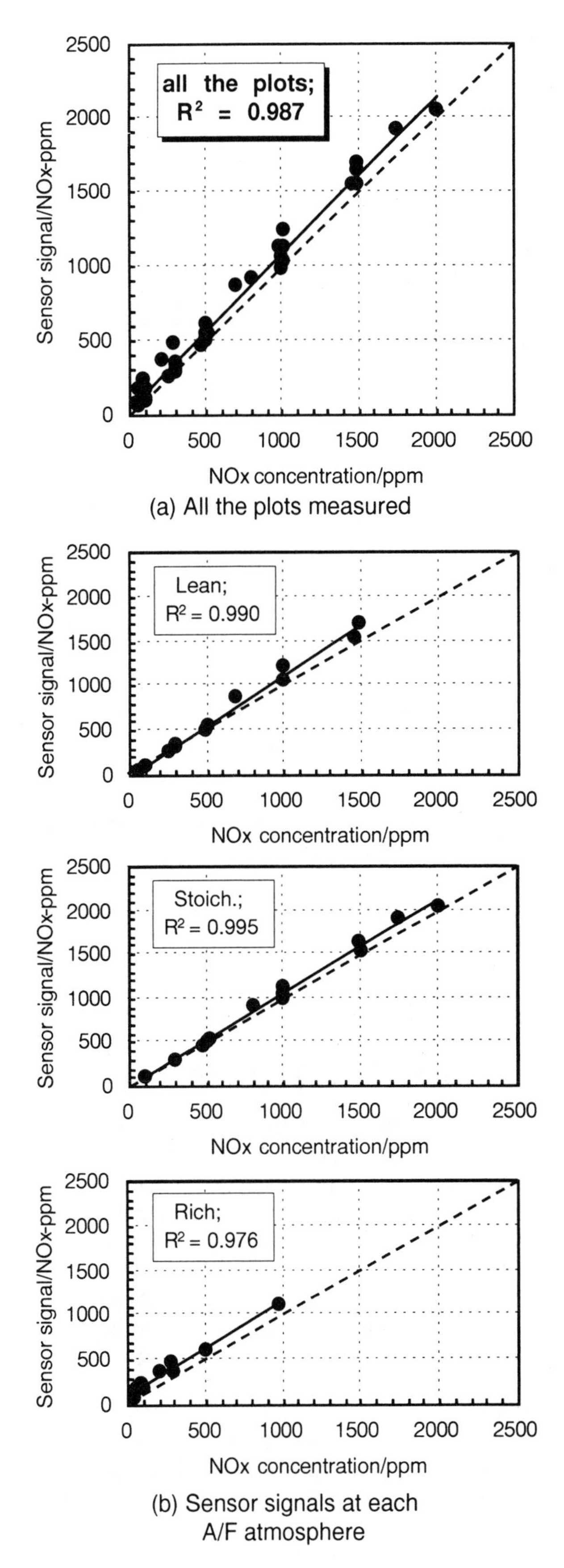

(a) All the plots measured

(b) Sensor signals at each A/F atmosphere

Figure 4; A/F dependency of NOx sensor signal on gasoline engine

(2) A/F dependency of sensor signal on gasoline engine

In order to verify an interference effect, A/F dependency of the sensor signal was measured on the 2.0-L gasoline engine. The sensor signal was measured in lean(A/F=16.3), stoichiometric point (A/F=14.7) and rich(A/F=13.2) conditions with gas temperatures ranging from 290℃to 720℃. In these various measuring conditions, the concentration of each gas component changed within the following wide range; CO; 0 to 3.2%, THC; 3 to 3900ppm, O_2; 0 to 2.3%, CO_2; 13.0 to 15.0%.

The test results are shown in figures 4 (a) and (b). The NOx sensor shows a stable linearity of its signal with a correlation coefficient of 0.987 for all the plots, as shown in figure 4 (a). The sensor signals in each A/F condition also indicate, in figure 4 (b), a good linearity with high correlation coefficient of 0.990 for lean, 0.995 for stoichiomentic point, and 0.976 for rich conditions. Therefore, it is presumed that the NOx sensor is feasible to use in practical A/F and temperature range of a conventional gasoline engine.

(3) Output characteristics on diesel engine

Temperature and A/F dependency of the NOx sensor signal was also evaluated on a diesel engine. A natural aspirated 7.4-L diesel engine was used and the NOx sensor was installed 1-meter downstream of an exhaust port and upstream of a diesel particulate filter(D.P.F.). The engine operating condition was changed from an idle to a full load, and the exhaust gas condition is changed widely from 90℃to 730℃ in gas temperature and from 18.1 to 100 in A/F.
It is considered that this wide range of A/F in this evaluation can cover a practical A/F range of a lean-burn gasoline engine as well.

Test results are shown in figures 5 (a) and (b) where the NOx sensor signal is converted to NOx concentration based upon a correlation between the sensor signal and NOx concentration in gas temperature range from 340℃ to 400℃ on the diesel engine. The NOx sensor indicates a good linearity of its signal for all the plots even in a wide range of A/F and a gas temperature. A correlation coefficient is 0.980. Therefore, it is presumed that the NOx sensor can register a stable signal in proportion to NOx concentration also in a practical condition on a lean-burn gasoline engine as well as a diesel engine.

In addition, however, there is a tendency of the sensor signal to shift-up a little with a gas temperature increase as shown in figure 5(b). Therefore, it is expected that the NOx sensor could show even higher accuracy with an additional temperature compensation for its signal also on a diesel engine.

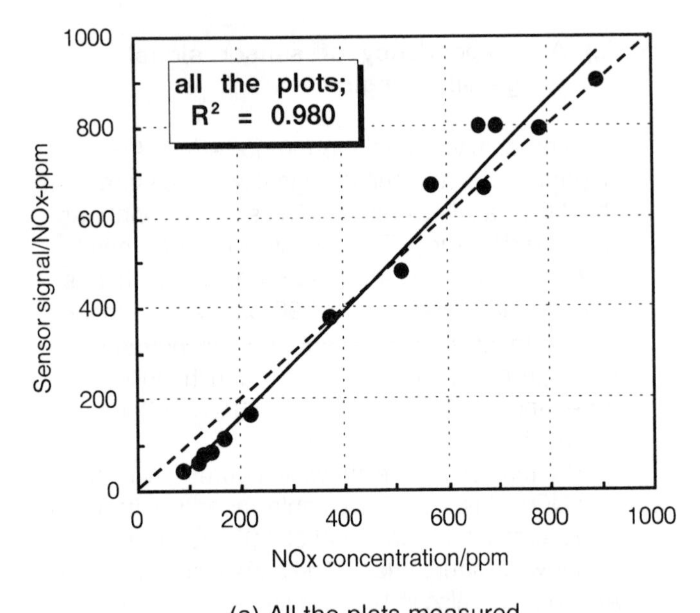

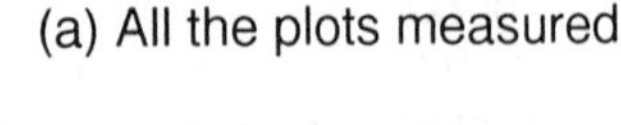
(a) All the plots measured

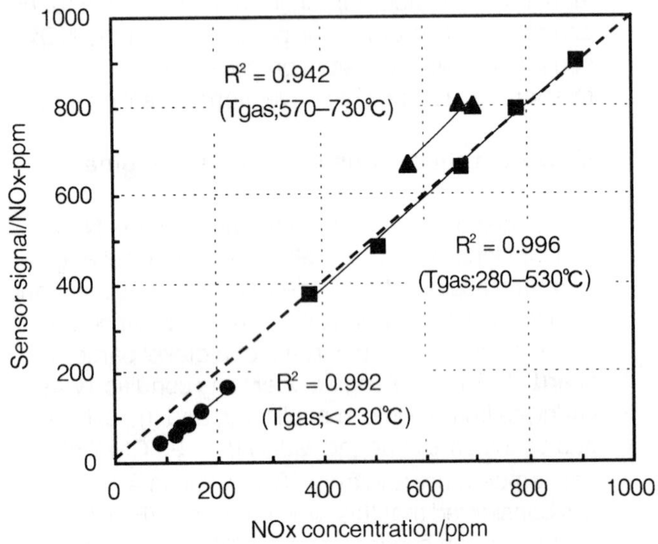

(b) Sensor signals classified by gas temperature

Figure 5; NOx sensor output characteristics on diesel engine

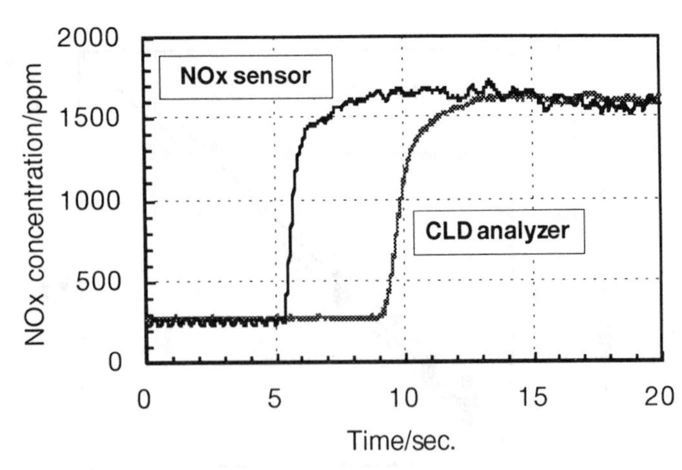

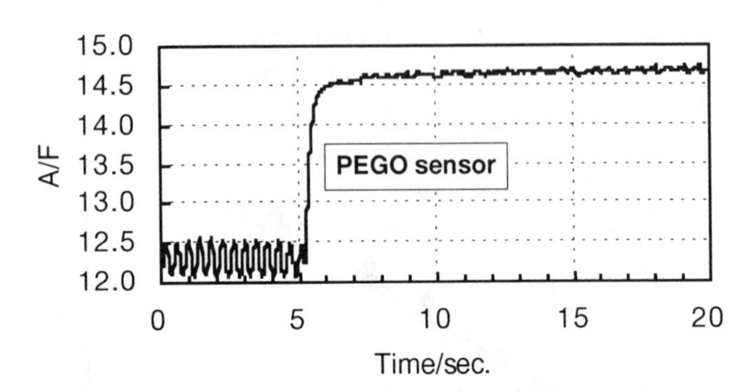

Figure 6; Response property of NOx sensor on gasoline engine

(4) Response property

Figure 6 shows the response property measured on a 2.0-L gasoline engine. The engine operated at a steady condition of 2400rpm/5.0kgm. A NOx sensor and PEGO(Proportional Exhaust Gas Oxygen) sensor were installed 1.5-meter downstream of an exhaust port and on the same circumference of the exhaust pipe upstream of the catalyst. When the A/F was switched from 12.3 to 14.7 by a step change of MAF(Mass Air Flow) sensor signal, each sensor signal were recorded. In order to confirm the signal level of the NOx sensor, a CLD gas analyzer (HORIBA/CLA-150) was also connected to the exhaust pipe at the same position as the sensors through a 6-meter sampling line and its signal was recorded.

The NOx sensor indicates 260msec. of 33 to 66% response time, although the PEGO sensor indicates 160msec. In addition, the NOx sensor registers almost the equivalent NOx indication to the CLD gas analyzer in this step-change period.

Also, FTP-75(HOT-505 mode) tests were conducted to check the response property in a transient state.

In the first test, the NOx sensor was installed downstream of a main catalyst at underfloor position on a 2.0-L gasoline vehicle. Figure 7 shows the NOx sensor signal in comparison with NOx indication by a CLD gas analyzer in direct sampling mode, an intake manifold pressure and a vehicle speed. The NOx sensor indicates quicker, sharper and narrower signal peaks than the CLD gas analyzer. Because the CLD analyzer was connected to a tailpipe with a 10m-long sampling line, so that an exhaust gas must mix up through the sampling line to get a broad signal and a time delay of its signal to some extent. Except for the broad signal with the time delay of the CLD analyzer, however, most of NOx sensor signal peaks is similar to that of the CLD gas analyzer.

It is generally said, in addition, that a NOx emission mainly generates when a vehicle is accelerated. The acceleration condition is represented by an increase in intake manifold pressure. In consideration of this point, most of timing of the NOx sensor signal peaks is quite consistent with the timing of the elevated intake manifold pressure.

Furthermore, in the second test, the NOx sensor was installed at the entrance to the CLD gas analyzer of a direct sampling mode to directly compare the NOx signals to each other. Figure 8 shows the test result. The NOx sensor indicates quite identical signal peaks with those of the CLD gas analyzer through the test. Therefore, it is supposed that the NOx sensor has a good response property to register its signal corresponding to NOx generation on a vehicle even in a transient state.

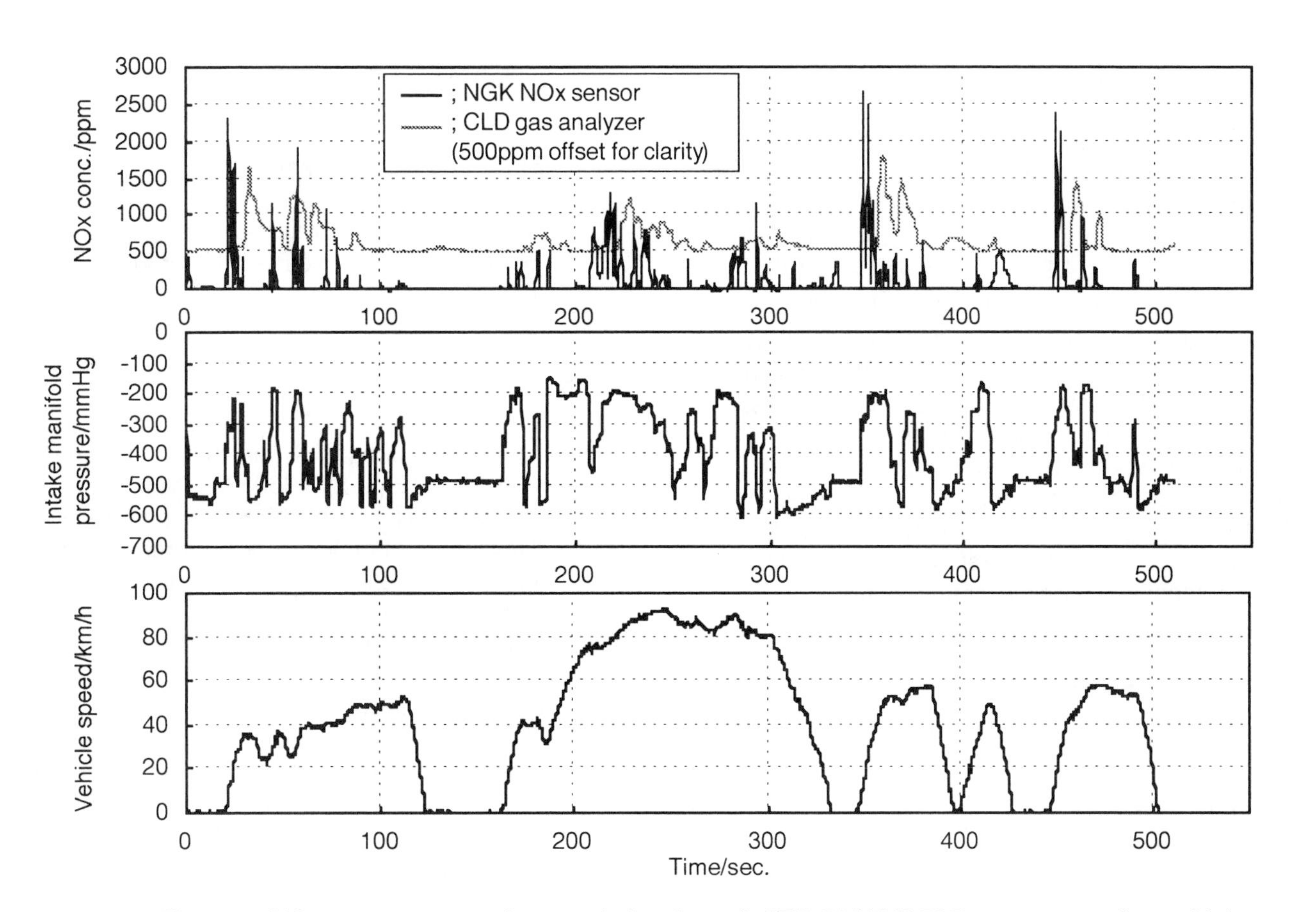

Figure 7; NOx sensor output characteristics through FTP-75(HOT-505) test on gasoline vehicle

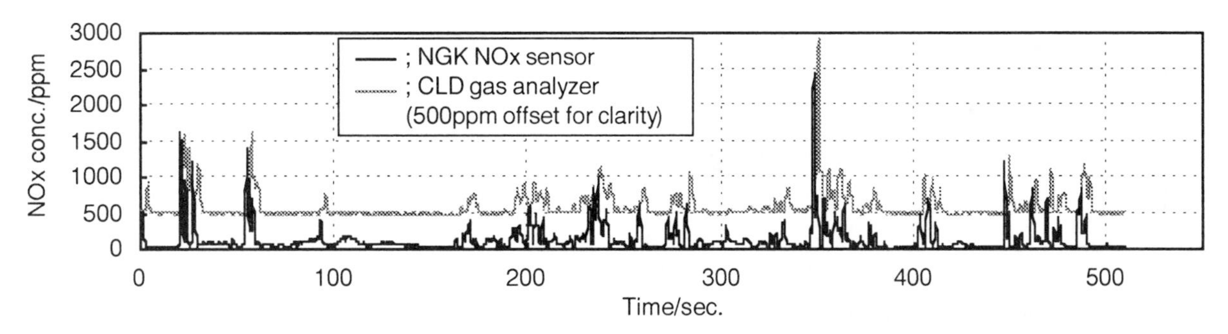

Figure 8; Signal comparison between NOx sensor and CLD analyzer in direct sampling mode through FTP-75(HOT-505) test on gasoline vehicle

NOx SENSOR FOR LOW NOx MEASUREMENT

The forementioned sensor, capable of measuring NOx by the pumping current, demonstrates characteristics feasible for on-board applications; a stable linear signal in proportion to NOx concentration and a quick response detecting NOx generation in a transient state as well.

Concerning a low NOx measurement, however, it is expected that an EMF-type is preferable for more sensitive detection in our sensing element design to the pumping current type.
Because the predetermined oxygen concentration in the second cavity is quite low and the EMF is based upon the Nernst equation. It means that a steep EMF signal change is expected in the NOx sensing cell even by low concentration change of NOx.

Figure 9 shows output characteristics of the EMF-type NOx sensor installed downstream of the catalyst on the 2.0-L gasoline engine. The NOx concentration was changed by the EGR ratio in lean(A/F=16.3), stoichiometric point(A/F=14.7) and rich(A/F=13.3), and in temperature range of 350℃ to 560℃. The sensor signal, the EMF voltage, decreases with NOx concentration increase in each A/F condition as an asymptote. Especially, in NOx concentration less than 300ppm which is a practical range for NOx catalyst and its control on lean-burn gasoline and diesel engines, the EMF sensor signal changes steeply. Therefore, it is presumed that the EMF-type NOx sensor would be preferable for low NOx measurement and possible for automotive application.

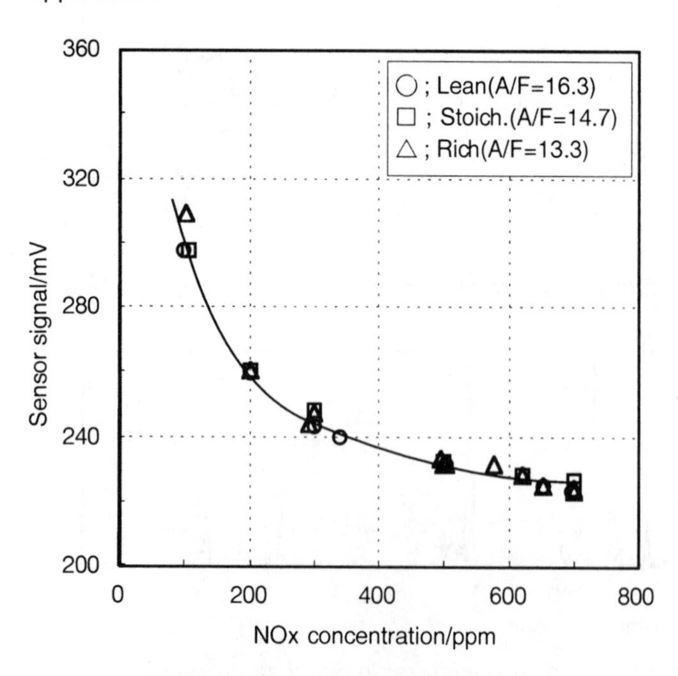

Figure 9; EMF-type NOx sensor output characteristics on gasoline vehicle

CONCLUSION

1. Pumping current type NOx sensor

(1) The NOx sensor indicates a stable linearity of its signal in proportion to NOx concentration in a wide range of temperature and A/F as a practical condition on both gasoline and diesel engines.

(2) The NOx sensor shows a good response in hundred msec. and a sharp signal peak following NOx generation in a transient state as well.

(3) The NOx sensor is feasible for on-board applications such as an engine management or catalyst control/monitor systems on both gasoline including lean-burn and diesel engines.

2. EMF-type NOx sensor

(1) The EMF-type NOx sensor is possible for on-board applications, especially for low NOx measurement.

REFERENCE

(1)N.Kato, K.Nakagaki, N.Ina, "Thick film ZrO2 NOx sensor", SAE paper #960334,1996
(2)N.Miyoshi, N.Takahashi, K.Kasahara et.al., "Development of new concept three-way catalyst for automotive lean-burn engines", SAE paper #950809, 1995
(3)A.Takami, T.Takemoto, H.Iwakuni et.al., "Development of lean burn catalyst", SAE paper #950746, 1995
(4)G.Smedler, A.Walker, D. Winterborn et.al., "High performance diesel catalysts for Europe beyond 1996", SAE paper #950750, 1995
(5)M.S.Brogan, W.Boegner et.al., "Evaluation of NOx storage catalysts as an effective system for NOx removal from the exhaust gas of leanburn gasoline engines", SAE paper #952490, 1995

961130

New Type of NOx Sensors for Automobiles

Yukio Nakanouchi, Hideyuki Kurosawa, Masaharu Hasei,
Yongtie Yan, and Akira Kunimoto
Riken Corp.

Copyright 1996 Society of Automotive Engineers, Inc.

ABSTRACT

New types of potentiometric NOx sensors suitable for use on automobiles were developed by using stabilized zirconia as a base solid electrolyte. It was found that the sensor with sensing electrodes of metal oxides ($CdMn_2O_4$ or $NiCr_2O_4$) showed excellent response to NOx in the concentrations between 20 and 4000 ppm at temperatures higher than 600 ℃. The electromotive force of the sensors was almost linear to the logarithm of the NOx concentrations with positive slope for NO_2 and negative slope for NO. Especially, the sensor fitted with $CdMn_2O_4$ gave excellent responses to NO at 600 ℃, while the sensor fitted with $NiCr_2O_4$ showed high sensitive to NO_2 at 650 ℃. The sensors were insensitive to CO, CO_2, CH_4, C_3H_6, O_2 and water vapor. The sensors were fabricated as planar types with the reference electrode exposed to the sample gas, so that the sensors were simple in structure and easy to manufacture.

INTRODUCTION

Air pollution from the exhaust of automobiles has been causing a serious problem in our modern society, and the regulations on the emission standards of pollutants from automobile are becoming more severe year by year. The studies about controllability of combustion for gasoline or diesel engines and the reduction of emissions by catalytic converters have been carried out to meet the emission standards [1-5]. The OBD-Ⅱ regulation from the U.S. Environmental Protection Agency (E. P. A.) now requires whether the catalyst used to clean the exhaust is in order. The monitoring system has been introduced by using the dual oxygen sensors [6][7], although this is not a direct method. If the combustion control and the catalyst monitoring are carried out with a direct detection of NOx (NO and NO_2) concentration, they will become more accurate and effective. For example, if the fuel ignition, EGR rate and so on are controlled with feedback information on NOx concentration in the exhaust, the engine-out emissions may be reduced. In addition, the deterioration of the catalyst can be easily judged by a NOx sensor behind the converter. From these reasons, a NOx sensor suitable for automobiles is urgently needed. The sensor needs to be compact, inexpensive, superior in thermal and chemical stability, and be able to in-situ and continuously detect NOx concentration at elevated temperatures.

While the engine-out NOx from a stoichiometric engine is mostly NO, the exhaust from diesel engines can contain significant amounts of NO_2 (up to 30 % of the NOx [8]). Also, catalytic converters can oxidize some of the NO to NO_2, particularly under lean conditions. Therefore, the ideal sensor would respond equally to NO and NO_2. Alternatively, a sensor sensitive to only NO could be used with a sensor sensitive to only NO_2 in order to determine the total NOx.

So far there have been many reports on NOx sensors based on semiconductive oxides such as In_2O_3-SnO_2 [9], Cr_2O_3-Nb_2O_3 [10], WO_3 [11]. Almost all of these sensors were limited to temperatures below 500 ℃, although the sensing material themselves were superior in thermal stability. On the other hand, the solid electrolyte NOx sensors using Na^+ conductor such as in NASICON ($Na_3Zr_2Si_2PO_{12}$) or β / β'' -alumina and a nitrate or nitrite auxiliary phase have been investigated [12][13]. The electromotive force of these sensors were linear to the logarithm of NO or NO_2 concentrations. We have also tried to develop the tubular sensor using stabilized zirconia and a binary nitrate sensing auxiliary phase [14], because stabilized zirconia is quite tough chemically and mechanically and has been extensively utilized for the A/F sensors equipped in automobiles. The tubular sensor was found to show good response to NOx. Furthermore, we have introduced the planar type sensor with the reference electrode exposed to the sample gas [15]. The planar type sensor gave the same NOx sensing behavior as the tubular one, and the dependency of electromotive force on oxygen concentration was mitigated considerably. Though the sensors based on solid electrolytes exhibited the Nernst's response to NOx, these sensors could only be operated below 500 ℃, because of the melting point limit of the nitrate or nitrite. Because of this, the NOx sensors reported up to now can not be used for combustion control for which the working temperature of the sensor is often required to be higher than 500 ℃.

Because of this situation, we attempted to explore a new type of NOx sensor operative at elevated temperature by combining a stabilized zirconia and a metal oxide. As a result, the sensors using Y_2O_3-stabilized zirconia and $CdMn_2O_4$ or $NiCr_2O_4$ gave excellent sensing properties to NO at 600 ℃ or NO_2 at 650 ℃, respectively. This paper deals with the NOx sensing characteristics and mechanisms of the novel sensors.

EXPERMENTAL

A schematic view of the fabricated NOx sensing device is shown in figure 1. An 8 mol% Y_2O_3-stabilized zirconia (YSZ) plate (4×4×0.3 mm) was used as a based solid electrolyte. For a sensing electrode, a metal oxide film (1×2.2 mm) was deposited on the YSZ plate by R.F. magnetron sputtering (JOEL, JEC-SP360R), and a Pt layer was prepared on top of the oxide film by a screen printing method. A Pt reference electrode was prepared on the same plate by the screen printing method, leaving a 1.5 mm-space between the two electrodes. Then the device was annealed in air for 1 hour at 1050 ℃. Pt wires were welded to the reference electrode and the Pt layer on top of the oxide film by electric welding (Unitek, UNIBOND-Ⅱ).

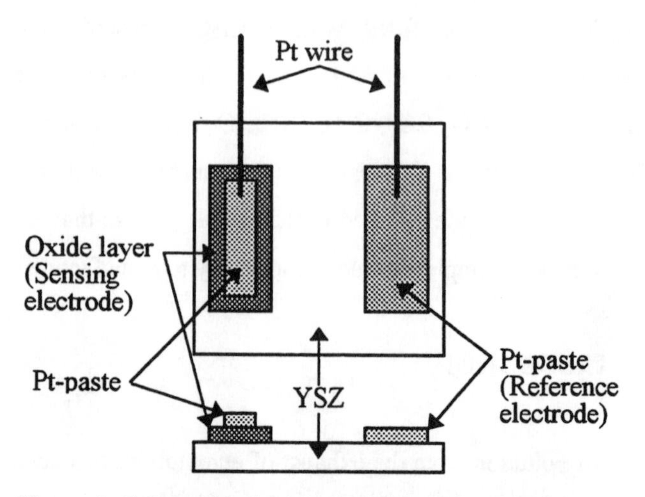

Figure 1. Schematic diagram of zirconia-based NOx sensing device using oxide electrode.

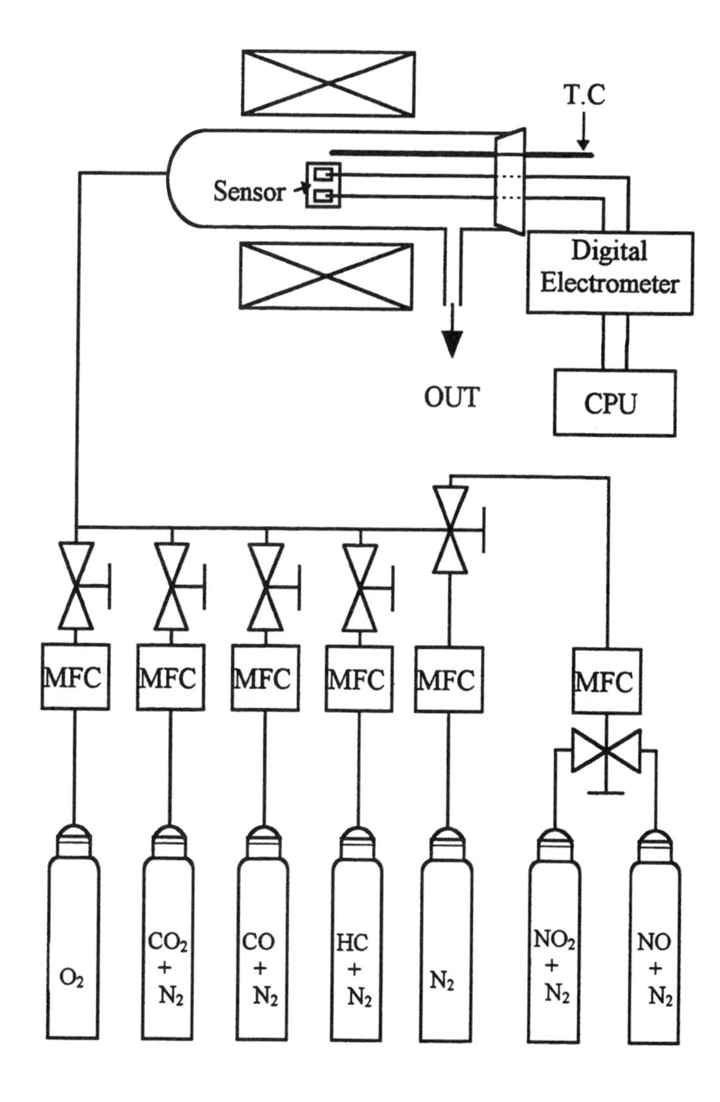

Figure 2. Measuring system for the sensing characteristics of NOx sensor.

The NOx sensing performance was examined in a gas-flow system equipped with a furnace, as shown in figure 2. The sample gases containing various concentrations of NO or NO_2 under a constant oxygen concentration of 4% were prepared by diluting parent gases with nitrogen and oxygen gases, with the total flow rate of 500 cm^3/min. The concentrations of CO, CO_2, CH_4 or C_3H_6 were varied by the same method. The electromotive force (EMF) of the device was monitored with a digital electrometer (Advantest, TR8652) while changing the concentrations of various gases at temperatures between 500 ℃ and 700 ℃.

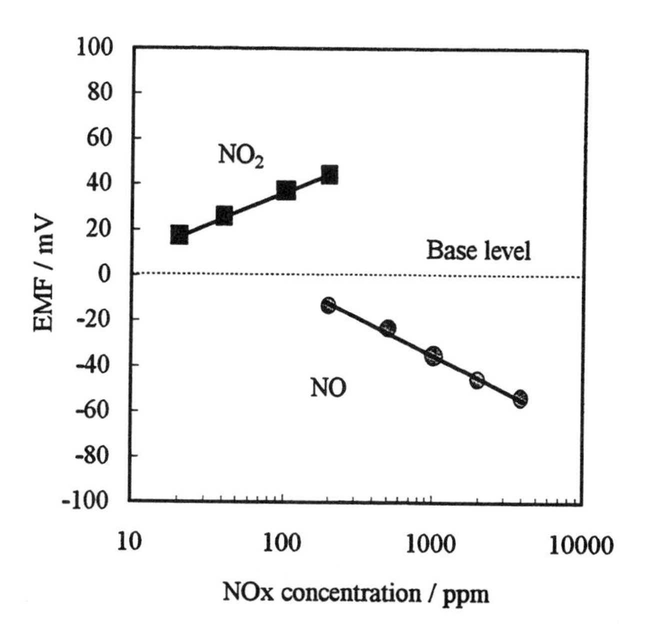

Figure 3. Dependence of EMF on NOx concentration for the sensor fitted with $CdMn_2O_4$ at 600 ℃.

RESULTS AND DISCUSSION

NO SENSOR - Figure 3 shows the EMF response of a device utilizing a $CdMn_2O_4$ layer upon exposure to several concentrations of NO_2 and NO at 600 ℃. The EMF was almost linear to the logarithm of NO or NO_2 concentrations. The EMF decreased with increasing NO concentration in the range 200~4000 ppm, with the slope of -31.8 mV/decade. The EMF increased with increasing NO_2 concentration in the range 20~400 ppm, with the slope of 27.3mV/decade. The sensor is highly sensitive to both NO and NO_2 at 600 ℃. The slope was slightly larger with NO than with NO_2. However, when compared at the same concentration, the sensitivity to NO was smaller than that to NO_2, where the sensitivity is defined as the difference between the EMF with the sample gas and the base level EMF, which usually approaches 0 V. In spite of this, the sensor will be expected to detect the NO concentration in the exhaust before the catalyst for a stoichiometric engine, where the NOx is almost entirely NO. Besides NOx, various

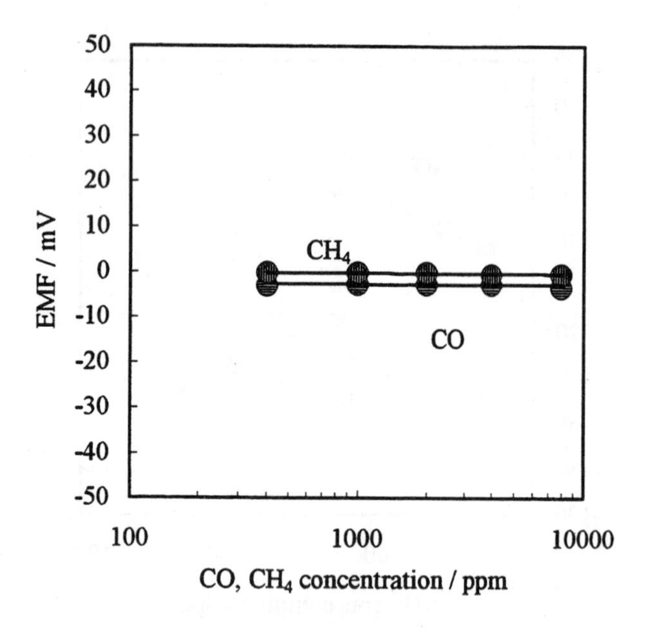

Figure 4. Dependence of EMF on concentration of CO or CH_4 for the sensor fitted with $CdMn_2O_4$ at 600 ℃.

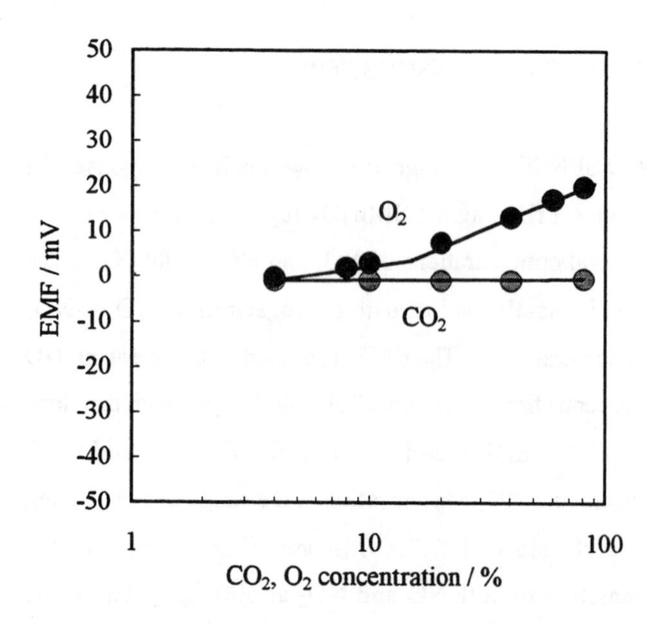

Figure 5. Dependence of EMF on concentration of CO_2 or O_2 for the sensor fitted with $CdMn_2O_4$ at 600 ℃.

gases such as CO, CO_2 and hydrocarbons exist in the exhaust from automobiles, and oxygen concentration changes with the condition of combustion, so for practical use the sensor should be resistant to interference from these gases. Figure 4 shows the EMF dependence of the sensor on the concentrations of CO or CH_4 at 600 ℃.

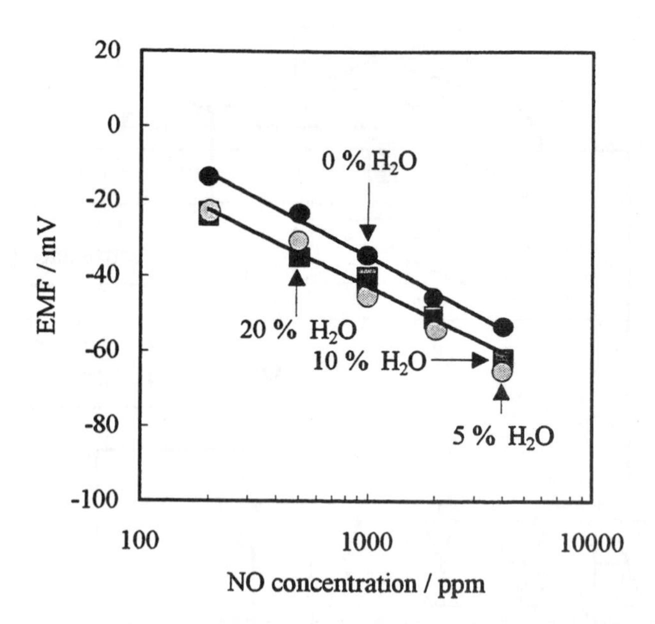

Figure 6. Influence of coexistent water vapor on NO detection for the sensor fitted with $CdMn_2O_4$ at 600 ℃.

Clearly, the EMF was not noticeably affected by introducing CO or CH_4 in the concentration range 400～8000 ppm. Figure 5 gives the shift of EMF with increasing the concentration of CO_2 or O_2. The sensor was insensitive to CO_2. In the case of oxygen, the EMF of the sensor was almost constant in the concentration range between 4 and 10 %, and increased with increasing oxygen concentration in the range 20～80%. Though the oxygen concentration varies with the condition of combustion, it is usually smaller than 10 % with stoichiometric or lean burn engines. Thus, for practical use, the influence of oxygen on the EMF of the sensor is not a serious matter.

The influence of water vapor on NO response was examined because water vapor is also present in the exhaust. Figure 6 shows the influence of coexistent water vapor on the EMF to NO for the sensor. It is found that the EMF of the sensor decreased slightly by introducing water vapor. However, the slopes for the dependence of the EMF on NO concentration remained unchanged in the water-vapor concentration of 0～20 %. So the sensor showed high resistance to water vapor, and can be used to detect NO in the exhaust because the concentration of water vapor is always lower than 20% there.

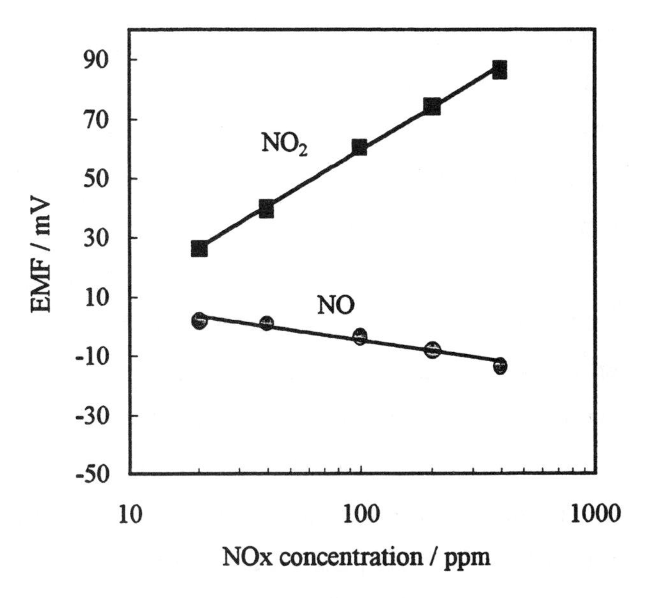

Figure 7. Dependence of EMF on NOx concentration for the sensor fitted with $NiCr_2O_4$ at 650 ℃.

NO_2 SENSOR - Figure 7 shows the dependence of the EMF for the device attached with $NiCr_2O_4$ on the NO_2 and NO concentrations at 650 ℃. The EMF values were linear to the logarithm of NO_2 and NO concentrations in the range 20～400 ppm, with slopes of 48.5 mV/decade for NO_2 and -13.9 mV/decade for NO, respectively. The slope and sensitivity to NO_2 were much larger than those for NO, though the EMF responses of the sensor to NO_2 and NO were in the opposite direction. These properties of the sensor were examined in the temperature range between 500 ℃ and 700 ℃. The largest slope occurred at 650 ℃. Figure 8 shows response transients to NO_2 for the sensor with $NiCr_2O_4$ at 650 ℃. The EMF was stable at a fixed concentration and responded quickly to a change in concentration. The times for the 90 % response and recovery was shorter than 15 s. This time included the gas-exchange time in the measuring system, which is estimated to be 8 s. The sensor with high sensitivity to NO_2 can be used to judge the deterioration of the de-NOx catalyst mounted on exhaust manifold. Because the NO_2

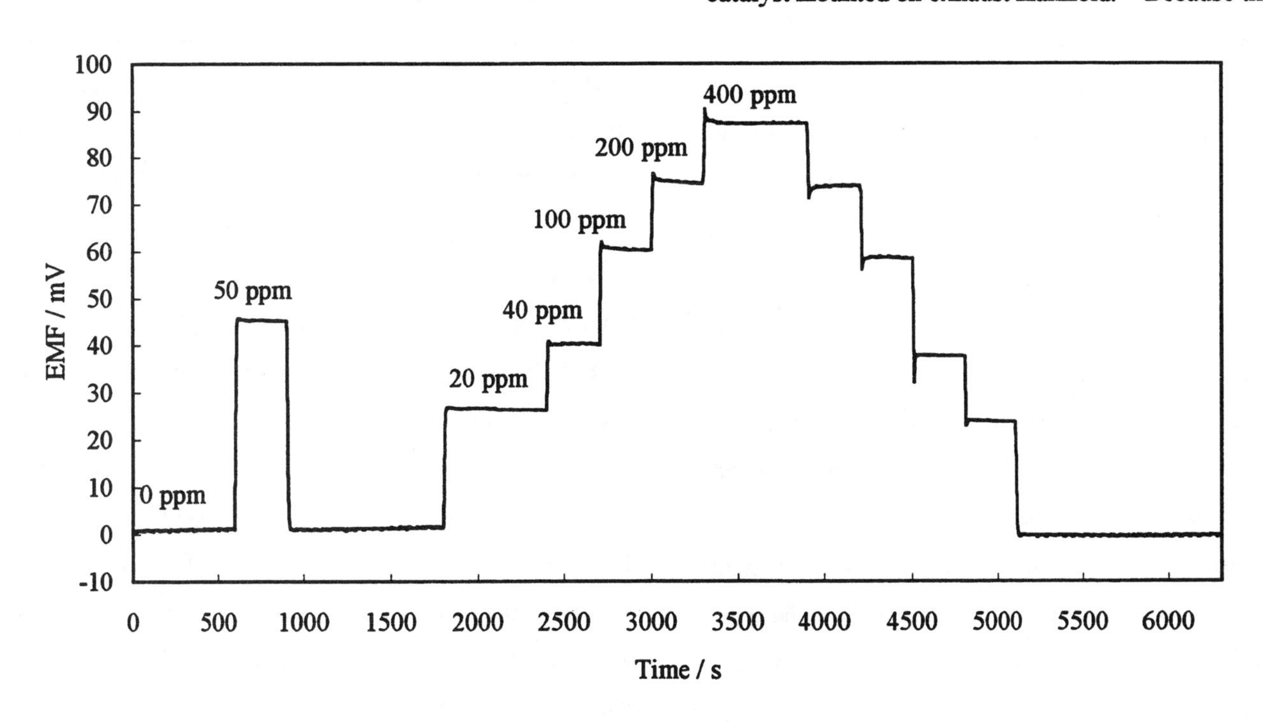

Figure 8. Response transients to NO_2 for the sensor fitted with $NiCr_2O_4$ at 650 ℃.

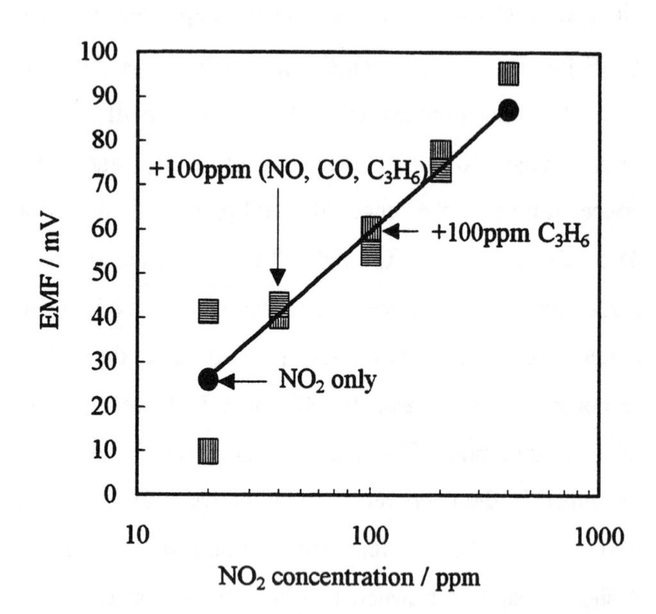

Figure 9. Influence of coexistent gases on NO_2 detection for the sensor with $NiCr_2O_4$ at 650 ℃.

will go up for a deteriorated de-NOx catalyst, where the lean environment would promote the oxidation of NO to NO_2. In addition, for a de-NOx system with hydrocarbon injection, the amount of hydrocarbon injection can be optimized by the feedback control with the sensor.

The influences of CO, CO_2, CH_4, C_3H_6 and O_2 on the EMF of the sensor were evaluated at 650 ℃, and the results revealed that the EMF was undisturbed by the presence of CO, CO_2, CH_4 up to 1000 ppm or O_2 up to 40%. The EMF of the sensor was changed by introducing C_3H_6, though the slope for the dependence of the EMF on C_3H_6 concentration was smaller than that in the case of NO. Furthermore, the influences of coexistent NO and C_3H_6 on NO_2-response were examined. Figure 9 gives the results obtained at 650 ℃. The EMF of the sensor to NO_2 was hardly affected by the coexistent 100 ppm CO, 100 ppm NO and 100 ppm C_3H_6, where the NO_2 concentration was higher than 40 ppm. Like the sensor fitted with $CdMn_2O_4$, the sensor also exhibited high resistance to water vapor.

MECHANISM OF NOx SENSING - The present sensor is constructed in the following cell structure.

Sample gas, Pt | Stabilized zirconia | Metal oxide, Pt,Sample gas
(Reference electrode)|(O^{2-} conductor)|(Sensing electrode)

Here the Pt layer on the metal oxide simply acts as a reservoir of electrons kept in equilibrium with the metal oxide. The potential of the Pt reference electrode, despite the fact that the electrode was exposed to the sample gas, is essentially determined by the O_2 concentration in the sample gas, because the Pt electrode is insensitive to other gases at elevated temperatures. On the other hand, the mechanism of potential generation on the oxide sensing electrode attached on a stabilized zirconia is somewhat complicated. The mechanism based on the mixed potential is widely accepted now, though further investigation is needed. The mixed potential has been discussed on the polarization curve of the sensing electrode in the cases of H_2 sensor and H_2S sensor [16][17]. Here a similar discussion is simply carried out without polarization curve measurements. At the sensing electrode, NOx and O_2 can undergo electrode reactions at the three-phase contact. In the gas containing NO, one can assume cathodic and anodic reactions involving oxide ions as follows.

$$O_2 + 4e^- \rightarrow 2O^{2-} \quad (1)$$

$$2NO + 2O^{2-} \rightarrow 2NO_2 + 4e^- \quad (2)$$

On the other hand, in the gas containing NO_2, one would tend to dissociate down to its equilibrium concentration just opposite to the above.

$$2O^{2-} \rightarrow O_2 + 4e^- \quad (3)$$

$$2NO_2 + 4e^- \rightarrow 2NO + 2O^{2-} \quad (4)$$

The reaction of (1) + (2) or (3) + (4), form a local cell and determines the mixed potential of sensing electrode with the cathodic and anodic reactions. The EMF of the sensor is the deference between the mixed potential of the sensing electrode and the potential of the reference electrode. Clearly, at a fixed concentration of O_2 the EMF increases with increasing NO_2 concentration, and decreases with increasing NO concentration.

SUMMARY AND CONCLUSIONS

New types of potentiometric NOx sensors were fabricated by using stabilized zirconia and metal oxide. The sensors exhibited good sensing performances to NOx at above 600 ℃. The electromotive force of the sensor was almost linear to the logarithm of NOx concentration with positive slope for NO_2 and negative slope for NO. Especially, the sensor fitted with $CdMn_2O_4$ gave excellent responses to NO at 600 ℃, while the sensor fitted with $NiCr_2O_4$ showed high sensitive to NO_2 at 650 ℃. Furthermore, the sensors are insensitive to CO, CO_2, CH_4, C_3H_6, O_2 and water vapor, and exhibited quick responses to NOx, with the time for 90 % response being shorter than 15 s.

The planar sensor with a reference electrode exposed to the sample gas is compact, and all of the constituent materials are superior in thermal stability. These features enhance the possibility of mounting the sensor directly on the exhaust manifold. Now the new sensor capable of self heating is being fabricated and tested. In the near future the sensor will be mounted on exhaust manifold, and its characteristics and durability will be examined for practical use.

REFERENCES

1. S. Shundoh, N. Komori and K. Tsujimura, "NOx Reduction from Diesel Combustion Using Pilot Injection with Pressure Fuel Injection" , SAE paper 920461.
2. T. Inoue, S. Matsushita, K. Nakanishi and H. Okano, " Toyota lean combustion system the third generation system" , SAE paper 930873.
3. N. Uchida, Y. Daisho, T. Saito and H. Sugano, "Combined effects of EGR and Supercharging on Diesel Combustion and Emissions" , ASE paper 930601.
4. M. Konno, T. Chikahisa, T. Murayama and M. Iwamoto, "Catalytic Reduction of NOx in Actual Diesel Engine Exhaust" , SAE paper 920091.
5. G. Muramatsu, A. Abe, M. Furuyama and K. Yoshida "Catalytic reduction of NOx in Diesel Exhaust" , SAE paper 930135.
6. J. W. Koupal, M. A. Sabourin and W. B. Clemmens, "Detection of Catalyst Failure On-Vehicle Using the Dual Oxygen Sensor Method" , SAE paper 910561.
7. J. S. Hepburn, D. A. Dobson, C. P. Hubbard, S. O. Guldberg, E. Thanasiu, W. L. Watkins, B. D. Burns and H. S. Gandi, "A Review of the Dual EGO Sensor Method for OBD- II Catalyst Efficiency Monitoring" , SAE paper 942057.
8. J. C. Hilliard and R. W. Wheeler, "Nitrogen Dioxide in Engine Exhaust" , SAE paper 790691.
9. G. Sberveglieri and S. Groppelli, "Radio Frequency Magnetron Sputtering Growth and Characterization of Indium-Tin Oxide(ITO) Then films for NO_2 Gas Sensors" , Sensors and Actuators, 15 p235-242, (1988).
10. T. Ishihara, K. Shiokawa, K. Eguchi and H. Arai, "Selective Detection of Nitrogen Monoxide by the Mixed Oxide of Cr_2O_3-Nb_2O_5" , Chemistry Letters, p997-1000, (1988).
11. M. Akiyama, Z. Zhang, J. Tamaki, N. Miura and N. Yamazoe, "Tungsten Oxide-based Semiconductor Sensor for Detection of Nitrogen Oxides in Combustion Exhaust" , Sensors and Actuators B, 13-14 p-619-920, (1993).
12. Y. Shimizu, Y. Okamoto, S. Yao, N. Miura and N.

Yamazoe "Solid Electrolyte NO_2 Sensors Fitted with Sodium Nitrate and/or Barium Nitrate Electrodes" , Denki Kagaku, 59, p-465-472, (1991).

13. S. Yao, Y. Shimizu, N. Miura and N. Yamozoe, "Using of Nitrate Auxiliary Electrode for Solid Electrolyte Sensor to Detect Nitrogen Oxides" , Chemistry Letters, p587-590, (1992).
14. H. Kurosawa, Y. Yan, N. Miura and N. Yamazoe, "Stabilized Zircinia-Based Potentiometric Sensor for Nitrogen Oxides" , Chemistry Letters, p1733-1736, (1994).
15. H. Kurosawa, Y. Yan, N. Miura and N.Yamazoe, "Stabilized Zirconia-Based Nox Sensor Operative at High Temperature" , Solid Ionics, 79 p338-343, (1995).
16. N.Miura and N. Yamazoe, Edited by T. Seiyama, Chemical Sensor Technology, 1 p123-139, Kodansha / Elsevier, Japan, (1988).
17. Y. Yan, N. Miura and N. Yamazoe, "Potentiometric Sensor Using Stabilized Zirconia and Tungsten Oxide for Hydrogen Sulfide" , Chemistry Letters, p1753-1756, (1994).

970459

Advanced Planar Oxygen Sensors for Future Emission Control Strategies

Harald Neumann, Gerhard Hötzel, and Gert Lindemann
Robert Bosch GmbH

ABSTRACT

This paper presents advanced planar ZrO_2 oxygen sensors being developed at Robert Bosch using a modified tetragonal partially stabilized zirconia (TZP) with high ionic conductivity, high phase stability and high thermo-mechanical strength.

Green tape technology combined with highly automated thickfilm techniques allows robust and cost effective manufacturing of those novel sensing elements. Standardization of assembling parts reduces the complexity of the assembly line even in the case of different sensing principles.

The sensor family meets the new requirements of modern ULEV strategies like fast light off below 10 s and linear control capability as well as high quality assurance standards.

High volume production will start in 1997 for European customers.

INTRODUCTION

After 20 years experience with closed loop controlled three way catalyst using a conventional thimble type ZrO_2 based oxygen sensor the world wide applied strategy of a two-step controller seems to be at its limit [1].

The main emissions appear today during warm up phase and at the points of maximum deviation from $\lambda = 1$ due to the two step controller behaviour (especially with aged catalysts).

Therefore new emission control strategies have been developed during the past years to meet the new LEV/ULEV requirements and to have further potential for future regulations. A linear λ-control strategy was presented by Honda [2].

The key point of this philosophy is to keep the conversion rate of aged catalysts high by reducing the two-step controller caused deviation from the ideal $\lambda = 1$ point. This requires a linear wide band sensor with high accuracy at $\lambda = 1$, new stage as well as aged. The functional principle of such sensors is already described [3].

A second strategy is proposed by various car manufacturers which keeps the wellknown two-step controller but starts the closed-loop mode during warm-up phase and is preferentially lean driven to reduce hydrocarbons. The challenge of this strategy is the fast light off capability of the upstream sensor which needs to be 2 to 4 times better than the capability of actual thimble type oxygen sensors.

Both strategies are supported by moving catalyst and sensor towards the engine outlet for further reduction of cut in times (close coupled catalyst). These measures increase temperature requirements at sensor tip to more than 1000 °C.

Furthermore customers want to reduce the total electrical power, size and weight of components (Example: V6-engine with 2 upstream and 2 downstream sensors
I. 4 high performance thimble type sensors 72 W
II. 4 high performance planar type sensors 28 W).

For all these reasons and considering the coming requirements of exhaust gas sensors BOSCH decided to apply planar technology (Fig. 1) [4] [5] [6] for the new sensor generation first because of better performance concerning the above mentioned requirements but secondly because of their potential for cost reduction and to be also a promising platform for the next generation of exhaust gas sensors for NO_x, HC, etc.

TECHNICAL SECTION

The most important supposition for a durable planar sensor generation is the development of high quality ZrO_2 substrates. Selection and processing of the ceramic ZrO_2 powder must follow sometimes contradictory aspects like

- low sintering temperature for good cofired electrode performance
- high bending strength and thermal shock resistance
- long term phase stability
- high ionic conductivity
- ability for cofiring to a monolithic structure together with alumina and platinum
- good performance in tape casting.

PREPARATION AND CHARACTERIZATION OF ZrO_2 ELECTROLYTES FOR PLANAR OXYGEN SENSORS

Tape casting - Since the Doctor-Blade-Process introduced by Howatt [7] in 1952 has proven itself suitable for the preparation of thin substrates [8] many efforts have been made for further improvements. For oxygen sensors an ultrafine grained prereacted tetragonal Zirconia (UPZ) with the addition of 8 weight % Y_2O_3 is used as electrolyte material because of its long term stability and improved thermo-mechanical stress ability.

We consider the pre-treatment of the starting powders as the most important step due to the ultra low grain size of the prereacted ZrO_2 particles. The ceramic raw materials are processed together with the organic components such as solvents, binders, and plasticizers to form a slurry. In order to prevent pore formation, the slurry is deaerated and then metered to a casting head.

The slurry is drawn out to form a tape between the two adjustable blades and a polyester film which is supported by a steel belt conveyor. During the process, the gap between the blades and the polyester film is monitored by an automatic film-thickness measuring and control unit. The steel belt conveys the cast tape into a tunnel-shaped drying section where the solvents are evaporated off by an adjustable, uniform flow of air. The drying temperature is regulated by controllable infrared radiators. The dry tape is removed at the end of the drying tunnel and roughly cut into single pieces. Electrolyte greensheet substrates are punched out of the dried flexible tape. The punching residue is redissolved and mixed in with the fresh slurry mixture. Production costs can be reduced considerably by recycling of up to 100 % of residue.

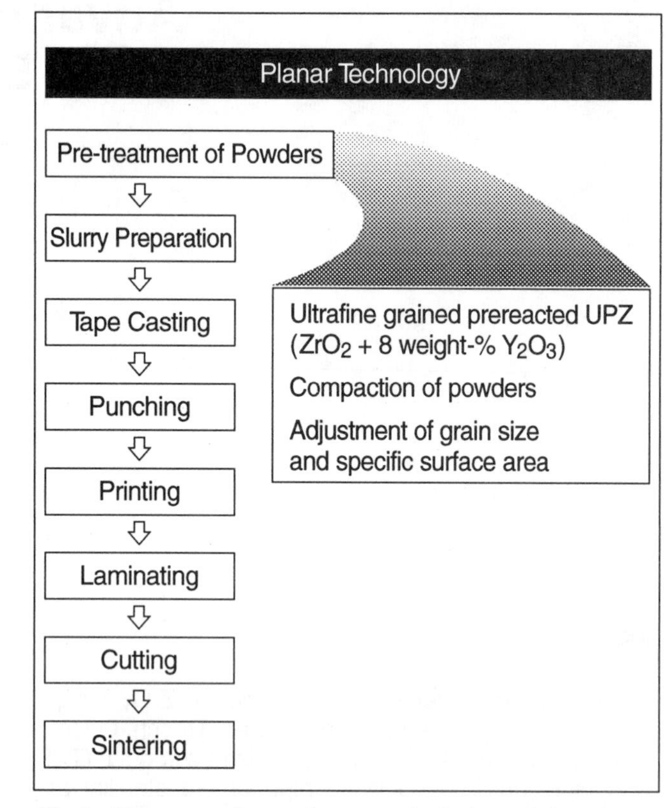

Fig.1: Process scheme for manufacturing of planar oxygen sensors.

Electrolyte characteristics - Green sheets are prepared using different starting materials and powder treatments - 30 mm disks are punched out and are sintered in an electrically heated laboratory furnace in air, ultrafine grained prereacted ZrO_2 (UPZ) and ultrafine grained mixed oxide ZrO_2 (UMZ) at temperatures below 1400 °C and fine grained mixed oxide ZrO_2 (FMZ) below 1450 °C. The samples show full density with very few pores see Fig. 2a, b.
The microstructure of UPZ shows better homogeneity compared to UMZ material, see Fig. 2a, b.

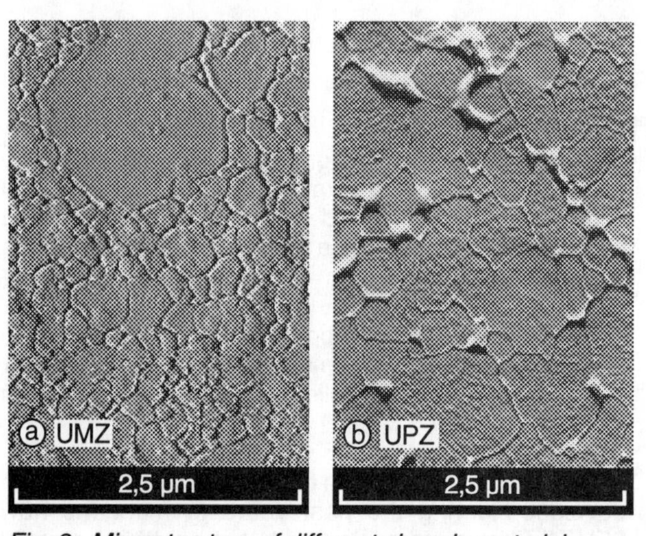

Fig. 2: Microstructure of different zirconia materials

The sintered test samples are used for the strength measurements before and after ageing in an autoclave under 190 °C and the corresponding water steam pressure.

At present there is no standardized test procedure for determining the strength of thin ceramic substrates. DIN 52 292 [9] defines a method for determining the flexural strength of disk-shaped glass and ceramic samples with thickness > 1 mm, see Fig. 3.

The calculation is valid provided that the flexure at the sample center is sufficiently small. The maximum local stress underneath the loading ring must be kept below 2 % of the center stress. In the case of substrate thickness < 1 mm, this value is soon exceeded using the radius relationships specified in DIN 52 292.

A test method for measuring the strength of ZrO_2 substrates with a thickness < 0.5 mm was developed based on the calculations made by Kao et al. [10]. Loading and support ring radii (r_1/r_2) are dimensioned such that ZrO_2 tapes with a thickness < 0.5 mm are subjected to a stress of maximum 1.02 x σb under the edge of the loading ring [11] when they reach their fracture stress σ_b at the sample center.

Fig. 4 shows the permissible fracture stress of ZrO_2 ceramics for various loading ring radii as a function of sample thickness. Because of the high strength values for a sample thickness of 0.4 mm only a very small diameter of loading ring of 1,88 mm is allowed.

Monoclinic phase content of the samples is detected by X-ray defraction analysis (XDA).

The results of bending tests and XDA as a function of the exposure time to 190°C and the corresponding water steam pressure are shown in Fig. 5.

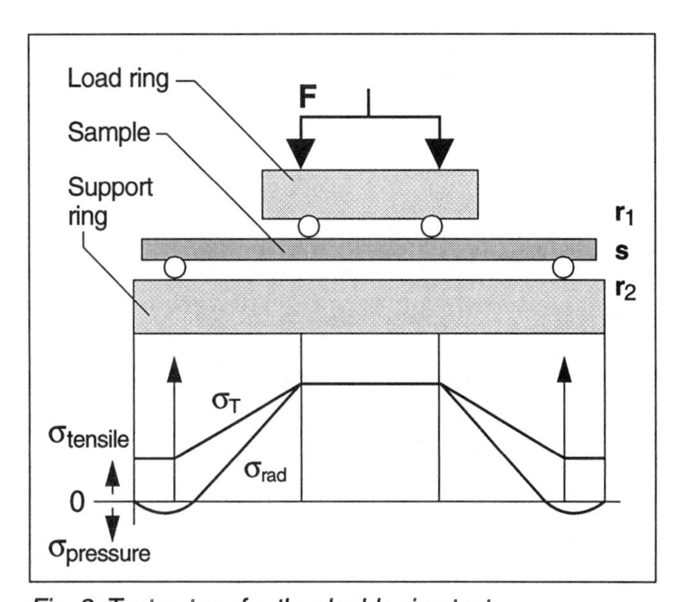

Fig. 3: Test set-up for the double-ring test.

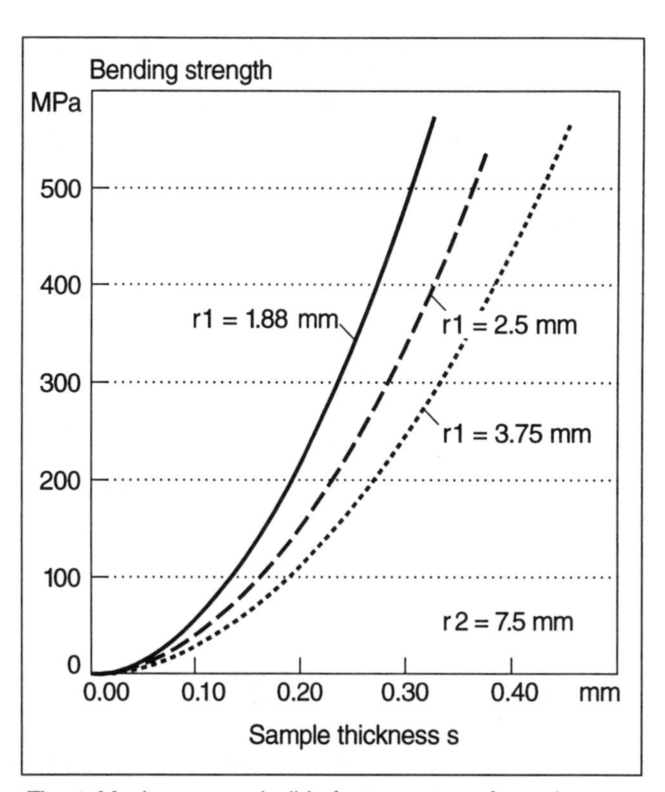

Fig. 4: Maximum permissible fracture stress for various loading ring radii r_1 as a function of sample thickness

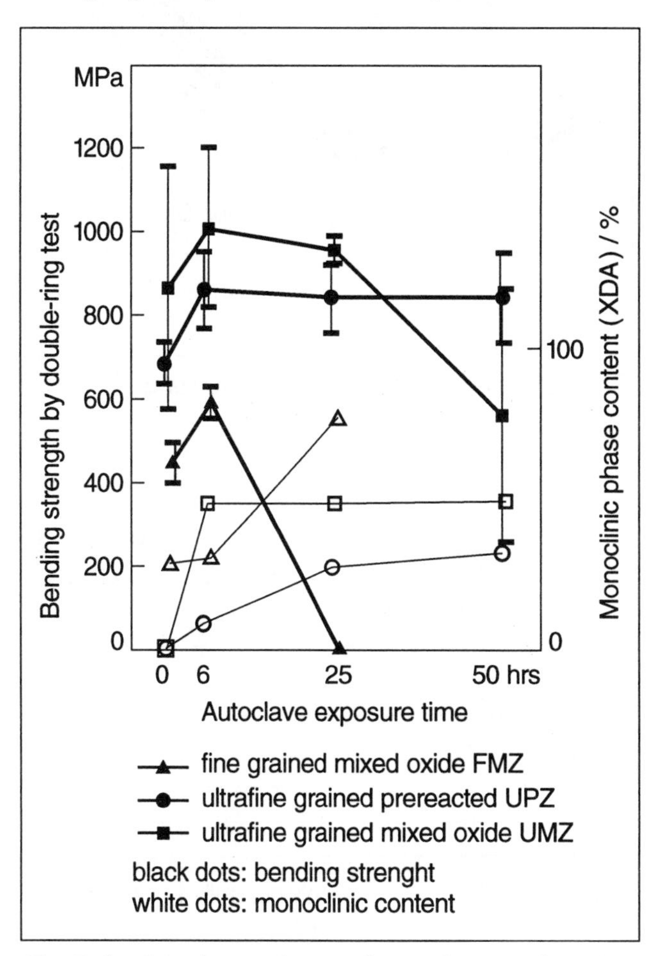

Fig. 5: Ageing of ceramic samples under 190 °C and corresponding water steam pressure

The use of the ultrafine grained prereacted zirconia (UPZ) does not reach the extremely high starting level of the fine grained mixed oxide material but shows lower scattering and better stability during ageing up to 50 hours. Phase analysis shows a strong increase in monoclinic content of the UMZ during the first 6 hours. The FMZ samples are completely desintegrated after 25 hours ageing.

SENSOR STRUCTURE

The ceramic layer structure of all planar sensing elements is achieved using consequently thickfilm technology. The robust (highly automated) screen printing process is applied for all functional layers.

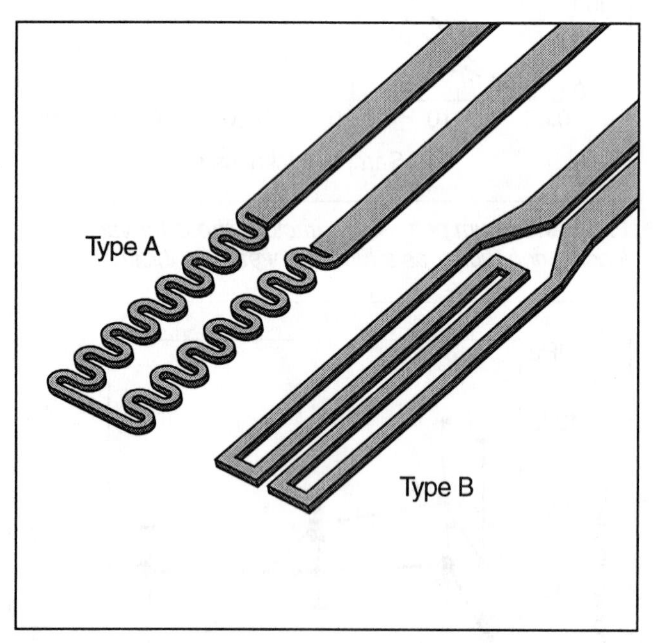

Fig. 6: Scheme of different heater structures

	T_{tip}, °C at 6 W ambient air	P, W heater at 13 V	R_{heater}, Ω at 20 °C
Type A	515	4.5	17
Type B1	512	5.5	13
Type B2	605	6.5	9
Type B3	500	only controlled mode	3

Table 1: Characteristics of different heater designs

The layouts are designed considering tolerances that the printing process is capable even for large green sheets (multiple printed panel).

Another aspect is to realize a modular sensor structure with the option to combine different green sheets (heater, air duct, etc.) to individual sensing elements. This is achieved by standardization of layouts, dimensions, vias and air ducts.

Together with subsequently similar assembling processes this results in a flexible high volume production with low set up times.

A further milestone is the development of a stable heater structure, being the basis for every fast light off sensor as well as for a heater integrated wide band sensor with its high operation temperature.

Former designs which often use a heater structure similar to Fig 6 (Type A) are limited to a heater power of some 5 W (heater not temperature controlled). Higher power causes ceramic cracks during engine runs. FEA calculations show that a longitudinally oriented heater design (Type B, Fig. 6) with individual lead width offers higher durability potential and allows about 50 % higher power without increasing the maximum tensile stress along the edges of the sensing element.

The tip temperature of Type B design at the same power can be further increased by reducing the thermal loss along the leads (high thermal coefficient, low resistance of lead material), which is realized with the fast light off heater (Type B2).

The $\lambda = 1$ fast light off sensor (LSF 4.7) is completed by laminating, cutting and sintering of three green sheets (heater sheet, reference air duct and sensing cell) as indicated in Fig 7.

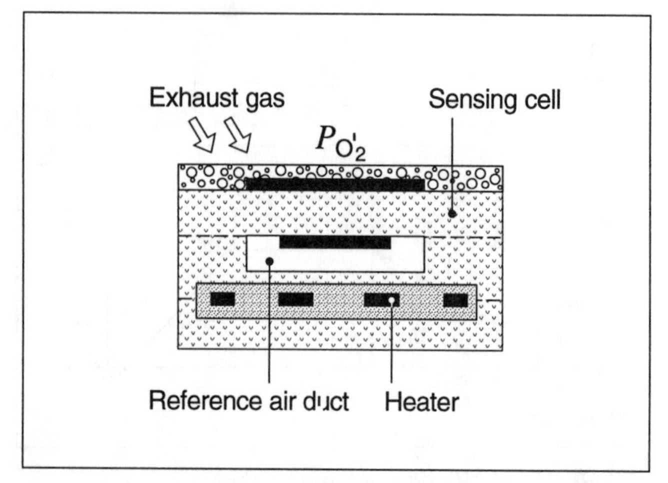

Fig. 7: Scheme of $\lambda = 1$ LSF 4.7 sensing element structure

The new wide band sensor design (LSU4.7, Fig. 8b) shows some additional improvements compared to the former UEGO design (Fig. 8a).

Major changes are the better heater coupling to the sensing cell because of the blind hole gas entrance instead of a thru hole, the porous diffusion barrier which allowes together with active cermet electrodes to leave out the insulation layer (luggin capillary) indicated in Fig. 8a. This results in a more robust ceramic structure and due to the perfect $\lambda = 1$ function of the sensing cell (good protection by the diffusion barrier, high temperature) high accuracy and long term stability for $\lambda = 1$ linear control applications.

For wide band sensors where a constant cell temperature has to be applied we use Type B heater structure but with low resistance (Type B3), see Table 1. Due to its high power this heater needs necessarily a closed loop control which also allows a smart heater cut in strategy. Good PID control behaviour is achieved by exposing the circular electrodes precisely to the hot area of the heater.

At least a special activation treatment of the cermet electrodes and a pigmentation process of the protective layer is applied to establish the full sensor function i. e. high catalytic activity and less electrode polarization at high current densities [12] [13].

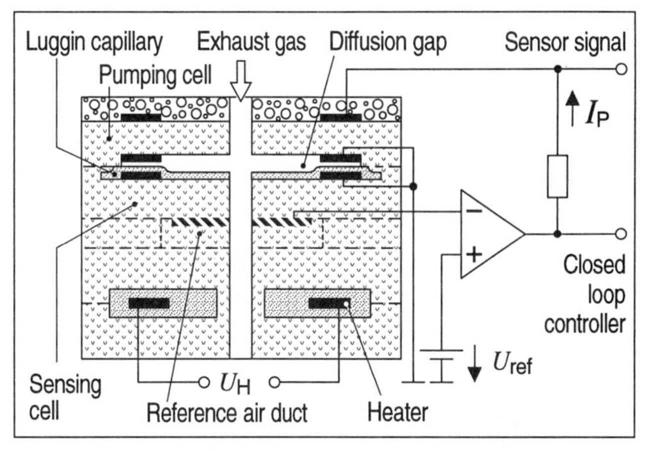

Fig. 8a: Old design of wide band sensor UEGO

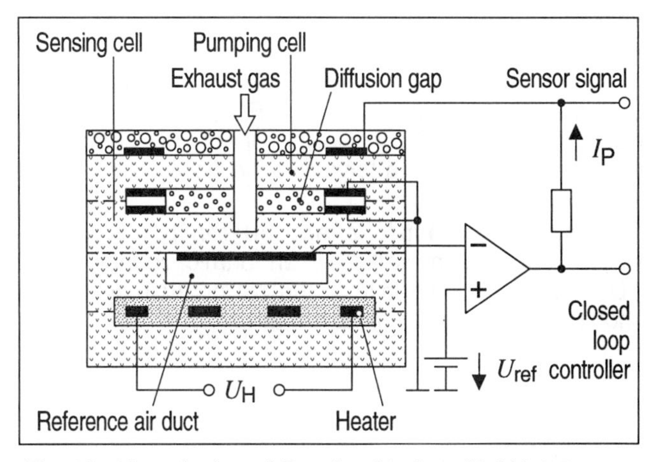

Fig. 8b: New design of Bosch wide band LSU 4.7

FUNCTIONAL ASPECTS

Fast light off behaviour of $\lambda = 1$ LSF sensor. Fig. 9 shows the influence of above described heater designs and activation/pigmentation treatment on the light off behaviour.

The experiments are carried out in a laboratory test bench with synthetic exhaust gas (PSG) at $T_{gas} = 20°C$.

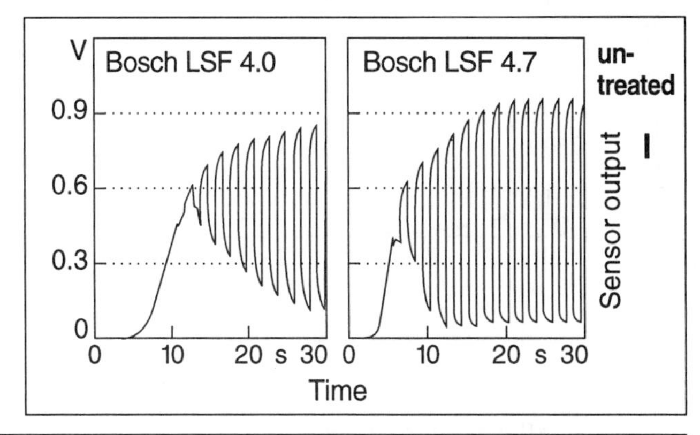

Fig. 9: Fast light off behaviour of $\lambda = 1$ LSF sensors

Test conditions:
$T_{gas} = 20°C$
$V_{gas} \approx 8 m/s$
$\lambda = 1.00 \pm 0.03$
f = 0.5 Hz

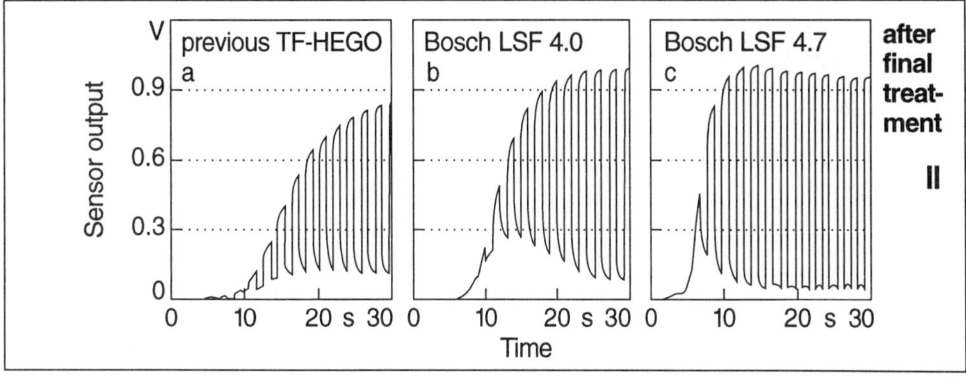

The sensor is exposed to a cold gas stream of about 8 m/s. The gas is switched between λ = 0.97 and 1.03 at a frequency of 0.5 Hz. At time t_0 = 0 the heater is powered with U_{heater} = 13 V. The light off time t_{lo} is defined by achieving a switching amplitude of U_s > 600 mV (rich phase) and U_s < 300 mV (lean phase).

A key parameter for the fast light off time t_{lo} is the internal sensor resistance R_s which is significantly lower for LSF 4.7 (Type B2 design), see Table 2.

Electrode polarization - If a current is forced to flow through the electrochemical cell an overpotential ΔU_{pol} caused by restricted transport of oxygen through the electrode/electrolyte interface influences the Nernst voltage.

In Fig. 10 this is shown for untreated planar lambda sensors of LSF 4.0 (P_{heater} = 6 W) and LSF 4.7 (P_{heater} = 4.8 W) in the upper row compared to the same sensors after activation (lower row).

A current of 1 mA is applied for 10 ms to the sensor causing immediately an ohmic voltage drop $\Delta U = R_s \cdot I$ additionally to the Nernst voltage which should be about zero (ambient air).

Heater/ Sensor type	t_{lo}, s at 20 °C	R_{sAC}, kΩ at U_{heater} = 10 V	R_{sAC}, kΩ at P_{heater} = 4 W
Untreated			
Type A	–	–	–
Type B1 (previous design LSF 4.0)	19.5 (fig. 9b I)	3	3.75
Type B2 (new fast light off LSF 4.7)	8 (fig. 9c I)	0.5	0.6
After final treatment			
Type A	18	9	2.4
Type B1 (previous design LSF 4.0)	13.5 (fig. 9b II)	2	2.5
Type B2 (new fast light off LSF 4.7)	6.5 (fig 9c II)	0.23	0.475

Table 2: Typical light off time (t_{lo}) and internal cell resistance R_s of the different heater designs

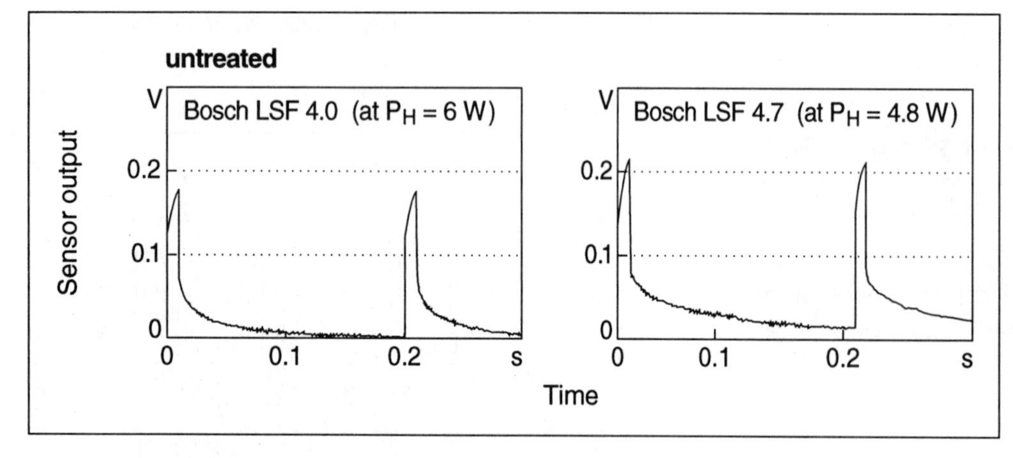

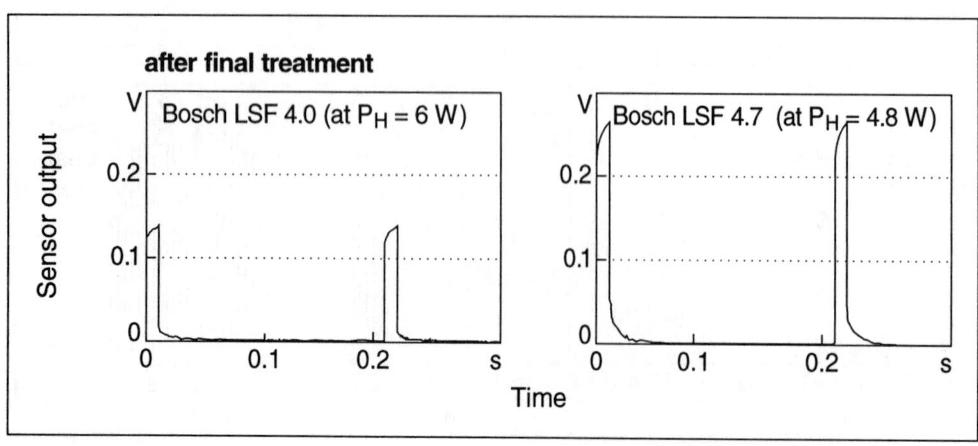

Fig. 10: Electrode polarization of planar λ = 1 sensors

Test conditions:
ambient air
I_p = 1 mA
for 10 ms
each 200 ms

The sensor output is additionally superimposed by the polarization ΔU_{pol} of the electrode, increasing the sensor output when the current is applied and decreasing (depolarization) when the current is switched off.

Untreated planar lambda sensors show higher polarized electrodes and longer depolarization times than activated sensors. This effect can be a problem for sensors downstream catalyst (cold applications) or for the sensing cell of the wide band lambda sensor.

Electrodes with high current flux capability are much more important for amperometric wide band lambda sensors (LSU 4.7, Fig. 8b). The p_{O2}-depending pumping current I_p is up to 1000 times higher than the measuring current of a galvanic $\lambda = 1$ sensor.

In lean atmosphere the inner pumping electrode pumps oxygen out of the diffusion gap till the opposite inner Nernst electrode (NE) is measuring a voltage of 450 mV ($p_{O2} \sim 10^{-9}$ atm resp.). Due to the p_{O2}-gradient along the height of the gap a significantly lower p_{O2}-value is necessary at the pumping electrode resulting in a high pumping voltage V_p.

To avoid blackening of the ceramic (reduction of ZrO_2) in this case both inner electrodes (NE and PE) have to be short circuited, see Fig. 8a, b. Therefore a part of the pumping current will flow from the inner Nernst electrode to the outer pumping electrode reducing p_{O2} at the inner Nernst electrode simultaneously, see Fig. 11.

However, to avoid polarization at the Nernst electrode of the previous design with less current flux capability, an additional insulating Al_2O_3-layer (Luggin capillary) is necessary to keep the part of pumping current flowing through the Nernst electrode at 2-5% of the total current, see Fig. 8a.

The LSU 4.7 design has as described above

- a porous filled diffusion gap
  (→ lower I_p-values)
- no additional insulation layer
  (→ more robust design)
- gas entry hole only in the upper ZrO_2-sheet)
  (→ heater layout focused to the electrodes)
  (→ Nernst-reference-electrode in hot spot region)
- highly activated cermet electrodes

Due to the slightly increased ceramic temperature (improved heater design) and a better alignment of hot area and electrodes the internal resistance of the sensing cell is significantly lower in the new design, see Table 3.

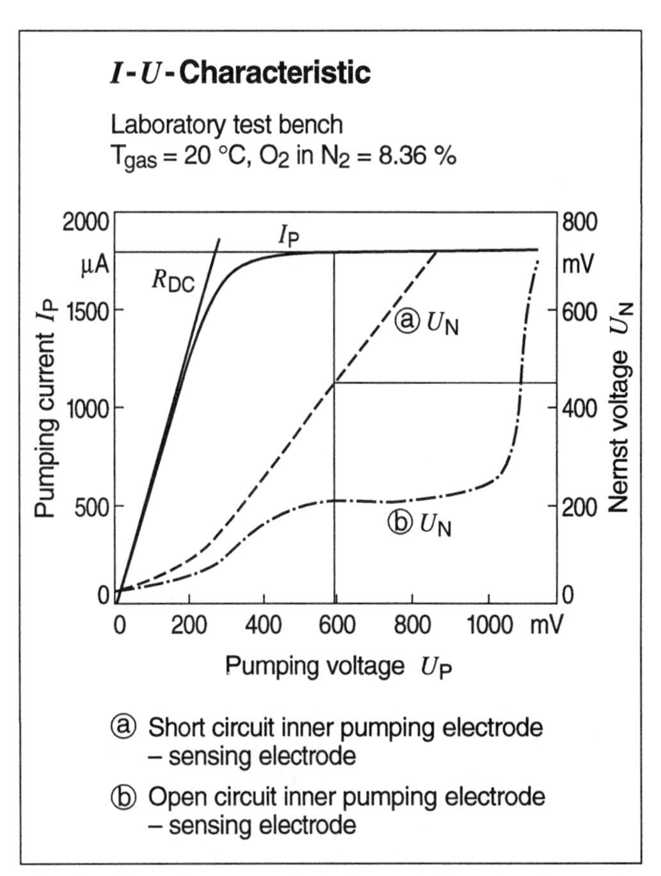

Fig. 11: Pumping characteristics and Nernst voltage of LSU 4.7, see Fig. 8b

	Previous UEGO	LSU 4.7
$I_{pumping}$ (total)	~ 10 mA	~ 4 mA
I_p (NE part)	< 0.5 mA (< 5%)	~ 1 mA (~ 25%)
V_p (typical)	< 4 V	< 1 V
R_i (pumping cell)	~ 200 Ω (AC)	~ 100 Ω (AC)
R_i (sensing cell)	~ 2 kΩ (AC)	~ 100 Ω (AC)
T_{cer}	~ 750 °C	~ 750 °C
P_{heater}	10 - 12 W	10 - 12 W
t_{lo}	< 30 sec	< 15 sec

Table 3: Typical characteristics of wide band sensor designs regarding Fig. 8a, b measured in ambient air

Together with the high current flux capable electrodes the sensing cell shows no polarization although the current is 4 times higher than in the previous design.

The more robust design (porous ZrO_2 barrier - elimination of the Al_2O_3 insulating layer within the ZrO_2-substrate) leads to negligible deterioration of the diffusion gap and the sensor signal over application time.

A further application of high current flux capable electrodes will be a pumped O_2-reference which can be used instead of an air reference. This is already done in a wide band λ-sensor, where a very small continuous current of about 20 μA is superimposed to the Nernst voltage. Because of the high operation temperature the current does not strongly influence the signal. If this principle is used in $\lambda = 1$-sensors at low temperatures (e. g. downstream catalyst) the signal would be significantly disturbed in case of electrodes with low activity.

An interesting approach could also be to install an intelligent pumping mode with nonpolarizing electrodes i. e. a pumping current of 1 mA for 10 ms followed by a 490 ms time period where the sensor is used like a conventional air reference sensor. After 500 ms the reference needs to be recharged again.

To develop gas sensors (NO_x, HC...) e.g. for catalyst monitoring the above described electrode and heater features will become a key.

These sensor types will need independently of the used measuring principle an improved heater with high power because of the fast light off requirements and exact temperature control. To eliminate O_2 cross sensitivity a high current flux capable electrode will be of further use for these sensors.

	New stage λ_{dyn}	2000 hrs hot/rich $\Delta\lambda_{dyn}$	150 hrs oil $\Delta\lambda_{dyn}$
Bosch fast light off LSF 4.7	1.009	< 0.005	< 0.006
Bosch wide band LSU 4.7	1.008	< 0.002	< 0.001

Table 4: λ_{dyn} characteristics of Bosch planar oxygen sensors after engine tests, measured in a 350°C laboratory test bench

Endurance Engine Tests - To prove the durability of the new sensor designs following special engine tests are performed

a) hot rich endurance run with 930°C tip temperature for 2000 hrs

b) an oil engine run with 0.5 weight-% in the fuel (heavy oil ash deposits) for 150 hrs, see Table 4.

The engine tests confirm the good quality of both designs especially the extremly low $\lambda = 1$ drift of the wide band sensor indicates that even heavy deposits around and in the gas entrance do not deteriorate the sensor signal.

CONCLUSION

This paper describes the advanced BOSCH planar sensor designs and demonstrates their high performance ready for the new ULEV-strategies. Their robustness has been proven in various design validation and endurance tests concerning functional, ceramic or mechanical aspects.

High volume production will start in '97 for MY '98. A significant increase is expected over the following years.

BOSCH believes that the planar technology is the promising platform for future NO_x and HC sensors. A need for such gas sensors is already announced. Implementation for catalyst monitoring and Diesel engines will take place within the next 5 years.

ACKNOWLEDGMENTS

The authors want to thank all members of the planar sensor team and the ceramic group for technical support. Special thanks to Elisabeth Stark, Ulrich Eisele, and Hans-Martin Wiedenmann for many helpful discussions.

REFERENCES

[1] H. -M. Wiedenmann, G. Hötzel, H. Neumann, J. Riegel, and H. Weyl, Exhaust Gas Sensors, in Automotive Electronics Handbook, Edited by Ronald Jurgen, Mc Graw Hill. Inc.
ISBN 0-07-033189-8 (1995)

[2] K. Antonius, A. Garner, S. Garrett: Honda first to have gasoline engine verified at ULEV exhaust levels, Honda News Press Release (Jan 1995)

[3] S. Soejima and S. Mase: Multi-Layered Zirconia Oxygen Sensor for Lean Burn Engine Application, SAE 850378 (1985)

[4] N. Higuchi, S. Mase, A. Lino, and N. Kato: Heated Zirconia Exhaust Gas Sensor Having a Sheet-Shaped Sensing Element, SAE 850382 (1985)

[5] T. Ogasawara and H. Kurachi: Multi Layered Zirconia Oxygen Sensor with Modified Rhodium Catalyst Electrode, SAE 880557 (1988)

[6] T. Yamada, N. Hayakawa, Y. Kami, and T. Kawai: Universal Air-Fuel Ratio Heated Exhaust Gas Oxygen Sensor and Further Applications, SAE 920234 (1992)

[7] G. N. Howatt: Method of producing high dielectric, high insulation ceramic plates, U. S. Patent 2,582,993 (1952).

[8] R. E. Mistler, D. J. Shanefield, and R. B. Runk: Foil casting of ceramics, in Ceramic processing before firing. Edited by G. Y. Onada and L. L. Hench (John Wiley and Sons, Inc., New York, Chichester, Brisbane, Toronto) 411-448 (1978).

[9] DIN 52 292 (1984): Testing of glass and glass ceramic material. Determining the flexural strength, double-ring test on disk-shaped samples with small test area.

[10] R. Kao, N. Perrone, and W. Capps: Large-deflection solution of coaxial-ring-circular-glass-plate flexure problem, J. Am. Ceram. Soc. 54 [11], 566-571 (1971).

[11] K. - H. Heußner, U. Eisele, and G. Lindemann: Flexural strength of ceramic substrates, Robert Bosch GmbH, In-house investigation. Not published (1991).

[12] German Patent DE 41 00 106 C1

[13] T. M. Gür, R.A. Huggins; Importance of electrode/ zirconia interface morphology in high temperature solid electrolyte cells; J. Appl. Electrochem. 17 (1987) 800-806.

960692

Chemical Sensors for CO/NO_x-Detection in Automotive Climate Control Systems

Kurt Ingrisch, Astrid Zeppenfeld, Michael Bauer, Botho Ziegenbein, Heiner Holland, and Bernd Schumann
Robert Bosch GmbH

Copyright 1996 Society of Automotive Engineers, Inc.

ABSTRACT

A new air quality sensor for climate control in automobiles has been developed. The sensor is designed for use in air conditioning systems with an air intake flap and a charcoal filter. The main indicators for the detection of air pollution are CO and NO_x.
The sensor elements consist of a SnO_2 layer made in thick film technology resulting in small sensor size and high sensitivity at reasonable cost. The sensor elements are packaged together with an evaluation circuit in a water resistant sensor housing for underhood installation. By means of specific dopants the sensor elements have been optimized in order to detect CO and NO_x with very low cross-sensitivity.
This paper describes the design and production of the sensor. The performance of the complete unit under typical field conditions is presented. The sensor is in serial production.

INTRODUCTION

The exhaust gases of diesel and gasoline engines are a mixture of many substances. The main component emitted by gasoline engines is CO and the main component emitted by diesel engines is

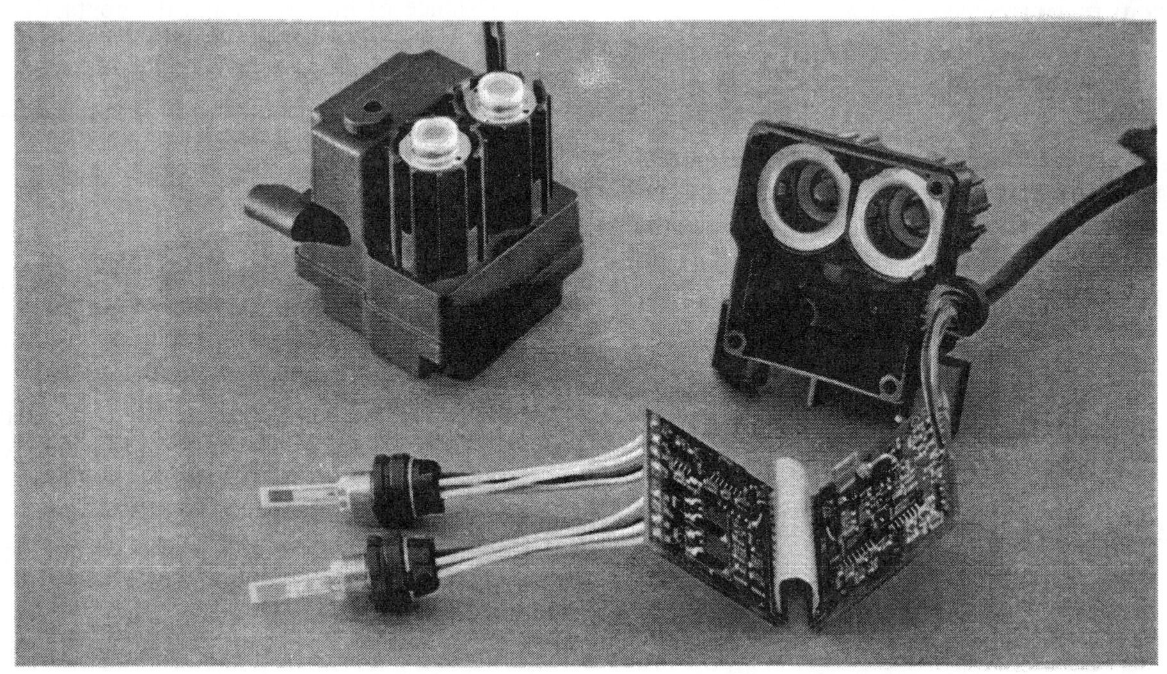

Figure 1:Photograph of the complete air quality sensor

NO_x.

When driving directly behind another car, there will be concentrations of CO ranging from 30 to 100 ppm whereas the concentrations of NO_x will be in a much lower range of 2 to 10 ppm.

Other components of the exhaust gases, some of which smell bad, are complex hydrocarbons in very low concentrations, which are difficult to detect. Therefore the ambient CO- and NO_x-concentrations are used as indicators for the degree of air pollution.

With a CO/NO_x-sensor it is therefore possible to control the air intake flap in air conditioning systems automatically in order to protect the passengers from short term air pollution peaks. In addition, the expensive charcoal filter, part of more sophisticated air conditioning systems, will have a much longer life.

This application requires a sensor for CO and NO_x respectively with high reliability and durability.

These advantages are offered by SnO_2-semiconducting sensors at reasonable costs [1], [2].

In the first part of the paper, we will describe the basic components, the sensor elements, and in the second part the whole device, the air quality sensor.

SENSOR ELEMENTS

Basic Design

The sensing principle is based on the conductivity change of the sensing layer when exposed to the target gases.

Figure 2 gives a schematic view of the sensor elements. All structures are screen printed on an alumina substrate. They consist of the heater at the back and the interdigitated electrodes and the sensitive layers on the front.

Electrodes and the heater are made of pure platinum, which is sintered at a temperature of about 1800 K. The sensing layer consists of tin oxide, specially doped for high sensitivity and selectivity to CO and NO_x respectively.

Operating Principle

The sensitive sensor layers consist of n-type doped semiconducting tin oxide. The structure of the layer is basically a porous material consisting of 20 to 60 nm diameter crystals (Fig.3). They grow together on their surface during the sintering process. The growth process is designed to form necks between the grains which are about one third of the total grain diameter. Conducting electrons move through the grains via the contact areas. The resistance of the layer is composed by the resistance of the grains and the contact areas [3], [4].

Due to the large surface of up to 60 m^2 per gram, a huge amount of gas can adsorb on the sensor surface. The size of the grains and necks is in the order of the mean free path of the electrons inside the oxide (Debye-length).

Chemical reactive gases have electron donor or acceptor properties. Therefore, the charge density in the whole material is drastically influenced by adsorbed gases, and the resistance of the material changes.

In clean air the sensor is mainly covered with oxygen. Since oxygen is an acceptor, electrons are drawn from the interior leading to a resistance higher than under vacuum conditions.

Reducing gases like CO react with the adsorbed oxygen thus lowering the surface coverage. Fewer electrons are trapped and the resistance decreases. On the other hand, NO_2 is a stronger acceptor than oxygen. Adsorption therefore leads to increased resistance [5].

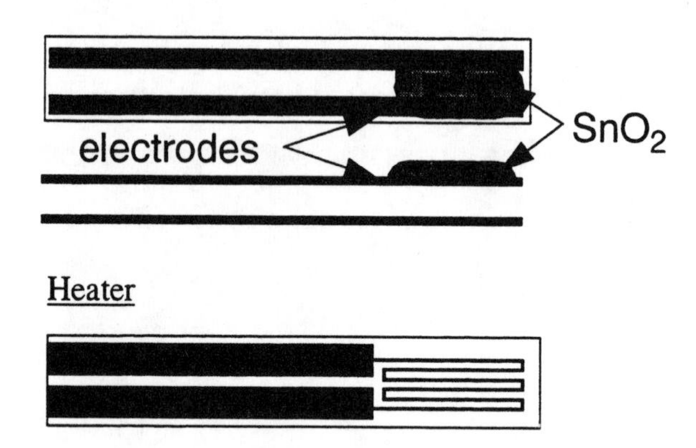

Figure 2: Drawing of the CO and of the NO_x sensor element

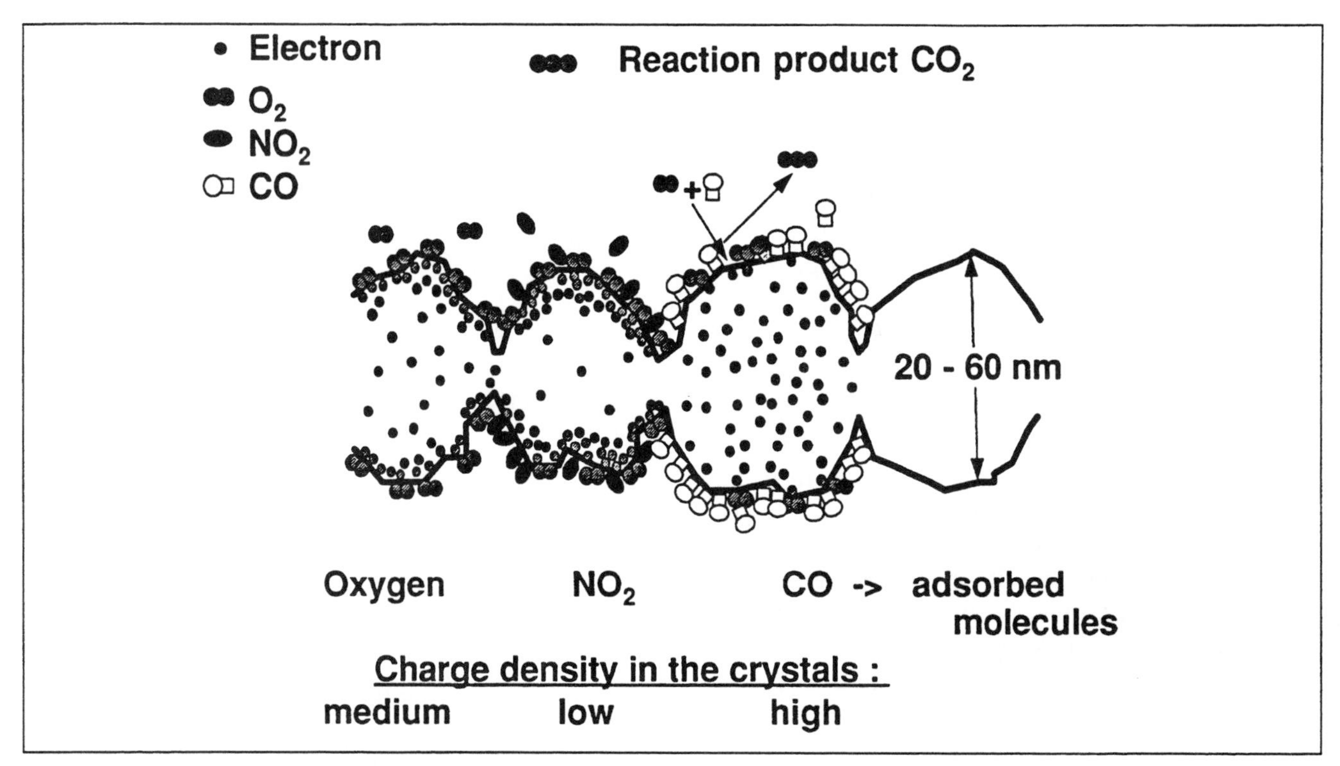

Figure 3: Principle behavior to different target gases

Sensor Materials

Wet chemical preparation methods are used to make the sensor materials. New nano technology precipitations are used to form 20-60 nm SnO_2-crystals which are coated after purification with additives like oxides of earth alkaline metals, rare earth metals and noble metal clusters.
After the chemical preparation the suspension is converted into a gel. It is dried, ground and mixed with organic binders to produce a screen printing paste. Rheology is adjusted to create optimized machine printable properties.
At the Robert-Bosch-Research Center several years of scientific research have been spent to develop materials with sufficient selectivity and long term stability.

Production

The sensor elements are produced using modern thick film technology. First, the platinum electrodes and heater are printed and subsequently fired at 1800 K. Then several printing and drying steps are performed on the sensing materials after which the sensor is annealed.

As reported, the sensor material is extremely sensitive even to very low amounts of additives (dopants). Therefore, some measures had to be taken in order not to contaminate it either during the manufacturing process or with any of the raw materials.
As a result, the sensors are produced in a separate new production line. The ceramic substrates are standard but specially handled and packaged by the supplier. The paste for the electrodes and the heater consists of pure platinum. Finally the annealing of the sensor material is performed in an oven with a quarz chamber. The last step is the testing of the electrical parameters and of the performance of the sensor element.

Characteristic of Sensor Elements

SnO_2-sensors are sensitive to many reducing and oxidizing gases like hydrocarbons, CO, NO, NO_2 as well as to humidity.
In many cases the dependence of the sensor resistance on the concentration of a target gas can be described by

$$R \propto c^{\beta}$$

whereby R is the sensor resistance, c *is* the concentration of the target gas and β is a characteristic parameter for the gas under investigation [3].
For practical use one often defines the sensitivity s by

$$s = \frac{R_0}{R_g} \quad \text{for } R_g < R_0$$

$$s = \frac{R_g}{R_0} \quad \text{for } R_0 < R_g$$

with R_0 the resistance in a carrier gas (synthetic air with 50 % humidity) and R_g the resistance at a special threshold concentration c of target gas, which is diluted in the carrier gas. Since for different gases R_g could be either smaller or greater than R_0, the definition of s is made in a way that the condition $s > 1$ is satisfied.
All resistances in this chapter are measured at a constant dc-voltage load of 2 V.

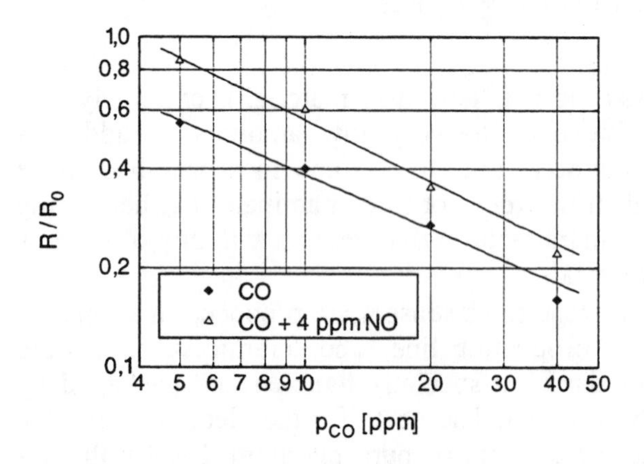

Figure 4: Relative resistance of the CO-element. Response to CO and (CO+4 ppm NO) in synthetic air.

Sensitivity and Cross-Sensitivity

Optimized for the application the CO-sensor shows a high sensitivity to CO and a low cross sensitivity to NO_x (Fig. 4).
The NO_x sensor element on the other hand has a high sensitivity to NO_x (with similar sensitivities to NO and NO_2) and a very low cross sensitivity to CO (Fig. 5).
The combination of the two sensor elements thus

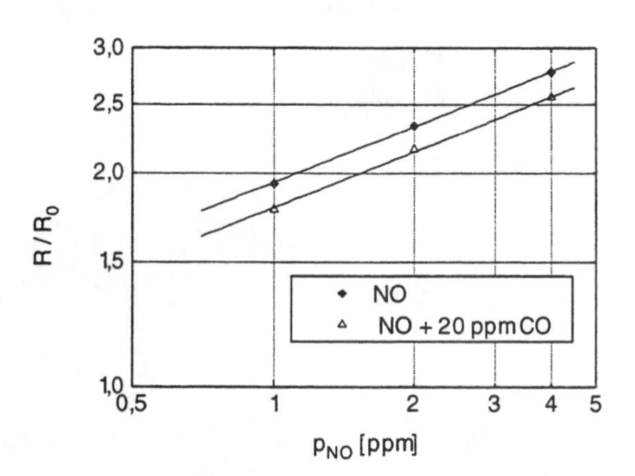

Figure 5: Relative resistance of the NO_x-element. Response to NO and (NO+20 ppm CO) in synthetic air.

offers the possibility to detect both CO and NO_x independently in a gas mixture. This is essential for the application as a traffic air pollution sensor. The resistance of both sensors also depends on the humidity (Fig. 6). Since the humidity normally changes slowly compared to CO and NO_x pollution peaks, this dependence can be compensated for mostly by the evaluation algorithm.

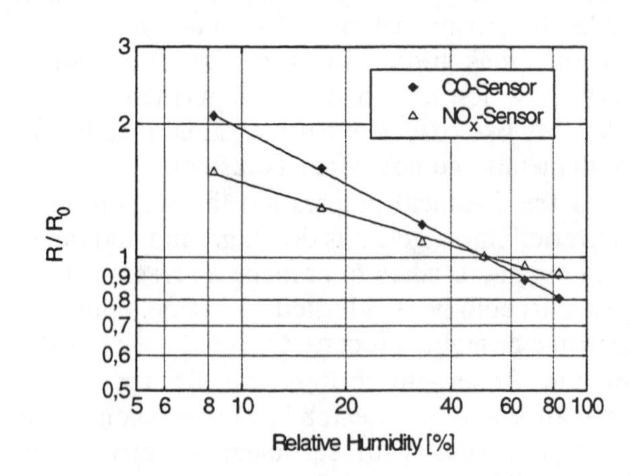

Figure 6: Relative resistance vs. humidity

Response Time

Both sensor elements show a response time to the target gas under 5 s at operating temperature (Fig. 7).

This is absolutely sufficient for the application. The response time is strongly dependent on the operation temperature. At 540 K it is, for example, five times longer than at 600 K.

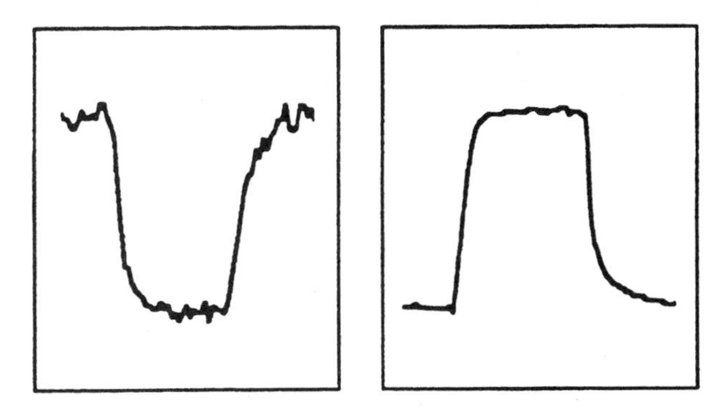

Figure 7: Response time of CO- and NO_x-element. The target gas is 15 s switched on

Operating Temperature

A very important parameter is the operating temperature. Pure SnO_2 is a n-type semiconductor with a low conductivity in the interesting temperature range of 400 - 700 K. To improve this the sensor element is doped with a donor-type element. Due to the semiconducting behavior the sensor resistance decreases with increasing sensor temperature. On the other hand the interactions of the gas molecules with the SnO_2-surface (adsorption, reaction, desorption) also show a distinct temperature dependence.

Figure 8 shows the dependence of the sensor resistance on the temperature for several target gas concentrations in the case of the CO-sensor.
The strong increase of resistance between 470 and 510 K can be attributed to the desorption of water molecules which adsorb at lower temperatures [1]. The NO_x-sensor shows a qualitatively similar behavior (Fig. 9).
From these measurements one can deduce the temperature dependence of the sensitivity, which is defined as

$$s_{CO} = \frac{R_0}{R_{40ppmCO}} \text{ for the CO-sensor and}$$

$$s_{NO} = \frac{R_{4ppmNO}}{R_0} \text{ for the } NO_x\text{-sensor.}$$

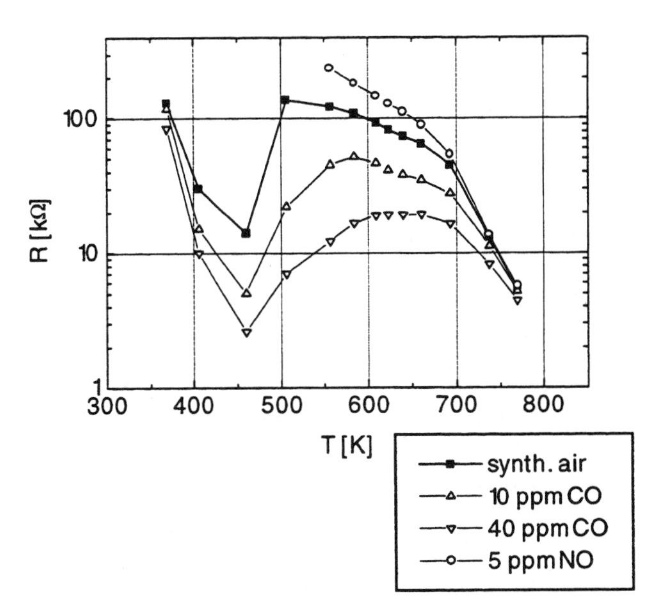

Figure 8: Resistance of CO-element vs. temperature for several concentrations

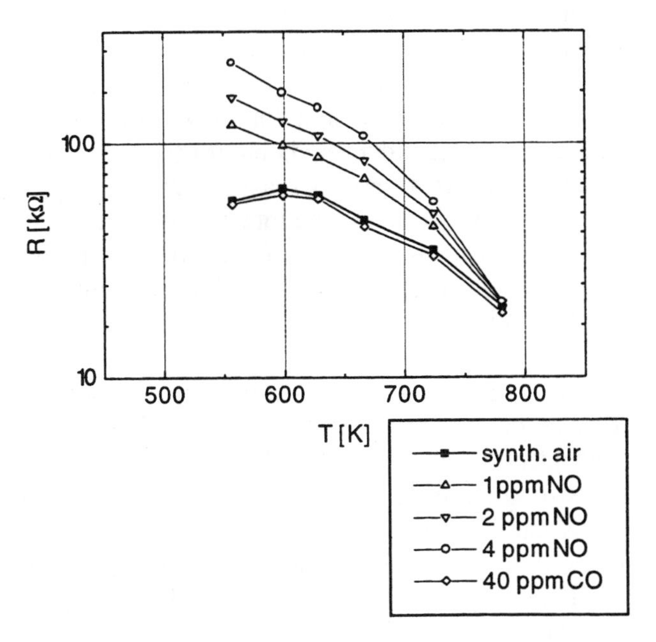

Figure 9: Resistance of NO_x-element vs. temperature for several concentrations

The sensitivity maximum is at 510 K for the CO-sensor. The NO_x-sensor shows decreasing sensitivity with increasing temperature (Fig. 10, 11). However, in order to define the optimum operating temperature, the dependence of the cross-sensitivities and response times also have to be considered. A temperature of 600 K for both sensors has been found to be a good compromise.

This is absolutely sufficient for the application. The response time is strongly dependent on the operation temperature. At 540 K it is, for example, five times longer than at 600 K.

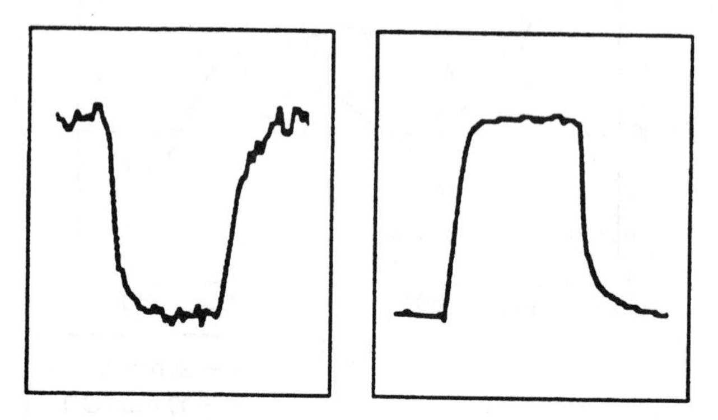

Figure 7: Response time of CO- and NO_x-element. The target gas is 15 s switched on

Operating Temperature

A very important parameter is the operating temperature. Pure SnO_2 is a n-type semiconductor with a low conductivity in the interesting temperature range of 400 - 700 K. To improve this the sensor element is doped with a donor-type element. Due to the semiconducting behavior the sensor resistance decreases with increasing sensor temperature. On the other hand the interactions of the gas molecules with the SnO_2-surface (adsorption, reaction, desorption) also show a distinct temperature dependence.

Figure 8 shows the dependence of the sensor resistance on the temperature for several target gas concentrations in the case of the CO-sensor.
The strong increase of resistance between 470 and 510 K can be attributed to the desorption of water molecules which adsorb at lower temperatures [1]. The NO_x-sensor shows a qualitatively similar behavior (Fig. 9).
From these measurements one can deduce the temperature dependence of the sensitivity, which is defined as

$$s_{CO} = \frac{R_0}{R_{40ppmCO}} \text{ for the CO-sensor and}$$

$$s_{NO} = \frac{R_{4ppmNO}}{R_0} \text{ for the } NO_x\text{-sensor.}$$

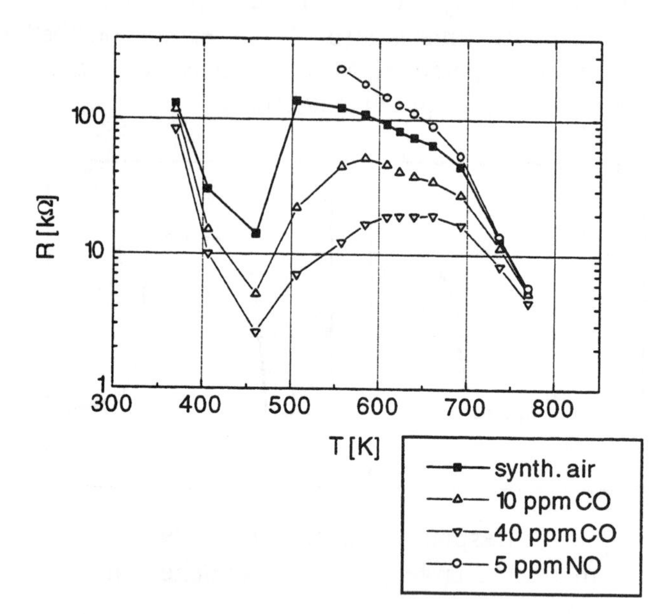

Figure 8: Resistance of CO-element vs. temperature for several concentrations

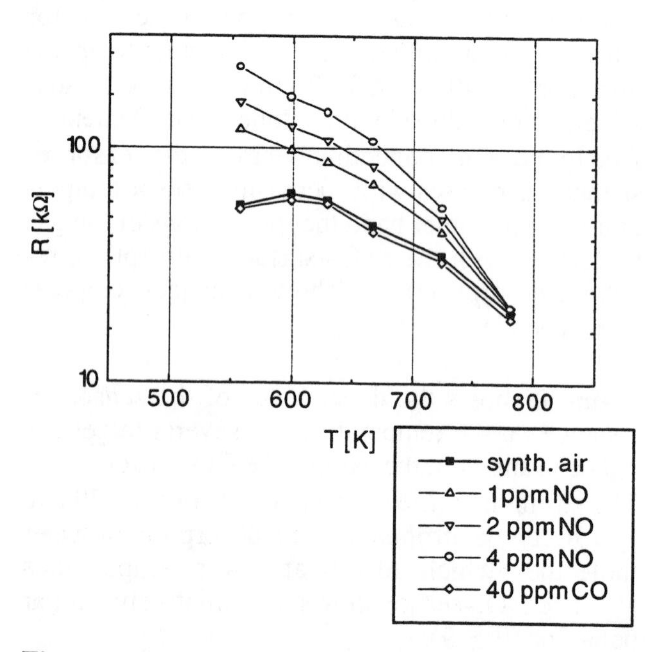

Figure 9: Resistance of NO_x-element vs. temperature for several concentrations

The sensitivity maximum is at 510 K for the CO-sensor. The NO_x-sensor shows decreasing sensitivity with increasing temperature (Fig. 10, 11). However, in order to define the optimum operating temperature, the dependence of the cross-sensitivities and response times also have to be considered. A temperature of 600 K for both sensors has been found to be a good compromise.

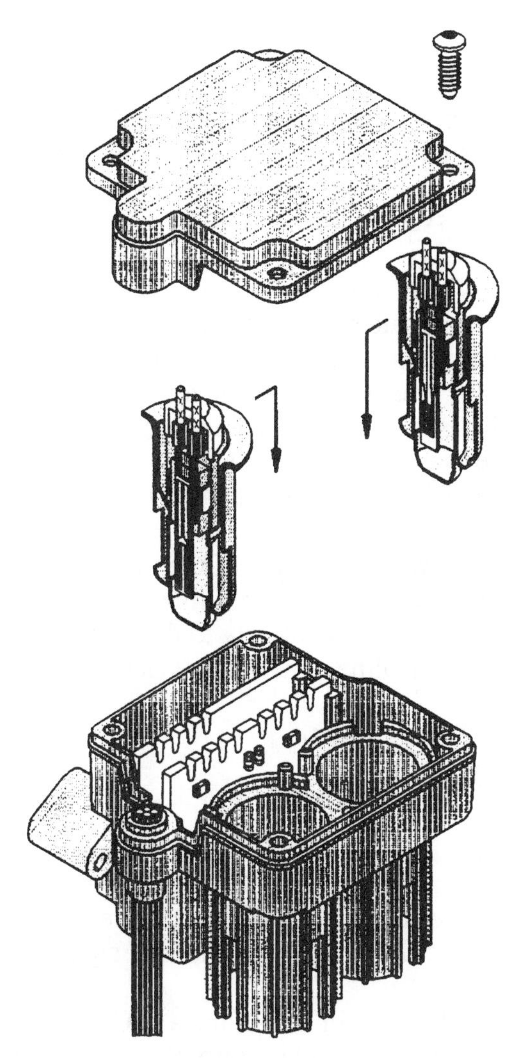

Figure 13: Assembly of the air quality sensor

Long Term Stability

Extensive durability tests are performed with the air quality sensor under various conditions.
Test A:
This test is performed in the laboratory under the following conditions:
The sensor is periodically switched on/off in the following intervals:
- 4 minutes on
- 1 minutes off

In the "On" period the sensor is working at operation temperature and connected with the evaluation circuit.

The tests give some indications about long term behavior of the gas characteristic due to the thermal aging of the sensitive layer and about the adhesion of the thick film layer and the alumina substrate.

Test B:
This test is performed like test A but installed directly at an intersection with high traffic load.
It provides information about the aging in extremly polluted air.

The results of tests A and B are similar. Therefore, only one typical result of test B will be discussed (Fig. 14).
It shows the basic level in synthetic air (50% relative humidity at 296 K) and the response to 2 ppm NO and 40 ppm CO, measured several times during the stability test of 1800 hours.
The basic level is stable and the sensitivity of the NO_x-element rises by about 20%. This stability is quite sufficient for the application.

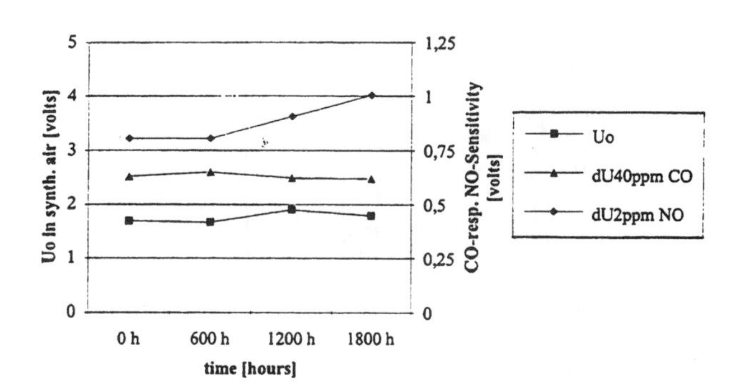

Figure 14: Durability test of the air qualitiy sensor (Test B)

Signal Conditioning and Application Aspects

The aspect most relevant to the application of the air quality sensor is an improvement of the climate in the interior of an automobile. CO- and NO_x-concentrations are correlated with the concentrations of bad-smelling and disturbing gases.
It is known that the relationship between the degree of perception of the human nose and the concentration of odours is not a linear one. Small variations at a low concentration level cause the

same perception as large variations at a high level. Therefore, the air quality sensor should demonstrate the same characteristic. This is achieved by taking the logarithm of the sensor element signals. The fresh air intake should be interrupted in situations where the CO-concentration, the NO_X-concentration or the sum of both concentrations show a certain increase within a fixed period of time. Thus, it is sufficient to generate only one output signal out of the two sensor element signals. Due to the fact that typical traffic emissions have a CO- to NO_X-concentration ratio of roughly ten to one and the ratio of sensitivities of the sensor elements is nearly inverse, a simple addition of the two logarithmic signals is sufficient.
Further evaluation of the sensor signal and generation of a logical signal for operating the air intake flap is made in the ECU of the air conditioning system.

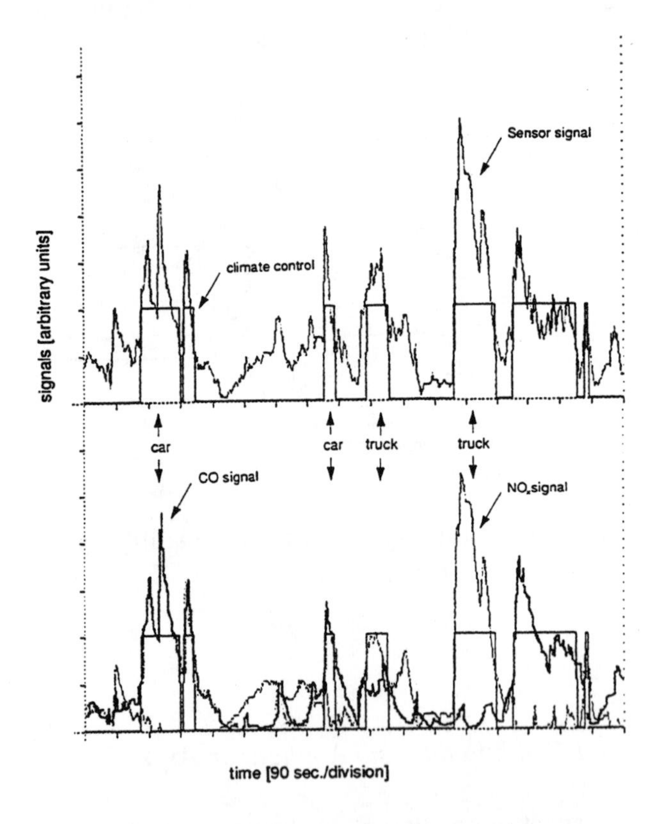

Figure 15: Test drive results with air qualitiy sensor

Upper diagram: sensor signal and logical output signal of the climate control system

Lower diagram: CO- and NO_X-signals [6]

The final part of the text describes one evaluation concept.
The basic level of air pollution in rural areas is normally significantly lower than in cities or in industrial districts. Furthermore, for safety reasons the flap can not be closed permanently. Consequently the system should only be used to detect pollution peaks. This behavior is also adapted to the sensitivity of the human nose. Therefore, it is necessary to create a floating zero level by averaging the actual signal over a defined period of time. At a certain threshold between zero level and the actual signal, the flap is opened or closed. This procedure has another advantage since it also compensates for zero level changes due to humiditiy cross-sensitivity.
Figure 15 shows an example of the function of the system in connection with a climate control system. The system reacts clearly to diesel emissions as well as to gasoline emission peaks [6].

CONCLUSIONS

A new air quality sensor for vehicle climate control is presented. The sensor extends the number of functions of the air conditioning system and increases the comfort which it provides significantly.
Using a CO- and, in addition, a NO_X-selective sensor element with an appropriate signal evaluation circuit, the masking effect of simultaneous CO emissions of gasoline engines and NO_X emissions of diesel engines has been solved.
The air quality sensor is in serial production and has proven its reliability in the field.
Additional efforts still have to be made with respect to an improved selectivity. The integration of additional sensing parameters (e.g. humidity, temperature) for automotive air conditioning systems which provide greater comfort can be realized in the future.

ACKNOWLEDGEMENT

A project of this complexity requires the contributions of many colleagues. We would like to thank H. Potthast and D. Elbe (for preparing sensor materials), G. Riehl (concept and coordination), T. Weigold (electronics), B. Wassmer (experiments), R. Maier (construction), A. Fischer (processing), H. P. Trah (calculations), M. Pfefferle, M. Baisch and T. Stumber (experiments).

REFERENCES

[1] W. Göpel, J. Hess, and J. N. Zemel (Eds.),
" Sensors - A Comprehensiv Survey"
Vol. 2: Chemical and Biochemical Sensors (Part 1 and 2), VCH Weinheim, New York, Basel, Cambridge, 1991

[2] W. Göpel,
"Chemosensoren mit elektrischen Transducern"
tm - Technisches Messen 62, 1995, Oldenburg Verlag

[3] John F. McAleer, Patrick T. Moseley, John O. W. Norris and David E. Williams
"Tin Dioxide Gas Sensors, Part 1.-Aspects of the Surface Chemistry revealed by Electrical Conductance Variations"
J. Chem. Soc.., Faraday Trans. 1, 1987, 83, 1323-1346

[4] John F. McAleer, Patrick T. Moseley, John O. W. Norris and David E. Williams
"Tin Dioxide Gas Sensors, Part 2.-The Role of Surface Additives"
J. Chem. Soc.., Faraday Trans. 1, 1988, 84, 441-457

[5] D. Kohl,
"Surface Processes In The Detection Of Reducing Gases With SnO_2-Based Devices",
Sensors and Actuators 18, 1989, pp. 71-114

[6] T. Stumber,
"Aufbau und Programmierung eines Auswertemoduls für chem. Sensoren auf Mikrocomputerbasis", Diplomarbeit (BA), R. Bosch GmbH, 95

SPEED AND ACCELERATION SENSORS

960758

Acceleration Sensor in Surface Micromachining for Airbag Applications with High Signal/Noise Ratio

M. Offenberg, H. Münzel, D. Schubert, O. Schatz,
F. Lärmer, E. Müller, B. Maihöfer, and J. Marek
Robert Bosch GmbH

Copyright 1996 Society of Automotive Engineers, Inc.

ABSTRACT

Employing novel surface micromachining techniques, a highly miniaturized, robust device has been fabricated. The accelerometer fulfills all requirements of state-of-the-art airbag systems. The present paper reports on the manufacturing and assembly process as well as the performance of the sensor.

The capacitive sensing element consists of a moveable proof mass of polysilicon on a single crystalline silicon substrate. A lateral acceleration displaces the proof mass and a capacitive signal is generated at a comb electrode configuration. An external IC circuit provides the signal evaluation and conditioning in a closed loop mode, resulting in low temperature dependency of sensor characteristics and a wide frequency response.

The sensor is fabricated by standard IC processing steps combined with additional surface micromachining techniques. A special deposition process in an epitaxial reactor allows the fabrication of moveable masses of more than 10 μm thickness.

As a result, working capacitances are up to 10 times higher as compared with conventionally fabricated surface micromachined accelerometers. Measurement ranges of $50g_n$ with mg_n resolution have been realized.

Secondly, the larger thickness results in an increased mechanical stiffness of the mechanical structure normal to the wafer surface. Hence, cross sensitivities of this sensing element are reduced by more than one order of magnitude. Also, fewer problems are encountered with the more rugged sensor structure during mounting and encapsulation.

INTRODUCTION

The market for automotive sensors is currently dominated by piezoelectric sensors or bulk micromachined sensors using piezoresistive or capacitive sensor elements. Recently, surface micromachined accelerometers with capacitive sensing using thin poly crystalline silicon layers have been transferred to volume production or are close to market introduction, respectively [1,2]. Two different product strategies are being followed: The monolithic integrated approach [1,3] and the two-chip concept using separate chips for sensing and evaluation [2]. There are numerous arguments for either solution. One main reason for an integrated approach is the extremely small signal produced by the capacitive sensing scheme. A capacitance resolution in the regime smaller than $1x10^{-18}$ Farad is regarded as extremely difficult. With a working capacitance of about 120fF and a proof mass of $0.28x10^{-9}$kg the noise floor is given by the Brownian noise limit [3,4] of about $0.12mg/\sqrt{Hz}$. To overcome this limit a larger proof mass is required. This can be achieved by larger geometries in either the vertical or the lateral dimension. Typically low stress poly layers are deposited in the amorphous state at or below 600°C with deposition rates under 10-20nm/min. Deposition times up to 10 hours for a 2μm thick layer are required. The fabrication of thicker layers than 2μm appears unfeasible due to exceedingly long process times. The lateral dimensions are restricted by residual stress gradients in the material resulting in bowing of the freed beams and eventually to mechanical contact to adjacent fixed elements of the structure.

Several approaches have been suggested to obtain a thicker material. Single-crystalline material patterned from SOI wafer [5] or by a combination of anisotropic and isotropic etching from the silicon substrate [6] have been proposed. This paper describes an approach for deposition and patterning of the sensor material that results also in larger geometries with the additional feature of a buried interconnection layer. Proof masses of well above 1×10^{-9}kg alleviate the effect of the Brownian noise limit. The process flow is suitable for the monolithic integration of the sensing element with on-chip electronics [7]. However sense capacitances close to or larger than 1pF also allow a simple two-chip realization that will be discussed in the following.

DESIGN OF THE SENSOR STRUCTURE

The capacitive sensing element consists of a comb arrangement as shown in Fig. 1.

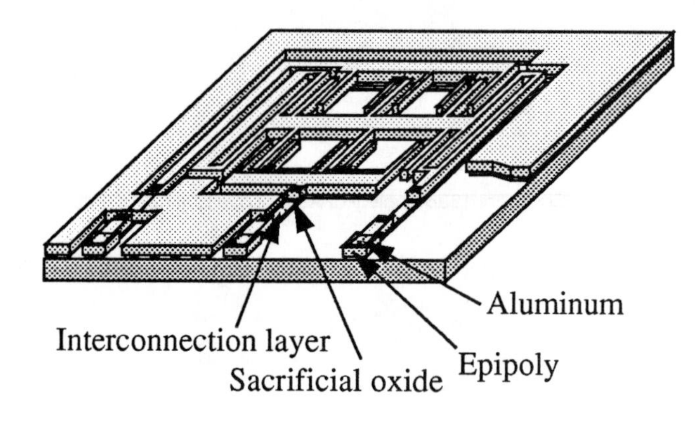

Fig. 1: Schematic drawing of the surface micromachined accelerometer

The proof mass is suspended by four tethers that are 11μm thick, 3μm wide and 270μm long. The mass is 900μm long and carries 60 electrodes that shift their position upon an external acceleration force versus two sets of fixed electrodes. The electrodes are 300μm long. They are connected by an additional poly interconnection layer underneath the structure. Hence, two capacitors C_1 and C_2 in a differential capacitor

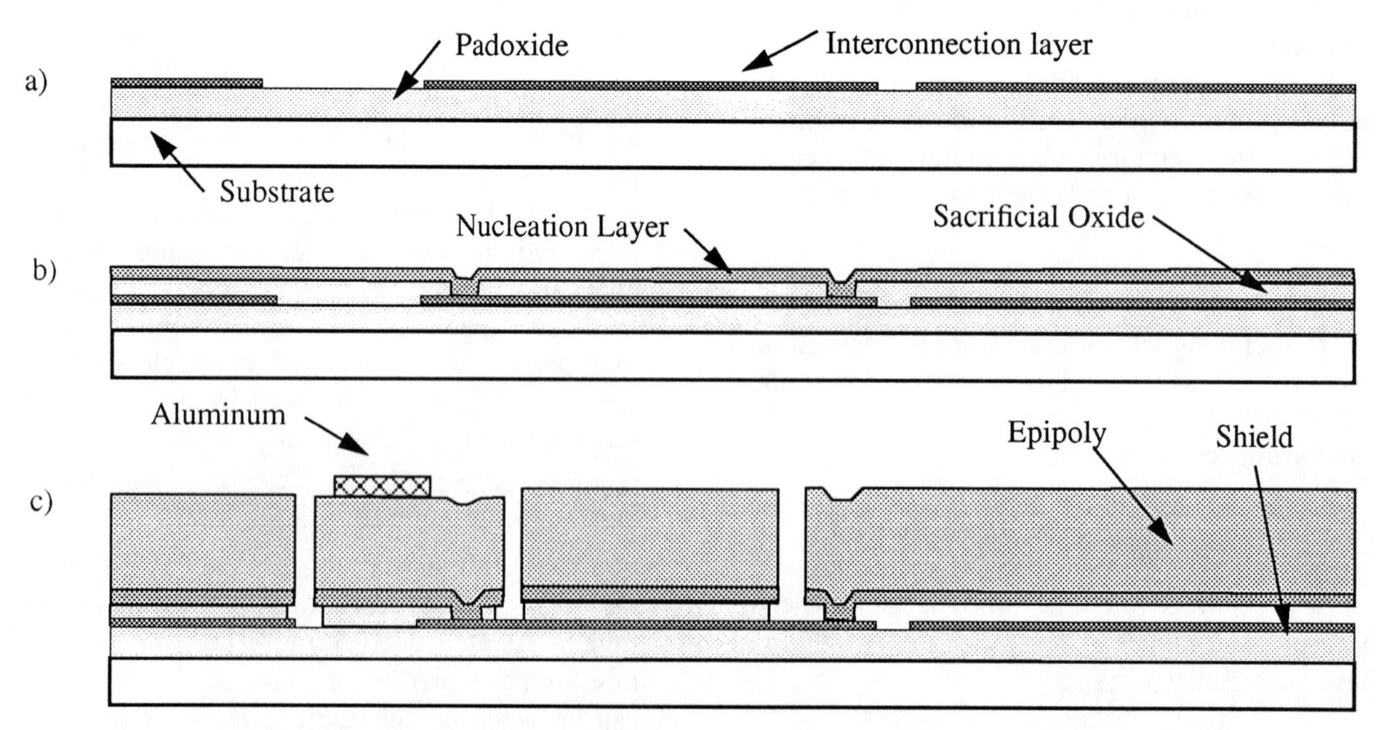

Figure 2: Processing sequence for the surface micromachined accelerometer

arrangement are formed. This second layer that is sandwiched between two oxide layers also provides an electrostatic shield under the proof mass and the connection to the aluminum bond pads. The lateral dimensions are optimized for a specific application but do not represent an upper limit. In principle also much larger structures can be realized.

SENSOR FABRICATION AND MATERIAL PROPERTIES

One feature of the fabrication process is the deposition of poly silicon for the sensor structure in an epitaxial reactor rather than in a LPCVD reactor. At elevated temperatures an 11μm thick polycrystalline silicon layer is deposited.

The sensing element is formed by three conductive and two isolating layers that are patterned with four lithographic mask steps. Fig. 2 shows the processing sequence that will be described in the following:

a) A 2μm thick thermal pad oxide 1 is grown on the silicon substrate.

A LPCVD polycrystalline silicon layer is deposited and patterned (mask 1) to form the interconnection layer.

b) A sacrificial oxide is deposited and patterned (mask 2) at low temperatures. Optionally via-holes through oxide 1 to the substrate can be formed by overlapping openings in mask layers 1and 2.

A blanket thin layer of LPCVD polycrystalline silicon is deposited.

c) Using the LPCVD poly silicon as a nucleation layer the 11μm thick poly crystalline layer is deposited in an epitaxial reactor at 1150°C. Since we use an epitaxial system we call this material epipoly. At 1150°C very large deposition rates on the order of 600 nm/min are obtained. The epipoly is doped and anisotropically etched (mask 3). A large aspect ratio of 5 with 11μm sidewalls and 2μm gaps can be realized using a fluorine based etch chemistry. Isotropic etching cycles and sidewall passivation cycles are alternated to obtain high anisotropy. The large selectivity versus photo resists eliminates the need of an oxide hard mask required for chlorine based processes. Subsequently aluminum is sputtered and patterned (mask 4) to form bond pads.
The sacrificial oxide underneath the sensing beam is removed to form freestanding structures.

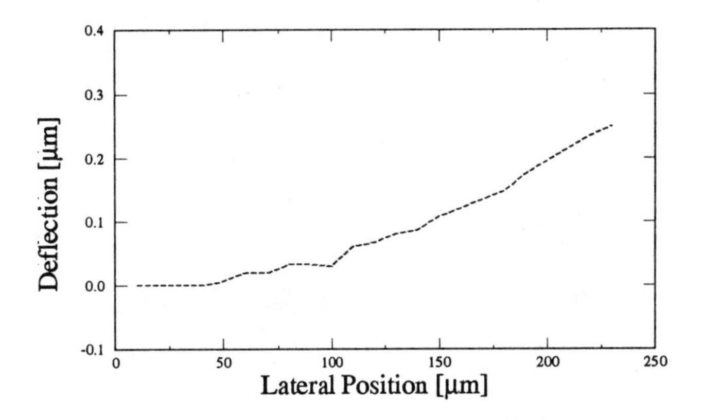

Fig 3. Surface profile of a released beam

Sacrificial oxides are usually removed in aqueous solution of hydrofluoric acid. Special care must be taken to avoid the stiction of the beam during the subsequent drying of the structure. Capillary forces might pull the movable parts to the substrate or to fixed electrodes. Additional means as freeze drying, supporting structures [8], or critical point drying [9] are necessary. We avoid the wetting of the wafer by the use of vapor HF for sacrificial oxide etch [10]. As a source for the etchant the vapor phase over a liquid HF solution is used. Since water plays an important role in the etch chemistry as a product and at the same time as an initiator of the process the etching process has to be carefully controlled. Condensation and water produced by the etch chemistry can lead to the formation of droplets, capillary forces and stiction. Increasing the wafer temperature above the temperature of the HF solution removes excess water from the wafer surface and avoids condensation. However the etch rate drops rapidly with increasing wafer temperature since water is needed as an initiator of the etching process. So the wafer temperature is slightly higher than the temperature of the HF solution to ensure proper etching without stiction.

Finally the wafer is diced and the chips are mounted. The movable structure is protected during these steps using a proprietary technique.

Fig. 4: SEM photograph of the sensing element

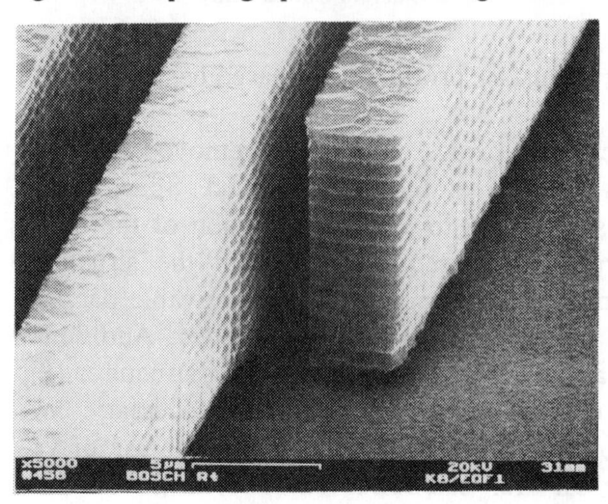

Fig. 5: SEM photograph showing a detail of the capacitive electrodes

The as-deposited epipoly shows almost ideal mechanical properties: The residual stress and the stress gradient in the structure are extremely small. However the subsequent doping of the polysilicon introduces a small compressive stress of about -5MPa. Long and thin tethers for the suspension of the proof mass could possibly buckle according to the Euler buckling law. Folded beams are usually used for compensation of mechanical stress. In this work a proprietary cantilever structure (patent pending) is used for the suspension of the proof mass that not only compensates compressive stress but converts it into tensile stress.

Concentration gradients of the dopant can lead to a curvature of the freed beams. Fig. 3 shows a typical surface profile of a released beam. The tip of a 1000µm long cantilever is bend up by approximately 2µm. This residual stress gradient presently restricts the maximum lateral dimension of the devices to 1000µm. Process optimization will decrease this residual stress gradient further.

Details of the completed sensor device are shown in Figs. 4 and 5. The trenching process assures excellent vertical dimensions of the capacitor plates.

EVALUATION CIRCUIT

The evaluation of the differential capacitive signal is performed on a separate chip using a closed loop scheme for force balancing. The block diagram and the corresponding clock scheme is shown in Fig. 6. On both sets of fixed electrodes complementary rectangular signals with a frequency of 200 kHz are applied. On and off voltages are ground (0V) and V_{dd} (e.g. 4 V), respectively. The voltage of the proof mass is set by a short pulse to 4Volt. The voltage remains constant during the subsequent switching cycle on the fixed electrodes if the proof mass is exactly centered (0g acceleration). Upon acceleration one capacitor, e.g. C_1, decreases in value. If the voltage on the fixed electrodes is switched from 4V to 0V, or vice versa, a charge transfer from C_2 to C_1 takes place. As the total amount of charge remains constant the voltage of the proof mass changes proportionally to the capacitance difference of C_1 and C_2. This voltage will be sampled and stored by a S&H unit. Subsequently a voltage on the proof mass will be restored to 4V. As a next step the voltage on the capacitor C_2 with an increased value will be switched from 4V to 0V. Consequently the voltage on the proof mass drops and is stored by a second S&H unit. The voltage of the proof mass will be restored to

4V and the next sampling cycle begins. Meanwhile the voltages stored in the S&H units will be subtracted from each other and added to a reference voltage of 1.7V. The result is used as an input to a comparator with the second input connected to a saw tooth voltage. The output of the comparator is the rectangular voltage that along with the inverted output is used to drive the fixed electrodes. Since the duty cycle is a function of the acceleration a larger mean-time averaged voltage difference is applied to the smaller capacitance or to the larger gap, respectively. As a consequence the proof mass is pulled back to the center position electrostatically. Hence, the loop for force balancing is closed. The output of the comparator can be used directly as a PWM signal for the acceleration or converted by a low pass filter to an analog output signal.

RESULTS AND DISCUSSION

The sensing element that is discussed in the following was specifically designed for the application in airbag deployment systems. In this case the working capacitance is equivalent to 800fF. It should be pointed out that the fabrication process allows also larger lateral geometries with higher capacitances.
The sensing element along with the evaluation circuit was mounted on a metal header of a DIL8 housing (Fig. 7)

Fig. 7: Chip photo of the mounted sensor and the evaluation circuit

The effect of mounting induced stress on the sensor output was investigated. The 0g output of sensors connected to the header by die attach and sensors that were only held by the bonding wires were compared. No significant offset shift was observed that could be linked to mounting induced stress. We attribute the immunity against external mechanical stress to the large thickness of the sensor material of 11μm and the stress compensation structures. The stiffness of the structure normal to the wafer surface is increased by two orders of magnitude compared to a 2μm thick poly silicon since it depends by the third power on the thickness.

Accelerations of different values were imposed

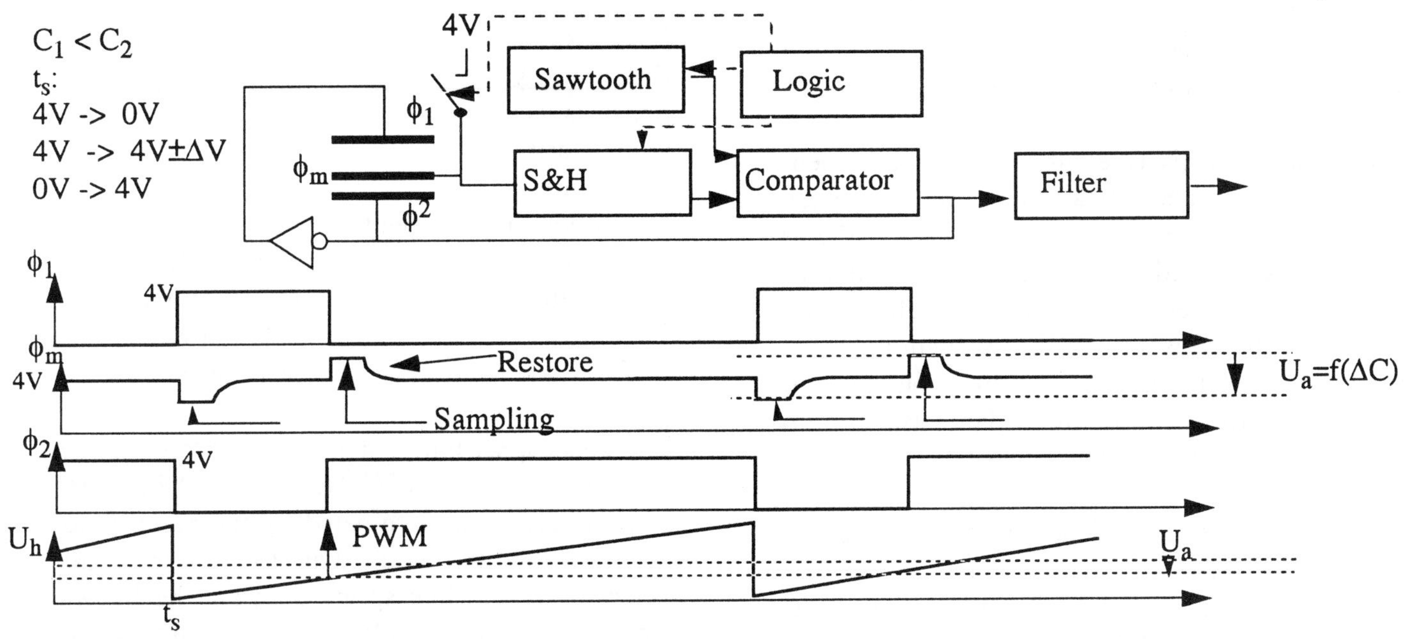

Fig. 6: Block diagram and clock sequence and for the evaluation of the differential capacitor

on the sensing element by a mechanical shaker. Fig. 8 shows the output characteristic of the device plotted versus the acceleration.

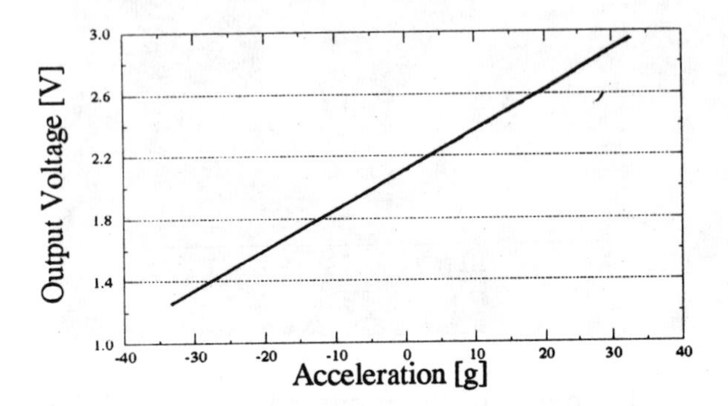

Fig. 8: Output characteristic of the accelerometer

The non linearity is below 1% in the range between -35g and +35g. This low value can be attributed to the force balancing scheme that limits the displacement of the proof mass at 35g to 30nm.

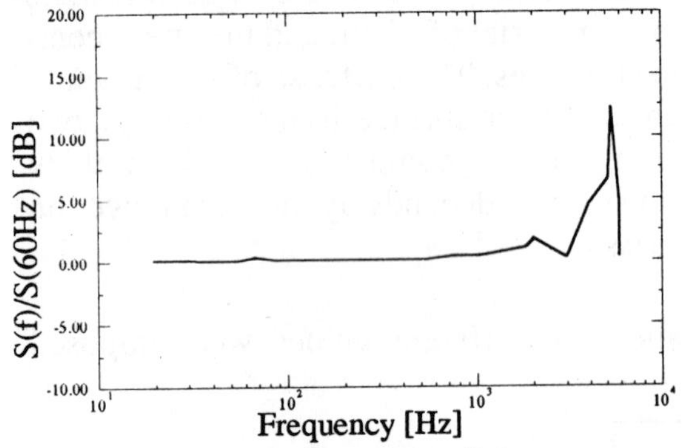

Fig. 9: Frequency response of the sensor system

The sensor sensitivity exhibits very little dependence on the mechanical excitation frequency below 1kHz. The peaks visible at 2kHz in Fig. 9 are partly generated by the attachment of the sensor to the mechanical shaker. The cross sensitivity was also tested by excitation of the sensor normal to the sensing direction. Due to the large stiffness caused by the large aspect ratio the cross sensitivity is well below 1%.

The temperature dependence of the devices has been studied. The temperature coefficient of the sensitivity (TCS) and the offset (TCO) are plotted in Figs. 10 and 11, respectively.

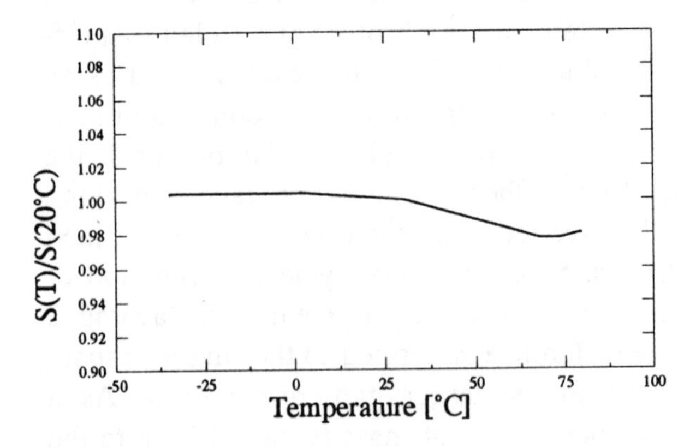

Fig. 10: Temperature coefficient of the sensitivity

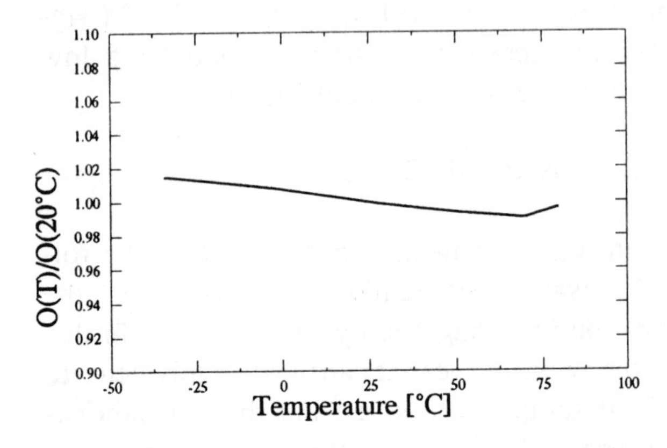

Fig. 11: Temperature coefficient of the offset

The following table summarizes the measured characteristics of the accelerometer:

Sensor size	0.8mm x 1.0mm
Sensor mass	5µg
Sense capacitance	~800fF
Nonlinearity	<1%
Sensitivity	24mV/g
Open-loop displacement	9nm/g
Noise floor	$0.8mg/\sqrt{Hz}$
Bandwidth	DC to 2.5kHz
TCS	200ppm/K
TCO	300ppm/K

The noise spectrum of the sensor system was determined to $19\mu V/\sqrt{Hz}$ or $0.8mg/\sqrt{Hz}$, respectively. It is dominated by the contribution of the evaluation circuit. Due to the comparably heavy mass and low spring constant the Brownian noise is negligible. The comparably low noise floor allows the use of this sensor not only for airbag application with a large measurement range but also for low g application such as ABS or dynamic vehicle control. If an acceleration range below 5g is addressed the sensor design can be adapted to obtain a larger sensitivity.

SUMMARY

A surface micromachining process is presented that allows the fabrication of long freestanding beams with larger thicknesses than obtained by conventional poly silicon techniques. A buried interconnection layer gives additional design flexibility for the sensing element. This type of surface micromachining allows the reduction of the chip area as compared to bulk micromachining. At the same time working capacitances around 1pF with small parasitic capacitances can be realized. Consequently the noise level as well as temperature and linearity errors are reduced. Characterization results demonstrate the potential for a large spectrum of applications ranging from crash sensing for airbag deployment systems to low g sensors.

ACKNOWLEDGEMENT

The authors would like to thank the team 'surface micromachining technology' for valuable contributions to the fabrication and the characterization of the device.

REFERENCES

[1] R.S. Payne and K.A. Dinsmore
"Surface micromechanical accelerometer: A technology update"
Soc.Auto.Eng.P-242, Detoit, MI, Feb. 25 - March1,1991, pp. 127-135.

[2] L.J. Ristic, R.J. Gutteridge, J. Kung, D. Koury, B. Dunn, H. Zunimo
"A capacitive type accelerometer with self-test feature on a double-pinned polysilicon structure"
Tech. Digest, 7th Int. Conf. Solid-State Sensors and Actuators (Transducer '93), p.810

[3] K.H.-L. Chau S.R. Lewis Y. Zhao, R.T. Howe, S.F. Bart, and R.G. Marcheselli
"An integrated force-balanced capacitive accelerometer for low-g- applications"
Tech. Digest, 8th Int. Conf. Solid-State Sensors and Actuatuators (Transducer '95), Stockhom, Sweden, June 1995, p.593

[4] T.B. Gabrielson
Mechanical-thermal noise in micromachined acoustic and vibration sensors"
IEEE Trans. on Electron Devices, vol. 40, pp. 903-909

[5] L. Zimmermann, J.Ph. Ebersohl, F. Le Hung, J.P. Berry, F. Baillieu, P.Rey, B. Diem, S. Renard, P. Caillat
"Airbag application: a microsystem including a silicon capacitive accelerometer, CMOS switched capacitor electronics and true self-test capability"
Sensors and Actuators A46-47 (1995) pp. 190-195

[6] K.A. Shaw, Z.L Zhang, and N.C. MacDonald
"SCREAM1: a single mask, single-crystal silicon, reactive ion etching process for micromechanical structures"
Sensors and Actuators A40 (1994) p.63

[7] M. Offenberg, F. Lärmer, B. Elsner, H. Münzel and W. Riethmüller
"Novel process for a monolithic integrated accelerometer"
Tech. Digest, 8th Int. Conf. Solid-State Sensors and Actuators (Transducer '95), Stockholm, Sweden, June 1995, pp. 882-592)

[8] R.L Alley, G.J. Cuan, R.T. Howe, K. Komvopoulos
"The effect of release etch processing on surface microstructure stiction"
Techn. Digest IEEE HiltonHead1992 p.202

[9] G.T. Mulhern, D.S. Soane, R.T. Howe
"Supercritical carbon dioxide drying of microstructures"
Tech. Digest, 7th Int. Conf. Solid-State Sensors and Actuators (Transducer '93), p.296

[10] M. Offenberg, B. Elsner, and F. Lärmer
"Vapor HF etching for sacrificial oxide removal in surface micro machining"
Electrochem. Soc. Fall-Meeting 1994, Ext. Abstr. No 671

950535

Angular Rate Sensor for Automotive Application

Toshihiko Ichinose and Jiro Terada
Matsushita Electronic Components Co.

ABSTRACT

The automotive industry has sought accurate, durable, and low cost mass produced angular rate sensors for chassis system control and navigational systems. An angular rate sensor based on the Coriolis force imposed on vibrating tuning forks has been developed for automotive applications.

This newly developed angular rate sensor uses tuning forks made with fabricated elinvar metallic material. Size reductions in these sensors have reduced their weight and allowed for mounting directly on the printed circuit board of the electronic control module.

Zero point temperature drift improvements have been developed through precision metal fabrication techniques.

Reduced weight improved the sensor's impact resistence.

PRINCIPLE TECHNOLOGIES

The study on the vibration gyro started in 1941 by using a tuning fork mechanism[1]. After that, many variations of the technology have been announced[2][3], however, the major technologies are categorized into two groups; one is a tuning fork design and the other is an acoustic bar design. The tuning fork design with orthonally arranged elements[4][5][6] is the more common technology trend.

The tuning fork type angular rate sensor is designed to convert the Coriolis force which appears on a moving object during rotation to an electric signal. The vibration is a way to get the moving object to detect rotation information. Therefore, the vibration mechanism technology is critical to the design of the sensor.

The advantages of the tuning fork design to the other design are;

1) From a stand point of scale models in various vibration modes, in other words interference among multiple partial resonance, the orthogonal tuning fork design has a simple structure. It is easy to design the tuning fork's natural vibration frequencies as well as separating its partial natural frequencies.
2) A stable suspension system for the tuning fork can be designed. It has a strong immunity to vibration and impact.
3) Compared with the acoustic bar design a large and stable amplitude can be obtained in the tuning fork design. This gives relatively high sensitivity without using a resonance mechanism.

In the orthogonal turning fork design, it is common to design a different natural frequency of the two top rectangular plates of the tuning fork than the entire natural frequency of the tuning fork. The top rectangular plates detect the Coriolis force as a result of an angular movement. We call this type of design a non-resonance mechanism. In the acoustic bar design, it technically is possible to design the sensor as a non-resonance mechanism by designing the section of the bar rectangle which has different natural frequencies between the driving direction and the detecting direction. However, this design will end up with very small sensitivity due to small gain. Therefore, most of the acoustic bar designs select a resonance mechanism.

The resonance mechanism has several advantages such as;

1) A couple of thousands times higher detection sensitivity can be obtained.
2) Since it has a large resonance Q value, it is little affected from outcoming noise except the natural resonance frequency.

However, the resonance mechanism has several significant disadvantages of the design.

1) Influence on the large resonance Q value caused by variation of the adhesive which attaches the piezoelectric devices is not negligible. A small deviation of the adhering conditions such as amount of glue and contact location can easily affect the entire sensor's characteristics significantly in a large Q condition.
2) The suspension method of the acoustic bar is not stable.
3) Sensor performance is affected by temperature changes due to the temperature coefficient of the piezoelectric devices and the adhesive as well the schift in structure.
4) The suspension system of the acoustic bar is not robust, so the reliability is insufficient.

The attached table 1. is a comparison analysis of the technologies between the turning fork design and the acoustic bar design.

1,1' : driving elements
2,2' : Sensing elements
3,3' : Tuning fork vibratory arm
4 : Support pin
5 : Supporting block
6 : Post
7 : Substrate

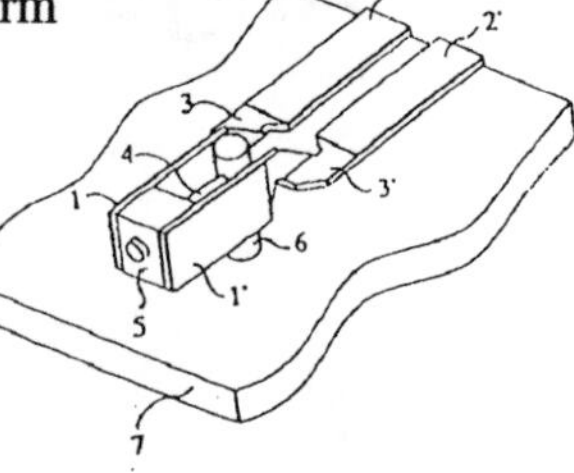

Fig.1 Perspective structural view of the angular rate sensor

Table 1 Comparative analysys between the tuning fork design and the accoustic bar design

	Tuning fork design	Acoustic bar design
Sensitivity	Poor	Good
Anti-vibration	Poor	Good
Anti-shock	Good	Poor
Response	Good	Poor
Suspention reliability	Good	Poor
Degree of freedom for frequency design	Good	Fair

STRUCTURE

Fig. 1 is a perspective structural view of the angular rate sensor developed recently.

The turning fork consists of

1) Right and left arms which are orthogonally folded elinvar (specific metal material designed for acoustic performance) plates of constant elastic metal at the middle. Rectangular piezoelectric devices of PZT group are adhered to these surfaces.
2) The supporting block which combines the above two pieces at their roots.

Each piece of the elinvar is designed to be separated to a detecting part as its top and a driving part as its root, and become as an arms of the tuning fork. Both of the arms are soldered onto the brass supporting block joints orthogonally to an iron post pin anchored through the brass base substrate.

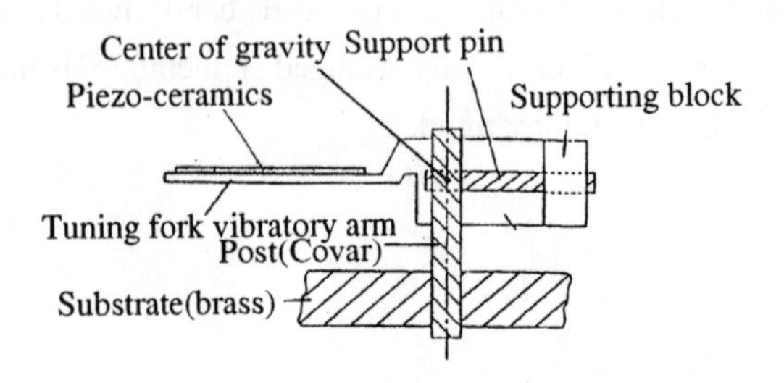

Fig. 2 Structural drawing of supporting method.

As shown in Fig. 2, the post pin is located inside of the tuning fork, and the length of the supporting pin is designed so that the center of the turning fork gravity exists on the post pin. Since the substrate is kept in vertical position during use and this construction makes the center of gravity for the tuning fork and hung center the same, the turning fork does not get any excess torque from vibration or outcoming shock. For example, it withstands a 1 meter drop test. Also, we could get its stable operation over a wide temperature range.

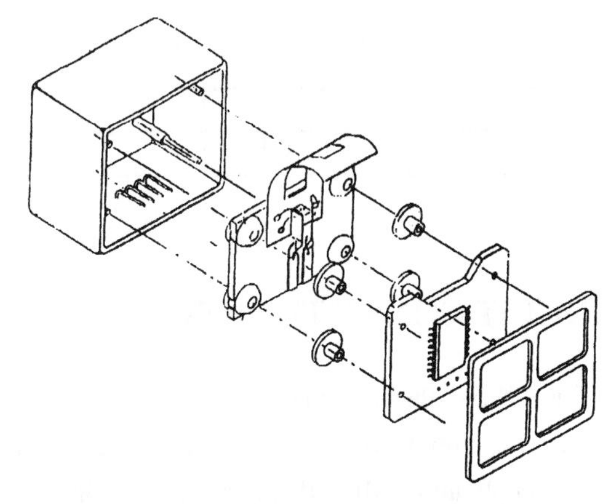

Fig 3 Case assembly layout

Fig. 3 shows the case assembly layout. The substrate on which the turning fork is mounted is settled into a plastic case with four rubber insulators.
The frequency response characteristic calculated from the rubber insulator and weight of substrate is shown in Fig. 4.

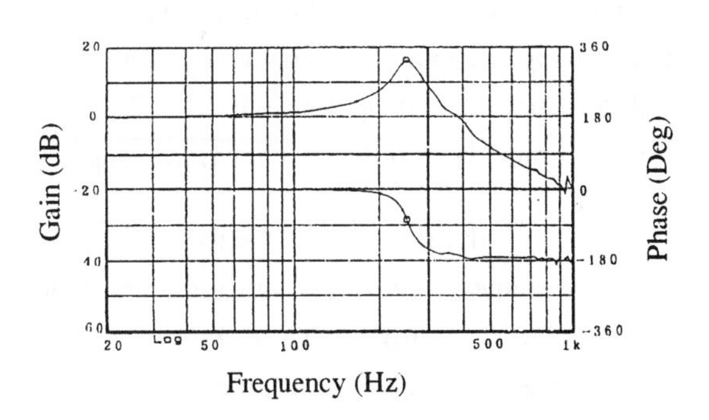

Fig 4, Frequency respons characteristic of rubber suspention

The driving circuit is installed opposite to the turning fork inside of the case. Its electric connection to the tuning fork is done through a flexible printed circuit board.

Inert gas sealed inside of the plastic case, prevents moisture condensation on the surface of the vibrating device which could affect the vibration condition when the temperature circumstance changes drastically.

The outside of the plastic case is metal shielded for EMC protection purpose, because the output signal from the piezoelectric device is small and easily disturbed by external noise.

DESIGN OF THE TUNING FORK

Regarding to design of a driving frequency of the tuning fork (f_{DR}), the resonance frequency f_{DR} is generally expressed as follows when the effective length of tuning fork vibration arm to be l (m), thickness to be t (m), elasticity to be E (N/m^2) and density to be P (Kg/cm^3) [6)]

Although this formula tells that only the length l and

$$2\pi f_{DR} = \frac{\alpha_{hv}^{2} \cdot t}{l^2}\sqrt{\frac{E}{P}} \quad (1)$$

where α_{hv}: eigenvalue

thickness t need to be designed properly in order to obtain a desired f_{DR}, actual f_{DR} sometimes differs due to various dimensional ratios. In fact, the formula (1) is not accurate enough in this case of structure shown in Fig, 1,
because of complication from many error factors such as thickness t, correction of elasticity E, difference of shape and material of the supporting pin and the post pin from simulation.

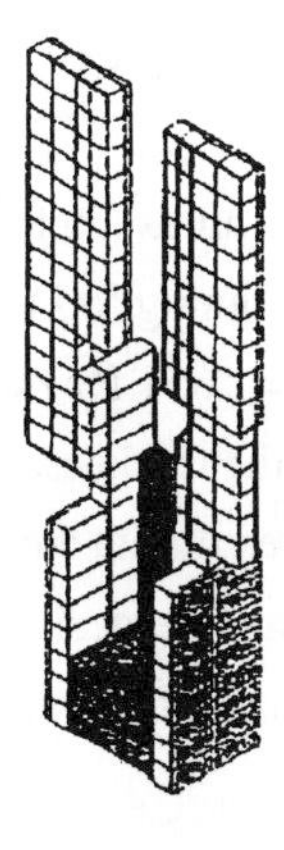

Fig.5 Analysis model for FEM.

In order to get more accurate frequency design of this tuning fork, as shown in Table 1, we applied an eigenvalue analysis by FEM, and designed the frequency both by a simulation result and by experiment. Fig. 5 is the analysis model of FEM we set up. It consists of three dimensions, 8 nodes, 6 planes, 380 elements, 5 planes, 6 elements and 721 nodes. For its boundary condition, we completely confined three direction of X, Y, Z, for the substrate, vertical and longitudinal direction for the supporting pin. Also, in order to avoid errors due to a bending force caused by reverse piezoelectric effect, we analyzed the vibration by short-circuiting the electrodes of the four piezoelectric elements.

Fig.6 Resonounce mode at f_{DR}

An example of analysis result in the so-called V shape vibration mode (see Fig. 6) of the tuning fork (f_{DR} = 1651 Hz), is shown in Table 2. It shows how the tolerance of each dimensions of the tuning fork (17 design parameters: W1 to 6, H1 to 7, D1 to 4) influence to the resonance frequency. The symbol W denotes the width of tuning fork, H is its length, and D is its thickness.

Table 2 shows, as an example, if the length H1 (one of the dimensions of the tuning fork) becomes extended by 1%, it influences the f_{DR} = 1651 Hz by 36Hz drop. On the other hand, 1% change of the thickness D2 influences it only 3Hz. This study tells as a result that those dimensions of W3, H1, H3, W2, and H2 must be designed and controlled by one digit more precisely than other dimensions in order to maintain the design quality.

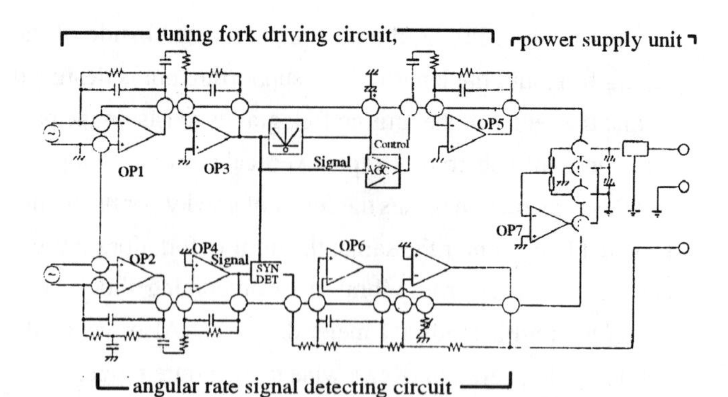

Fig 7, Circuit construction

Table 2 Effect of size parameters on f_{DR} Unit: %

	D1	D2	D3	H1	H2	H3	H4	H5	W2	W3	W6
Effect	8	3	7	36	15	19	11	8	19	50	8

CIRCUIT CONSTITUTION

Fig. 7 shows the circuit is divided into three blocks; tuning fork driving circuit, angular rate signal detecting circuit, and power supply unit. All of them are included into two bipolar IC chips. The voltage regulator IC outputs a constant volts and it is a low drop voltage type. It can operate at 7.5 volts power supply which is calculated by subtracting the drop voltage drop of a reverse connection preventive diode and absorption circuit of load dump surge from a battery's minimum voltage of 9 volts.

The metallic part of the tuning fork is biased at 3.5 volts, and the driving circuit of the tuning fork and the angular rate signal detecting circuit operate with this 3.5 volts as an imaginary ground level (GND) of two power supplies.

The tuning fork driving circuit activates one of the two piezoelectric elements adhered to both arms of the tuning fork, and does the AGC control so that the output electric charge from the other piezoelectric elements keeps constant. Generally speaking, piezoelectric elements have large deterioration of piezoelectric efficiency both by ambient temperature change and aging. However, by applying this driving method, the amplitude of the tuning fork is controlled to maintain as constant sensitivity as possible, to reduce the temperature drift of the sensitivity and aging change. In addition, in order to shorten the start up time, the amplifier gain is set to maximum right after power up until the amplitude reaches to a specific level.

The angular rate signal detecting circuit amplifies the AC electric charge produced from the detection element. A DC output signal is provided from this signal. The electric charge produced from the detection element is feeble. For example, when the sensor rotates by 1° per second (1 °/sec), the number of electrons appear on the device is only approximately 100 pieces per second. To the contrary, the electric charge not related to the Coriolis force appears at a level of 100 times more than the angular rate signal, which means the noise is 100 times bigger than the true signal. The majority of this non-related electric charge is from a bending motion of the detecting element caused by a leakage of the tuning fork vibration energy as shown in Fig. 8. However, since this signal has a 90 degree phase shift from the genuine signal of angular rate, it can be canceled by synchronous detection.

We introduced a double balance style in the synchronous detection circuit, so that the switching speed of the transistor is hardly depend on temperature changes.

The original source signal is so feeble that all of the input of the circuitry applied current amplifying with imaginary short construction but not voltage amplifying. This construction prevented voltage noise from the surrounding circuitry going into the input circuitry block.

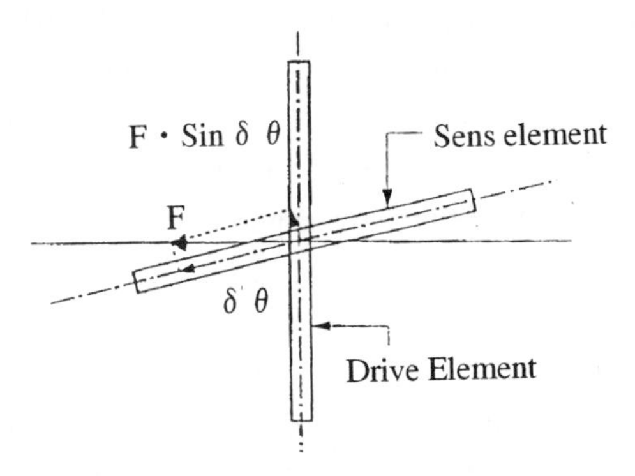

Fig 8. Principle of Angle Error

ZERO POINT DRIFT

In the car navigation system application, the sensor output is integrated to provide is angular displacement. So if the sensor output with no angular rate input, which we call zero point voltage or null voltage, drifts due to environmental change, it could cause a large error even if the drift to be converted is very small.

Some of the causes of this zero point drift are the value of undesired electric charge component (Q) created by the driving, change ΔQ of undesired charge component (Q) caused by temperature changes, error value of phase (Y) in synchronous detection, and phase error ΔY by temperature changes. These phase error will cause an offset voltage by an undesired charge component (Q) due to temperature change, resulting in a zero point voltage drift due to the through rate of the operational amplifier and others.

Therefore the temperature drift D of zero point voltage due to undesired charge component and phase error in synchronous detection is expressed as follows.

$$D_t = A \cdot (\Psi \cdot \Delta Q + Q \cdot \Delta\Psi) \qquad (2)$$

where A: constant.

The directional sensitivity component of detection element of driving inertial force (F) for generating the undesired charge component (Q) is expressed below. This assumes the resonance angular frequency to be ω_0, deviation angle of orthogonality to be $\delta\theta$, velocity to be $V = V_0 \sin(\omega_0 t)$ and mass of detection elements to be m,

$$F \cdot \sin(\delta\theta) = m \cdot V_0 \cdot \omega_0 \cdot \cos(\omega_0 t) \cdot \sin(\delta\theta) \qquad (3)$$

The above is be converted as the following by comparing with Coriolis force.

$$|F \cdot \sin(\delta\theta)| \,/\, |Fc| \propto (\omega_0 / \Omega) \cdot \sin(\delta\theta) \qquad (4)$$

Therefore, to decrease the undesired charge component (Q) responsible for sensitivity of Coriolis force, it is needed to decrease the deviation angle of orthogonality ($\delta\theta$) or to lower the resonance angular frequency of tuning fork (w_0). As the deviation angle of orthogonality ($\delta\theta$) increases, as shown in Fig. 9, the undesired charge component increases.

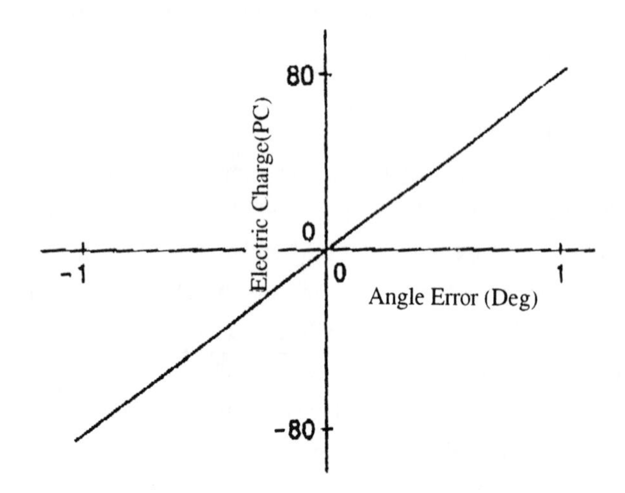

Fig.9 Angle Error and null signal

By introducing in the manufacturing process the measurement and correction technology into the orthogonal folding of the elinvar by precision press, the undesired charge component can be decreased to 1/5 of the conventional level. The zero point temperature drift can be improved significantly by this procedure.

RESPONSE

It does not require a high speed response in the resonance mechanism. Because the frequency response characteristic of the sensor output to the input angular velocity is determined almost completely by the Q value (mechanical sharpness) of the resonance element.

However, the angular rate sensor we developed is of "non-resonance mechanism". It uses the different resonance frequencies between the detection unit and driving vibration unit. Therefore it has a flat frequency responsiveness and theoretically it is possible to cover from DC to nearly the driving frequency (f_{DR}).

In order to specify the sensor for the application of the frequency usage, a five pole low pass filter behind the synchronous detection circuit is installed to attenuate the noise caused by detection. The filter constant is determined to match the required frequency band of the control object. The filter constant depends on the application's demand. For example,

(1) In a four wheel steering (4WS) application of the angular rate sensor, high speed responsiveness is required. The response speed is set as 3.6ms and the band width is set from DC to 100 Hz.

(2) For the navigation application, the response time is set as 20ms, and the band is set from DC to 7Hz.

As shown in the above, this angular rate sensor has the flexibility of being able to set up and change its response time easily by only changing the circuitry constant. Modifying the tuning fork vibration element is not required.

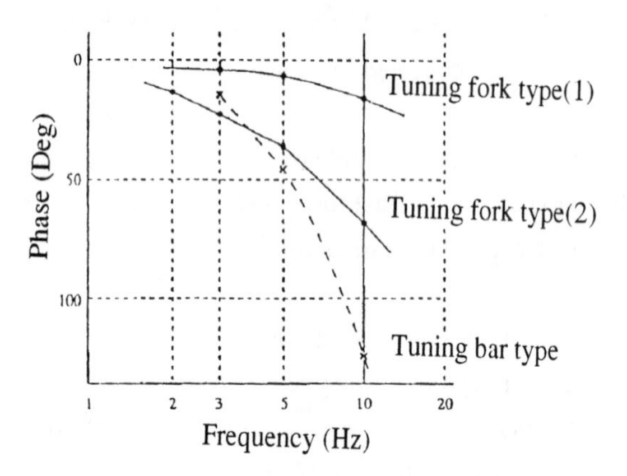

Fig 10 Phase Delay of Angular Rate Sensor

Fig. 10 shows the result (representative values) of phase delay from the output signal due to the applied angular velocity caused by oscillating the sensor module around the detection shaft. For an acoustic bar design, we used one angular rate sensor available on the market. The tuning fork (1) reprensents one of our design for 4WS application, and the tuning fork (2) is from our another of our designs for navigation application.

In this Fig. 10. assuming on oscillation frequency inputted in to the angular rate sensors to be Ωx = 10 Hz, the phase delay is about 13° in the 4WS angular rate sensor (1) and about 72° in the navigation angular rate sensor (2). The delay time τ appears as the following.

$$\tau_{(1)} = (1/10) \times 13° / 360° = 3.6 \text{ [ms]}$$

$$\tau_{(2)} = (1/10) \times 72° / 360° = 20 \text{ [ms]}$$

CONCLUSION

Table 3 summarizes the key specifications of the angular rate sensor developed by the above study.
The angular rate sensor is designed as a tuning fork with one-point suspension system. The driving elements and the sensing elements are aligned orthogonally in the tuning fork.
As shown in Photo 1, 2, 3 is an instance of angular rate sensor.

The improvements in the design and manufacturing process are summarized below;

1) In addition to the precise orthogonal fabrication technology of the elinvar material, we introduced measurement techniques and adjusting techniques in to the sensor's production. As a result, significant improvement of the zero point temperature drift has been achieved.

2) The impact immunity (for example, 1m drop durability) has been improved drastically by locating the tuning fork's center of the gravity in the post from the sensor's substrate.

3) The disadvantage of the non-resonance type, which is low sensitivity, was resolved by the unique circuitry designs such as current amplification with imaginary short at the input circuitry, double balance style implementation in the synchronous detection circuit and EMC protection.

Table 3 Key specifications of the angular rate sensor

Item	Specification
Operation temperature range	-40 to 80°C
Supply voltage	+8 to +15 Volts
Supply current	20mA MAX>
Sensitivity	25 ± 2.5mV/°/sec Vcc = 12V, T = -40 to 80°C
0 point output temperature drift	2.5 ± 0.1V, Vcc = 12V, T = 25°C 2.5 ± 0.6V, Vcc = 12V, T = -40 to 80°C
Output	0.5 to 4.5V Vcc = 12V, T = -40 to 80°C
Dynamic range	±50 deg/sec Vcc = 12V, T = -40 to 80°C
Output ripple	less than 10 mV Vcc = 12V, T = 25°C
Response	less than 9.0deg 5Hz, T = 25°C

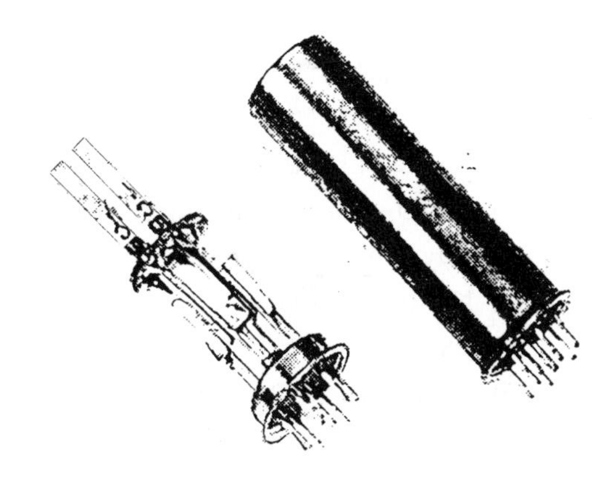

Photo 1 Angular rate sensor of ordinary type

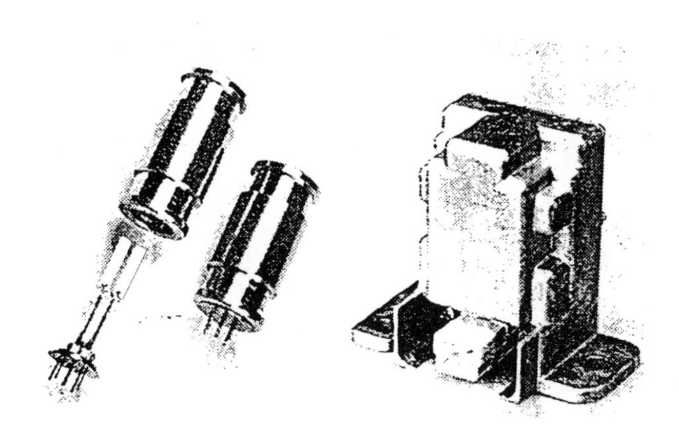

Photo 2 Angular rate sensor of four wheel steering (4WS) application

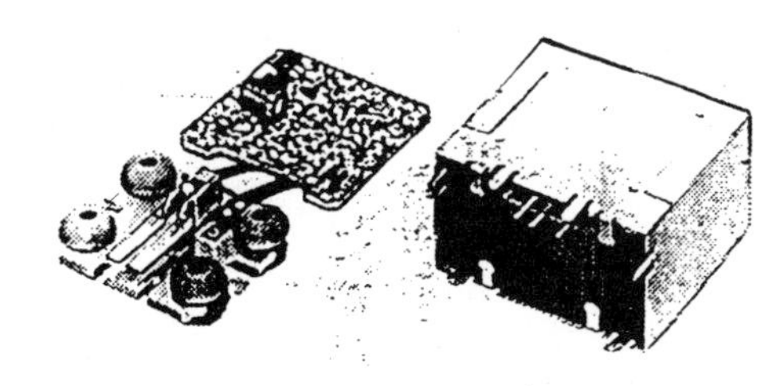

Photo 3 Angular rate sensor of navigation application

REFERENCE

1) USP 2,247,960
1) USP 2,247,960
2) USP 2,513,340

3) W. D. Gates, "Vibrating angular rate sensor may threaten the gyroscope", Electronics, June No. 10 130 (1968)

4) USP 4,628,734

5) Tanaka et al., "Development of the Active 4WS SystemUsing Yaw Rate Feedback control for the 1991 SOAROR", Toyota Technical Review, 19.50 (1991)

6) J. Terada, T Ichinose et al., "Development of angular rate sensor", JSAE Autum Convention Proceedings. No. 936. 1993

940833

Servo Acceleration Sensor for Automotive Applications

Sumio Masuda, Noboru Watanabe, Norio Miyahara, and Hiroshi Iiyama
Jeco Co., Ltd.

ABSTRACT

This paper describes a low-cost servo acceleration sensor suitable for automotive applications such as the antilock braking system or the traction control system. The sensor comprises a weight, a pair of flat springs, a magnetic circuit, an optical system and an electric circuit. The weight consists of a torquer coil and a pair of damper plates.

In order to reduce the cost, we developed a unique construction suitable for mass production.

The results of the performance tests show an excellent linearity and fine temperature characteristics over the operating temperature range of -40 to +85° C.

INTRODUCTION

Recent popularization of automotive control technology such as the anti-lock braking system (ABS) or the traction control system (TCS), increases a demand for acceleration sensors which generate analog output voltage proportional to acceleration. These sensors are required to detect acceleration range of ±1.5G ($1G=9.8ms^{-2}$) and frequency response from D.C. to approximately 10Hz. In addition, they are also required high reliability and low-cost.

In the present work, we developed a servo acceleration sensor which satisfies above requirements. Advantages of our servo acceleration sensor over various type acceleration sensors such as the piezoresistive sensor[1,2] are as follows:

(1) It is suitable for D.C. response.
(2) It has an excellent linearity.
(3) Expensive facility investments are not necessary. And,
(4) Our mechatronics technology can be effectively applied.

Although the conventional servo acceleration sensors have superior accuracy, linearity and frequency response characteristics, their complicated structure makes their cost extremely high. Therefore, they are applied only in limited areas such as aircraft control. Our mechatronics technology has made it possible to design a low-cost, high performance and high reliability servo acceleration sensor for automotive applications.

Table 1 Target Specifications of Servo Acceleration Sensor.

Item	Specification
Detecting Range	-1.5 to +1.5 G (-14.7 to $+14.7\ ms^{-1}$)
Operating Voltage	8 to 16 V(DC)
Operating Temperature	-40 to +85 °C
Storage Temperature	-40 to +105 °C
Frequency Response	DC to 30 Hz
Offset Voltage (-40 to +85° C)	2.50 ± 0.07 V
Gain (-40 to +85° C)	1.333 ± 0.067 V/G ($0.1360 \pm 0.0068\ Vm^{-1}s^{2}$)

DESIGN OF SERVO ACCELERATION SENSOR

TARGET SPECIFICATION - Table 1 shows the target specifications of the acceleration sensor for automotive applications. It is also needed to pass a series of environmental examinations for automotive use.

STRUCTURE - Figure 1 shows a mechanical structure of the servo acceleration sensor. The sensor consists of a weight, a pair of flat springs, a magnetic circuit, an optical system and an electric circuit.

The weight, which consists of a torquer coil, a coil bobbin and a pair of aluminium damper plates, is suspended by the flat springs which are also used as conductors to apply current to the torquer coil.

Two yokes and a pair of permanent magnets make a magnetic circuit which crosses the torquer coil and the damper plates. The magnetic circuit has two roles. One is to generate Lorentz force proportional to the current through the torquer coil. The other is to generate eddy current in the damper plates and thereby to damp the motion of the weight.

The block-diagram of the electric circuit and the servo system are shown in Figs. 2 and 3, respectively.

When the feedback loop shown in Figs. 2 and 3 is open, the equilibrium between the displacement x of the weight from the neutral point, x=0, and the acceleration α is shown as

$$m\alpha = kx, \quad \text{or,} \quad x = \frac{m}{k}\alpha, \tag{1}$$

where m and k are mass of the weight and spring modulus, respectively.

Displacement of the weight is detected by the optical system which consists of a light emitting diode (LED), an position sensitive light detector (PSD) and an optical slit on the weight.

When the feedback loop is closed, a differential amplifier and an error amplifier supply current,

$$I = \mu(x - x_0), \tag{2}$$

to the torquer coil, where μ is a constant. x_0 denotes the position of the weight where the torquer coil current I=0: as described later, the

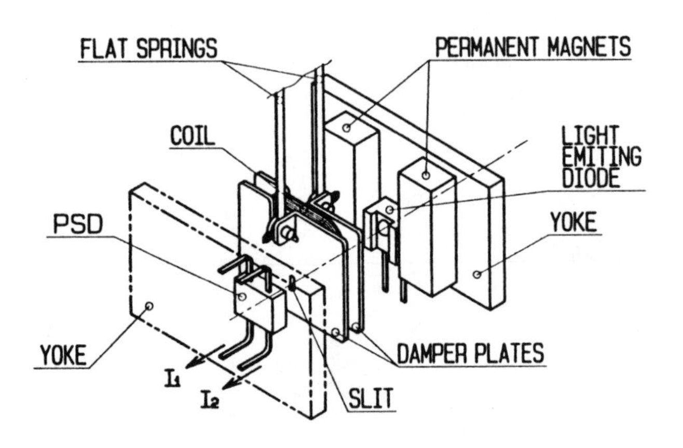

Fig. 1 Schematic structure of the sensor element.

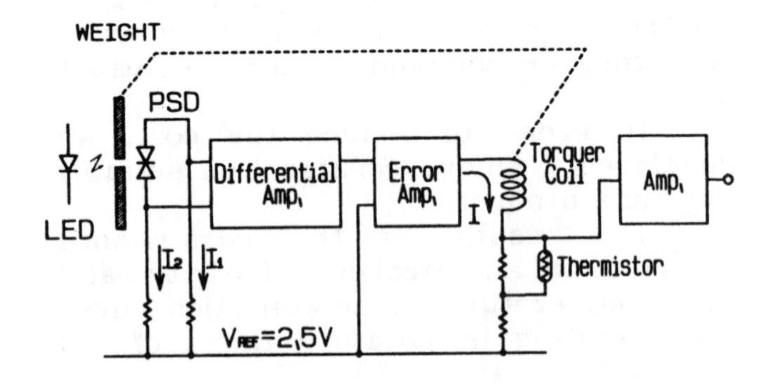

Fig. 2 Block diagram of the electric circuit.

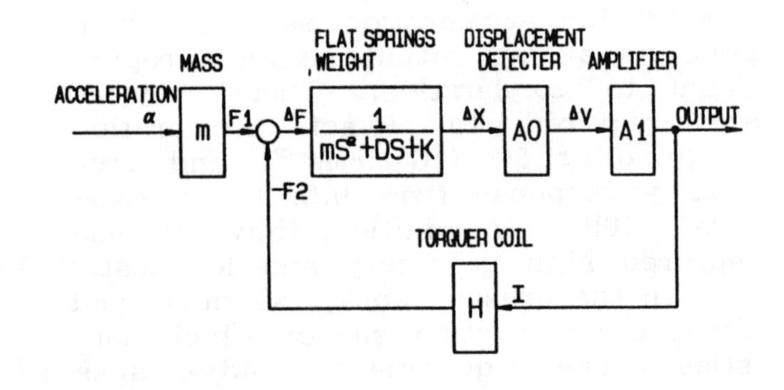

Fig. 3 Servo system.

position of the weight is restricted near x_o by the servo system. The equilibrium equation becomes

$$ma = kx + \kappa IB, \quad (3)$$

where κ and B are a constant and flux density, respectively. From EQs (2) and (3),

$$I = \frac{\mu m}{k+\mu\kappa B}a - \frac{\mu k}{k+\mu\kappa B}x_o = Ga + I_o, \quad (4)$$

where $G=\mu m/(k+\mu\kappa B)$ and $I_o=-\mu k x_o/(k+\mu\kappa B)$ are called current gain and offset current, respectively. EQ (4) indicates that I is proportional to the acceleration a. The sensor output voltage proportional to the acceleration is obtained by an I-V conversion circuit.

The position of the weight also obtained from EQs (2) and (3).

$$x = \frac{m}{k+\mu\kappa B}a + \frac{\mu\kappa B}{k+\mu\kappa B}x_o. \quad (5)$$

Because we designed the sensor as $k<<\mu\kappa B$,

$$x \simeq \frac{m}{\mu\kappa B}a + x_o. \quad (5')$$

The slope, $m/\mu\kappa B$, is far smaller than the slope, m/k in EQ (1). Therefore, the working position of the weight is restricted within a narrow range near x_o.

TEMPERATURE COMPENSATION - In EQ (4), all constants except B and x_o are not significantly influenced by temperature. EQ (4) indicates that the temperature dependence of B causes temperature dependence of both the gain and the offset of sensor output voltage, whereas, the temperature dependence of x_o causes a drift of the output offset voltage. In order to design an acceleration sensor for automotive use, it is necessary to compensate these temperature influences.

The magnetic flux density B is influenced by temperature because the flux from the permanent magnets decreases as temperature increases. We compensate this temperature dependence using a thermistor as shown in Fig. 2.

The position x_o is affected by temperature because of the temperature characteristics of the LED and the PSD. The PSD provides photocurrents I_1 and I_2 according to both the position and the intensity of incident light spot. These photocurrents are influenced by temperature.

Figure 4 shows an experimental result of temperature dependence of the relation between I_1-I_2 and the displacement of the weight. In this experiment, displacement is applied by inclining the sensor whose feedback loop is opened on purpose, and, I_1-I_2 is obtained from the output voltage of the error amplifier. As is obvious from Fig. 4, the least temperature dependence of I_1-I_2 is obtained when such the position of the weight is chosen as $I_1-I_2=0$.

As mentioned previously, the weight is restricted within a narrow range near x_o. Therefore, if x_o is so adjusted as $I_1-I_2=0$, then the offset current I_o is constant regardless of temperature so that the output offset voltage V_o may not be influenced by temperature. Such condition can be realized by adjusting the position of the weight as $I_1-I_2=0$ under no acceleration and no torquer coil current.

In addition to the temperature characteristics of the components described above, mechanical stresses due to thermal expansion of mechanical elements may become causes of the fluctuation of the sensor output. In order to

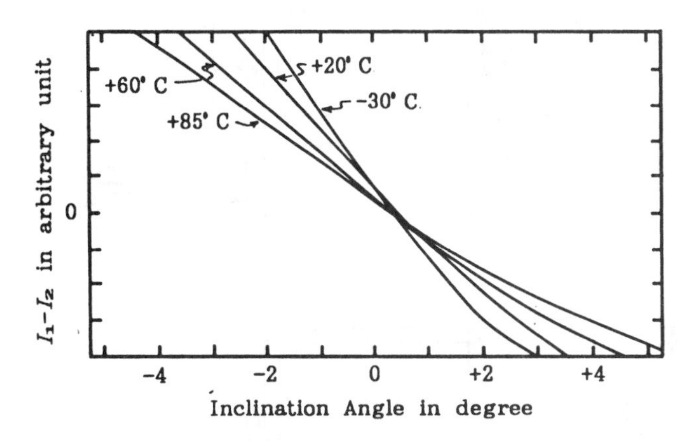

Fig. 4 A temperature dependence of the relation between the difference of photocurrents, I_1-I_2, and the displacement of the weight, x. Displacement is applied by inclining the sensor whose feedback loop is open. Abscissa indicates inclination angle which is corresponding to the displacement.

diminish the effects of the mechanical stresses we carefully designed the mechanical construction of the sensor (e.g. the structure of the weight which does not apply a stress on the flat springs) so that thermal expansion can be easily liberated.

CONSIDERATION TO DESIGN THE LOW-COST SENSOR - Our mechatronics technologies are of great use for designing the low-cost, high reliability acceleration sensor. Our designing policies to reduce cost and maintain performance and reliability are:

(1) to simplify the structure,

(2) to reduce the number of components,

(3) to use molded plastic or pressed metal instead of shaved metal as much as possible, and,

(4) to design the structure suitable for mass-production and automatic production.

An external view of the acceleration sensor is shown in Fig. 5.

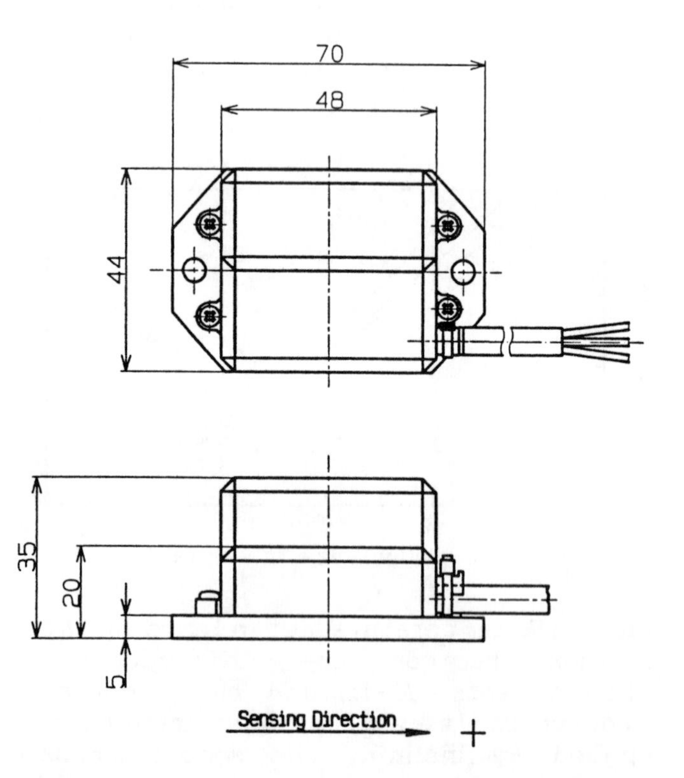

Fig. 5 External appearance of the servo acceleration sensor.

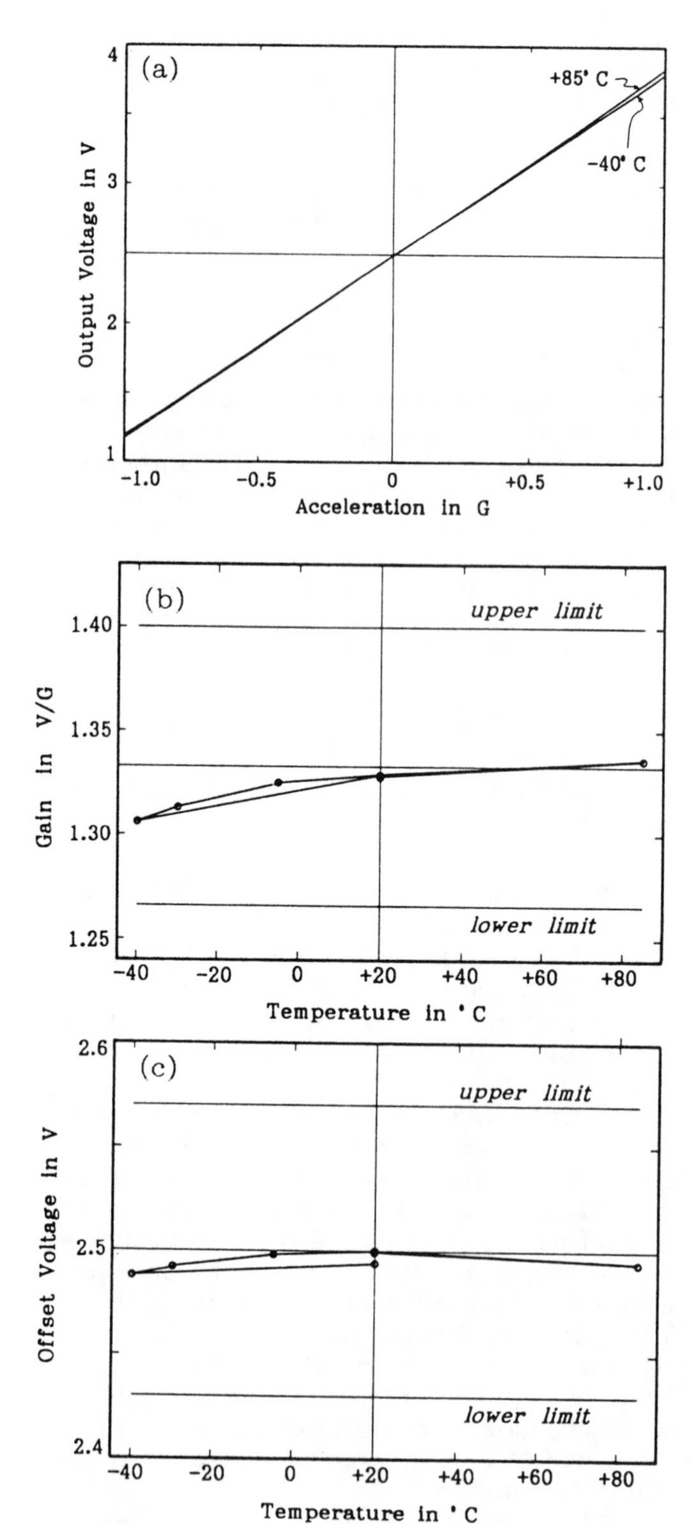

Fig. 6 Typical temperature characteristics of the servo acceleration sensor. (a) Output voltage vs acceleration at -40° C and +85° C. (b) Gain vs temperature. (c) Offset voltage vs temperature. Measurements are executed in order of +20° C, -40° C, -30° C, -5° C, +20° C, +85° C and +20° C.

RESULTS OF THE PERFORMANCE TEST AND DISCUSSIONS

Fifteen sensors were manufactured on an experimental basis. Figure 6 shows a typical temperature characteristics of one of the samples. In the temperature range -40 to +85° C, the sensor output does not influenced significantly by the temperature. Temperature characteristics of the other fourteen samples also well satisfied the target specifications. These results indicate that our temperature compensation technique and stress-free structure effectively fulfill the intended functions.

Measured frequency characteristics showed flat sensitivity in the frequency range of D.C. to 30Hz.

Reliability tests are now being performed and no significant problem has been experienced so far. Typical reliability test items are shown in Table 2.

CONCLUSION

We have developed a low-cost servo acceleration sensor suitable for automotive applications such as the antilock braking system. Temperature compensation, stress-free structure, low-cost design and reliability which are required for the automobile sensor are especially considered. Fifteen samples produced on an experimental basis satisfy the target specifications through temperature range -40 to +85° C.

We are convinced that our sensors will soon be brought to practical use.

Table 2 Typical Reliability Tests under Operation.

Endurance Reliability Test
High Temperature Test (storage)
Low Temperature Test (storage)
High Temperature Marginal Test
Low Temperature Marginal Test
Thermal Shock Test
Dew Formation Test
Temperature and Humidity Cycle Test
Vibration Test
Shock Test
High Voltage Marginal Test
EMI Test
Electrostatic Test

ACKNOWLEDGEMENT

The authors wish to express their gratitude to Mr. H. Takagi, Managing Director, Jeco Co., Ltd, for his encouragement and discussions for this work. They also thank to Mr. K. Tamura, Jeco Co., Ltd., for his assistance in experiments.

REFERENCES

[1] J. Bryzek, K. Petersen, L. Christel and F. Pourahmadi (1992): New technologies for Silicon Accelerometers Enable Automotive Applications, SAE Technical Paper, 920474.

[2] M. Tsugai, M. Bessho, T. Araki, M. Onishi and T. Sesekura (1992): Piezoresistive Acceleration Sensor for Automotive Applications, SAE Technical Paper, 920476.

930235

A New Type of Magnetic Motion Sensor and Its Application

Ivan J. Garshelis
Magnetoelastic Devices, Inc.

ABSTRACT

A new type of sensor capable of sensing the direction and velocity of motion of smooth surfaced metallic targets is described. The sensor consists of a small permanent magnet and a magnetic field sensor displaced from each other along the line of motion, each at a small fixed distance from the target surface. Operation is based on the motion dependent location, strength and polarity of magnetic field sources created within proximate target regions via their passage through the field of the magnet. The fields arising from these target regions are detected by the magnetic field sensors, typically Hall effect or magnetoresistive devices. With ferromagnetic targets, a non-volatile memory of the direction of last occurring motion is also provided. Utility of these sensing capabilities is illustrated by descriptions of applications for automatic turn signal canceling and antitheft devices, back-up alarm activation and for engine misfire detection via crankshaft speed variation.

INTRODUCTION

The advantageous use of magnetic field sensors (FS) together with permanent magnet (PM) field sources for sensing relative motion of mechanical parts is well established. In automotive applications such devices are routinely applied in a variety of position and rotational speed sensing applications. Both passive (wherein an emf is induced in a coil by a time varying magnetic field [1, 2]) and active (e.g. Hall effect or magnetoresistive [3, 4]) FSs find use in these applications. Advanced designs employing purposefully shaped magnets to linearize throttle position sensing [5], in addition to the recent integration of both PMs and FSs within ball bearings [2, 4, 6], attest to the acceptance of this basic technology for use in the automotive environment. Further additional applications have included steering wheel position sensing, distributorless ignition, tachometers, and ABS systems. While the application areas expand and the specific device constructions become ever more imaginative, all of these motion sensors apply the same physical phenomenon and thus share a similarity in certain constructional features. This is illustrated in Fig. 1.

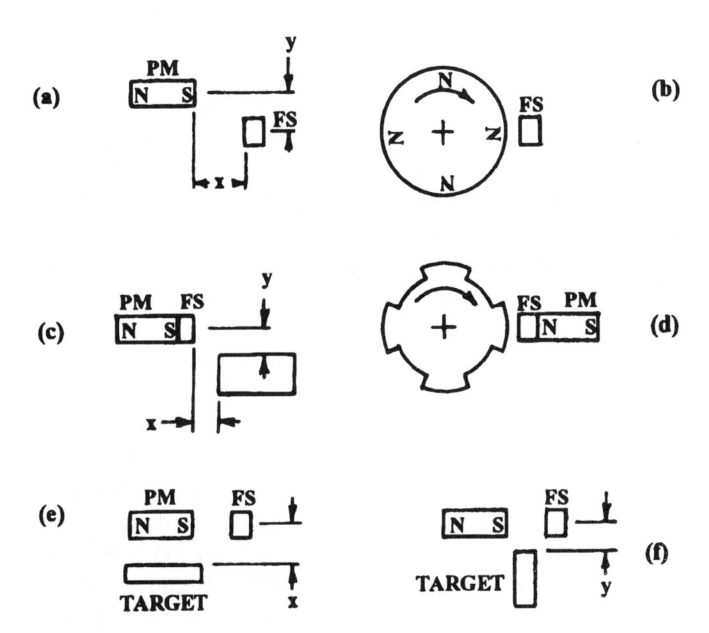

Fig. 1. Arrangements of elements in conventional types of magnetic motion sensors. Motion is sensed by the changing flux through the FS brought about by:
(a), (b) Varying proximity to a PM pole region.
(c), (d) Varying proximity to a flux enhancing ferromagnetic target.
(e), (f) Varying proximity to a flux bypassing ferromagnetic target.

Operation of these sensors specifically depends on the spatial gradient in the field intensity of the PM, i.e., the field is strongest at the poles and falls off rapidly with distance from a pole. Thus in the simplest case shown in Fig 1 (a), the FS detects a stronger field as the distance, x or y, separating it from the magnet pole decreases. It follows that the FS in (b) will experience periodic increases, decreases and reversals of field direction as the multipolar magnet rotates. The changing position of a "soft" ferromagnetic "target" (i.e., a material having relatively high permeability and low coercivity) within the sensed field region can also alter the field intensity at the FS location. Thus, decreasing x or y in (c) will increase the field intensity at FS and rotation of the toothed target wheel in (d) will cause periodic fluctuations in the field intensity at FS. "Flux shunting" arrangements, illustrated in (e) and (f), act oppositely, i.e., decreasing target distances x or y decreases the field intensity at FS.

In all variations of these arrangements, the motion being sensed affects the field intensity at the FS location by altering a dimension of an air gap in the magnetic circuit, thereby varying the relative permeance of those flux paths that are common to both the PM and the FS.

Many applications use digital techniques wherein the motion of interest is divided into a discrete number of uniform portions. This is typically accomplished either by patterning a "target" member with periodic physical features, e.g., teeth as shown in Fig. 1 (d), or with magnetic pole pairs as shown in Fig. 1 (b). In either case, the motion of the target (clearly not limited to rotation) is mirrored in a corresponding pattern of magnetic flux changes through the FS. Motion of the target is thus detected in distance increments equal to the pitch of its magnetically salient features. The varying output signal of the FS is often itself digitized to allow only one or another of two discrete values, switching between these at critical rising or falling flux densities. Position of the target is tracked by counting these output level switchings while speed is determined from either the time interval between switchings or from the switching frequency. The direction of motion is determined either from asymmetrical features of the output level switchings or from the phase sequences of signals from two (appropriately spaced apart) field sensors.

Using these conventional methods, the unambiguous detection of motion requires at least one switching event. The determination of speed requires at least two such switchings (more often three, to determine a full period). Acceleration cannot be determined until after two speed measurements have been made and direction of motion cannot be ascertained with a single FS until it can be assured that acceleration is not the source of an observed asymmetry. For rapid determination of start-up direction, tracking of oscillatory motions and monitoring changing accelerations, a target pattern with fine pitch is required. This increases the complexity, required precision, and cost of the sensing system.

A NEW TYPE OF MOTION SENSOR

A different type of magnetic device for sensing both the speed and the direction of motion (i.e., the velocity) of a target surface that requires neither a pattern of periodic salient physical features nor of magnetic poles has been described [7, 8]. In addition to operating with smooth targets, the sensor output signal at zero speed can, with ferromagnetic targets, also indicate the direction of last occurring motion.

Figure 2 shows the basic elements of this new motion sensor (collectively called "DMS" from Direction of Motion Sensor) and their functional relationships. The DMS is seen to consist of a PM and a FS, each mounted rigidly at small distances from a target surface and displaced from each other along a line parallel to the target motion. The polarization of the PM and the sensing axis of the FS are generally oriented normal to the target surface (other orientations of the FS are

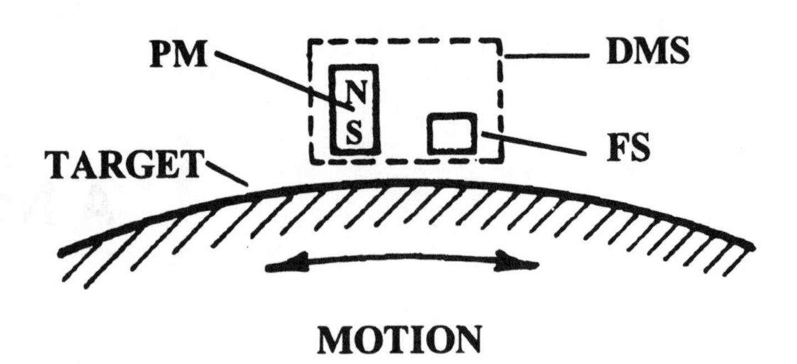

Fig. 2. Arrangement of elements comprising a DMS motion sensor.

sometimes preferred). The FS is typically a Hall effect or magnetoresistive element but, as will become understood, may for certain applications, be a passive type of field sensing element. In general, a suitable target will have a smooth, uninterrupted (over the range of motion of interest), flat or uniformly curving surface. The unvarying separating distances between the PM, FS, and target, clearly preclude the conventional (i.e., variations in permeance) modes of operation. The lateral displacement of the PM relative to the FS, while seemingly a trivial constructional variant nevertheless makes possible two entirely new and different modes of operation. It is appropriate to separately examine the underlying principles and operation of each since one derives from the magnetic remanence of ferromagnetic targets while the other depends on asymmetrical eddy currents induced in conductive targets.

OPERATION

A. FERROMAGNETIC TARGETS - Operation of the DMS with ferromagnetic targets requires that the target first be magnetically "initialized" by a single movement past the PM over the full active range of motion. As shown in Fig. 3, this initial movement through the locally strong field of the PM will convert surface regions (depth will depend on target characteristics and the strength of the PM relative to the target size) of an initially unmagnetized target (a) into the uniformly remanently magnetized condition shown in (b). (Note that this magnetization "points" toward the PM. If the N and S poles of the PM are interchanged, the remanent magnetization would still be parallel to the line of motion, although now "pointing" away from the

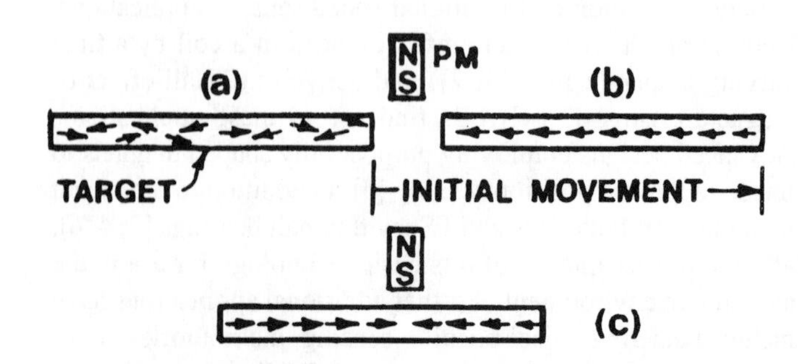

Fig. 3. Unmagnetized target (a) is remanently magnetized by initial movement to (b). Subsequent to and fro movements divide target into oppositely polarized regions on either side of PM (c).

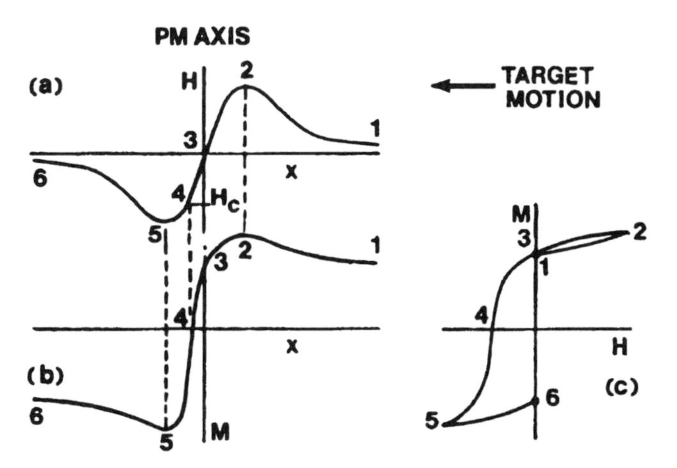

Fig. 4. Sequence of magnetization reversal events by passage of target from *right to left* through longitudinal field of PM.

PM.) Subsequent movement in the opposite direction (c) reverses the polarity of the remanent magnetization in those portions of the target which again pass through the strong field region of the PM. Further forward or reverse motion of the target will always leave it divided into two oppositely polarized regions separated by a transition zone located at the current PM position. The sequence of events by which the magnetization polarity is reversed by passage through the field of the PM is illustrated in Fig. 4.

Curve (a) in this figure simply shows how the longitudinal component of the PM field (H) varies with distance (x) from its axis. Both H and x are relative values since the absolute value of H depends on the PM material and size as well as on its distance from the target surface, against which, x is normalized (Only the longitudinal component of field need be considered since the axial component is the same at (x) as at (-x). Moreover, the relatively low coercivity of steels used for mechanical parts (typically 5 to 50 Oe) will, by self demagnetization, preclude significant remanent magnetization normal to the target surface.) Curve (b) shows how the magnetization M at points x along the target varies in response to H This curve is developed by combining curve (a) with the hysteresis loop characteristic of the target material shown in (c).

A point on the target, initially far from the PM axis (where H = 0), remanently magnetized in condition (1), is seen to encounter an increasingly intense field as it moves towards the PM (1 - 2 in (a)). Since this field has the same polarity as M, it tends to increase M along the path 1 - 2 in (c). Thus, when the point on the target reaches the position of maximum H in (a), it is also at its peak value of M, as shown in (b) and (c). As this point continues to move toward the PM axis, H decreases toward zero (2 - 3 in (a)) and M relaxes to its remanent value (the same as 1) along path 2 - 3 in (c). After crossing the PM axis, the point encounters an increasingly negative field and this acts first to diminish the remanent magnetization until becoming zero where $H = H_C$, (the coercive field) at 4. With further movement, M again starts to increase, now with reversed polarity, and continues to increase until reaching a peak negative value when the point reaches the position of maximum H at 5. As the target continues to move, the point being followed encounters a continuously diminishing H, dropping effectively to zero at 6 where M has again relaxed to its remanent value but now reversed in polarity from its starting condition at 1.

While Fig. 4 has been used to describe the sequence of magnetization altering events at a single point in the target, it should be clear that curve (b) is effectively a "snapshot" of M at all points in the target that are within the 1 - 6 transition zone at any instant in time. This being the case, it can be seen that by far the greatest *change* in M per unit distance along the target is in the 3 - 5 region and this, in fact, is many times larger than the *change* in M within the equal length 2 - 3 region. The *changes* in M within the 1 - 2 and 5 - 6 regions are seen to be small, gradual and not significantly different from each other. Thus target motion to the left establishes a magnetization gradient of significant steepness only within the target region to the left of the PM axis.

In similar fashion, target motion to the right will develop the steepest magnetization gradient on the right side of the PM axis. This is shown in Fig. 5 The creation, and directional shift in location, of a steep magnetization gradient are fundamental to the operation of the DMS with ferromagnetic targets. The magnetic hysteresis of the target material is clearly necessary for the creation of these gradients..

Since spatial gradients of magnetization act as sources of magnetic fields, such fields will arise from the gradients created in the target. The field of greatest strength arises from the region having the steepest gradient (e.g., 3 - 5 in Fig. 4, 7 - 9 in Fig. 5), and this field can have significant strength even in the space outside the target. Since for both directions of motion the steep gradient is created on the "downstream" side of the PM axis, the field is naturally stronger in that region of space than on the "upstream" side. It should thus be clear that a FS, displaced to one side of the PM axis (as in Fig. 2), will experience a different field intensity depending on whether, by the motion of the target, it is located upstream or downstream of the PM. It should also be clear that the magnetization distribution created within the target by its leftward (Fig. 4(b)) or rightward (Fig. 5(b)) movement is maintained by the field of the PM which is fixed in the stationary frame of the DMS. Thus, although either of these magnetization distributions may exist anywhere along the target, they always reside within the transition zone at the PM location. As a consequence, the asymmetry in the fields on the two sides of the PM continues to exist even after the target stops

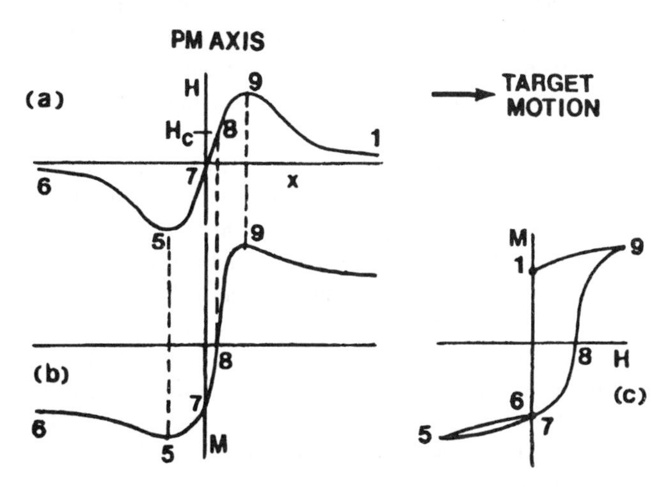

Fig. 5. Sequence of magnetization reversal events by passage of target from *left to right* through longitudinal field of PM.

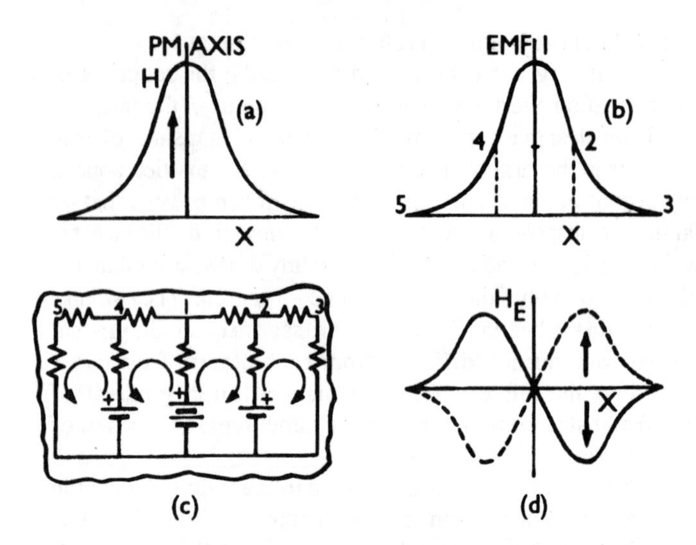

Fig. 6. (a) Variation in the field intensity H with distance x (in the line of motion) from the PM axis.
(b) Variation in the emf induced in a transverse target element moving right to left at constant velocity through the field shown in (a). (This emf distribution would be of opposite polarity for left to right motion.)
(c) Circuit (with the simplification of lumped elements rather than the actual, continuous distribution) showing the resulting eddy currents flowing in a target plane normal to the P< axis. (The directions of current flow would all be reversed for left to right motion. The resistance elements shown are in reality complex impedances having inductive components.)
(d) Variation with distance x in the field intensity and polarity of the field created by the eddy currents. The dashed line shows the field resulting from left to right motion.

moving. In fact, the field sensed by the FS is alterable *only* by a reversal in the direction of target motion. The DMS therefore provides a non-volatile indication of the direction of last occurring target motion.

A somewhat smaller additional magnetization gradient will be created on the *upstream* side of the PM in continuously rotating targets (or those rotating more than one full revolution). Details of operation of the DMS under these conditions are explained in [7]. It is sufficient to indicate here that the DMS provides the same non-volatile indication of the direction of last occurring motion with rotating as with oscillating targets.

B. CONDUCTIVE TARGETS - In moving, electrically conductive targets, ferromagnetic or not, velocity dependent emfs are induced in those regions passing through the normal field of a PM. Due to the gradient in the magnetic field intensity, these emfs vary with position and thus cause local electric currents, e.g., eddy currents, to circulate. Fig. 6 provides a simplified illustration of this effect. As with all currents, these eddy currents have associated magnetic fields. In this case, since they originate within areas of the target proximate to the PM, they can be readily sensed by the nearby FS. The elementary conclusion of Fig. 6, namely, that the eddy currents will circulate in opposite directions on either side of the PM, is supported by consideration of Lenz's Law. Thus, since points on the target which are *approaching* intersection with the PM axis are passing through a region of increasing field intensity, the reaction field created by the circulating eddy currents will oppose this increase. In contrast, after passing the PM axis, points on the target move through a region of diminishing field intensity and thus the reaction field on the downstream side will now oppose a decrease. Clearly then, the eddy current field on the upstream side of the PM axis will be of opposite polarity to the PM field but will have the same polarity on the downstream side. This is indicated in Fig. 6(d).

The symmetrical PM field distribution shown in Fig 6(a), which exists in the presence of a stationary target, is thus seen to be altered by the asymmetrical contribution of eddy current fields created by the target's motion. The magnitude of the eddy current contribution to the total field depends on the eddy current density, which in turn, depends on the intensity and gradient of the PM field, the electrical conductivity of the target material and on the target velocity. Simple theory would predict a linear dependence on all of these factors. Previous studies [9,10] of the eddy current fields arising from non-smooth, e.g., toothed, targets rotating past a permanent magnet, indicate a linear dependence on velocity and target conductivity. These studies neither considered the altering effect of the eddy current fields on the generated emfs (analogous to armature reaction in electrical machinery) nor the inductances of the current paths (which limit the rate of change of current) nor non-linear effects of magnetization in ferromagnetic targets. In any case, the asymmetry has obvious directionally dependent characteristics. In this regard it is analogous to the situation previously described with ferromagnetic targets. Thus it should again be clear that a FS, displaced to one side of the PM axis (as in Fig. 2), will experience differing field intensities depending on whether, relative to the ongoing motion of the target, it is located upstream or downstream of the PM. In this case, the magnitude of the difference in intensities will vary directly with the speed of the target, dropping to zero at zero speed. This is the basis for operation of the DMS with moving conductive targets.

Most ferromagnetic target materials are also electrically conductive and thus operation of the DMS with *moving* ferromagnetic targets will expectedly show a combination of both magnetization gradient and eddy current effects. The expected

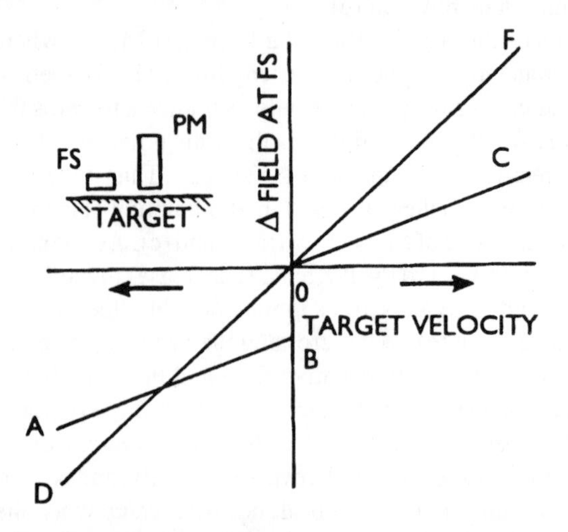

Fig. 7. Transfer functions of DMS for both ferromagnetic conductive targets (ABOC) and for nonferromagnetic, conductive targets (DOF).

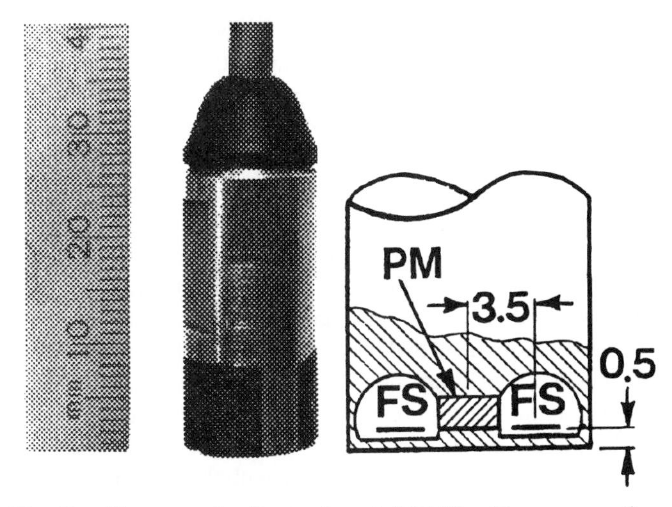

Fig, 8. Photograph of experimental DMS with cross-section showing internal details.

changes in field intensity at the FS for both ferromagnetic and non-ferromagnetic, conductive targets as a function of velocity, are shown in Fig. 7. The fact that, for any one direction of motion, the field changes associated with both the magnetization gradient and the eddy currents have the same polarity, is clearly a fortuitous circumstance.

RESULTS WITH EXPERIMENTAL SENSORS

Fig. 8 shows the details of construction of an experimental DMS. A 2.5 mm square by 1.5 mm thick Nd-Fe-B PM with two Hall effect FSs (Texas Instruments Inc. TL 173C) were mounted on an aluminum frame and encapsulated in epoxy. As shown in the cutaway, the chips in the IC package and the face of the magnet were approximately in the same plane, 0.5 mm back from the active face of the DMS. The tubular body and the flat were for convenience in mounting in a variety of experimental fixtures. Two FSs, one on either side of the PM were used to reduce sensitivity to ambient fields, including rejection of the signal due to the common field of the PM itself and to provide symmetrical output signals of opposite polarity for the two directions of motion. Each FS had a zero field output signal of 6.0 V when powered by 12 Vdc and a polarity dependent sensitivity to field of 1.5 mV/Oe. The difference in the two FS outputs was taken as the DMS output signal, V_0.

The effect of varying magnetic characteristics (by heat treatment) of a simple high carbon steel wire target are shown in Fig. 9. While the three targets exhibit similar maximum sensitivities and eventual fall off rates with increasing air gap between the DMS face and the wire surface, their respective maximum sensitivities are seen to occur at different gaps. Insensitivity to air gap variation over a modest range is seen to be possible. For any one combination of PM and target characteristic, this desirable feature can usually be optimized at a specific lateral displacement of the FSs from the PM axis.

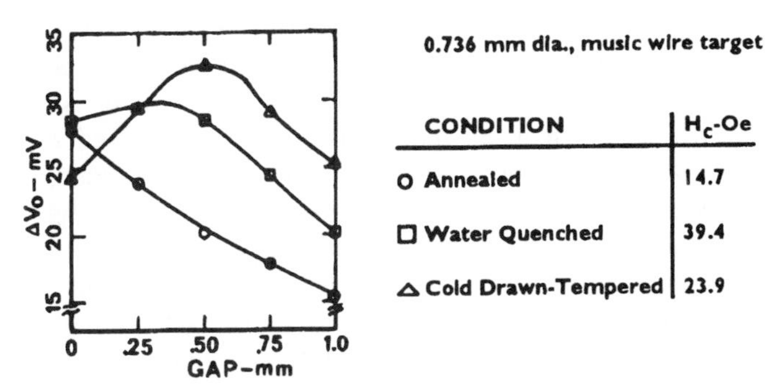

Fig. 9. Variation in output signal of the DMS shown in Fig. 8 with distance from music wire targets in the metallurgical conditions indicated.

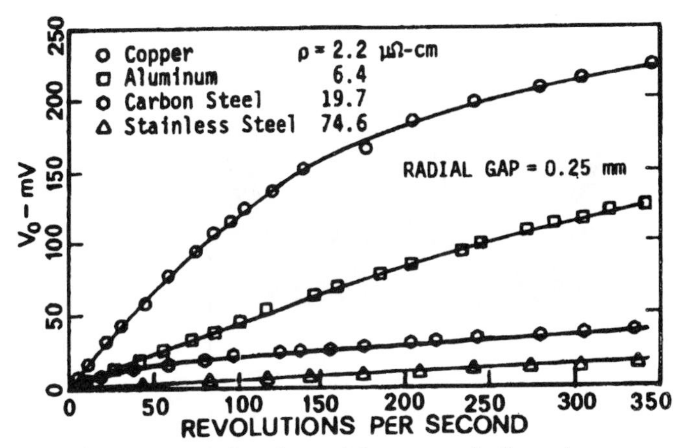

Fig. 10. V_0 vs rotational speed for targets indicated.

Speed dependent signals for a variety of target materials are shown in Fig 10. The data for all materials except the high resistivity stainless steel show a non-linearity with speed characterized by a downward curvature. This tendency is most pronounced in the high conductivity copper and in the ferromagnetic carbon steel targets. The data also clearly shows the expected decrease in sensitivity with increasing target resistivity.

The ability to sense rapidly changing eddy currents is illustrated in Fig. 11. The three, narrow chordal slits in the target surface serve to increase the resistance of the eddy current paths. The resulting decrease in eddy current as these regions pass the PM is manifested by the large drops in signal apparent in the oscillogram. The three small holes, being further from the target surface show a much smaller, though still visible effect. Radial runout is also indicated by the slowly varying signal baseline. The "DMS" used to obtain this oscillogram was a standard differential magnetoresistive sensor (Siemens FP 212 L 100-22) designed for the conventional modes of sensing illustrated in Fig. 1. Nevertheless, the device contains, in one small package, two spaced apart (1.6mm) magnetoresistors (the FSs) behind which is mounted a samarium-Cobalt biasing magnet (the PM). While this is not the preferred location for the PM (since not designed for this new mode of use), its performance is nevertheless adequate for some applications. Other commercially available devices containing two spaced apart Hall effect sensors unfortunately have switching rather than linear outputs (e.g. Siemens TLE 4920 and Allegro Microsystems Type 3059), thus making them unsuitable for most DMS applications.

APPLICATIONS

The unique modes of operation of this motion sensor together with its novel capabilities clearly recommend its consideration for use in new application areas. Also, as is often the case following the introduction of any new capability,

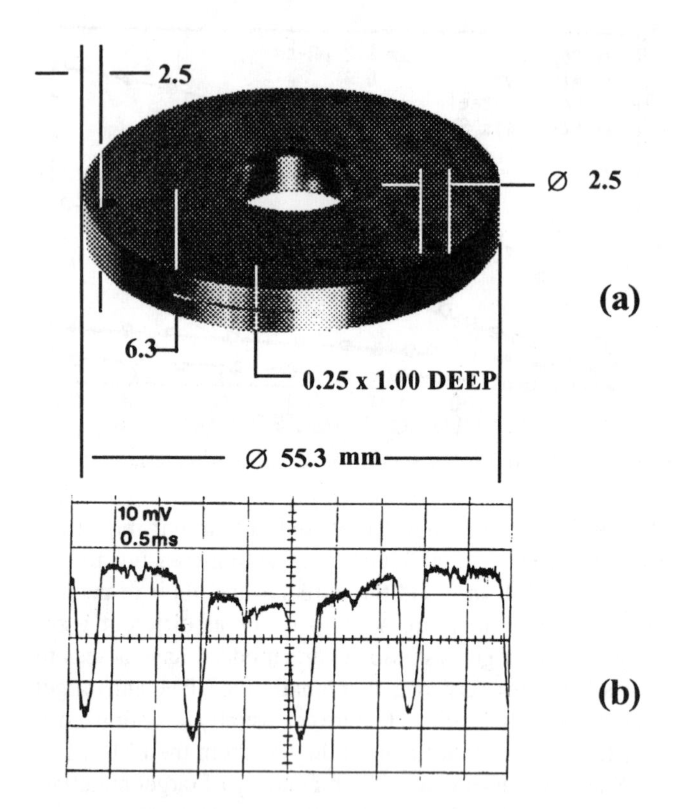

Fig. 11. (a) Cast aluminum target disc having 3 holes and 3 slits as shown.

(b) Oscillogram showing the output signal from a dual magnetoresistive DMS in a bridge circuit excited with 5V dc. The face of the sensor was 0.25mm from the rim of the disc which was rotating at 16,400 RPM.

previously unrecognized, or marginally solved problems become seeable in a new light, and this invites their reconsideration. In this section, a few such applications, related to the safe and efficient operation of engine driven motor vehicles, are identified. They are chosen to illustrate the utility of this new sensor's non-volatile direction memory capability and its real time velocity sensing ability.

1. TURN SIGNAL CANCELING DEVICE - Few, if any, drivers of road vehicles have not encountered, either by observations of other vehicles or by experiences with their own vehicle, failures of the automatic turn signal canceling feature incorporated in conventional turn signal switches. Apprehension, confusion and annoyance are common reactions to driving behind a vehicle flashing a false signal of intention to turn, as is the feeling of embarrassment upon becoming aware of one's own such errant signaling. Considering the important safety implications of proper indications of turn intention, these are justifiable reactions.

The problem of imperfect reliability in the operation of automatic canceling mechanisms in turn signal switches has been acknowledged and various schemes have been suggested or employed to alert the driver of the offending vehicle to the need to manually cancel false turn signal indications. For example, Chrysler [11] has, in recent models, installed a system which sounds a chime if the vehicle is driven further than 0.8 km at or above some modest speed with the turn signal on. Cadillac [12] offers a system that increases the volume of the turn signal under similar conditions. Considering that the sound associated with the turn signal has already escaped the driver's notice (possibly due to road noise, loud music, hearing problems, etc.) a chime or volume increase may be only little more successful in drawing attention to the false signaling. Considering also the maturity of the designs of presently used mechanical canceling mechanisms, and their excellent but still dangerously imperfect reliability, it becomes clear that some other approach (e.g., a non-mechanical device, either alone or as a redundancy) might significantly reduce turn signal canceling failures.

When it is realized that the cams, detents and springs used in mechanical canceling mechanisms function to detect the reversing rotation of the steering shaft at the completion of a turn, the potential utility of the DMS for this application is immediately understood. In fact, since with the DMS there are no mechanical parts to encircle the steering shaft, it becomes more than merely an alternative to conventional canceling mechanisms. The ramifications of this are more far reaching than immediately apparent since it opens a host of new design options for turn signal operation, including those associated with the location and mode of operation of the turn signal initiating switch itself. With a DMS providing the automatic canceling function there is no longer a need for parts of the canceling mechanism of these switches to be installed axially over the top of the upper steering shaft. Replacement or maintenance, if needed, will no longer require removal of the steering wheel, which in many modern vehicles also involves removal of the air bag and associated hardware.

There is in fact, for this application, no functional necessity to locate the DMS on or near the switch used to initiate the turn signal. Since there is a correlated motion between all portions of the steering system, from the steering wheel to the steerable, road contacting wheels, all parts of this system undergo a coordinated reversal in their motions during the return to straight ahead travel. Thus there is an option to locate the DMS away from areas already burdened with a high density of control and/or other devices. The requirements for a satisfactory target surface (e.g., steel, smooth, having a single degree of freedom, accessible) are clearly met anywhere along the upper steering shaft. Less obvious surfaces such as the smooth side of the rack are also possible. Optimal location for the DMS can be determined by wiring harness routing and other considerations.

If the signals are to be operated by a three position switch of the conventional (minus the mechanical latching feature) type, optional canceling by the driver can still be accomplished with the customary action. A circuit diagram illustrating how a turn signal system might incorporate a DMS together with a conventional type of initiating switch is shown in Fig. 12, with a detailed explanatory caption. The left or right turn signal, initiated by a momentary movement in the appropriate direction of the turn signal lever, is electronically latched "on" until canceled by the DMS as it senses the start of the steering system's return toward the straight ahead position. A momentary movement of the turn signal lever in the opposite direction will also serve to cancel the turn signal. Since motion reversal detection sensitivity of the DMS can be tailored by appropriate choice of the PM, PM-FS spacing, target coercivity and air gap, no special provisions for "lane changing" signals would be required.

With only a modest electronic addition, the same DMS

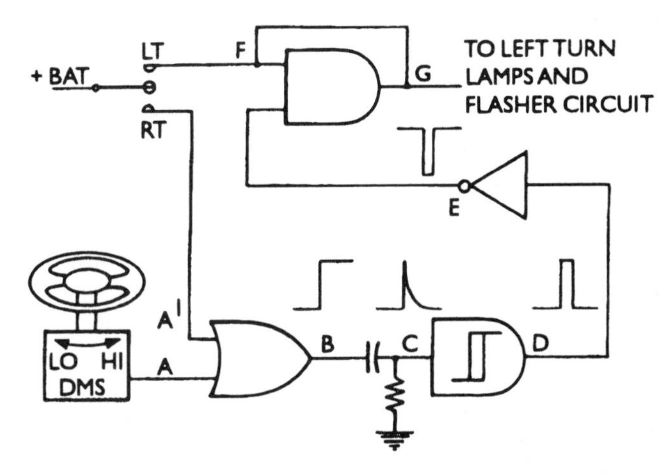

Fig. 12. Schematic illustration of left turn signal operation using the DMS.

Prior to initiating a left turn signal, the output of the DMS may be either HI or LO depending on whether the last movement of the steering wheel was to the right or to the left, respectively.

If it is HI when the left turn (LT) switch is momentarily actuated, B will be HI, else B will be LO. In either case C will be low as will be D. E will therefore be HI for either condition of the DMS output.

When the LT switch is momentarily closed (F and E now HI), G goes HI, starting the "intent to turn left" indication in the conventional manner. G is latched HI through the feedback connection G - F.

During the left turning maneuver, B will either remain or become LO. In either case E remains HI. As the steering wheel is turned to the right in order to bring the vehicle back to straight travel, the DMS output becomes HI. The transition to HI appears at C, is squared at D and inverted to a LO (of short duration) at E, unlatching the HI at G, thereby canceling the turn signal. Momentarily actuating the RT switch will cancel a LT turn signal in the same way via A' going HI. The RT turn signals operate analogously.

used to cancel the turn signal indication may also function as an antitheft device. Each time movement of the steering wheel is reversed, whether or not turn signals are used, the DMS output signal level switches. A preset counter can, unless disabled by a knowing driver, be used to trigger an alarm system in the vehicle, or otherwise indicate that the vehicle is being unlawfully operated. While such a system does not prevent a theft, it may well increase the likelihood of apprehending the suddenly thwarted thief.

2. BACKUP ALARM - As a matter of safety (for pedestrians and following vehicles) and of convenience (when driving at night) modern road vehicles are equipped with backup lamps [13] which are turned on whenever the transmission control mechanism is in the "reverse" position. Horns and other audible alarms on mobile, construction and industrial machinery and on school buses are similarly activated [14]. Forward warning horns are activated in analogous fashion [15]. The value of the backup alarm as a safety enhancing feature in automobiles is also recognized [16].

Nevertheless, all of these travel direction signaling devices only indicate the driver's *intended* direction of *powered* travel. They fail to detect, and thus can provide no warning if, e.g., the vehicle is (deliberately or accidentally) allowed to roll in a direction not in congruity with the selected transmission position. There are many instances, especially involving backward rolling, where such anomalous motion constitutes a hazard. A DMS sensing the rotation directions of any wheel, or rotationally connected part, together with the transmission position switches, can be used to provide a visual and/or audible signal to alert both the driver and persons who may unwittingly be in harm's way. By using a toothed or otherwise periodically inhomogenous target, the DMS signal can also be processed to distinguish past from present action.

3. ENGINE MISFIRE DETECTION AND DIAGNOSTICS - Exhaust emissions of motor vehicles constitute a major source of air pollution. Concern for air quality has a considerable impact on the design of engines, engine controls and on fuel and exhaust systems. It is common practice for regulatory agencies to require periodic (typically annual) testing of exhaust emissions to insure continued compliance with statutory standards. Means to more promptly alert the driver to engine dysfunctions affecting exhaust emissions would further benefit air cleanliness. Engine misfires, too subtle for the driver to notice, have been identified as causing a significant increase in exhaust pollutants. Overheating of the catalytic converter coating by ignition of unburned fuel, discharged from an engine when a cylinder misfires, diminishes its efficiency and can cause permanent damage. Diagnostic systems to detect misfires have been mandated for inclusion, by 1997, on all light and medium duty vehicles sold in California [17]. Similar regulations in other states and by federal agencies is following.

Misfire detection systems based on a variety of different strategies are being developed in response to this mandate. Common to all of these approaches is the identification of a misfire by the appearance of an anomaly in the normal operation of the engine. Since, by definition, a misfire is a power stroke event characterized by a reduction in combustion efficiency, a misfire can be recognized by a diminished (relative to normal) effective cylinder pressure during the power stroke [18], a reduction in the resulting torque applied to the crankshaft [19], or even a temporally related fluctuation in the composition of the exhaust gas. Since these are challenging parameters to measure directly, especially with low cost, reliable and minimally intrusive sensors, the presently most well advanced misfire detection systems rely instead on the more easily measured anomalous fluctuations in instantaneous crankshaft speed [20-22]. These speed fluctuations reflect the varying angular acceleration imposed on the crankshaft by the net generation or utilization of torque associated with the cycle of events *normally* occurring in each cylinder. A missing or reduced torque impulse during the power stroke of any cylinder is mirrored in the acceleration pattern and this is evidenced by an anomaly in the speed fluctuation waveform. Common practice with these systems is to digitize the rotational position of the crankshaft with toothed wheels and measure (in clock periods) time intervals for successive, intertooth rotations of the crankshaft. The already present flywheel gear is often used for this purpose to avoid additional parts and related costs. Conventional magnetic gear tooth sensors are used to provide the signals to

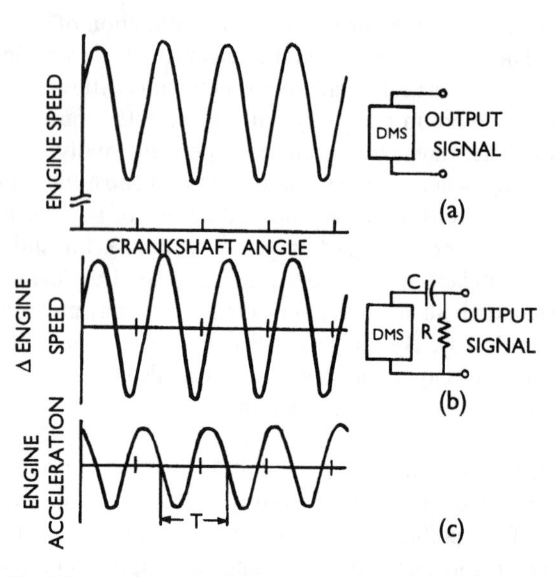

Fig. 13. (a) Raw output signal of the DMS varies with the instantaneous engine speed (simplified waveform shown).
(b) Capacitive coupling (high pass filtering) eliminates the average component of velocity.
(c) Differentiating (RC<<T) provides the acceleration waveform.

start and stop the clock period counters. Variations in clock count per tooth at constant average speed (constant clock count per revolution) manifest the fluctuations in instantaneous speed. Misfires are indicated by anomalies in the "normal" clock count fluctuation waveform. Various schemes that involve filtering, engine modeling, and computation are currently used to reliably identify the occurrence of an anomaly over the full range of engine speed, load and road conditions.

The DMS provides an attractive alternative to the above described method of sensing crankshaft speed fluctuation. As previously explained, the DMS provides an output signal proportional to target velocity when used with a smooth, conductive target. Thus, by mounting a DMS proximate to an appropriate surface of the crankshaft (or opposite a face or rim of a plate or disc of a more conductive material, e.g., aluminum, mounted on the crankshaft), a signal proportional to instantaneous velocity becomes directly available without computation. This is shown in Fig. 13(a). The fluctuating component is readily separated from the average by coupling the DMS output to its load through a capacitor; this is illustrated in Fig. 13(b). (It should be noted that there is no need for two, differentially connected field sensors if only the fluctuating component of velocity is required, since the steady component of the signal due to the field from the permanent magnet is not passed through the capacitor.) If the product of RC is small relative to the signal period, the voltage across R will be proportional to $d\omega/dt$, i.e., to the instantaneous crankshaft acceleration, α. This is illustrated in Fig. 13 (c). If the Hall effect field sensor is replaced with an inductive sensor, the raw DMS output signal will be directly proportional to α, since the voltage induced in the coil is proportional only to the time rate of change of the eddy current field (which is itself proportional to velocity).

Unlike the digital speed sensing methods in which clock counts per tooth diminish with increasing engine speed, the output signal of the DMS is a true analog of the speed or of the acceleration. At the least, this makes the processing of the raw sensor signals to extract a fluctuation waveshape computationally less intensive. It is not inconceivable that additional, analog signal processing and comparison methods based on waveshape features such as crest factor, harmonic content or phase relationships could provide sufficient information to reliably detect the required level of misfires. Thus, the DMS mode of sensing may enable the construction of a small and economical, stand alone, misfire detector.

SUMMARY

A simple motion sensing device operating on unconventional magnetic principles has been described. Although seeming to be only casually different in construction from previously known types of magnetic motion sensing devices, this new sensor nevertheless has an entirely different mode of operation. Unlike conventional sensors which detect target movement from the effect of such movement on the relative permeance of flux paths linking the permanent magnet field source and the field sensor, the output signal of this new sensor derives from the detection of fields from *new sources* created within the target by its motion. The location, strength and polarity of two additional field sources carries the information concerning the direction and speed of the target motion. Localized eddy currents induced in any conductive target provides a velocity dependent source of additional field. In ferromagnetic target materials, another field source (associated with the steep magnetization gradient in the transition zone between remanently magnetized regions of opposite polarities) also provides a non-volatile memory of the direction of last occurring motion. The device operates with smooth surfaced targets and is not dependent on specific surface features. This combination of unusual capabilities has been shown to have utility over a broad spectrum of applications in motor vehicles.

REFERENCES

1. B.M.F. Bushofa, T. Meydan, *Passive non-contact amorphous speed sensor*, IEEE Trans. Magn. Vol. MAG-23, No. 5, p. 2197, 1987.

2. B. Oswald, *A new generation: ABS capable wheel bearings*, SAE Paper No. 910699, 1991.

3. R. Podeswa, U. Lachmann, *Differential Hall ICs for gear tooth sensing in hostile environments*, Sensors, Vol. 6, No. 1, p. 34, 1989.

4. R. Lugosi, M. Brauer, A.J. Santos, J.A. Hilby, *The design of speed and position sensors in conjunction with bearings in automotive applications*, SAE Paper No. 900206, 1990.

5. R. Bicking, G. Wu, J. Murdock, D. Hoy, R. Johnson, *A Hall effect rotary position sensor*, SAE Paper No. 910270, 1991.

6. S.F. Brown, C. Rigaux, *Speed sensor integration in wheel bearing hub units*, SAE Paper No. 910899, 1991.

7. I.J. Garshelis, W.S. Fiegel, *A motion sensor having non-volatile direction memory*, IEEE Trans. Magn., Vol 27, No. 6, p. 5429, 1991.

8. Patents applied for.

9. A. Bastawros, *A simplified analysis for eddy-current speed transducer*, IEEE Trans. Magn., Vol. MAG-23, No. 3, p. 1905, 1987.

10. J.D. Rickman, Jr., *Eddy current turbocharger blade speed detection*, IEEE Trans. Magn., Vol. MAG-18, No. 5, p. 1014, 1982.

11. Chrysler 1992 Service Manual: Electrical, Heating-AC, and Emissions. p. 8J-1.

12. J.K. Chalsma, '93 cars speed to the showroom, Ma chine Design, Vol. 64, No. 19, p. 12, 1992.

13. SAE J1076 MAR90, Standard.

14. SAE J994 MAR85, par. 6.2, Recommended Practice.

15. SAE J1105 SEP89, Recommended Practice.

16. Hammacher Schlemmer, Wireless back-up alert, Cat. No. 45952W, p. 46, 1991.

17. State of California Air Resources Board, Mail Out No. 92-30, p. A-8, 1992.

18. G.W. Pestana, *Engine control methods using combus tion pressure feedback*, SAE Paper No. 890758, 1989.

19. R.D. Klauber, E.B. Vigmostad, J. V. Gerpen, D.V. Meter, F.P. Sprague, F. Reiter, Jr., *Miniature magnetos trictive misfire sensor*, SAE Paper No. 920236, 1992.

20. W.B. Ribbens, G. Rizzoni, *Applictions of precise crankshaft position measurements for engine testing, control and diagnosis*, SAE Paper No. 890885, 1989.

21. S.J. Citron, J.E. O'Higgins, L.Y. Chen, *Cylinder by cylinder engine pressure and pressure torque wave form determination utilizing speed fluctuations*, SAE Paper No. 890486, 1989.

22. Pat. #4,697,561, Citron, Assigned to Purdue Research Foundation, 1987.

The sensored bearing

Integration of the ABS sensor in hub unit bearings

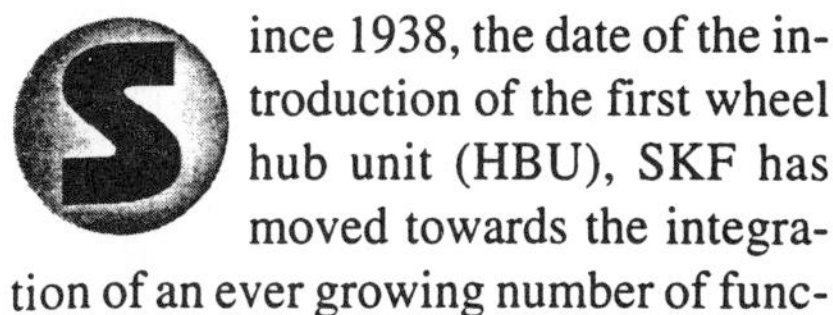

Since 1938, the date of the introduction of the first wheel hub unit (HBU), SKF has moved towards the integration of an ever growing number of functions in car wheel bearings.

In its highest evolved version (HBU III), the hub unit bearing is greased and sealed for life. It is usually integrated as a structural element in the suspension and represents the limiting factor of the wheel and disc brake.

Even ABS (Antilock Braking System) is not a recent idea; the first patent by Bosch was seen in 1936 with the first finalised designs and applications being seen around the 60s and 70s.

Over the last 20 years the system has progressively evolved beyond mechanics to include sophisticated electronics and ever improving performance and reliability. The official market launch was seen in the late 70s as an optional extra for "top of the range" cars.

Today, ABS systems are used on 30% of vehicle production in Europe and 40% of vehicle production in North America. It is forecast that by the year 2000, these figures for class "B" and "C" vehicles will rise to 60%-65% in Europe and 75%-80% in North America.

The biggest consequence of these increases in volumes will be a progressive fall in the price of the entire ABS system. (Qualified sources foresee that by the year 2000, the OEM price of the entire ABS system will be equal to 20%-25% of that in 1980.)

Traditional (Passive) ABS sensors and their limitations

The wheel speed sensor in most widespread use today is of the passive type; the generated output signal is an alternating voltage with a frequency and amplitude proportional to the rotational speed of the wheel. An electronic "black box" control system processes these signals and determines whether it is necessary to modulate the braking pressure on each wheel to avoid locking.

In the most traditional configuration, the velocity/speed sensor is usually positioned on the suspension mounting facing the impulse ring.

In the case of driven wheels, it is normally mounted on the CV joint while for non-driven wheels it is commonly mounted on the wheel hub (Figure 1,2).

Both of these solutions are not without their drawbacks - for example:

- The need to maintain very tight tolerances on the position of the sensor and impulse ring.
- Automatic mounting on assembly lines is difficult and costly.

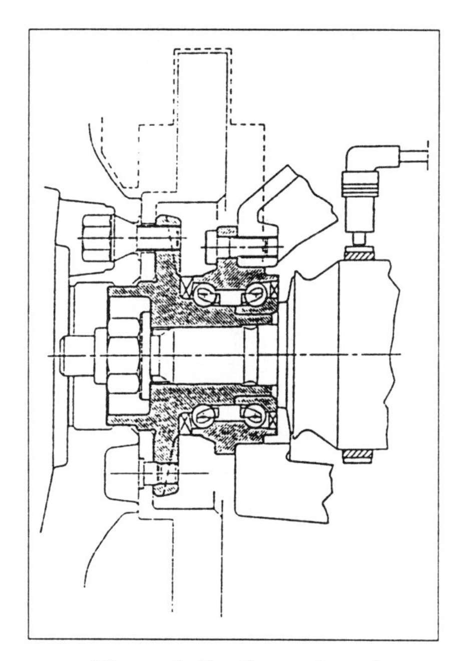

Figure 1:Configuration of traditional ABS sensor on a driven wheel (HBU HIE)

- Verification of the signal output can only take place at the final stage of assembly of components.
- They are exposed to risks of environmental damage during operation on the vehicle.
- Low output signal under a velocity of 2-3 km/h - with the consequences of not being able to run other systems such as traction control and navigation systems.

As we will see, these limitation are overcome by integrating the passive sensor in the bearing.

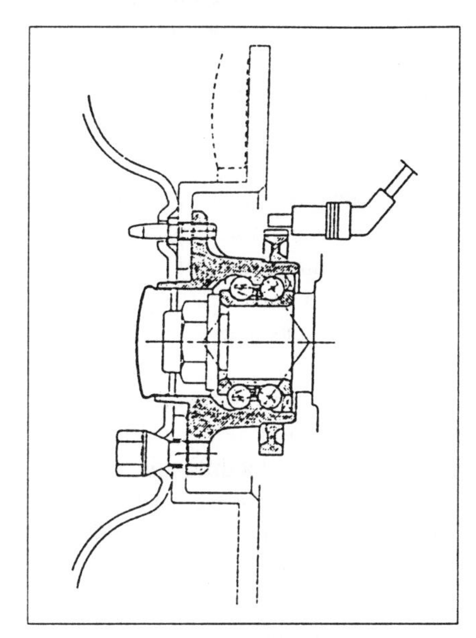

Figure 2: Traditional ABS sensor configuration on "free" wheels (HBU II)

Alternative to the traditional ABS sensor (active sensor)

Staying with sensors that are not integrated in the bearing, there exist at least two alternatives to the traditional passive sensor: the Hall cell and magnetoresistance. Both are described as being "active" in as much as that it is necessary to have an external source of power supply for them to be able to function. They generate a signal of constant amplitude for the entire range of velocities - consequently they can still be used by control systems even when the velocity is close to zero.

This characteristic renders them particularly interesting so that the signal may also be used for functions other than ABS such as traction control, tachometry and navigation systems. Notwithstanding these potential advantages, active sensors are actually of little use on normal production vehicles.

Integration of the ABS sensor in the wheel hub unit

The use of ABS on an ever increasing range of vehicles and the integration of the ABS sensor in the wheel hub unit renders the developments in ABS particularly interesting both in technical and economic terms.

Thanks to their position and precision, wheel bearings are in fact the ideal place to effect the measurement of rotational velocity. The integration of the sensor in the bearing overcomes the limits of the traditional sensor previously illustrated.

At SKF, this concept has been around for a long time and demonstrates our conviction that a wider diffusion of ABS systems would not have been possible without an improvement (in terms of reliability, efficiency and cost) in all the component parts, including the sensor. It was in the early exploratory phase of the development of ABS in the 1980s that the advantages and disadvantages of the traditional solution and levels of integration were defined as:

- Integration of the impulse ring only
- Integration of a miniature traditional sensor
- Integration of a miniaturised active sensor.

At the same time the possibility was hypothesised for developing a new concept -passive sensor with improved performance and being more suited to miniaturisation and integration in the bearing, (project "MAPS"). The conclusion drawn from this exploration convinced us to concentrate our efforts on the more ambitions and advanced level of integration, MAPS.

In this configuration, the sensor is completely integrated into the bearing - it is enclosed in a plastic body connected to the non-rotating ring. The sensor is then connected via a connecting plug to the electronic control system. (Figures 3 and 4)

As has been already acknowledged, MAPS did not result in a simple miniaturisation of the traditional passive sensor - it works according to a different physical principle which has better characteristics compared to traditional sensors:

- Lowest measurable velocity 0.8-1 km/h
- Maximum radial and axial compactness
- Greater precision and regulatory of signal.

Full compatibility with current electronic control systems is maintained throughout. Thanks to the collaboration of qualified technical partners

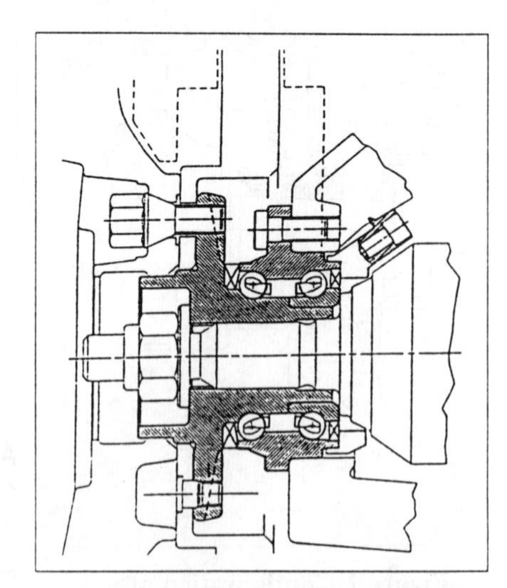

Figure 3: Configuration of ABS sensor integrated on the bearing for driven wheels (HBU HIE)

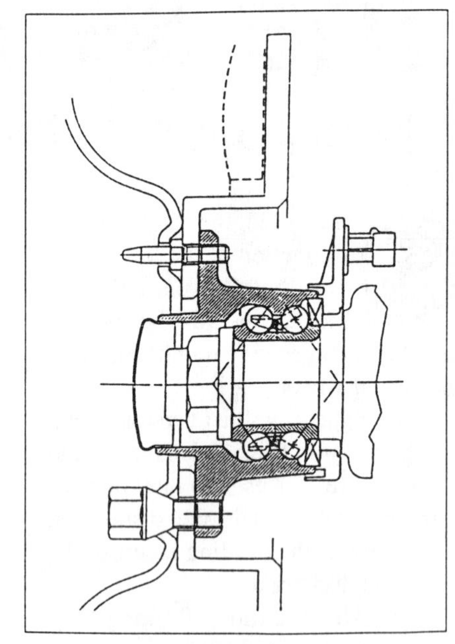

Figure 4: Configuration of ABS sensor integrated on the bearing for non-driven wheels

(TRW-SIPEA), the support and encouragement of Bosch and the support of some auto makers (FIAT, GMHA, Opel, SAAB) the transition of the development phase to production was achieved in a short time and without any particular problems.

Today, for SKF the integration of the sensor does not represent a research activity any more - it represents a productive reality.

In 1995 we have, in fact, produced more than 1,500,000 units (HBU II and HBU III) with integrated sensor for the European and overseas markets.

Further production increases are forecast throughout the 1990s especially when considering the by now proven advantages offered by the integrated sensor:

- Lower total cost
- Fewer parts to manage
- Lower assembly time
- Lower precision required on suspension components
- Better signal characteristics which can be used for traction control, navigation systems or for tachometry.■

ENGINE KNOCK SENSORS

962103

Spark Ignition Engine Knock Detection Using In-Cylinder Optical Probes

Zhihong Sun
Argonne National Lab

Perry L. Blackshear and David B. Kittelson
University of Minnesota

ABSTRACT

Two types of in-cylinder optical probes were applied to a single cylinder CFR engine to detect knocking combustion. The first probe was integrated directly into the engine spark plug to monitor the radiation from burned gas in the combustion process. The second was built into a steel body and installed near the end gas region of the combustion chamber. It measured the radiant emission from the end gas in which knock originates. The measurements were centered in the near infrared region because thermal radiation from the combustion products was believed to be the main source of radiation from a spark ignition engine. As a result, ordinary photo detectors can be applied to the system to reduce its cost and complexity.

It was found that the measured luminous intensity was strongly dependent upon the location of the optical sensor. However, good correlations existed between the luminosities measured from two probes and cylinder pressure quantities in terms of crankangle and magnitude of peak pressure of normal combustion cycles. Knock can be detected by the spark plug optical probe as a high frequency ripple similar to that observed on the pressure waveform. The knock intensity was quantified by the peak-to-peak band-pass filtered luminosity with the center frequency set at the resonant frequency of the combustion chamber. This luminous knock intensity was found to correlate well with the integral of the root-mean-square (rms) of the band-passed cylinder pressure signal.

Filtered peak-to-peak oscillations in luminosity measured by an inexpensive infrared detector sighting across the combustion chamber offer an alternative to pressure or acceleration measurements in the sensing and control of engine knock.

INTRODUCTION

Knock in spark ignition (SI) engines is believed to be associated with the spontaneous ignition of the end gas before the normal turbulent flame front reaches it [1,2]. The end gas autoignition produces rapid and spatially non-uniform energy release and pressure rise which sets up a pressure wave or shock wave propagating back and forth across the combustion chamber, exciting the acoustic resonant frequency modes of the chamber. Knock not only puts a constraint on the engine efficiency since it limits the compression ratio but also restricts the maximum indicated mean effective pressure (IMEP). Therefore it must be eliminated or controlled in the entire operating range. At present the most common methods of avoiding knock are improving fuel quality, optimizing engine design and operating parameters, and retarding spark advance when knock is detected by an accelerometer. Engine-mounted accelerometers have been adopted in many production engines and have worked well at low engine speed. However, the signal-to-noise ratio rapidly diminishes at higher speed due to the mechanical noises superimposed on the acceleration signal. As a result, the engine feedback control is either disabled or uses retarded timing, causing the engine to lose its efficiency and power. The future of knock sensors holds many significant challenges associated with reducing measurement error and complexity [3]. Thus, in-cylinder direct measurement techniques are desirable, as they directly measure the combustion event corresponding to knock.

The application of the in-cylinder optical combustion sensors for knock detection has been proposed by few researchers [4,5]. These sensors apparently are not affected by any mechanical and electrical disturbances, which is a definite advantage in detection accuracy over the accelerometer based sensors, and may meet the sensing challenges for production engines. However, since most of the detection systems reported so far used expensive devices, such as monochromators and photomultipliers, they are not economically practical for use in the low cost engine control systems. Nutton and Pinnock developed a prototype in-cylinder optical combustion measurement system utilizing fiber optic probes and low cost silicon photodetectors [6]. The system was successfully applied to the production engines to measure combustion performance, but very little work was reported on knock characterization and measurement.

In this study, a practical and low cost SI engine optical combustion sensing system was applied for knock detection.

The measurements were centered in the near infrared region because thermal radiation from the combustion products produces the strongest source of radiation from a spark ignition engine, strong enough that inexpensive photodetectors can be employed.

COMBUSTION RADIATION IN SI ENGINES

In general, the spectra of the turbulent flame front for SI engines burning gasoline and other liquid hydrocarbon fuels show the bands largely from radicals of C_2 and CH in the visible region and OH in the ultraviolet region [7]. Lavoie et al. conducted a time-resolved spectroscopic measurement from the ultraviolet to the near infrared region on a single cylinder L-head CFR. engine [8]. The spectrum obtained consisted of the chemiluminescent radiation of $CO+O \rightarrow CO2+h\upsilon$ and $NO+ON \rightarrow O2+h\upsilon$ continua along with the familiar OH, CH, C_2, and H_2O bands.

Studies of the infrared region beyond 1.0 μm have also been made in engine flames. Nearly all the emission from engine normal flames in the 1.0 to 3.0 μm region results from H_2O and CO_2 formed as products of combustion [7]. Most recently, Remboski et al. measured the luminous emission in the near infrared region in a single cylinder SI engine using an in-cylinder optical probe coupled with a monochromator [9]. The wavelength band used in the study was chosen as 927.7 nm, one of the H_2O bands existing between 900 and 1000 nm, because a strong sensitivity to burned-gas conditions was obtained without apparent effects from the surface of the combustion chamber or flame emission. Results showed good correlation between the measured radiant intensity and cylinder pressure quantities such as maximum pressure, crank angle of maximum pressure, and IMEP. Based on this study, Yong et al. developed a monochromatic gas radiation model and concluded that the measured radiation energy is proportional to the mass of burned gas which radiates and the burned gas temperature through an exponential function, and is inversely proportional to the total surface area of the combustion chamber which receives the radiation energy [10]. Further evaluation of the model led to the conclusion that thermal radiation from the main combustion products is the main source of radiation from an SI engine, while chemiluminescent radiation and thermal radiation involving NO_x can be a minor part of the total radiance [11].

The onset of knock is always accompanied by a rapid increase in luminous intensity, after which the luminous emission fluctuates in unison with the cylinder pressure [4,6,12]. The light fluctuation is attributed to the compression wave propagating back and forth across the combustion chamber. The high frequency knocking pressure waves cause the burned gases to reilluminate by adiabatic heating. The emission spectrum of knocking combustion sometimes shows a heavy continuous portion in the red. That is believed to be the blackbody radiation from soot formed by cracked lubricating oil at the higher temperatures attained when knock occurred [7].

EXPERIMENTAL TECHNIQUE AND APPARATUS

SINGLE-CYLINDER TEST ENGINE - A single cylinder CFR SI engine with a cylindrical combustion chamber was used in this research because the optical accesses to the combustion chamber could be obtained without any structural modification. In addition, its strong structure can resist severe knock and makes it a perfect device for knock research. The resonant frequency of the combustion chamber has been found to be 6.2 KHz. The engine was coupled to a synchronous AC motor maintaining a constant operating speed of 900 RPM. A Chromalox air heater with 1000 Watt capacity is situated in the intake manifold right before the intake valve. There is no throttle in the intake manifold. The compression ratio (CR) was set at 8.5:1 for most of the experiments conducted. The regular unleaded gasoline was used throughout the study. A simple PC-based open-loop electronic ignition and fuel injection system was designed and constructed to replace the original ignition and fueling systems for more precise control of ignition timing and fuel metering of such an aged engine.

The cylinder pressure was measured by using a KISTLER model 601B1 piezoelectric pressure transducer with a natural frequency of 300 KHz. The transducer was mounted on an adapter which replaced the original thermal plug assembly in the cylinder head near the end gas region. The charge yielded by the transducer was then converted to a proportional voltage by a KISTLER model 5004 dual mode charge amplifier for further processing. The air-to-fuel ratio was determined by measuring the intake air and fuel flow rates separately.

OPTICAL SENSORS - The combustion luminosity was detected simultaneously by two optical combustion sensors with different formats. The first sensor (probe 1) is an optical probe constructed in the standard spark plug (Champion model UD-16) of the engine as shown in Figure 1, top. An optical access to the combustion chamber was obtained by inserting a 1 mm diameter, 3 meter long fused quartz optical fiber into a hole drilled all the way through the center electrode and the terminal stud of the spark plug. The quartz fiber provides excellent transmission in visible and near infrared spectral regions. The numerical aperture of the optical fiber is 0.20, that is the transmitted light is collected by the light gathering end from within a 23^0 included-angle acceptance core. One critical requirement of the optical probe is that the light gathering end has to stay free of deposits from combustion products and cracked lubricant in order to transmit flame radiation successfully. To meet this requirement, the end of the optical fiber was maneuvered to extend beyond the wall of the center electrode into the combustion chamber. This design was found to be effective in reducing the amount of deposits on the end face [6]. In addition, the spark generated by the spark plug in each working cycle of the engine also cleans the deposits on the optical fiber. Since the plastic jacket of the optical fiber can not withstand the temperature reached at the center electrode, the jacket embedded in the electrode was removed. The optical fiber was permanently bonded in place using a high temperature adhesive which also served to prevent the escape of combustion products via the interface between the fiber and the center electrode. A small section of the ground electrode that completely blocked the front end of the center electrode was machined off to free the light gathering face of the optical probe, while a 0.7 to 0.9 mm gap between two electrodes was still guaranteed for normal operation. This modified spark plug does not alter its ignition characteristics according to the information provided by its manufacturer. The high tension connection was rebuilt to allow both the supply of ignition voltage to the spark plug and the passage of the optical fiber which is terminated with a standard connector.

The second optical sensor (probe 2) is a steel body probe incorporating a 1 mm diameter, 3 meter long fused quartz optical fiber, the same way as with the first sensor, as illustrated in Figure 1, bottom. The probe was mounted next to the pressure transducer on the pressure transducer adapter situated in the end gas region of the cylinder as shown in Figure 2. The intention was to measure the radiation from the end gas.

LIGHT FILTERS AND PHOTODETECTORS - The narrowband interference band-pass filters were used to transmit the desired wavelength intervals of the broadband combustion luminosity signals collected by the optical probes. The filter has a center wavelength of 900 nm, a bandwidth (Full Width at Half Maximum) of 20 nm and a minimum peak transmission of 40%.

Two silicon photodiodes, each with a peak responsivity of 900 nm, were used to accept filtered or unfiltered combustion luminosities and to convert them into electrical currents. They were carefully selected with low noise levels and high sensitivities to detect low light signals. The effective area of each photodiode is 1 mm^2. The application of these relatively large area photodiodes reduced the necessity for highly accurate alignment of the fibers. The electrical signals were amplified by the amplifiers for further processing.

SIGNAL CONDITIONING AND DATA ACQUISITION - To measure knock intensities, the pressure and optical signals were first high-passed by RC filters to remove DC components. The signals were then amplified to increase the resolution and reduce noise for the highest possible accuracy of data acquisition. Finally, Band-pass filters were applied to isolate the knocking pressure and luminosity oscillations from other high frequency components. The center frequency of the band-pass filter for pressure signal processing was set at 6.3 KHz which is the resonant frequency of the CFR engine combustion chamber.

Data were collected via a National Instruments AT-MIO-16F-5 interface board that has a maximum sampling rate of 200 KHz. A falling edge of the TDC signal of the intake stroke was applied to commence a data acquisition sequence. Individual A/D conversions were triggered externally by the digital crankangle pulses at every 0.125 crankangle degree (CAD), which offers the effective sampling rate for each data input channel of 43.6 KHz. This sampling rate was chosen as a compromise between accuracy and limitation of the instrumentation. For knock measurement, the

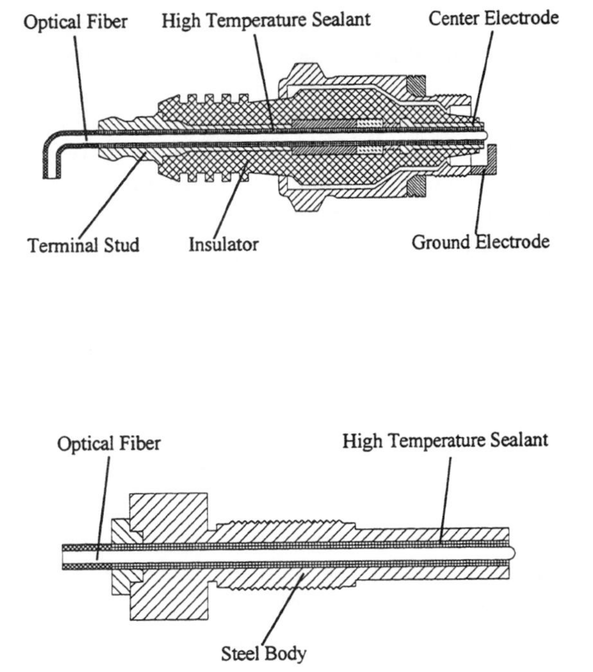

Figure 1: Optical combustion sensors. top: optical spark plug sensor, bottom: optical combustion probe

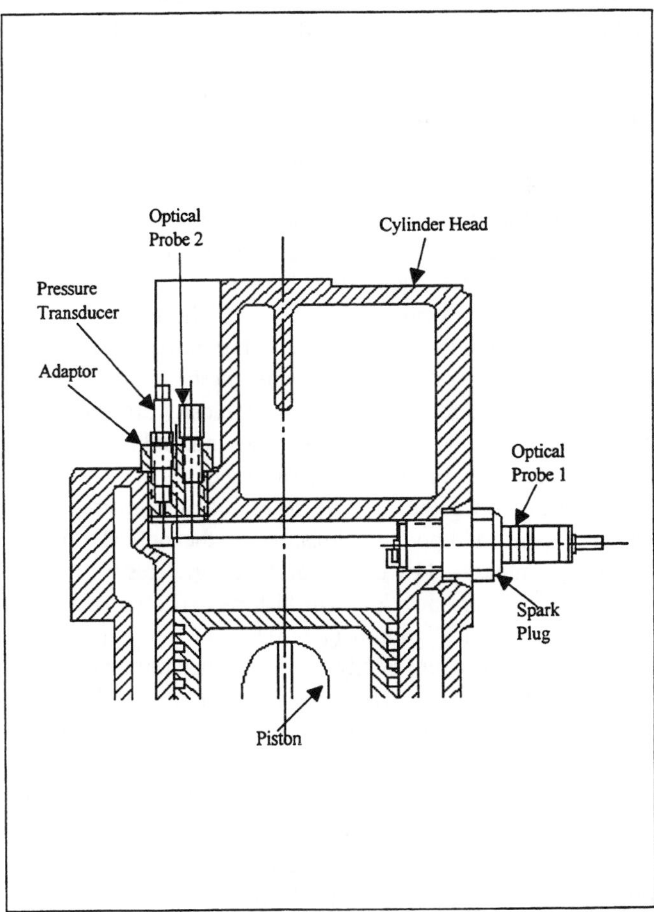

Figure 2: Cylinder head assembly with pressure transducer and optical sensors

signals were digitized from -50 to 100 CAD ATDC during which knock is most likely to occur at the spark timings considered. It not only minimized the effects of valve closings and other engine vibrations which often show up on the cylinder pressure signal, but also significantly reduced the computer's memory and storage requirements. Data acquisition programs were developed using a graphical programming language in National Instruments LabVIEW for Windows 2.5.1 software development environment.

RESULTS AND DISCUSSION

CHARACTERISTICS OF THE MEASURED RADIANCE FOR NORMAL COMBUSTION - Two optical combustion probes were installed at different locations of the combustion chamber as described in previous section. There are two reasons to mount the optical probes in those particular locations:

1. Both spark plug and pressure transducer adapter are the most ideal devices available on the CFR engine for incorporating the optical probes without any modification of the combustion chamber. Once installed, the probes do not disturb the flow field and flame process in the chamber.

2. Integrated into the spark plug, optical probe 1 detects radiation mainly from burned gas behind the flame front throughout the combustion process. Optical probe 2, situated near the cylinder wall farthest from the spark plug, monitors emission from unburned gas near the end gas region before the flame front reaches the probe close to the end of the combustion process. This arrangement makes it possible to study the effect of optical sensor location on the combustion luminosity measurement with the least number of optical probes.

Figure 3 shows cylinder pressure and luminous emission intensities, luminosities, measured by two optical probes as a function of the crankangle degree for a typical normal combustion cycle. After conversion to electrical signals, the luminosity signals were amplified by the same factor before being digitized. For comparison purposes, the instantaneous pressure and luminosity levels are normalized by their corresponding peak values and presented in the bottom graph. Note that the units for pressure and luminosity are arbitrary throughout this paper unless otherwise indicated since no comparison between engines will be attempted.

As shown in Figure 3, the measured luminosities are strongly dependent upon the location of the optical probe. Luminosity 1 measured by optical probe 1 correlates well with the cylinder pressure development histories. The sharp spike at the location of the ignition (7 CAD BTDC) is apparently due to the spark. The luminosity reaches a maximum when the combined effects of cylinder temperature and H_2O concentration are optimum. The maximum luminosity occurs around the location of maximum cylinder pressure at 28 CAD ATDC (about 0.5 CAD after the crankangle of maximum pressure for this particular cycle).

Throughout most of the combustion process, the level of luminosity 2 picked-up by optical probe 2 is significantly lower than that of luminosity 1 as shown in Figure 5, top. This is because the probe can not detect the radiation from

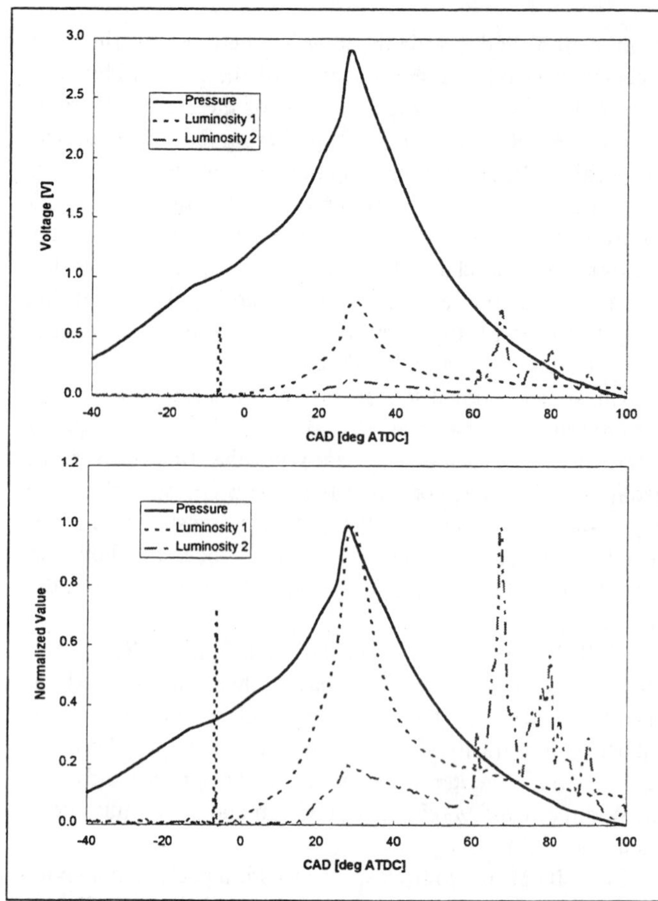

Figure 3: Pressure and luminosities for a typical normal combustion cycle. CR=8.5, SA=7 CAD BTDC, A/F=14.6, WOT. top: original signals, bottom: normalized signals

unburned gas directly until the arrival of the flame front towards the end of the combustion due to its location and orientation (refer to Figure 2). The luminosity reaches a peak at around the crankangle of maximum cylinder pressure as does luminosity 1.

It was observed that more and more cycles had only one peak in luminosity 2 when the air-to-fuel ratio was made leaner than stoichiometric, while other operating conditions remained unchanged. This is due to the fact that leaning out the intake charge lengthens all stages of the combustion process and slows down the flame propagation speed. In some cycles, the flame develops and propagates so slowly that there is insufficient time to complete combustion prior to the exhaust valve opening. The flame may even be extinguished before the exhaust valve opens and before the flame has propagated across the combustion chamber. All of the cycles recorded had one peak when air-to-fuel ratio was adjusted to 17.5.

In Figure 4, the crankangles of maximum luminosity 1 and peak cylinder pressure from normal combustion cycles are correlated by varying ignition timing from 10 CAD to 25 CAD BTDC with a 5 CAD interval at stoichiometric air-to-fuel ratio conditions. The compression ratio was set at 6:1 so that knock did not occur in any of the cycles within the ignition timing range examined. At each running condition, data from 100 consecutive engine cycles were sampled to account for the effect of cycle-to-cycle variations. It can be seen from the figure that two crankangles correlate well on

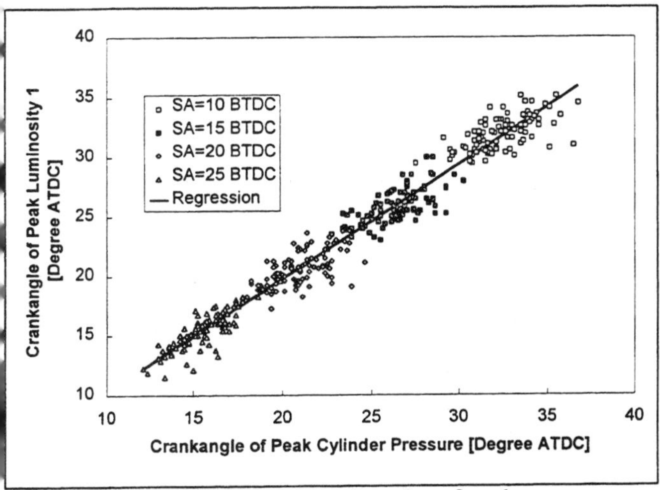

Figure 4: Correlation of crankangles of peak pressure and peak luminosity 1 for several levels of spark advance, CR=6:1, A/F=14.6, WOT, 87 ON fuel, 100 consecutive cycles for each spark advance. r^2=0.97

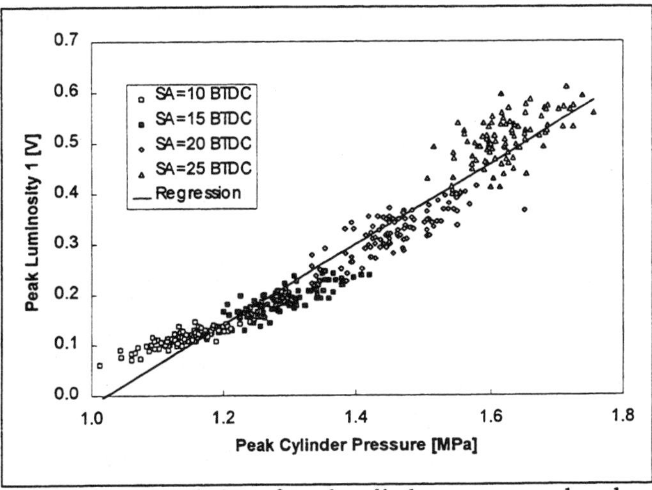

Figure 5: Correlation of peak cylinder pressure and peak luminosity 1 for several levels of spark advance, CR=6:1, A/F=14.6, WOT, 87 ON fuel, same engine cycles as in Figure 7. r^2=0.94.

cycle-to-cycle basis with the square of the correlation coefficient (or the coefficient of determination r^2) of 0.97. The correlation is consistent with the result reported by Remboski et al. who measured flame luminosity in a single cylinder engine using one optical probe [9]. In that work, the probe was mounted near the exhaust valve and situated about half way in the flame traveling path, that is from the spark plug to the cylinder wall opposite to the spark plug. The correlation, however, disagrees with the report from Nutton and Pinnock who found the luminosity peaks at the location of 90% mixture burned [6]. Since detailed information regarding the location of the optical probe in that paper was not given, it is hard to explain why Nutton and Pinnock's result is different from that of this study. Accurate determination of the crankangle of peak cylinder pressure is useful for feedback engine control because this crankangle is directly related to the occurrence of maximum indicated mean effective pressure (IMEP).

Linear relationship was also found between maximum luminosity 1 and peak cylinder pressure on cycle-to-cycle bases as illustrated in Figure 5.

CHARACTERISTICS OF THE MEASURED RADIANCE FOR KNOCKING COMBUSTION - The ultimate goal of this research is to apply the optical combustion sensors to engine knock detection. Since a strong correlation exists between measured combustion luminosities and cylinder pressure, it is expected that the luminosity signals could be applied to sense knocking combustion in a similar way as with the pressure signal. Because the cylinder pressure transducer is by far the most accepted knock sensor, the cylinder pressure histories of knocking cycles serve as the references for knock identification throughout the study. Knock may be characterized in terms of knock intensity and knock occurrence crankangle.

Knock intensity (KI) has been quantitatively characterized in number of ways. In this study, they were quantified by the following two parameters based on the band-pass filtered pressure signal: the maximum peak-to-peak amplitude of the filtered signal (KI p-p), or the integral of the root-mean-square (rms) amplitude of the filtered signal within a time domain which was chosen from the time of ignition until the end of ringing of the signal (KI rms). Both parameters were determined mathematically from digitized band-pass filtered pressure data. The KI rms definition characterizes both the magnitude and duration of the knock signal. The relationship of the two is shown in Figure 6. It can be observed that for low and moderate knock intensities the two definitions agree very well as claimed by Leppard [13]. However, the magnitudes of KI rms are larger than those of KI p-p when knock is heavy, indicating that the recorded pressure oscillations following knock not only have large amplitudes but also have long duration. Knock intensity depends on the mass of the end gas at the time of autoignition and the energy release rate in the end gas during knock. It has been shown that knock intensity increases as autoignition occurrence crankangles get progressively closer to TDC and the mass of unburned charge increases [1]. In this case, autoignition occurs at multiple exothermic centers in the end gas for at least some of the cycles. Therefore, heavy knock is

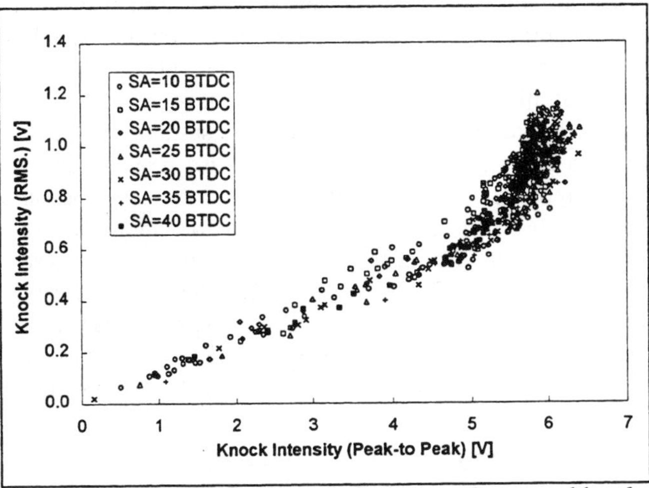

Figure 6: Correlation of KI p-p and KI rms for several levels of spark advance, CR=8.5:1, A/F=14.6, 87 ON fuel, 100 consecutive cycles for each spark advance. r^2=0.83.

expected to produce high amplitude and long duration pressure oscillations due to the large amount of end gas and the fast energy release rate involved. Since engine damage from knock depends on its intensity and duration, the importance of applying KI rms definition to quantify knock severity is apparent.

Knock occurrence crankangle was defined as the moment when an initial abrupt increase in cylinder pressure trace occurred as indicated in Figure 7, top. It was determined from the calculation of the rate change of the pressure with respect to the crankangle (dP/dCAD) between two successive sample points. Based on the examination of a large number of pressure traces of knocking cycles at various operating conditions, knock was assumed to occur if (dP/dCAD) exceeded a threshold value of 1.02 MPa/CAD. The accuracy was judged to be ± 0.13 CAD at the data sampling rate used.

For most of the knock experiments conducted, the compression ratio was chosen as 8.5:1, and the air-to-fuel ratio as 14.6, stoichiometric, unless otherwise specified. Audible knock started at 12 CAD BTDC in this running condition. At least 100 cycles of data were sampled for every running condition whenever a correlation was desired. Shown in Figure 7 are luminosity 1, luminosity 2, and cylinder pressure signals plotted against crankangle respectively for a typical knocking combustion cycle; the spark advance was set at 35 CAD BTDC. Note that the amplification of raw luminosity 2 signal was made 10 times larger than that of luminosity 1 signal in order to accentuate the portion of luminosity 2 corresponding to the knock process. This results in signal saturation later in the cycle which is not shown in this figure.

Visual inspection of Figure 7 shows that luminosities 1 and 2 exhibit distinct features, which is in contradiction to the result from Ohyama el al. [4] who reported that the measured luminosity of a knocking combustion cycle is independent of the viewing field of the optical sensor. There is a very clear similarity between the luminosity 1 and the cylinder pressure curves. For this particular cycle, knock occurred at 5.6 CAD BTDC as determined from the pressure trace. The sudden rise in pressure at this crankangle indicates an increased rate of heat release at the time of autoignition which causes a steep increase in radiant emission as well. A pronounced change in slope of luminosity 1 around the knock occurrence crankangle is apparent as illustrated in this Figure. After knock onset, the pressure oscillation in the first tangential mode, or the lowest frequency mode, results in the gas charge sloshing back and forth across the cylinder [2]. In addition, the propagation of large amplitude pressure waves across the combustion chamber originated from the end gas autoignition process causes changes in local density of the cylinder content. These lead to oscillations in radiant emission since its intensity depends on the density of emitters. The changing gas radiation in the vicinity of the spark plug is manifested on luminosity 1 as a high frequency ripple similar to that noted in the pressure trace.

The luminosity 2 signal also rises sharply at knock onset as shown in Figure 7 due to rapid burning in the end gas region that is close to optical probe 2. It then fluctuates in unison with the pressure signal.

In general, the noise level of luminosity 2 is high due to

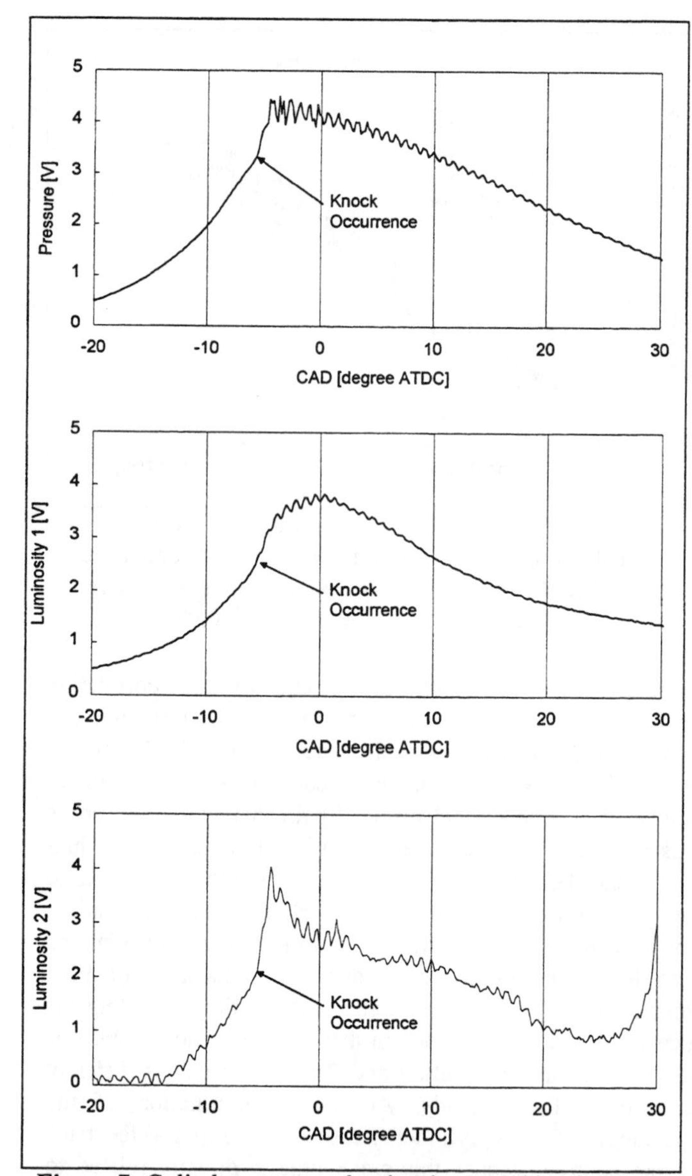

Figure 7: Cylinder pressure, luminosity 1, and luminosity 2 signals of a knocking cycle, CR=8.5:1, A/F=14.6, SA=35 CAD BTDC, WOT. 87 ON fuel.

its low signal level. Even after knock onset, the signal level is significantly lower than that of luminosity 1. This was unexpected because the optical probe 2 was placed near the chamber wall opposite the spark plug where one would expect the end gas to be located. Upon the occurrence of autoignition, the pressure and temperature in the end gas region rise substantially due to the rapid release of the unburned charge's chemical energy, which may cause a strong increase in luminosity in the region. A possible explanation for the low luminosity signal detected is that the location of autoignition sites varies considerably with engine operating conditions and from cycle-to-cycle. The engine used in this study was equipped with a shrouded intake valve which induced a counter clockwise swirl. Liiva et al. determined the autoignition locations in the end gas of a CFR engine with such a swirl motion by using a two dimensional triangulation scheme and four pressure transducers positioned orthogonally to each other on the cylinder periphery in the clearance space [14]. The autoignition sites were found in the neighborhood of the exhaust valve. Thus, they may not be directly included in the acceptance core of the optical probe 2. In addition, temperature gradients exist across the burned gas with the highest temperature in the neighborhood of the spark plug, which is also confirmed in a knocking disk-chambered engine by a most recent computational study [2]. At the end of combustion, the temperature of the gases in the vicinity of the spark plug could be as much as 300 to 400^0 K higher than that at the opposite end of the chamber. This effect may also contribute to the higher luminosity 1 signal due to its exponential dependence on burned gas temperature [10].

Figure 8 shows only part of the knocking portion from the luminosity and pressure traces of a knocking cycle. Note that the evidence of knock was picked up first simultaneously by the pressure transducer and optical probe 2 because they were mounted on an adapter positioned nearest to the end gas region. The initial discontinuity in slope appears on the luminosity 2 trace less than 0.2 CAD later than that of the pressure. The subsequent pressure oscillation is approximately

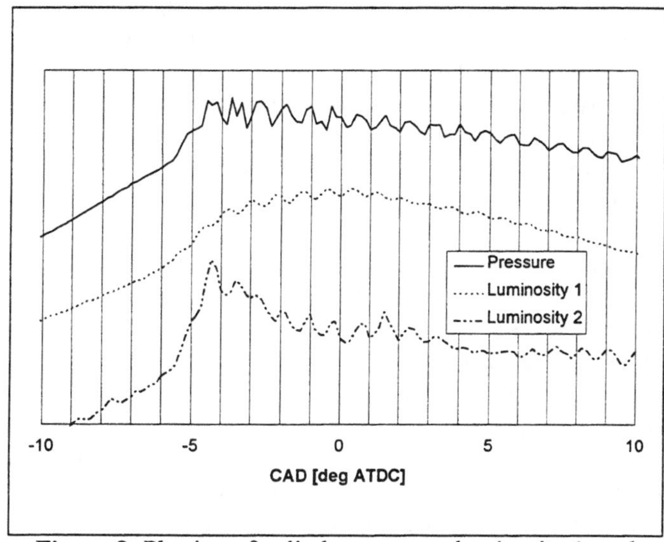

Figure 8: Phasing of cylinder pressure, luminosity 1, and luminosity 2 signals for a knocking cycle, CR=8.5:1, A/F=14.6, SA=35 CAD BTDC, WOT, 87 ON fuel.

in phase with that of luminosity 2, while the phase difference between the pressure and luminosity 1 traces is about 180^0. Therefore, the propagation of pressure waves between the location of the spark plug and that of the pressure transducer adapter, corresponding to the first tangential mode, is evident. The initial wave speed is about 800 m/s, which was estimated from the cylinder bore and the time interval between the first peaks of pressure and luminosity 1. The speed of sound is predicted to be 895 m/s by using the mean cylinder gas temperature over the crankangle range of -5 to 5 CAD ATDC. It can also be observed from the figure that the amplitudes of oscillations of luminosity signals, especially the luminosity 1 signal, decay faster than that of the pressure trace.

CORRELATIONS BETWEEN LUMINOSITY AND PRESSURE SIGNALS IN KNOCKING COMBUSTION - As discussed in the previous section, knock onset is indicated by a sudden increase in luminosity. Therefore, the knock occurrence crankangle may be identified from the first derivative of the luminosity signals with respect to the crankangle due to their strong similarity to the pressure signal. Computation of the first derivative of luminosity 1 signal was carried out from two successive sample points starting from 20 CAD after ignition crankangle to eliminate the effect of the sharp spike from the spark. Knock was defined to occur when the first derivative exceeded the value of 1.0 Volt/degree. In Figure 9 the knock occurrence crankangle measured from the pressure signal is plotted versus the knock angle measured from luminosity 1 signal on an individual cycle basis. The two results compare well with one another. The difference between these two measurements is less than 0.5 CAD.

The biggest problem in the calculation of the first derivative of luminosity 2 to identify the knock occurrence was the high signal noise. The signals are very noisy due to the low signal levels until about 5 degrees before they reach the first peak. Fortunately knock consistently occurred at less than 5 degrees before the crankangle of peak cylinder pressure for all of the measurements examined. Thus the signal differentiation was only performed in this narrow 5 CAD domain in order to avoid high signal noise region, that is, it started at 5 degrees before the first peak crankangle for every operating condition. The threshold value used to judge knock onset was found to be 1.1 Volt/degree. Excellent correlation between the knock occurrence crankangle measured from the pressure signal and the knock angle measured from luminosity 2 signal was obtained as displayed in Figure 10.

The luminosity 1 and luminosity 2 at knock occurrence crankangle are shown plotted against the cylinder pressure at knock onset in Figures 11 and 12 respectively for the same data sets used in Figures 9 and 10. As expected, the luminosities increase with the pressure and good correlations were achieved in both cases. The correlation coefficient is higher for luminosity 2 than for luminosity 1, which is simply because the location of optical probe 2 was closer to that of the pressure transducer. However, the luminosities depart from the linear relationship when ignition timing was advanced to 35 CAD BTDC.

KNOCK IDENTIFICATION - Checkel and Dale developed a knock indicator based on the third derivative of the cylinder pressure trace in the peak pressure region to quantitatively measure the knock intensity on a CFR engine [15]. There was a strong interest in applying this algorithm to combustion luminosity signals due to its potential for predicting end gas autoignition at wide range of knock intensity levels. However, attempts to identify knock by this method was not successful.

Attention was then focused on the effect of the characteristic pressure fluctuations on the luminosity signals. The frequency spectra of the knocking luminosity signals were first examined in order to identify knock components among other background noises. Figure 13 depicts the frequency spectra of luminosity 1 for three different spark advances. Each spectrum is the average of 100 consecutive cycles. It contains some noise components at the lower frequency region (less than 5 KHz) and an outstanding spectral peak at around 6.4 KHz that is in consistent with the fundamental frequency of the combustion chamber (within

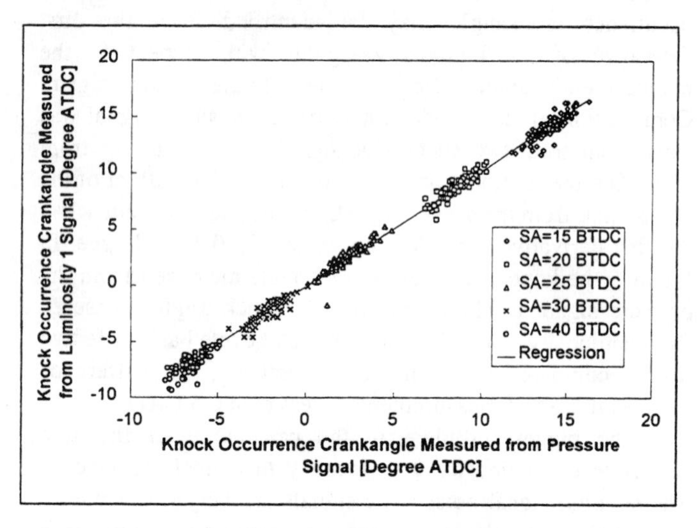

Figure 9: Correlation of KI p-p and KI rms for several levels of spark advance, CR=8.5:1, A/F=14.6, 87 ON fuel, 100 consecutive cycles for each spark advance. r^2=0.83.

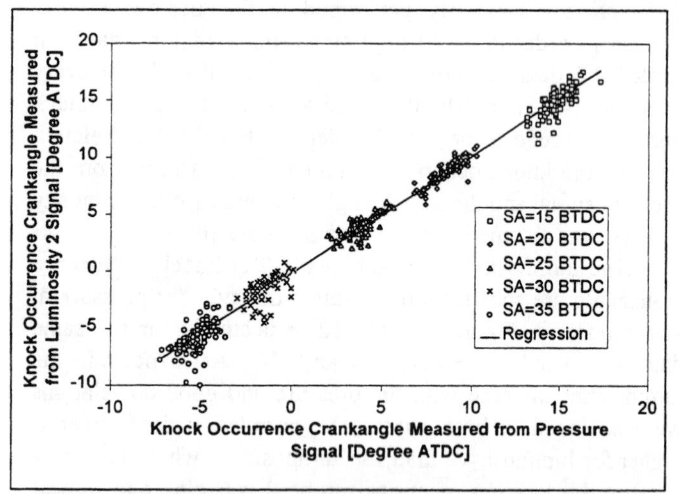

Figure 10: Comparison of knock occurrence crankangle measured from pressure signal and from luminosity 2 signal, CR=8.5:1, A/F=14.6, 87 ON fuel, 100 consecutive cycles for each spark advance, r^2=0.99.

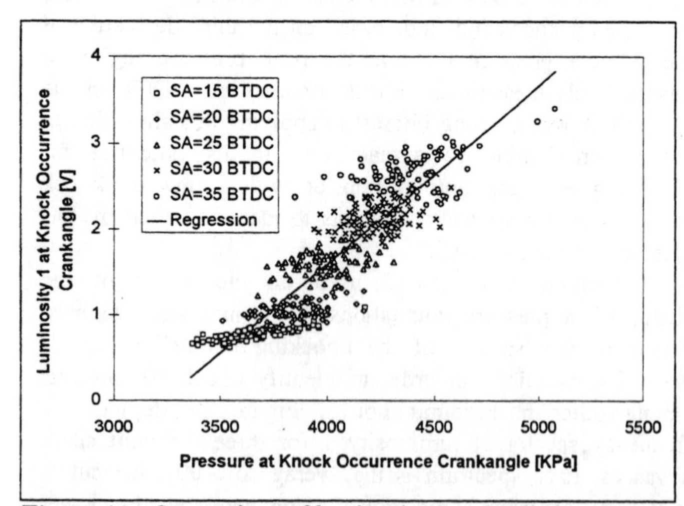

Figure 11: Comparison of luminosity 1 and cylinder pressure at knock occurrence crankangle, CR=8.5:1, A/F=14.6, 87 ON fuel, 100 consecutive cycles for each spark advance. r^2=0.81.

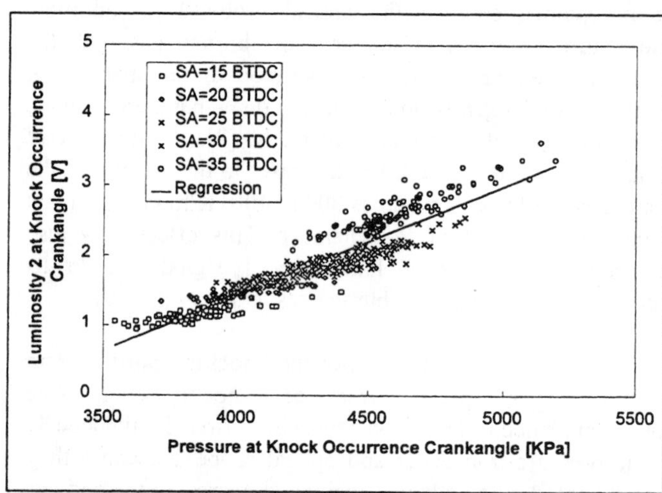

Figure 12: Comparison of luminosity 2 and cylinder pressure at knock occurrence crankangle, CR=8.5:1, A/F=14.6, 87 ON fuel, 100 consecutive cycles for each spark advance, r^2=0.86

1.6%). The spectral peak of 10.1 KHz, the second tangential mode, also appears consistently for all of these running conditions. The noise components even existed when the engine was not knocking. In addition, the power magnitudes of the peaks are higher when ignition timing is advanced due to higher average knock intensity. Further study of a large number of spectra showed that the detected fundamental frequency did not shift with spark advance and knock intensity.

The frequency analysis on luminosity 2 signals were performed from 5 degrees before the knock occurrence crankangle to 25 degrees after the peak pressure for each cycle in order to accentuate the knocking portion and to avoid noisy low signal region. Figure 14 plots the average frequency spectrum of luminosity 2 of 100 consecutive cycles with spark advance of 35 CAD BTDC. Although the spectrum contains a prominent peak at 6.4 KHz, it also includes high magnitude noise components especially in the lower frequency region.

From this analysis, it is concluded that knock is detectable by simply processing the high frequency

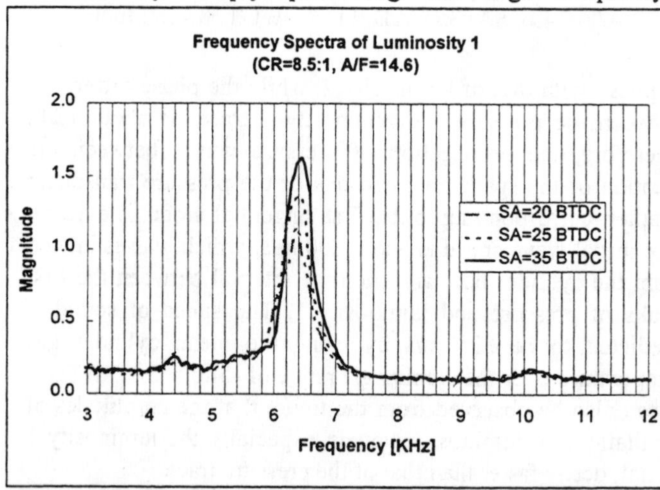

Figure 13: Frequency spectra of luminosity 1, average from 100 knocking cycles for each spark advance, CR=8.5:1, A/F=14.6, 87 ON fuel.

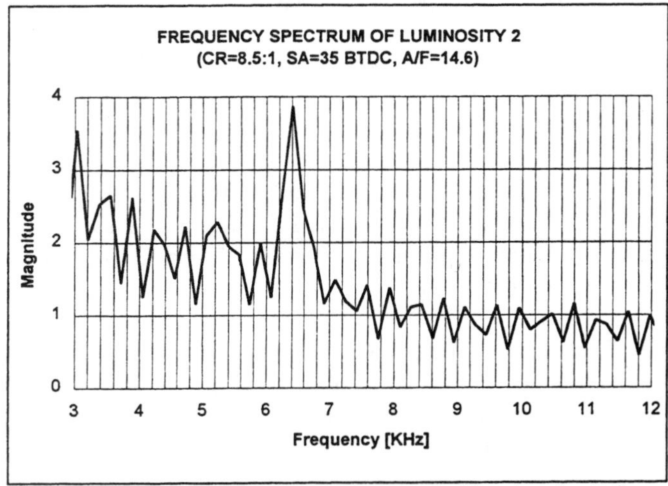

Figure 14: Average frequency spectrum of luminosity 2 from 100 consecutive cycles, CR=8.5:1, SA=35 CAD BTDC, A/F=14.6, 87 ON fuel.

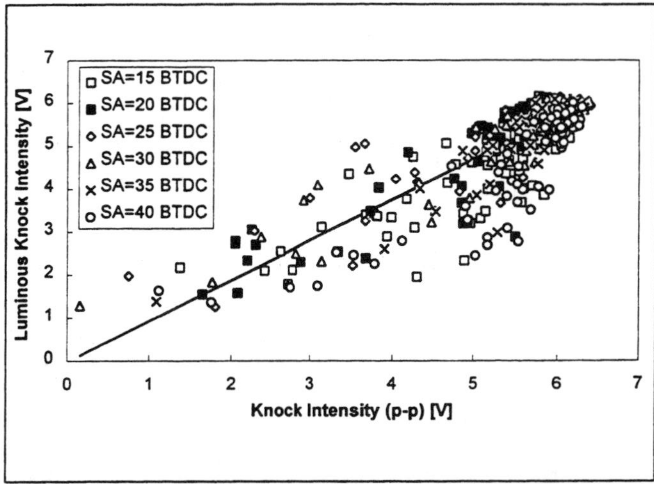

Figure 15: Correlation of KI (peak-to-peak) and luminous knock intensity with spark advance variation, A/F=14.6, CR=8.5:1, 87 ON fuel, 100 consecutive cycles for each spark advance. r^2=0.73.

component of the luminosity 1 signal in the same way as for the cylinder pressure signal sensing. A "luminous knock intensity" was thus defined as the peak-to-peak amplitude of the band-pass filtered luminosity signal. In order to simplify the measurement system, the optical interference filters were removed.

A comparison between luminous knock intensity and peak-to-peak pressure knock intensity with spark advance variation on a cycle-to-cycle basis in Figure 15 shows that the two agree reasonably well in predicting the level of knock with r^2 of 0.73. However, a better correlation is achieved when the luminous knock intensity is compared with rms pressure knock intensity as illustrated in Figure 16 for the same sets of data (r^2=0.84). As mentioned earlier, heavy knock is usually associated with the autoignition of a large amount of end gas and faster heat release rate. In this case, the luminous knock intensity should be high due to the high cylinder gas temperature produced. The correlation was further investigated under different air-to-fuel ratio (Figure 17) and inlet air temperature (Figure 18) running conditions. The correlation is consistent in both cases as shown.

Attempts to detect knock by using the luminosity 2 signal were not successful. The difficulty was associated with the noise presented in the signal.

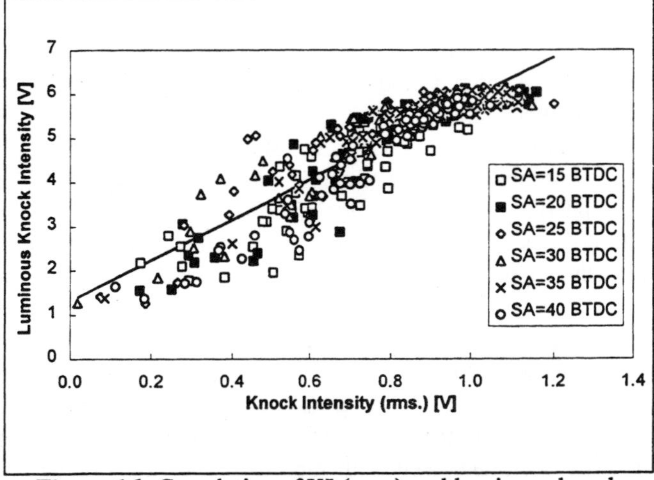

Figure 16: Correlation of KI (rms.) and luminous knock intensity with spark advance variation, A/F=14.6, CR=8.5:1, 87 ON fuel, 100 consecutive cycles for each spark advance, r^2=0.84.

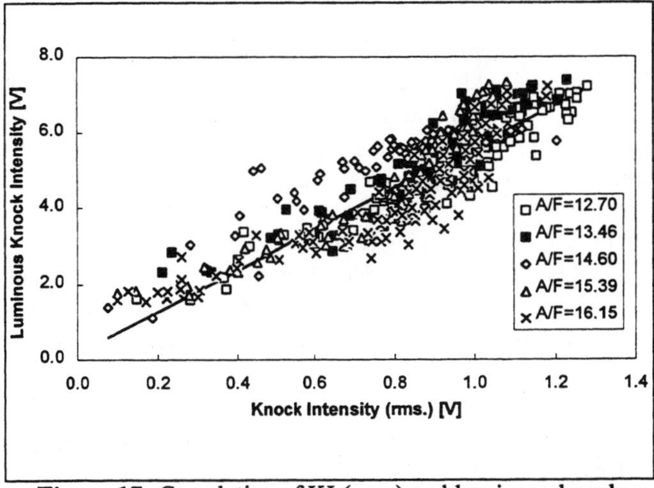

Figure 17: Correlation of KI (rms.) and luminous knock intensity at different air-to-fuel ratio, A/F=14.6, CR=8.5:1, 87 ON fuel, 100 consecutive cycles for each air-to-fuel ratio, r^2=0.84.

CONCLUSIONS

A simple and low cost SI engine optical combustion sensing system has been demonstrated to have potential applications to the field of knock detection. Two optical probes were applied to the system and the measurements were centered in the near infrared region. The results can be summarized as follows:

1. The measured luminosities were strongly dependent upon the location of the optical sensor. However, good correlations exist between the luminosities measured from the two probes and the cylinder pressure quantities in terms of crankangle and magnitude of peak pressure of normal combustion cycles.

2. Knock was measured in terms of knock intensity and

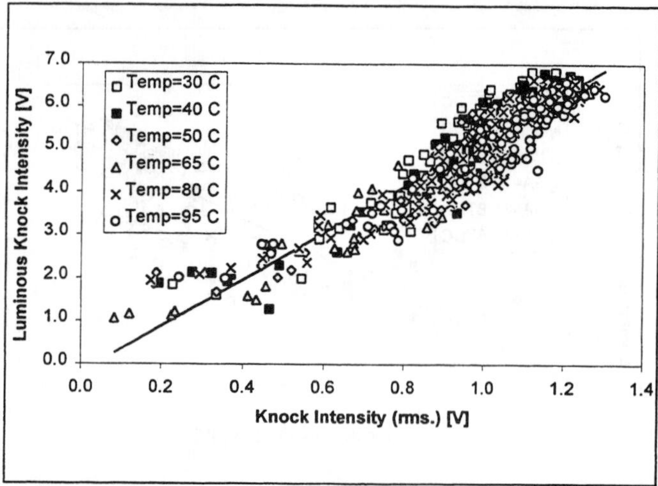

Figure 18: Correlation of KI (rms.) and luminous knock intensity at different inlet air temperatures, A/F=14.6, CR=8.5:1, 87 ON fuel, 100 consecutive cycles for each inlet air temperature, r^2=0.86.

knock occurrence crankangle. Based on the cylinder pressure measurement, it was found that the rms. knock intensity correlated well with the peak-to-peak knock intensity when knock was light and moderate; while the former is larger than the later when knock was heavy.

3. Knocking combustion was shown to cause oscillations of the luminosity signals of similar phase and frequency to the associated pressure oscillation. At knock onset, the luminosities were found to correlate with the cylinder pressure in both magnitude and knock occurrence crankangle.

4. The spark plug optical sensor can be used to detect knocking combustion. The knock intensity can be quantified by the peak-to-peak band-pass filtered luminosity, with center frequency of the filter set at the resonant frequency of the combustion chamber. This luminous knock intensity was found to correlate well with the integral of the rms of the band-passed cylinder pressure signal having the same pass band.

ACKNOWLEDGMENT

The authors gratefully acknowledge Dr. Matthew L. Franklin and Mr. Darryl Thayer of the University of Minnesota, Dr. David L. Haugan, and Dr. Ruonan Sun of John Deere Product Engineering Center for valuable discussions and suggestions related to this work, Professor Jimei Sun and Dr. Jianhua Zhang of Jilin University of Technology for providing some optical components.

This work was conducted at the Internal Combustion Engine Research Laboratory, University of Minnesota. It was not funded through Argonne National Laboratory.

REFERENCES

1. Konig, G. and Sheppard, C. G.W., "End Gas Autoignition and Knock in a Spark Ignition Engine", SAE Paper 902135, 1990.

2. Blunsdon, C. A. and Dent J. C., "The Simulation of Autoignition and Knock in a Spark Ignition Engine with Disk Geometry", SAE Paper 940524, 1994.

3. Dues, S. M., Adams, J. M., and Shinkle, G. A., "Combustion Knock Sensing: Sensor Selection and Application Issues", SAE Paper 900488, 1990.

4. Ohyama, Y., Ohsuga, M., and Kuroiwa, H., "Study on Mixture Formation and Ignition Process in Spark Ignition Engine Using Optical Combustion Sensor", SAE paper 901712, 1990.

5. Spicher, U., Spiegel, L., Reggelin, B., and Heuser, G., "Investigation into the Applicability of an Optical Fiber Sensor for Knock Detection and Knock Control System", SAE Paper 922370, 1992.

6. Nutton, D. and Pinnock, R. A., "Closed Loop Ignition and Fueling Control Using Optical Combustion Sensors", SAE Paper 900486, 1990.

7. Gaydon, A.G., The Spectroscopy of Flames, 2nd ed., Champman and Hall Ltd., London, 1974.

8. Lavoie, G. A., Heywood, J. B., and Keck, J. C., "Experimental and Theoretical Study of Nitric Oxide Formation in Internal Combustion Engines", Combustion Science and Technology, Vol. 1, 1970.

9. Remboski, D. J., Plee, S. L., and Martin, J. K., "An Optical Sensor for Spark-Ignition Engine Combustion Analysis and Control", SAE Paper 890159, 1989.

10. Yang, J., Plee, S. L., Remboski, D. J., Jr., and Martin, J. K., "Comparison Between Measured Radiance and a Radiation Model in a Spark-Ignition Engine", Transactions of the ASME, Journal of Engineering for Gas Turbines and Power, Vol. 112, July, 1990.

11. Yang, J., Plee, S. L., and Remboski, D. J., Jr., "Relationship Between Monochromatic Gas Radiation Characteristics and SI Engine Combustion Parameters", SAE Paper 930216, 1993.

12. Affleck, W. S. and Fish, A., "Knock: Flame Acceleration or Spontaneous Ignition", Combustion and Flame, Vol.12, 1968.

13. Leppard, W. R., "Individual Cylinder Knock Occurrence and Intensity in Multicylinder Engines", SAE Paper 820074, 1982.

14. Liiva, P. M., Cobb, J. M., and Acker, W. P., "Swirl, Fuel Composition, Localized Hating, and Deposit Effects on Engine Knock Location", SAE Paper 932814, 1993.

15. Checkel, M. D. and Dale, J. D., "Computerized Knock Detection from Engine Pressure Records", SAE Paper 860028, 1986.

951237

Engine Knock Control Via Optimization of Sensor Location

James W. Forbes, Kevin R. Carlstrom, and William J. Graessley
Ford Motor Co.

ABSTRACT

This paper describes a procedure used to aid in the control of IC engine knock, an autoignition phenomenon that results in customer annoyance, loss of power, and potential engine damage. Since a control system can only function as well as the signal it is provided, input signal optimization is critical to the robustness of the system. Optimization of the input signal starts with a properly located physical transducer on the engine block. The locating process begins with laser holometry to evaluate compliant regions of the block. Holographic data, block vibration spectra and empirical engine data are then used to identify the most promising sensor locations. These locations are then verified with a broadband accelerometer mounted on a dynamometer engine. This process allows the highest available signal to noise locations to be found in a systematic and efficient manner.

INTRODUCTION

As engine designers strive for powerplants with higher specific output (power/displacement) and decreased tailpipe emissions, finding methods to increase engine operating efficiencies becomes paramount. As a result, accurate knock detection and spark control allows spark advance to be maintained at maximum brake torque timing (MBT) providing maximum power while avoiding knock which can cause customer annoyance and/or engine damage. At its lightest levels, knock is heard as a high pitch metallic sound resulting from rapid combustion of the end gas in the knocking cylinder[1]. This combustion sound is transmitted as vibration through the walls of the block where it can be used as an early indication of excessive knock[2].

A number of knock detection and knock control algorithms are currently being used in production applications[3], but the success rate of all these systems is limited by the accuracy of their input signal. In a piezo-electric element based knock detection systems, proper placement of the sensor is one of the most important steps in providing an accurate signal. Many other events in the engine cycle can generate frequencies in the knock region. The sensor, therefore, must be located where knock engine vibrations are significant with respect to background engine vibrations.

While analytical methods exist to identify knock sensor locations and the frequency(ies) at which knock will occur[4][5], these methods are approximations and do not account for all factors that contribute to engine block vibration. A purely hit-or-miss method of placing knock sensors in prospective locations on an engine and testing the signals is time consuming, costly, and must be done late in the development process since any engine component change can change signal response at a location.

A method was developed to identify optimal knock sensor mounting locations that is more accurate than analytical methods and more efficient than the hit-or-miss method. This allows the knock sensor location to be optimized and the knock spectral region to be quickly identified in the early stages of knock system design, and can be rapidly re-evaluated if the engine design is changed in a way that affects the transmissibility of knock signals.

PROBLEM DEFINITION

The key to a successful knock detection system is the proper placement of the knock sensor(s) on the engine block or cylinder head. Ideal mounting would be in a location where pressure oscillations in the cylinders due to knock are transmitted as vibrational energy, but other sources of vibration (valve noise, piston slap, etc.) are rejected - a location where knock from all cylinders can be detected with a suitable signal-to-noise ratio. The ideal situation would involve situating the sensor early in the engine design stage, casting knock transmission paths into the block and/or cylinder head, locating the sensor to balance the signal-to-noise from all cylinders so each can be detected equally, and providing the required mounting boss in an optimal orientation.

An engine is generally not designed with the knock system in mind. Instead, a knock sensor is typically added to an existing design, and to keep costs down, the sensor's mounting location is generally confined to one of a few packagable locations. As a result, sensor placement is usually the result of locating an existing, unused mounting boss that will allow the sensor to package without interference from other components and will hopefully provide an acceptable signal-to-noise ratio for knock.

Sensor packaging is one of the largest obstacles to designing the knock system. Aerodynamic hood lines and modern control systems result in extremely crowded engine compartments. Components such as engine mounts, alternator, manifolds, etc. have mounting priority due to the physical orientation needed for proper operation. Other components tend to have higher packaging priority because of their bulk, maximum operating temperature requirements, and air flow requirements. Items such as wiring harnesses; vacuum and fuel lines; and hydraulic and coolant hoses take up additional space.

After components with higher priority are packaged, there are usually very few mounting locations available on the engine, and of these locations a limited number have the clearance necessary to package the knock sensor. When the mounting restrictions for the sensor itself are taken into account, (maximum sensor temperature, maximum vibration amplitude, electromagnetic interference concerns, etc.) the available locations decrease even further.

Another factor in the knock sensor mounting process is the type of sensor used. If the frequency at which the block vibrates during knock events is consistent through the entire engine speed range, a resonant type sensor may be used. This type sensor has an internal resonance that has a relatively narrow bandwidth (~ 100-700 Hz) and requires no additional circuitry to filter the knock signal. A linear sensor, on the other hand, has flat response up to a frequency that is typically above that used for knock detection (~ 20-25 kHz). The sensor output is then band-passed by the knock detection circuitry, which allows the system the flexibility to detect knock over a wider frequency range.

In regards to the mounting process, the type of sensor used must be taken into account since the space requirements are different for each sensor. A resonant sensor is typically larger than the linear sensor and requires more clearance for installation and wire harness connection.

The subject of this paper was an existing engine design to which a knock system was added. Choice of mounting locations was limited to four existing bosses, including one that was used by a motor mount bolt. All of the chosen locations would package a linear sensor, but only two locations would package a resonant-type sensor without interference.

Although the number of available locations may be limited, it is still important to choose the location that will provide the best knock signal. The evaluation process to determine the final sensor location on few locations can still be quite involved. If the sensor is being packaged on a new engine design where the mounting location is open, the task can become extremely time consuming. An efficient and reliable process is needed to narrow down the available mounting choices to the one location that will provide the best signal-to-noise ratio from all cylinders. An optimal process would also identify the frequency of engine block vibration at that location throughout the range of operating conditions, would be efficient enough to save valuable development time, and would be robust enough to be used on a variety of engine configurations.

LABORATORY PROCEDURE

To achieve a systematic process for sensor location, a laboratory procedure has been developed and validated that allows a robust sensor location to be found with a minimum of effort. The procedure centers around the use of laser holographic interferometry, or holometry (a full-field vibration measurement technique) [6][7], to provide a systematic method of determining all engine surfaces that transmit the effects of engine knock efficiently. The method allows sensor location to be accomplished in a rapid, non-destructive manner that minimizes the use of costly and time consuming dynamometer testing.

DESCRIPTION OF HOLOMETRY - Laser holometry is a photographic method of observing vibrations and deformations over an entire object. A schematic of the test setup used in this procedure is shown in Fig. 1. In conventional photography, light reflected from the object exposes the photographic emulsion, creating an image. With holometry, however, light from the object combines with a reference beam to capture a three dimensional picture of the object. Two forms of holometry used as part of this procedure are real-time holometry and time-average holometry.

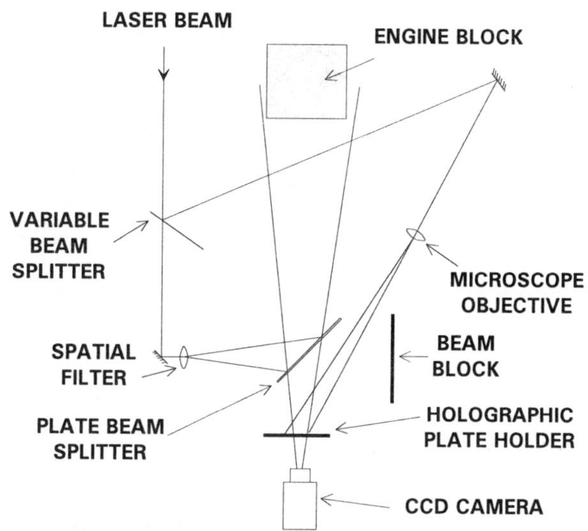

Figure 1: Schematic of holographic setup showing the path of the laser beam and the positions of the optics

Real-time holometry involves exposing and developing a single hologram of the undeformed object and replacing it in the holographic plate holder. The test object is then viewed through the plate while both the object and plate are illuminated with laser light. In this configuration, any motion of the part relative to the initial exposure will be observed in real time as a fringe pattern. The motion will be seen as bands of light and dark (fringes) varying in intensity according to the square of the zero order Bessel function of the first kind, with each fringe representing motion of less than a micrometer.

Time-average holometry is also used in this study. This technique involves recording a single, long exposure hologram of an object vibrating at a predetermined resonant mode. The resulting fringe pattern will highlight the nodal lines of the mode as bright areas with the antinodes banded by fringes of diminishing contrast. These fringes are also of the Bessel function variety and result from the fact that as the object vibrates, antinodes will dwell most often at their extreme states of motion where velocity is zero while nodes will remain motionless.

THE HOLOMETRY PROCEDURE - The holometry process for knock sensor location can be broken down into the following steps: setup and preparation of the engine, frequency determination, holometry, and interpretation of interferograms & location selection as detailed below.

Setup and Preparation - One of the advantages of this experimental method is that a full powertrain is not needed to locate the sensor. A bare engine block with only the major structural elements can be used for analysis which minimizes testing cost and reduces complexity. The block is then painted with a retro-reflective paint to improve the efficiency of the holographic imaging. The painted block is fixtured to a vibration isolated optics table with the rear-face-of-block on the table and excited with an electromagnetic shaker.

Frequency Determination - Prior to holographic testing the resonant frequencies of the block are determined. Slow frequency sweeps are used to excite the block and the block response is observed via real-time holometry. From this procedure major out-of-plane modes are seen. To find more subtle modes a laser vibrometer, which functions as a non-contact accelerometer, is used. Vibrometry has the advantage of having no internal resonance, so it can measure very high frequencies without undo distortion of the signal. In addition, the vibrometer can be quickly repositioned at different locations. Using the vibrometer, frequency spectra were taken at a number of locations on the block (a typical spectrum is shown in Fig. 2). Frequencies where amplitude peaks occur are tabulated along with the real-time holometry results. Based on these techniques, frequencies for time-average holometric imaging were chosen.

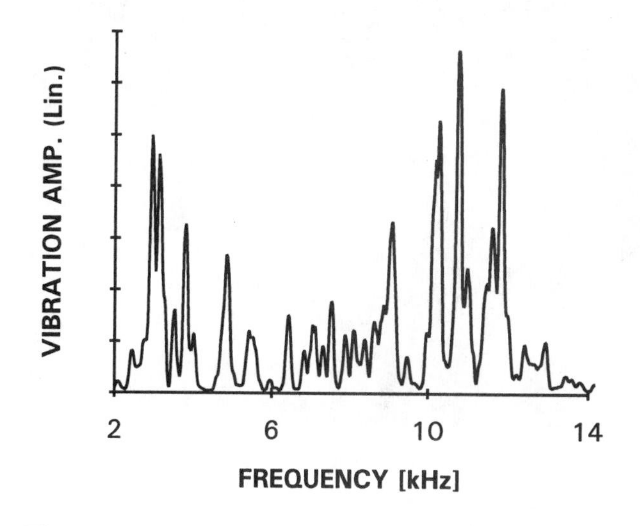

Figure 2: Typical vibrometer power spectrum showing block resonances in the knock frequency range

Holometry - Based on the frequencies determined in the previous step, time-average interferograms were generated for the driver and passenger sides of the engine block, one side of the block is shown in Fig. 3. Interferograms at two of the more active frequencies are shown in Figs 4 and 5 for the view in Fig. 3. These images contain the full field mode shape at the given frequency.

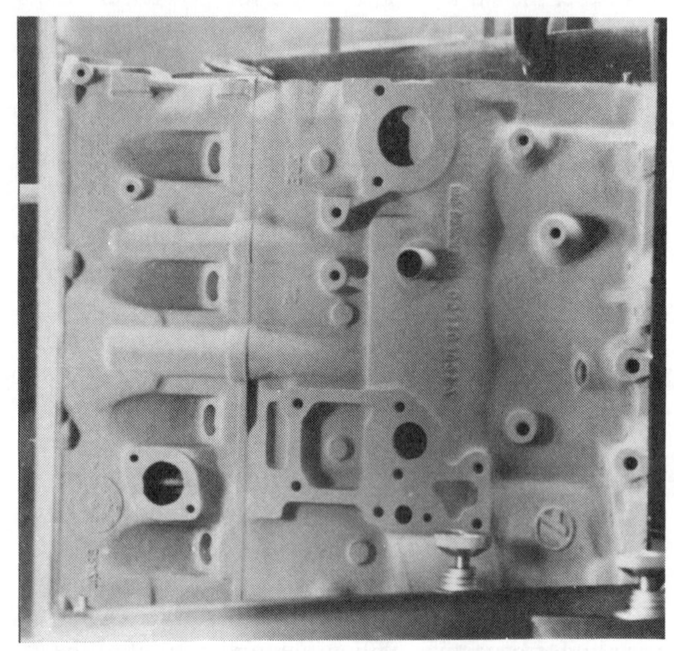

Figure 3: Holometric setup view of the side of the engine block

Figure 4: Interferogram of a vibration mode at a typical knock frequency near 5 kHz.

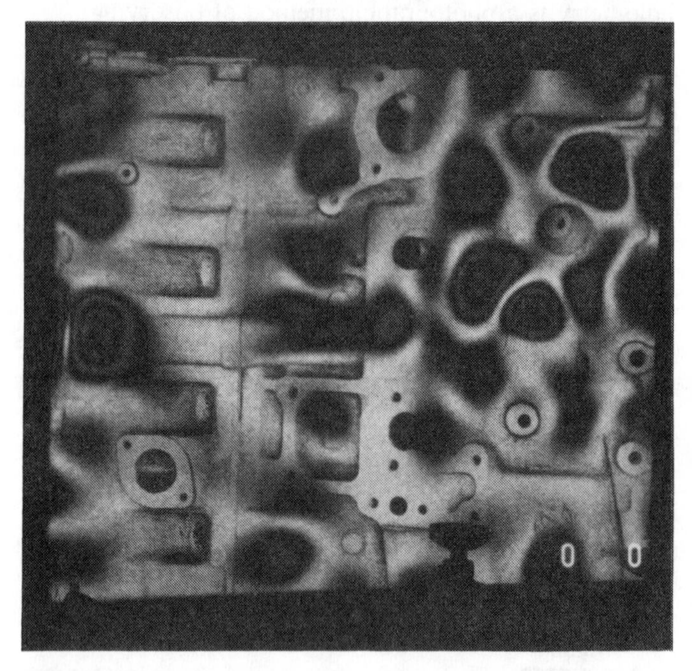

Figure 5: Interferogram of a vibration mode at a typical knock frequency near 11 kHz

Interpretation of Interferograms and Location Selection - Based on the fringe patterns (alternating light and dark bands) seen in the interferograms, potential locations were chosen to yield the highest signal-to-noise ratio at a given frequency. An ideal sensor location will respond well to knock frequencies and be insensitive to input to other excitation sources. The two frequencies shown represent modes at lower and upper ends of a typical knock spectral region. The high modal density seen in these images indicates that sensor signal quality will be highly spatially dependent.

RESULTS AND PROCESS VERIFICATION

To determine the actual signal to noise ratio of a given location, a sensor needs to be evaluated on a running engine. Using an engine dynamometer, accelerometers, and a dynamic signal analyzer, the vibration signal at locations with high potential were measured at high engine load. Vibration frequency spectra were recorded at baseline engine timing to establish a background noise level without knock and with controlled knock induced via spark advance. The difference between the baseline and knocking vibration levels gives the signal to noise ratio of the knock sensor seen at that particular location.

Vibration data for all locations were acquired simultaneously for the baseline condition and then for the knock condition. This minimized the effect of the variable severity of knock from run to run. Using this method also made the calibration of the accelerometers less critical since both knock and no knock runs are taken with the same transducer and then only relative changes observed

With this method, data were taken at various load/speed conditions with typical resulting power spectra shown in Fig. 6 for the case of a predetermined good location and Fig. 7 for a location that would have been desirable from an assembly standpoint. These traces were generated with the analyzer not synchronized to the firing of the cylinders so that a few knock events are averaged in with many samples of background noise. This method simulates one type of engine control strategy. To simulate different strategies, the analyzer could be set up to gate to specific portions of the combustion cycle, and it can also be gated to the firing of each cylinder to ensure all are being detected from a specific location.

Frequency and bandwidth of the knock system can also be established from the test data. The background vibration data taken at baseline spark timing includes valve noise, piston slap, accessory vibration, and noise from any other source during normal engine operation that can corrupt the knock signal. The power spectra will indicate frequencies where knock signal-to-noise is highest throughout the engine operating range.

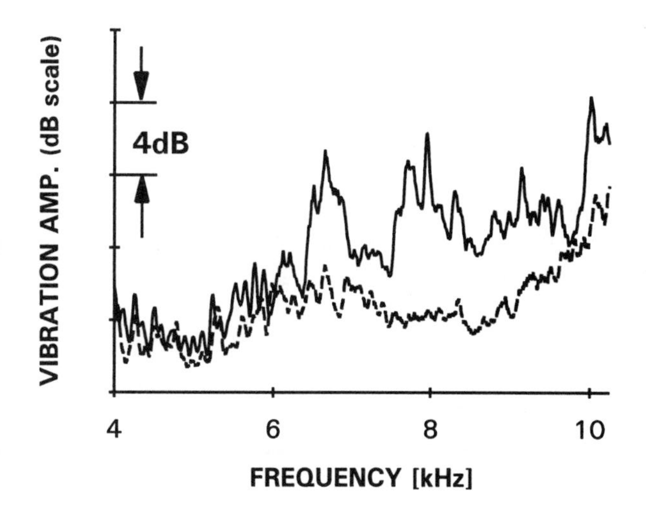

Figure 6: Power spectra of knock (solid line) and no-knock (dashed line) vibration at a high S/N location

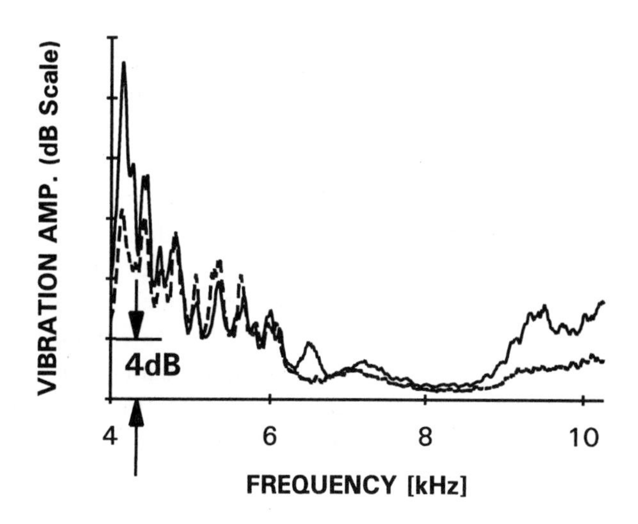

Figure 7: Power spectra of knock (solid line) and no-knock (dashed line) vibration at a low S/N location

CONCLUSIONS AND FUTURE WORK

Many methodologies may be employed to find an optimal transducer location to detect engine knock. The method presented here represents a systematic approach to reveal locations with the highest potential to detect knock. Through the use of laser holometry a greater understanding of the high frequency dynamics of the engine structure can be used to determine a robust location with little sensitivity to variations in the block structure or sensor tolerances. Locations with both high signal to noise ratio (S/N) and a broad frequency range of detection help determine the robustness of the location. The dynamometer mapping of high potential locations has allowed the method to be verified and refined based on running engine results.

These refinements have challenged some initial assumptions of the nature of knock detection. One example of this is that in early studies locations on the cylinder head were not considered because it was assumed that impactive motions of the valves would reduce the detectability of any knock signal. Through empirical studies this was found not to be the case and that if the proper location was used, the knock could be discriminated from other high frequency sound events.

Another conclusion determined with the aid of this process is that changes to the structure of an engine block made to improve sound quality characteristics can minimize the availability of good knock detection locations. As engine structure is tuned to reduce the high frequency radiated noise the vibration modes that prove useful for sensor location are broken up and the need for a systematic process becomes more critical.

Based on the foundations presented, in this paper future work to tune the sound quality and knock response of the engine is underway. By beginning work early in the product development process, engine structure that is favorable to the detection of knock and to the packaging of a knock sensor can be incorporated readily. This minimizes the chance of being constrained to a less than ideal location later in the development process. Early work on knock control also allows new strategies for control to be employed. If these strategies are based on certain quality of type of input signal, then the method presented in this paper can be employed to find that signal.

REFERENCES

[1] K.M. Chun, J.B. Heywood, "Characterization of knock in a spark-ignition engine," SAE Paper 890156, 1989.

[2] T. Priede and R. K. Dutkiewicz, "The effect of normal combustion and knock on gasoline engine noise," SAE Paper 891126, 1989.

[3] P.V. Puzinauskas, "Examination of methods used to characterize engine knock," SAE Paper 920808, 1992.

[4] E. F. Obert, "Internal combustion engines and air pollution," Harper & Row, New York, 1973, pg. 107.

[5] R. Hickling, D. A. Feldmaier, F. H. K. Chen, J. S. Morel, "Cavity resonances in engine combustion chambers and some applications," J. Acoust. Soc. Am., 73(4), pp. 1170-1178, April 1983.

[6] G. M. Brown and R. R. Wales, "Vibration analysis of automotive structures using holographic interferometry," Proc. SPIE, 398, pp. 19-22, 1983.

[7] G. M. Brown, J. W. Forbes, M. M. Marchi, R. R. Wales, "Comparison of holographic interferometry to other test methods in automotive testing," SPIE, 2004, 1993.

922370

Investigation into the Applicability of an Optical Fiber Sensor for Knock Detection and Knock Control System

U. Spicher, L. Spiegel, and B. Reggelin
FEV Motorentechnik GmbH & Co. KG

G. Heuser
Technical University of Aachen

ABSTRACT

A new fiber optic sensor has been used to detect knocking combustion. With this sensor it is possible to detect high frequency signals which are free from electrical and mechanical disturbance. By using the maximum signal rise of the detected optical signals for each combustion cycle, it is possible to clearly seperate knocking and non-knocking cycles. The detected maximum signal rise was used in a preliminary test as the input of a knock control system.

1 INTRODUCTION

Due to fuel limitations and the increase in oil prices, the improvement of engine efficiency is one of the most important goals of research and development work. The part load efficiency of engines can be enhanced by increasing the compression ratio. The increase in compression ratio is, however, limited by a phenomenon occurring at high loads - engine knock.

In automotive technology, the knock phenomenon has always been one of the major limitations to the improvement of the engine efficiency. Knocking combustion may lead to different kinds of engine damages. For example, erosion of the chamber walls, cylinder head gasket failure and piston failure e.g. melting of the piston [1-5]. The investigations of Woschni et al. [6] and Essig et al. [7] show, that the surface temperatures of the chamber walls do not distinctly increase during knocking combustion. They state, that therefore the thermal stress of the chamber walls could not be the reason for the damage. The investigations of Maly et al. [8] show, that these failures are mainly caused by the mechanical stress indicated by the pressure waves which occur during knocking combustion. On the other hand some authors [9,10] measured increasing surface temperatures when knocking occurs. In their opinion knock damage is caused by extreme high local pressure combined with temperature peaks weakening the material of the surface.

During the investigation of knocking combustion many theories have been created. Three of these theories have often been quoted in recent literature [11-13]:

- Auto-ignition theory: the unburned gas ahead of the normal flame front reaches a condition such that a rapid self-ignition process at one or more points may occur.

- Detonation theory: the normal spark-ignition flame front undergoes a transition from a subsonic deflagration to a supersonic detonation coupled with a shock wave.

- Rapid-entrainment theory: the normal spark-ignition flame front is accelerated to high, but still subsonic, speeds causing rapid rates of pressure rise.

König et al. [14,15] observed, during their measurements, only autoignition, which generally had multiple initiation points. They characterized three different modes of autoignition:

1) deflagration
2) thermal explosion
3) developing detonation

These modes are described as follows [14,15]:

- Deflagration: low mean end-gas temperature and steep temperature gradients. With such gradients, a weak pressure wave immediately propagates away from the center. Combustion undergoes a gradual transition to deflagration and continues as a normal flame. Knock may be non-existent to moderate. Flame speeds are similar to the main flame.
- Thermal explosion: high mean end-gas temperature and small temperature gradients. A near simultaneous chemical reaction follows ignition at a center in the almost isothermal gas. This might be characterized as a thermal explosion, or homogeneous autoignition. Knock may be moderate. Apparent flame speeds are very high. Due to the heterogenity of the engine charge no pure thermal explosions were observed.
- Developing detonation: Intermediate mean end-gas temperature and temperature gradient. When the gradient is smaller than a critical value, a strong shock is created with an intensity high enough to initiate and sustain intense chemical reaction. This would ultimately lead to a steady state detonation. Knocking can be violent. Flame speeds are very high.

Due to the heterogenity of the engines charge, no pure thermal explosion was observed [15]. Engine failure is mainly caused by strong pressure waves. Therefore in the case of autoignition the developing detonation will be the most important case for engine damage.

To avoid knock damage, knocking combustion must be eliminated over the entire engine operating range. To avoid knocking combustion under high load conditions, production engines are normally operated with ignition timings which are adjusted with a safe clearance distance to the knock limit and/or for best efficiency. To compensate for this disadvantage while avoiding engine knock, a practical knock control system is desirable. It would allow engine operation close to the knock limit and/or at best efficiency without knocking combustion under all conditions.

The functioning of a knock control system depends on reliable detection of the knock characteristics by the knock sensor. Modern sensors are generally based on the detection of the knock effects. Most systems include an acceleration pick-up fixed at the cylinder head or on the engine block which registers the mechanical vibrations produced by the knocking combustion [16-21]. This does not allow a safe detection since the typical knock frequencies are often superposed by valve train noises or other engine vibrations, particularly at high engine speeds. These drawbacks can be avoided by detecting the knock phenomenon directly within the combustion chamber of the engine.

The in-cylinder pressure as well as the light emission of the combustion process are the main parameters which seem to be suitable for a good extraction of knock information by detecting the knock phenoenon within the combustion chamber. When using the in-cylinder pressure, the input signals of a knock control system may be disturbed by other signals e.g. the noise of valve closing. It is therefore not suited for production engines. Extensive studies [13] have shown that the light emissions of a knocking combustion and a non-knocking combustion are clearly different from each other. In addition, the flame propagation velocity is much higher when engine knock occurs. Therefore, the detection of knocking combustion with an optical sensor seems to be the best way to develop a suitable knock control system for engines. When using an optical knock sensor in production engines photo-electric cells may be used instead of photomultipliers, to reduce costs.

In this paper a knock sensor using optical fibers is presented. This sensor is able to detect high frequency signals free from electrical and mechanical disturbance. This allows simple evaluation procedures which nevertheless yield better results for knock detection than using conventional sensors.

The optical fiber sensor, signal processing and method of evaluation of the detected signals for individual combustion cycles of varying knock intensity are discussed in detail. A number of alternative knock intensity evaluation procedures are examined for the optical sensor. The feasibility of using the optical sensor for knock control systems is demonstrated.

2. EXPERIMENTAL RESULTS

2.1 TEST RIG AND MEASUREMENT TECHNIQUE - To determine the combustion process in engines, a multi-optical fiber technique has been developed and is used for investigations on combustion chamber studies, knocking combustion, and influences of engine parameters on flame propagation [22-27]. Analyses of the flame propagation measurements with this optical fiber technique have shown that during knocking combustion, which will damage the engine, the flame propagation of the combustion zone (reaction zone) is completely different from the propagation during non-knocking combustion. Therefore, this technique seems to provide a good prospect for use in a knock control system.

Thus, to detect the knocking combustion phenomenon a specially developed knock sensor using optical fibers is used. With this knock sensor one is able to detect high frequency light

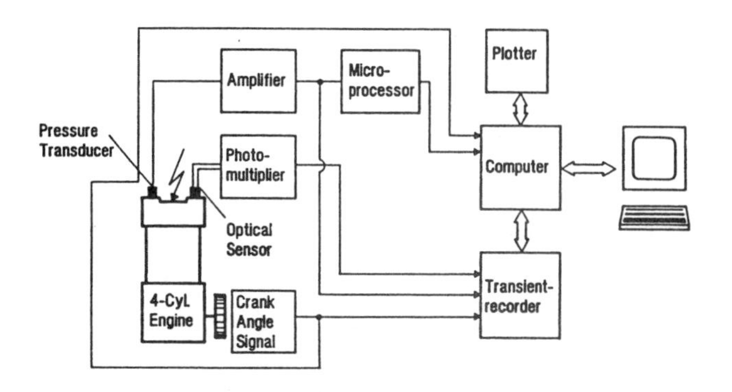

Figure 1 : Test Rig and Measurement Equipment

emission signals free from electrical and mechanical disturbance. Figure 1 shows a schematic representation of the test rig and the measurement equipment.

The optical sensor is installed in the cylinder head. The radiation of the flame is detected by this sensor and transmitted by optical fibers to photomultipliers, which convert the optical radiation signals of the flame into electrical output. To detect the radiation of the most important gas components involved in the engine combustion process, optical fibers with a spectral range from 250 nm to 800 nm and photomultipliers with a spectral range from 250 nm to 650 nm were used. The radiation of the flame is detected by three separate optical fibers in the sensor and three corresponding photomultipliers. The signals from the photomultipliers are recorded by a transient recorder. A sample rate of 500 kHz was used and time window began 20° BTDC with a duration of 70° CA. The transient recorder signals are stored and analysed by a computer after the measurement and can be printed on a plotter or on a monitor. The in-cylinder pressure is obtained simultaneously with the flame radiation measurement by using a pressure transducer (Kistler 6061, resonance frequency 100 kHz). The pressure signal from the transducer is amplified by the pressure amplifier and stored additionaly to the transient recorder temporarily in a microprocessor high speed data acquisition system over a complete cycle of 720° CA [28]. In addition to the optical flame signals the in-cylinder pressure signals can be transfered to the computer and be analysed in the same way as the optical signals. Thus, a comparison of both measurements concerning the identification of the knock behavior is possible.

The measurements presented in this paper were performed in a water-cooled, four-cylinder, four-stroke engine. The engine specifications are:

Swept Volume	: $V_H = 1.588\ 10^{-3}\ m^3$
Bore	: $D = 79.5\ 10^{-3}$ m
Stroke	: $S = 80.0\ 10^{-3}$ m
Compression Ratio	: CR = 9.3
Combustion Chamber	: Heron

The investigations were carried out using a test rig with an electronically controlled direct-current-machine. The temperature of oil and water could be regulated externally. The fuel was injected into the inlet port. The heron combustion chamber had a squish area of about 47%. The design of the combustion chamber is shown in Figure 2. The location of the exhaust valve, the intake valve, and the spark plug as well as the position of the pressure transducer is indicated in the figure. The position of the optical sensor with the three optical fibers can also be seen.

The following specifications were required for the design of the optical sensor:

- the combustion chamber shape should not be changed by the sensor
- the optical sensor should be located close to the region where knocking combustion occurs, because the closer the sensor is positioned to the knock onset location the clearer knocking can be detected
- the volume elements observed by the optical fibers should be relatively small

Under these specifications the optical sensor, shown in Figure 3 was designed. The three optical fibers are mounted in three specially designed optical probes, which are screwed into an adapter. The angles between the probes are 120 degrees. To protect the optical fibers from the influence of the in-cylinder pressure and the combustion temperature, sapphire rods are glued inside the optical probes. The flame radiation is detected through small bores with a diameter of approximately 1 mm. The observed volume elements are cone-shaped with a view angle of about 10.4 degrees. The size of one volume element is about 0.01% of the compression volume of the cylinder at the crank angle interval when knock normally occurs. The whole volume created by the three small bores in front of the sapphire rods is less than 0.01% of the compression volume. Experience with flame propagation

Optical Sensor
Optical Fiber 1
Optical Fiber 2
Optical Fiber 3
Exhaust Valve
Intake Valve
Spark Plug
Pressure Transducer

Figure 2 : Design of the Combustion Chamber

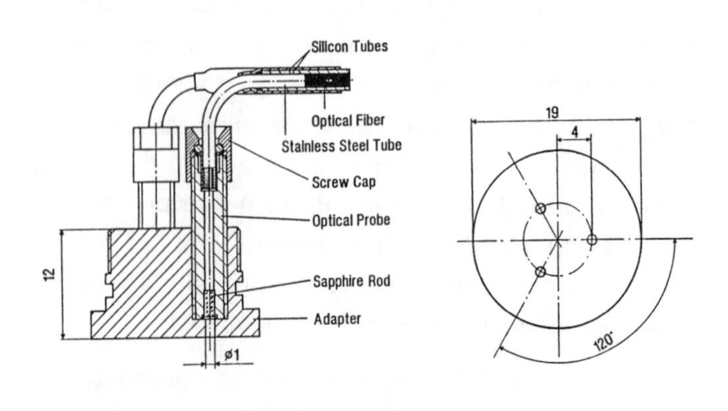

Figure 3 : Optical Sensor

investigations under knocking and non-knocking conditions have shown, that these small bores have no influence on the knocking process. The optical fibers are connected to the optical probes by screw caps and are accommodated in stainless steel tubes which provide protection from the high temperature water in the water-cooled engine. To seal the optical fiber against the water, two additional silicon tubes are glued on the stainless steel tube.

2.2 DEFINITION OF KNOCK INTENSITY - To quantify the knock phenomenon, it is helpful to introduce a variable which enables the characterization of the knock intensity. This can be done either by processing the pressure trace and/or by analysing the light signals from the optical fiber sensor.

Figure 4 shows a pressure trace of a knocking combustion cycle for the time interval in which knock occurs. The knock is indicated by the high frequency oscillations in the pressure trace. This high frequency pressure trace $p_{(t)}$ can be smoothed resulting in a low frequency signal $L_{(t)}$, represented by the dashed line. By using a formula from the field of electronics to calculate the energy of an alternating signal it is possible to calculate knock intensity $E_{\Delta p}$. By changing this formula to use it for calculation within the high frequency pressure oscillations it yields the knock intensity $E_{\Delta p}$ of one combustion cycle within a time window (t_1, t_2):

$$E_{\Delta p} = \int_{t_1}^{t_2} (p_{(t)} - L_{(t)})^2 \, dt$$

The knock intensity $E_{\Delta p}$ shows good correlation to the maximum pressure oscillation Δp_{max} [13], especially at low knock intensities. Increasing the knock intensity leads to scattering of $E_{\Delta p}$. Due to the calculation of the knock intensity within a time window the knock intensity comprises the pressure oscillations during the expansion. In the case of using Δp_{max} this part of the signal is completely ignored. The knock intensity $E_{\Delta p}$ from the pressure trace is used as a comparator to the knock intensity value detected by the optical fiber sensor. To define such a knock intensity the differences between the optical signals of a non-knocking combustion and the optical signals of a knocking combustion shall be discussed.

Figure 5 shows the cylinder pressure and the intensity of the flame radiation for one non-knocking combustion cycle (Ignition Timing 167° CA ABDC). The signals are shown in a crank angle range from 170 to 230° CA after bottom dead center (ABDC). The cylinder pressure reaches the maximum value of approximately 19 bar at about 210° CA ABDC. At about 203° CA the

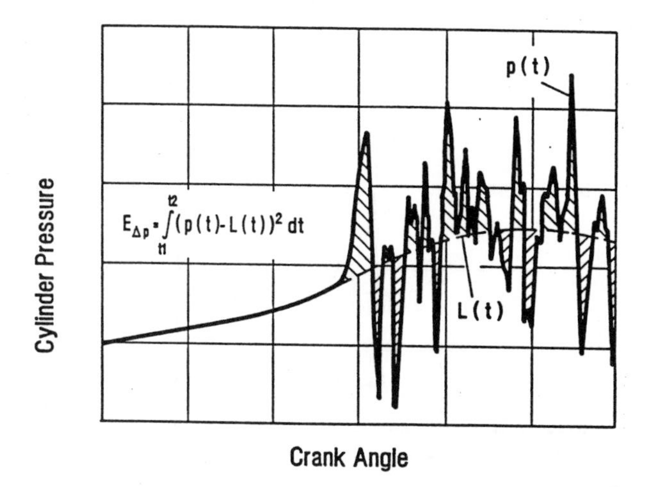

Figure 4 : Determination of Knock Intensity from the Pressure Trace

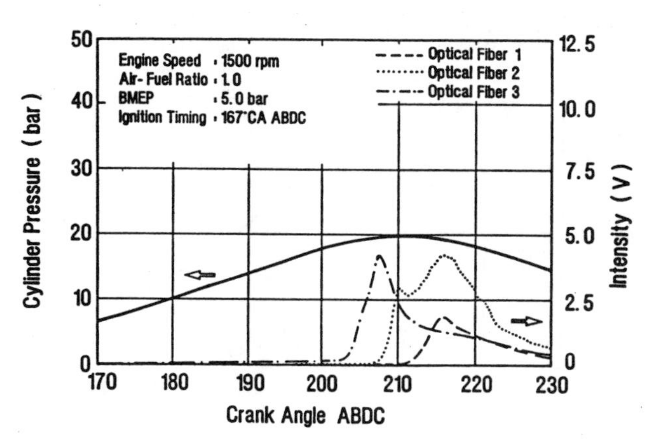

Figure 5 : Time Slopes of Cylinder Pressure and Optical Fibers for a Normal Cycle

flame front reaches the optical fiber 3 first then fiber 2 was reached at about 208° CA and at last fiber 1 was reached at about 212° CA. The difference between the beginning of the intensity and the maximum of the intensity is between 3 to 5° CA. The maximum of the intensity for optical fiber 1 is about 2 Volts, for the optical fibers 2 and 3 about 4 Volts. Due to the non-spherical flame propagation and the cyclic variation the optical fibers 1 and 2 are not reached by the flame at the same time.

Figure 6 shows the cylinder pressure and the intensity of the flame radiation represented by the three optical fibers for a knocking combustion cycle. To produce knock the ignition timing was adjusted to an earlier spark advance (157° CA ABDC). The cylinder pressure shows a significantly higher level than for the non-knocking combustion cycle. The high frequency oscillations which indicate knocking combustion can clearly be seen.

A strong difference in the flame radiation can be noticed between the non-knocking combustion cycle shown in Figure 5 and the knocking combustion cycle. The maximum of the intensity is much higher for knocking combustion than for non-knocking combustion. The intensity of the optical fiber 3 increases due to the normal flame propagation at about 190° CA. At about 194° CA an additional strong increase in the intensity can be observed. This strong increase in the intensity is caused by knocking combustion. It can be noticed right before the knock effect can be seen in the pressure trace. This behavior can be explained by the fast flame propagation during knocking combustion, because the knocking reaction zone moves very fast through the small volume element and induces this steep increase in the light signals.

For the optical fibers 1 and 2 a strong increase in the intensity from the zero level can be noticed, because the knocking flame propagation reaches the volume elements of these probes directly. The steep gradients in the light signals of the flame radiation during knocking combustion can be used as an indicator for knocking combustion and the definition of the knock intensity.

Figure 7 shows a definition of the knock intensity from a detected light signal.

$$KI = \tan\alpha$$

The velocity of the knocking flame can be calculated by the timing when the strong increase in the light signals occurs. In this preliminary test the time difference between the increase in the signals of the different optical fibers should be used instead of the velocity as another indicator for knocking combustion by a given size of the optical sensor. For a non-knocking combustion

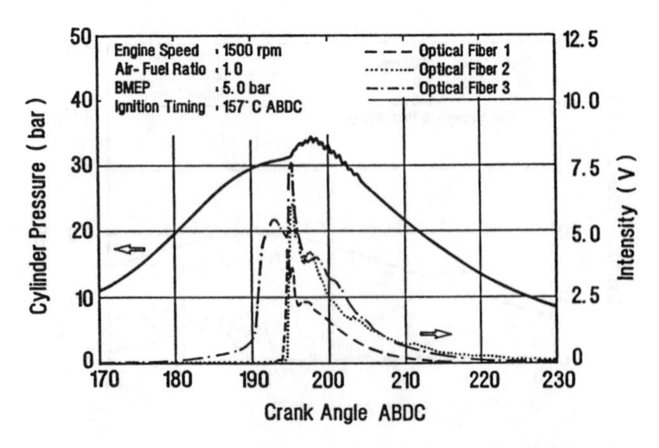

Figure 6 : Time Slopes of Cylinder Pressure and Optical Fibers for a Knocking Cycle

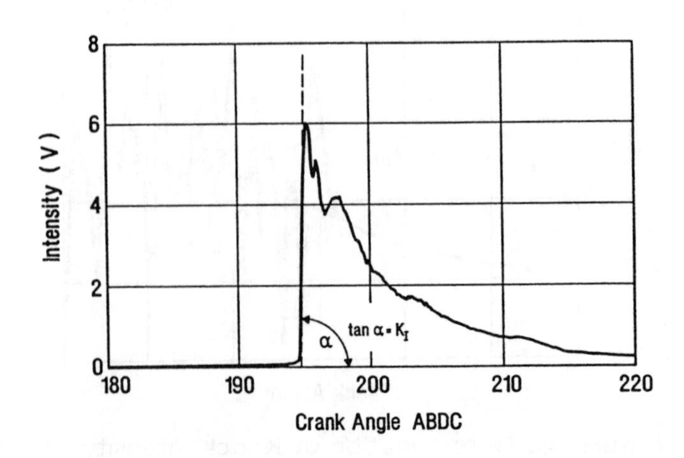

Figure 7 : Definition of the Knock Intensity KI

cycle (figure 5) the difference in signal increase of fiber 1 and 2 is about 4° CA. For the knocking combustion cycle in figure 6 the difference is less than 1° CA.

2.3 RESULTS OF KNOCK INTENSITY - The differences between normal and knocking combustion were investigated for the following engine operating conditions.

Engine speed (n)	: 1500 rpm
Air-Fuel Ratio (AFR)	: 1.0
Brake Mean Effective Pressure (BMEP)	: 5 bar

The ignition timing was changed from 167° CA to 147° CA in steps of 5 degrees. At all measurements, both the cylinder pressure and the light emission of the flame radiation of 16 combustion cycles were detected. The knock intensity $E_{\Delta p}$ was used to decide between normal and knocking combustion. The maximum increase in the flame radiation signal was used as a magnitude for the knock intensity KI.

Figure 8 shows the maximum signal rise from optical fiber 1 versus knock intensity $E_{\Delta p}$ in a double logarithmic diagram. The borderline between non-knocking and knocking combustion can be set at a knock intensity of about 10^{-5} bar^2 s. The non-knocking combustion cycles are plotted as circles and knocking combustion cycles are plotted as closed points. Most of the non-knocking combustion cycles have a signal rise of about 0.5 to 1 V/°CA. Just a few of the non-knocking combustion cycles have a signal rise more than 1 V/°CA. The signal rise of the knocking combustion cycles varied from 4 to 80 V/°CA. A clear differentiation between knocking and non-knocking cycles by the light signal rise is possible.

A classification of the signal rise is shown in figure 9 with ten classes and a class width of 4 V/°CA. The lowest class contains all the non-knocking combustion cycles while the classes 2 to 10 contain the knocking combustion cycles.

To avoid knocking combustion in an engine, ignition timing must be adjusted to later crank angle degrees when knock occurs. Thus, to determine the correlation between ignition timing and knocking combustion, respectively, the

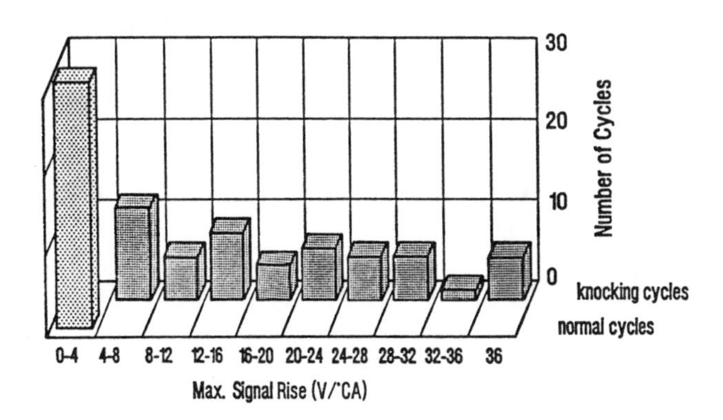

Figure 9 : Classification of the Signal Rise of Optical Fiber 1

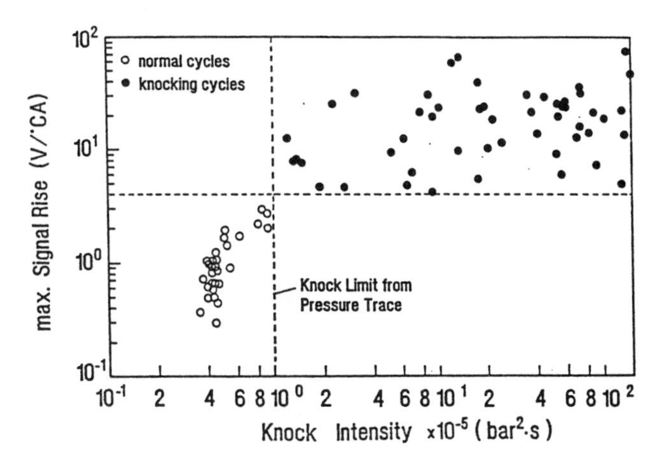

Figure 8 : Maximum Signal Rise of Optical Fiber 1 versus Knock Intensity $E_{\Delta p}$

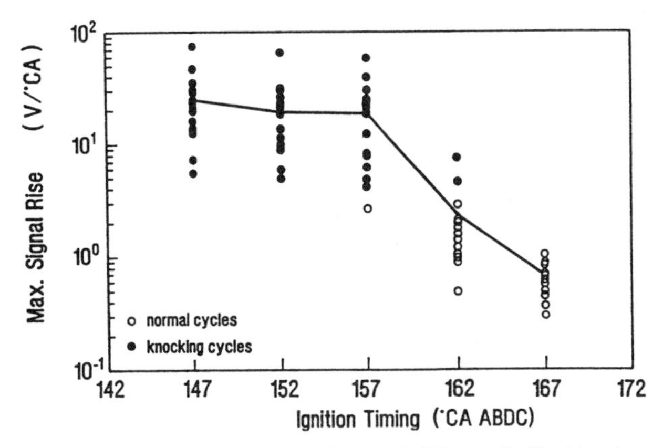

Figure 10 : Maximum Signal Rise of Optical Fiber 1 versus Ignition Timing

knock intensity must be known for each engine and engine operating condition. Figures 10 shows the maximum signal rise (knock intensity KI) versus the ignition timing. At an ignition timing of 162° CA the first two knocking cycles can be noticed. With an ignition timing of 157° CA there is only one normal combustion cycle while the others are knocking cycles. The full line shows the average of the cycles for each of the investigated ignition timings. A significant increase between the ignition timing of 162° CA and 157° CA can be observed. This clear influence of ignition timing on the maximum signal rise can be used for an anti-knock control system.

The time differences in signal increase between fibers 1 and 2 is shown in figure 11. The time differences for the most knocking cycles are lower than 1° CA. The time differences for non-knocking cycles varied from 0 to 5° CA. In the case of non-knocking cycles, cyclic variations in the flame propagation cause the spread of the crank angle difference between optical fiber 1 and 2. A clear separation of normal and knocking cycles is not possible using the time differences in signal increase alone.

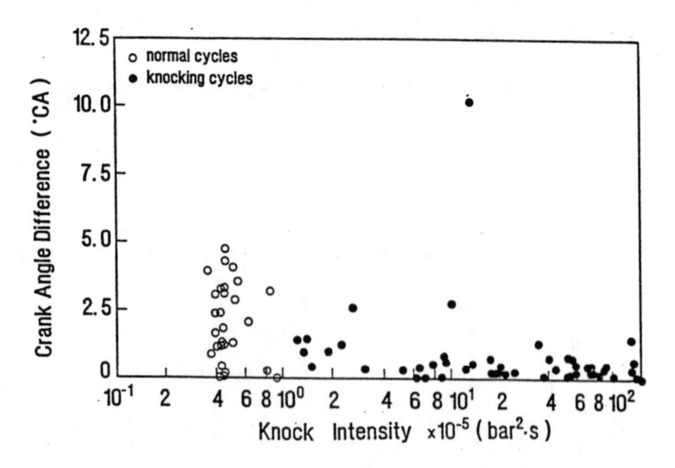

Figure 11 : Crank Angle Difference in Light Signal between Optical Fiber 1 and 2

2.4 KNOCK CONTROL SYSTEM - Based on the optical flame signals focused on knock identification, a knock control system with a suitable electronic signal processing system was developed and tested under the same engine conditions. The knock controller is a closed-loop regulator system consisting of four units which are: one optical fiber, the signal processing unit, the regulator itself and the ignition timing unit.

THE SIGNAL PROCESSING UNIT - The increase of the flame signal is measured by differentiating the analog input light signal. The peak detector samples the maximum signal rise for each cycle. The analog output voltage is held during one cycle untill the next signal peak. The peak detector must be capable of detecting a peak of less than 500 nanoseconds with adequate accuracy. As the system is measuring the signal rise, the output also depends on the signal gain. Deposits on the sensor for example cause a reduced signal output. For long term stability an automatic gain control (AGC) is necessary. The prototype operated without AGC and produced useful results. With an automatic gain control it is possible to use an optical knock sensor within a road performance of about 30 000 km [29].

THE REGULATOR - The analog PI regulator consists of an integrating component and a proportional component with a very low gain. The integrating component completely compensates the static error and the proportional component serves for faster settling of the system. All coefficients can be adjusted manually. A differential component of the regulator was not useful due to the high noise level of the input signal. Dependent on the input voltage of the knock sensor the regulator changes the ignition timing. The new ignition timing is set for the next combustion cycle.

THE IGNITION TIMING UNIT - An analog input voltage controls the ignition timing within a 10 degree CA range in addition to an externally adjustable offset. A high resolution crank angle decoder allows to change the ignition timing in 1/10° CA steps.

The signal processing unit was tested at various engine speeds, different BMEPs and by changing the ignition timing.

Figure 12 shows a typical pattern of the optical knock signal (maximum signal rise) for an engine test conducted under operating conditions near knock limit. It can be seen from the results that some of the combustion cycles show a knocking behavior (higher than 4V) while most of the combustion cycles show a non-knocking behavior.

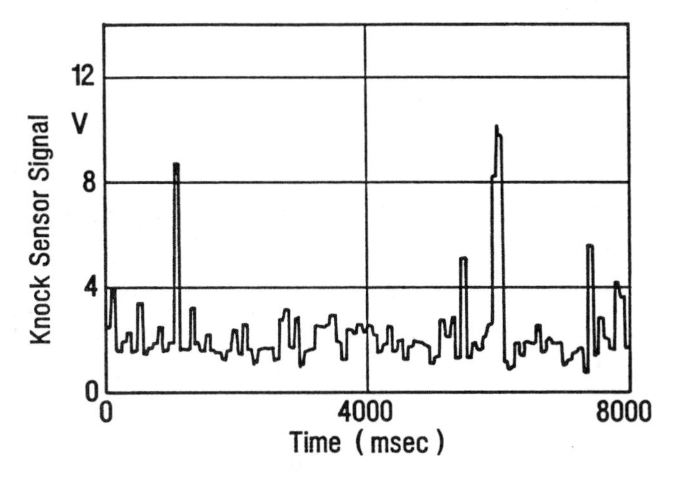

Figure 12 : Knock Sensor Signal
(IT= 160° CA ABDC, speed= 1500 rpm, λ= 1.0 (AFR), BMEP= 5.0 bar)

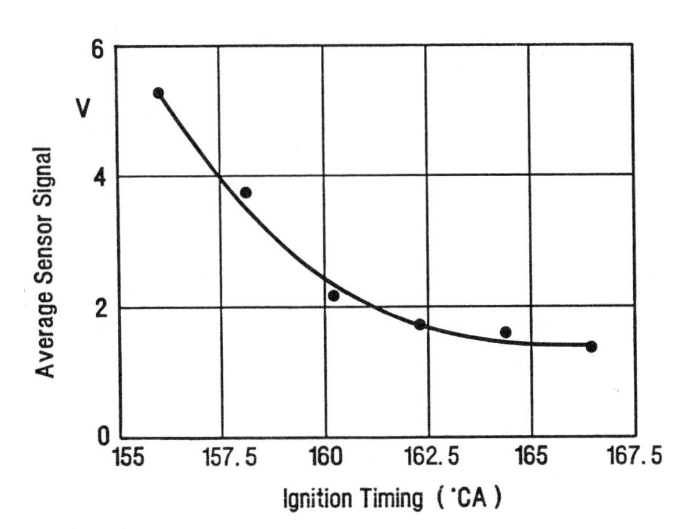

Figure 13 : Knock Sensor transfer Function
(speed= 1500 rpm, λ = 1 (AFR), BMEP= 5 bar)

Figure 13 shows the knock sensor transfer function. The ignition timing was changed in steps of two degree crank angle. To receive the knock sensor transfer function an average over a period of 8 seconds is used. The rise time varies from about 10 μsec to 200 μsec. This requires a wide dynamic range for the measuring unit. At an ignition timing of 162° CA ABDC (speed= 1500 rpm, λ = 1 (AFR), BMEP= 5 bar) the mean value of the peak amounted to 48.2 V/msec and a rise time of 52 μsec (2V input signal amplitude). The standard deviation of the output signal was about 50%, because of the cyclic variations of the combustion process. The sensitivity to ignition timing variation at an ignition timing of 160° CA was 0.6 V per degree CA.

The knock sensor signal is the input of the analog PI regulator. The output of the regulator is connected to the ignition delay unit performing in a closed-loop with the combustion engine. The system shows stable operation for all regulator gains and time constants.

Figure 14a shows the closed-loop regulator's response to a 5 degree step, created by changing the offset of the ignition timing. Figure 14b shows the regulator input signal. Each

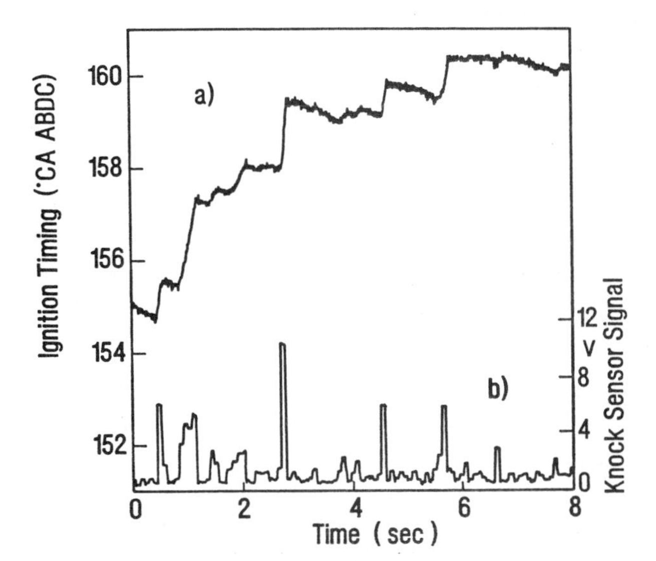

Figure 14 : Regulator Step Responce
(5 Degree Step)

time when knocking occurs the ignition timing changes in steps to later timings until the ignition timing is adjusted for non-knocking combustion again. At the adjusted parameter the knock regulator has a time constant of about three seconds. This is the maximum speed if allowing a ignition timing jitter of 0.5 degree CA. To increase the response time of the knock regulator a higher jitter would need to be allowed.

As shown before, the maximum signal rise of the flame radiation spread within a wide range. This means a poor signal to noise ratio (SNR). If operating the regulator with a high gain and small time constant, the ignition timing would change in a range of several degrees. A slow regulator or low pass filter has an averaging effect and high accuracy can be achieved. This means the SNR is the limiting factor for system speed. A better S/N ratio may be achieved by weighting the signal with a nonlinear transfer function.

Due to the periodicity of the ignition and combustion process, the flame signal is a time discrete signal. The output of the maximum signal rise unit can be compared to a signal of a sample and hold unit. The maximum delay is one combustion cycle. The phase delay caused by the engine is negligible compared to the regulator and has no influence on the regulator's stability.

As a prototype, the controller operated without compensating the influence of speed and load. Due to the influence of speed and load on flame propagation an adjustment of the knock sensor signal is necessary. Furthermore, a pre-control of the ignition timing is necessary to achieve a fast transient response. Other parameters influencing the combustion process, such as air-fuel-ratio and air temperature, can be compensated by the error feedback of the closed loop controller.

A microprocessor controlled version of the regulator would have several advantages, mainly faster response and higher accuracy. A digital version would include pre-compensation and knock limit compensation. The signal increases because the knock value is dependent on engine speed. The knock limit is not constant. Air fuel ratio and load are less significant but not without influence. Dynamic regulator gain and time constants would allow a faster response of the system.

As there is an error feedback signal, it is possible to develop an adaptive version. The system would be able to adapt its parameter for pre-compensation to the current parameters of the engine, teaching itself by the error signal. Engine parameters may change by time due to decreasing compression, etc. The permanent memory of the engine management system would always contain the actual engine parameters.

3 CONCLUSIONS

Based upon the favorable results of the tests described in this paper, it appears that the newly developed optical fiber sensor and the corresponding knock control system can be applied to the detection and control of knocking combustion.

With this system it is possible to determine knocking combustion for each single combustion cycle by using the maximum gradient in the signal rise of the detected optical signal. Only one optical fiber per cylinder is necessary as the input signal of this knock control system. To make this system less costly for production engines, photoelectric cells should be used instead of photomultipliers.

Knocking combustion and non-knocking combustion can be clearly separated by the optical flame signals, detected with one optical fiber. A comparison of the knock intensity detected by the optical fiber sensor and from the pressure trace has shown that there is a good agreement between both measurement concerning the decision whether the cycle is a knocking one or not.

Using the knock intensity detected by the optical sensor, the cyclic variations of the combustion process concerning knock can be determined. A mean knock intensity of several combustion cycles (for example 25 cycles) can be determined. This mean knock intensity can be used as an input for the regulator to adjust the ignition timing with the ignition timing unit.

The maximum delay of the complete knock control system is just one combustion cycle. Thus, the system is fast enough to adjust the ignition timing directly when knock occurs.

To use a optical knock control system in a production engine further investigations concerning deposits and long time stability must be carried out.

4 References

1. W.E. Betts, "Avoiding High Speed Knock Engine Failures", V.W. Internal Symposium on Knocking of Combustion Engines, 1981
2. Kolbenschmidt, "Technisches Handbuch"
3. H. Kornprobst, G. Woschni and K. Zeilinger, "Erkenntnisse über das Verhalten der Kolbenringe bei Klopfbetrieb", AUTEC 1988
4. F. Renault, "A New Technique to Detect and Control Knock Damage", SAE paper 820073
5. G. Betz and J. Ellermann, "Knock-Related Piston Damage in Gasoline Engines, Knock Measurement Technique, Aspects of Piston Failure Prevention", V.W. Internal Symposium on Knocking of Combustion Engines, 1981
6. G. Woschni and J. Flieger, "Experimentelle Untersuchungen zum Wärmeübergang bei normaler und klopfender Verbrennung im Ottomotor", MTZ 43, 1982
7. G. Essig, K. Schellmann and E. Wacker, "Kolbenbeanspruchung bei klopfender Verbrennung", MTZ 44, 1983
8. R.R. Maly et al., "Theoretical and Experimental Investigation of Knock Induced Surface Destruction", SAE paper 900025
9. W. Lee and H.J. Schaefer, "Analysis of Local Pressures, Surface Temperatures and Engine Damages under Knock Conditions", SAE paper 830508
10. H. Zhao et al., "Characterization of Knock and Its Effect on Surface Temperatures", SAE paper 920514
11. D.H. Cuttler and N.S. Girgis, "Photography of Combustion During Knocking Cycles in Disc and Compact Chambers", SAE paper 880195
12. P.V. Puzinauskas, "Examination of Methods Used to Characterize Engine Knock", SAE paper 920808
13. H.P. Kollmeier, "Untersuchungen über die Flammenausbreitung bei klopfender Verbrennung", Dissertation RWTH-Aachen, 1987
14. G. König and C.G.W. Sheppard, "End Gas Autoignition and Knock in a Spark Ignition Engine", SAE paper 902135
15. G. König et al., "Role of Exothermic Centres on Knock Initiation and Knock Damage", SAE paper 902136
16. S.M. Dues et al., "Combustion Knock Sensing: Sensor Selection and Application Issues", SAE paper 900488
17. K. Obländer et al., "Antiklopf-Regelung am Ottomotor", ATZ 82, 1980
18. R. Fritz and A. Richter, "Klopfregelung am Porsche-944-Turbomotor", MTZ 46, 1985
19. H. Brüggemann et al., "Der neue Sechszylinder-Vierventilmotor mit 3.0 l Hubraum für den neuen Mercedes-Benz 300 SL-24", MTZ 50, 1989
20. P. Hensler et al., "Der Porsche-Motor 944 S2", MTZ 50, 1989
21. K.P. Schmillen and M. Rechs, "Different Methods of Knock Detection and Knock Control", SAE paper 910858

22. U. Spicher, "Untersuchungen über die räumliche Ausbreitung und das Erlöschen der Flamme bei der motorischen Verbrennung", Dissertation RWTH-Aachen, 1982

23. U. Spicher and A. Velji, "Measurements of Spatial Flame Propagation and Flow Velocities in a Spark Ignition Engine", XX. Symposium on Combustion/ The Combustion Institute 1984/ pp. 19-27

24. U. Spicher and H.P. Kollmeier, "Detection of Flame Propagation During Knocking Combustion by Optical Fiber Diagnostics", SAE paper 86153

25. U. Spicher et al., "An Experimental Study of Combustion and Fluid Flow in Diesel Engines", SAE paper 872060

26. U. Spicher et al., "Application of a New Optical Fiber Technique for Flame Propagation Diagnostics in IC Engines", SAE paper 881637

27. U. Spicher and H. Kröger, "Flame Propagation Diagnostics in Engines by Multi-Optical Fiber Technique", 2. International Conference on Methodology and Innovations in Automotive Experimentation, Florence, November 1988

28. G. Schmitz et al., "A Fast Intelligent VMEbus System for Combustion Analysis in Engines", 19th ISATA, Monte Carlo, 24th - 28th October 1988

29. W. Moser, "Der Lichtverlauf bei der ottomotorischen Verbrennung", XX FISITA CONGRESS Wien, Das Automobil in der Zukunft, 1984

900488

Combustion Knock Sensing: Sensor Selection and Application Issues

Steven M. Dues, Joseph M. Adams, and George A. Shinkle
Delco Remy Div.
General Motors Corp.

ABSTRACT

Knowledge learned through the successful application of millions of knock sensors on automotive engines is reported. An explanation of the basic characteristics of the knock phenomena and their relationship to sensing capabilities is given. Popular conceptions and misconceptions concerned with engine knock and the variables affecting it are examined. Sensing methods are described, with the emphasis being on vibration sensitive devices. Application methodologies and issues are discussed, with a review of potential pitfalls in sensor selection, sensor location, and systemization.

INTRODUCTION

Combustion control systems, when engineered properly, can increase customer satisfaction with his automobile by improving performance and fuel economy. Minimizing spark knock is a major aspect of many engine control systems of today because knock is both detrimental to the engine and objectionable to the customer. Knock control allows the engine to be calibrated closer to its optimum operating point with deviations only when knock actually occurs as opposed to designing conservatively to avoid knock under all operating conditions. Also, controlling knock reduces the vehicle octane requirement (1)*, allowing the use of lower octane fuels.

Many of today's knock control systems are limited by the sensing device in the system. The purpose of this paper is to review knock sensor selection criteria through discussion of the knock phenomenon, types of sensors, application issues, popular misconceptions, and sensor performance issues. The information is based on experience gathered through hundreds of knock sensing applications with over twenty million units on production vehicles.

BASIC KNOCK PHENOMENA

Comprehension of the knock phenomena begins with an empirical understanding of the combustion process itself. For explanation purposes, divide the combustion chamber volume into infinitesimally small three-dimensional cube elements containing oxygen and fuel molecules. The combustion process of each element is a chemical reaction which produces heat and chemical products as explained by the theories of chemical kinetic reactions (2). The molecules have distinct probabilities of reaction and reaction rates as a result of each element's mixing rate, heat input, and time history. The combustion process begins as a spark is introduced into several of these elements by the spark plug.

* Numbers in parentheses designate references at end of paper.

The energy input raises the temperature of the elements near the spark plug until the above-mentioned chemical reactions occur. As the element reacts, it produces heat and expands. A portion of the heat is transferred into the next element and the expansion compresses that next element. With the proper amounts of temperature, pressure, and time, as defined by the surface shown in Figure 1, the element will spontaneously combust. As long as the element is below the combustion surface it has not "totally reacted." Whenever the temperature-pressure-time surface is exceeded the element has "totally reacted" or burned. It should be noted that at some period in time this element's reaction will become luminescent. If this phenomena was examined from a more global scale, it would appear as if a flame front was passing through the element.

Normal combustion can be thought of as a controlled propagation of the reactions from element to element. However, under certain conditions, as the propagation of reactions approaches the end-gas region of the combustion chamber, the temperature, pressure, and time histories of the elements comprising the end-gas are "simultaneously" approaching the combustion surface shown in Figure 1. This results in an extremely fast propagation of reacting elements which releases tremendous amounts of energy.

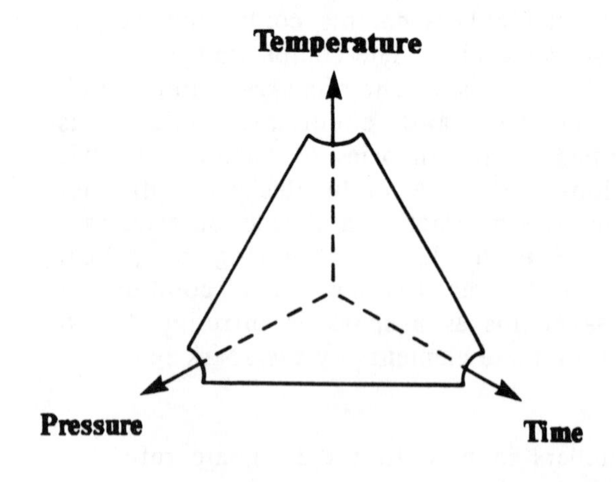

Figure 1 - Combustion Pressure-Time-Temperature Surface

The amount of energy released by the end-gas is a function of the number of elements (amount of unburned fuel mass) contained in the end-gas and how fast the reaction occurred. This sudden release of energy increases the pressure of the end-gas. The pressure rise then expands across the combustion chamber generating the phenomena called knock. An empirical method of explaining this phenomena is as follows.

The sudden pressure rise sets up a shock wave in the combustion chamber whose strength is proportional to the pressure differential across the shock wave. This differential will be greater for faster energy release rates and for larger amounts of unburned fuel mass in the end-gas. This explains the statistical probability for cycle-to-cycle variation of knock intensity. Each combustion cycle is statistically different but the mean changes of the probability are repeatable. Therefore, any change to the combustion process which increases the average end-gas mass or the combustion rate will increase knock intensity.

The shock wave traverses the combustion chamber in a complex manner due to the varying boundary conditions that the wave crosses. This wave is analogous to an oblique shock wave which generates vibrations whose frequency content is dependent upon its initial conditions and the cylinder dimensions. The fundamental frequency of the shock wave is a function of the distance it travels (bore diameter) and the local speed of sound (average bulk temperature of the combustion chamber). The relationship can be expressed by Equation 1 (3,4).

Equation 1: $$f_n = \frac{k_c * c}{B} = \frac{k_t \sqrt{T}}{B}$$

Where:

f_n = Fundamental Knock Frequency

k_c, k_t = Constants

c = Local Speed of Sound

B = Cylinder Bore Diameter

T = Bulk Combustion Chamber Temperature

As noted by other authors (3,4), acoustic theory for disc-shaped chambers states that the previous equation can be multiplied by a mode shape factor to identify the frequencies of higher vibrational modes. The factors for mode shapes two through four have been documented as approximately 1.65, 2.07, and 2.26. Again, realizing the knock phenomena is caused by an oblique shock wave, the amount of energy which is transferred into each mode shape resonance will be determined by the initial conditions and propagation of the shock wave.

Practical experiences can be combined with the above theory to explain many knock phenomena.

A. While many factors influence the occurrence knock, its frequency is controlled primarily by the combustion chamber temperature (or local speed of sound) and the cylinder bore diameter, as given in the preceding equation. Because of this temperature dependence, the knock frequency is subject to the same statistical variation as combustion. Deviations from the mean fundamental knock frequency as high as +/- 400 hertz have been observed for a given engine model. The engine speed only affects the knock frequency through how it changes the average combustion chamber temperature. Experience has confirmed that the fundamental frequency is inversely proportional to the bore diameter, as exemplified in Figure 2.

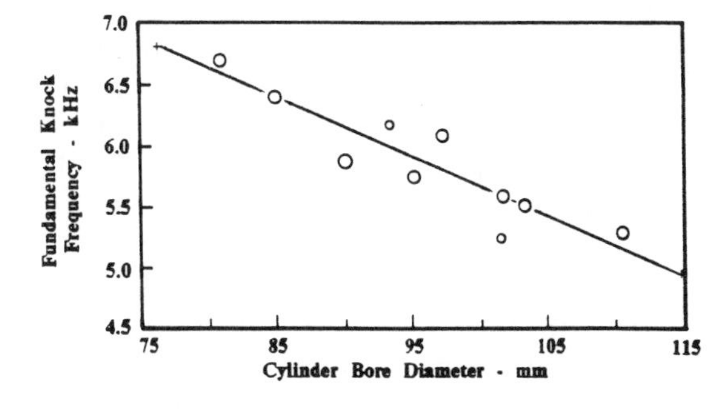

Figure 2 - Knock Frequency as a Function of Bore Diameter

If an engine's mean knock frequency is above the predicted frequency by several hundred hertz, the average combustion chamber temperature is most likely higher than normal. Conversely, if the knock frequency is several hundred hertz below the predicted value, the average combustion chamber temperature is lower than normal. These abnormal combustion temperatures may have adverse affects on emissions.

B. Other researchers (3,4) have discussed the fact that, if a combustion chamber is not properly designed, the faster the combustion rate in that chamber, the more likely it is to knock. This can be explained by the fact that the pressure rise of the end-gas is faster and, therefore, heat of the end-gas elements cannot be transferred to the quenching layer of the combustion chamber as in slower burning combustion chambers. Also, after knock has been initiated the chamber is more likely to knock on the next cycle, and the knock is likely to be more severe (5). This phenomena has been observed in many engines tested and has been described as "thermal runaway."

C. High swirl combustion chambers have been introduced in recent years to drastically increase the combustion rate. These engines are more likely to knock than conventional engines for the reasons stated above. Also, these engines have introduced a new knock frequency which is not explained by conventional acoustic theory. Experience has shown that the predominant frequency is approximately 1.25 to 1.30 times the calculated fundamental frequency. An exact theory explaining this new frequency has not been formulated, but it may be caused by the swirl interfering with the oblique shock wave.

D. The number of elements in the end-gas which will spontaneously combust is dependent, along with the items previously mentioned, upon their composition. Part throttle operation leaves more residuals in the end-gas which slows the chemical reaction, reducing the probability of knock occurring. Wide open throttle operation, on the other hand, reduces the residuals in the combustion chamber, increasing the chances for knock occurrence. Rapid throttle angle changes also increase the probability to knock by minimizing residuals. This phenomena is called "tip-in" knock.

Any application of a knock sensing device to an engine must include a thorough testing of all operating environments. Several knock frequency modes can occur in the combustion chamber and the excitation and travel of the shock wave can transfer its energy into any and all of these modes. Therefore, when choosing a knock vibration mode to sense, one must be careful to select a mode which is most prevalent under all operating conditions, not just a single test point.

SENSOR TYPES

Numerous methodologies have been proposed for sensing engine knock in combustion control systems. The most prevalent approach in the automotive industry today is sensing vibrations induced by knock. Most knock sensors utilize a piezoelectric element to transform mechanical vibrations into electrical signals and they are regularly categorized by the usable range of their response, as shown in Figure 3. Flat sensors have a high resonant frequency, resulting in a relatively flat response in the fundamental knock frequency range, five to eight kilohertz. The resonant frequency of Spike Resonant and Broadband Resonant sensors is centered on the mean knock frequency of the engine, capitalizing on the built-in mechanical amplification and filtering characteristics of the sensors. The difference between these two groups is the bandwidth of the resonance. Spike sensors have a narrow bandwidth, on the order of one hundred hertz, while Broadband Resonant sensors have bandwidths approaching one thousand hertz.

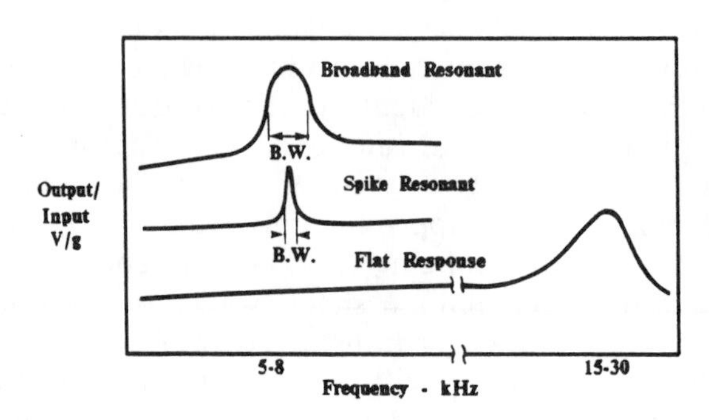

Figure 3 - Sensor Transfer Function Comparison

Flat frequency response sensors offer the advantage of commonalty since one sensor design can be used on several engine models with different knock frequencies. However, controller electrical filtering must be optimized for each engine model in order to distinguish knock from noise. Spike sensors are tuned to the frequency of knock for a particular engine but are limited by manufacturing variations, either in the sensor or the engine, which may cause misalignment in the knock frequency and their response, resulting in nondetected knock events. Broadband Resonant sensors such as the type produced by Delco Remy combine the benefits of the Flat and Resonant sensors. Their mechanical amplification and filtering of the signal over a broad bandwidth in the knock range provides high signal levels while allowing for normal engine variations.

Vibrational knock sensors measure a secondary response to the knock event in the combustion chamber. From a pure engineering standpoint, a more direct measurement technique is certainly desirable to reduce measurement error and complexity. Cylinder pressure, combustion temperature, and in-cylinder flame measuring devices are being developed, using technologies such as piezoelectric, optical, fiber optic, micro-machined silicon, and chemically sensitive materials. Each of the currently available devices has advantages and disadvantages as compared to present knock sensors, and should be evaluated on the basis of system requirements, packaging needs, and cost to benefit analysis as mentioned later in this paper to justify their implementation.

APPLICATION ISSUES

The majority of knock control systems in production today provide global engine control by retarding the timing based on the input of a single knock sensor. However, some systems in use and under development offer either individual cylinder knock control or utilize multiple sensors. Individual cylinder control is most beneficial if the engine is not uniform such that some cylinders are more prone to knock. A global control system would continually retard the timing of all cylinders to correct the few that knock easily, whereas an individual cylinder control system would adjust only the troublesome ones. Single sensors can be used with either type of control system. Alternatively, multiple sensors may be used to improve knock detectability or, if necessary, to increase reliability. Many issues, including control strategy, desired knock control level, engine characteristics, and a cost to benefit analysis should be considered when defining the total control system. The ensuing section assumes a single sensor application, yet many of the ideas apply to multiple sensor systems as well.

Figure 4 outlines the critical knock sensor application process and serves as a reference for this discussion. The level of control desired from the system should be defined, with consideration of target customers, before application work begins. For instance, do customers desire total elimination of audible knock, or would they rather have improved performance with some limited amount of audible knock? Referring to Figure 5, this step involves determining how close to the knock limit the engine will operate, and how far it will be allowed to stray into the knock region before the control system retards the spark. Another preliminary step is to determine the probable knock frequency. For a given bore diameter, experience based relationships such as that displayed in Figure 2 can predict the frequency of the desired mode. However, factors such as turbochargers and high swirl combustion chambers may alter the knock frequency, as mentioned previously. This step allows initial selection of a resonant sensor; later data analysis will confirm the actual mean knock frequency and its variation.

When determining where to locate knock sensors, whether designing a new engine or adapting an existing one, the primary concern should be to find locations which are mechanically stiff. Stiffer locations allow transmission of the knock-induced vibrations while limiting the level of noise input to the sensor. The authors' experience and the work of others (4) has shown that locations which react to vibrations with significant motion have also been excessively prone to noise from sources other than knock. Locations low on the engine block and centered axially are recommended because of their stiffness and remoteness from other noise generating components, such as valve trains and accessory drives. It is recommended that several locations be identified for further analysis because knock vibrations do not always obey common engineering intuition. Table 1 gives further location selection considerations.

Studies should be conducted to rate prospective locations based on their signal amplitude and ability to respond to knock vibrations from all cylinders equally. Instrument grade accelerometers and, if proper mounting is possible, actual knock sensors can be used to evaluate locations under knock and no-knock conditions. Graphical representations of complex engine mapping techniques create a visual comparison of potential mounting locations. As shown in Figure 6, location A would be selected as the better mounting position because it has the highest signal amplitude with the least variation.

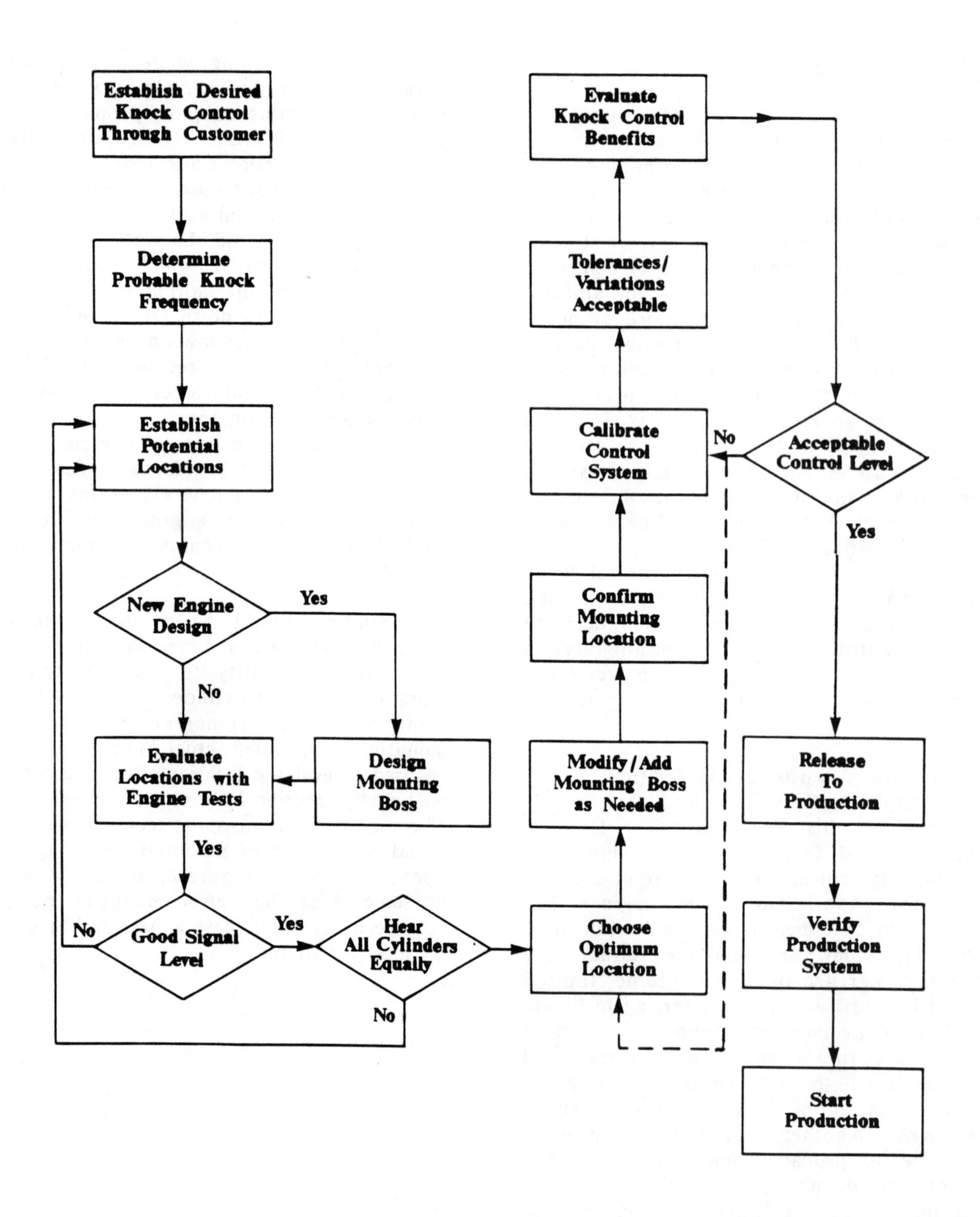

Figure 4. Application Process Flowchart

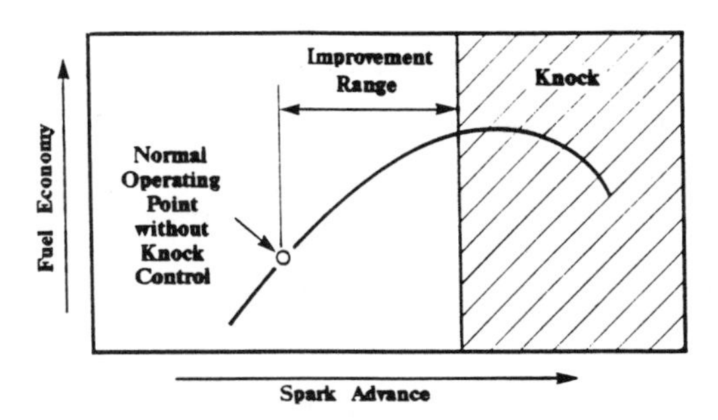

Figure 5 - Knock Control Calibration Improvements

Table 1 - Location Selection Recommendations

1. Choose a mechanically stiff location centrally located on the lower portion of the engine block.

2. Position away from noise generating components, such as the valve train and accessory drives.

3. Locations between two cylinders are better than on cylinder center lines because of increased stiffness.

4. Limit exposure to harsh environments and high temperatures.

5. Consider manufacturability, process flow, and serviceability.

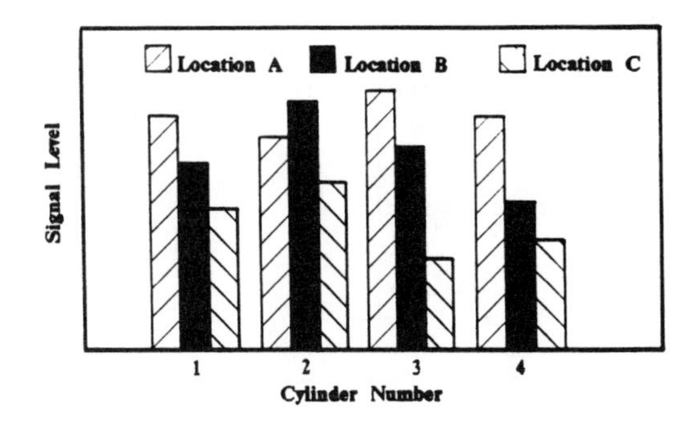

Figure 6 - Location Rating Example

After the location has been selected provisions must be made for mounting the sensor if they are not already present. The modification often involves casting a boss into the block to provide material for the sensor threads to engage and to increase mounting stiffness. When designing a new engine, bosses can often be cast into the block at prospective sensor locations to aid in the selection process. Once proper mounting conditions are present the sensor location can be confirmed and the system calibrated. Also, if the sensor is particularly frequency sensitive, such as a Spike sensor, additional testing will be necessary to insure that all manufacturing tolerances are acceptable. After the calibration has been determined, the level of control available should be compared to that desired and modifications made if necessary. Final verification of the production system is recommended before the start of production.

SENSOR SELECTION CRITERIA

Sensor selection criteria should include factors other than performance, as shown in Table 2. While many of these factors are manufacturer dependent, several general comments can be made about the three sensor categories. Spike and Broadband Resonant sensors generally have higher output levels in the knock frequency range than Flat sensors because of the amplification caused by the resonant response. Figures 7 and 8 demonstrate the signal levels and signal to noise ratios of a Broadband Resonant sensor and a Flat sensor mounted next to each other and responding to the same input.

Table 2 - Sensor Selection Criteria

Performance
- Signal Level
- Signal to Noise Ratio (at knock frequency)
- Performance Degradation at Temperature
- Sensor Orientation Requirements

Environment
- Temperature Capability
- Ruggedness - Vibration and Handling
- Sealed Connections
- Environmentally Sealed (Humidity, Salt Spray, etc.)
- Compatibility With Automotive Fluids

Reliability
- Durability
- Proven Reliability
- Serviceability

Manufacturability
- Quality Level
- Robustness to Manufacturing Variation
- Assembly Tolerance Control Requirements

Systems Issues
- Wiring Harness Interface (No. of Wires)
- EMC - Electromagnetic Compatibility
- Packageability (Size)
- Total Systems Cost Impact Driven by Sensor

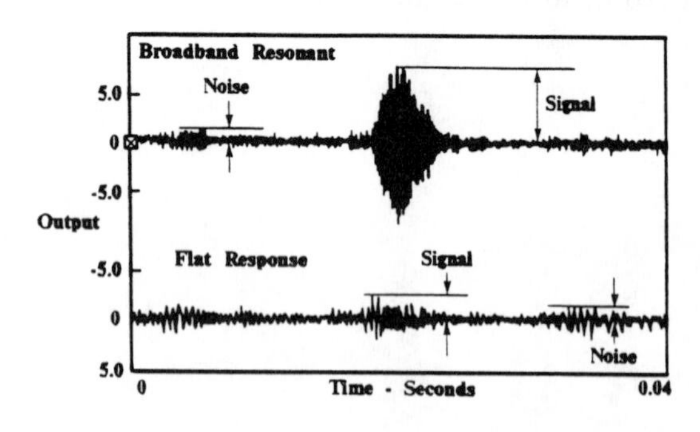

Figure 7 - Time Based Signal to Noise Ratios

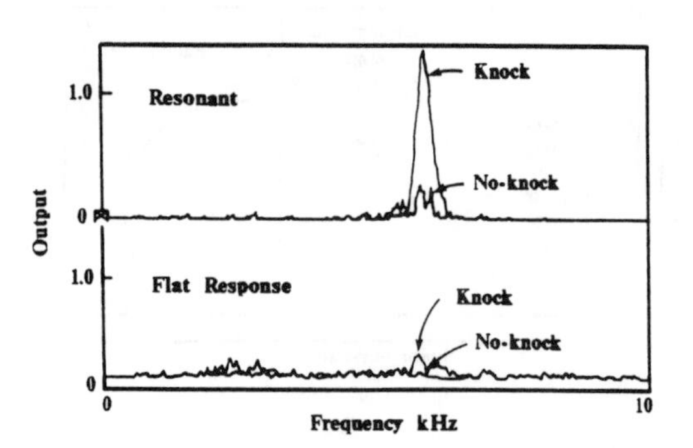

Figure 8 - Frequency Based Signal to Noise Ratios

As noted earlier, spike sensors commonly require holding tight manufacturing tolerances to control the resonant frequency, and may raise application costs. Flat sensors typically increase the cost of the knock system because of the need for high quality electrical filters and amplifiers in the controller to distinguish knock from background noise. They also often raise costs because their low outputs generally require a second lead to provide a stable ground and protect against radio frequency and electromagnetic interference. While some Broadband and Spike sensors also need a second wire because of low output, many do not. The selection matrix below, Table 3, highlights the major decision criteria for selection of the sensor type.

Table 3 - Sensor Comparison Matrix

PARAMETER	BROADBAND RESONANT	SPIKE RESONANT	FLAT
SIGNAL LEVEL	5	3	1
SIGNAL TO NOISE RATIO	4	4	3
EASE OF APPLICATION	4	3	5
IMPACT ON SYSTEM COST	4	4	2
SENSITIVITY TO VARYING KNOCK FREQUENCY	4	1	4
HIGH SPEED SENSING (>6000 RPM) CAPABILITY	2	2	2
TOTAL	23	17	17

Ratings: 1 - 5, With 5 Being Best

POPULAR MISCONCEPTIONS

The field of knock sensing has been plagued by misinformation. The true cause of this is unknown, but it is at least partially driven by the following forces:

1. Knock sensing typically utilizes vibrations in the five to eight kilohertz frequency range. These frequencies tend to generate very small displacements and thus are difficult to "visualize" and comprehend.

2. Knock sensing is highly tied to the particular company's engine control strategy and algorithm development, so systems philosophy has controlled many technical decisions.

3. Vibration sensitive knock sensors measure a secondary phenomenon, thus knock-like vibrations generated from other engine components are sometimes interpreted as knock.

Several of the popular misconceptions deserve additional discussion.

A. Flat sensors have better signals than Resonant sensors at high engine speeds.

At high engine speeds all of the vibration type sensors have reduced signal to noise ratios. This is due to the inherent vibration "noise" of the engine in the knock frequency range increasing exponentially with engine speed such that it begins to mask the knock signal. Since each engine and sensor location has its own signal characteristics, engine parameters and sensor placement affect the signal to noise ratio as much as the sensor itself.

B. Resonant sensors can drive the block to vibrate.

This misconception can best be addressed by visualizing a simple mechanical engineering model consisting of two masses. The dynamic mass of the worst case resonant sensor is roughly 75 grams while the dynamic portion of the engine block (where the sensor mounts) is several orders of magnitude greater. Thus the potential amount of vibration input to the engine from the sensor has to be negligible.

C. Flat sensors are newer technology than Resonant sensors.

Both types of sensors have been around for some time. The performance of Flat sensors is very similar to laboratory accelerometers which have been used for decades, and resonant knock sensors were used on some aircraft engines in the 1940's. All mechanical systems have a natural resonant frequency; the Flat sensor's resonance is just moved beyond the knock frequency range, whereas resonant devices take advantage of free mechanical amplification and filtering of the signal rather than doing it with electronics.

D. Resonant sensors cause a time delay due to ringing while Flat sensors do not.

This is an interesting perspective since the resonant sensors are similar to a "bell" which rings at its natural frequency. However, sufficient internal dampening exists at these high frequencies that the sensors basically respond as a forced response system only. That is, they create a signal only when excited by the engine. Perhaps there is some infinitely small period of time required to damp the vibrations, but no concerns of this sort have ever proven out in the authors' experience. The above statements are corroborated by Figure 7, which shows the signals generated by a knock event for a Resonant sensor and a Flat sensor that are mounted side by side.

E. Resonant type sensors require more application work.

As discussed earlier, the application process is nearly the same for Resonant and Flat sensors. Resonant sensors require selecting the appropriate frequency sensor while the correct filter frequency must be chosen for Flat sensors. The spike sensor certainly requires much engineering effort to insure that sensor frequency tolerances, engine variation, and controller tolerances are compatible. However, the Broadband Resonant sensor is very tolerant of these variations, as is the Flat sensor.

F. The best sensor location is where the engine tends to be most compliant.

The general feeling seems to be that the location on the engine structure which responds to vibration with the most motion will provide more input to the sensor than a stiff location, thus allowing it to detect knock easier. While compliant locations will input larger amplitude vibrations to the sensor, it must be remembered that they will be more sensitive to lower frequency vibrations, and will therefore amplify noise more than the higher frequency knock. As noted earlier, a stiff area that responds well to the forced vibration of the knock event offers less sensitivity to noise (4), yielding greater sensing capability, than a compliant area.

G. Flat sensors offer better filtering capabilities than Resonant sensors, allowing the elimination of more noise frequencies.

The application of electronic filtering is independent of the sensor type. Similar filters can be used with either type of sensor to remove unwanted frequencies, but the properties of the Resonant sensor often make them unnecessary. Caution must be used in specifying "tight" filters to insure that the variation in the knock frequency does not move it outside of the passband.

H. Resonant type sensors cannot be used in individual cylinder knock control systems.

Any of the three types of sensors can be used with either a global or an individual cylinder knock control system. The primary difference between these two control systems is how they react to knock once it is detected. Both systems need sensors which give the best probability for knock detection (high signal to noise ratio) and match other sensor selection criteria established for the application. As mentioned previously, Broadband Resonant offer benefits over the other two types of sensors.

SUMMARY

This paper has presented a logical, experience based review of vibration type knock sensors and criteria for selection of an optimum sensor for automotive engines. Knock sensors are a vital portion of a complex control system. A total system analysis was recommended to select a sensor which yields best system performance, highest reliability, lowest cost, and highest customer satisfaction. The future of combustion sensors holds many significant challenges associated with developing more direct measurement techniques. The next generation of combustion sensors are beginning to emerge from the engineering laboratories of the world, and they must compete with the existing combustion control sensors and systems from a cost versus benefit standpoint to be successful in the marketplace.

The keys to an effective combustion/knock control system are a full understanding of the combustion process, application issues, customer desires, and system needs. While this may appear an arduous task, proper attention to detail can yield sub-audible knock control with technology that is available today. It is hoped that this paper will further the state of knowledge in the complex field of engine knock sensing.

REFERENCES

(1) J. H. Currie, D. S. Grossman, J. J. Grumbleton, "Energy Conservation with Increased Compression Ratio and Electronic Knock Control," SAE Paper 790173, 1979.

(2) E. F. Obert, Internal Combustion Engines and Air Pollution, New York: Harper & Row, Inc., 1973.

(3) K. Sawamoto, et al., "Individual Cylinder Knock Control by Detecting Cylinder Pressure," SAE Paper 871911, 1987.

(4) N. Nakamura, et al., "Detection of Higher Frequency Vibration to Improve Knock Controllability," SAE Paper 871912, 1987.

(5) K. M. Chun and J. B. Heywood: "Characterization of Knock in a Spark-Ignition Engine," SAE Paper 890156, 1989.

ACKNOWLEDGEMENTS

The authors wish to acknowledge the efforts of many GM personnel in several divisions who helped make this paper possible by supplying background information. Significant assistance was provided by Barry Owens, Allen Atkinson, Reb Gooding, and Gerald Fattic of Delco Remy Division, GMC.

TORQUE SENSORS

970605

Development of a Magnetoelastic Torque Transducer for Automotive Transmission Applications

Ivan J. Garshelis
Magnetoelastic Devices, Inc.

Jonas A. Aleksonis and Christopher A. Jones
Methode Electronics, Inc.

Robert M. Rotay
Chrysler Corp.

Copyright 1997 Society of Automotive Engineers, Inc.

ABSTRACT

The development of a transducer for sensing the torque on the output shaft of a four speed rear wheel drive automatic transmission is described. Magnetoelastic polarized ring technology was selected based on its independence from shaft properties and its non-contact mode of sensing. The ring and several intermediate sleeves were attached by press fits onto an experimental shaft. The magnetic field arising from the ring with the application of torque was sensed by flux gate sensing elements. Ability to accurately measure the output torque of an engine driven transmission over its full range of torque and speed was demonstrated by dynamometer tests.

INTRODUCTION

Automotive transmissions serve to match the available torque and speed of the vehicle's engine (or other on-board prime mover) with the tractive effort required at the wheels to attain or maintain the speed and direction of travel desired by the driver. With a *manual* transmission, the driver, on the basis of experience and mastery of the required skills, willfully selects the gear ratio deemed most suitable for the specific driving circumstance (e.g., drive-away, acceleration, hill climbing). With an *automatic* transmission, the driver's desires, relative to the speed of the vehicle, are primarily signaled by manipulation of the throttle. The gear selection lever is repositioned only for directional control and to accommodate special driving conditions, e.g., descending long or steep hills. Based on the engine's response to motion and position of the throttle, as well as to incline and other conditions of the road, the transmission must be automatically guided to quickly select the *most appropriate* gear ratio. The determination of when best to shift from one gear ratio to another and the control of the complex series of events that actually effect the shift require a variety of sensory inputs and a stored knowledge (map) of the engine's speed-torque-throttle position characteristics. The driver's non-involvement with the actual mechanics of shifting results in a heightened awareness of accompanying sounds, vibrations and irregular accelerations. Thus it is the burden of the designer of an automatic transmission to not only ensure that the ratio best able to provide the efficient transfer of power from the engine to the wheels is selected, but that the actual shifting events will take place in a manner that is also smooth and maximally transparent to the driver.

The problem is complicated by variations in the engine map with engine temperature, barometric pressure (altitude), engine wear and between individual engines. It would appear that these problems could be simplified if the actual tractive effort was factored into the control of the transmission since this is the quantity most actively involved in executing the driver's speed demands. In the absence of braking, and except for the rotational inertia of intervening parts, the tractive effort at the wheels has a fixed (for any one vehicle) numerical relationship with the torque on the output shaft of the transmission. Thus, knowledge of the actual value of the instantaneous torque on this shaft should enable the development of improved shift control systems.

The potential for improving shift quality and other aspects of vehicle performance from signals representative of the torque on transmission shafts is well recognized [1-5]. Appreciation of the value of such information has stimulated many investigations into *estimating* this torque from indirect sources, e.g., computations involving engine and torque converter maps [4] or changes in shaft rotational speed [5]. Nevertheless no reports on the development of a *direct* torque sensing means that is sufficiently accurate, reliable and economical for use on production vehicles have yet appeared. This paper reports on the details of such a development.

DEVELOPMENT PROGRAM

A program aimed at developing a torque sensor to a stage sufficient to allow determination of its suitability for the intended use was outlined as follows:

1. Establish a *tentative* specification for the torque sensor. Consider the requirements in sufficient detail to identify preferred sensor locations, expected measurement range, operating environment and performance targets.
2. Critically examine known torque sensing technologies and select the one considered most appropriate for this application. (During this phase it was deemed important to maintain a constant awareness of the goal: a suitably accurate, extremely reliable, economically justifiable and mass producible torque sensor.)
3. Design experimental hardware.
4. Fabricate and assemble necessary parts.
5. Test experimental assemblies under static (non-rotating) loads and under unloaded rotational conditions.
6. Load test an assembled transmission on a dynamometer.

In-vehicle testing was not considered to be part of this development phase since such would clearly require the kind of preparation that would not be undertaken without encouragement from step 6 results. Because of the need to schedule dynamometer testing well in advance, static and unloaded spin testing of the shaft and skeletal sensor constructions were to be relied on to avoid seriously unpleasant surprises during dynamometer testing of the assembled transmission.

SPECIFICATION

The output shaft of a four speed automatic transmission used in front engine, rear wheel drive, light trucks and utility vehicles was selected for installation of a first generation torque sensor. A photograph of the targeted shaft as it normally exists without a sensor is shown in Fig. 1. The sensor was to be located in the region, forward of the speedometer take-off gear, indicated as A A on the photo.

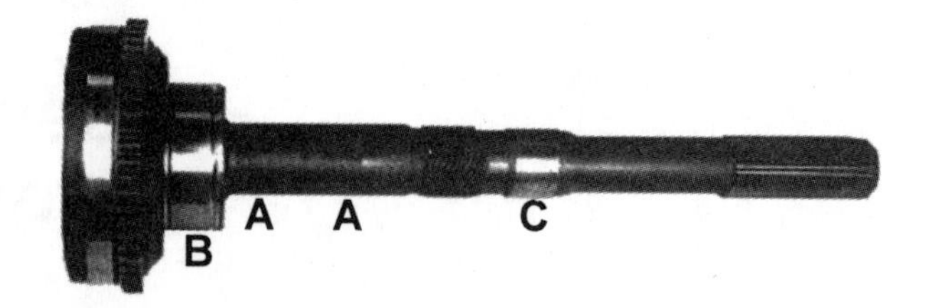

Fig. 1. Transmission output shaft. A A indicates torque sensing region. The shaft is supported in the overdrive housing by ball bearings at B and C.

Tentative target specifications were as follows:

- Location: Overdrive unit output shaft of Chrysler Model 44RE/42RE transmission.
- Shaft Diameter: 35 mm at sensor location.
- Shaft Material: Forged AISI/SAE 8620, Carburized, OQ & T (0.040/0.070" case, HRC 58/62) [6].
- Torque Range: 0-2700 N·m.
- Overload: 4700 N·m forward, 2700 N·m reverse.
- Speed Range: 0-5000 rpm (highest accuracy expected to be required in middle ranges).
- Resolution: <7 N·m (1 part in 400).
- Linearity: Desired, monotonicity required.
- Hysteresis: TBD (rezeroing in neutral acceptable).
- Output Signal: 0.2 - 4.8 Vdc corresponding to 0-2700 N·m. CW or CCW directionality to be independently determined.
- Environment: Cast aluminum enclosure. Exposure to transmission fluids. Temperature range: -40 to +150 °C.
- Response: 30 Hz bandwidth (opening estimate).
- Power Supply: 12 Vdc system, as per SAE - J1211.

SELECTION OF TORQUE SENSING TECHNOLOGY

BROAD CLASSIFICATION - A wide variety of sensing technologies are embodied in the many available instruments for measuring the torque being transmitted by a rotating shaft. By one means or another these *torque transducers* all serve to provide an electrical signal that is a linear analog of the torque. All operate by sensing some manifestation of the shear stress which effectively transmits the torque along the shaft. While these devices are more commonly categorized according to their underlying operating principles, for the purposes here we find it more suitable to divide all torque transducers into only two, constructionally distinct categories, namely: those that utilize shaft mounted electrical components and those that do not. In consideration of the rigors of the present application, including circumstances that may arise during manufacture, assembly and repair of the transmission as well as the conditions of actual use, technologies utilizing shaft mounted electrical components were felt to be insufficiently robust. It was also felt that designs utilizing shaft mounted electrical components would be too labor intensive to be cost competitive. In fact, constructions relying on adhesively bonding any components to the shaft were deemed inappropriate. Thus, although strain gauges have been suggested for other automotive applications [7] and capacitive elements have been investigated for use on automotive axles [8], these technologies were not considered further. For similar reasons, as well as the absence of reports on their use in automotive applications, SAW devices [9] were also excluded from more detailed examination.

TORSION BAR - Technologies based on measurement of shaft twist angle were briefly considered. The angle of twist, θ, for a 35 mm diameter (*d*), SAE 8620 steel shaft (G = shear modulus = 80 GPa) under the full range torque, T = 2700 N·m, was found from

$$\theta = \frac{32LT}{\pi d^4 G} \tag{1}$$

to be only 0.0131° per mm of shaft length. Even if a clear gauging region 100 mm in length (*L*) could be made available, the angle measurement system would need to discriminate displacements as fine as 0.00328° in order to provide the desired 1 part in 400 resolution. Such resolution challenges even instrument grades of "torsion bar" torque transducers, e.g., those based on the light passing through axially spaced optical shutters [10]. It was judged as unlikely that variable reluctance systems, such as those developed for automotive power steering applications (wherein typical twist angles are between 4° and 7° [11]), could be sufficiently refined for use with shafts having the required torsional stiffness. While this difficult angular resolution requirement could be relaxed by reducing the shaft diameter in the gauging region, this would risk plastic yielding of the shaft under peak torque conditions. Also, increasing shaft compliance would alter the spectrum of torsional oscillations in the drive train, with possibly serious adverse consequences. For these and other reasons associated with the manufacturability of needed high precision components, this approach was not considered further.

MAGNETOELASTIC (PERMEABILITY) - Detailed attention was directed to magnetoelastic (often called, "magnetostrictive") torque sensing technologies. It is well known that stress acts to alter the orientation of the *spontaneous magnetization,* the presence of which characterizes a material as ferromagnetic [12]. (Magnetomechanical effects due to stress manifest an intrinsic magnetoelastic interaction generally present in this class of materials.) Among the consequences of this stress induced magnetization reorientation are changes in the permeability of a stressed sample from measured values when in the unstressed condition. The utility of this phenomenon for the measurement of torque on a rotating shaft can be readily appreciated since magnetic coupling enables the determination of properties such as permeability without any need to physically contact the stressed body. The further attractiveness of utilizing changes in the magnetic properties *of the shaft itself* in order to determine the torque being transmitted is obvious. Fleming [13] has studied and compared three common configurations of torque transducers that develop their output signals from stress induced permeability changes.

Although well understood and seeming to be ideally appropriate, reported efforts to use these methods on powertrain shafts nevertheless fail to demonstrate suitable accuracy over the required torque and temperature ranges. The underlying problem stems from the fact that powertrain shafts are designed to fulfill demanding mechanical functions.

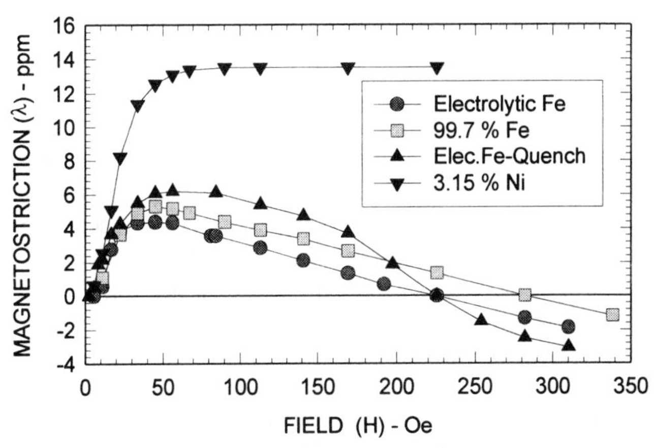

Fig. 2. Effects of composition and thermal history on the magnetostriction of iron and dilute alloys.

Except for those instances when torque sensing is the overriding concern, e.g., in commercial torque transducers, the choices of shaft size, material, fabrication methods, heat treatments and other processing steps are all aimed at meeting mechanical performance and cost criteria. The influences of composition and processing conditions on magnetic and magnetoelastic properties of the shaft are only rarely considered [2, 14]. One generally ignored property, *magnetostriction* (λ), plays an essential role in determining both the direction and amplitude of changes in permeability with stress [15]. While *saturation* magnetostriction (λ_s) [16] is an intrinsic material property, the variation of permeability with stress follows from the variation of magnetostrictive strain ($\delta l/l = \lambda$) with applied field (H), and this is a sensitive function of microstructure as well as composition. Data plotted in Fig. 2 (taken from Schulze [17]) for some dilute alloys of iron clearly shows these dependencies. The reversal in sign of λ seen in all but the 3.15% Ni-Fe alloy is reflected in a variable *direction* of change in permeability with stress in these materials (the *Villari* reversal [18]). Shaft materials exhibiting a Villari reversal are clearly not good choices for this means of torque sensing. SAE 8620 [6] has too little Ni, Si, Al or Co to squelch the non-monotonic λ vs H characteristic found in iron, its principal constituent.

The effective permeability of a shaft surface, even in the absence of stress, varies significantly with temperature as well as with composition and microstructure. Moreover, the term, "permeability" does not define a single, material dependent property, since its value, and the manner and range of its variability with stress, depends on the amplitude and frequency of the measuring field (as well as on other, more incidental influences on shaft magnetization, e.g., the earth's field). Thus, such performance characterizing features as linearity, hysteresis, dynamic range and temperature effects are all dependent on the excitation field parameters in this class of torque sensors.

Additional complications pertinent to this specific application include the potential for drifting calibration caused by *irreversible* changes in permeability associated with stress excursions beyond the endurance limit of the shaft material during peak overloads. Similar drifts would accompany even subtle microstructural changes associated with prolonged time at temperatures in the tempering range. These, in addition to concerns with normal shaft to shaft variations in characteristics affecting calibration, were considered serious enough to eliminate "permeability" forms of magnetoelastic torque transducers from further consideration. (The existence of the problems associated with the inappropriate and inconsistent magnetoelastic properties of typical shaft materials are well recognized [1, 2, 3, 7, 14, 19, 20, 21]. Persistent attempts at solving these problems, by e.g., characterizing individual shafts [19], ad hoc material development [20], attachment of sleeves of more suitable materials [21] or even remote variations on this basic sensing theme such as inducing eddy currents in Cu conductors electroplated on the shaft [3], have yet to demonstrate adequate performance for this demanding application.)

MAGNETOELASTIC (POLARIZED RING) - By avoiding dependence on permeability, inconsistent or unstable magnetic properties of the shaft do not affect the operation of magnetoelastic torque transducers employing magnetically polarized ring constructions [22]. Figure 3 shows the arrangement of functional elements comprising this type of transducer. The transducer consists only of a ring of magnetoelastically active material attached to the shaft in such manner as to assure that they act as a single mechanical unit and a proximate magnetic field sensor. (Mechanical attachment of the ring to the shaft is most often accomplished by a simple press fit.) As indicated in the figure, the ring is magnetically divided into two regions having oppositely directed circumferential polarizations. When torque is applied to the shaft these polarizations become more or less steeply helical; the ring effectively develops magnetic "poles" uniformly distributed around its central band (indicated in gray) and, of opposite polarity, at its ends. The strength and polarity of the magnetic field associated with these poles is linearly analogous to the magnitude and CW or CCW

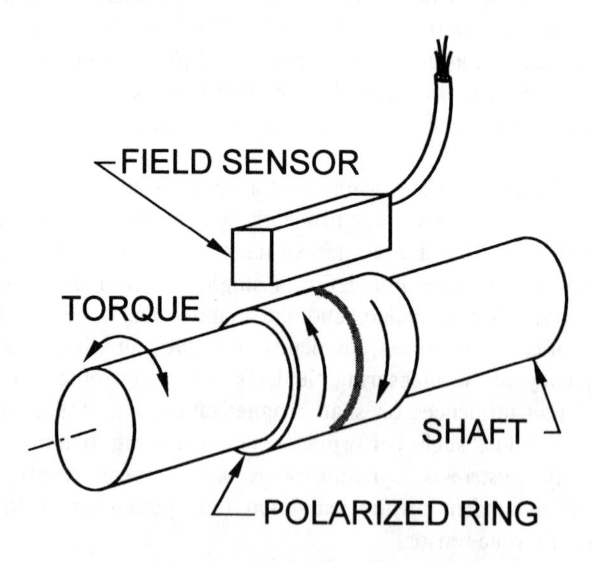

Fig. 3. Functional elements of a polarized ring torque transducer. The arcuate arrows indicate the directions of remanent magnetization (i.e., polarization) in each ring region. When torque is applied, magnetic poles develop along the gray band and at each of the ring ends.

directionality of the applied torque. The output signals from one or more magnetic field sensors thereby provide the measure of the torque. (Further explanation of the operating principles and related details can be found in the referenced technical and patent literature [22-25].) The sensitivity, i.e., the field intensity per unit of torque in this type of transducer is based primarily on the size and *intrinsic* material properties of the polarized ring. This, in combination with other features, including the complete absence of any form of excitation field, greatly simplifies (or may even eliminate) individual transducer calibration. A properly constructed transducer of this type will, by its nature, generate no field (i.e., zero) with zero applied torque, independent of its temperature. Hence, temperature compensation, if needed at all, would be limited to transducer sensitivity (i.e., the "span") and would expectedly be generic for any one design. Calibration drift due to thermally induced microstructural changes in the ring is unlikely since typical rings undergo final heat treatments at 483 °C (for 3 hours), far above the specified maximum operating temperature. The simplicity of the construction portends a reliable and economical device. All of these clearly desirable features, together with advances in the technology development and experience in its use since first suggested for electric power steering applications [25], directed its selection for this application

TRANSDUCER DESIGN AND CONSTRUCTION

GENERAL - Although the functional elements shown in Figure 3 comprise a complete torque transducer, two factors associated with the existing transmission shaft (Fig. 1) necessitated the use of additional components, at least at this experimental stage of development. Firstly, as with all applications involving shafts made of ferromagnetic steels, it is found desirable to interpose a non-magnetic spacing ring between the magnetically polarized ring and the shaft. Spatial separation of the field generating ring from the shaft allows more of the magnetic flux to follow paths through the space on the outside of the ring where the field sensor(s) are located. If the ring is mounted directly on a typical ferromagnetic shaft, nearly 90% of the generated flux will take lower reluctance paths through the shaft and thus not be available for sensing. Besides reducing this loss, spatial separation minimizes any effects that incidental shaft magnetization may have on the operation of the transducer. It has been found in some applications that an austenitic stainless steel "insulating" ring having a radial thickness as small as 1 mm reduces the flux loss to less than 25%. Greater thickness will naturally be even more efficient.

The size (diameter in the sensing region) of the presently available transmission shaft made it necessary to add still another component (ring) to the basic construction. Since the polarized ring generates a magnetic field in proportion to the torsional (shear) stresses over its cross section, it is important to ensure that these stresses accurately reflect the *applied* torque and *only* the applied torque. Should there be any plastic yielding within the shaft or ring, or any slip at their interface under any conditions of applied torque, the stresses will not uniformly relax to zero when the torque is removed.

Under such inelastic conditions there will be a *residual* stress distribution over the cross section with the ring stressed in the opposite direction to that which existed when the high torque was applied. The magnetic field arising as a result of this residual stress will be sensed as a torque even though the applied torque is zero. The field sensor output signal at zero applied torque will thus be different (i. e., there will be *hysteresis*) depending on whether inelastic events occurred during CW or CCW applied torques. Hysteresis will also be manifested as differing values of output signal for the same value of torque, depending on whether the torque is increasing to, or decreasing from, a torque high enough to cause inelastic events. While zero shifts at zero torque can be electronically eliminated (by resetting the signal when in neutral) and minor hysteresis is of little practical concern, it is clear that yielding of the shaft and interfacial slip would best be avoided.

MECHANICAL DETAILS - The extent of this problem and its solution are readily understood from the following analysis. The shear stress associated with the transmission of a torque, *T*, through a shaft of diameter, *d*, varies from zero at the center to a maximum value (τ_m) at the surface, found from

$$\tau_m = \frac{16T}{\pi d^3}. \qquad (2)$$

For d = 35 mm, under the peak overload torque of 4700 N·m, τ_m will reach 558 MPa. While no torsional yield strength data for carburized 8620 steel has been found, considering that the tensile yield strength (0.2% offset) for the core of a bar of this size is almost certainly $\leq$ 800 MPa [6], and that the torsional yield strength of steels is typically about 57% of the tensile value (i.e., $\tau_y \leq$ 456 MPa), yielding of some portion of the cross-section under overload conditions seems likely. An obvious solution is to enlarge d in the region where the ring is to be installed and such would expectedly be the eventual course of action. Presently limited by the existing shaft dimensions, it would appear that d could be effectively increased by simply increasing the wall thickness of the spacing ring. After a moment's reflection, it will be realized that the shear stress in each ring is transferred to it via frictional forces arising from the contact pressure at all underlying interfaces. Without unlimited, independent control of friction coefficients, high interfacial contact pressures are required. Except for some exotic (read: expensive) Ni based alloys, non-magnetic materials have insufficient strengths to act elastically under the required contact pressures. Hence it was considered necessary to build up the effective diameter of the shaft in the sensing region by pressing on a heat treated steel sleeve. The problem was further complicated by the fact that the shaft in region AA (Fig. 1) was not sufficiently larger than the bearing journal at C to obtain the desired interference. The shaft was thus built up to the desired diameter by means of *two* hardened steel sleeves, sized such that mounting,, by pressing on the outer one, increased the contact pressure between the inner one and the shaft. To avoid distortion under the heavy press-on forces required, and to provide better control of the alignment of the parts during their assembly, each interface was machined to a shallow taper (typical included angle ≈ 1°). Figure 4 shows the details of the final assembly design.

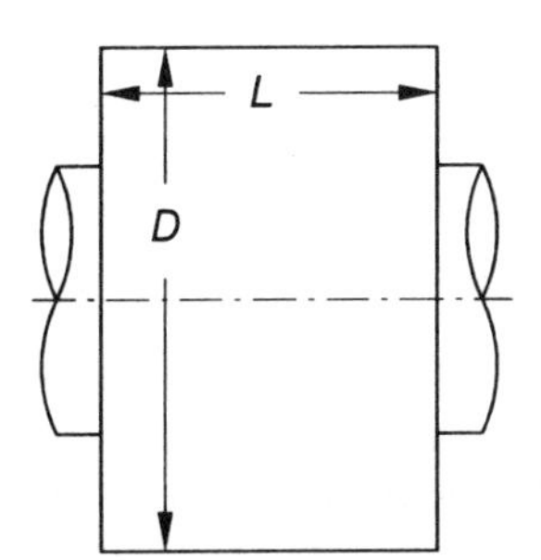

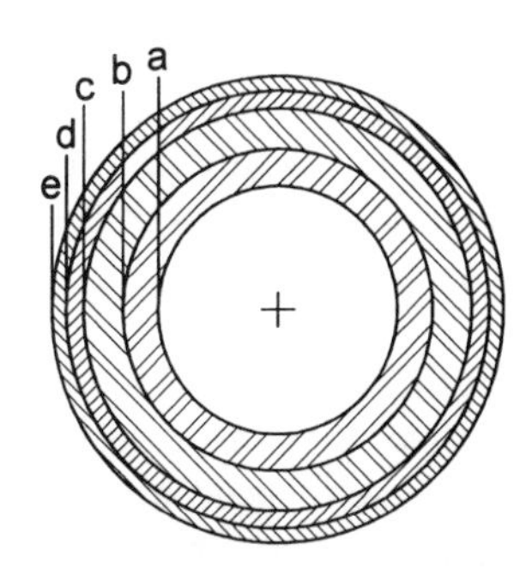

Fig. 4. Details of experimental ring assembly. Inner two rings - OQ & T AISI 4340; "Insulator" ring - Carpenter 22 Cr 13 Ni 5 Mn; Outer ring - Teledyne Vasco C-250, aged 3 h @ 483 °C. Axial length L = 45.7 mm. Outside diameter = 66.0 mm. Nominal radii at each indicated interface: a = 17.6; b = 22.9; c = 28.7; d = 31.2; e = 33.02.

It is of course, only the outermost ring of the four ring set that has an active role in the torque *sensing* function. The use of an interference fit for this ring served the dual purposes of assuring the mechanical integrity of its attachment to the underlying "shaft" and instilling a circumferential tensile (hoop) stress. This is needed to establish the circumferential direction as the stable orientation for the remanent magnetization. An 18% Ni maraging steel was chosen for this ring on the basis of its high strength and desirable magnetic properties [26]. Although it was not considered necessary for the magnetically polarized ring to be any longer than 20-25 mm in order to fulfill its transducing function, it was thought best at this stage to keep all rings the same length in order to reduce the possibility of slippage at any of the interfaces.

MAGNETIZATION - The two, oppositely polarized regions in the outer ring were instilled by simply holding a North-South/South-North pair of bonded ferrite magnets close to the ring surface while rotating the shaft (for about 100 revolutions) [22]. The polarizations were circumferentially homogenized by passing an approximately 300 peak ampere, 60 Hz current, axially through the shaft.

FIELD SENSOR - In general, the magnetic field that arises when torque is applied to the shaft can be sensed by any of several suitable means. Both Hall effect and "flux gate" sensors [27] were used for static load and unloaded spin testing of the experimental transducer shaft. Since the output signals from presently available Hall effect devices often drift with temperature, a flux gate design, having an inherently stable and temperature independent zero field output signal, was chosen for installation into the transmission housing. At this stage in the development it was thought prudent to avoid possible problems from shaft runout, bending stresses or other sources of circumferential inhomogeneity by combining the signals from two, diametrically opposed, sensor pairs. The actual sensors consisted of small diameter, solenoidal windings of high temperature insulated magnet wire, each encircling a single strand of near zero λ_s amorphous wire

(Unitika DC2T). These core/coil "assemblies" were cemented into grooves machined into an aluminum ring that had been turned to fit an axially concentric bore already existing in the overdrive housing. The radial air gap between the ring and sensor surfaces was 0.5 mm.

ELECTRONICS - Figure 5 shows a block diagram of the circuit used with the core/coil assemblies to obtain a field (hence torque) dependent signal. Details of its operation are described in the (lengthy) caption.

TESTING

STATIC TESTS - An apparatus for applying controlled and accurately measurable torques to the stationary shaft was constructed. The shaft was firmly clamped at the forward (large) end, while, through a system of links and levers and a yoke having a matching internal spline, torque in the form of a force couple was applied to the free end. The push and pull forces comprising this couple were applied through tension/compression strain gauge load cells. The forces were

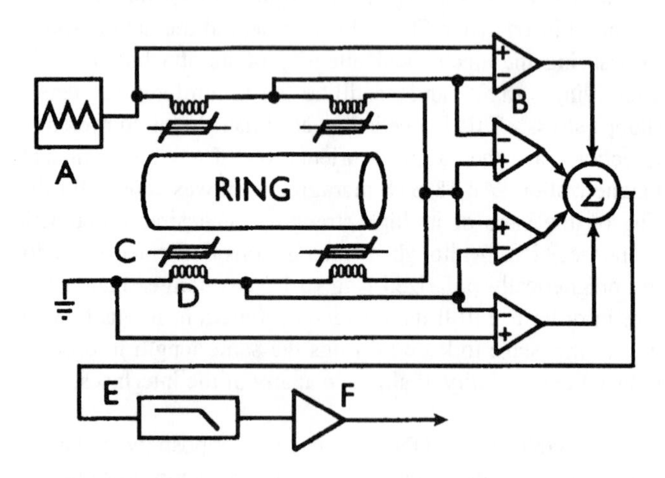

A - Triangle Wave Oscillator
B - Comparators
C - Square loop wire cores
D - Coils
E - Low Pass Filter
F - Output Buffer
Σ - Summing Amplifier

Fig. 5. Block diagram of electronic circuit used with the flux gate field sensors. The triangle wave current (≈ 13 kHz) through the coils sweeps the core wires through major hysteresis loops. In the absence of torque, the magnetization within each of the cores reverses at the same amplitude of coil current. When torque is applied, the associated magnetic field adds to the excitation field on the cores adjacent to one region of the ring while subtracting on those adjacent to the other. Since all core magnetizations reverse at one characteristic value of *effective* field, the cores adjacent to each region now reverse at different current amplitudes, i.e., at different times in the triangle wave. The comparators, which in the absence of torque spend equal (half) periods with high and low outputs, now have asymmetrical high and low periods. Their average outputs, after summing, filtering and buffering, thereby vary up or down depending on the field amplitude and polarity, hence on the torque amplitude and direction.

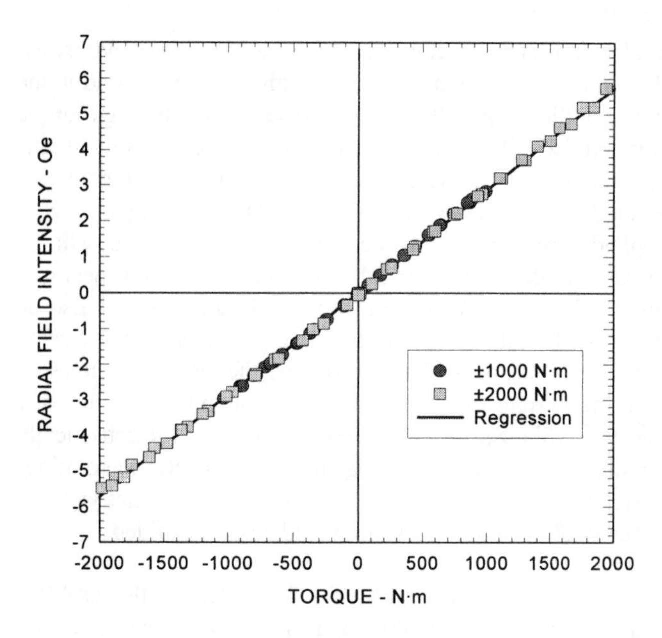

Fig. 6. Variation in the radial field intensity appearing near the axial center of the ring with applied torque. (Distortion of the yoke and load cell limitations prevented accurate exploration at much higher torques.)

produced by a hydraulic "jack" which, by its placement at one or the other end of a lever, provided for the application of CW or CCW torques. A commercial gaussmeter with Hall effect probes was used to measure the actual radial and longitudinal field intensities in the space around the ring. Typical data obtained for complete cycles of increasing and decreasing torques in each direction are plotted in Fig. 6. The excellent linearity and minimal hysteresis are clearly indicated.

DYNAMIC TESTS - After assembly of the overdrive components to the shaft and further assembly into the housing, the overdrive unit was attached to a Chrysler 44RE transmission. The transmission, coupled to a 4.0 liter 6 cylinder in-line engine, was installed in a fully instrumented

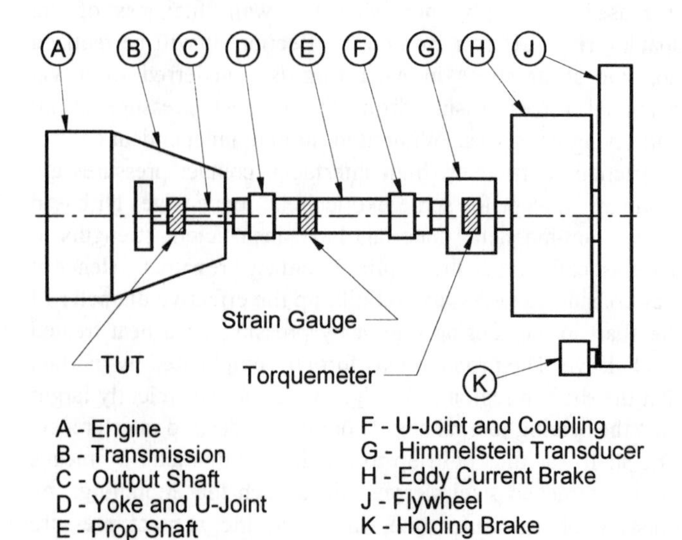

Fig. 7. Arrangement of apparatus used for dynamic tests. The shaded areas indicate the three torque measuring devices.

dynamometer test cell. The driveline arrangement is schematically illustrated in Fig. 7. In addition to the measurement of the output signal from the transducer under test (TUT), torque was measured by a custom built and calibrated strain gauge bridge installed on the prop shaft and by a Himmelstein Model MCRT 8-02TA Torquemeter (rated for 50,000 lb.·in. up to 5000 RPM). Other measured parameters included: engine speed, throttle position, output shaft speed and transmission oil temperature.

All measurement data was sampled at rates appropriate to the specific test and acquired by computer. All details of the test regimens were under computer control except for the selection of the transmission gear ratio. This was either allowed to change automatically by the normal operation of the transmission or else continuous operation in first gear was manually selected. Testing followed several different regimens. For example, while the transmission was kept in first gear, targeted values of speed and output torque were established by interactive adjustments of the throttle position and dynamometer brake current. While some data was collected in this way at various output shaft speeds and torques, one limitation or another prevented the attainment of steady state operation at high enough torques to fully test the transducer. Automatic test cycles provided more interesting data. Prior to the beginning of each such cycle, the output shaft was held stationary by the holding brake on the flywheel while the engine idled. The test was launched by simultaneously releasing the brake and snapping the throttle to a preestablished opening. The engine would accelerate more or less rapidly depending on throttle position and the transmission stepped through its gear ratio changes as the flywheel reached appropriate speeds. After reaching the limiting output speed, the throttle was cut and the flywheel was brought to a halt by the eddy current brake. Tests of this nature were performed for approximately 10% increments in throttle position. Data (at 100 samples/s) from one test run at full throttle (nominally 100%) are shown in Figures 8-12.

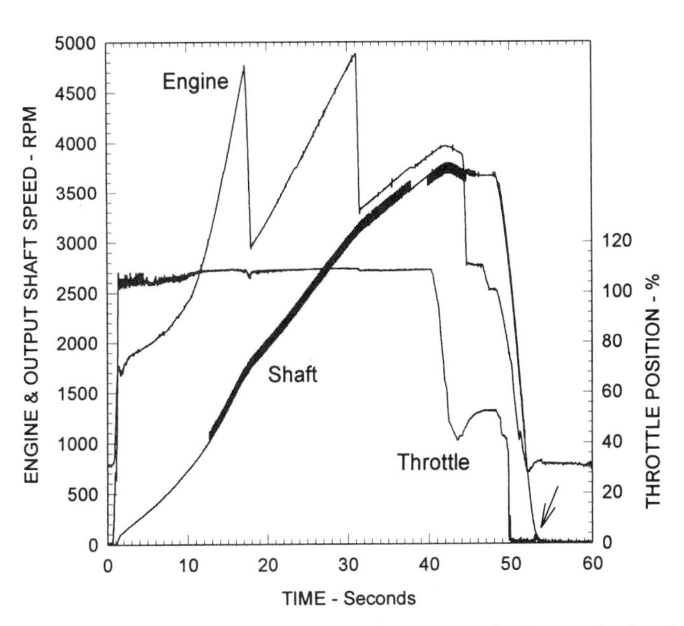

Fig. 8. Variations in engine and output shaft speeds (and throttle position) during 100% throttle automatic test cycle.

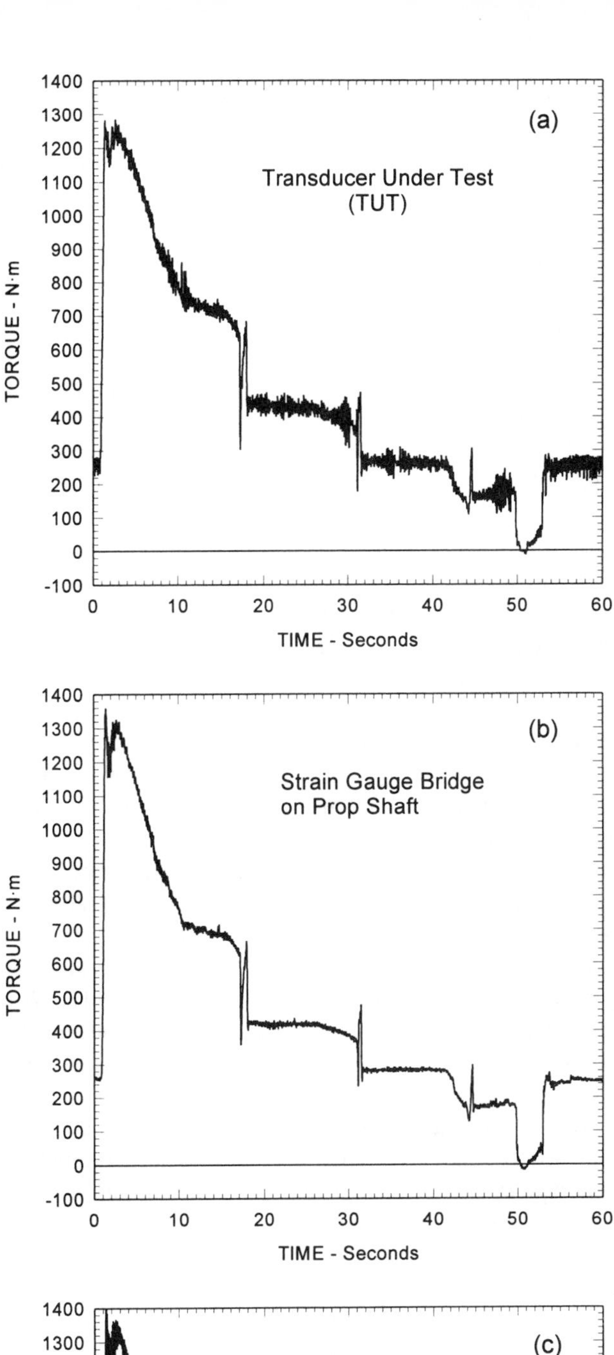

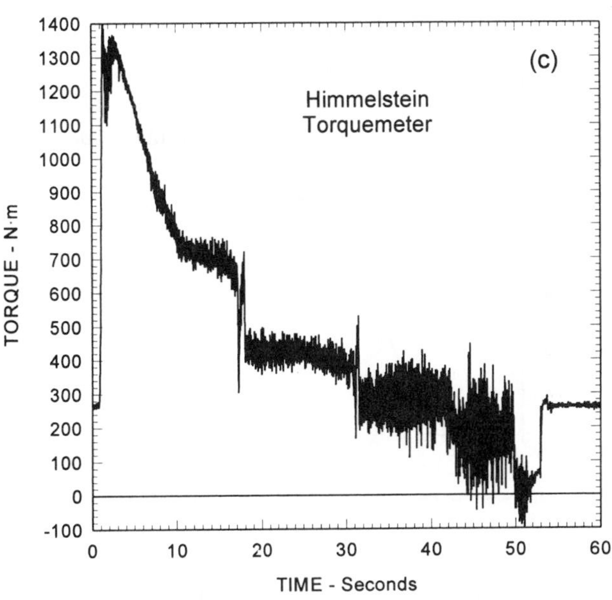

Fig. 9. Torque as measured by: (a) the TUT; (b) the strain gauges on the prop shaft; (c) the torquemeter.

The progression of active events throughout the 60 second duration of the automatic test cycle is shown in Fig. 8. Following the opening of the throttle, which effectively starts the test, the output shaft is seen to undergo a continuous acceleration as the transmission steps between first and second and second and third gears, while decelerating slightly prior to the shift into overdrive. The throttle was cut to about half open just before the shift into overdrive, near 45 seconds into the test, after which the eddy current brake was applied and the throttle shut (near 49 seconds) bringing the shaft to a halt slightly after 54 seconds.

Figure 9 shows the variation in the output torque as indicated by each of the three measurement devices shown in Fig. 7. This figure shows that the measurement of torque under the dynamic conditions found in engine/transmission drivelines is not a simple matter. The precise answer to the question, "What is the magnitude of the torque?" is seen to depend on the type and location of the sensor. While this complicates the performance assessment of any torque measuring device under study, it also shows that there is a wealth of information that torque measurement can provide.

The "noise" seen to be present in varying degrees in these three figures is not noise in the classical sense but represents actual torque variations experienced by each sensor; different in each due to torsional oscillations. It is quite normal for the dynamic torques on a distributed system containing elements of varying compliances and inertial masses to vary from location to location. This is readily seen by examining the expanded data shown in Figs. 10-12. While the TUT is seen to provide an average signal that for the most part accurately follows the prop shaft strain gauge data, a superposed oscillatory signal of $\approx$ 9 Hz is seen to be pervasive. This signal component is believed to have arisen from oscillations of the engine on its mounts, a conclusion supported by its presence even when the output shaft was stationary, e.g., as seen in Figs. 9 and 10 during the first one second of the test and in Fig. 9 (and the right end of Fig. 12) during the last 6 seconds of the test cycle. These oscillations are seen to be stimulated to substantial amplitudes, both in particular engine speed ranges and during rapid changes in engine acceleration accompanying shifting events and throttle changes. Chaotic and severe bouncing of the engine and transmission was observed during the tests but no records were made to correlate the timing and extent of such events.

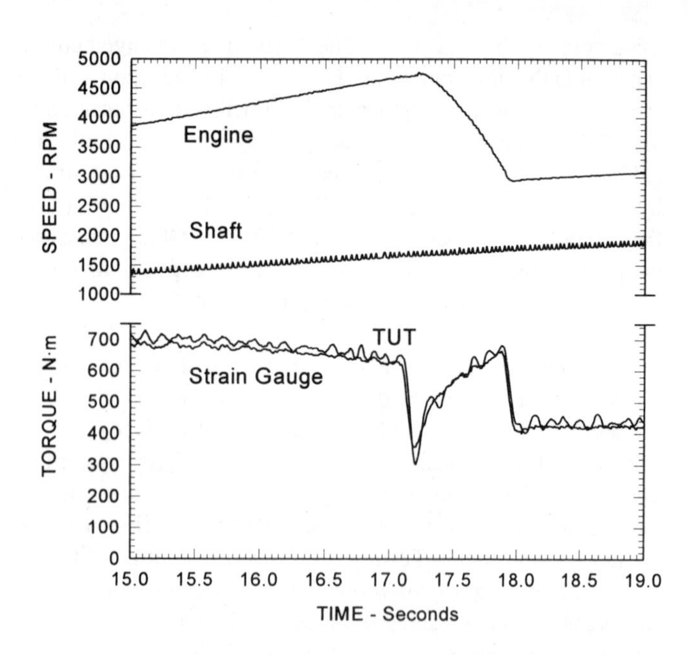

Fig. 11. Variations in engine RPM, shaft RPM and torque, as indicated by both the strain gauges on prop shaft and the TUT, during shift from first to second under full throttle conditions.

Torsional oscillations in the torquemeter records are seen in Fig. 9(c) to be even more severe, but these were of a higher frequency ($\approx$ 13.5 Hz), reflecting a resonant association of the torquemeter compliance with the flywheel and dynamometer

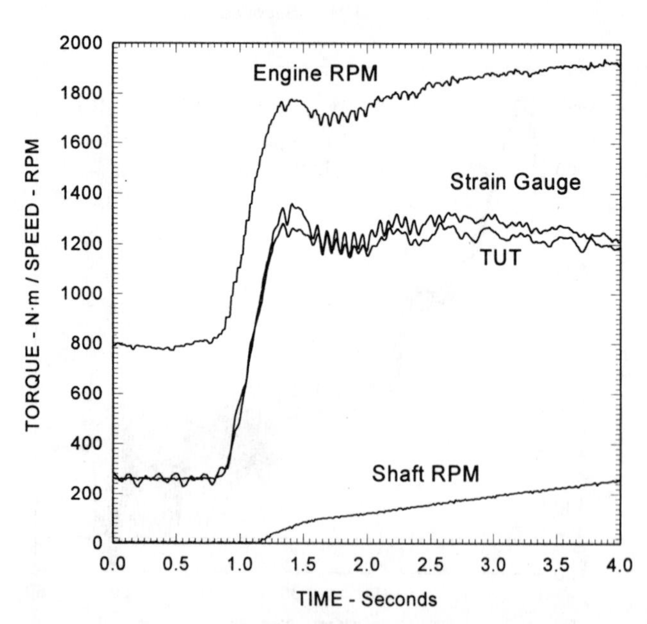

Fig. 10. Variations in engine RPM, shaft RPM and torque, as indicated by both the strain gauges on the prop shaft and the TUT on the transmission output shaft, during the first 4.0 seconds of the automatic test at full throttle.

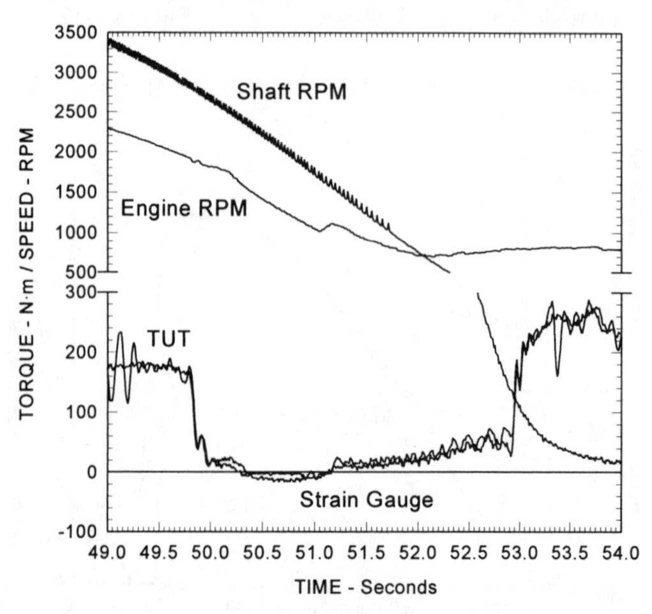

Fig. 12. Variations in engine RPM, shaft RPM and torque, as indicated by both the strain gauges on the prop shaft and the TUT, during final throttle shut down. As seen by the arrow in Fig. 8, the throttle "bounced" to almost 5 % near 53 seconds.

rotor inertias. These oscillations are seen to have been absent (as would be expected) when the output shaft was stationary at the beginning and end periods of the test cycle. The prop shaft is seen to be largely free of the large oscillatory torques found at both ends of the driveline. This too is expected since this shaft is insulated by the compliance of the U-joints and the non-linearity of the torque and velocity transfers associated with the angularity of the coupled shafts.

Other than for the additional oscillatory torque, the TUT signal is seen to track the prop shaft data quite accurately at all torque levels except those over about 1200 N·m. The torque indicated by the TUT is clearly seen, by the expanded data in Fig. 10, to be consistently low from 1.2 - 4 seconds into the test. Since no such fall-off in sensitivity is seen in the static data plotted in Fig. 6, it would appear that a saturation problem may exist in the transfer function of the flux gate sensor. While this certainly needs to be investigated, it is also interesting to note that, as indicated in Fig. 9, the peak torque indicated by the torquemeter *exceeds* that shown by the prop shaft by a similar amount in the same region.

The expanded data in Figs. 10-12 provide an opportunity to examine both the ability of the TUT to follow transient torques and to observe the large amplitude that the oscillatory torques can reach when stimulated, as well as the rapidity with which these are damped.

During the many tests that were performed, over a period of several hours, the transmission oil temperature varied only from RT (≈ 20 °C) to 76 °C, only a small fraction of the specified operating range. Nevertheless, within this limited range no significant changes in the performance of the TUT were observed.

CONCLUSION

It is recognized that the developments reported here are but the first steps towards the ultimate application of torque transducers in automotive transmissions. Even at this early stage of development the described transducer clearly shows the ability to accurately measure the actual transmitted torque using simple and reliable hardware. As the technology advances, it can be reasonably expected that performance will be even further improved. A possible additional benefit from the use of this polarized ring technology is the reduction in overall sensor count, since adaptations of the basic technology have already been shown to provide rotational speed information in the transducer signal [28].

Next steps in the development will include further dynamometer tests utilizing a larger engine and more sophisticated loading means. This will allow dynamic testing up to the full specified CW and CCW torques, including excursions into the overload range. Following this, there will undoubtedly be extensive in-vehicle testing to provide the data necessary for the development of suitable control algorithms. Only then can it be determined whether actual torque measurement on automotive transmission shafts can provide the anticipated benefits.

ACKNOWLEDGMENTS

We would like to express our thanks to Hong Liu for her invaluable assistance with the FEA design and stress analysis of the ring system and to David Cripe for his design and construction of the field sensors and associated electronics. We would also like to thank the managements of our respective organizations for their support and encouragement.

REFERENCES

[1] William J. Fleming, "Magnetostrictive Torque Sensors-Derivation of Transducer Model", SAE Paper No. 890482, (1989).

[2] Hiroyuki Aoki, Junichi Maruyama, Munekatsu Shimada, Katsuji Tanizaki, Shinichiro Yahagi, and Takanobu Saito, "Torque Sensor with Shape Anisotropy", Tech. Digest of the 7th Sensor Symposium, 83-86, (1988).

[3] Jarl R. Sobel, Jan Jeremiasson, and Christer Wallin, "Instantaneous Crankshaft Torque Measurement in Cars", SAE Paper No. 960040, (1996).

[4] Masahiko Ibamoto, Hiroshi Kuroiwa, Toshimichi Minowa, Kazuhiko Sato and Takeshi Tsuchiya, "Development of Smooth Shift Control System with Output Torque Estimation", SAE Paper No. 950900, (1995).

[5] Toshimichi Minowa, Kenichirou Kurata, Hiroshi Kuroiwa, Masahiko Ibamoto, and Masami Shida, "Smooth Torque Control System Using Differential Value of Shaft Speed", SAE Paper No. 960431, (1996).

[6] AISI/SAE 8620, Composition: C- 0.18/0.23; Mn-0.70/0.90; P- 0.035 max; S- 0.040 max; Si- 0.15/0.30; Ni-0.40/0.70; Cr- 0.40/0.60; Mo- 0.15/0.25; Fe- balance. For recommended heat treatments and mechanical properties, see, for example, *Engineering Properties of Steel*, Philip D. Harvey, Editor, American Society for Metals, pp 189-193, (1982).

[7] Erich Zabler, Frieder Heintz, Anton Dukart, and Peter Krott, "A Non-Contact Strain Gage Torque Sensor for Automotive Servo Driven Steering Systems", SAE Paper No. 940629, (1994).

[8] J. D. Turner, "The development of a thick-film non-contact shaft torque sensor for automotive applications", J. Phys. E: Sci. Instrum., 82-88, (1989).

[9] anon., Machine Design, 48, Description of a torque sensor based on the stress dependence of surface acoustic waves (SAW) developed by Sensor Technology Ltd., Banbury, UK (Jan. 11, 1996).

[10] Series I and Series II Torque Transducers, manufactured by Vibrac Corporation of Amherst, NH, and described in company literature, are claimed to obtain 1% accuracy with 0.5° full scale deflection using optical shutter techniques.

[11] See for example: U.S. Patent No. 5,394,760 (1995), E. K. Persson, P. K. Webber, and D. L. Perry, "Torque Sensor for a Power Assist Steering System", Assigned to TRW Inc.

[12] Richard M. Bozorth, *Ferromagnetism*, D. Van Nostrand, Princeton, NJ (1951), p 12.

[13] William F. Fleming, "Magnetostrictive Torque Sensors - Comparison of Branch, Cross, and Solenoidal Designs", SAE Paper No. 900264, (1990).

[14] Munekatsu Shimada, "Magnetostrictive torque sensor and its output characteristics", J. Appl. Phys., 73, (10) 6872-6874, (1993).

[15] R. M. Bozorth, op. cit. (Ref. 12), pp 595-627.

[16] R. M. Bozorth, op. cit. (Ref. 12), p 632.

[17] Alfred Schulze, "Magnetostriction I.", Z. Physik, 50, 448-505, (1928).

[18] R. M. Bozorth, op. cit. (Ref. 12), 602.

[19] U.S. Pat. No. 5,495,774 (1996), R. D. Klauber, E. B. Vigmostad and F. P. Sprague, "Magnetostrictive Torque Sensor Air Gap Compensator", Assigned to Sensortech L. P.

[20] U.S. Pat. No. 5,107,711 (1992), H. Aoki, S. Yahagi and T. Saito, "Torque Sensor", Assigned to Nissan Motor Co. Ltd.

[21] U. S. Pat. No. 4,506,554 (1985), "Magnetoelastic Torque Transducer", K. Blomkvist and J. O. Nordvall, Assigned to ASEA Aktiebolag.

[22] I. J. Garshelis and C. R. Conto, "A torque transducer utilizing a ring divided into two oppositely polarized regions", J. Appl. Phys. 79 (8), 4756-4758, (1996).

[23] I. J. Garshelis, "A Torque Transducer Utilizing a Circularly Polarized Ring", IEEE Trans. Magn., 28, 2202-2204, (1992); I. J. Garshelis, "Investigations of Parameters Affecting the Performance of Poarized Ring Torque Transducers", IEEE Trans. Magn., 29, 3201-3203, (1993).

[24] U. S. Pat. Nos. 5,535,555 (1994); 5,465,627 (1995) and 5,520,059 (1996), I. J. Garshelis, "Circularly Magnetized Non-Contact Torque Sensor and Method for Measuring Torque Using Same", Assigned to Magnetoelastic Devices, Inc.

[25] Ivan J. Garshelis, Kristen Whitney and Lutz May, "Development of a Non-Contact Torque Transducer for Electric Power Steering Systems", SAE Paper No. 920707, (1992).

[26] Ivan J. Garshelis, "Magnetic and Magnetoelastic Properties of 18% Ni Maraging Steels", IEEE Trans. Magn, Vol. 26, 1981-1983, (1990).

[27] Daniel I. Gordon and Robert E. Brown, "Recent Advances in Fluxgate Magnetometry" IEEE Trans. Magn., MAG-8, 76-82, (1972).

[28] Ivan J. Garshelis, Christopher R. Conto and Wade S. Fiegel, "A Single Transducer for Non-Contact Measurement of the Power, Torque and Speed of a Rotating Shaft", SAE Paper No. 950536, (1995).

960040

Instantaneous Crankshaft Torque Measurement in Cars

Jarl R. Sobel, Jan Jeremiasson, and Christer Wallin
ABB Industrial Systems AB

Copyright 1996 Society of Automotive Engineers, Inc.

Jarl R Sobel, Jan Jeremiasson, Christer Wallin
ABB Industrial Systems AB

ABSTRACT

The instantaneous torque of an internal combustion engine has been measured in front of the flywheel, using a solenoidal magnetoelastic torque sensor. The measured torque in a four cylinder engine has been compared with pressure signals during bench tests, and excellent correlation with values of the mean effective pressure for each cylinder have been obtained. Road tests have been performed in a car with a five cylinder engine and automatic transmission. Results with respect to engine control and misfire detection are presented, which show only minor disturbances from driving on rough roads.

INTRODUCTION

This paper deals with measurement of torque in cars using a magnetoelastic non-contact torque sensor. Magnetoelasticity is a property of all ferromagnetic materials which causes the magnetic permeability of the material to change when it is subjected to mechanical stress. This effect is closely related to magnetostriction, i.e. the effect which causes an unstressed material to change shape upon magnetization. When stress is small there is a thermodynamic relation between the change of induction with stress and the magnetostriction [1]. The principle of non-contact torque measurement using magnetoelasticity is to let a magnetic field penetrate the shaft, and detect changes in flux caused by the torsional stress in the shaft. Most magnetoelastic transducers use a high frequency magnetic field, which is measured by the voltage it induces in a coil. This gives the sensor a low output impedance, high signal level, and high immunity to electrical interference. The high frequency output signal is demodulated to give a rapid step response.

Since the sensor shaft is manufactured from steel, it can be made to replace the component where the torque is to be measured. In other cases it can be integrated with other parts using conventional joining methods such as beam welding. However, close control of the material is required to ensure the right magnetic properties. In principle the design of the non-contact torque sensor is relatively simple, and it can be mass produced at a competitive price. It is particularly well suited for automotive use, where it could open up many new applications.

To illustrate the potential of this sensor this paper deals with measurement of the instantaneous torque between the crankshaft and the flywheel, or converter in automatic transmissions. The most important measurement is the compliance with the On Board Diagnostics II (OBDII) Regulation by the California Air Resource Board to detect misfire[2].

In this application the sensor was mounted in a vehicle to study influences on the measured torque when driving on rough roads.

The measurement of torsional stress in the crankshaft has been performed before by several authors e.g.[3] and [4]. However, the transducer utilized by these authors is based on a magnetoelastic torque transducer called the Cross Torductor [5] [6] invented by O Dahle of our company in 1953. In this design the mechanical stress is measured locally on the surface of the shaft. Therefore, due to stress being distributed irregularly around the circumference of the shaft, the output signal will fluctuate as the shaft rotates, even when the shaft is unloaded. The output signal of such a transducer does not reflect the true instantaneous torque in the crankshaft. Instead, conclusions have to be drawn based on a comparison between the output of the sensor at the same angular position but at different revolutions [4].

Alternative methods to estimate the crankshaft torque are indirect, via pressure sensors, or via accurate measurements of engine speed variations [7] [8]. Another method is by measuring the twist of the crankshaft between the flywheel and front end of the engine [9].

SOLENOIDAL MAGNETOELASTIC TORQUE SENSORS

The torque transducer used in this paper is classified by Fleming [10] as a solenoidal torque sensor. The basic design of this type of torque sensor was first invented by the Russians L.N Tseytlin, A.K Shimokhin and F.B Kim in 1979 [11] and 1981 [12], and it exists in different forms.

The main characteristic is that the shaft is magnetized with a solenoidal and rotationally symmetric magnetic field using a primary coil concentric with the shaft. The surface of the shaft is prepared with some kind of anisotropy in axially separated annular zones. The anisotropy forces the field in each zone to depart from its natural direction parallel to the axis of rotation of the shaft, and follow either of the principal stress directions of the torsional stresses to be measured. These stress directions are directed at angles ±45° from the axial direction. In the zone where magnetic field lines follow the compressed direction, the magnetic permeability will decrease due to the magnetoelastic effect. Consequently, the output voltage of the secondary coil surrounding this zone will decrease. In other zones where the magnetic field follows the stretched direction, the magnetic permeability and the corresponding output voltage will increase. The sensor output is then obtained as the difference of these voltages simply by connecting the secondary coils in series.

A significant property of this type of sensor is that the electrical output is insensitive to rotation of the transducer shaft. In practice the error is less than ±0.2%. This is especially important when measuring torque variations with the crank angle.

The distinguishing feature between different solenoidal magnetoelastic torque sensors is how the necessary anisotropy is formed.

In the sensor described in [11] the anisotropy is geometric, and formed by cutting parallel grooves in the surface of the shaft, inclined at the appropriate ±45° angles.

This was subsequently improved in [12], where the anisotropy is both magnetic and geometric. Here the zones are formed by creating a knurl in the shaft surface with the troughs of the knurl at the required angles. The magnetic anisotropy is created by residual stresses due to the plastic deformation of the material. This type of sensor has been further developed by I Garshelis [13].

Other ways to create a magnetic anisotropy using amorphous ribbons is followed by Sasada and others. Sasada describes a sensor where the amorphous ribbons are glued to the surface of the shaft in a ±45° chevron pattern[14].

The torque sensor referred to in this report has been under development at our company since 1985. The anisotropy is in this case neither geometric nor magnetic. Instead, an anisotropic surface conductivity is formed by electroplating thin copper strips on the surface of the shaft. Due to eddy currents in the strips, the magnetic field is confined to the free surface between the strips. The strips can be made very thin, on the order of 50 microns, depending on the frequency of the magnetic field.

The primary advantage with this method is that the surface of the shaft is smooth, without grooves or knurls that cause stress concentrations, which in turn could cause failure from fatigue or overload.

A particular complication with the measurement of crankshaft torque was the influence of temperature gradients on the zero signal of traditional two zone solenoidal sensors. This temperature gradient is caused as the crankshaft side of the sensor has a different temperature than the flywheel side. Since magnetic permeability is temperature dependent, this gives rise to a difference between the zones which is not caused by torsion but shows up as a zero shift of the sensor output. The problem was solved by using a special three zone configuration, as described in [15].

Figure 1 outlines such a transducer. In this case the housing which holds the concentric coils is centered around the shaft using a ball bearing at each end.

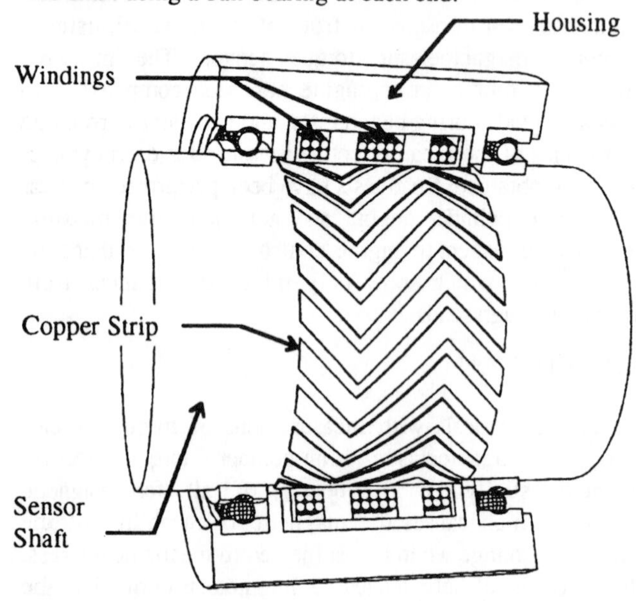

Figure 1: Three zone magnetoelastic torque sensor

The influence of temperature, in particular on the slope of the output signal with respect to the torque has also been solved [16]. In addition, the influence on measurement from static magnetic fields has been eliminated [17].

CRANKSHAFT TORQUE MEASUREMENT

The torque of the crankshaft in front of the flywheel is a very important parameter which could be widely used to control the condition and efficiency of a combustion engine. The torque measured at this position gives the actual output from the engine. The rapid response from the transducer is much faster than any torque variation in the crankshaft. An important advantage gained by measuring the torque is that information of the instantaneous torque from all cylinders can be obtained by using just one sensor. This can make it possible to design adaptive functions which compensate for aging of engine components, manufacturing tolerances, different fuel qualities and local or global environmental variations for the entire life time of the engine. It would also be possible for an engine diagnostic system to make more extensive checks of the engine status, which may prevent a possible engine breakdown.

In large diesel engines for example, the cylinder balancing is very important for reducing exhaust emissions, vibrations and increasing fuel economy. An on-board crankshaft torque sensor can be used to continuously monitor and control engine performance.

MECHANICAL INTEGRATION OF THE TORQUE SENSOR WITH THE CRANKSHAFT - A solenoidal torque transducer inevitably requires a certain length to perform its function. However, it is possible to integrate it with a crankshaft without extending the total length of the engine.

Figure 2 shows a design where the sensor is built into a 2.3 liter, in-line, 5-cylinder, naturally aspirated engine.

In the installation as shown by Figure 2, the measuring shaft and the housing for the concentric coils are located in a cylindrical cavity at the rear end of the crankshaft. The housing with the coils is fixed to the engine block. The torque sensor has a measuring length of 35 mm, and is connected to the crankshaft through a standard spline joint. The other side is welded into the carrier plate.

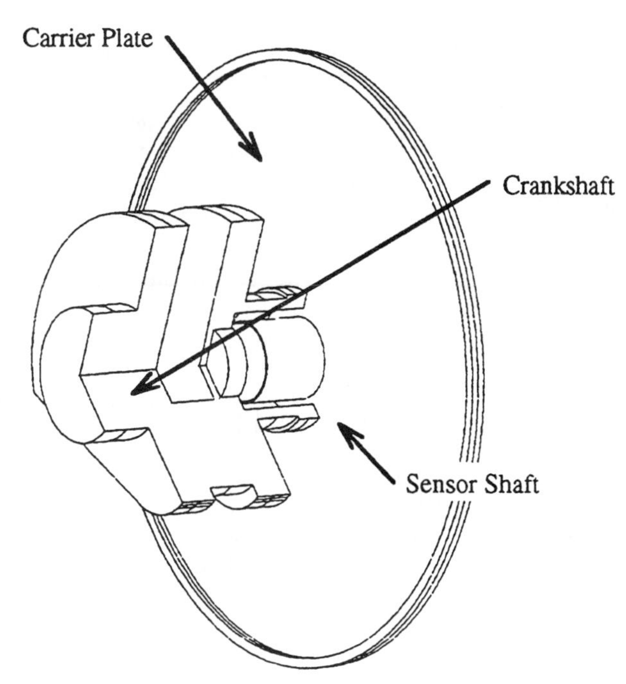

Figure 2: Crankshaft integrated solenoidal torque sensor

Calculations using FEM modeling of the crankshaft with a reduced thickness of material underneath the last base bearing has shown that neither the bearing function nor the crankshaft strength is seriously affected.

The introduction of the measuring shaft will cause a slight decrease in the stiffness of the crankshaft and a subsequent reduction of its resonant frequency. The torsional oscillations at maximum speed can be kept at acceptable amplitudes by conventional methods, and the engine lifetime will not be affected. In fact, engine life can be increased by using a torque signal to control the throttle and the speed of the engine, thus protecting it and other parts of the drivetrain from overload.

TORQUE CALCULATED FROM PRESSURE - The following measurements were performed on a four cylinder Opel engine (2 liters, 16 valves) in a test cell at IMH – Institut Für Motorenbau Prof. Huber GmbH, in Munich, Germany.

In this case the engine was extended, so that the torque sensor could be bolted between the crankshaft and the flywheel. Cylinder pressure was measured with Kistler Quartz Pressure Transducer Type 601A.

To calculate the torque from the cylinder pressure measurements, each pressure signal was multiplied by the corresponding lever and piston area. The four torque contributions thus obtained are then added together.

The results shown in Figure 3 are from an interesting study at low engine speed, namely at 1200 rpm and 10 Nm load, where the natural roughness of the engine – "cyclic variability" – gives rise to large variations in the pressure produced. Figure 3 shows the torque calculated from pressure during one cycle (two revolutions) as a function of the Crank Angle (CA).

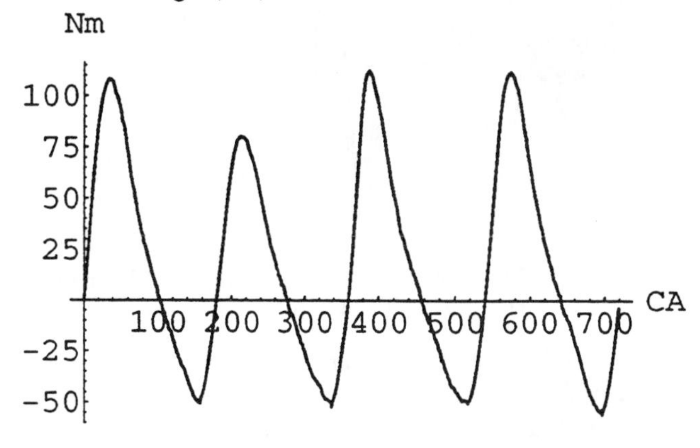

Figure 3: Torque on the crankshaft calculated from cylinder pressure. 1200 rpm, 10 Nm

INERTIAL FORCES - In every reciprocating engine the forces on the pistons are balanced by inertial forces needed to accelerate the pistons and the piston rods. These forces can be calculated using a simple model with point masses and constant rotational speed, found in elementary textbooks on the subject.

Inertial forces calculated using this simple model under the same conditions as above yield a result shown below in Figure 4.

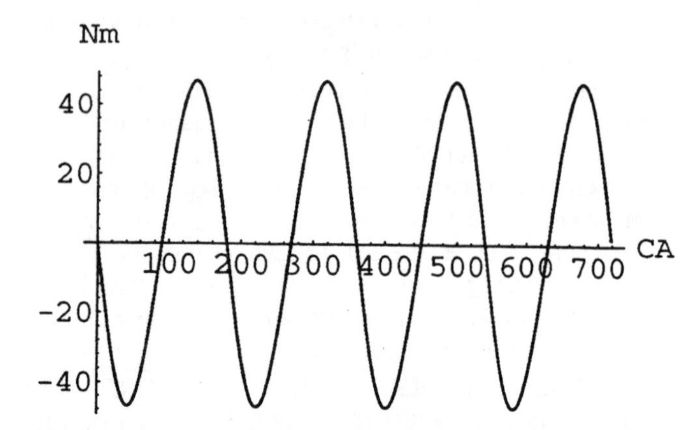

Figure 4: Torque from inertial forces. 1200 rpm

COMPARISON WITH MEASURED TORQUE - In order to calculate the level of the mean torque, friction must be taken into account, which is outside the scope of this study. Instead the mean torque was eliminated from both the calculated and the measured torque.

Figure 5 shows the calculated torque and the measured torque during one cycle (two revolutions) as a function of the Crank Angle (CA). The rapidly oscillating signal is the measured torque. The cause of the rapid oscillation is a torsional resonance of the crankshaft.

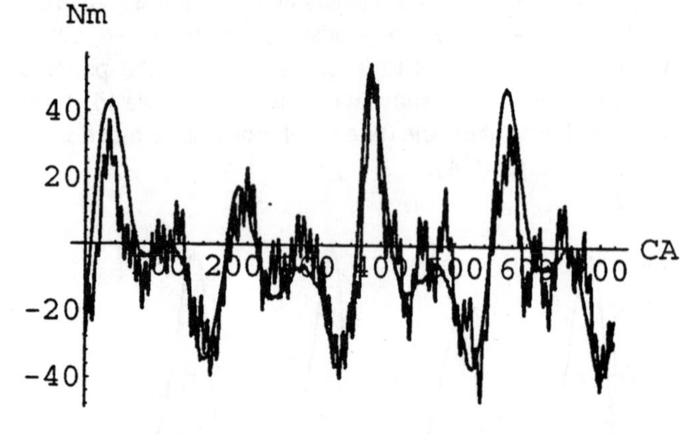

Figure 5: Measured and calculated torque at 1200 rpm and low load in a 4 cylinder engine.

The result presented in Figure 5 indicates that the torque signal could be used to calculate the shape of the pressure curve, at least when friction and influence of torsional resonances are negligible. However, since the torque signal takes inertial forces, resonances and friction into account, it should be better suited for engine control than a pressure signal. Instead of controlling the engine depending on what happens inside the cylinders, you are able to control it depending on what comes out of the engine.

CORRELATION BETWEEN THE AVERAGE TORQUE PER STROKE AND THE INDICATED MEAN EFFECTIVE PRESSURE

The Indicated Mean Effective Pressure (IMEP) from each combustion is important in many ways.

In the case of misfire detection, this parameter can be used as a reference variable, quantifying the quality of the combustion process, and thus indicating the possible occurrence of a misfire.

In the case of engine control, the objective could be to adjust the injection in each cylinder separately, and make each cylinder produce as much work as possible for a certain amount of injected fuel. This could also be used to obtain optimum balance between the cylinders.

The Indicated Mean Effective Pressure is proportional to the work performed by each cylinder during one cycle. By dividing this work with π radians (for a four cylinder four-stroke engine), we obtain a scaled IMEP-value corresponding to the average torque per stroke.

The average torque per stroke is in this case obtained by averaging the measured torque over an interval of 180 degrees containing the combustion. Note that such an integration of the torque due to inertial forces equals zero.

Thus, in order to analyze whether the torque signal can be used for misfire detection or for engine control purposes, we compare the average torque per stroke with the corresponding work produced by each combustion divided by π.

The results were obtained from the same study on the four cylinder engine as referred to above.

LOW SPEED AND LOW LOAD - Average torque per stroke and scaled IMEP-values during 76 combustions (19 complete engine cycles) at 1200 rpm and 10 Nm load are plotted in Figure 6 below. The true separation of the two curves would be due to friction forces which are not considered in this context. The curves are separated to improve clarity.

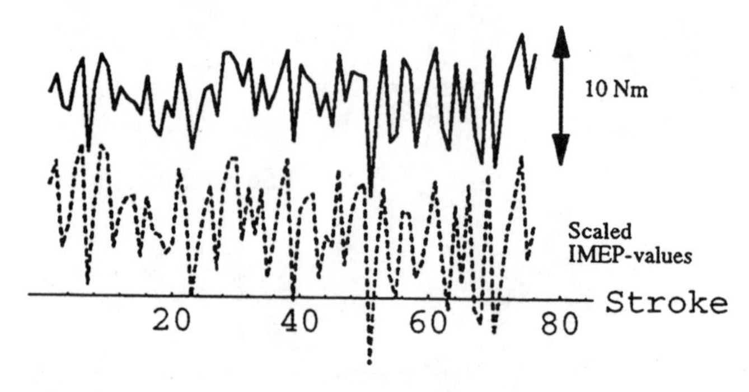

Figure 6: Average torque per stroke and scaled IMEP-values at 1200 rpm, 10 Nm load.

The good correlation in Figure 6 shows that the average torque per stroke transmitted to the flywheel during a combustion in a four cylinder engine is only slightly influenced by pressure in the other cylinders, or by differences in work due to internal engine friction.

Note the large variation in scaled IMEP and average torque per stroke due to cyclic variability at this low speed and load.

HIGH SPEED AND LOW LOAD DURING MISFIRE - At high engine speeds, the crankshaft is much more influenced by torsional resonances and inertial forces. However, the addition from these components to the work is ideally equal to zero, as noted above. Therefore the correlation between work calculated from pressure and torque remains very good, even at low loads.

The case of 76 ignitions at 5000 rpm and 10 Nm load is shown in Figure 7 which also contain the effect of three consecutive misfires of cylinder 1. Again, the curves are separated to improve clarity.

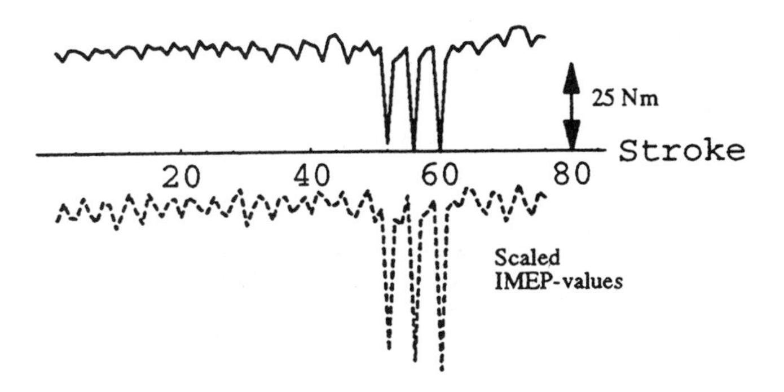

Figure 7: Average torque per stroke and scaled IMEP-values at 5000 rpm and 10 Nm load during three consecutive misfires of cylinder 1

The unevenness produced by a cylinder misfire, is quite obvious in both curves.

ON BOARD MISFIRE DETECTION

The following measurements were performed on a five cylinder engine, mounted in a car with front wheel drive and automatic transmission. The torque transducer was integrated with the crankshaft as described in Figure 2 above. Misfire was induced by turning off the voltage to the spark plug at regular intervals.

INTERMITTENT MISFIRE - Measurements were performed at several loads and speeds. In all cases the change in the average torque per stroke produced at misfire is larger than six times the standard deviation of the torque signal in the absence of misfire.

Some typical examples are presented here.

Torque during misfire at Low Speed - Detection at low speed is by far the easiest case, due to the absence of influence on the torque by any crankshaft torsional resonances. Therefore the absence of an ignition is readily visible in the instantaneous torque as shown below in Figure 8 as a function of the Crank Angle (CA) in degrees.

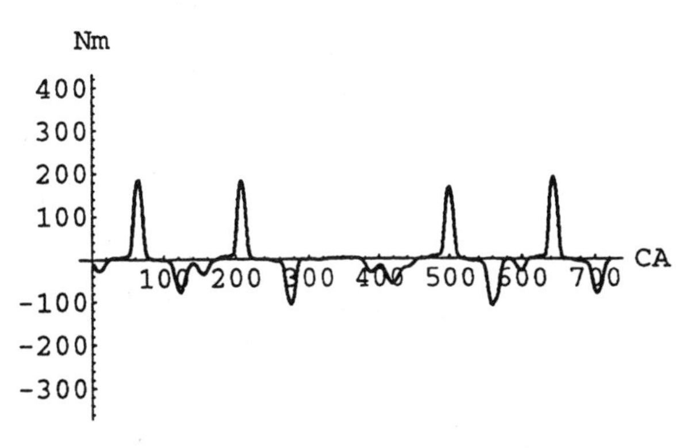

Figure 8: Instantaneous torque for one cycle at 1500 rpm in neutral gear during misfire

The average torque per stroke is also significantly reduced as seen in Figure 9 below. Here it is shown over a period of 50 revolutions during 5% misfire.

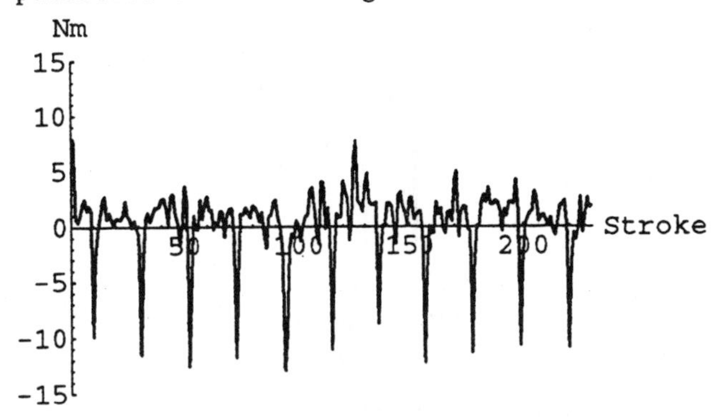

Figure 9: Average torque per stroke at 1500 rpm in neutral gear. 5% misfire

In this case, when ignition is absent the average torque per stroke is reduced by about 10 Nm. At higher loads this torque reduction at misfire increases, as could be expected.

Torque at High Speed - As rotational speed increases, the torque resulting from the ignition in each cylinder is dominated by the torque produced by unbalanced inertial forces, as well as the torsional resonance of the crankshaft.

This is illustrated in Figure 10 below, a typical case at 4000 rpm and neutral gear.

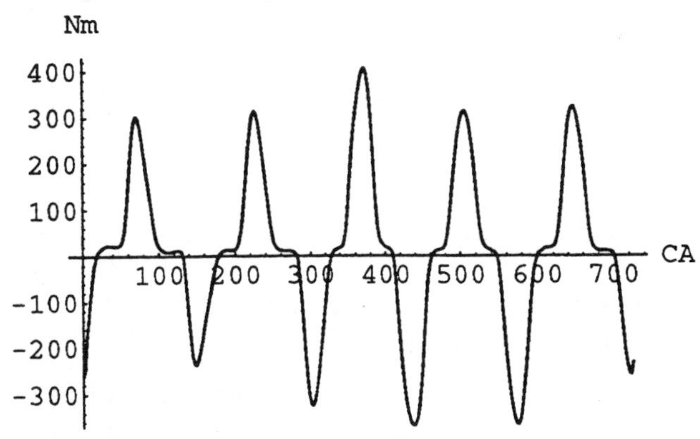

Figure 10: Instantaneous torque at 4000 rpm in neutral gear during misfire

In this case it is difficult to see that cylinder 5 (corresponding to the fourth peak in the graph) is misfiring, unless we compare this measurement with the curve without misfire, measured during the same conditions.

However, when properly averaged, the mean torque produced by inertial forces amounts to zero, as well as the torsional oscillations induced by the firing of the cylinders.

Therefore, when the average torque per stroke is studied, the same significant reduction is found in the output from a misfiring cylinder as found above.

In Figure 11 below, the average torque per stroke is calculated under the same conditions as above. Misfire occurs here in 5% of the ignitions

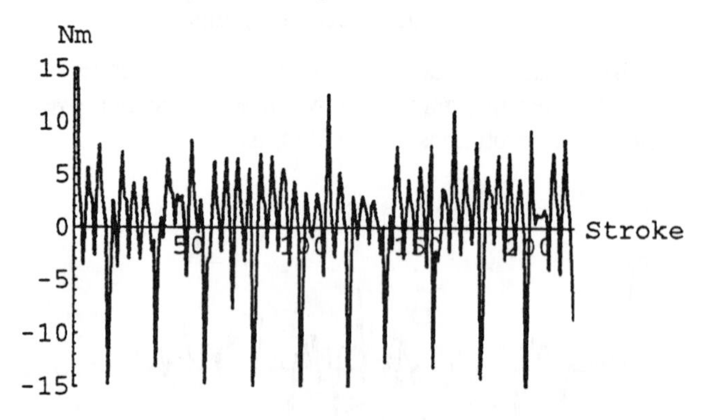

Figure 11: Average torque per stroke at 4000 rpm in neutral gear. 5% misfire

The misfires are clearly visible.

Note that in Figure 9 and Figure 11 above, the torque contributions from all cylinders are shown in sequence.

Summary - The results of our measurements can be summarized in Figure 12, which describes the mean drop in average torque per stroke for an induced misfire at engine speeds ranging from 1500 to 4000 rpm for engine running in neutral gear, and a load of about 50 Nm.

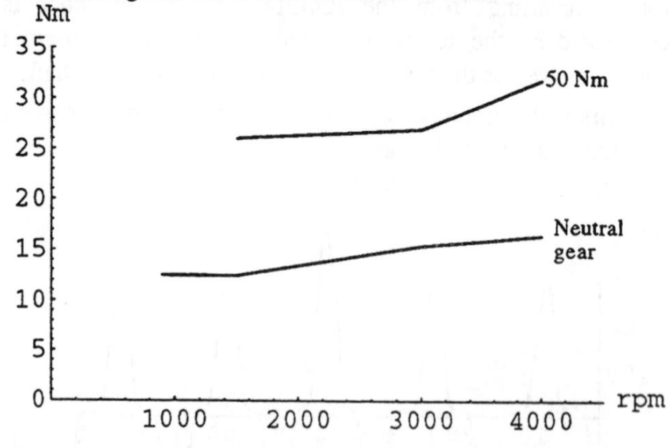

Figure 12: Mean drop in average torque per stroke for an induced misfire

The drop in torque is about 10 Nm when driving in neutral gear, and increases with speed and load. The increase with speed is probably coupled to the increase in load on the engine due to increased internal friction losses in the case of the engine running in neutral gear. In the case with a load of 50 Nm it is coupled to the increase in power output.

Misfire at engine speeds higher than 4000 rpm were not measured in the car due to problems with the flywheel tooth sensor at these speeds. However this has been measured in bench tests with good results.

DETECTION OF CONTINUOUS MISFIRE - Since the torque contributions can be compared between cylinders, detecting continuous misfire in one cylinder is the same as detecting a single misfire.

Cases where two or more cylinders are continuously misfiring are shown below.

As an example, in Figure 13 below the situation at 4000 rpm and a load of 60 Nm, during continuous misfire in cylinder 4 and 3 is shown.

The firing order for this engine is 1-2-4-5-3.

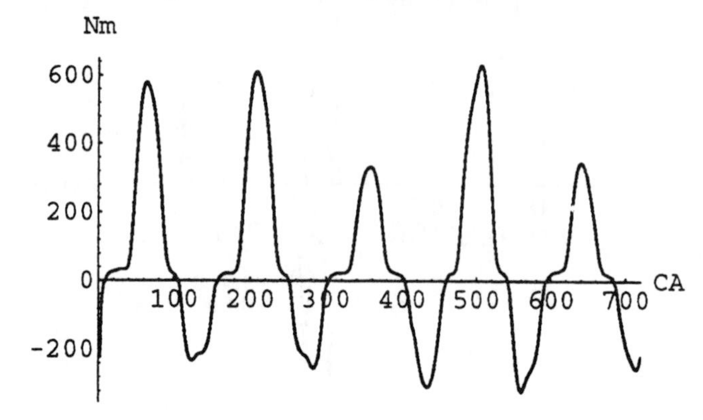

Figure 13: Continuous misfire in two cylinders at 4000 rpm and 60 Nm

The misfire gives rise to a clear drop in the maximum torque.

Even continuous misfire at three cylinders can be detected, as shown below in Figure 14 where cylinders 2,4 and 3 are continuously misfiring.

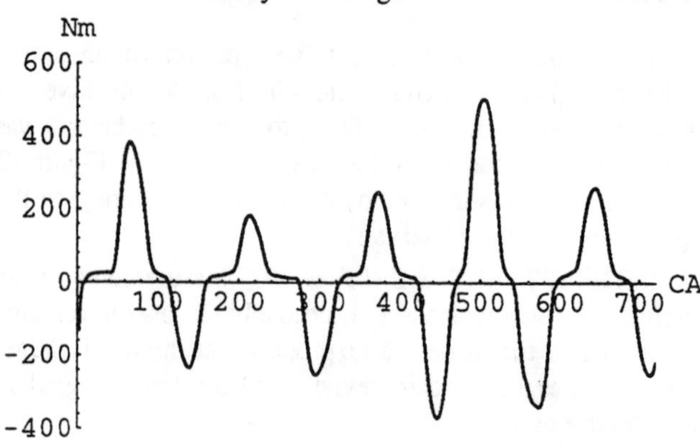

Figure 14: Continuous misfire in three cylinders at 4000 rpm and 15 Nm

INFLUENCE FROM DRIVING ON ROUGH ROADS - Theoretically there is little affect upon the torque measured in front of the flywheel or converter when driving on rough roads. This is due to the fact that this torque is reduced by the gearbox, and also to the large moment of inertia of the flywheel/converter compared with the crankshaft.

The measurements performed are not intended to be by any means exhaustive. Instead, the measurements indicate the magnitude of road disturbances, and demonstrate the main advantages obtained by measuring torque in front of the flywheel.

Belgian Pavé - Measurements were performed on so called Belgian Pavé at different speeds and in different gears.

Figure 15 below shows the torque noise, defined as the standard deviation of the average torque per stroke with respect to the mean torque, for speeds between 40 km/h and 60 km/h. To facilitate the comparison with typical signal levels, the scale on the y-axes is the same as in Figure 12 above.

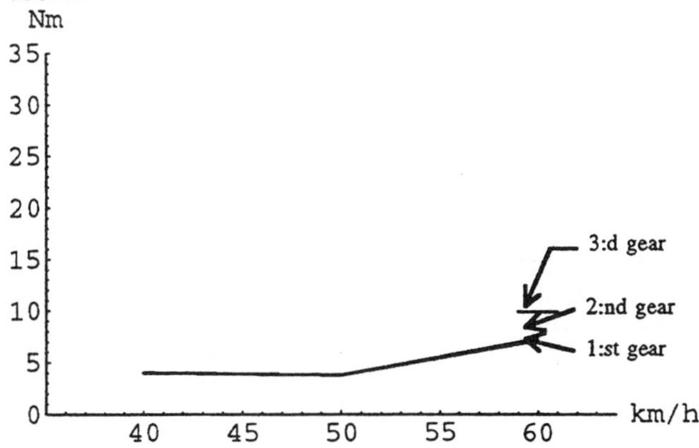

Figure 15: Torque noise levels at different speeds and in different gears when driving on Belgian Pavé

The noise levels are very similar when driving in first and second gear. When driving in third gear the noise levels increase, which is expected due to the reduced gear ratio.

Comparing with Figure 12 we see that the noise levels are well below the average signal due to a misfire, even at zero torque output (neutral gear).

In Figure 16 below, a case of 2% induced misfire was measured while driving on Belgian Pavé at 40 km/h in 1:st gear.

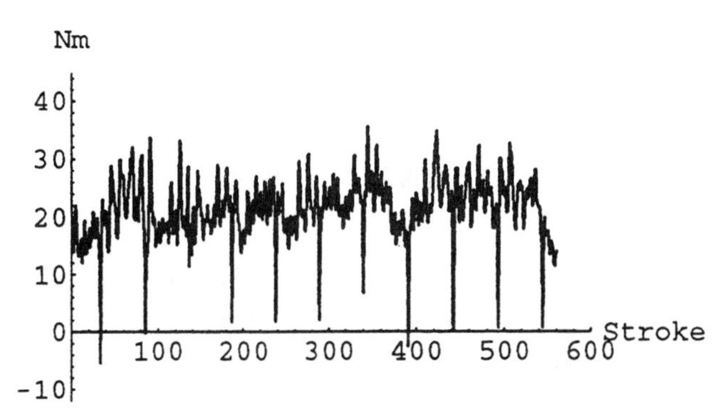

Figure 16: Average torque per stroke.
Belgian Pavé, 40 km/h, 1st gear, 2% induced misfire.

Washboards - When driving on the washboard track at 40 km/h, oscillations in a range of frequencies from 6 Hz to 22 Hz are excited. These frequencies are intended to excite drivetrain oscillations.

When driving in first gear, the major oscillations occur at 11 and 15 Hz which is illustrated between strokes 1500 and 2000 in Figure 17 below

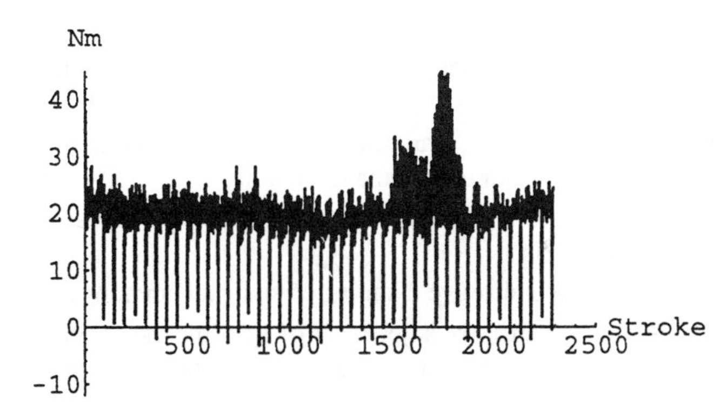

Figure 17: Average torque per stroke.
Washboard, 40 km/h, 1st Gear, 2% induced misfire.

A detail of the figure above is shown in Figure 18 below

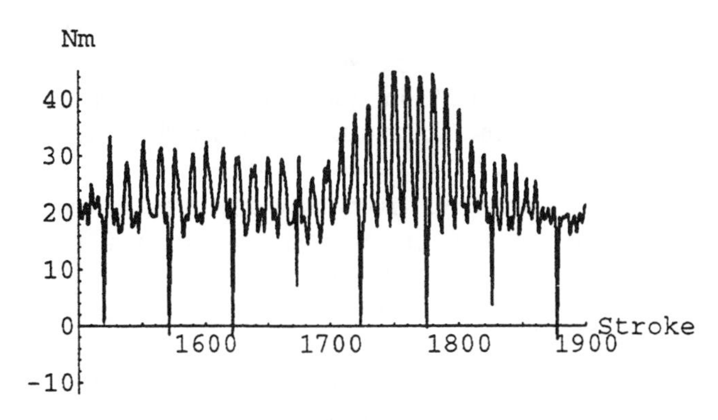

Figure 18: Average torque per stroke.
Washboard, 40 km/h, 1st Gear, 2% induced misfire.

The drivetrain torque resonance gives rise to a torque signal in front of the flywheel with an amplitude of about 20 Nm. Still, misfires are clearly seen.

OTHER AUTOMOTIVE TORQUE SENSOR APPLICATIONS

Our sensor has also been tested in other automotive applications, some of which are mentioned below.

CRANKSHAFT TORQUE SENSOR - Other examples of possible applications for a crankshaft torque sensor are:

- Cylinder balancing in diesel engines
- Turbo pressure control
- Lean combustion control
- Automatic optimization of ignition
- Idle control
- Electronic throttle control
- Garage diagnostic tool

In other applications it may be preferable or required that the torque sensor is mounted at other positions in the drivetrain away from the crankshaft.

GEARBOX TORQUE SENSOR - If the torque sensor is integrated in the gearbox, the torque output , together with other signals, can be used for:

- Determination of gear shifting to obtain maximum efficiency or minimum exhaust emission
- Line pressure control in continuously variable transmissions
- Monitoring and active damping of torsional vibrations
- Gearshift control by adjustment of ignition and throttle towards a desired torque level to obtain smoother gearshifts
- Reduction of drivetrain noise
- Elimination of vehicle speed oscillations

The measured torque has a very fast step response of less than 1 millisecond. Furthermore it does not deteriorate with time or when the engine is operating under abnormal conditions.

Mean torque estimated from mass flow sensors and other sensors, can only be recalculated every 10 milliseconds with an accuracy of about 5% in steady state. During dynamic processes such as gear shifting and acceleration, the accuracy of estimated torque is about 20%.

A more accurate and faster torque measurement could make the gear shifting process smoother and faster, thus improving comfort and operating life of the gearbox. In addition, it would be possible to monitor and compensate for clutch wear.

DRIVETRAIN - To integrate a torque sensor in the drivetrain gives advantages especially in 4-wheel drive systems, where the main aim would be to develop an intelligent system to improve performance, simplify the 4-wheel drive transmission, and save costs. One important feature is to improve the driving performance and stability of vehicles on roads with a low friction coefficient on their surface. A system with torque sensors could be used to optimize and distribute the driving force to the front and rear wheels, according to the running condition of the vehicle. The braking performance may also be improved.

ELECTRICAL POWER STEERING TORQUE SENSOR - Electrical Power Steering requires an inexpensive and highly reliable sensor to provide the electrical control signal indicating the torque exerted by the driver on the steering wheel [18]. The high reliability, as well as the ability to compensate for external static magnetic fields, makes the Solenoidal Magnetoelastic Torque sensor a good candidate to fulfill the specification for a non-compliant steering column torque sensor, although the extreme demands on overload capability in combination with high demands on zero-stability are still hard to meet.

CONCLUSIONS

This report describes the use of an inexpensive transducer which measures the true instantaneous torque, and is suitable for automotive applications.

To demonstrate its capabilities, the torque sensor has been evaluated in engine bench tests, and integrated with the crankshaft in a car with automatic transmission.

Close correlation with the instantaneous torque computed from cylinder pressure signals is shown at low engine speeds. Furthermore, induced misfires are clearly visible in the torque signals, even when driving on rough roads.

This sensor can already be used for various purposes such as engine control, drivetrain management and control of an automatic transmission in order to improve economy, handling, and reduce environmental pollution. There is also the possibility of a reduced cost, since this sensor may eliminate the need for certain other sensors.

REFERENCES

[1] Richard M Bozorth, "Ferromagnetism", p. 732, IEEE Press 1993 (written in 1951)

[2] State of California, AIR RESOURCES BOARD, MAIL OUT #95-03

[3] Y. Nonomura, J. Sugiyama, Ksukada, & M. Takeuchi, Toyota Motor Corp. "Measurements of Engine Torque with Intra Bearing Torque Sensor", SAE Technical Paper 870472

[4] R. Klauber et al., "Miniature Magnetostrictive Misfire Sensor", SAE Technical Paper 920236

[5] O Dahle, "Device For Indicating and Measuring Mechanical Stress within Ferro-magnetic Material", US Patent 2,912,642, assigned to Asea Allmänna Svenska Elektriska Aktiebolaget, issued Nov 1959.

[6] O Dahle, "Measuring Means for Measuring a Torsional Stress in a shaft of Magnetostrictive Material", US Patent 3,011,340 assigned to Asea Allmänna Svenska Elektriska Aktiebolaget, issued 1961

[7] Stephen J. Citron et al, "Cylinder by Cylinder Engine Pressure and Pressure Torque Waveform Determination Utilizing Speed Fluctations", SAE Technical Paper 890486.

[8] H-M Koegler et al, "Method and Apparatus for Diagnosing Internal Combustion Engines", US Patent 5,157,965 assigned to AVL, Austria, issued 1992

[9] Georg F Mauer et al, "On-Line Cylinder Diagnostics on Combustion Engines by Noncontact Torque and Speed Measurements, SAE Technical Paper 890485.

[10] W.J Fleming, "Magnetostrictive Torque Sensors - Comparison of Branch, Cross and solenoidal Designs", SAE Technical Paper 900264.

[11] A K Shimokhin, L N Tseytlin, "Magnetoelastic Torque Sensor", USSR Patent 667836, issued June 1979

[12] L N Tseytlin, A K Shimokhin, F B Kim, "Magnetoelastic Torque Transducer", USSR Patent 838448, issued June 1981

[13] I Garshelis, "Magnetoelastic Torque Transducer", European Patent Specification 0 270 122 B1, issued Sept 1991.

[14] Sasada et al, "Torque Transducers with Stress-Sensitive Amorphous Ribbons of Chevron-Pattern", IEEE Trans. on Magn., Vol. MAG-20, No. 5, Sept 1984

[15] J R Sobel, "Magnetoelastic Torque Transducer", US Patent 4,873,874, assigned to Asea Brown Boveri, issued Oct 1989

[16] J R Sobel, D Uggla, H Ling, "Magnetoelastic Noncontacting Torque Transducer", PCT Patent Application 9500289, March 1995

[17] J R Sobel, J Palmquist et al , "Detection of Static and/or Quasistatic Magnetic Fields in Magnetoelastic Force and Torque Transducers", US Patent 5,122,742 , assigned to Asea Brown Boveri, issued 1992.

[18] I Garshelis et al, "Development of a Non-Contact Torque Transducer for Electric Power Steering Systems", SAE Technical Paper 920707.

940629

A Non Contact Strain Gage Torque Sensor for Automotive Servo Driven Steering Systems

Erich Zabler, Frieder Heintz, Anton Dukart, and Peter Krott
Robert Bosch GmbH

ABSTRACT

Tapping of one or more torques (ranges 10 Nm and 60 Nm) on the steering column for the purpose of servo control must satisfy high accuracy requirements on the one hand and high safety requirements on the other hand. A suggestion for developing a low-cost solution to this problem is described below: Strain gages optimally satisfy both these requirements: However, for cost reasons, these are not applied directly to the steering column but to a prefabricated, flat steel rod which is laser welded to the torque rod of the steering column. The measuring direction of the strain gages is under 45° to the steering column axis. The strain gages are either vacuum metallized onto the support rod as a thin film or laminated in a particularly low-cost way by means of a foil-type intermediate carrier.

The relatively low output signals of the measuring resistor bridge are amplified on site (on the steering column) with an integrated electronic evaluation circuit and converted into a frequency-analog, highly interference-proof signal. Signal output and power supply may be connected to the fixed parts of the body neither by collector rings nor by cables. Both connections are made totally non-contact by means of a very simple and low-cost rotary transformer which surrounds the steering column concentrically and has two separate winding chambers. In this case, it is practically impossible for air gap fluctuations and similar interferences to falsify the frequency-analog measuring signal.

OVERVIEW OF THE TORQUE MEASURING METHODS

Torques are measured by recording the mechanical stress σ of the test shaft at the point of stress using suitable instruments. This problem is far more difficult to solve on a rotating shaft than on a fixed one (Fig. 1); and the problem of taking accurate measurements is compounded by that of transmitting them with as little error as possible.

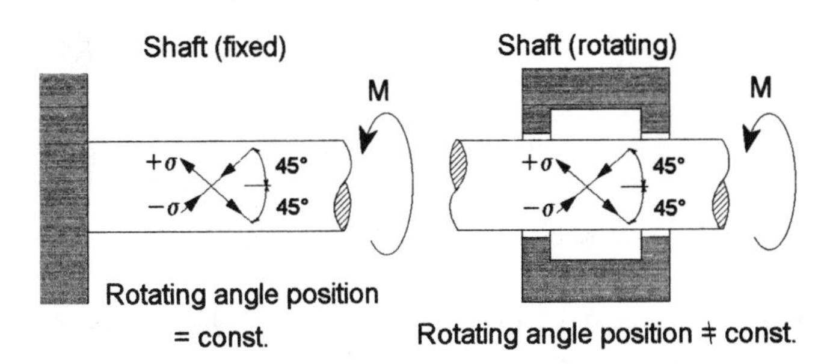

Fig. 1 - Torque measurement on a fixed and a rotating shaft

Many solutions to these problems have been proposed, but only the most recent and most important will be discussed below:

MEASUREMENT OF THE TORSION ANGLE - A shaft subjected to a moment M is twisted over an axial length l through a torsion angle $\Phi = \text{const} \cdot l \cdot M$ (Fig. 2). There are two non-contact methods of measuring this torsion angle:

Two identical speed taps at interval l - If two incremental speed sensors (e.g. inductive, magnetic or optoelectronic) are mounted at an axial distance l, then under load M there is a time shift of $\Delta t = \text{const} \cdot M$ in

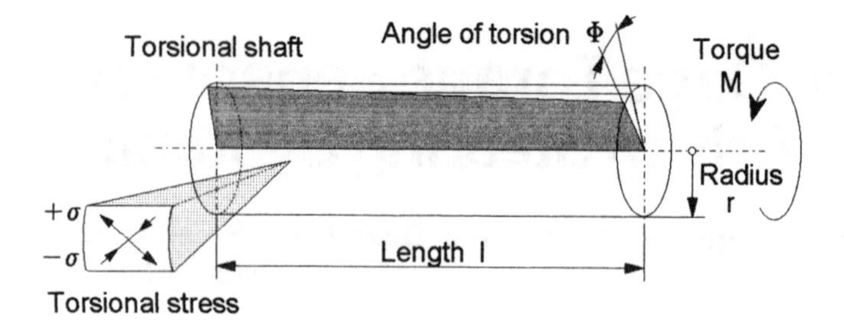

Fig. 2 - Deformation of an elemental cylindrical shaft due to torsion

the two sensor signals (Fig. 3). Apart from the fact that in this method extremely high demands have to be made on the speed sensors, the measurement signal is only available at discrete points in time.

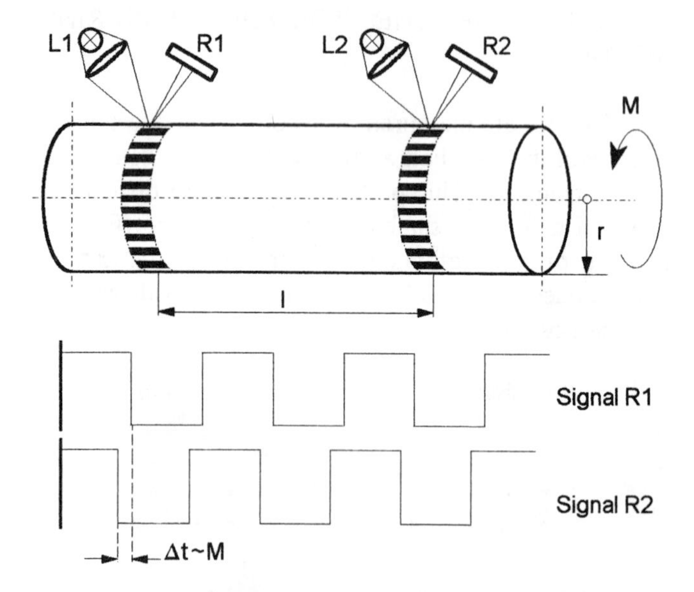

Fig. 3 - Torque measurement by means of phase shift in two (photoelectric or otherwise) speed sensors

Eddy current sensor - As described in [1], this method involves mounting two slotted sleeves manufactured from a good electrical conductor (Cu or Al) at a distance l on the test shaft:
The relative angular position of these sleeves is registered with the aid of a fixed high-frequency coil (1..3 MHz) which encircles the shaft (Fig. 4): Minimum attenuation exists when the slots in both sleeves are in congruence; maximum attenuation when the slots in one of the sleeves are just obscured by the segments in the other, i.e. when the coil detects a quasi-closed attenuator sleeve. Although this provides a continuous signal, the air gap between the sleeves and the coil must be kept extremely narrow (it may even be necessary to have the sleeves brush against the coil) and the system must be manufactured and operated with very narrow tolerances

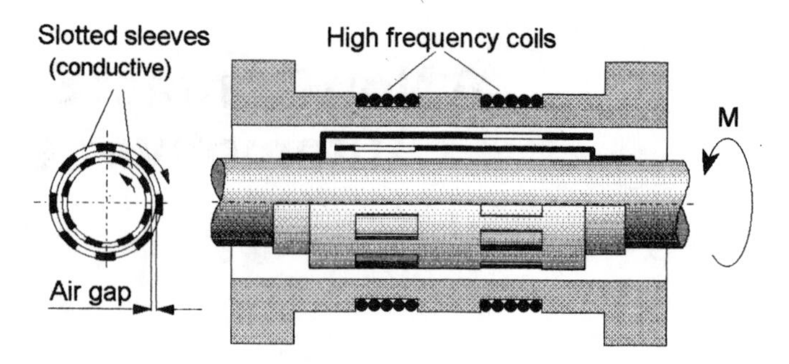

Fig. 4 - Eddy current sensor

if the measurement is to be accurate. For technical reasons, the measurement effect can only be readily evaluated at close to the coils' natural resonance frequency; the attenuation change due to measurement in this range amounts to only a small percentage of the basic attenuation.

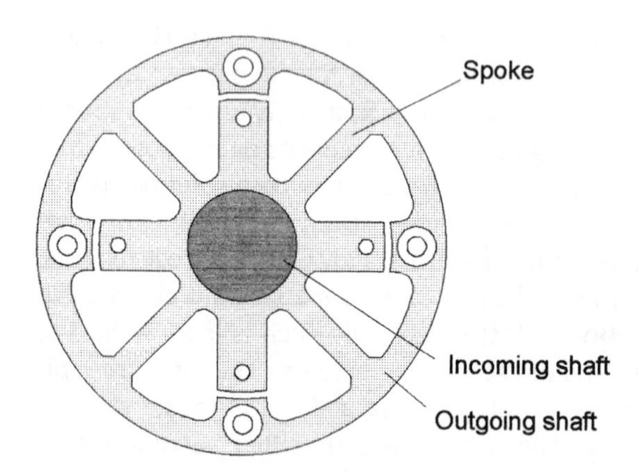

Fig. 5 - Radial torsion spring

The structural length of such a sensor based on measurement of the torsion angle can be shortened by using a radially acting spoked spring (Fig. 5). The radial enlargement, however, means that required volume for installation remains approximately the same.

The common disadvantage of these methods is the overall length l of the structure essentially required to attain useful measurements of torsion angles of $\Phi = 0.1 \dots 1°$; in technical realizations, this typically lies somewhere in the order of 10 cm.

MEASUREMENT OF THE TORSIONAL STRESS - In contrast to tapping a torsion angle, methods which directly measure the material stresses due to torsion, do not essentially require a certain length of structure; they can provide measurements practically at a single point on any length of shaft.

Magneto-elastic sensors - Transductors operating inductively directly utilize the magneto-elastic properties of the shaft; this method exploits the anisotropy of the permeability μ_r occurring under load. Two forms of transductors exist: One which is arranged concentrically around the shaft (Fig. 6), and the other which is mounted radially on the shaft.

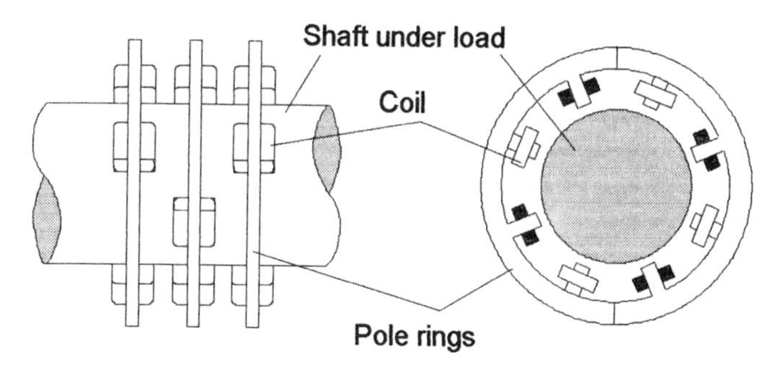

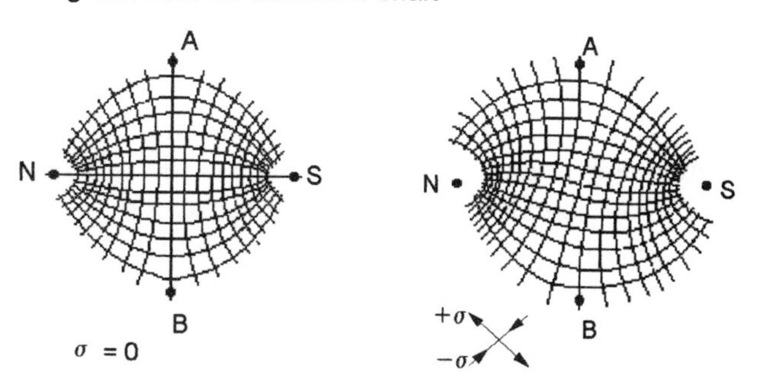

Fig. 6 - The magneto-elastic principle

Apart from its complexity and its sensitivity to air gap fluctuations, this method always requires a compromise between the mechanical and magnetic properties of the shaft. After all, most of the materials that can be used display considerable magnetic hystereses which accordingly falsify the measurement results.

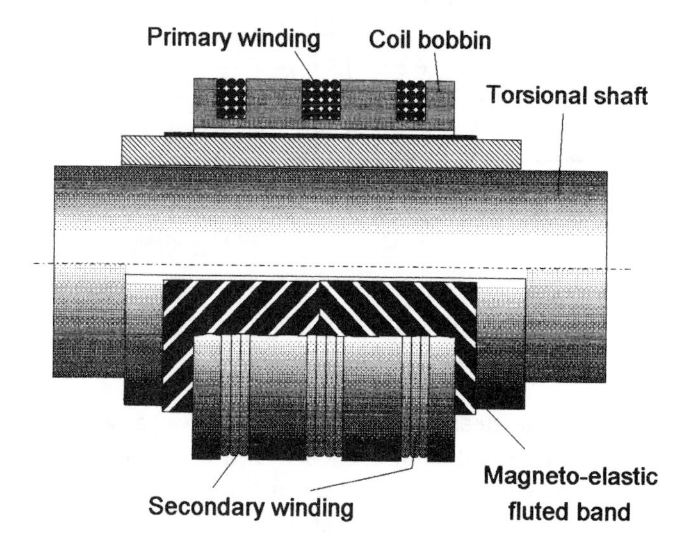

Fig. 7. Coaxial transductor

The disadvantages of the transductors are largely obviated if the surface is friction-coated with amorphous metal (Fig. 7). This material, which is manufactured in foil form and is approximately 30 μm thick, displays excellent magnetic properties as well as - when specially made - excellent magneto-elastic properties. However, a method of fastening the foil onto the surface of the shaft with long-term reproducible adhesion and without impairing the properties of the foil material, has yet to be publicized.

<u>Strain gages (SGR)</u> - In contrast to amorphous metal foils, various forms of strain gages can be mounted on the surface of the shaft with good adhesion. Although the measurement effect of conventional strain gages is only weak (relative resistance change $\Delta R/R < 0.3$ %), this method can be regarded as the most accurate; here, however, there are the additional problems of non-contact signal transmission and power supply.

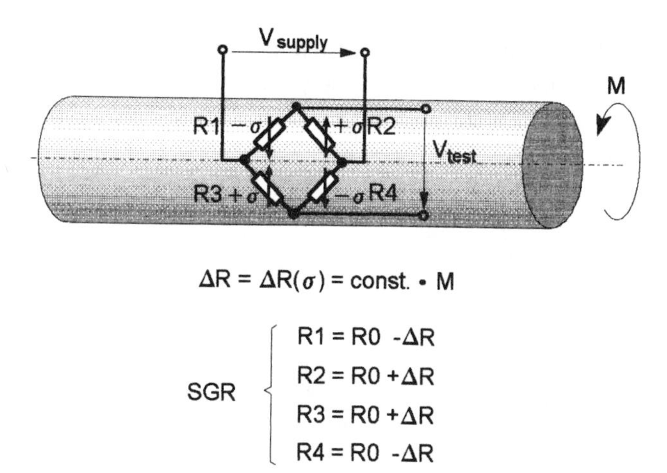

Fig. 8 - Arrangement of strain gages on the torsion shaft

Since the stresses due to torsion produced in the shaft are aligned at under ± 45° to the longitudinal axis, the strain gages are mounted on the shaft so as to form a full bridge in the same alignment (Fig 8). With the strain gages in this configuration, tensile/pressure forces or bending stresses result in no misalignment of the bridge, and thus no output signal is generated. The supply and output voltages must be transmitted contact-free to and from the bridge. If the bridge is supplied with alternating current, the voltages can be transferred by means of coaxial coupling transformers arranged concentrically around the shaft (Fig. 9). However, not only the low level of the output signal, but also the influence of air gap fluctuations and tolerances, can easily lead to considerable errors if no appropriate - and usually sophisticated - countermeasures are taken.

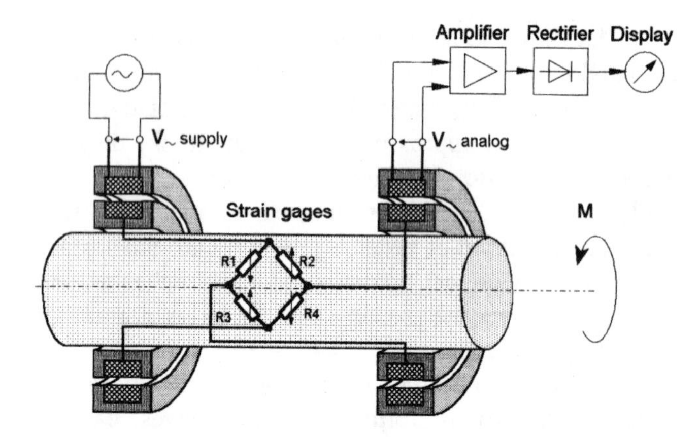

Fig. 9 - Strain gage sensor torque measurement with non-contact transfer via transformer and without electronics on the shaft

Much more profitable is a configuration such as that illustrated in Fig. 10, with electronic components on the shaft in addition to the strain gages: Any kind of a.c. supply voltage can be transferred and its amplitude need not be kept constant. This supply voltage is converted on the shaft into a stabilized d.c. voltage for supplying the bridge.

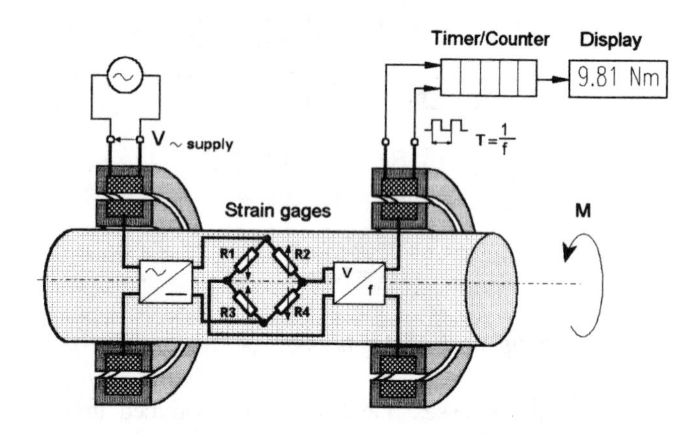

Fig. 10 - Strain gage sensor torque measurement with non-contact transfer via transformer and with electronics on the shaft

Another electronic unit amplifies the bridge's output signal and converts it into a frequency-analog form in which it can be easily decoupled in turn and transmitted without contact and practically error-free via a second transformer. Frequency-analog output signals are often referred to as quasi-digital signals, since they permit extremely straightforward digitalization of measurement results, including any form of further digital processing, transfer (bus), storage and display, through the use of a counter at the fixed end.

SPECIAL REQUIREMENTS IN THE AUTOMOBILE; SELECTION OF A MEASUREMENT PRINCIPLE

There is an obvious need in several cases for non-contact transmission of measurement signals, particularly forces and torques, from rotating parts in the automobile to the electronic control units located in the chassis: For example, the transmission of information from the road wheels, drive shaft and steering column. This paper was prompted by the design of an electronically controlled, servo driven power steering system as illustrated in Fig. 11. Such a system requires the measurement first and foremost of the manual steering torque applied by the driver at the wheel, and, if necessary, of the total steering torque as the sum of manual and servo moments.

The entire steering system, and in particular the steering column, is of major significance in the safety of a vehicle. Although it does not rotate constantly, it can usually be turned through more than two full revolutions in either direction, which means that any electrical transmission of a torque signal via a rolling winding must be regarded as risky and too susceptible to error. Even collector rings are incapable of providing the accuracy and reliability required for such a sensor.

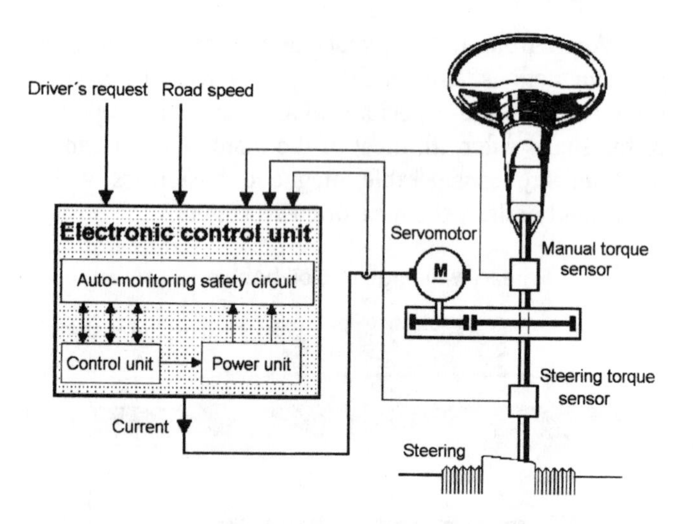

Fig. 11 - Servo driven power steering system with torque tapping

Since the steering column itself is physically much too large, a special shaft section with a correspondingly tapered cross section is inserted for the measurement. It must be ensured, however, that the measurement points can withstand oversteering by the driver (panic moment) or an attempted theft with the steering lock engaged. In sensor view of the very limited space available for

Measuring range, manual torque :	± 10 Nm (M_{Nm})
Measuring range, total torque:	± 60 Nm (M_{Nt})
Overload resistance M_{mmax}, M_{tmax}:	± 120 Nm
Operating temperature:	- 40 ... +125 °C
Permissible measurement time:	< 1 ms
Endurance: 2,000,000 load changes ± 2.5 M_N	
Dimension:	
steering column diameter:	30 mm
length of measuring point:	25 mm
Permissible dimensions of the	
whole device: diameter:	50 mm
length:	60 mm
Characteristic / Tolerances:	as per Fig. 12

Table 1 - The technical requirements of the sensor

installation, the volume of the sensor is also restricted. The and electronics on the shaft are operated in a lubricant such as ATF fluid which is required for the servo gear mechanism. Because the sensor is a component of an extremely fast control system, a measuring time of 1 ms must not be exceeded. However, the most important requirement in a motor vehicle is low-cost large-scale manufacture. The system developers foresee a cost of less than DM 30 for a straightforward measurement system. The main technical requirements of the sensor are summarized in Table 1.

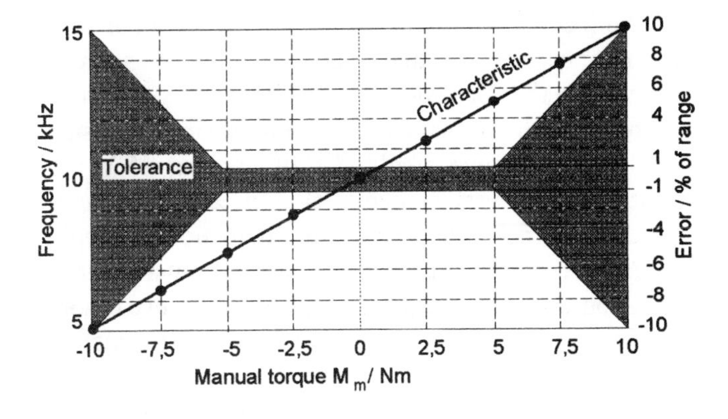

Fig. 12 - Characteristic curve and permissible deviation of the steering moment sensor

In view of the low volume required for the structure, methods based on torsion angle measurement appear to have their disadvantages. Of the stress-measuring, i.e. quasi-elemental, measuring methods, the strain gage method is by far the most accurate and reliable; in the form as illustrated in Fig. 10, it is the method that best satisfies the above requirements.

REALIZATION OF A FREQUENCY-ANALOG STRAIN GAGE TORQUE SENSOR WITH NON-CONTACT TRANSMISSION

Since the measuring principle selected is already known [2] and implemented in sophisticated measuring instruments, the main problem lays in finding ways and means of making the goal of extremely low-cost manufacture for use in the automobile seem attainable.

PRINCIPLE OF THE PRESS-FITTED / WELDED STRAIN GAGE TORQUE SENSOR, MEASUREMENT SHAFT - Manually affixed, conventional foil-type strain gages are not acceptable for use in the automobile. More rational methods of laminating on the strain gages and thin or thick film techniques, however, are only suitable for small components which have to be absolutely smooth for this type of attachment; i.e. the strain gages must not be mounted in a recess in the part. For this reason, there have in the past been problems in using such high-quality strain gages on larger machine parts which as a whole are not suited to the more rational attachment methods described above.

One way out of this dilemma is offered by press-fit and welded techniques used by us and illustrated in Fig. 13. A small conical or cylindrical specimen, so to speak, is cut from the large part, fitted with strain gages in the most convenient way and then re-inserted into the large machine part with as much adhesion and long-term stability as possible. Of course, the cutting of the bore and the manufacture of the inserted measuring section are two separate processes, but it is advisable to always use the same material as for the large part so as to obtain as homogeneous a bond as possible, free from harmful side-effects.

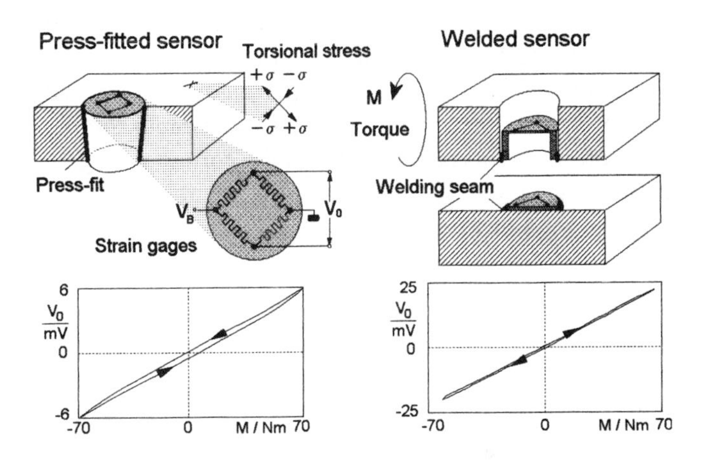

Fig. 13 - Principle and application of press-fitted and welded strain gages for torque measurement

Conical, usually solid, press-fitted parts have the advantage of being subsequently removable using suitable extractor tools, but even when pressed in as securely as possible, display a small hysteresis in their characteristic curve due to the residual micro-friction. Cylindrical welded sensors, partially hollow like pressure sensors and thus virtually pot-shaped, do not display this hysteresis, as is illustrated in the comparison in Fig. 13. Laser welded joints, if they are properly done as in the cross section in Fig. 14, possess adequate long-term stability.

If the form of the torsion element is level, a flatter sensor carrier can be welded onto the surface without the need for a bore. For one realization, sputtered thin film sensors with thin or thick film glass insulation were first considered; however, thin film sensors on laminated plastic support rods appear to be far more cost-effective.

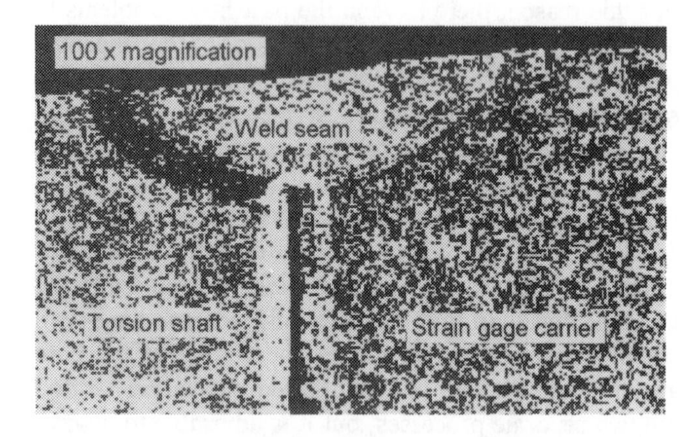

Fig. 14 - Welded connection between a potshaped strain gage carrier and the torsion shaft

Both rectangular and circular cross sections of measuring shaft were considered. In order to avoid having to provide an additional safeguard against overload for any weaker measuring point for the manual torque, the shaft was dimensioned uniformly corresponding to the greatest range of the total moment; i.e. the shaft was designed for a measuring range of 60 Nm:

Rectangular cross section:	*10 x 5.5 mm²*
Circular cross section:	*16 mm diameter*

In each case the transition to the full cross section of the steering shaft of 30 mm is tapered. In both the simulation and practical experiments, only the circular cross section has provided the required long-term stability.

SIGNAL AND POWER TRANSFER - As already illustrated in diagrammatic form in Fig. 10 and more accurately in Fig. 15 below, the rotary transformer for the transfer of signals and power is composed of two chambers each arranged concentrically around the shaft: A smaller chamber for signal transfer and a larger one for the power supply. A double-U profile is used so as to make both chambers as magnetically resistant as possible to mutual interference. Additional shielding is provided by an aluminium partition which, because of eddy current effects, has a screening, magnetically insulating effect.

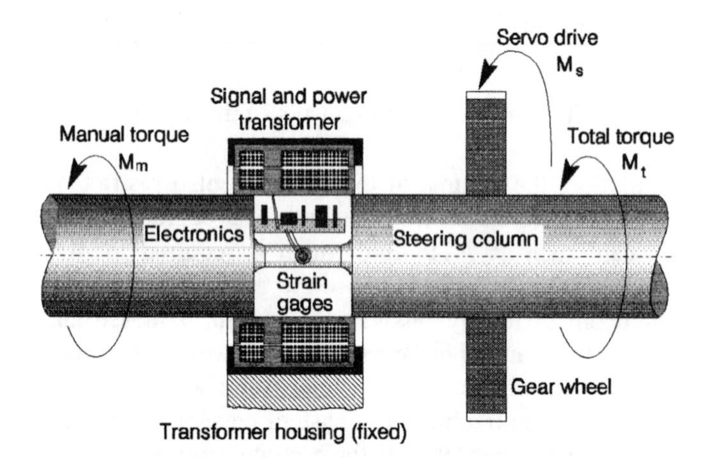

Fig. 15 - Construction of the sensor and rotary transformer

The fixed and rotating parts of the transformer core are concentrically arranged; the air gap of approx. 0.3 mm that this provides guarantees such good coupling between the primary and secondary sides that sufficiently high voltage amplitudes of approx. 15 V are available, even if very simple steel is used in the manufacture of the core components for both power supply and signal transfer. The efficiency of the transformer is 50 % at an operating frequency of 1 kHz. An even more cost-effective alternative is to manufacture the transformer core using plastic-bonded ferrite material (injection moulding). In fact, the higher permeability and lower eddy current losses with this material provide a considerable improvement in electrical efficiency.

SIGNAL ELECTRONICS - For a fully active power steering system, i.e. one that is effective at all road speeds, it is not just a matter of measuring the manual and total moments, but for safety reasons, each sensor should really have its own backup and the signals be independently amplified and transferred with two independent transformer windings according to the redundancy principle.

For the preliminary realization, however, operation was restricted as illustrated in Fig. 16 to the low speed range encountered during parking (parking aid), so that the system could be constructed with just one torque measuring point and no redundancy. This approaches very closely the prototype construction shown here which is built using mainly discrete components.

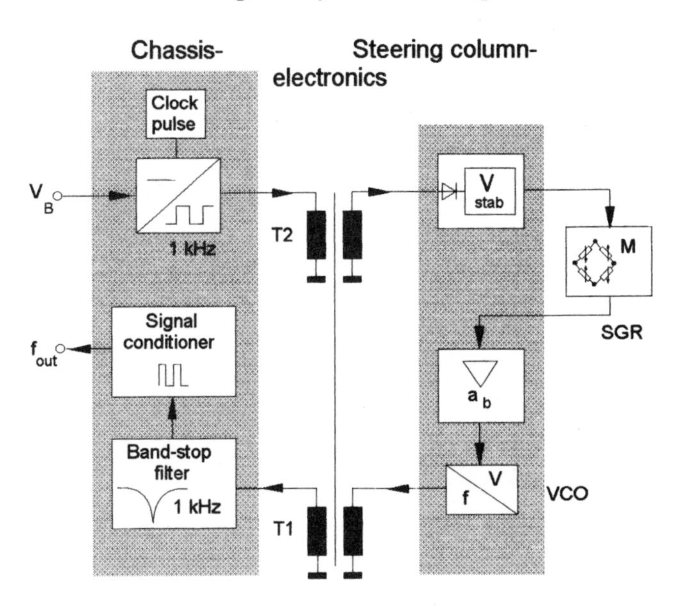

Fig. 16 - Block diagram of a measuring system for a simple parking aid

The entire electronic circuitry is divided into one part fixed in the chassis and one part that rotates on the steering column:

The steering column circuit amplifies the low output voltages from the sensor bridge (max. 20 mV) to a level of approx. 3 V, which is sufficient to actuate an integrated conventional voltage/frequency converter (VCO). It also contains an integrated voltage stabilizer which converts the rectified output voltage of the transformer into a highly stable d.c. voltage (15 V) for supplying the sensor bridge and the other circuit components (total current consumption: 17 mA). For series manufacture, this circuitry could be integrated into a single low-cost IC which could if necessary be encapsulated on the steering column together with the sensor system. However, it is also adequate to simply coat the chip with fluorsilicon gel.

The chassis circuit uses the simplest means available to generate a square-wave voltage at 1 kHz which supplies the primary winding of the power transformer. It is also advisable to incorporate the receiver circuit for the frequency-analog measuring signals which are passed through a 1 kHz band-stop filter prior to processing in order to eliminate the residual crosstalk from the supply transformer. This circuit need not be located directly beside the transformer, but could be accommodated at some distance where the operating conditions are more favorable or even in the corresponding system control unit.

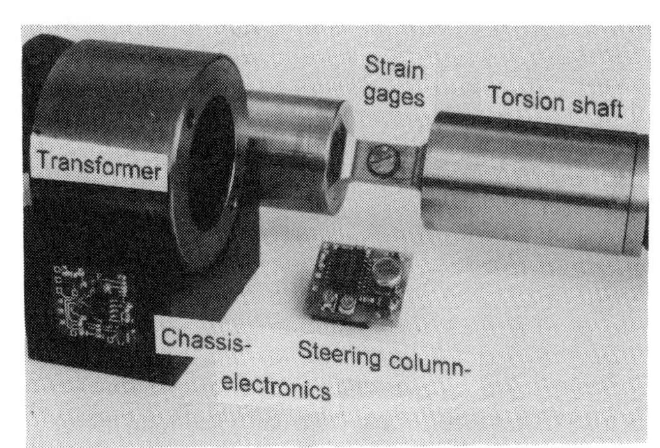

Fig. 17 - Components of the sensor system for the parking aid

Fig. 17 illustrates a disassembled sensor system for a measuring shaft with rectangular cross section. For round shafts, the discrete steering column circuitry was mounted on a flexible backing material which can be wrapped around the shaft to save space and can therefore be accommodated entirely inside the transformer.

MEASUREMENT RESULTS

The measuring electronics are designed and tuned so as to provide the desired characteristic curve illustrated in Fig. 12. The torque tapping of the torsion bars designed for a nominal moment of 60 Nm and possessing a cylindrical torsion rod (diameter 16 mm) displayed a measurement sensitivity of 0.3 mV/Nm and - according to Table 2 - only very minor error deviations, even referred to a measuring range of 10 Nm:

Offset drift:	0.01 %	of range / K
Temperature hysteresis:	0.10 %	of range
Load hysteresis:	0.13 %	of range

Table 2 - Sensor deviations

As illustrated in Fig. 18, the zero point temperature error of the amplifier circuit and downstream voltage/frequency converter amounted to only ± 0.5 % of range in the temperature range -20 ... +120 °C, and even less than ± 0.1 % of range between 0 °C and 120 °C. The same graph also shows the total error of the sensor and electronics.

As expected, the transfer of the frequency-analog signal caused no additional error. The variations due to the rotation, angular position and eccentricity of the steering column are also negligible.

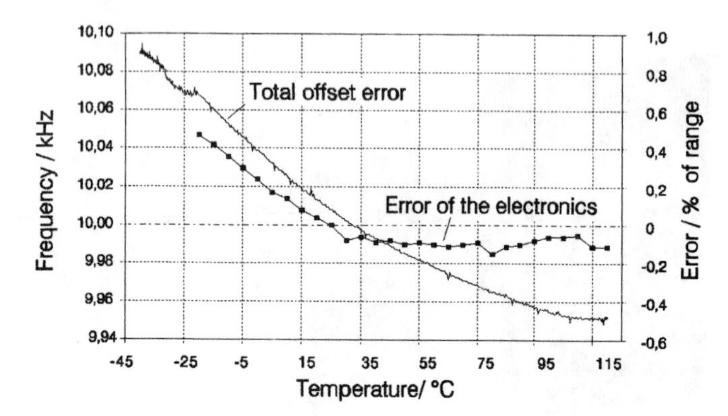

Fig. 18 - Offset error of the electronics and total offset error referred to the measuring range of 10 Nm

CONCLUSION

The method of equipping small, usually cylindrical press-fitted or weldable material specimens with highly reliable and accurate strain gages in a very low-cost process means that these measuring elements, which have already proven their measuring capability, can be used inexpensively for series implementation in larger force and torque-carrying mechanical components in the automobile, such as the steering column, axles, etc. The electronics can be easily integrated directly at the measuring point. In conjunction with the low-cost manufacture of the transformer this permits highly accurate, reliable and absolutely non-contact measurement of torques or other forces, even in rotating parts. The estimated costs lie in a range that automobile manufacturers will accept.

REFERENCES

[1] Dobler, K. et al.: "Drehmomentmessung mit Hilfe von Wirbelstromeffekten" (torque measurement with the aid of eddy current effects), Sensoren und Sensorsysteme Handbook for Engineers, 5th Edition, Expertverlag 1991

[2] Baldauf, W.: "Frequenzanaloge Drehmomentmessung mit Oberflächenwellen-Resonatoren" (frequency-analog torque measurement by surface-acoustic-wave resonators), Technisches Messen 58 (1991), Oldenburg Verlag

900264

Magnetostrictive Torque Sensors – Comparison of Branch, Cross, and Solenoidal Designs

William J. Fleming
Seat Restraints Research & Development
TRW Vehicle Safety Systems Inc.
Washington, MI

ABSTRACT

Intense worldwide activity is currently focused on development of magnetostrictive torque sensors. The sensors are both non-contact and provide high sensitivity in combination with robustness. They are therefore prime candidates for use in torque-feedback closed-loop controls of automotive engines and transmissions.

Previously, both linear and nonlinear analyses of the branch-design of magnetostrictive torque sensors were given. This paper goes beyond prior work to include cross-design and solenoidal-design sensors (designs that are more commonly used).

For each sensor design: general models are derived, equivalent electrical and magnetic circuits are developed, and equations governing signal outputs are given. A comparison is done of magnetic circuit operating behavior for sensors designed to fit into the same space on a shaft made of maraging steel.

MAGNETOSTRICTIVE TORQUE SENSORS have been the object of extensive development over the past few years. For example, Table 1 of a 1989 paper [1]* listed 44 published references on worldwide developments of magnetostrictive sensors appearing during the years, 1982 thru 1988.

An additional thirty magnetostrictive sensor publications have appeared this past year. These new publications provide an update to Table 1 of the 1989 paper [1], and are listed in Table 1 of the present paper. The publications/patents cited in these tables demonstrate that intense sensor development activity is currently taking place in North America, Japan, and Europe.

* Numbers in brackets designate references at end of paper.

Table 1
1989-1990 Worldwide Magnetostrictive Sensor Development Activity*

Corporation	Development Activity Level	Refs.
North America:		
-Mag Dev Inc./Kubota Ltd.	high	[4-6]
-United Technologies/ Spectrol Electronics	medium	[7]
-Purdue University	low	[8]
-U.S. Navy	low	[9]
Japan:		
-Nissan Motor/Daido Steel	high	[10-13]
-Toyota Motor/Toshiba	high	[14-16]
-Matsushita Electric/ Panasonic	high	[17-19]
-Kubota Ltd./Mag Dev	high	[5,20]
-Honda Motor	medium	[21,22]
-Kyushu University	high	[23,24]
Europe:		
-ASEA/Brown Boveri	high	[25-28]
-Daimler-Benz Motor	low	[29]
-Sensglas Terfenol/Edge Technologies	medium	[30]
-Delft University	low	[31]
-Przemyslowy Inst. Auto.	low	[32]
-Academy Sciences, USSR	medium	[33]

* These are new publications, appearing this past year

There are many applications where torque measurement greatly benefits vehicular performance, and these therefore are the reason for the current level of intense sensor development activity. Applications such as: passenger car powertrains, agricultural tractors, truck/off-highway vehicles, stationary powerplants, and manufacturing machines were previously outlined in Table 2 of Ref. [1].

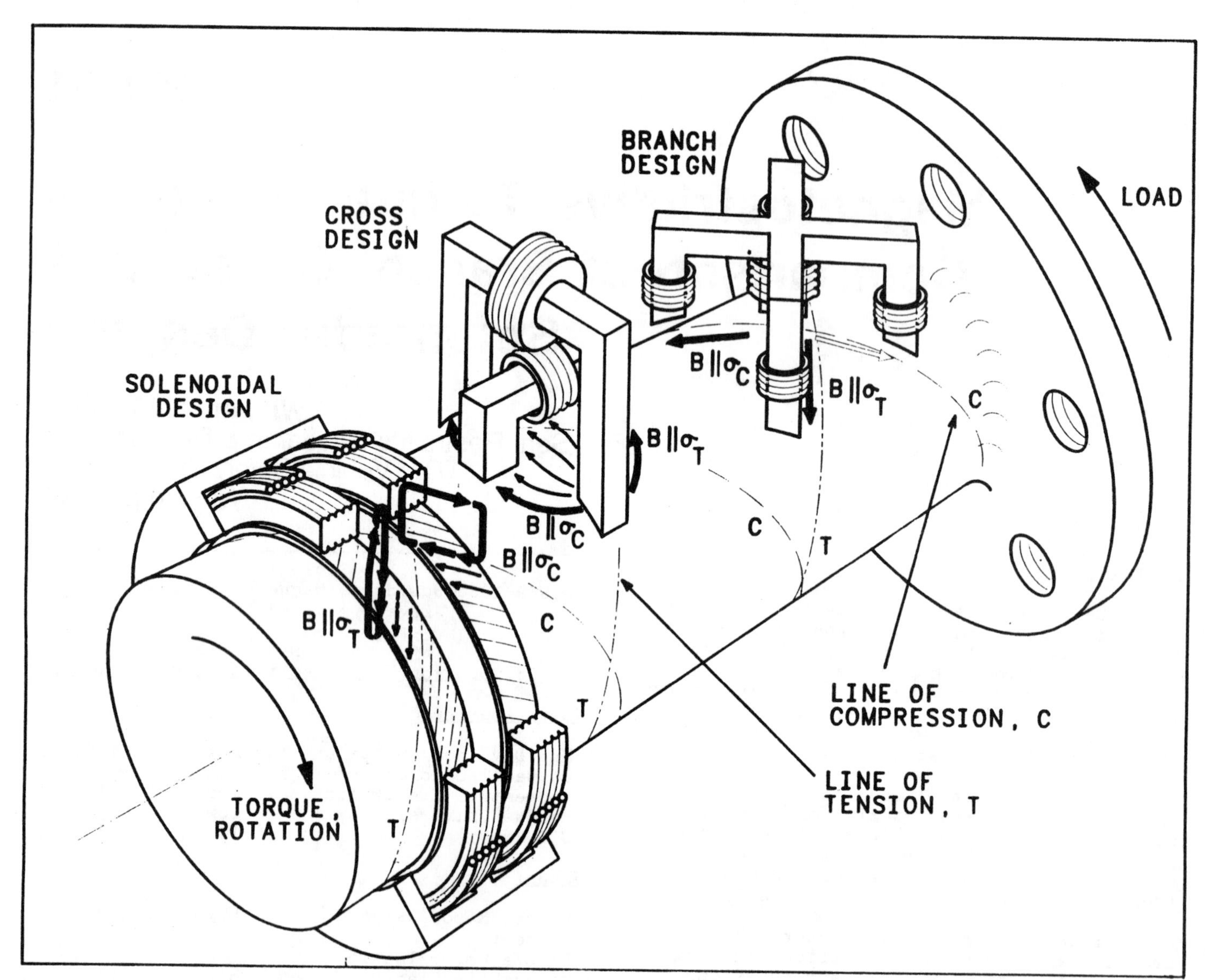

Fig. 1. Branch, Cross, and Solenoidal Design Torque Sensors -- Sensing Torque In the Same Shaft.

Magnetostrictive torque sensors are used in these applications because they feature: noncontact measurement, rapid response, stability, good accuracy, and high sensitivity in combination with capacity to withstand overloading.

A general, linear, model for magnetostrictive torque sensors was previously derived [1,2]. Subsequently, that model was extended to include the nonlinear effects of magnetic saturation in the sensing member material [3].

BRANCH, CROSS, AND SOLENOIDAL SENSOR DESIGNS

Prior analyses [1-3] focused on development of a single-branch sensor element that was a fundamental building block for branch-design sensors. This paper goes beyond the prior work and includes analyses of both cross-design and solenoidal-design sensors (designs that are more commonly used). All three designs, sensing torque in the same shaft, are shown schematically in Fig. 1.

In the present paper, equivalent electrical and magnetic circuits are developed, equations governing signal outputs of each sensor design are given, and calculations are made to compare magnetic circuit operating behavior for sensors designed to fit into the same space on a shaft made of maraging steel.

SENSOR MODEL DEVELOPMENT

Analytical models for each of the three major types of sensor design -- branch, cross, and solenoidal designs shown in Fig.1 -- are derived below.

BRANCH DESIGN -- It was shown in Appendix A of Ref. [1] that: "output signals of two-branch, four-branch, and multi-branch magnetostrictive sensors are given in terms of output signals corresponding to a single-branch sensor element."

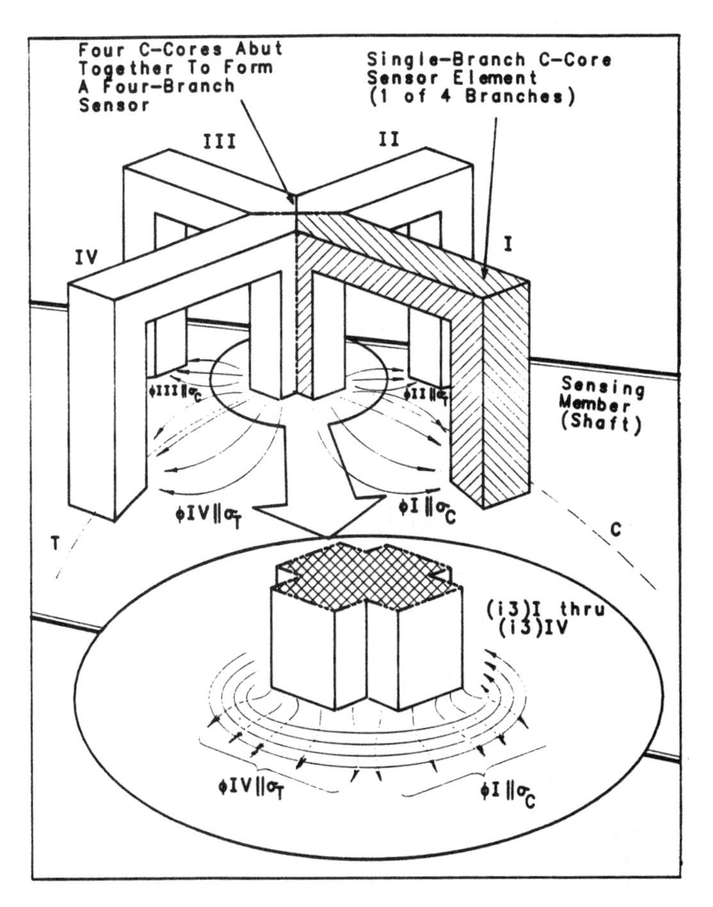

Fig. 2 Multi-Branch Sensor Made Up of Individual Single-Branch Sensor Elements, Showing Division of Magnetic Flux Flow Φ, And Detail of Associated Eddy Current Flow i_3.

As shown in Fig. 2, the multi-branch sensor (in this case a four-branch sensor) has its branches joined in parallel to a common center (excitation) pole. The single-branch sensing elements consist of separate "C-core" -shaped segments, abutted and joined together in Fig. 2 to form the center pole of the multi-branch sensor.

A single-branch sensor element, highlighted in Fig. 2, consists of a single C-core element, scaled appropriately to represent the pole it represents in the four-branch sensor.

Also shown in Fig. 2 are the approximate flow lines of magnetic flux Φ induced in the sensing member (shaft) by the sensor. Flux flowing into the shaft splits into four nearly equal parts while passing over to the corner detection poles. The flow paths of flux, associated with each of the four branches, are aligned predominately parallel along principle lines of stress in the shaft (due to applied torque).

Fluxes for each path are identified in Fig. 2 as components: $\Phi_{I}\|\sigma_C$, $\Phi_{II}\|\sigma_T$, $\Phi_{III}\|\sigma_C$, and $\Phi_{IV}\|\sigma_T$; where the subscripts I thru IV correspond to branches (quadrants) I thru IV of the sensor. Stresses σ_C and σ_T correspond, respectively, to principle components of compressive and tensile stress in the shaft that are due to applied torque.

In the inset view of Fig. 2, detail is shown of how two of the flux components, $\Phi_{I}\|\sigma_C$, and $\Phi_{IV}\|\sigma_T$, flow away from the excitation pole. The inset also shows how eddy currents associated with these fluxes, $(i_3)_I$ and $(i_3)_{IV}$, flow in accordance with Lenz's Law concentrically around the excitation pole in counter-clockwise directions so as to create flux that opposes the generated flux components.

For clarity, other flux components, $\Phi_{II}\|\sigma_T$ and $\Phi_{III}\|\sigma_C$; and eddy currents, $(i_3)_{II}$ and $(i_3)_{III}$, were omitted in the inset of Fig. 2. Note: Physical processes depicted in Fig. 2 were previously shown in Figs. 4, C.3, and C.4 of Ref. [1].

Each one of the four branches in Fig. 2 is representable by the single-branch C-core element shown in Fig. 3. The C-core element is located in proximity to a magnetostrictive sensing member. It is separated from the sensing member by an air gap of typical size of 0.5-to-1.0 mm. This arrangement permits magnetic flux to be coupled into a sensing member in a noncontact manner.

In prior work [1-3], branch-design sensors were analyzed in terms of transfer impedance Z, which is the ratio of sensor output voltage to input current.

When sinusoidal fixed-frequency excitation is assumed (as was done in Ref. [1]), and rms values of voltage and current are used (as denoted by capital letter symbols); transfer impedance is defined as:

$$Z := \frac{V_2}{I_1}, \qquad (1)$$

where V_2 is the detection-coil output voltage and I_1 is the excitation-coil input current, both shown in Fig. 3.

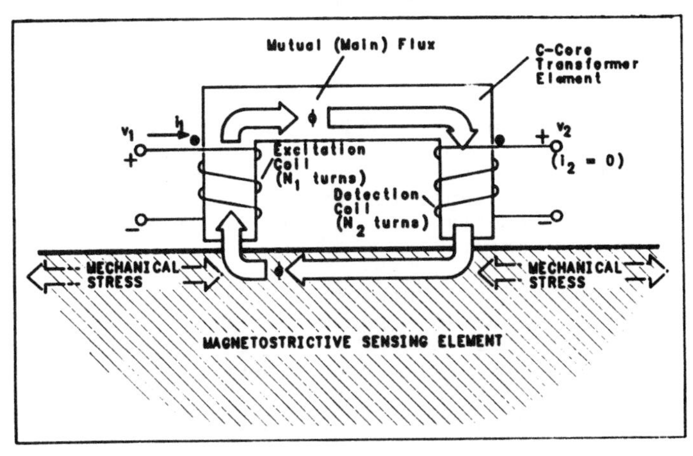

Fig. 3 Simplified Representation of a Magnetostrictive Single-Branch Sensor (Cross Sectional View).

Recall that voltage V_2, when coil 2 terminals are open circuit, is defined by Faraday's Law given in Eqs. (2) and (5) of Ref. [1], and is given as follows:

$$V_2 := j\omega \cdot N_2 \cdot \left[\Phi + \Phi_{12}\right] , \qquad (2)$$

where $\omega = 2\pi f$ is the radian excitation frequency, N_2 is the number of turns in the detection coil, Φ is the main flux (shown in Fig. 3), and Φ_{12} is leakage flux crossing between C-core pole pieces (shown in Fig. 4 below).

Transfer Permeance -- In the present paper, it is convenient to analyze sensors in terms of transfer permeance P_Φ, which is defined as the ratio of output main flux Φ to input magnetomotive force $N_1 \cdot I_1$. Thus, transfer permeance is given as::

$$P_\Phi := \frac{\Phi}{N_1 \cdot I_1} , \qquad (3)$$

where Φ is the main flux shown in Fig. 3 (the flux that passes through the subsurface of the sensing-member), and $N_1 \cdot I_1$ is the input magnetomotive force of the excitation coil (number of turns N_1 times excitation current I_1).

When Eqs. (1) thru (3) are combined, the following result is obtained:

$$Z := j\omega \cdot N_1 \cdot N_2 \cdot \left[P_\Phi + \frac{P_{12}}{3}\right] , \qquad (4)$$

where transfer permeance P_Φ is given by Eq. (3), and cross-leakage permeance P_{12} is given by Eq. (C.7) of Appendix C in Ref. [1]. The complete equivalent magnetic circuit for the single-branch sensor, and its simplified form containing the two permeance elements P_Φ and P_{12} of Eq. (4), are shown in insets (a) and (b) of Fig. 4.

Multiplication of P_{12} in both Eq. (4), and also shown in Fig. 4, by the factor 1/3 accounts for the distributed nature of leakage flux along the lengths of adjacent C-core poles. The factor of 1/3 is appropriate because only one of the poles includes an excitation coil [1].

Also note that the use of the transfer permeance variable P_Φ in Fig. 4(b) has replaced four permeances and the magnetomotive force generator $N_3 \cdot i_3$ in Fig. 4(a), therein markedly simplifying the equivalent magnetic circuit.

Equation (4) shows that transfer permeance P_Φ is a basic parameter that determines the influence of the sensing member on the electrically measured impedance Z and output voltage V_2 of the sensor. In Appendix A of this paper, it is shown that P_Φ is given as follows:

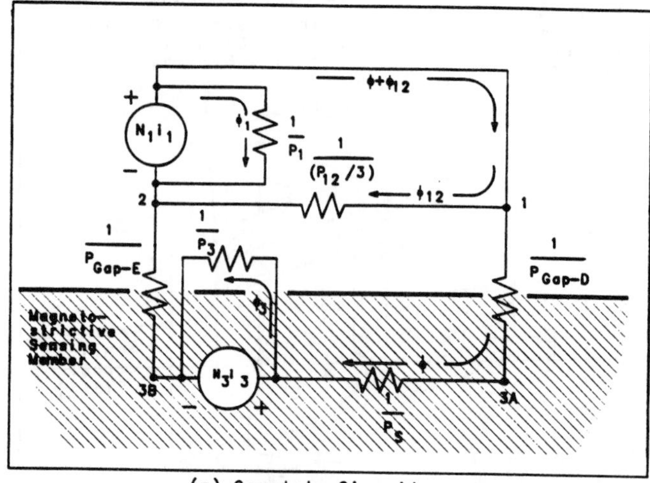

(a) Complete Circuit

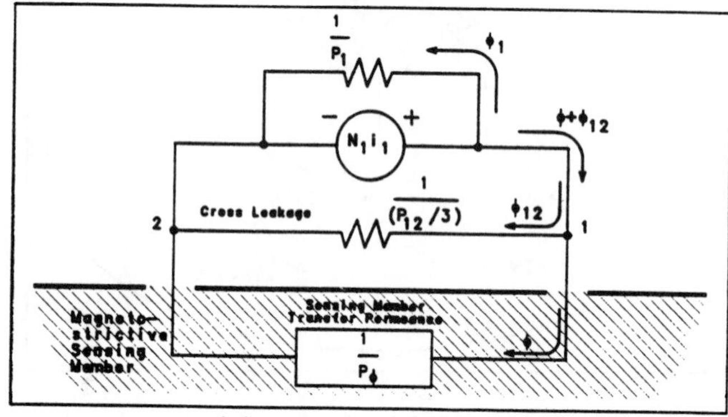

(b) Simplified Circuit

Fig. 4 Equivalent Magnetic Circuits For the Single-Branch Sensor -- (a) Complete Circuit (taken from Fig. 6 of Ref. [1]), and (b) Simplified Circuit Utilizing Transfer Permeance P_Φ Defined in Eq. (3). Here:
- i = instantaneous value of current
- Φ = flux
- P = Permeance
- N = coil turns
- D = Detection Pole
- E = Excitation Pole.

$$P_\Phi := \frac{1}{\left[\frac{1}{P_{GAPS}}\right] + \left[\frac{1}{P_{S\Phi}}\right]} , \qquad (5)$$

where:

$$P_{GAPS} := \frac{1}{\left[\frac{1}{P_{GE}}\right] + \left[\frac{1}{P_{GD}}\right]} , \qquad (6)$$

is the series combination of air gap permeances P_{GE} and P_{GD} associated with the excitation-pole E and the detection-pole D, and:

$$P_{S\Phi} := P_S \left[1 - \alpha_S\right] \quad , \tag{7}$$

where

$$\alpha_S := \frac{1}{1 + \dfrac{P_3}{P_S}\left[1 - \dfrac{j}{\epsilon}\right]} \quad , \tag{8}$$

and $P_{S\Phi}$ is an effective value of sensing-member permeance that includes both the effects of change of permeability due to applied stress [i.e., the magnetostrictive effect used to sense applied torque], and the effects of eddy-current-induced opposing flux (Lenz's Law).

Note: Appendix A shows that both P_Φ and $P_{S\Phi}$ in Eqs. (5) and (8) include effects of eddy current loss. Consequently, both quantities have real and imaginary parts which are introduced via the operator $j = \sqrt{-1}$ in Eq. (8).

Specifically: P_S and P_3 in Eq. (8) are the sensing-member permeance and the Lenz's Law opposing-flux permeance, j equals $\sqrt{-1}$, and ϵ is a field strength parameter defined by Eq. (16) of Ref. [1] and described in Appendix B. Air gap permeance P_{GAPS} in Eq. (6) is computed using procedures given by Eqs. (C.1) thru (C.6) of Ref. [1]

Although it is not very obvious, it can be shown that when Eqs. (4) thru (8) are combined, exactly the same expression for transfer impedance, as previously given by Eqs. (17) thru (25) in Ref. [1], is also obtained here. This is true because of the identity relationship given by Eq. (A.10) in Appendix A.

Significance of Transfer Permeance Analysis -- Transfer permeance P_Φ is defined by Eqs. (5) thru (8), as derived above. In the present study, where three different sensor designs are considered, it is simpler to analyze sensors in terms of transfer permeance. This is the reason therefore that the transfer permeance approach is used instead of the transfer impedance approach which was previously used in Refs. [1-3].

Nonlinear Analysis -- Effects due to nonlinear properties of magnetic materials associated with sensing members were previously analyzed [3]. The method of solution for the nonlinear problem is described in Appendix B.

Sensor Output Signal -- Sensor output signal can be determined in a straightforward manner. For example, consider signal detection method A (constant mmf control, using constant current excitation), which was previously analyzed in Appendix A of Ref. [2].

When method A is used, normalized signal (defined as the signal-to-operating-point ratio) s_A is, from Eq. (A.19) of Ref. [2], given as:

$$s_A := \frac{V'_2 - V_2}{V_2} := \frac{\Delta |Z|}{|Z|} \quad , \tag{9}$$

where V'_2 is the value of detection coil output voltage when stress is applied, V_2 is the voltage when there is no applied stress, and $\Delta|Z|$ is the change in the magnitude of the transfer impedance Z created by applied stress (due to torque or force).

Note: in the derivation of Eq. (9), a full-wave bridge detector circuit was assumed. Thus, Eq. (9) only provides signal magnitude, and does not utilize phase information that is also available [3].

In Eq. (4), the transfer impedance Z equation above, at a fixed operating condition, all factors are constant except the permeance terms. Thus, if Eq. (4) is substituted into Eq. (9), one obtains:

$$s_A := \frac{\Delta \left| P_\Phi + \dfrac{P_{12}}{3} \right|}{\left| P_\Phi + \dfrac{P_{12}}{3} \right|} \quad , \tag{10}$$

and, when Eq. (3) above and Eq. (10) of Ref. [1] are used, Eq. (10) above reduces to:

$$s_A := \frac{\Delta \left| \Phi + \Phi_{12} \right|}{\left| \Phi + \Phi_{12} \right|} \quad , \tag{11}$$

Since cross-leakage terms P_{12} and Φ_{12} do not change when stress is applied, Eqs. (10) and (11) show that cross leakage permeance has an attenuating effect, and it diminishes the output signal of the sensor (as described previously in Ref. [2]).

On the other hand, as seen from Eqs. (10) and (11) above, and for practical sensors with small values of P_{12}, transfer permeance P_Φ basically governs the output signal of the sensor. Transfer permeance P_Φ therefore plays an important role in analyzing performance of sensors, and for this reason is the object of discussion in the present work.

This approach will be used in the following engineering analyses of branch, cross, and solenoidal design sensors. Branch sensors will be described first.

Multi-Branch Sensors -- All equations up to this point apply for the basic single-branch sensor element shown in Fig. 3. However, it was previously shown in Appendix A of Ref. [1] that output voltage V_{2m} for detection coil "m" is given as:

$$V_{2m} := I_1 \cdot Z_m \quad , \tag{12}$$

where I_1 is the center-pole excitation current (common to all branches) and Z_m is the transfer impedance of the m-th branch in the multi-branch sensor. For example, index "m" goes from I thru IV for the four-branch sensor shown in Figs. 1 and 2.

In the present paper, Z_m is given by Eq. (4). Permeance parameters for multi-branch elements, shown in Fig. 2, are scaled appropriately to represent single-branch sensor elements corresponding to individual branches of the multi-branch sensor. Equation (4) is evaluated using the appropriately scaled parameters to determine Z_m for branch "m."

If the flux Φ passing through a given branch is wanted, it is obtained by substituting, separately into Eqs. (3) and (5) thru (8); permeance parameters associated with each single-branch element for each of the four branches.

The equivalent magnetic circuit for the four-branch sensor, and its simplified form corresponding to use of transfer permeance, as defined by Eqs. (3) and (5) thru (8), are shown in Fig. 5.

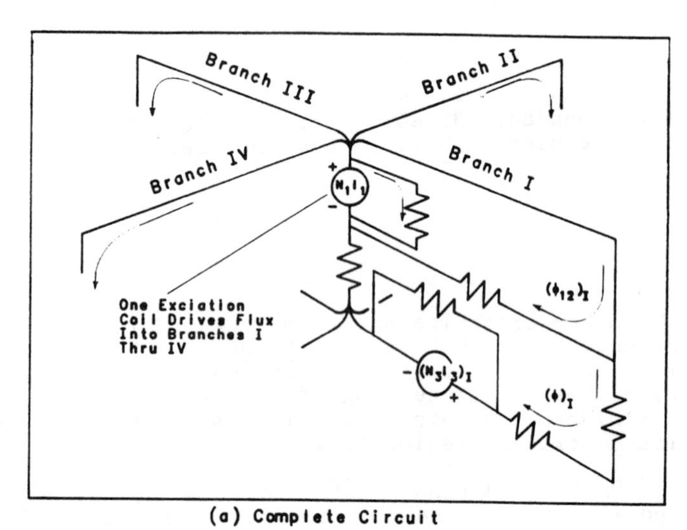

(a) Complete Circuit

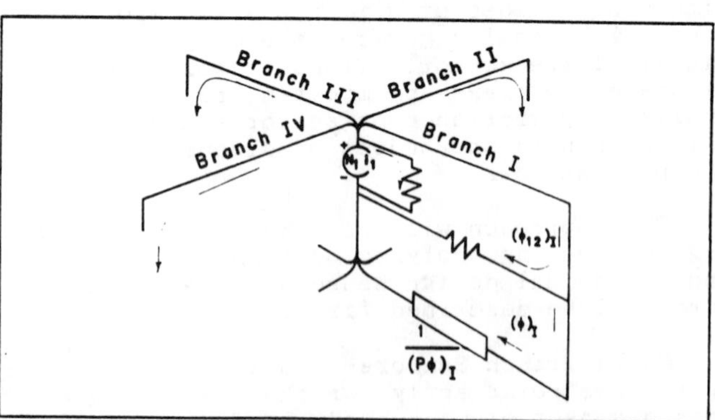

(b) Simplified Circuit

Fig. 5 Equivalent Magnetic Circuits For A Four-Branch Sensor -- (a) Complete Circuit (taken from Fig. A.2 of Ref. [1]), and (b) Simplified Circuit Using Transfer Permeances $P\Phi$.

CROSS DESIGN -- Shown in Fig. 1, the cross design sensor, also called a Torductor(R), was first developed by Dahle of the Swedish ASEA company -- see Refs. 20-22 of Ref. [34].

A cross design sensor is described in Ref. 11 of the present paper. Previously, there was considerably more development of this design. For example, Refs. 2, 9-11, 20-23, 26, 37, 45 and 48 in Ref. [1] all describe cross sensors developed from 1982 to 1988; whereas Refs. 11, 13, 14, and 20-22 in Ref. [34] describe developments prior to 1982.

The cross sensor is shown in more detail in Fig. 6. Only the excitation coil and associated C-core pole piece are shown. The excitation pole directs flux across an air gap and into the shaft (sensing member) where it spreads out and then converges to return via the opposite pole of the C-core.

Flux flow between poles of a C-core was previously modeled for the branch design sensor using results of magnetostatic field theory. The flux-flow field, shown in Fig. C.3 of Ref. [1], is the same as that shown in Fig. 6 here. For analysis purposes, this

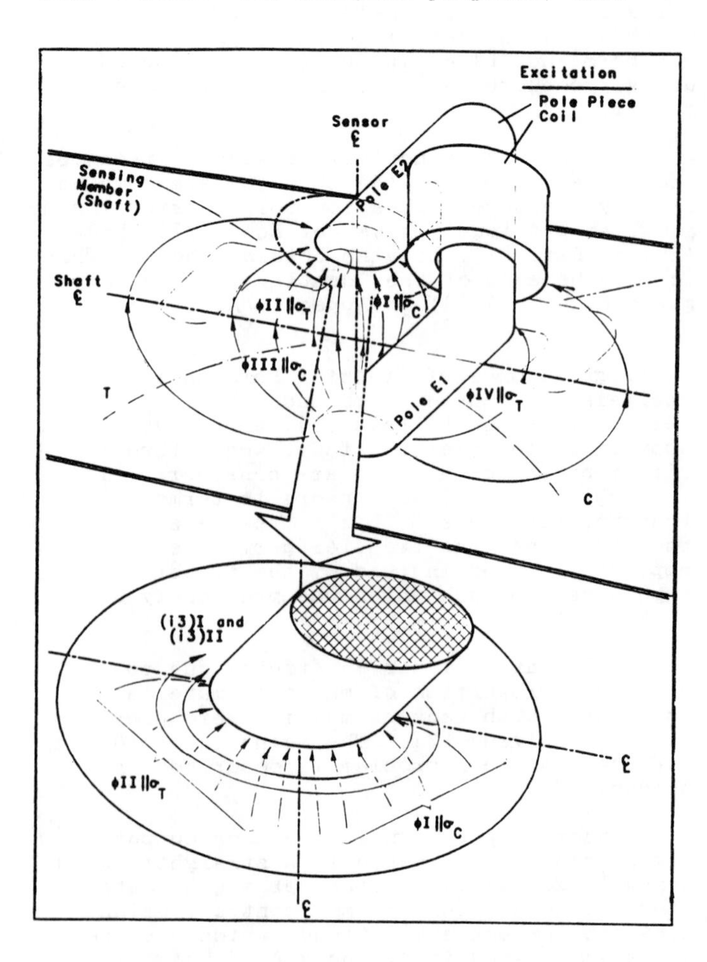

Fig. 6 Cross-Design Sensor With Division of Magnetic Flux Flow Into Quadrants Defined By Sensor and Shaft Centerlines.

flux field is subdivided into four quadrants, I thru IV, as identified in Fig. 6. The quadrants are delineated by the intersection of the symmetry centerlines for the sensor and the shaft, as seen in Fig. 6.

Principle lines of stress due to applied torque are also shown in Fig. 6. Note that flux in quadrant I (in the upper right-hand corner of the top illustration) flows predominately parallel to compressive lines of stress. Similarly, fluxes in quadrants II, III, and IV flow, respectively, predominately parallel to: tensile, compressive, and tensile lines of stress.

Fluxes for each path are identified in Fig. 6 as components: $\Phi_I \| \sigma_C$, $\Phi_{II} \| \sigma_T$, $\Phi_{III} \| \sigma_C$, and $\Phi_{IV} \| \sigma_T$; where the subscripts I thru IV correspond to quadrants I thru IV. Stresses σ_C and σ_T correspond, respectively, to principle components of compressive and tensile stress in the shaft.

In the inset of Fig. 6, detail is shown of how two of the flux components, $\Phi_I \| \sigma_C$, and $\Phi_{II} \| \sigma_T$, flow towards the excitation pole, and how eddy currents associated with these fluxes, $(i_3)_I$ and $(i_3)_{II}$, flow in accordance with Lenz's Law concentrically around the excitation pole in clockwise directions so as to create flux that opposes the generated flux. (Physical processes depicted in Fig. 6 were previously described in conjunction with Figs. 4, C.3, and C.4 of Ref. [1]).

Equivalent Magnetic Circuit -- In Figs. 3 and 4 above, it was shown that a single-branch C-core sensor element could be described by the simplified equivalent magnetic circuit shown in Fig. 4(b). However, for the branch sensor, each flux was assumed to flow predominately parallel to principle lines of stress.

The situation is much different for the cross sensor's C-core element, shown in Fig. 6. As already discussed, in two quadrants flux flows are predominately parallel to lines of compressive stress and in the other two quadrants, flows are along lines of tensile stress. For this reason, the transfer permeance of the cross sensor must be decomposed into four parts, with one part for each of the four quadrants shown in Fig. 6.

Total transfer permeance is still defined by Eqs. (3) and (5) thru (8) above, but it is split into four parts in accordance with the equivalent magnetic circuit shown in Fig. 7. Hence, transfer permeance components: $(P\Phi)_I$, $(P\Phi)_{II}$, $(P\Phi)_{III}$, and $(P\Phi)_{IV}$ in Fig. 7 will combine to equal $P\Phi$, as defined by Eqs. (3) and (5) thru (8). Note: the circuit in Fig. 7 is essentially the same as the one shown in Fig. 5 of Dahle's 1960 paper, Ref. [35].

[Note: one might argue that a shunt permeance element, unaffected by stress, should be added in Fig. 7 to account for flux passing directly along the sensor centerline axis. Along this centerline, effects due to compressive and tensile stresses would have compensating effects and would cancel. Addition of such a shunt permeance to make a more exact equivalent circuit for the cross sensor is, however, unwarranted for two reasons. First, evaluation of the shunt permeance parameter itself would be very difficult. And secondly, a more exact model of this aspect of the sensor would not be consistent with the many approximations already made in this analysis [1-3]].

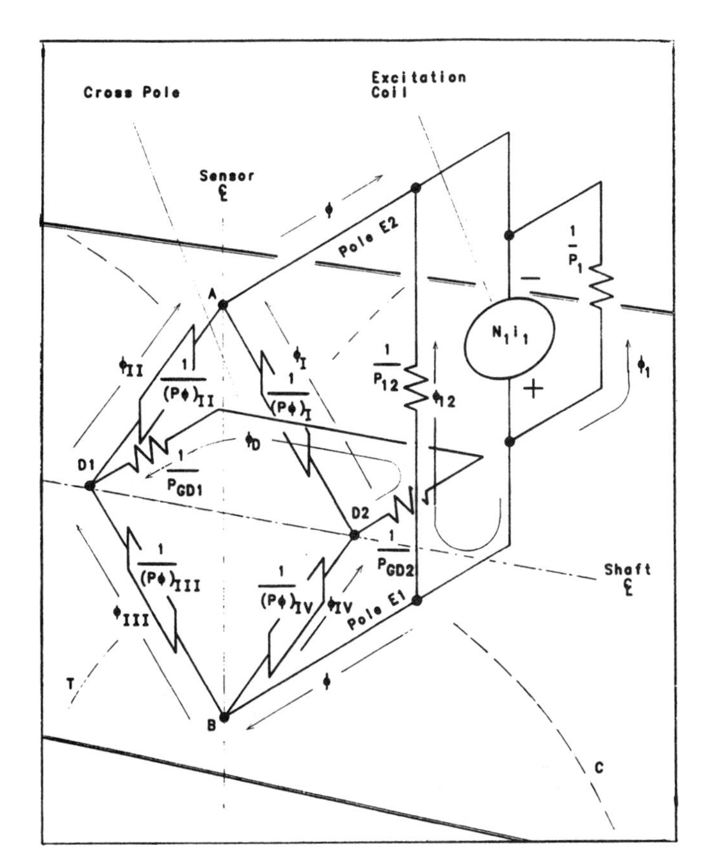

Fig. 7 Equivalent Magnetic Circuit For the Cross-Design Sensor -- Derived From the Simplified Circuit Shown in Fig. 4(b).

In Fig. 7, flux Φ_D through the cross-pole is shown in the sensor magnetic circuit. The cross pole was previously shown in Fig. 1. It detects magnetic scalar potential-difference flux Φ_D. In the absence of applied torque, and for symmetric sensor geometry and uniform (homogeneous) sensing-member material properties, the four components of transfer permeance $P\Phi$ in Fig. 7 are equal and no flux Φ_D flows in the cross pole.

This ideal condition of balance will be designated as the "symmetric-uniform-unstressed" condition, or simply as the "SUU-condition." For this condition, fluxes flowing in each quadrant of the cross sensor -- Φ_I, Φ_{II}, Φ_{III}, and Φ_{IV} -- are all equal.

However, when torque is applied, and for a positive-magnetostrictive sensing material; transfer permeances $(P\Phi)_I$ and $(P\Phi)_{III}$ in the compressive-stress quadrants decrease (reluctance values increase), whereas transfer permeances $(P\Phi)_{II}$ and $(P\Phi)_{IV}$ in the tensile-stress quadrants increase (reluctance values decrease). As a consequence, an unbalance of the four-element magnetic-bridge permeance circuit of Fig. 7 takes place. The SUU-condition no longer applies, and flux Φ_D therefore flows through the cross pole.

Permeances P_1, P_{12}, P_{GD1}, and P_{GD2} in Fig. 7 represent, respectively, permeances associated with: self-leakage flux Φ_1, cross-leakage flux Φ_{12}, and detected flux Φ_D. Flux Φ_D crosses air gaps from the shaft to the detection poles D1 and D2. Note: cross leakage permeance P_{12} is not multiplied by the factor 1/3 in Fig. 7, as it was in Fig. 4. This is because there are no excitation coils on the pole-piece legs of the cross sensor and no magnetomotive force (mmf) gradient exists [36].

Transfer permeances $(P\Phi)_I$ thru $(P\Phi)_{IV}$ are defined by Eq. (5) above, but with the following changes required to match the cross sensor geometry.

Air gap permeance P_{GAPS} in Eq. (5) is set equal to one-half the total gap permeance of its associated excitation pole, and all detection-pole gap permeance is included in the elements P_{GD1} and P_{GD2} seen in Fig. 7. For example, when evaluating permeances $(P\Phi)_I$ thru $(P\Phi)_{IV}$; gap permeances in Eq. (5) are given, respectively, as:

$$\left[P_{GAPS}\right]_I := \frac{P_{GE2}}{2}, \quad (13)$$

$$\left[P_{GAPS}\right]_{II} := \frac{P_{GE2}}{2}, \quad (14)$$

$$\left[P_{GAPS}\right]_{III} := \frac{P_{GE1}}{2}, \quad (15)$$

and

$$\left[P_{GAPS}\right]_{IV} := \frac{P_{GE1}}{2}, \quad (16)$$

where poles E1 and E2 are identified in Fig. 7, and gap permeances P_{GE1} and P_{GE2} are computed using procedures given by Eqs. (C.1) thru (C.6) of Ref. [1].

Gap permeances P_{GD1} and P_{GD2} are similarly computed using procedures given by Eqs. (C.1) thru (C.6) of Ref. [1].

Under SUU-conditions, total sensing-member permeance P_S and Lenz's Law opposing-flux permeance P_3 are both measured between excitation poles E1 and E2. The total values of these parameters are divided into four components, corresponding to the four transfer-permeance elements shown in Fig. 7. The components are defined such that, when combined back into a single equivalent value, the same values of P_S and P_3 for permeance between poles E1 and E2 are obtained as when they are split into four components.

Because sensing-member permeance P_S and the Lenz's Law opposing-flux permeance P_3 follow common flux paths to those for the transfer permeance $P\Phi$, they can be subdivided the same way as was done for transfer permeance.

As seen in Figs. 6 and 7, transfer permeances in quadrants I and IV, and those in quadrants II and III, are both connected in series. The series combinations are then connected in parallel, therein forming a magnetic analog of a Wheatstone bridge.

It is easily shown that for a balanced bridge, under SUU-conditions, components of permeance in the bridge will have individual values identical in value to that of the total value measured at the excitation terminals (poles E1 and E2) of the bridge circuit.

This is true because the series combinations of component I plus component IV, and components II and III, will be both double the total value. And since the series combinations are then connected in parallel, the resultant total value becomes exactly the same as the individual component values.

For example, under SUU-conditions, sensing-member permeances P_S and the Lenz's Law opposing-flux permeances P_3 -- used in Eq. (8) to evaluate transfer permeances $(P\Phi)_I$ thru $(P\Phi)_{IV}$ in Fig. 7 -- are given as follows:

$$\left[P_S\right]_I := \left[P_S\right]_{II} := \left[P_S\right]_{III} := \left[P_S\right]_{IV} := P_S, \quad (17)$$

and

$$\left[P_3\right]_I := \left[P_3\right]_{II} := \left[P_3\right]_{III} := \left[P_3\right]_{IV} := P_3, \quad (18)$$

where P_S and P_3 are, respectively, the sensing-member permeance and the Lenz's-Law opposing-flux permeance for the total (undivided) sensor. Values of both P_S and P_3 are computed using procedures given by Eqs. (C.10) thru (C.22) of Ref. [1].

Thus, even though the magnetic circuit for the sensing-member transfer permeance of the cross design is segmented into four parts -- $(P\Phi)_I$ thru $(P\Phi)_{IV}$ in Fig. 7 -- these parts, when combined in accordance with Eqs. (13) thru (18), will reduce to the single-element $P\Phi$ transfer permeance shown in Fig. 4(b).

To summarize, the reason transfer permeance is split (segmented) into four parts is the following. As illustrated in Fig. 6, flux flows predominately parallel to lines of compressive stress in two quadrants, and it flows along lines of tensile stress in the other two quadrants. Consequently, to depict the opposing magnetostrictive effects of compressive and tensile stresses, transfer permeance is segmented into four parts using the method described by Eqs. (13) thru (18).

If torque is applied to the shaft, and for a positive-magnetostrictive sensing material; permeances $(P_S)_I$, $(P_3)_I$, $(P_S)_{III}$, and $(P_3)_{III}$ in compressive-stress quadrants decrease, whereas permeances $(P_S)_{II}$, $(P_3)_{II}$, $(P_S)_{IV}$, and $(P_3)_{IV}$ in tensile-stress quadrants increase. Equations (5) thru (8) are used to compute transfer permeance, and the following conditions hold:

$$\left[P_{\Phi}\right]_{I} < \left[P_{\Phi}\right]_{II} , \qquad (19)$$

$$\left[P_{\Phi}\right]_{I} := \left[P_{\Phi}\right]_{III} , \qquad (20)$$

and

$$\left[P_{\Phi}\right]_{II} := \left[P_{\Phi}\right]_{IV} , \qquad (21)$$

Hence, due to applied torque, an unbalance in the transfer permeance components -- defined by Eqs. (19) thru (21) -- of the magnetic circuit bridge occurs. This means that detected (output) flux Φ_D flows in the permeance bridge circuit of Fig. 7.

Magnetic Circuit Solution -- An exact solution of the magnetic circuit for the cross sensor, shown in Fig. 7, is difficult to obtain. For example, the circuit in Fig. 7 has three independent mmf loops which lead to three independent equations that must be solved simultaneously [37]. Closed-form symbolic solution of the circuit is difficult. Moreover, exact circuit solution is again unwarranted considering the many approximations already made in this analysis [1-3].

A reasonable assumption is that the detected flux Φ_D in Fig. 7 is negligibly small compared to other fluxes Φ_I, Φ_{II}, Φ_{III}, and Φ_{IV}. When this assumption is made, there are only two independent circuit loops and solution is greatly simplified. Specifically, this assumption implies that in the circuit of Fig. 7, even when there is an applied torque (and SUU-conditions do not hold); the following conditions hold:

$$\Phi_{I} := \Phi_{IV} , \text{ and } \Phi_{II} := \Phi_{III} . \qquad (22)$$

When Eqs. (22) and Ampere's Law are used, mmf loop equations, corresponding to the right-hand and left-hand branches of the circuit in Fig. 7, can be written as follows:

$$\left[\Phi_{I}\right]_{Cross} := \frac{N_1 \cdot I_1}{\dfrac{1}{\left[P_{\Phi}\right]_{I}} + \dfrac{1}{\left[P_{\Phi}\right]_{IV}}} , \qquad (23)$$

and

$$\left[\Phi_{II}\right]_{Cross} := \frac{N_1 \cdot I_1}{\dfrac{1}{\left[P_{\Phi}\right]_{II}} + \dfrac{1}{\left[P_{\Phi}\right]_{III}}} , \qquad (24)$$

where $N_1 I_1$ is the mmf ampere-turns provided by the excitation coil (the capital "I" symbol denotes that rms values of current are used), and transfer permeances $(P_\Phi)_I$ thru $(P_\Phi)_{IV}$ are defined by Eqs. (5) and (13) thru (18).

Equations (23) and (24) give functional relationships for fluxes $(\Phi_I)_{CROSS}$ and $(\Phi_{II})_{CROSS}$ in terms of flux density B of the shaft material. This is true because the equations are functions of permeances P_S and P_3 that are both B-field dependent.

As an example, in Eq. (23) which describes magnetic-circuit flux $(\Phi_I)_{CROSS}$: transfer permeance $(P_\Phi)_I$ is evaluated using field-dependent permeability μ_I which corresponds to effects of compressive stress acting on shaft material in quadrant I. Similarly, $(P_\Phi)_{IV}$ is evaluated using permeability μ_{IV} which corresponds to effects of tensile stress acting in quadrant IV. Hence, the magnetic circuit equation (23) depicts the combined effects of both compressive and tensile stresses; i.e., the combined effects of quadrant-I compressive stress, and quadrant-IV tensile stress.

When applied torque exists, Eqs. (19) thru (21) show that permeances $(P_\Phi)_{II}$ and $(P_\Phi)_{IV}$ will increase (reluctances decrease) due to tensile stress, while permeances $(P_\Phi)_I$ and $(P_\Phi)_{III}$ decrease due to compressive stress. Thus, the mmf drop in Fig. 7 across one permeance -- e.g., $(P_\Phi)_I$ -- will increase, while the mmf drop across the other -- e.g., $(P_\Phi)_{IV}$ -- will decrease. A unbalance condition is therefore created in the permeance-bridge, causing detected flux Φ_D to flow in the sensor.

Sensing member permeance P_S is a nonlinear function of field strength; and, as discussed in Ref. [3], is defined in terms of flux density B. Equations for nonlinear sensing-member permeance are simultaneously solved with the circuit flux equations -- either Eqs. (23) or (24). Different solutions for field strength are obtained for each quadrant -- one solution for the compressive-stress quadrant, and one for the tensile-stress quadrant. Further description of the solution method is given in Appendix B.

Magnetic permeabilities of sensing-member materials increase due to application of tensile stress and decrease due to compressive stress (as true for positive magnetostriction materials, includes most every material of engineering interest). Thus, solutions of Eq. (23) give a value of flux density B'_{IV} in quadrant IV that is larger than B'_{I} in quadrant I. This behavior is shown in Fig. B.3 of Appendix B.

Recall that magnetic-circuit equations (22) thru (24), which were used to obtain the above solution, were based on the assumption that detected flux Φ_D was negligibly small compared to fluxes Φ_I thru Φ_{IV}. Therefore, once flux solutions for Φ_I thru Φ_{IV} are obtained, validity of this assumption must be checked.

The detected flux Φ_D in Fig. 7 is determined using Ampere's Law that states that the mmf's and magnetic potential drops in a magnetic circuit loop will sum to equal zero. Thus, for circuit loop A-D1-D2 in Fig. 7, one finds that:

$$\Phi_D := \frac{\dfrac{\Phi_{II}}{\left[P_{\Phi}\right]_{II}} - \dfrac{\Phi_{I}}{\left[P_{\Phi}\right]_{I}}}{\dfrac{1}{P_{GD1}} + \dfrac{1}{P_{GD2}}}, \qquad (25)$$

where fluxes Φ_I and Φ_{II} are found using the solution procedure described in Appendix B. Permeance parameters $(P\Phi)_I$, $(P\Phi)_{II}$, P_{GD1}, and P_{GD2} are evaluated using procedures described above.

Note: if Ampere's Law equation is written for circuit loop B-D2-D1, one obtains exactly the same result as in Eq. (25) because of the symmetry relationships in Eq. (22).

Nonlinear Analysis -- A method of nonlinear solution was previously analyzed [3], and is described in Appendix B.

There is, however, one main difference between nonlinear solutions for the branch-design sensor and the cross-design sensor. For the branch sensor, there is only one sensing-member permeance element in the magnetic circuit, as defined by Eq. (3); whereas for the cross sensor there are two elements in series, as defined by Eqs. (23) and (24). The ramifications of this difference are described in Appendix B.

Sensor Output Signal -- When sinusoidal fixed-frequency excitation is assumed, as was done in Ref. [1], and a detection coil is used to measure the detected flux Φ_D; output voltage V_D, when coil 2 terminals are open circuit, is given by Faraday's Law as follows:

$$V_D := j\omega \cdot N_2 \cdot \Phi_D, \qquad (26)$$

where $\omega = 2\pi f$ is the radian excitation frequency, N_2 is the number of turns in the detection coil, and Φ_D is the detected flux defined by Eq. (25).

The output signal, defined by Eq. (26), includes both magnitude and phase information. If a full-wave bridge detector circuit were used, the sensor would not be able to discriminate torque reversal because the output signal (with magnitude information only) would always be positive. For this reason, phase-synchronous detectors must be used with the cross sensor to provide torque sign-reversal information by detecting output signal phase change with respect to phase of the excitation (reference) signal.

Note: cross-leakage flux Φ_{12} does not affect the output signal V_D in Eq. (26) because it flows perpendicular to the axis of the detection coil -- see Figs. 1 and 7. In other words, in accordance with the generalized form of Faraday's Law [38], the vector dot product of flux density B_{12} and coil flux-sensing-area A is therefore zero.

SOLENOIDAL DESIGN -- Shown above in Fig. 1, the solenoidal design sensor, also called a Torkducer(R) [5], was first developed in 1977 by Tseitlin, A.K. Shimokhin and F.B. Kim of the Soviet Union [39].

Recent reports on solenoidal design sensors are described in Refs. 4-6, 10-12, 16-18, 25-28 and 30 of the present paper. These thirteen publications/patents show the intense activity on development of this sensor design. Prior developments are described in Refs. 8, 12-16, 31-32A, 36, 40 and 41 in Ref. [1] -- which describe solenoidal sensors developed during the years, 1982 through 1988.

The solenoidal sensor of Fig. 1 is shown in greater detail in Fig. 8. As its name implies, solenoidal coils encircle the shaft. There are a total of four coils, divided into two coil pairs. Each coil pair consists of an excitation coil and a detection coil. For best coupling to the sensing member, and for reduced cross leakage, excitation coils are wound in the space closest to the shaft. Companion detection coils are then positioned concentrically on top of the excitation coils, as illustrated in Fig. 8.

Also shown in Fig. 8 are grooved or knurled bands that are formed on the surface of the shaft [5,6]. The grooves are aligned parallel to principle lines of stress in the shaft. In this way, interfering effects of residual stresses in the shaft are diminished -- see Refs. 8, 32, 40, 41 and 46 of Ref. [1] for further information. The addition of grooves greatly reduces unit-to-unit variability of sensing-member magnetostrictive material properties, thereby permitting interchangeable use of sensors and sensing shafts without need for extensive recalibration.

In accordance with the analysis approach used in previous work [1], probable flux path modeling is again used to determine permeances associated with the flux compo-

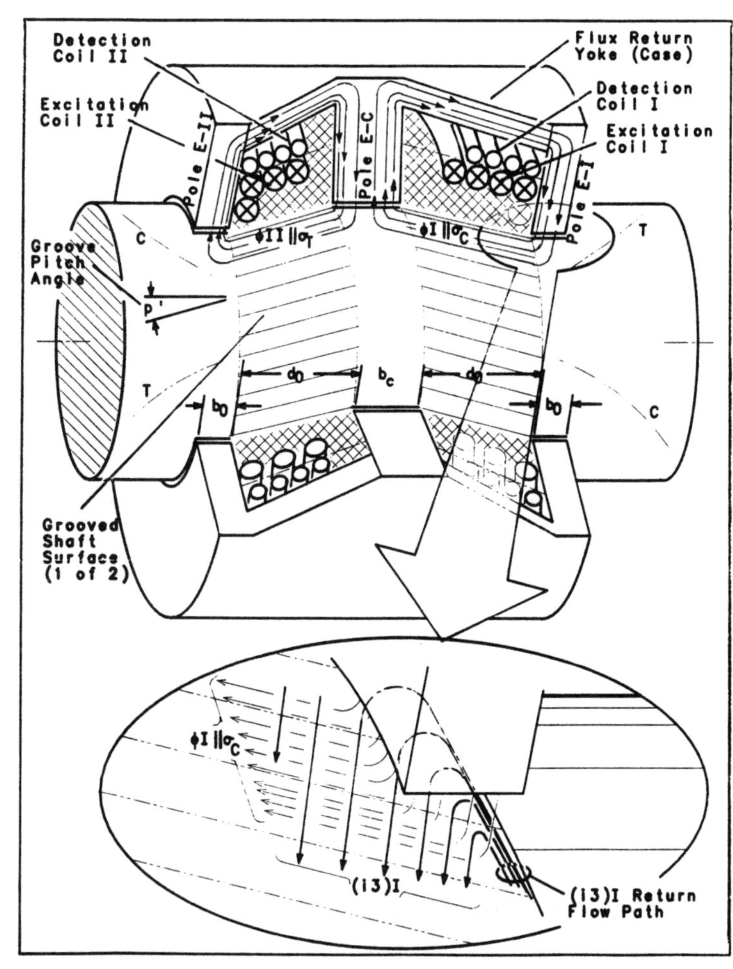

Fig. 8 Solenoidal-Design Sensor (Cut-Away View) Showing Magnetic Flux Paths.

nents shown in Fig. 8. Excitation coils I and II create two main fluxes, $\Phi_I \| \sigma_C$ and $\Phi_{II} \| \sigma_T$; where the subscripts I and II correspond to fluxes created by coils I and II.

Polarities of the excitation coils in Fig. 8 cause the fluxes to both circulate in clockwise directions. Excitation polarities that produce co-circulating flux components, as shown in Fig. 8, have been found experimentally to provide best torque measurement sensitivity [40].

Stresses σ_C and σ_T in Fig. 8 correspond, respectively, to lines of principle compressive stress and principle tensile stress in the shaft. By design, the grooves formed on the shaft are made parallel to the lines of principle stress.

In the enlargement of Fig. 8 (inset), detail is shown of how the main flux component, $\Phi_I \| \sigma_C$, flows out of excitation pole I; while eddy current created by this flux, $(i_3)_I$, flows orthogonal to the main flux. This behavior is in accordance with that required by electromagnetic field boundary conditions [41].

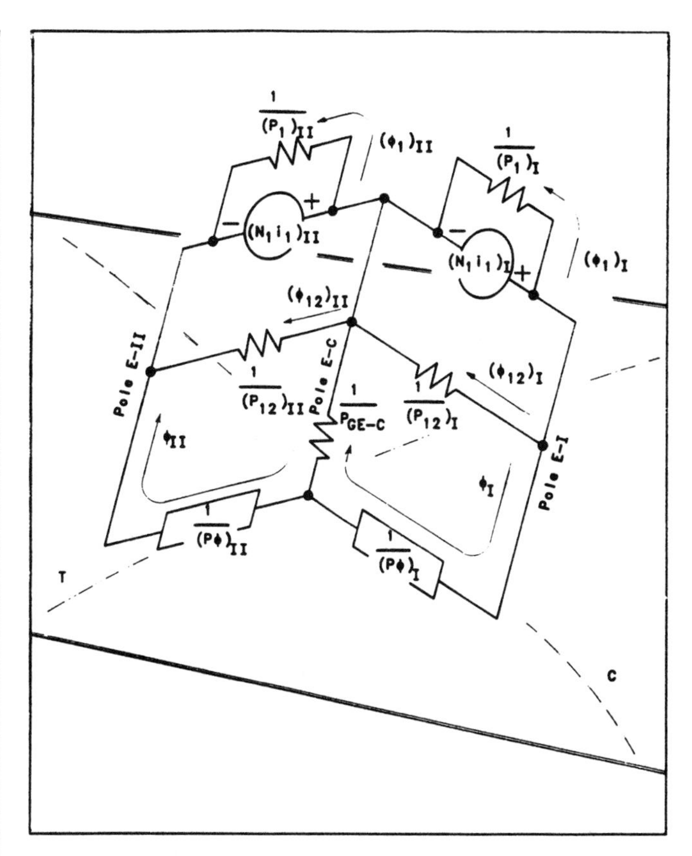

Fig. 9 Equivalent Magnetic Circuit For the Solenoidal-Design Sensor.

Also shown in the inset of Fig. 8 is an eddy current flow $(i3)_I$ return path. This return path is predicted by probable field considerations. Using this line of reasoning, it also follows that periodic, repeating, flow patterns of eddy current and Lenz's-Law opposing flux must exist around circumferences of sensing-member grooved bands. This must be the case because, if at least two repeating patterns of eddy current did not exist, eddy current would not be able to return from one side of a band to the other.

Finally, probable field considerations lead to the conclusion that the number of repeating patterns of eddy current flow will depend on the diameter of the shaft, the width of the grooved band, and the angle that the flux flows (in the grooves). However, further discussion of this behavior requires either electromagnetic field modeling, or field measurements, and in any case is beyond the scope of this paper.

Equivalent Magnetic Circuit -- The equivalent magnetic circuit for the solenoidal-design sensor is shown in Fig. 9. This circuit is derived from the simplified circuit shown in Fig. 4(b). At first glance, the circuit looks simply like that for a two-branch sensor, something like that shown in Fig. 5(b), but with two branches instead of four.

There is a major difference, however, between a two-branch magnetic circuit and the solenoidal circuit in Fig. 9. Namely, the circuit for the two-branch sensor has a single excitation mmf source, whereas the solenoidal sensor circuit has two mmf sources (two excitation coils), designated as mmf-$N_1 i_1$ sources I and II in Fig. 9. These mmf sources generate flux in the parallel-connected circuit loops of Fig. 9. These fluxes, Φ_I and Φ_{II}, flow through the grooved subsurfaces of the shaft, as shown in Fig. 8.

In Fig. 9; permeances P_1, P_{12}, and P_{GE-C} represent, respectively, permeances associated with self-leakage fluxes Φ_1, cross-leakage fluxes Φ_{12}, and fluxes Φ_I and Φ_{II} flowing from the shaft to the excitation E-C through the center-pole air gap GE-C.

Note that cross leakage permeances P_{12} in Fig. 9 are not multiplied by the factor 1/3, although they were for the branch sensor. This is because the excitation coils of the solenoidal sensor are wound around the shaft and no magnetomotive force (mmf) gradient exists along the side (cross leakage) walls of the sensor case [36].

Transfer permeances $(P\Phi)_I$ and $(P\Phi)_{II}$ are still defined by Eq. (5) above, but with the following changes required to match the solenoidal sensor geometry.

Air Gap Permeance -- the term P_{GAPS} in Eq. (5) is set equal to the gap permeance of its associated excitation pole. All of the center pole gap permeance is included in the element P_{GE-C} seen in Fig. 9. Whereas, for permeances $(P\Phi)_I$ and $(P\Phi)_{II}$; gap permeances in Eq. (5) are given, respectively, as:

$$\left[P_{GAPS}\right]_I := P_{GE-I} \quad , \qquad (27)$$

and

$$\left[P_{GAPS}\right]_{II} := P_{GE-II} \quad , \qquad (28)$$

where poles E-I and E-II are identified in Figs. 8 and 9.

Gap permeances P_{GE-I}, P_{GE-II}, and P_{GE-C} are again computed using the probable flux path procedures of Appendix C in Ref. [1]. However, due to the different geometry of the solenoidal sensor, the following changes are made.

First, a circumferential length Lc is defined as:

$$L_C := 2 \cdot \pi \cdot \left[R + \frac{G}{2}\right] \quad , \qquad (29)$$

where R is the shaft radius and G is the shaft-to-sensor air gap (assumed uniform and equal for all poles). Next, area A in Eq. (C.2) of Ref. [1] for direct crossing flux is given as:

$$A := L_C \cdot b \quad , \qquad (30)$$

where b is the width of the pole which, as shown in Fig. 8, is given as: $b = b_O$ for poles E-I and E-II, and $b = b_C$ for the center pole E-C.

For fringing flux path 1 in Fig. C.1 of Ref. [1], the term $2\pi(r + G/2)$ in the P_1 permeance term of Eq. (C.3) of Ref. [1] is replaced by:

$$\text{Total Peripheral Length} := 2 \cdot L_C \quad , \qquad (31)$$

where Lc is given by Eq. (29), and the factor of 2 takes into account both sides of the pole edge.

For fringing flux path 2 in Fig. C.1 of Ref. [1], the P_2 permeance term of Eq. (C.4) of Ref. [1] is replaced by:

$$P_2 := \frac{2 \mu_0 L_m}{\pi} \ln\left[1 + \frac{g_2}{G}\right] \quad , \qquad (32)$$

where

$$L_m := 2 \cdot \pi \cdot \left[R + G + \sqrt{G \cdot \left[G + g_2\right]}\right] \quad , \qquad (33)$$

Here: μ_0 is the permeability of air, g_2 is the fringing dimension defined by Eq. (C.6) of Ref. [1], and Lm is a mean circumferential length derived by using probable flux path analysis methods -- see Refs. 73-76 of Ref. [1].

The expressions given in Eqs. (30) thru (33) are substituted into Eq. (C.1) of Ref. [1], thereby permitting evaluation of gap permeances P_{GE-I}, P_{GE-II}, and P_{GE-C}.

Sensing Member And Lenz's-Law Opposing Flux Permeances -- To complete the evaluation of transfer permeances $(P\Phi)_I$ and $(P\Phi)_{II}$ using Eq. (5) above, sensing member permeances $(P_S)_I$ and $(P_S)_{II}$ and Lenz's-Law opposing flux permeances $(P_3)_I$ and $(P_3)_{II}$ -- which enter into Eqs. (7) and (8) -- must be determined. Methods for evaluating these permeance terms for the solenoidal sensor are given here.

In the solenoidal sensor, flow of main flux Φ through the sensing member is straightforward. Figure 8 shows that fluxes Φ_I and Φ_{II} simply flow in a sheet-like layers from one circumferential pole-ring to the other.

The thickness of the flux sheet is given by the penetration depth δ', defined by Eq. (C.11) of Ref. [1]. The cross-sectional breadth of the flux sheet is to good approximation equal to the shaft circumference, $2\pi R$, where R is the radius of the shaft. The distance d' that the flux travels parallel to the grooves in the band is given as:

$$d' := \frac{d_0 + b_0 + \frac{b_C}{2}}{\cos(p')} , \qquad (34)$$

where pole dimensions d_0, b_0 and b_C are shown in Fig. 8; and p' is the pitch angle of the grooves (normally equal to 45 degrees).

When the above expressions are combined, the following relationship for sensing member permeance P_S, analogous to Eq. (C.10) of Ref. [1], is obtained:

$$P_S := \mu \cdot \delta' \cdot \frac{2 \cdot \pi R}{d'} , \qquad (35)$$

where μ is the permeability of the sensing member material, δ' is the flux penetration depth given by Eq. (C.11) of Ref. [1], R is the radius of the shaft, and d' is the flux travel distance given by Eq. (34).

Sensing member permeances $(P_S)_I$ and $(P_S)_{II}$ are both evaluated using parameter values appropriate to paths I and II in Eq. (35). Typically, geometric parameters R and d' are the same for paths I and II. But sensing material properties μ and δ' will exhibit opposite types of magnetostrictive behavior, therefore requiring separate evaluations of Eq. (35) for paths I and II.

Once sensing-member permeances are determined using Eq. (35), Lenz's-Law opposing flux permeances $(P_3)_I$ and $(P_3)_{II}$ are determined using Eq. (C.21) of Ref. [1]. However, no reliable probable-flux-path method of computing P_{3A}, the above-surface air permeance in Eq. (C.21) of Ref. [1], is available. This is an inherent limitation of the probable flux path model used here.

To overcome this limitation, permeance P_{3A} was evaluated in the present study by setting its value to that necessary to make the computed value of sensor transfer permeance match experimentally measured values.

Empirically, it was found that P_{3A} could be estimated using the following relationship:

$$P_{3A} := 2 \cdot P_S \left[\mu_{max}, f_{REF} \right] , \qquad (36)$$

where the factor of 2 gave the best match with experiment; and $P_S(\mu_{MAX}, f_{REF})$ equals the value of the sensing-member permeance P_S of Eq. (35), evaluated at the point of maximum shaft-material permeability μ_{MAX} (for zero applied stress), and at reference frequency f_{REF} set equal to 1000 Hz (the value which empirically gave the best match). Because permeance P_{3A} is associated with opposing flux flow through air, it's value once evaluated using Eq. (36) remains constant, independent of changes in frequency and/or flux density.

Magnetic Circuit Solution -- The object here is to solve for flux components Φ_I and Φ_{II} which circulate through the sensing member transfer permeances $(P_\Phi)_I$ and $(P_\Phi)_{II}$ in the magnetic circuit shown in Fig. 9.

Ampere's Law is used to write mmf loop equations, corresponding to the circuit loops I and II in Fig. 9, as follows:

$$\left[N_1 \cdot I_1\right]_{II} - \Phi_{II} \cdot \left[\frac{1}{P_{GE-C}} + \frac{1}{\left[P_\Phi\right]_{II}} \right] + \Phi_I \cdot \left[\frac{1}{P_{GE-C}} \right] := 0 , \qquad (37)$$

and

$$\left[N_1 \cdot I_1\right]_{I} - \Phi_{I} \cdot \left[\frac{1}{P_{GE-C}} + \frac{1}{\left[P_\Phi\right]_{I}} \right] + \Phi_{II} \cdot \left[\frac{1}{P_{GE-C}} \right] := 0 , \qquad (38)$$

where $(N_1 I_1)_I$ and $(N_1 I_1)_{II}$ are the mmf (ampere-turns) provided by excitation coils I and II, the capital "I" symbol denotes that rms values of current are used, $(P_\Phi)_I$ and $(P_\Phi)_{II}$ are the transfer permeances defined by Eqs. (5) and (27) thru (35), and P_{GE-C} the center pole gap permeance is defined by Eqs. (29) thru (33).

Since mmf sources $(N_1 I_1)_I$ and $(N_1 I_1)_{II}$, transfer permeances $(P_\Phi)_I$ and $(P_\Phi)_{II}$, and the center pole gap permeance P_{GE-C} are known quantities; flux components Φ_I and Φ_{II} can therefore be computed by simultaneous solution of Eqs. (37) and (38). Solutions for Φ_I and Φ_{II} are found in this manner, and are written as follows:

$$\Phi_I := \frac{\left[N_1 \cdot I_1\right]_{II} + \left[N_1 \cdot I_1\right]_{I} \cdot \left[1 + \frac{\frac{1}{\left[P_\Phi\right]_{II}}}{\frac{1}{P_{GE-C}}} \right]}{\frac{1}{\left[P_\Phi\right]_{I}} + \frac{1}{\left[P_\Phi\right]_{II}} \cdot \left[1 + \frac{\frac{1}{\left[P_\Phi\right]_{I}}}{\frac{1}{P_{GE-C}}} \right]} , \qquad (39)$$

and

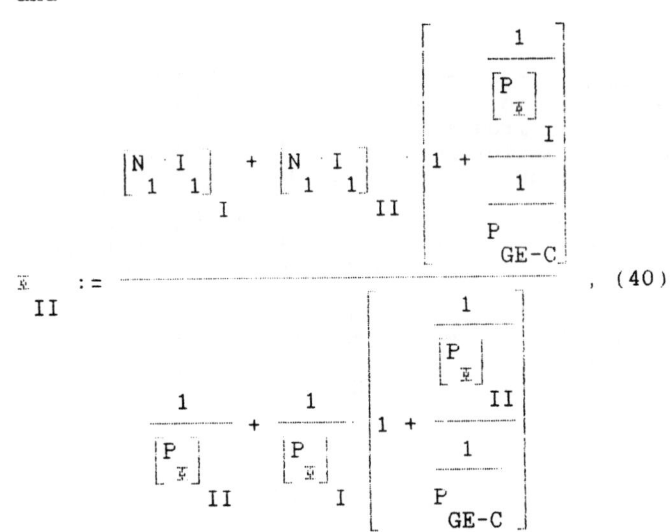

$$\Phi_{II} := \frac{\left[N_1 I_1\right]_I + \left[N_1 I_1\right]_{II} \cdot \left[1 + \frac{\frac{1}{\left[P_\Phi\right]_I}}{\frac{1}{P_{GE-C}}}\right]}{\frac{1}{\left[P_\Phi\right]_{II}} + \frac{1}{\left[P_\Phi\right]_I}\left[1 + \frac{\frac{1}{\left[P_\Phi\right]_{II}}}{\frac{1}{P_{GE-C}}}\right]}, \quad (40)$$

where terms used here were previously defined in conjunction with Eqs. (37) and (38).

Under SUU-conditions (with no applied torque), and with equal mmfs $(N_1 I_1)_I = (N_1 I_1)_{II}$; transfer permeances $(P\Phi)_I$ and $(P\Phi)_{II}$ are equal, and Eqs. (39) and (40) show that the circulating flux components Φ_I and Φ_{II} will also be equal.

However, if torque is applied to the shaft, and for a positive-magnetostrictive sensing material; transfer permeance $(P\Phi)_I$ associated with the direction of compressive stress decreases, whereas permeance $(P\Phi)_{II}$ associated with the tensile stress direction increases. In this case, flux components Φ_I and Φ_{II} will be unequal and the nonlinear analysis methods described below are used to obtain the solution.

Nonlinear Analysis -- There is, again, another important difference between nonlinear solutions for the branch-design sensor in Ref. [3] and that for the solenoidal-design sensor. For the branch sensor, there was only one sensing-member permeance element in the magnetic circuit, as defined by Eq. (3) above; whereas for the solenoidal sensor there are two simultaneous solutions, one for circulating flux Φ_I and one for Φ_{II}, as defined by Eqs. (39) and (40). The ramifications of this difference are described in Appendix B.

Sensor Output Signal -- Sinusoidal fixed-frequency excitation of both coils I and II in the illustrated sensor of Fig. 8 is assumed. Both coils are also assumed to be excited by a common power source; i.e., at the same frequency. Detection coils I and II in Fig. 8 are used to measure flux components Φ_I and Φ_{II}, and are assumed to have equal numbers of turns N_2.

The output signal V_D of the solenoidal sensor is simply given by the difference of output voltages of detection coils I and II. Output voltages of each coil, when terminals are open circuit, are given by Faraday's Law -- Eq. (26) above with Φ_D replaced, respectively, by Φ_I or Φ_{II}. When Eq. (26) is used, together with the above assumptions, one obtains the result:

$$V_D := \omega \cdot N_2 \left[\left|\Phi_{II}\right| - \left|\Phi_I\right|\right], \quad (41)$$

where $\omega = 2\pi f$ is the radian excitation frequency, N_2 is the number of turns in detection coils I and II, and Φ_I and Φ_{II} are the detected flux components determined by nonlinear solution and defined by Eqs. (39) and (40).

The output signal, defined by Eq. (41), includes only magnitude information because it was assumed that separate full-wave bridge detector circuits were used to measure voltages associated with each detection coil. Hence a desirable sign-reversal of output signal with reversal of applied torque direction is obtained here without using phase-synchronous detector circuits.

Note: because of symmetry of the solenoidal sensor, seen in Fig. 8, cross-leakage fluxes Φ_{12} will have have nearly equal effects on voltage output signals from each detection coil. Effects of cross-leakage flux will therefore mostly cancel out in the differential output signal of the sensor, defined by Eq. (41).

MODEL SUMMARY

Equivalent magnetic circuits for the branch, cross, and solenoidal design sensors have been developed and were shown, respectively, in Figs. 5(b), 7, and 9. The equivalent circuit models are grouped together and are shown on a common shaft in Fig. 10. A comparison of Fig. 1 and Fig. 10 shows how equivalent magnetic circuits of the sensor designs correspond to their physical configurations.

Figure 10 also illustrates that the three sensor designs are all modeled in terms of various combinations of a basic single-branch sensing element shown in Fig 4(b). It was for this reason that great emphasis was placed on its understanding in this, and in prior work [1-3].

BEST EXCITATION

Previous study had shown that at low frequencies of excitation (below approximately 5000 Hz), large values of excitation current I_1 were required to achieve maximum normalized output signal s. For example, this behavior was shown in Fig. 9 of Ref. [3]. The value of excitation current required to reach this point approximately corresponded to that needed to operate at the knee of the d-c magnetization curve for the sensing-member material.

[Note: normalized signal-to-operating-point ratio s_A for constant current sensor excitation is computed using Eqs. (9) thru (11). As discussed in Appendix A of Ref. [2], normalized signal $s = s_A$ is a key

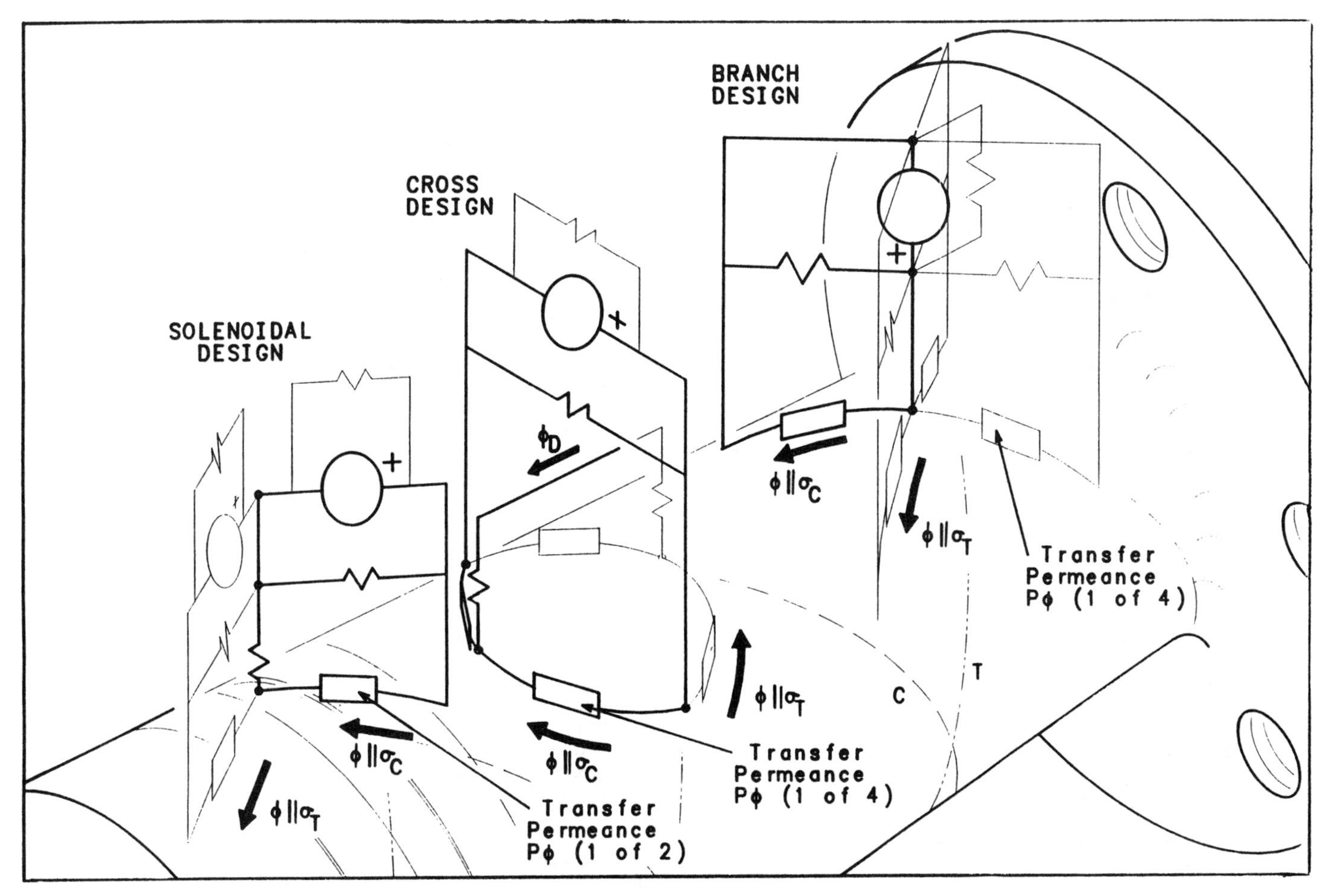

Fig. 10 Side-By-Side Comparison of Equivalent Magnetic Circuits For The Branch, Cross, and Solenoidal Design Torque Sensors -- Mounted On the Same Shaft.

output parameter because both sensor signal-to-noise performance and stress-measurement sensitivity are proportional to it].

An unexpected, but very important result, was found during the course of this study. The discovery was made that significantly higher values of normalized signal s are obtained by reducing excitation current, while at the same time increasing excitation frequency. Normalized signal strengths as much as 5-to-10 times stronger than those obtained at low frequencies were found when excitation frequencies were increased into the range, 50,000 Hz-to-500,000 Hz.

This behavior is shown in Fig. 11, where normalized output signal s is plotted vs. excitation current I_1 for a single-branch sensor operated at both excitation frequencies ranging from 500 Hz up to 5,000,000 Hz. The behavior is also exhibited in analysis of cross and solenoidal sensor models.

Experimental observations of this behavior have been made. In the words of Sasada [42]: "signal sensitivity can be increased by increasing the exciting frequency, and .... It should be better to use a higher exciting frequency for realization of the torque sensor of sufficient sensitivity .... with reduced exciting current." In fact, best signal strength at high excitation frequencies has been reported by many investigators, as seen in Table 2.

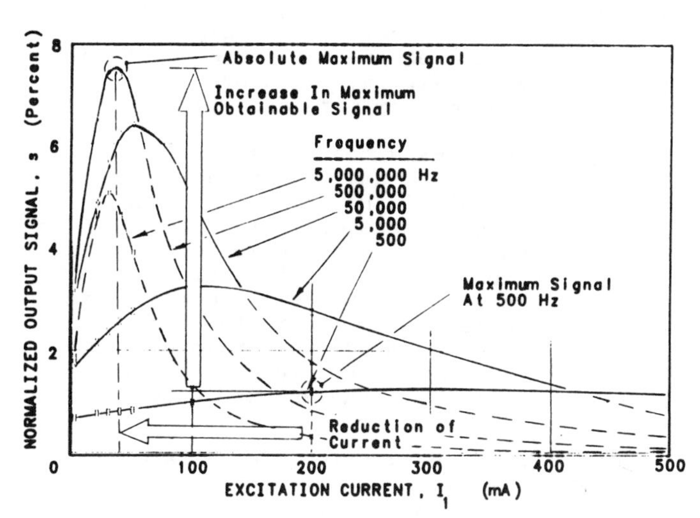

Fig. 11 Effect of Increasing Excitation Frequency On Normalized Output Signal As a Function of Excitation Current.

Table 2
Reported Best Excitation Frequencies
For Magnetostrictive Sensor Operation

Reported Best Excitation Frequency (Hz)	Investigator (Affiliation, Year)	Ref.
50,000	Ishino (Kubota, 1989)	[5]
99,000	Sasada (Kyushu Un., 1989)	[24]
30,000	Aoki (Nissan, 1988)	[11]
50,000	Kobayashi (Toshiba, 1988)	[14]
10,000	Garshelis (Mag Dev, 1988)	[44]
10,000-to-100,000	Kobayashi (Toshiba, 1988)	[28]*
20,000	Sugiyama (Toyota ,1987)	[22]*
50,000	Sasada (Kyushu Un., 1987)	[42]
5,000	Himmelstein (Himmel, 1986)	[16]*
100,000	Sasada (Kyushu Un., 1986)	[35]*
170,000	Winterhoff (AEG-Tel, 1985)	[42]*
6,000	Scoppe (Avco Lyc., 1984)	[6]*
20,000-to-50,000	Iwasaki (Aisen Seiki,1983)	[31]*
20,000	Harada (Kyushu Un., 1982)	[36]*
2,000	Fleming (Gen Motors, 1982)	[3]*
5,000	Dorman (Mech Tech, 1980)	[13]*
60	Dahle (ASEA, 1960)	[2]*

* Asterisks denote references taken from Ref. [1].

Experimental reports listed in Table 2 are explained in terms of the present model as follows.

Greater signal output at higher frequencies occurs due to reduction of sensing-member permeance P_s. This is caused by reduced flux penetration (i.e., skin depth) δ'. Note: flux penetration is defined by Eq. (C.10) of Ref. [1] and Eq. (B.7) of the present paper.

At sufficiently high frequencies, sensing-member permeance eventually becomes small -- its reluctance becomes large -- compared to that of the air gaps. This means that attenuating effects of air gap reluctance in the equivalent magnetic circuit of, for example, Fig. 4(a) are overcome because sensing-member permeance is low.

On the other hand, output signal ultimately falls off at still higher frequencies (above 1,000,000 Hz) as a consequence of increased cross leakage of flux between poles. In this case, sensing-member permeance P_s becomes extremely small, to the point where it becomes small compared to cross leakage permeance P_{12} in the equivalent magnetic circuit of, for example, Fig. 4(a). In this case, cross leakage permeance therefore shunts (attenuates) the output signal.

A further condition for achieving maximum signal is that excitation current must be adjusted such that flux density B in the sensing member -- given by Eq. (B.6) of Appendix B -- is approximately equal to that value $B = B_{MAX}$ in Eq. (15) of Ref. [3] where maximum stress-induced permeability change occurs for the sensing material. Figures 2 and 3 of Ref. [3] show magnetization curves depicting this behavior, whereas Eqs. (B.1)-(B.3) of the present paper define the behavior.

If excitation current is reduced below the point where sensing-member flux density B can no longer achieve maximum permeability change, rapid drop-off of signal takes place. Below this point, signal drop off occurs regardless of how high one increases the excitation frequency. Hence, there will be one particular combination of excitation current and frequency at which absolute maximum signal occurs. This point is labeled in Fig. 11. However, absolute maximum strength signal s is not achieved without incurring some operating disadvantages, and these will be discussed below.

A full set of operating parameters for the sensor analyzed in Fig. 11 are plotted in Fig. 12 as a function of excitation current. At each value of current, frequencies were found that gave relative maximum output signal for that value of current. Maximum signal operating-conditions in Fig. 12 were obtained numerically using solution procedures described for the branch sensor in Appendix B. At each combination of excitation current I_1 and frequency f, the maximum obtainable normalized signal s and signal strength S are graphed in inset (a).

Figure 12(a) shows that an absolute maximum normalized signal s of -7.5 percent and maximum signal strength S of -58.5 V are predicted at fairly high frequencies of 200,000 Hz and 500,000 Hz, respectively. [The minus signs indicate that compressive stress was applied to the sensing member, where values of v_{1C} used in Eq. (B.2) of Appendix B were taken from Table B.2].

However, Fig. 12 also shows that:

- for excitation current above about 100 mA, the frequency for maximum signal, shown in inset (a), rapidly decreases to values well below 50,000 Hz
- power dissipation P_D (due to frequency-dependent eddy current losses), shown in inset (b), is 3-to-5 times greater than for low frequency operation
- flux penetration depth δ', shown in inset (b), is reduced to 0.02 mm (800 microinch) for frequencies above 500,000 Hz
- excitation voltage V_1, shown in inset (b), is 10-to-15 times greater than for low frequency operation, and becomes as large as 790 Vrms at 500,000 Hz

Obviously high frequency and high voltage levels not only are potential EMI (electromagnetic interference) problems, but also necessitate use of expensive excitation electronics. Of greater significance, however, is the limited flux penetration of only 0.02 mm.

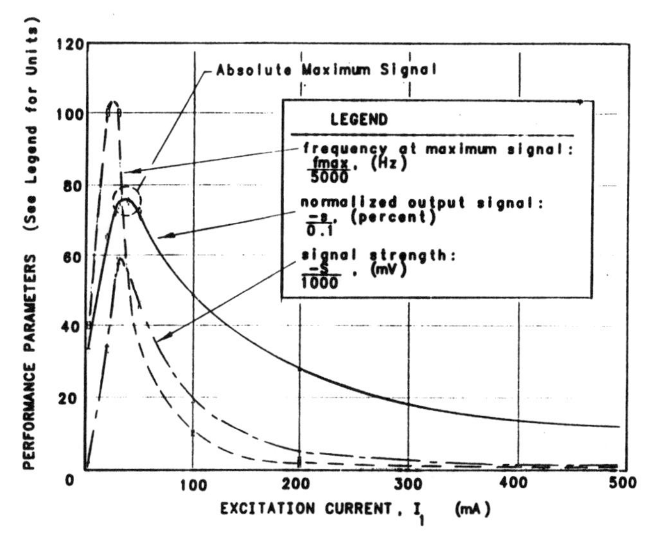

(a) Output Signals s and S, and Excitation Frequency At Which Maximum Signals Occur

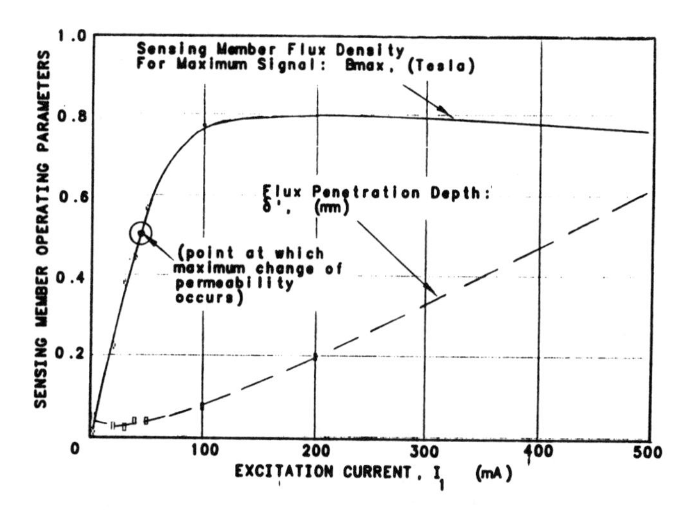

(c) Flux Density and Flux Penetration In Sensing Member

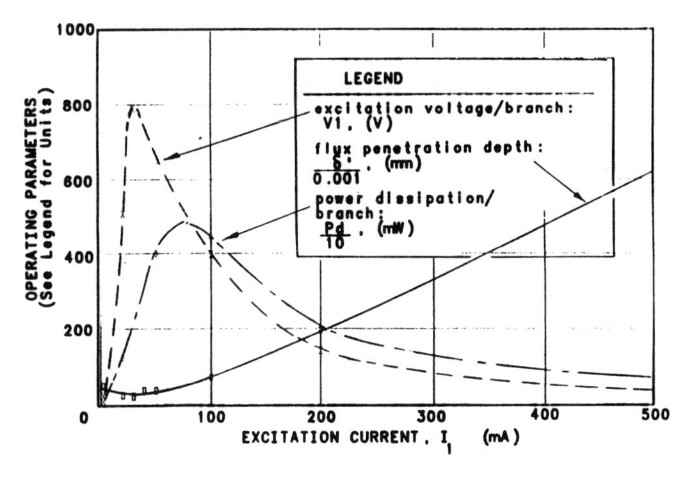

(b) Required Excitation Coil Voltage, Flux Penetration, and Power Dissipation in Sensing Member

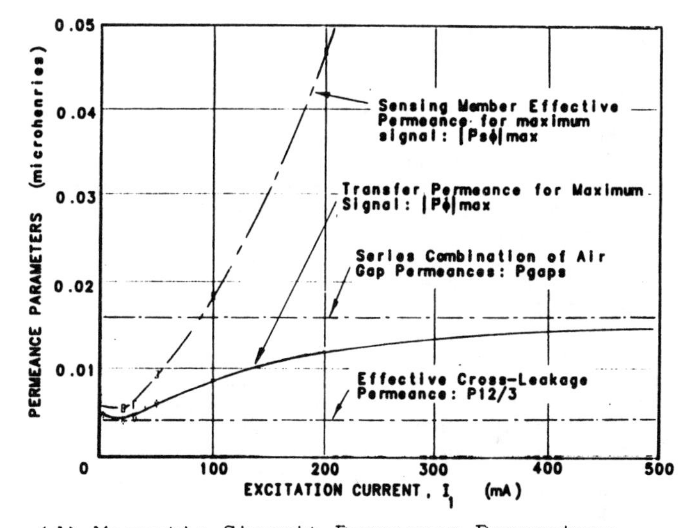

(d) Magnetic Circuit Permeance Parameters

Fig. 12 Effect of Increasing Excitation Frequency On Normalized Output Signal and On Other Operating Parameters of the Sensor. At Each Operating Point, Output Signal Was Maximized for That Combination of Excitation Current and Frequency.

The 0.02-mm depth of penetration is roughly equal to the surface roughness of No. 1 dull-finish sheet steel [43]. When flux penetration is this small, imperfections and inhomogeneities of material permeability near the surface of a rotating-shaft sensing member will create potentially severe signal noise.

Hence, if high excitation frequencies were used, magnetic inhomogeneities in a rotating shaft material would be accentuated. A torque sensor mounted over the rotating shaft would, in effect, function like an endless-loop tape recorder. A repeating noise pattern would be created with repetition rate equal to the frequency of shaft rotation. Primarily for this reason, sensors are therefore normally operated at frequencies no higher than approximately 50,000 Hz.

Insets (c) and (d) of Fig. 12 show behavior of the magnetic parameters of the sensor as excitation current and frequency are varied to maintain maximum relative output signal. The particular shaft material chosen (maraging steel -- see Table B.2 in Appendix B) has its greatest change of permeability with applied stress at B = 0.5 Tesla. Figure 12(c) shows that the value of flux density for maximum signal, the B_{MAX}-curve, passes through the 0.5-Tesla point at I_1 = 45 mA. As expected, Fig. 12(a) shows that this indeed is exactly the point at which absolute maximum output signal occurs.

Figure 12(d) shows the magnetic-circuit permeance parameters associated with the sensor. At large values of excitation current, sensing-member permeance $|P_{S\Phi}|$ greatly exceeds air gap permeance P_{GAPS}. For this condition, inset (a) shows that relative maximum signal levels are consequently low. This behavior is analogous to a voltage-divider circuit, where the small permeance (high reluctance) of series air gaps attenuate the effects of P_{sg}. [Note: the equivalent magnetic circuit for the sensor was shown in Fig. 4(a)].

For excitation current below about 45 mA, further decrease of P_{sg} in Fig. 12(d) is no longer advantageous because the sensor signal is begins to experience attenuation due to the shunting effects of cross-leakage permeance P_{12}. In finding solutions of relative maximum signal, avoidance of these shunting effects is evidenced by the upturned tail of the P_{sg}-curve for $I_1 < 45$ mA seen in inset (d). At the same time, avoidance of shunting effects also accounts for the downturned tail on the excitation frequency curve seen in inset (a).

SUMMARY AND CONCLUSIONS

Equivalent magnetic circuits for the branch, cross, and solenoidal design sensors have been developed and are shown, respectively, in Figs. 5(b), 7, and 9. These equivalent circuit models were grouped together and are shown on a common shaft in Fig. 10. Comparison of Figs. 1 and 10 above showed how equivalent magnetic circuits of the sensor designs correspond to their physical configurations.

Figure 10 also illustrated that the three sensor designs were all modeled in terms of various combinations of a basic single-branch sensing element, shown in Fig. 4(b). It was for this reason that great emphasis was placed on its understanding in this, and in prior work [1-3].

Analysis revealed that significantly higher values of normalized sensor output signal are obtainable by reducing excitation current while increasing excitation frequency. Normalized signal strengths as much as 5-to-10 times stronger than those obtained at low frequencies are predicted when excitation frequencies are increased to above 50,000 Hz.

However, stronger output signal is not achieved without incurring disadvantages of: potential EMI problems, requirement for more expensive electronics, and introduction of increased signal noise.

In Appendix B, calculations are made to compare magnetic circuit operating behavior for sensors designed to fit into the same space on a shaft made of maraging steel.

ACKNOWLEDGMENTS

Use of computer resources and the support of Barney Bauer at TRW Vehicle Safety Systems, Washington, MI, are much appreciated.

REFERENCES

PRIOR MODEL DEVELOPMENT

1. W.J. Fleming, "Magnetostrictive Torque Sensors -- Derivation of Transducer Model," SAE Paper 890482, companion paper (part 1) to Ref. [2] below, presented at the International Congress, Detroit, MI, February 27-March 3, 1989.

2. W.J. Fleming, "Magnetostrictive Torque Sensors -- Analysis of Performance Limits," SAE Paper 890483, companion paper (part 2) to Ref. [1] above, presented at the International Congress, Detroit, MI, February 27-March 3, 1989.

3. W.J. Fleming, "Magnetostrictive Torque Sensor Performance -- Nonlinear Analysis," IEEE Trans. on Vehicular Technology, Vol. VT-38, Issue 3, (to be published), November, 1989.

WORLDWIDE MAGNETOSTRICTIVE SENSOR DEVELOPMENT ACTIVITY DURING THE PAST YEAR

North America

4. "Mag Dev Eyes Uses For Sensing Technology," Automotive Electronics News, p.10, July 17, 1989.

5. R. Ishino and R. McConnell, "A Noncontact Torque Sensor," Conference Proceedings, Paper 109C, presented at Sensors Expo International, Cleveland, OH, September 12, 1989,.

6. "Special Steel and Knurls Perfect Torque Sensor," Engineering and Technology Guide, Machine Design, p.48, November 23, 1989.

7. "Torque Sensor For Powertrain Applications," Sensors, product brochure, p.5, United Technologies/Spectrol Electronics Div., La Puente, CA, SAE Exhibitor literature, February 1989.

8. A. Hossain and M. Rashid, "Force Transducer Using Amorphous Metglas Ribbon," 1988 IEEE Industry Applications Meeting, Conference Record, pp. 1815-1822, Vol. 2, October 2, 1988.

9. H. Savage et al., "Magnetostrictive Torque Sensor," Patent Application 7-374-112, Department of the Navy, Washington, D.C., NTIS Report No. USGAD-DO14-169/7, June 21, 1989.

Japan

10. S. Edo et al., "Magnetostrictive Device for Measuring Torsional Torque," U.S. Patent 4,823,620, assigned to Nissan Motor Co., issued April 25, 1989.

11. H. Aoki et al., "Torque Sensor With Shape Anisotropy," 7th Sensor Symposium, Technical Digest, pp. 83-86, 1988.

12. M. Todoroki et al., "Magnetostrictive Stress Measurement Apparatus," U.S. Patent 4,833,926, assigned to Nissan Motor Co., issued May 30, 1989.

13. H. Aoki et al., "Torque Detecting Device, U.S. Patent 4,840,073, assigned to Nissan Motor Co. and Daido Steel, issued June 20, 1989.

14. T. Kobayashi et al., "Built-In Amorphous Torque Sensor for an Induction Motor," IEEE Transl. J. Magnetics Japan, Vol. 3, pp. 226-235, March 1988.

15. Y. Nishibe et al., "Torque Detecting Apparatus," U.S. Patent 4,811,609, assigned to Toyota Motor Co., issued March 14, 1989.

16. Y. Nonomura et al., "Torque Detecting Apparatus," U.S. Patent 4,803,885, assigned to Toyota Motor Co., issued February 14, 1989.

17. H. Hase et al., "Magnetically Operated Non-Contact Magnetic Torque Sensor For Shafts," U.S. Patent 4,780,671, assigned to Matsushita Electric, issued October 25, 1988.

18. H. Hase et al., "Torque Sensor," U.S. Patent 4,823,617, assigned to Matsushita Electric, issued April 25, 1989.

19. M. Wakamiya et al., "Stress Sensitive Sensors Utilizing Amorphous Magnetic Alloys," U.S. Patent 4,785,671, assigned to Matsushita Electric, issued November 22, 1988.

20. "Mag Dev Licenses Its New Magnetoelastic Noncontact Sensor Technology to Kubota Ltd.," Sensors & Instrumentation News, July 1989.

21. T. Yagi et al., "Torque Sensor," U.S. Patent 4,817,444, assigned to Honda Motor, issued April 4, 1989.

22. N. Kimura et al., "Mechanical Quantity Sensor Element," U.S. Patent 4,784,003, assigned to Honda Motor, issued November 15, 1988.

23. I. Sasada et al., "Detection of Instantaneous Torque Using the Magnetostrictive Effect In a Practical Ferromagnetic Shaft," IEEE Trans. on Magnetics, Vol. MAG-24, No. 6, pp. 2886-2888, November 1988.

24. I. Sasada et al., "Generation of Zero-Level Fluctuation and Its Reduction In the Magnetic Head Type Torque Sensor With L-R Bridge Configuration," International Magnetics Conference, 1989, Conf. Proc.

Europe

25. J. Nordvall, "Magnetoelastic Force Transducer," U.S. Patent 4,825,709, assigned to ASEA Aktiebolag, issued May 2, 1989.

26. J. Sobel, "Magnetoelastic Force Transducer," U.S. Patent 4,823,621, assigned to ASEA Aktiebolag, issued April 25, 1989.

27. J. Nordvall, "Magnetoelastic Force Transducer," U.S. Patent 4,802,368, assigned to ASEA Aktiebolag, issued February 7, 1989.

28. J. Sobel, "Magnetoelastic Torque Transducer With Double Sleeve," U.S. Patent 4,845,999, assigned to ASEA Brown Boveri, issued July 11, 1989.

29. E. Schiessle et al., "Device for the Contactless Indirect Electrical Measurement of the Torque at a Shaft," U.S. Patent 4,805,466, assigned to Daimler-Benz, issued February 21, 1989.

30. "The Magnetoelastic Breakthrough," product brochure, Sensglas Terfenol AB, Lund, Sweden, and Edge Technologies, Ames, IA, USA, 1989.

31. R. Wolffenbuttel (Delft University of Technology, The Netherlands), "Non-Contact Torque Sensing In a Steel Axle," Paper 89030, Conf. Proc., presented at ISATA, Florence, Italy, May 30, 1989.

32. J. Justat (Przemyslowy Inst. Auto., Warsaw, Poland), "Magnetostrictive Torque Transducers," Elektronika, Vol. 29, No. 3, pp.13-19, 1988.

33. V. Franyuk and L. Ivan'kovich, (Academy of Sciences of the Belorussian SSR), "Eddy-Current Transducer Based On Flat Spiral Coils," Soviet J. Nondestructive Testing (English Translation), Vol. 24, No. 4, pp. 235-238, December 1988.

ADDITIONAL REFERENCES

34. W.J. Fleming, "Engine Sensors: State of the Art," SAE Paper 820904, presented at Convergence '82, the International Congress on Transportation Electronics, Dearborn, MI, October 4, 1982.

35. O. Dahle, "The Ring Torductor -- A Torque-Gauge Without Slip Rings, for Industrial Measurement and Control," ASEA Journal, Vol. 33, No. 3, pp. 23-32, 1960.

36. R. Peek, Jr. and H. Wagar, "Magnetic Circuit With Distributed Constants," Section 9.2, Switching Relay Design, pp. 317-322, D. Van Nostrand Co., Inc., New York, 1955.

37. H.H. Skilling, "Topology of Networks," Chapter 8, Section 13, Electric Networks, pp.197-198, John Wiley & Sons, New York, 1974.

38. R. Eisberg and L. Lerner, "Faraday's Law: The Crucial Role of Changing Magnetic Flux," Physics Foundations and Applications, Vol. II, pp. 1177-1179, McGraw-Hill Book Co., New York, 1981.

39. L.N Tseitlin, A.K. Shimokhin and F.B. Kim (Moscow Machine-Instruments Institute), "Magneto-Elastic Torque Converter," Soviet Union Patent 0838448,, published June 17, 1981, patented December 26, 1977.

40. Hiroyuki Aoki, "Operating Polarities of Solenoidal Torque Sensor Excitation Coils," private communication, Nissan Motor, Central Engineering Laboratories, Yokosuka, Japan, May 22, 1989.

41. S. Ramo et al., "Application of Maxwell's Equations: Penetration of Electromagnetic Fields Into a Good Conductor," Section 4.12, pp. 249-254, Fields and Waves In Communication Electronics, John Wiley & Sons, Inc., New York, 1965.

42. I. Sasada et al., "Characteristics of Chevron-Type Amorphous Torque Sensors Constructed By Explosion Bonding," IEEE Trans. on Magnetics, Vol. MAG-23, No. 5, pp. 2188-2190, September 1987.

43. "Cleaning and Finishing of Stainless Steel," Metals Handbook, 8th Edition, Vol. 2, Heat Treating, Cleaning and Finishing, pp. 599-606, American Society for Metals, Metals Park, OH, 1964 (eleventh printing, July, 1982).

44. I.J. Garshelis, "Magnetoelastic Torque Transducer," U.S. Patent 4,760,745, assigned to Mag Dev Inc., issued August 2, 1988.

45. "Maraging Tool Steels," Metals Handbook, 9th Edition, Vol. 3, Properties and Selection: Stainless Steels, pp. 446-447, Tool Materials and Special-Purpose Metals, American Society for Metals, Metals Park, OH, 1980.

46. I.J. Garshelis, "Magnetostrictive Properties Data For Maraging Steel -- Unimar 300, Supplied by Universal-Cyclops Specialty Steel, Pittsburgh, PA," (private communication), Mag Dev Inc., Pittsfield, MA, July 27, 1989.

APPENDIX A: DERIVATION OF TRANSFER PERMEANCE EQUATIONS

It is shown here that the sensing-member loop between nodes 1 and 2 of the equivalent magnetic circuit for the basic single-branch sensor element in Fig. 4(a) can be reduced to the simplified circuit shown in Fig. 4(b). This simplification is made possible by the following derivation of transfer permeance equations.

First, note that the complete magnetic circuit in Fig. 4(a) has three circuit elements in the sensing member, and these are located between nodes 3A and 3B. The three elements are: P_S the sensing-member permeance, P_3 Lenz's Law opposing-flux permeance, and mmf source $N_3 I_3$ associated with eddy current i_3. These elements were described previously in Eqs. (1) thru (10) of Ref. [1]. The objective here is to first reduce these three elements to a single element that will in itself completely represent the sensing-member.

In accordance with Ampere's Law, the magnetomotive force, mmf_{AB}, between nodes 3A and 3B in Fig. 4(a) is given as:

$$mmf_{AB} := N_3 \cdot i_3 + \frac{\Phi}{P_S} , \qquad (A.1)$$

where N_3 is the single-turn short-circuited winding associated with induced eddy current i_3, Φ is the mutual (main) flux which passes through the sensing member, and P_S is the sensing-member permeance.

When Faraday's Law, Kirchhoff's Law, and Ampere's Law -- given by Eqs. (3), (6), and (9) of Ref. [1] -- are combined, and sinusoidal fixed-frequency excitation is assumed; one obtains the result:

$$i_3 \cdot \left[R_3 + j\omega \cdot N_3^2 \cdot P_3 \right] := j\omega \cdot N_3 \cdot \Phi , \qquad (A.2)$$

where resistance R_3 accounts for eddy current losses, P_3 is Lenz's Law opposing-flux permeance, $\omega = 2\pi f$ is the radian excitation frequency, and terms i_3, N_3, and Φ were defined in Eq. (A.1).

Recall that ϵ, a field-dependent impedance parameter, was defined by Eqs. (14) and (16) of Ref. [1], and is given as:

$$\epsilon := \frac{\omega \cdot N_3^2 \cdot P_3}{R_3} , \qquad (A.3)$$

where all parameters have already been defined above.

Equations (A.1) thru (A.3) are combined, terms i_3 and R_3 are eliminated, and the factor ωN_3^2 is canceled out; then, after rearrangement, one obtains the expression:

$$mmf_{AB} := \left[\frac{\Phi}{P_S} \right] \cdot \left[1 + \frac{\dfrac{P_S}{P_3}}{1 - \dfrac{j}{\epsilon}} \right] , \qquad (A.4)$$

By definition, permeance is equal to the ratio, Φ/mmf. Using this definition, an effective value of sensing-member permeance is obtained from Eq. (A.4) as follows:

$$P_{S\Phi} := \frac{\Phi}{mmf_{AB}} := P_S \cdot \left[1 - \alpha_S \right] , \qquad (A.5)$$

where

$$\alpha_S := \frac{1}{1 + \frac{3}{P_S} \cdot \left[1 - \frac{j}{\epsilon}\right]} , \qquad (A.6)$$

These expressions are used in Eqs. (7) and (8) of this paper.

It is now straightforward to obtain the desired relationship for transfer permeance. From the magnetic circuit in Fig. 4(a). In accordance with Ampere's Law and using the effective value of sensing-member permeance $P_{S\Phi}$ in Eq. (A.5); the magnetomotive force mmf_{12} between nodes 1 and 2 in Fig. 4(a) is given as:

$$mmf_{12} := N_1 i_1 := \frac{\Phi}{P_{GD}} + \frac{\Phi}{P_{S\Phi}} + \frac{\Phi}{P_{GE}} , \qquad (A.7)$$

where $N_1 i_1$ is the mmf generated by the excitation coil, P_{GD} and P_{GE} are air gap permeances associated with the detection-pole D and the excitation-pole E, and $P_{S\Phi}$ is the effective value of sensing-member permeance given by Eqs. (A.5) and (A.6).

By definition, transfer permeance is given by Eq. (3) of this paper, which when combined with Eq. (A.7) gives the result:

$$P_{\Phi} := \frac{\Phi}{N_1 i_1} := \frac{1}{\frac{1}{P_{GAPS}} + \frac{1}{P_{S\Phi}}} , \qquad (A.8)$$

where

$$P_{GAPS} := \frac{1}{\frac{1}{P_{GE}} + \frac{1}{P_{GD}}} , \qquad (A.9)$$

The term P_{GAPS} is the series combination of air gap permeances associated with the excitation-pole E and the detection-pole D, and $P_{S\Phi}$ is the effective value of sensing-member permeance given by Eqs. (A.5) and (A.6).

Five elements between nodes 1 and 2 of the complete magnetic circuit in Fig. 4(a), in the sensing member, can now be replaced by a single transfer-permeance element, defined by Eq. (A.8). The resulting, simplified, magnetic circuit for the sensor is shown in Fig. 4(b) of the paper.

To compare with prior analysis of Ref. [1], it can be shown mathematically that the following identity holds:

$$P_{\Phi} := P \cdot (1 - \alpha) , \qquad (A.10)$$

where P_{Φ} is the transfer permeance given by Eq. (A.8), P is the series permeance term defined in Eq. (11) of Ref. [1], and α is given by Eq. (C.17) of Ref. [1].

The α-term, as discussed in conjunction with Eq. (C.16) of Ref. [1], is useful in physically depicting the sensor. It gives the ratio of eddy-current mmf, $N_3 I_3$, to excitation mmf, $N_1 I_1$; thereby resulting in the single independent-mmf source equivalent-magnetic circuit shown previously in Fig. A.2 of Ref. [1].

When air gaps are reduced to zero, the α-term of Eq. (C.16) in Ref. [1], and the α_S-term in Eq. (A.5) here become equal. Thus, these terms differ primarily in that the α-term of Ref. [1] includes effects of air gaps, while the α_S-term here does not.

APPENDIX B: SOLUTION OF NONLINEAR EQUATIONS

The solution for sensor operating point is nonlinear due to nonlinear behavior of the sensing material characteristic curves, such as those defined by Eqs. (B.1)-(B.4) below and shown in Figs. 2 and 3 of Ref. [3]. In brief, the magnetic flux predicted by the circuit equation of the sensor must equal that corresponding to the nonlinear field characteristic curve of its sensing material in order to obtain a solution point. Further discussion is found in Ref. [3].

The same approach is also followed here, and nonlinear solution procedures for each of the three sensor designs are outlined below.

BASIS USED FOR COMPARISON -- The three sensor designs, previously illustrated in Fig. 1, are shown in cross-sectional detail in Fig. B.1. The branch sensor has five coils -- a central coil for excitation and four corner coils for detection purposes. The cross sensor has two coils -- one coil for excitation and one coil for detection. And the solenoidal sensor has four coils -- two coils for excitation and two coils for detection.

For comparison purposes, the sensors were all assumed to detect torque in an identical 38.10-mm (1.500-inch) diameter shaft. Although this shaft diameter was arbitrarily selected, it is nevertheless thought to fit well within the range of shaft sizes of current engineering interest. The shaft material was assumed to be maraging steel, which is known to have excellent magnetostrictive properties [6,44,45]. The three sensors are all assumed to be separated from the shaft by identical 0.50-mm (0.020-inch) air gaps.

In the present analysis, all performance comparisons for the three designs will be normalized to the same shaft length. This is done because shaft space is at a premium in production powertrains, and therefore the sensor design with the best performance, per unit of shaft length, is obviously more attractive.

Figure B.1 shows that all three sensor designs were limited to the same length of shaft space, which in the present study was chosen equal to 15.56 mm. This choice corresponds to the length required to accommodate a pole separation of 10.00 mm in the branch design sensor.

Within the same 15.56-mm shaft space, cross and solenoidal designs in Fig. B.1 provide pole separations that are larger than the 10.00-mm separation of the branch design. Specifically, the cross and solenoidal designs provide pole separations of, respectively, 12.33 mm and 14.86 mm.

Larger pole separation translates to better torque measurement performance. So, based solely on packaging considerations illustrated in Fig. B.1, cross and solenoidal designs have an advantage over the branch design. Packaging advantages of cross and solenoidal sensors are primarily arise from the freedom to mount coils in the interior of the sensors, away from magnetic pole outer perimeters.

Note, however, that since the cross design is configured like a detachable sensing head, it can be simply installed and removed without disconnecting the driveshaft on which it measures torque. It therefore has this advantage over the solenoidal design that requires disconnection of the driveshaft for sensor installation and/or removal.

Table B.1
Parameter Values Used In Sensor Performance Calculations

Parameter	Units	Value Used: Branch Design	Value Used: Cross Design	Value Used: Solenoidal Design	Comment
N_1	turns	1000	1000	500	[a]
N_2	turns	1000	1000	500	[a]
G	mm	0.50	0.50	0.50	[b]
r_E	mm	2.640	2.640	--	[b]
r_D	mm	1.320	1.320	--	[b]
do_E	mm	9.105	18.86	6.380	[b]
do_D	mm	--	9.686	6.380	[b]
bo	mm	--	--	0.700	[b]
bc	mm	--	--	1.400	[b]
h	mm	8.425	10.87	6.910	[b]
R	mm	19.05	19.05	19.05	[b]
p'	degrees	--	--	45	[c]

[a] Numbers of coil turns are dependent on amounts of space available for coils in sensor envelopes shown in Fig. B.1, and/or are limited by resistance of wire (i.e., for the solenoidal sensor).
[b] Values of: air gap G, radii of poles r, separations of pole do, widths of ring poles b, heights of poles h, and radius of shaft R -- were taken from a CAD drawing corresponding to Fig. B.1.
[c] Pitch angle p' of grooves on shaft.

PARAMETERS USED IN CALCULATIONS -- Parameter values used in calculation of performances of the three sensors in Fig. B.1 are listed in Table B.1.

NONLINEAR MODEL FOR MAGNETOSTRICTIVE MATERIALS PROPERTIES -- Equations used in the present study were previously derived in Ref. [3], and are given here for reference purposes.

Appropriate depiction of the knee-shaped characteristic B-H curve of magnetostrictive materials was obtained using Eq. (12) of [3], which is as follows:

$$\mu_R := \mu_1 \cdot \left[d + \frac{B}{B_S}\right] \cdot \left[1 - \frac{B}{B_S}\right]^2 , \quad \text{for} \quad B \geq 0 , \quad (B.1)$$

where μ_1 and d are constants, B is the peak value of flux density in the sensing material, B_S is the value of B at very high field strength H, and μ_R is the relative permeability of the sensing material.

Change of permeability $\Delta\mu/\mu$ due to applied stress is, as discussed in [3], given by the same functional relationship -- Eq. (B.1) -- used to describe permeability; and is given as follows:

$$\frac{\Delta\mu}{\mu} := \frac{1}{\mu_1} \cdot \mu_R^{\ v} , \quad \text{for} \quad B \geq 0 , \quad (B.2)$$

where v_1 is constant, and μ_1 and μ_R are defined in Eq. (B.1).

Due to applied stress, permeability changes from μ to μ'; where, as taken from Eq. (10) of [3], μ' is given as:

$$\mu' := \mu \cdot \left[1 + \frac{\Delta\mu}{\mu}\right] , \quad (B.3)$$

where $\Delta\mu$ is the change of sensing member permeability due to applied stress (due to applied torque or force).

Sensing member impedance factor ϵ, a field strength parameter that appears in Eq. (8) of the present paper, was defined by Eq. (18) of [3] which is as follows:

$$\epsilon := \frac{1 + \left[\frac{B}{B_S}\right]^2}{1 + 3 \cdot \left[\frac{B}{B_S}\right]^2} , \quad (B.4)$$

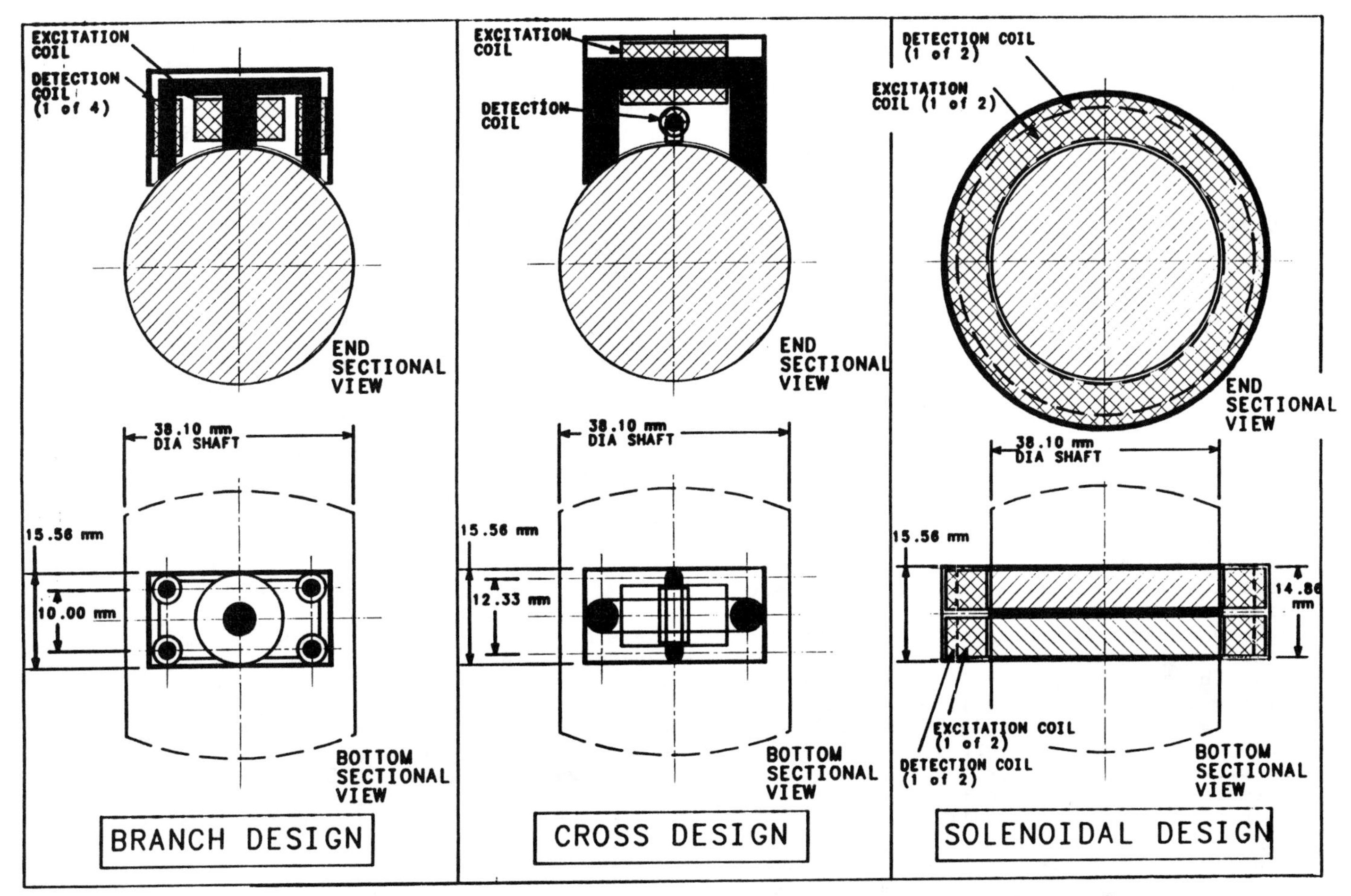

Fig. B.1 Cross-Sectional Views of Branch, Cross, and Solenoidal Torque Sensors Designed To Fit Into the Same Space On A Shaft Made of Maraging Steel. See Tables B.1 and B.2 for Further Detail.

where B and B_S have the same meaning as in Eqs. (B.1) and (B.2).

SELECTION OF SHAFT MATERIAL -- As previously mentioned, the shaft was assumed to be made of maraging steel. This material was chosen because of its known excellent magnetostrictive properties [6,44,45]. Appropriate properties of maraging steel, provided by Garshelis [46], are given in Table B.2. When these parameter values are substituted into Eqs. (B.1) and (B.2), characteristic curves for shaft permeability and stress-induced permeability change $\Delta\mu/\mu$ are computed.

Curves for μ_R', the perturbed value of permeability due to application of stress, are computed from μ_R and $\Delta\mu/\mu$ using Eq. (B.3); whereas ϵ, the field strength parameter, is computed using Eq. (B.4).

SELECTION OF OPERATING POINT -- Based on the survey of experimental results given in Table 2, and the model prediction of best signal at higher frequencies shown in Figs. 11 and 12, an excitation frequency of 50,000 Hz was selected in this study. As discussed above, frequencies higher than this introduce noise problems, whereas frequencies lower than this result in reduced output signal. Thus, the choice of 50,000 Hz is a compromise based on these engineering considerations.

With frequency fixed at 50,000 Hz, excitation current was varied until the point of maximum relative signal strength was found. All magnetic circuit calculations shown below for the three sensor designs in Figs. B.2 thru B.4 were made at these operating conditions.

MAGNETIC CIRCUIT SOLUTIONS -- Nonlinear solutions for each of the three sensor designs are outlined next.

BRANCH DESIGN SENSOR -- Individual branches of the multi-branch sensor, shown in Figs. 1, 2, and 5 are solved separately. Permeance parameters for each branch are scaled appropriately to represent single-branch sensor elements that correspond to individual branches of the multi-branch sensor.

Values of fluxes Φ_I thru Φ_{IV} passing through the branches are obtained by substituting, separately into Eqs. (3) and (5) thru (8); and using the permeance parameters associated with each single-branch element. As seen in the equivalent circuit of Fig. 5(b) and discussed in Appendix A of Ref. [1], each

Table B.2
Parameter Values That Describe Magnetostrictive Properties of Maraging Steel [6, 44-46]

Parameter	Value Used	Units	Comment
Nonlinear Magnetic Parameters			
μ_1	1019.6	--	[a]
d	0.09808	--	[a]
B_S	1.8660	Tesla	[a]
v_{1C}	-1.5294	--	[b]
v_{1T}	0.1912	--	[b]
μ_0	$4\pi \times 10^{-7}$	Henry/m	[c]
ρ	38.5×10^{-8}	ohm-m	[d]
P_{3A}-B	0.4063	uH	[e]
P_{3A}-C	0.3543	uH	[e]
P_{3A}-S	2.6999	uH	[e]

[a] The constants μ_1, d, and B_S were chosen to match the conditions: initial relative permeability = 100 at B = 0, and maximum relative permeability = 200 at B = 0.50 Tesla. Matching of conditions was done using Eqs. (13) thru (15) of Ref. [3]. The permeability values are appropriate for maraging steel, annealed at 815°C and then aged at 480°C [44,46].

[b] The constants v_{1C} and v_{1T} for, respectively, compressive and tensile stresses of -75 MPa and +75 MPa were chosen to match the conditions: $(\Delta\mu/\mu)_{MAX}$ = -0.3000 and +0.0375 at B = 0.50 T. These values represent effects due to a constant 30 N·m applied torque acting on a 12.7-mm O.D. shaft [44]. Figure 9 of Ref. [1] depicts effects of stress anisotropy for compressive and tensile applied stresses. These effects account for the widely different values of v_1 given here. The constants d and B_S are the same for both Eqs. (B.1) and (B.2).

[c] This is the constant for the permeability of vacuum.

[d] Electrical resistivity ρ of maraging steel was taken from Ref. [46].

[e] Values of above-surface air permeance P_{3A} for the branch-B, cross-C, and solenoidal-S, sensors were computed using Eq. (36), and the physical data given in Tables B.1 and B.2.

circuit branch is driven by the identically same mmf, $N_1 I_1$, of the center excitation coil.

For constant MMF (constant current) operation the sensor, the magnetic-circuit flux curve for each sensor branch is governed by Eq. (C.19) of Ref. [1] which, when combined with Eq. (A.10) above, can be expressed as:

$$\Phi_I\left[B,f,v_1\right] := \sqrt{2}\cdot N_1\cdot I_1\left|P_\Phi\left[B,f,v_1\right]_I\right| , \quad (B.5)$$

where Φ_I is the peak value of the fundamental-harmonic component of the mutual (main) circuit flux for branch I which is a function of flux density B, excitation frequency f, and applied stress defined by v_1; $\sqrt{2}$ converts from rms to peak values for the assumed sinusoidal excitation, $N_1 I_1$ is the rms value of mmf generated by constant-current excitation, and P_Φ is the transfer permeance of the sensor branch given by Eq. (5) of this paper.

In the P_Φ function of Eq. (B.5), the argument $v_1 = v_{1C}$ is used for sensor branches I and III that are aligned along lines of compression in the shaft, whereas $v_1 = v_{1T}$ is used for branches II and IV, aligned along lines of tension. Hence, when Eqs. (B.1)-(B.3) are used, together with v_1 values in Table B.2, appropriate relationships for branch fluxes Φ_I thru Φ_{IV} are obtainable using Eq. (B.5).

The sensing-material nonlinear permeance characteristic for each branch of the sensor is governed by Eq. (C.15) of Ref. [1], which is as follows:

$$\Phi_{SM}\left[B,f,v_1\right] := B\cdot\delta'\left[B,f,v_1\right]\cdot L\cdot g_u , \quad (B.6)$$

where Φ_{SM} and B are peak values of the fundamental-harmonic components of the flux and flux density signals inside the sensing material, L is a span dimension of the sensor measured between pole centerlines (shown in Fig. C.3 of Ref. [1]), g_U is a function of sensor geometry given by Eq. (C.8) of Ref. [1]. Note: the ratio, Φ_{SM}/H, where H = B/μ is the applied field strength, defines the nonlinear permeance corresponding to Eq. (B.6).

In Eq. (B.6), flux penetration depth δ' (Eq. (C.11) of Ref. [1]) is given as:

$$\delta'\left[B,f,v_1\right] := \sqrt{\frac{\rho}{2\cdot\mu_R\left[B,v_1\right]\cdot\mu_0\cdot\pi\cdot f}} , \quad (B.7)$$

where ρ is the resistivity of the sensing material, μ_R is the relative permeability value defined by Eq. (B.1), and f is the excitation frequency.

When permeability curves like those defined in Eqs. (B.1)-(B.3) -- as shown in Fig. 2 of Ref. [3] -- and the data of Tables B.1 and B.2 are used; one can compute how fluxes Φ_I, Φ_{II}, and Φ_{SM} vary as a function of flux density B. Results of these calculations -- made using Eqs. (B.5) and (B.6) -- are shown in Fig. B.2, where the parameters for the branch sensor of Table B.1 were used. The excitation frequency of 50,000 Hz was used, and an excitation current I_1 rms = 85 mA was found to produce maximum signal strength at this frequency and was therefore also used.

Functionally, flux density B enters into the Φ_{SM} expression of Eq. (B.6) directly as a linear factor, while field-dependent permeability μ enters Eq. (B.6) via the flux penetration term δ'. The Φ_I and Φ_{II} expressions of Eq. (B.5) are more involved. First, transfer permeance P_Φ is a complex variable; second, P_Φ includes effects of field-

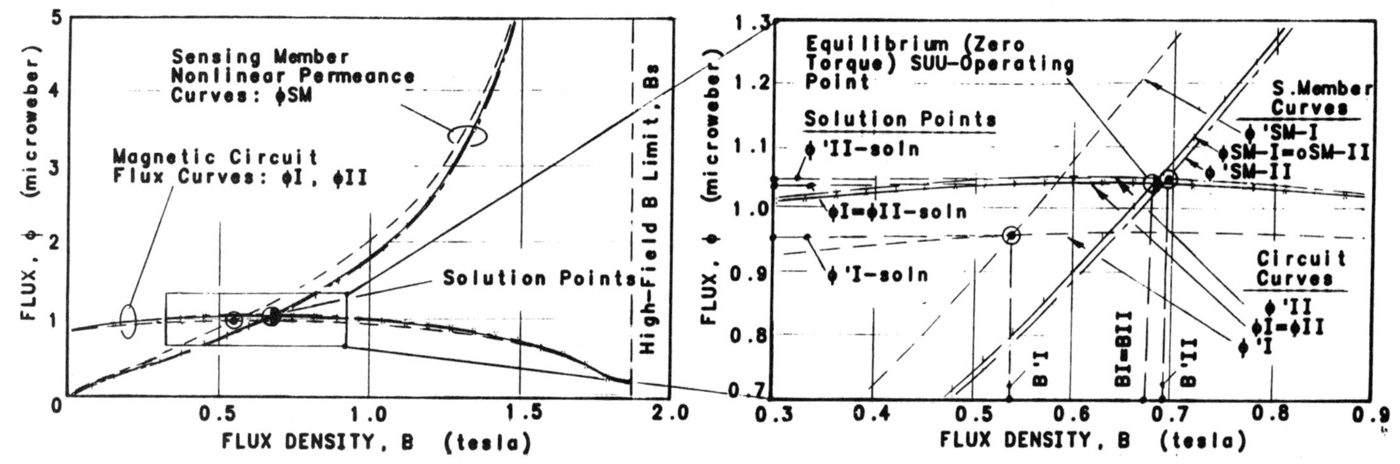

Fig. B.2 Nonlinear Field-Dependent Circuit Flux and Sensing-Member Nonlinear Permeance Curves Corresponding To the Point of Maximum Signal Strength For The Branch Sensor, Computed Using Eqs. (B.5) and (B.6), With Excitation: f = 50,000 Hz, and I_1rms = 85 mA.

dependent permeability in both the P_S and P_3 terms [see Eqs. (A.5) and (A.6)]; and thirdly, the factor α_S includes effects of the field-dependent impedance parameter ϵ [see Eq. (A.6)].

Figure B.2 shows that flux curves Φ_I, Φ_{II}, and Φ_{SM} change from initial values to perturbed values (as designated by the prime notation). This occurs because permeability changes from its unperturbed μ_R curve to its perturbed value μ_R' (due to application of stress). Solution points where magnetic fluxes Φ_I and Φ_{II} equal Φ_{SM} are shown in the enlarged inset on the right-hand side of Fig. B.2.

In Fig. B.2, the perturbed value of flux density B'_I, which is aligned with compressive stress, is decreased due to decrease of permeability, and there is also a decrease in flux Φ'_I-solution. On the other hand, the perturbed value of flux density B'_{II}, aligned with tensile stress, is increased due to increase of permeability, and there is an increase in flux Φ'_{II}-solution. This behavior occurs because maraging steel, defined in Table B.2, exhibits positive magnetostrictive material properties.

Summary -- In the nonlinear analysis, peak values of magnetic variables such as Φ and B are used, whereas rms values of electrical variables such as I and V are used. All variables are assumed to have sinusoidal waveforms and only the fundamental-harmonic components of the variables are considered. In addition, all solutions are obtained for constant applied torque acting on a sensing member shaft.

Numerical Methods -- Simultaneous solutions of the nonlinear sensor model for the intersecting Φ-B curves -- given by Eqs. (B.5) and (B.6) -- are shown in Fig. B.2. In prior work [3], these solutions were obtained using computer-spreadsheet Lotus 1-2-3(c) software, together with the numerical methods of interval halving and false position.

Much faster, more efficient, solutions have since been obtained with computer programs that utilize solve-block vector functions available using MathCAD(c) software. For example, solutions published in Figs. 7 and 9 of Ref. [3], that took 1-to-2 hours to complete, can be completed in less than a minute when MathCAD(c) is used instead.

CROSS DESIGN SENSOR -- The general method of nonlinear solution for the cross-design sensor is basically the same as for the branch sensor. There is, however, an important difference between nonlinear solutions for the two sensor designs.

For the branch sensor, there is only one sensing-member permeance element in each of its magnetic circuit branches, as defined by Eq. (B.5) above. However, for the cross sensor there are two elements in series, as defined by Eqs. (23) and (24) of this paper.

The cross sensor magnetic circuit, shown in Fig. 7 of this paper, is solved for each quadrant -- one solution for each compressive-stress quadrant, and one for each tensile-stress quadrant. Under constant current excitation, magnetic-circuit flux in Eq. (23), is given as:

$$\Phi_{Cross}\left(B, f, v_1\right) := \frac{\sqrt{2}\,\left[N_1 I_1\right]}{\left|\dfrac{1}{P_\Phi\left(B, f, v_{1C}\right)_I} + \dfrac{1}{P_\Phi\left(B, f, v_{1T}\right)_{IV}}\right|}, \quad \text{(B.8)}$$

where Φ_{CROSS} is the magnitude of the peak value of the fundamental-harmonic component of mutual (main) flux shown in the right-hand side of the circuit in Fig. 7, $\sqrt{2}$ converts from rms to peak values for the assumed sinusoidal excitation, $N_1 I_1$ is the rms value of mmf generated by constant-current excitation, and $(P\Phi)_I$ and $(P\Phi)_{IV}$ are the transfer permeances defined by Eqs. (19) thru (21) of this paper.

Equation (B.8) gives magnetic-circuit flux Φ_{CROSS} in terms of transfer permeance $(P\Phi)_I$, evaluated using field-dependent constant v_{1C} which corresponds to effects of compressive stress acting on shaft material in quadrant I; and $(P\Phi)_{IV}$, evaluated using constant v_{1T} which corresponds to effects of tensile stress acting in quadrant IV. Hence, this equation depicts the combined effects of both compressive quadrant-I stress and tensile quadrant-IV stress.

Similar to the branch-design sensor, nonlinear magnetic permeance in quadrant I predicted by the circuit equation (B.8), must equal that predicted by the sensing material equation, Eq. (B.6), in order to obtain a solution for Φ. To obtain a solution in quadrant I, it is assumed that the nonlinear permeance for quadrant I is equal to one-quarter of that obtained if the entire shaft material were subjected to compressive stress.

This assumption must be made because no electromagnetic field solution, consistent with other assumptions made in this study, is available to describe permeance in all quadrants due to combined effects of compressive stress in quadrants I and III, and tensile stress in quadrants II and IV.

For compressive stress acting in quadrant I, one therefore uses Eq. (B.6) to obtain the result:

$$\Phi_{SM\text{-}I}\left[B,f,v_{1C}\right] := \frac{B}{2}\cdot\delta'\left[B,f,v_{1C}\right]\cdot L_E\cdot g_u \quad \text{(B.9)}$$

where $\Phi_{SM\text{-}I}$ and B are peak values of the fundamental-harmonic components of the flux and flux density signals in sensing-material quadrant I. <u>Note</u>: the ratio, $\Phi_{SM\text{-}I}/H$, where $H = B/\mu$ is the applied field strength, defines the nonlinear permeance corresponding to Eq. (B.9).

The factor of 1/2 in Eq. (B.9) accounts for division of total flux into two equal parts. (As shown in Fig. 6, fluxes Φ_I and Φ_{II} enter pole E2. It is therefore assumed here that Φ_I is one-half of the total flux).

Flux penetration depth δ' in Eq. (B.9) is evaluated using the field-dependent constant v_{1C} which corresponds to effects of compressive stress acting on shaft material in quadrant I. Parameters L_E and g_u are functions of the sensor geometry which are evaluated using Eqs. (C.8) and (C.9) of Ref. [1].

To solve for nonlinear permeance in quadrant IV, it is again assumed that all quadrants of the shaft material are subjected to the same stress, but in this case they are subjected to tensile stress. Hence, for tensile stress, one uses Eq. (B.6) to obtain the result:

$$\Phi_{SM\text{-}IV}\left[B,f,v_{1T}\right] := \frac{B}{2}\cdot\delta'\left[B,f,v_{1T}\right]\cdot L_E\cdot g_u \quad \text{(B.10)}$$

where $\Phi_{SM\text{-}IV}$ and B are peak values of the fundamental-harmonic components of the flux and flux density signals inside sensing-material quadrant IV. The factor of 1/2 again accounts for the division of total flux into two equal parts, flux penetration depth δ' in Eq. (B.10) is evaluated using the field-dependent constant v_{1T} which corresponds to effects of tensile stress acting on shaft material in quadrant IV. Parameters L_E and g_u are functions of the sensor geometry which are evaluated using Eqs. (C.8) and (C.9) of Ref. [1].

When permeability curves like those defined in Eqs. (B.1)-(B.3) -- as shown in Fig. 2 of Ref. [3] -- and the data of Tables B.1 and B.2 are used; one can compute compute how fluxes Φ_{CROSS}, $\Phi_{SM\text{-}I}$, and $\Phi_{SM\text{-}IV}$ vary as a function of flux density B. Results of these calculations -- made using Eqs. (B.8) thru (B.10) -- are shown in Fig. B.3, where the parameters for the cross sensor of Table B.1 were used.

The excitation frequency of 50,000 Hz was used, and an excitation current I_1rms = 73 mA was found to produce maximum signal strength at this frequency and was therefore also used. Other parameters are given in Tables B.1 and B.2

Figure B.3 shows that under SUU-conditions, when no torque is applied, flux solutions $\Phi_{SM\text{-}I}$ and $\Phi_{SM\text{-}IV}$ are equal. When torque is applied to the shaft, flux curves Φ_{CROSS}, $\Phi_{SM\text{-}I}$ and $\Phi_{SM\text{-}IV}$ change

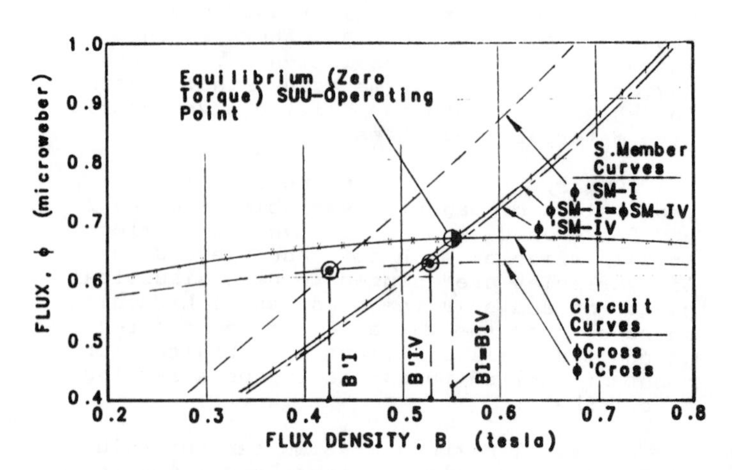

Fig. B.3 Nonlinear Field-Dependent Cross Sensor Flux and Sensing-Member Nonlinear Permeance Curves -- Computed Using Eqs. (B.8) thru (B.10), and Corresponding To the Point of Maximum Signal Strength.

from initial values to perturbed values (as designated by the prime notation).

Note that perturbed values of flux density B'_I and B'_{IV}, shown in Fig. B.3, are associated with each solution. Both B'_I and B'_{IV} are decreased with respect to the zero-torque SUU-operating point solution points.

Normally, the positive magnetostriction properties of tensile stress would increase B'_{IV} above the SUU-operating point, as was the case for the branch sensor in Fig. B.2. However, due to the pronounced, 8-to-1, anisotropy in field constants v_{1C} and v_{1T} for ferromagnetic metals like maraging steel, given in Table B.2, and the fact that compressive permeance is connected in series with tensile permeance in the cross sensor flux equation (B.8), compressive stress effects dominate those of tensile stress.

Thus, Fig. B.3 shows that the negative-direction-shifting effects of compressive stress dominate in the cross sensor flux curve, given by Eq. (B.8), such that the B'_{IV} solution point is pulled downwards, opposite its behavior if it were by itself.

<u>Note</u>: in accordance with symmetry relationships given in Eq. (22), solutions for remaining fluxes Φ_{SM-II} and Φ_{SM-III} are, respectively, simply equated to solutions for fluxes Φ_{SM-I} and Φ_{SM-IV}. And finally, output signal of the cross-design sensor is determined using Eqs. (25) and (26) of this paper.

SOLENOIDAL DESIGN SENSOR -- The general approach to solving this nonlinear problem is basically the same as for the branch-design and cross-design sensors.

There is, however, another important difference between the nonlinear solutions described above, and those for the solenoidal-design sensor. In both the above solutions, flux-carrying branches in each magnetic circuit need only be solved one branch at a time; whereas for the solenoidal sensor there are two flux-carrying branches which require simultaneous solution -- one for circulating flux Φ_I and the other for circulating flux Φ_{II}, as defined by Eqs. (39) and (40). The ramifications of this difference are described here.

Values of circulating fluxes Φ_I and Φ_{II} in Fig. 9 are each defined by simultaneous solutions given in Eqs. (39) and (40) of this paper. As before, the circulating magnetic flux Φ_I in loop I is governed by a circuit equation, and it must equal flux governed by the sensing-material nonlinear permeance characteristic for band I of compressive-stress sensing material. Similarly, circulating flux Φ_{II} in loop II must equal flux governed by the permeance characteristic of the tensile-stress sensing material in band II.

Fluxes Φ_I and Φ_{II} are defined as:

$$\left|\Phi_I\right| := \sqrt{2}\,\left|\Phi_I\left[B,f,v_{1C}\right]\right| , \qquad (B.11)$$

and

$$\left|\Phi_{II}\right| := \sqrt{2}\,\left|\Phi_{II}\left[B,f,v_{1T}\right]\right| , \qquad (B.12)$$

where $|\Phi_I|$ and $|\Phi_{II}|$ are peak values (magnitudes) of the fundamental-harmonic components of the circulating fluxes shown in the magnetic circuit in Fig. 9; the functions $\Phi_I[B,f,v_{1C}]$ and $\Phi_{II}[B,f,v_{1T}]$ are given by Eqs. (39) and (40); and $\sqrt{2}$ converts from rms to peak values for the assumed sinusoidal excitation.

In the grooved-surface band I of the shaft, shown in Fig. 8, one uses Eq. (B.6) to obtain:

$$\Phi_{SM-I}\left[B,f,v_{1C}\right] := B\,\delta'\left[B,f,v_{1C}\right]\,2\,\pi\,R , \qquad (B.13)$$

where Φ_{SM-I} and B are peak values of the fundamental-harmonic components of the flux and flux density signals in sensing-material band I, and flux penetration depth δ' is evaluated using the field-dependent constant v_{1C} which corresponds to effects of compressive stress acting on shaft material in grooved-surface band I.

The term, $2\pi R$, in Eq. (B.13) accounts for the fact that flux uniformly flows from the circumference of grooved-surface band I of the solenoidal sensor; where R is the shaft radius.

In grooved-surface band II of the shaft, one can use Eq. (B.13) above, but changed to reflect the fact that it now applies to the tensile-stress band II of the solenoidal sensor. The desired sensing-member band II equation is:

$$\Phi_{SM-II}\left[B,f,v_{1T}\right] := B\,\delta'\left[B,f,v_{1T}\right]\,2\,\pi\,R , \qquad (B.14)$$

where Φ_{SM-II} and B are peak values of the fundamental-harmonic components of the flux and flux density signals inside grooved-surface band II, and flux penetration depth δ' is evaluated using the field-dependent constant v_{1T} which corresponds to effects of tensile stress acting on shaft material in grooved-surface band II. As before, the term, $2\pi R$, in Eq. (B.14) accounts for the fact that flux uniformly flows from the circumference of grooved-surface band II of the solenoidal sensor; where R is the shaft radius.

When permeability curves like those defined in Eqs. (B.1)-(B.3) -- as shown in Fig. 2 of Ref. [3] -- and the data of Tables B.1 and B.2 are used; one can compute how fluxes Φ_I, Φ_{II}, Φ_{SM-I}, and Φ_{SM-II} vary as a function of flux density B. Results of these calculations -- made using Eqs. (B.11) thru (B.14) -- are shown in Fig. B.4, where the parameters for the solenoidal sensor of Table B.1 were used.

The excitation frequency of 50,000 Hz was used, and an excitation current $I_{1\,rms}$ = 48.5 mA was found to produce maximum signal strength at this frequency and was therefore also used. Other parameters are given in Tables B.1 and B.2

Figure B.4 shows that under SUU-conditions when no torque is applied, flux solutions Φ_{I} and Φ_{II} are equal. These solutions are at the point where curves 2 and 5 intersect. When torque is applied to the shaft, flux curves Φ_{I}, Φ_{II}, Φ_{SM-I} and Φ_{SM-II} change from unperturbed values to perturbed values (as designated by the prime notation). When torque is applied, solution points for magnetic fluxes are shown in Fig. B.4.

Note that perturbed values of flux density B'_{I} and B'_{II}, shown in Fig. B.4, are associated with each solution. B'_{I} is decreased with respect to the SUU-operating point solution for the band-I compressive-stress solution, and B'_{II} is increased for the band-II tensile-stress solution. This behavior occurs because maraging steel, defined in Table B.2, exhibits positive magnetostrictive material properties.

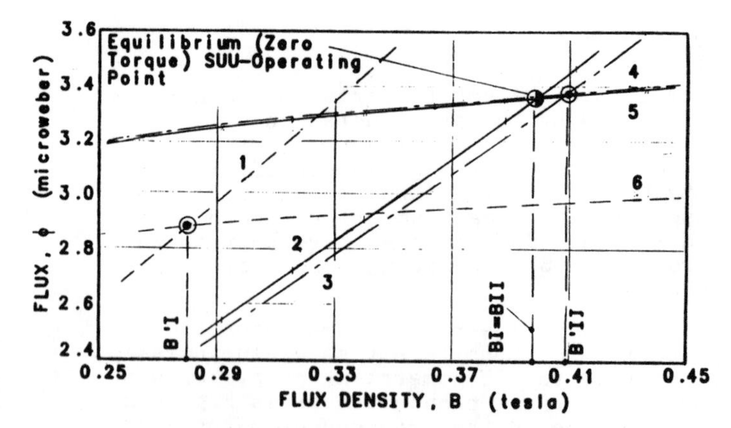

Fig. B.4 Nonlinear Field-Dependent Solenoidal Sensor Flux and Sensing-Member Nonlinear Permeance Curves -- Computed Using Eqs. (B.11) thru (B.14), and Corresponding To the Point of Maximum Signal Strength.

Curve	Description
1	Φ'_{SM-I}
2	$\Phi_{SM-I} = \Phi_{SM-II}$
3	Φ'_{SM-II}
4	Φ'_{II}
5	$\Phi_{I} = \Phi_{II}$
6	Φ'_{I}

AIR-FUEL RATIO SENSORS

970856

Local Air-Fuel Ratio Measurements Using the Spark Plug as an Ionization Sensor

Raymond Reinmann, André Saitzkoff, and Fabian Mauss
Lund Institute of Technology

Copyright 1997 Society of Automotive Engineers, Inc.

ABSTRACT

The influence of variable air-fuel ratio inside a spark ignition engine is examined by the use of an ionization sensor. The measured ion currents are used for predicting the local air-fuel ratio in the vicinity of the spark plug. In order to support the results, a theoretical analysis has been made. An instationary chemical kinetic model burning a mixture of iso-octane and n-heptane is used for the calculations. The results are used to reconstruct the crank angle resolved ion current that has been measured in an engine. This technique has been developed in order to offer a supplementary low-cost facility of controlling the air-fuel ratio within the combustion chamber of an engine.

INTRODUCTION

The local air-fuel equivalence ratio or excess air factor λ, present at the time of the early flame kernel development in an internal engine, is of major interest for the total understanding of the combustion process. In a spark ignition engine it is the gas mixture close to the spark plug that is of outermost importance. Local variations in λ do exist, since the combustible mixture has an inhomogeneous distribution prior to ignition. These variations are often referred to as one of the main contributors to the phenomenon called cycle-to-cycle variations in maximum cylinder pressure [1]. The flame initiation process is relatively long and will therefore significantly affect the total combustion process. If the local λ is far from the preferred and pre-set stoichiometric value, the combustion will suffer from lower reactivity and hence slower initial combustion and eventually leading to a lower engine efficiency. Consequently, the emission of air pollutants will increase beyond the regulated limits. Another disadvantage of the cyclic variations, if they are too large, is the undesirable rough engine performance that causes less comfortable driving. Finding a way of measuring and ultimately controlling the local λ would therefore definitely be of interest to the engine manufacturers.

In order to control the air-fuel ratio, a reliable technique for measuring the λ-value is needed. The oxygen sensor that most modern cars use for controlling and regulating the λ-value can not give any information about the local conditions around the spark plug at ignition, since it gives a global mean value from all cylinders. Due to this fact, the oxygen sensor can not yet detect the individual cylinder performance regarding the air-fuel ratio. One extreme possibility that can not be detected by the oxygen sensor would, for instance, be if half of the cylinders burn lean mixtures and the other half burn rich mixtures. The oxygen sensor is a rather costly part, therefore the use of one sensor per exhaust port is avoided although it would have been the most desirable solution. By creating a closed loop for regulating the strategy of fuel injection by an air-fuel sensor inside the cylinder close to the spark plug, the overall engine performance can be increased. This capability might become very useful in the most modern direct injected spark ignited engines.

Alternative ways of measuring the air-fuel ratio in the individual cylinders with low extra costs have been investigated with some different approaches during the last years. The light emitted by the cyanogen radical (CN) from the spark channel and initial flame kernel is somewhat dependent on the λ-value [2]. This technique demands a fiber optic connection to the cylinder, via e.g. the spark plug, and some additional optical filters and light detectors in order to work properly. Another successful way has been to measure the time for the residual charge, in the ignition system, to pass through the spark gap when the discharge has ended [3]. This charge has to be released by the ions and electrons that have been created by the spark process and the initial flame kernel. By measuring the spark voltage with a capacitive clamp-on probe on the high voltage cable, a signal that correlates well with the air-fuel ratio has been found [4].

This paper will however deal with the strong correlation that exists between the ion current, measured by the spark plug as a sensor, and the oxygen sensor [5]. A typical change of the ion current as a function of the crank

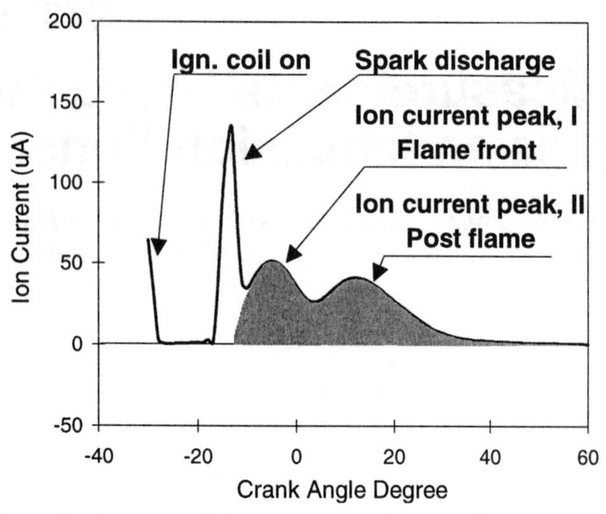

Figure 1. *A measured ion current with an explanation to the different occurring peaks. This is an averaged signal from 500 cycles at 2500 rpm and at a medium load. The measurements were done in an Opel 1600 cm^3 4-cylinder engine.*

angle, taken from our measurements, is shown in figure 1. The part of the ion current signal which correlated best with the λ-value measured by the oxygen sensor, was found by using a neural network [5]. This trained network found the strongest correlation in the first peak of the ion current signal. Encouraged by these results, we have made a detailed analysis of this correlation with an emphasis on why the correlation exists and what makes it that strong. The first peak in the ion current originates from the excessive chemi-ionization in the reaction zone of the propagating flame. The flame front remains only for a limited time in the region where it can be detected. When the reaction zone has progressed into areas that can not be measured by the spark plug, the ion sensor reflects the conditions of the recombining ions that was formed during the combustion. As the ion current is decreasing due to the recombining ions, the products of the combustion process are approaching their equilibrium concentrations. At this point there will also be a contribution from the thermal ionization that takes place in the post flame zone, i.e. the burned gases [6,7]. This is what generates the second ion current peak.

The main topic of this paper is however an explanation of the dependence between the first peak of the ion current and the local air-fuel ratio in the vicinity of the spark plug. In order to do this we make use of a chemical kinetic model that incorporates the electro-chemical reactions necessary to calculate the excessive ionization responsible for the charge production in the flame front.

FORMATION OF THE ION CURRENT

The technique of measuring ion currents in flames has been well known for a long time. It is almost two centuries, since Volta in the year 1801 discovered the electrical conductivity of a freely propagating flame. In the year 1906, Tufts was the first scientist arguing for an ionization process that had something to do with the chemistry in the flame front. In the early 50´s to 60´s this process was thoroughly investigated as it was considered to be one of the explanations for the high flame speed of burning hydrocarbons [8-10]. Recently, this area has been the subject for a newborn interest as the spark plug of an SI engine has become used as the ion current sensor to probe the physical and chemical properties inside the combustion cylinder. The success of sensing a gas volume of high temperature with violent chemical reactions has been used by many groups, but its application in an internal combustion engine is still rarely reported. Most of the work that has been published deals with continuous burners at pressures between 10^2-10^5 Pa [8-10].

The ionization process of molecules and atoms requires a considerable amount of energy. For the most frequently appearing species found in combustion systems the required energy is in the range of 4-16 eV, see table 1. In 1957 Calcote published a comprehensive survey of the ionization process that is present in the reaction zone of a flame [8]. He stated that the ionization was not accomplished by thermal ionization, but rather by a chemi-ionization process and a cumulative excitation process of molecules, atoms or radicals which can lead to ionization. The chemi-ionization process is caused by the energy released from an elementary chemical reaction together with the thermal collision energy, i.e. in the presence of exothermic reactions the ionization rate of the reaction products is enhanced. This can be explained by the crossing of the specific potential energy surfaces of the involved reacting molecules [9,11]. As a result of such a reaction, one of the products can actually end up in an excited or an ionized state.

Species	Ionization energy (eV)	Species	Ionization energy (eV)
N_2	15.5	C_2	12.0
N	14.53	CH	11.13
O_2	12.2	CH_2	11.82 & 10.4
O	13.614	CH_3	9.84
H_2	15.427	CH_4	12.61
H	13.595	C_2H_2	11.41
H_2O	12.6	C_3H_3	8.67
OH	13.18	C_3H_8	11.14
H_3O		CHO	9.88
CO	14.05	CH_3O	9.2
CO_2	13.84	Na	5.138
NO	9.25	K	4.339

Table 1. *Ionization energies of some commonly occurring combustion related species together with some minor species having a low ionization potential.*

The most commonly reported ions in the reaction zone of flames are CH_3^+, CHO^+, $C_3H_3^+$, and H_3O^+. The CH_3^+-ion is usually found in concentrations of less than one thousandth of CHO^+ and $C_3H_3^+$ [12]. The H_3O^+-ion is formed later than the others, since it is not formed direct from an elementary reaction but rather with a charge transfer reaction, see equation 2. This leaves us with CHO^+ and $C_3H_3^+$ as the primary ions, meaning that they control the formation of all other present ions. Three specific reactions have been considered the most probable for governing the major part of the ion formation and the total ion concentrations because of a favorable thermochemistry [8-10,12,13], see equations 1-3. Actually, the CHO^+ production can be accomplished by two different reactions starting from either the ground state CH or an excited state of the same radical. CH is present in much lower concentrations in the electronically excited states than in the ground state, therefore we neglect the contribution of this reaction in our present study. In addition, the temperatures in the reaction zone is high enough in order to produce CHO^+ from the ground state of CH.

$$CH + O \xrightarrow{k_1 = 5\times10^{-14}\ cm^3/s} CHO^+ + e^- \quad (1)$$

$$CH(^2\Sigma^+) + O \rightarrow CHO^+ + e^- \quad (1')$$

$$CHO^+ + H_2O \xrightarrow{k_2 = 7\times10^{-9}\ cm^3/s} H_3O^+ + CO \quad (2)$$

$$CH\left(A^2\Delta\right) + C_2H_2 \rightarrow C_3H_3{}^+ + e^- \quad (3)$$

The dominating flame ion has by most groups been found to be H_3O^+, using a mass spectrometer or a Langmuir probe. This can also be seen from the reaction rate coefficients shown in the reactions above. The reaction rate of reaction 2 is much higher than for reaction 1, therefore almost all CHO^+ are transformed into H_3O^+. The removal of this ion is obtained by the dissociative recombination with an electron to form water and a hydrogen atom.

$$H_3O^+ + e^- \xrightarrow{k_4 = 2.3\times10^{-7}\ cm^3/s} H_2O + H \quad (4)$$

This reaction is very fast but the collision probability for these two species is quite small. The electron density will also be reduced by the competing electron attachment to electronegative species such as O and OH. Due to these facts the H_3O^+ ion can be present quite far into the post flame region.

In the post flame zone, NO increases its importance because of its low ionization energy. The governing ionization process is thermal ionization since most of the violent exothermic chemical reactions have ended and the fast radicals have reached their equilibrium concentration. In a previous paper [6] we have presented results showing that an ionization equilibrium model can predict the ion current in agreement with our experiments. It follows that the thermal ionization is the governing ion formation process in the post flame gases of an engine. In a continuos atmospheric burner however the temperature of the post flame gases is lower and therefore the thermal ionization gets less dominant. Other papers have concluded that the recombination of ions, formed in the reaction zone of the propagating flame, predicts the post flame ion concentration and their contribution to the current through the post flame region [8,14-16].

As stated above, the strongest and best correlation between the ion current and the air-fuel ratio was found in the parts of the ion current that is related to the reaction zone. Hence, a more detailed analysis is necessary for the interaction between physics and chemistry in the propagating flame front after ignition of the hot flame kernel produced by the spark. In the next section a brief description of the used chemical kinetic model is given together with a steady-state assumption concerning the ion concentrations.

CHEMISTRY

Large efforts have been made for the description of the high and the low temperature oxidation of large aliphatic hydrocarbons like n-heptane and iso-octane [17]. On the basis of the reaction schemes compiled in reference [18] a simplified mechanism has been developed that is applicable for instationary Navier-Stokes equations 19-21]. A combination of these mechanisms leads to a RON dependent kinetic scheme, applicable for engine conditions. This scheme consists of 268 elementary reactions and 61 chemical species. A fully implicit backward differential scheme of variable order is used to solve the correspondent second order discretised Navier-Stokes equation.

The chemical kinetic model that is briefly described above is used to determine the species concentrations as a function of time or crank angle degree. Especially important are the concentrations that exist in the flame front, i.e. at ignition. For an accurate calculation of these concentrations a refined grid is used that is moving with the flame front. The pressure that is measured in the engine is used as input to the model. The obtained concentrations are used in the following steady-state assumptions that describes the emerging ion concentrations due to the chemi-ionization process.

The reaction rates (k_1, k_2 and k_4) for the reactions 1, 2 and 4 are given in references [10,22]. By assuming that the involved species in these reactions have reached a steady-state concentration in the flame front, we can estimate the ion concentrations. This assumption is not bad since the time-scales to reach these steady-state concentration are in the order of nano- to microseconds. The concentrations are given by

$$\left[CHO^+\right] = \frac{k_1[CH][O]}{k_2\left[H_2O\right]} \quad (5)$$

$$\left[H_3O^+\right]_{max} = \frac{k_2\left[CHO^+\right]\left[H_2O\right]}{k_4\left[e^-\right]} = \frac{k_1[CH][O]}{k_4\left[e^-\right]} \quad (6)$$

since $[H_3O^+]_{max} \cong [e^-]$ in the reaction zone, equation 6 can be rewritten as

$$\left[H_3O^+\right]_{max} = \left(\frac{k_1[CH][O]}{k_4}\right)^{1/2} \quad (7)$$

An evaluation of equation 7 leads to a steadily increasing ion concentration as the relative air-fuel ratio is decreased on the lean side of stoichiometry. The reason for this is that the amount of fuel is increased and thus the CH concentration increases. The oxygen concentration is not affected in the same drastic way since the fuel is not yet fully oxidized, meaning that most of the oxygen that was present prior to ignition is still available. As the gas mixture gets even richer, below λ=1.0, the pyrolysis and oxidation of the fuel will be incomplete. The oxygen concentration will start to decrease at a higher rate and the CH concentration will start to increase at a slower rate. As a consequence the H_3O^+ concentration gets oxygen controlled leading to a maximum at approximately λ=0.8 - 0.9 and then for even richer conditions the concentration decreases. For λ-values larger than 0.8 we can express the H_3O^+ concentration.

$$\left[H_3O^+\right]_{max} = \frac{const}{\sqrt{\lambda}} \tag{8}$$

The chemical calculations gives the distribution of the species concentrations along the cylinder radius. The reaction zone thickness that is calculated by the model is smaller than the actual thickness inside the engine since the calculations do not consider the influence of the gas flow and its inherent turbulence. A typical thickness of the flame inside an engine cylinder is in the literature given to be in the order of 0.1 mm [23]. This thickness includes the wrinkling of the flame front due to the turbulence. From the number of ions and electrons, the ion current can be calculated by applying the equations for the drift velocities of the ions [6].

The reaction rate for the $C_3H_3^+$ formation is at this time still not found in any literature. Therefore, as a first order approximation of the problem at hand we neglect the influence of this ion. A second order approximation can be obtained by using the amount of thermal ionized C_3H_3 at the temperature given from the chemical calculation. The third order approximation, that can be made, is to reduce the ionization potential with the heat of formation for reaction 3 [12-13]. These assumptions will give $C_3H_3^+$ concentrations that are lower than the real concentration for the first two approximations. The last approximation will result in too large concentrations, since not all of the released energy from the exothermic reaction, 3, will be transferred into excitation energy that will reduce the ionization potential.

EXPERIMENTS

An Opel engine was equipped with an ion current sensor in order to probe the combustion inside one of the cylinders. The engine is a straight 4 cylinder engine, with 16 valves, a compression ratio of 10.5 and a displacement volume of 1600 cm^3. The maximum torque is 150 Nm and is achieved at 3800 rpm. The maximum brake power is 80 kW at 6000 rpm. The engine is available in the Opel models: Astra, Corsa and Tigra.

The igniting sparks were produced by a non standard one coil per cylinder ignition system which provides a short intense discharge. The short discharges are necessary since the spark currents are large in comparison to the obtained ion currents and will therefore dominate the measured current if it is present during the measuring of the ion current, see figure 1. The measured spark current that is shown in figure 1 is not correct since what can be seen in this figure is only the electrical interference emerging from the spark. In addition, the spark current is flowing in the opposite direction and hereby completely destroying the ion current signal.

The ion currents are sampled together with the pressure data and oxygen sensor output by an AVL Indimaster 670. The pressure measurements are done via a piezoelectric crystalline Kiestler pressure transducer mounted inside the cylinder. The errors made in the measurements are in the order of 5 %. The pressure measurements were acquired in order to perform the chemical calculations, when evaluating the current contribution from the different species in the flame front. It can also be used for calculating the heat release.

The measuring series were organized in the following manner:

- the engine velocity ranged over 1300 - 4000 rpm
- the torque was varied between 19 and 85 Nm
- the λ was varied between 0.9 and 1.1.

Two series of the measurements mentioned above were performed. In the first one, the ignition timing was kept constant for each parameter setup while varying the λ and in the second series, the crank angle for maximum pressure is kept within the range of 12-15° ATDC, in order to achieve a MBT condition, while the λ is varied. For each series, 500 individual cycles were collected. These 500 cycles were used to produce a mean ion current that was used to find the overall correlations between the ion current and the λ-value. In figure 2, the recorded ion current is shown for different mean air-fuel ratios. The most obvious behaviors upon

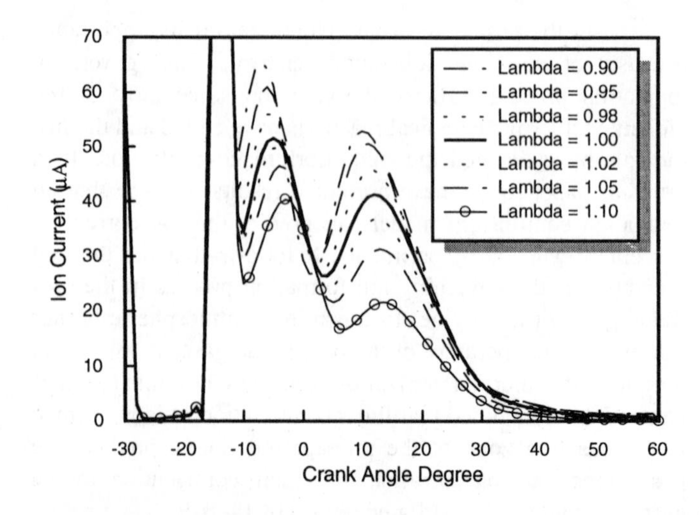

Figure 2. *Ion current waveforms as a function of the crank angle degrees for different λ-values. The relative air-fuel ratio is indicated in the legend. The engine operated at 2500 rpm with a torque of 85 Nm.*

λ-changes is examined for all our cases in the next section.

In the correlation graphs shown below, the legends need some clarifying information of what they represent. The first part of the legend refers to which one of the two series is being used, "io" is the one with constant ignition times and "ion" is the one with regulated peak pressure position. The last part of the legend gives the information about the engine speed. "13" refers to 1300 rpm, etc. There is one exception to this; "24", "25" and "26" are all referring to 2500 rpm. In these experiments the load conditions were varied. The number of appearing legends in the graphs is a measure of the applicability or quality of the method, since some engine conditions lead to ion current signals that could not be evaluated by the chosen technique.

EVALUATION OF MEAN ION CURRENTS

The well known fact that the first peak of the ion current contains information about the air-fuel ratio is in this section examined in more detail. For this preliminary analysis the averaged ion current curves were used in order to determine what could be expected from the individual ones. Furthermore, the behavior of the averaged ion current curves with the air-fuel ratio will be used for predicting the individual cycle variability of the local air-fuel ratios existing in the vicinity of the spark plug. From the graphs shown in this section it will be possible to express the air-fuel ratio as a function of different ion current parameters.

$$\lambda = f_1(\text{parameter}_1) + f_2(\text{parameter}_2) + \ldots \quad (9)$$

The best correlation between the ion current and the air-fuel ratio exists for the maximum value of the first ion current peak, see figure 3. In addition to the air-fuel ratio dependence the peak value seems to be direct proportional to the engine speed. The scattering from this relation, that can be seen in figure 3, is explained by the difference in engine load conditions and hence differing gas densities. The position of the first peak also slightly changes for the different

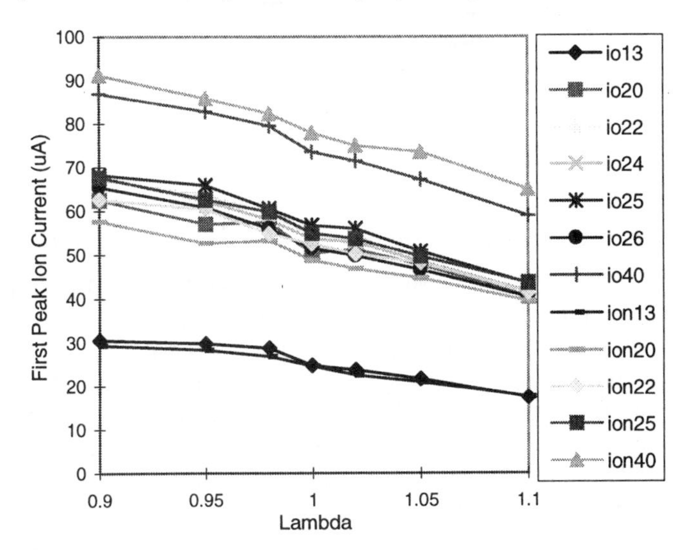

Figure 3. *First ion current peak amplitude dependence on the air-fuel ratio.*

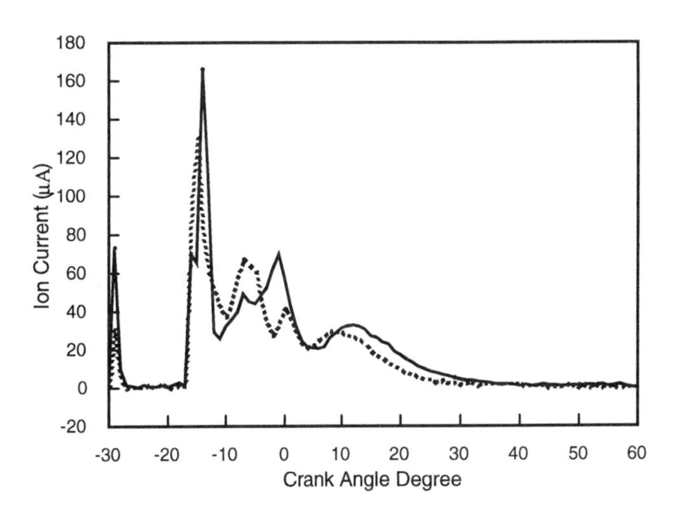

Figure 4. *Two examples of noisy ion currents with a subdivided first peak. Most of the noise in the figure has been removed by reducing the sample speed of the ion current signal by a factor of 10. The sample speed for all figures is 1 sample/CAD. The engine conditions were 2500 rpm, 65 Nm and λ=1.0.*

engine conditions which gives an additional change in the gas density.

In the individual curves, however a problem of finding the maximum current emerges as the first ion current peak can divide into several peaks and can also be very noisy, see figure 4. This phenomena occurs when the irregular gas flow inside a cylinder moves the flame kernel in the same irregular way. For this reason, we have to process the ion current data by an appropriate method that will not fail due to these kind of disturbances. Filtering, averaging and integration techniques were tried with varying success. An additional demand for the evaluation process is its apparent need of speed, since the time between two subsequent cycles is limited. Filtering and curve fitting processes usually takes too long time on the microcomputers available today. Therefore, we looked for a quick, simple and reliable method. The method of locally averaging the ion current peak was found to be both accurate and fast without loosing the strong correlation that exists between the peak ion current and the λ-value, see figure 3. The averaged electrical charge is calculated by first finding the maximum current in a pre-defined crank angle window (CAW), e.g. -12 - 5 would be appropriate for the signals shown in figure 2. When the peak current is found we integrate the current in a ± 5 CAW. The integrated value is then divided by the length of this window in order to get the mean current. From figure 5, it can be seen that for higher engine speeds (4000 rpm) the mean ion current is relatively higher than for the lower engine speeds. This behavior is explained by the effect from the current of the spark. At high engine speeds the first ion current peak occurs too close to the interference signal originating from the spark.

The ignition delay of a hot ignitable gas mixture is dependent on the stoichiometric conditions of the gas as well as the gas temperature. For lean conditions, the amount of available specific energy is decreased and thereby the flame temperature is reduced. The reduced flame temperature will

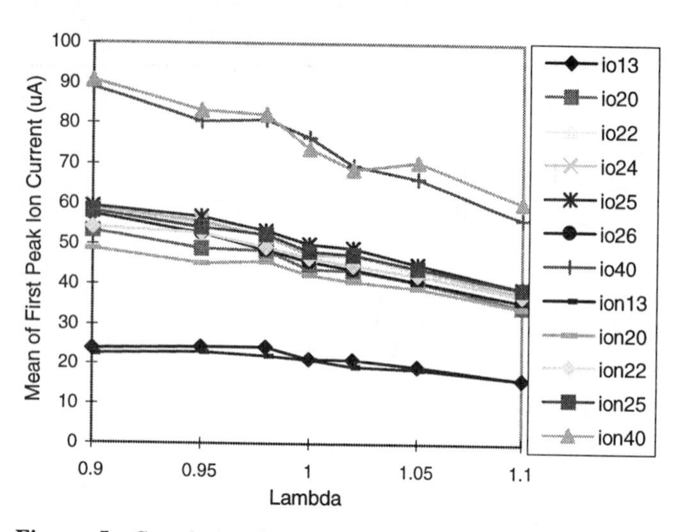

Figure 5. *Correlation between the integration of the first ion current peak divided by the peak duration.*

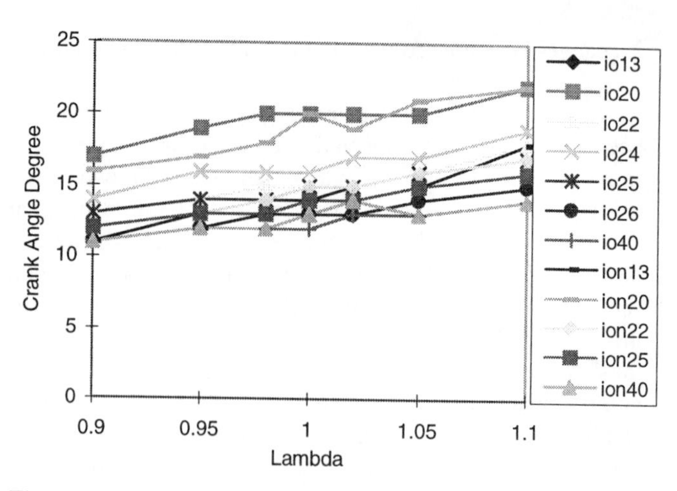

Figure 6. *Correlation of the distance in CAD between the ignition and position of the first ion current peak with the air-fuel ratio.*

result in a lower reactability of the gas, thus leading to a slower combustion as well as a longer ignition delay. The position of the first ion current peak in reference to the ignition time is directly proportional to the ignition delay and can therefore be used in order to visualize this parameter. This dependence is shown in figure 6 where the increase in ignition delay clearly can be seen for lean mixtures. The two uppermost curves are referring to very low load conditions of the engine, therefore the engine temperature is lower leading to lower gas temperatures. In addition, the ignition timing is set early which also reduces the gas temperature at the ignition time.

In figure 7 the correlation between the duration of the first ion current peak and λ is shown. This change of duration can also be explained by the change of reactability of the gas mixture. In the air-fuel ratios that has been investigated the flame speed is highest for the richest conditions and gradually decreases with increased λ-value. As can be seen from the scattering of the curves, the flame speed is not a function of air-fuel ratio only. The initial gas temperature is another parameter that affects the flame speed to large extent whereas the influence of changing pressure is small. The high load and high temperature engine conditions have a shorter first peak duration than the low load and low temperature conditions. This is verified by the curves in figure 7.

An ion current parameter that is connected with those shown in figures 3 and 6 is the rising edge derivative of the first ion current peak. The ion current was not derivated with respect to the crank angle degrees, instead the derivative was approximated by the first peak value divided by the distance between ignition and the first peak position. This parameter contains information about both the total number of ions, i.e. the peak value, and the ignition delay. These two parameters varies in strictly opposite directions as the air-fuel ratio is changing, therefore the dependence increases by dividing the two parameters. The need for a strong dependence is evident, if the ion current is to be used for extracting information about the gas mixture. In figure 8 this strong correlation is shown.

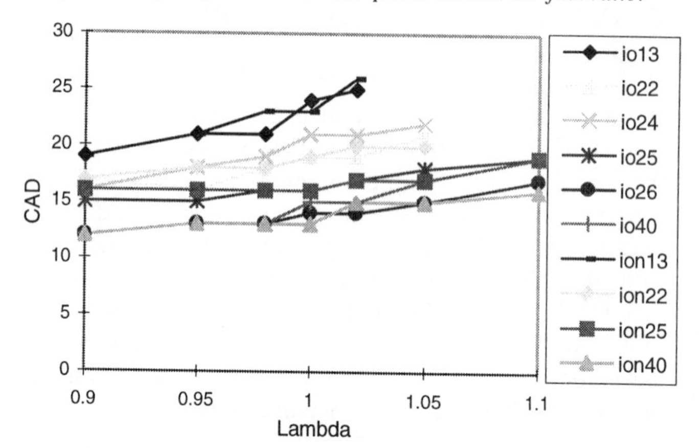

Figure 7. *Correlation between the duration of the first ion current peak and the air-fuel ratio.*

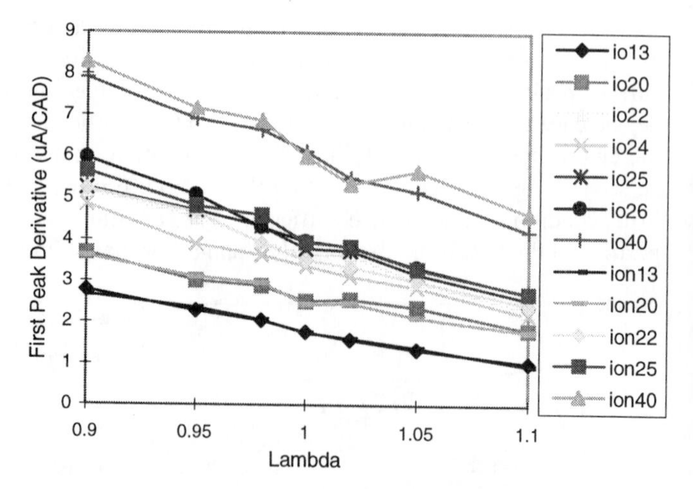

Figure 8. *The rising edge derivative of the first ion current peak.*

Although, we have earlier stated that the strongest correlations between the ion current and the air-fuel ratio exist in the reaction zone, correlations can also be found in the post flame region. A clear dependence for the second ion current peak on the air-fuel ratio is shown in figure 9. The explanation of this dependence is not evident, since we have explained the current flowing through the post flame gases by the thermal ionization process of NO [6-7]. In the λ-range that we have examined, the NO concentration should increase with air-fuel ratio according to the Zeldowich mechanism. The same facts that explains the higher NO concentrations for

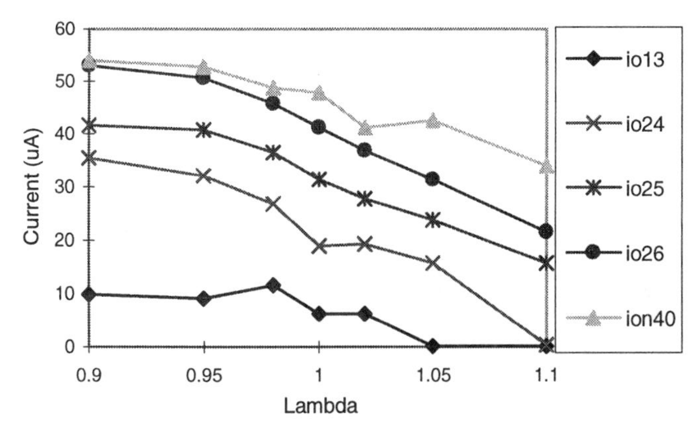

Figure 9. *Dependence of the second ion current peak value on air-fuel ratio.*

the leaner conditions also explains the lower ion currents. The explanation for the above changes is the oxygen concentration. Higher oxygen concentrations yield higher NO concentration within the given limits. As the oxygen atom is very electronegative, it will also be very likely to attach a free electron. When the electron is bound to the oxygen atom the mobility of the charge is highly reduced, therefore yielding lower ion currents. Note that this dependence only can be used for the medium to high load conditions. In the low load conditions the second peak is hard to find.

In the correlation graphs that are described above (3,5,8 and 9) a clear dependence on the engine speed can be noticed. This fact is by no means astonishing, since the ion concentration and thereby the amount of electrical charge is independent of the engine speed. The current, I, that arises due to moving electrical charges, Q, is expressed by the simple equation

$$I = \frac{dQ}{dt} = \frac{dQ}{d\theta}\frac{d\theta}{dt} \tag{10}$$

where $dQ/d\theta$ is constant but $d\theta/dt$ is not. At higher engine velocities the crank angles, θ, per time, t, is increased and therefore the current has to increase. This explanation is valid as long as the drift velocity of the ions and electrons is much larger than the gas flow in the cylinder. For all normal engines this is the case. There is also a clear dependence of the ion current on the engine load. For higher loads the density of the gas increases and therefore the total ion density increases.

The relations that has been presented in this section can all be used for determining the air-fuel ratio locally around the spark plug. In the next section only the derivative is used for such an evaluation. In addition, there is now enough evidence that a correlation should be found when performing a theoretical analysis of the problem. The results obtained from this analysis is given below.

EVALUATION OF INDIVIDUAL CYCLES

The ion current behavior that was extensively evaluated as a function of air-fuel ratio in the previous section is used in order to predict the local λ-value close to the spark plug. These variations have previously been investigated in an engine by the use of Laser Induced Fluorescence (LIF) [24-25]. In figure 10 an example of the pictures obtained by this technique is shown. The variation in air-fuel ratio is visualized by the changing gray scales of the picture. However, this picture is taken in a completely different engine and is used just for visualization of the actual change in air-fuel ratio that can exist locally around the spark plug.

The derivative, shown in figure 8, is used for the evaluation of the individual cycles. In order to exemplify the procedure one specific engine condition is chosen for a more detailed description of the evaluation process. The chosen engine condition is 37.5 Nm torque at 1300 rpm.

The first thing that has to be made is to form a function where the λ-value is a function of the derivative. This is accomplished by fitting a linear curve to the specific experimentally deduced curve in figure 8. From this linear equation the searched function can be derived.

$$\lambda = \frac{24.45 - \dfrac{dI}{d\theta}}{18.17} \tag{11}$$

Where the current I has to be inserted in μA. Secondly, the 500 individual cycles have to be processed in order to calculate their individual derivative. As mentioned above in the previous section the problem with multiple "first" peaks has to be taken into account. The procedure of locally averaging the maximum ion current peak was found to be unnecessarily complicated. The finally chosen search algorithm was even simpler; all peaks within a given CAW were localized. The mean value of the absolute ion current value and the mean value of its position was successfully used in the evaluation. The result of the evaluation can be seen in figure 11.

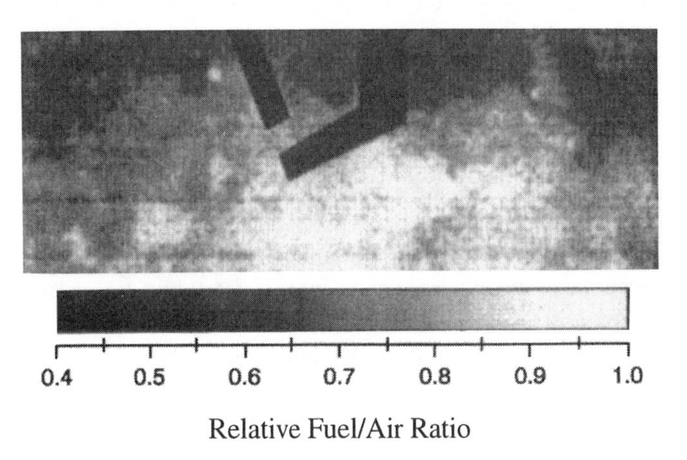

Figure 10. *A two-dimensional LIF picture of the fuel distribution 1 crank angle prior to ignition. The preset ϕ-value is 0.65 which corresponds to λ=1.54. The field of view is 13 × 30 mm^2 in which the area in front of the electrodes has been removed [24-25].*

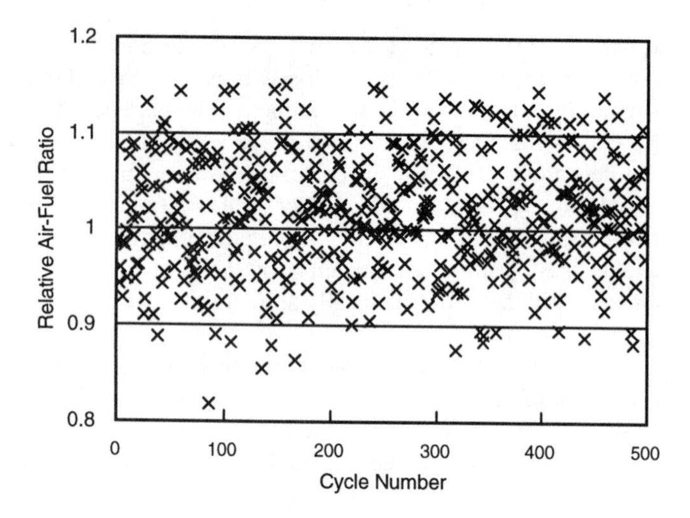

Figure 11. *The individual local λ-values for the 500 cycles of the ion13100 case, i.e. the mean λ-value is set to 1.00.*

The mean value of the predicted air-fuel ratios together with the standard deviation is statistically calculated. The mean value of λ taken from figure 11 is equal to 1.0178 and is not far from the preset value of 1.0. The standard deviation of the predicted λ-value is 0.0611, this is very close to that obtained by the LIF measurements [25]. It is worth to note that the distribution of the λ-values, as indicated in figure 11, reflects the true distribution of the real local air-fuel ratio and is not an error distribution.

In order to validate the prediction ability for various engine conditions this procedure is repeated for some different engine conditions. The results are presented in figure 12 and 13. In these plots the location of the circular bubbles refer to the engine load and rpm in figure 12 and rpm and λ-value in figure 13, whereas the radius of the bubbles are related to the standard deviations. The standard deviations in λ ranged from 0.0611 to 0.1183. The numbers given above the bubbles for the different conditions are the mean values of the individually predicted relative air-fuel ratios. The maximum error in the predicted mean air-fuel ratio is 3.1 %.

All the evaluations that are done in figure 12 are referring to the λ-value of 1.00. The reason for this is that the engines of today use a three-way catalyst and therefore have to be run under stoichiometric conditions. If the evaluation is performed for the available data at λ=0.9 and 1.1, approximately half of the cycles have to be evaluated by extrapolation, since they fall outside the investigated region. In spite of this fact the evaluation was performed anyway for the two conditions mentioned above. The results can be seen in figure 13. At λ=0.9 the predicted mean value is too high since the derivative can not be approximated by a linear curve below λ=0.9 [26], therefore the decrease in derivative in this area will be evaluated by a too high λ-value.

Additionally, it can be seen that the standard deviation increases with decreasing air-fuel ratio as well as for increasing engine speeds. In figure 13 the standard deviations ranged between 0.0427 and 0.1744. The maximum error made by predicting the air-fuel ratio partly outside the investigated region is 1.8 % at λ=1.1 and 6.2 % at λ=0.9.

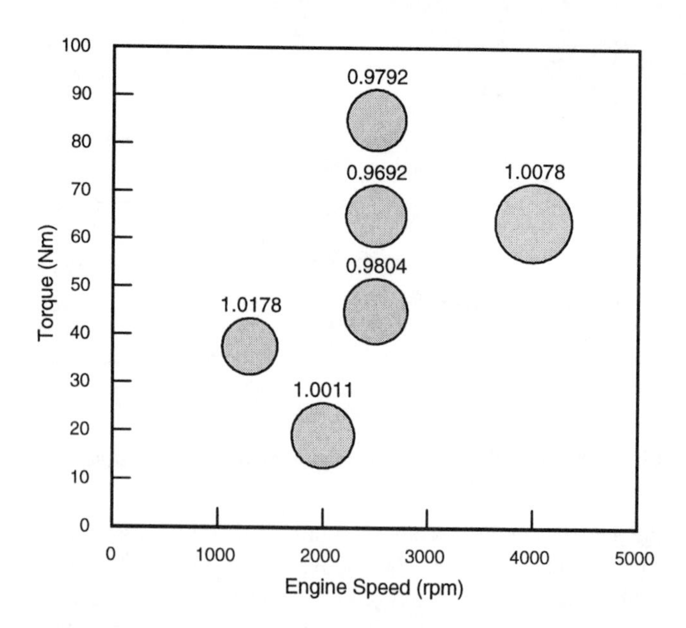

Figure 12. *Bubble chart of the individual engine conditions. The values above the bubbles are the mean values of the air-fuel ratio and the radius of the bubbles represent the standard deviations.*

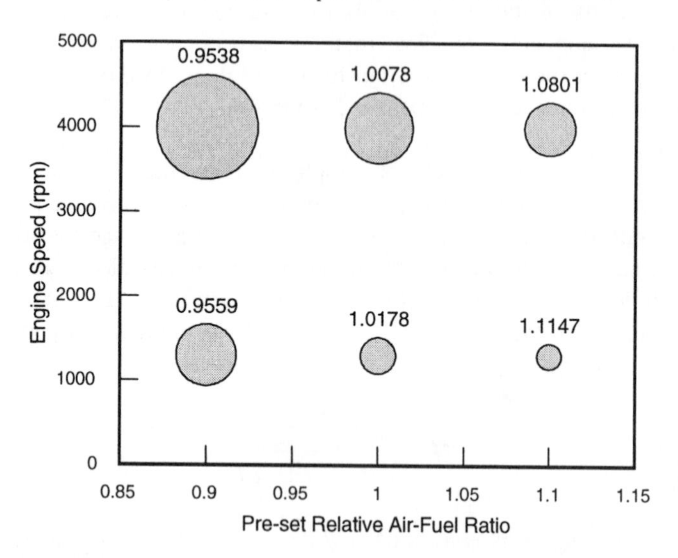

Figure 13. *Bubble chart of the individual evaluations of two engine conditions at the border of our investigation.*

THEORETICAL AND CHEMICAL EVALUATION

In order to test the above hypothesis that describes the first ion current peak generation, a brief numerical theoretical analysis is conducted along the lines described above. The theoretical analysis will be performed on three different engine conditions in which the load is kept constant and the λ-value is changed from 0.9 to 1.1 in steps of 0.1. The ion concentrations are calculated from the species concentrations according to equation 5 and 7. The concentrations of CHO^+ have shown to be only a small fraction of the concentrations of H_3O^+, therefore this ion is not being further used in this evaluation. Actually, all CHO^+ is transformed into H_3O^+ because of the reaction rates that control their formation and consumption.

The contribution from $C_3H_3^+$ is also investigated although the reaction rate is not known at this moment. By calculating the ionization degree of C_3H_3 it can be concluded that the contribution from this ion can be neglected. The contribution to the current was approximately 5 orders of magnitude lower than the contribution from H_3O^+. If on the other hand the ionization potential of C_3H_3 is reduced by the energy released from reaction 3, the contribution to the current is of the same order as the contribution from H_3O^+. This assumption is however over predicting the ion concentrations. This analysis of the $C_3H_3^+$ contribution to the ion current leads to the conclusion that this ion can be neglected for the moment, at least until the reaction rate is found.

The ion concentrations will contribute in different ways depending on their location in the engine cylinder. There is basically only two parameters that will affect the generated current for a given ion concentration. These two parameters consist of the electrical field, E, and the volume of the reaction zone, V_{rz}, of the ions. The current can be formulated

$$I = \frac{n_e V_{rz} v_d}{r} = \frac{\varphi_s n_{tot} V_{rz} E\mu}{r} \tag{12}$$

where n_e is the ion concentration, v_d is the drift velocity, r is the distance between the reaction zone and the center of the electrode gap, φ_s is the molar fraction of ion s, n_{tot} is the total gas density and μ is the ion mobility. In the reaction zone there is no doubt that the main contribution of the ion current is generated by the free electrons.

The electrical field, generated by the DC voltage across the spark gap, in and around the spark plug gap, is very complex because of the geometry of the spark plug electrodes, the piston and the cylinder top. A theoretical analysis of the existing field has to be made, yielding a simplified picture that with fair agreement corresponds to this field. The electrical field from an arbitrary electrically charged surface is given by the standard equation

$$E = -\nabla V = \frac{1}{4\pi\varepsilon_0\varepsilon}\int \frac{\sigma\, dS}{r^2}\frac{\bar{r}}{r} \tag{13}$$

where V is the electrical potential, and ε is the permittivity, σ is the charge density on the electrode surface S, r is the distance between the charged element to the position in which the field is sought and $\bar{r}$ is the direction orthogonal to the surface. The electrical field within the electrode gap can be viewed as approximately constant within a radius of 1.25 mm from the cathode axis. Far beyond this radius, the field is decreasing as a function of the inverse square of the radius according to equation 13. In the intermediate regions the field is smoothly approaching the behavior of the distant regions.

The high concentration of ions are mainly existing in the flame front. The total number of ions are increasing as the square of the radius since it is the surface of the increasing flame kernel that gives the number of ions. Beyond a radius of 10 mm the number of ions will increase as a linear function of radius since most of the ions are present in areas that are effectively shielded by the spark gap geometry. These facts also imply that the measurements done with an ionization sensor are quite local.

The ion mobility used in equation 12 for the calculation of the ion drift velocity irrespective of sign is given by

$$\mu = \frac{e\lambda_{mfp}}{mv_r} \tag{14}$$

where e is the unity charge, λ_{mfp} is the mean free path, m is the mass of the ion and v_r is the random velocity of the ion. The mobility of the ions is dependent on their location but this influence can be neglected without loosing the general description behavior.

In figure 14 the calculated ion currents can be seen. The absolute values of the currents maximum are too large and the relative changes are too high. The results are however accurate enough in order to perform further investigations improving the approximations made in the models used for this calculations. As can be seen in figure 14 the current flowing through the post flame region is not calculated by this model. A detailed description of how this can be made is found in reference [7].

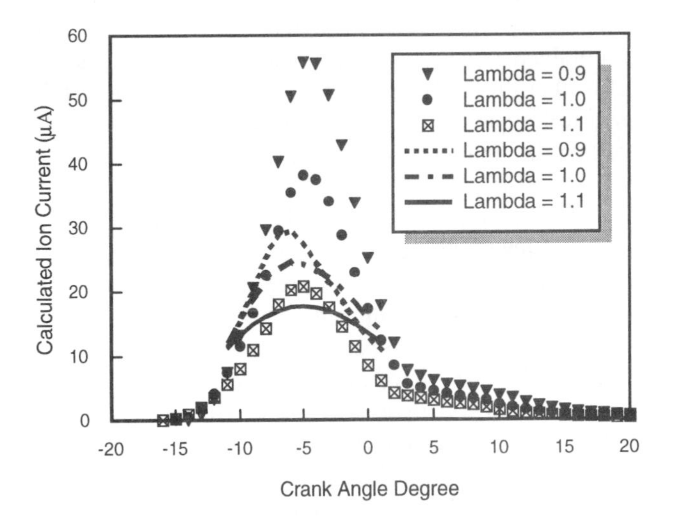

Figure 14. *The ion currents that was calculated from the ion concentration achieved from the chemical kinetic model are indicated by the lines and the measured ion currents are represented by the symbols according to the legend. The engine condition was 37.5 Nm torque at 1300 rpm. The symbols represent the experimentally recorded ion currents.*

CONCLUSION

It has been shown that the ion current that is measured through the spark gap can be used for measurements of the local λ close to the spark plug with a high accuracy, ±3 %. This accuracy could have been improved even more by the use of a combination of all the ion current parameters that has been presented in this paper. The scattering of local λ can be quite far from the preset λ measured by the lambda sensor, since the lambda sensor is measuring a global mean value over the cycle and also over the different cylinders. This makes the ion sensor for λ measurements a good complement, especially for stratified charge engines where differences in the mixture concentrations are desirable and has to be well controlled. If the results from the ion current system can be fed back into the fuel injector controller system, it will offer the possibility of an almost ultimate engine control system.

Not only can this ion sensor technique be used for monitoring the air-fuel ratio for the feedback system that controls the fuel injection. It can also be used as a sensitive tool for engine mapping in order to obtain the maximum driving conditions at the specific engine condition of interest. Stratified charge engines will benefit largely from such a system since it reveals the true condition at the spark plug location at ignition. Other on-board techniques have been investigated but the "ion-sense" technique seems to be the most easily applicable one.

Considering the conclusions drawn one decade ago regarding the use of ion probes, the potential of using the ion current has gone from a very low to a very high potential. In reference [27] the main conclusions were rather negative and can be summarized by: the large variance in the ion probe data, especially at high air-fuel ratios together with the poor and inconsistent correlation with the peak pressure and mass fraction burned implies a negligible potential for use in control of the air-fuel ratio. One of our main conclusions then has to be that this technique has now progressed into a fully mature technique that could successfully be used for controlling the air-fuel ratio in a spark ignition engine.

The results from the chemical analysis seems promising although the curves in figure 14 did not completely match. The errors made in the theoretical analysis can not be neglected, especially regarding the chemical kinetic model. The use of a mixture of iso-octane and n-heptane instead of a standard gasoline facilitated the calculations to a large degree but also imposed an unavoidable error. In conclusion, the theoretical analysis was without doubt successful in predicting the ion current in the first peak of a typical engine ion current curve. The anomaly shown in figure 14 can be reduced by improving this first order model.

ACKNOWLEDGMENT

This work was financially supported by the Swedish National Board for Industrial and Technical Development, (NUTEK).

REFERENCES

[1] Keck, J. C., J. B. Heywood and G. Noske, "Early Flame Development and Burning Rates in Spark Ignition Engines and Their Cyclic Variability" SAE 870164

[2] Merer, R. M. and J. S. Wallace "Spark Spectroscopy for Spark Ignition Engine Diagnostics" SAE 950164

[3] Miyata, S., Y. Ito and Y. Shimasaki "Flame Ion Density Measurement Using Spark Plug Voltage Analysis" SAE 930462

[4] Shimasaki, Y., M. Kanehiro, S. Baba, S. Maruyama, T. Hisaki and S. Miyata "Spark Plug Voltage Analysis for Monitoring Combustion in an Internal Combustion Engine" SAE 930461

[5] Auzins, J., H. Johansson and J. Nytomt "Ion-Gap Sense in Misfire Detection, Knock and Engine Control" SAE 950004

[6] Saitzkoff, A., R. Reinmann, T. Berglind and M. Glavmo "An Ionization Equilibrium Analysis of the Spark Plug as an Ionization Sensor" SAE 960337

[7] Saitzkoff, A., R. Reinmann, F. Mauss and M. Glavmo "In-Cylinder Pressure Measurements using the Spark Plug as an Sensor" SAE Paper, 1997

[8] Calcote, H. F. "Mechanisms for the Formation of Ions in Flames" *Combustion and Flame*, Vol. **1**, p. 385, 1957

[9] Calcote, H. F. "Ion Production and Recombination in Flames" Eigth Int. Symposium on Combustion, p. 184, 1960

[10] Green, J. A. and T. M. Sugden "Some Observations on the Mechanism of Ionization in Flames Containing Hydrocarbons" Ninth Int. Symposium on Combustion, p. 607, 1965

[11] Levine, R. D. and R. B. Bernstein Molecular Reaction Dynamics and Chemical Reactivity Oxford University Press, 1987

[12] Lawton, J. and F. J. Weinberg Electrical Aspects of Combustion Clarendon Press, Oxford 1969

[13] Kistiakowsky, G. B. and J. V. Michael
"Mechanism of Chemi-Ionization in Hydrocarbon Oxidations"
J. Chem. Phys., Vol. **40**, p. 1447, (1964)

[14] Collings, N., S. Dinsdale and D. Eade
"Knock Detection by Means of the Spark Plug"
SAE 860635

[15] Wenzlawski, K. und D. Heintzen
"Ionenstrommessung an Zündkerzen von Ottomotoren als Klopferkennungsmittel"
MTZ Motortechnische Zeitschrift **51** (1990)

[16] Krämer, M. and K. Wolf
"Approaches to Gasoline Engine Control Involving the Use of Ion Current Sensory Analysis
SAE 905007

[17] Nehse, M. and J. Warnatz.
"Kinetic Modelling of the Oxidation of Large Aliphatic Hydrocarbons"
26th Symposium (International) on Combustion, The Combustion Institute, 1996

[18] Chevalier, C., P. Louessard, U. C. Müller and J. Warnatz
"A Detailed Low Temperature Reaction Mechanism of n-Heptane Auto-Ignition"
Int. Symp. on Diagnostics and modeling of combustion in internal engines, COMODIA 90, Kyoto The Japan Society of Mechanical Engineers, 19908

[19] Müller, U. C.
Reduzierte Reaktionsmechanismen für die Zündung von n-Heptan und iso-Oktan unter Motorrelevanten Bedingungen, Ph.D. Thesis RWTH-Aachen 1993

[20] Müller, U. C., N Peters,. and A. Liñán
"Global Kinetics for n-Heptane Ignition at High Pressures"
24th Symposium (International) on Combustion, The Combustion Institute, 1992

[21] Pitsch, H., N. Peters,. and K. Seshadri
"Numerical and asymptotic studies of the structure of premixed iso-octane flames"
26th Symposium (International) on Combustion, The Combustion Institute, 1996

[22] Miller, W., J.
Ionization in Combustion Processes
in Oxidation and Combustion Revs, Vol **III**, p. 97, Edited by C.F.G Tipper, Elsevier Press, Amsterdam, 1968

[23] Heywood, J. B.
Internal Combustion Engine Fundamentals.
McGraw-Hill, 1988, ISBN 0-07-100499-8

[24] Neij, H., B. Johansson and M. Aldén
"Development and Demonstration of 2D-LIF for Studies of Mixture Preparation in SI Engines"
Combustion and Flame, Vol. **99**, p. 449-457, 1994

[25] Johansson, B., H. Neij, M. Aldén and G. Juhlin
"Investigations of the Influence of Mixture Preparation on Cyclic Variations in a SI-Engine, Using Laser Induced Fluorescence"
SAE 952463

[26] Clements, R. M. and P. R. Smy
"The Variation of Ionization with Air/Fuel Ratio for a Spark-Ignition Engine"
J. Appl. Phys., Vol. **47**, No. 2, February 1976

[27] Anderson, R. L.
"In-Cylinder Measurement of Combustion Characteristics Using Ionization Sensors"
SAE 860485

DEFINITIONS, ACRONYMS, ABBREVIATIONS

CAD - Crank Angle Degree.
CAW - Crank Angle Window
TDC - Top Dead Center (0 CAD.)
ATDC - After Top Dead center.
BTDC - Before Top Dead Center.
RON - Research Octane Number
MBT - Maximum Brake Torque
WOT - Wide Open Throttle.

970869

An Optical Sensor for Measuring Fuel Film Dynamics of a Port-Injected Engine

Timothy L. Coste
Control Devices, Inc.

Lawrence W. Evers
Michigan Technological University

Copyright 1997 Society of Automotive Engineers, Inc.

ABSTRACT

Increasingly stringent emissions regulations and customer demands for high efficiency and smooth performance demand highly accurate control of the air-fuel ratio of automotive spark-ignition engines. Electronic port fuel injection provides the necessary control by adding a precise quantity of fuel for a given amount of air drawn in by the engine. Ideally, the metered fuel will consist only of fine droplets and vapor. In reality, the fuel spray impinges upon the walls of the intake port, creating a liquid fuel film. The fundamentally different transport mechanisms of the liquid fuel compared to vapor or fine droplets greatly complicates the analysis of the fuel delivery system. Past research has provided models of fuel film dynamics in intake ports of port-fuel-injected engines, yet to date no practical method of measuring fuel films has been presented. This research attempts to fill this gap with the design and development of a method of measuring port wall wetting in production, port-fuel-injected engines. This work describes the design and installation of multiple optical sensors [1] flush in the engine's intake port and engine test results for two of the installed sensors.

INTRODUCTION

The fuel film thickness measurement system presented here is designed for the specific constraints of measuring fuel film in an intake port. The sensor requires access only to the underside of the fuel film and measurements are taken optically. In addition, the sensing face is small and can be mounted flush with the surface of the intake port. Optical measurement and flush mounting of the sensors minimizes intrusion into fuel film behavior. The single sided access requirement eases the problem of installing the sensors in the tight confines of the intake port region of the cylinder head. An earlier version of the film sensor system is presented in a paper entitled "Fuel Film Dynamics in the Intake Port of a Fuel Injected Engine." [2]

BASICS OF THE SENSOR SYSTEM

Bundles of optical fibers carry light from lasers to the port-mounted sensors where transmitting components in each sensor guide light into the fuel film where it is internally reflected by the fuel-air interface. Receiving components of each sensor guide the reflected light to other optical fiber bundles. The reflected light is sent to photodetectors and conditioning electronics, physically isolated from the running engine. The intensity of the reflected light varies with varying film thicknesses; the intensity variation is converted into a voltage; and a calibration curve converts the voltage to a depth. Measurements are taken in an operating engine, unaltered except for installation of seven film sensors, an in-cylinder pressure transducer, and a universal exhaust gas oxygen sensor, installed in a dynamometer test facility. Engine operation is controlled by the factory engine control computer. The engine is operated at various steady-state and transient conditions, and dynamic film depths measured are examined and compared to measured engine data.

Results show the sensors are capable of measuring dynamic changes in fuel films in operating engines. Dynamic fuel film signals were correlated with intake port events, such as injector firing, on a cycle-by-cycle basis. Fuel film was found to vary, both in magnitude of film and in the manner port events effect measurements, with changes in engine conditions. Thus, a viable technique has been developed for measuring fuel films in an operating engine, and measurements provided may be used to validate film dynamics models and further understanding of the relationships between fuel films and engine performance.

OPTICAL SENSOR OPERATION

The basic concept of the film sensor is to use the liquid/vapor interface, formed between fuel film and the port environment, to act as a reflector of a light beam. A conical light beam is directed at the interface and reflected back into the liquid film using the principal of total internal reflection. Figure 1 (a) shows light refraction as it passes through an interface while Figure 1 (b) shows light beyond the critical angle being totally internally reflected.

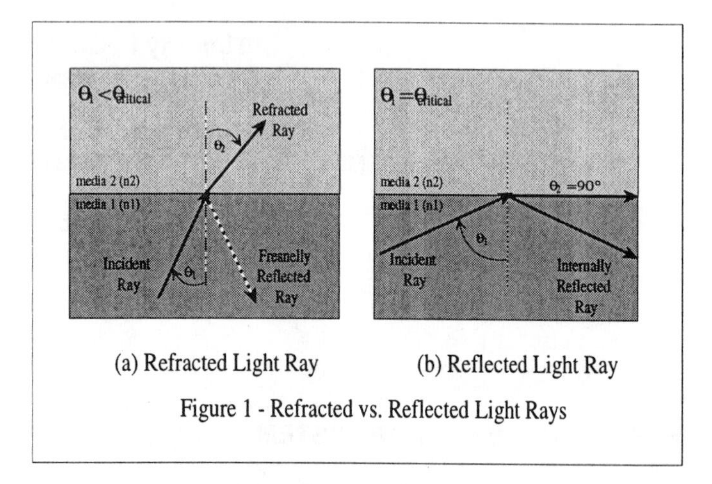

Figure 1 - Refracted vs. Reflected Light Rays

A receiver element is sized such that it is the same size as the reflected light beam. Thus, depending upon the distance the interface is from the flush mounted film sensor (i.e. the film depth), different proportions of the light beam strike the receiver element. Figure 2 shows a simplified two dimensional version of the sensor at three different film thicknesses.

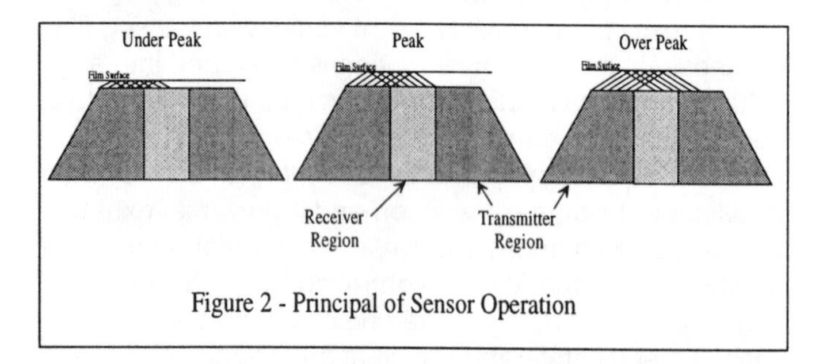

Figure 2 - Principal of Sensor Operation

The middle graphic of Figure 2 shows the sensor at the film thickness which produces the maximum voltage output. At this thickness, the reflected light beam exactly fills the receiver element, producing the maximum signal. The left graphic shows a film thickness less than the designed thickness where the reflected light beam does not fill the receiver element, producing a smaller than maximum signal. In this case, some quantity of the light beam re-enters the transmitter element it emanated from and is lost. Finally, at thicknesses greater than the designed peak thickness, as in the graphic on the right of Figure 1, the light beam has passed beyond the point of filling the receiver element and the signal is again less than the maximum. Here the symmetric shape of the film sensor can be noted as the lost light in this case actually enters the transmitting element on the opposite side of the sensor.

The design of the sensor used in the engine tests reported here is shown in Figure 3. This design includes several features intended to better control the path of the light rays both as they leave and as they return to the optical sensor.

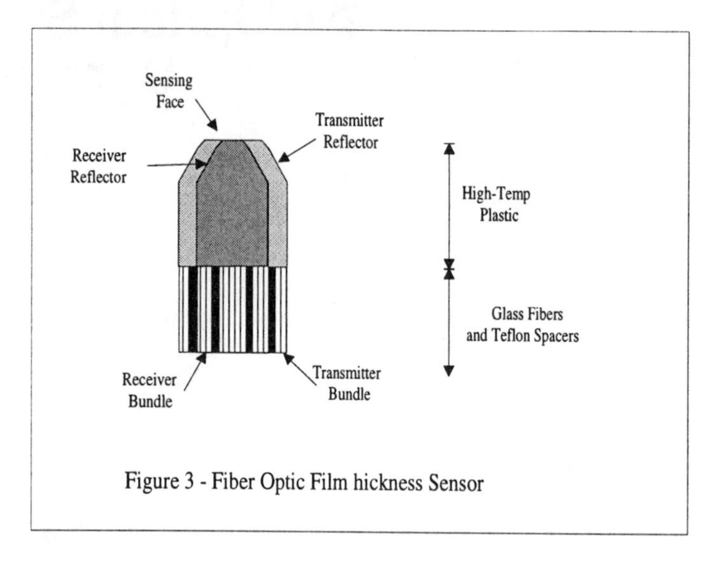

Figure 3 - Fiber Optic Film hickness Sensor

SENSOR INSTALLATION

ENGINE - The engine used in the tests was a 1.9 liter four cylinder port-fuel-injected engine. The engine uses a single overhead cam and two valves per cylinder. The intake port contains a high degree of swirl, and the fuel injectors are aimed directly at the intake valve and fire on a closed valve under most engine conditions. The modifications to the engine were restricted to the installation of an in-cylinder pressure analysis system, a high speed UEGO sensor and emissions testing equipment in the exhaust, and the optical sensors themselves. Figure 4 depicts the engine test facility.

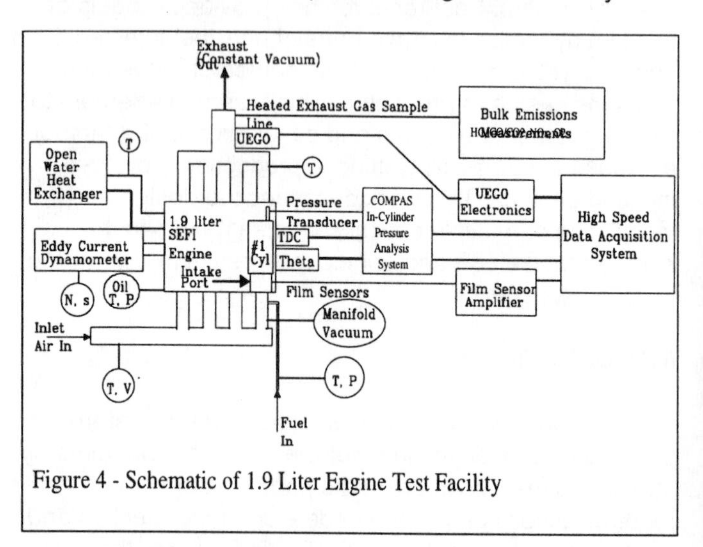

Figure 4 - Schematic of 1.9 Liter Engine Test Facility

FUEL INJECTORS - The fuel injectors used consisted of three different styles of injector. One style was the standard production injector supplied with the engine. This injector style created a pyramidal fuel spray with a Sauter Mean Diameter (SMD) of approximately 125 µm. The other two styles of injectors used a Compound Silicon Micro-Machined (CSMM) orifice to create a conical spray with SMD's of approximately 50-65 µm.

One style of CSMM injector has a spray diameter approximately equal to the production injectors (referred to as the narrow CSMM injector in this paper) and the second style of CSMM injector had a spray cone diameter approximately 25% larger in diameter than the production injectors (referred to as the wide CSMM).

SENSOR LOCATIONS - Seven optical film sensors were installed in the intake port of the engine. Figure 5 depicts the locations of each sensor on a rubber mold made of the intake port geometry. Figure 5 (a) depicts the basic port shape and relative position of the fuel spray. Figure 5 (b) depicts the sensor positions on the rubber mold, as viewed from the front of the engine. Four of the sensors are arrayed around the intake valve seat in approximately 90° increments. Sensor One is at the front of the engine. Ninety degrees clockwise (as viewed from above the engine) from sensor one is sensor two. This sensor is directly in-line with and facing the fuel spray from the fuel injector. Figure 5 (c) depicts the sensor positions on the rubber mold, as viewed from the left side of the engine. This view is looking back into the intake manifold and is facing the fuel spray. Figure 5 (d) depicts the sensor positions on the rubber mold as

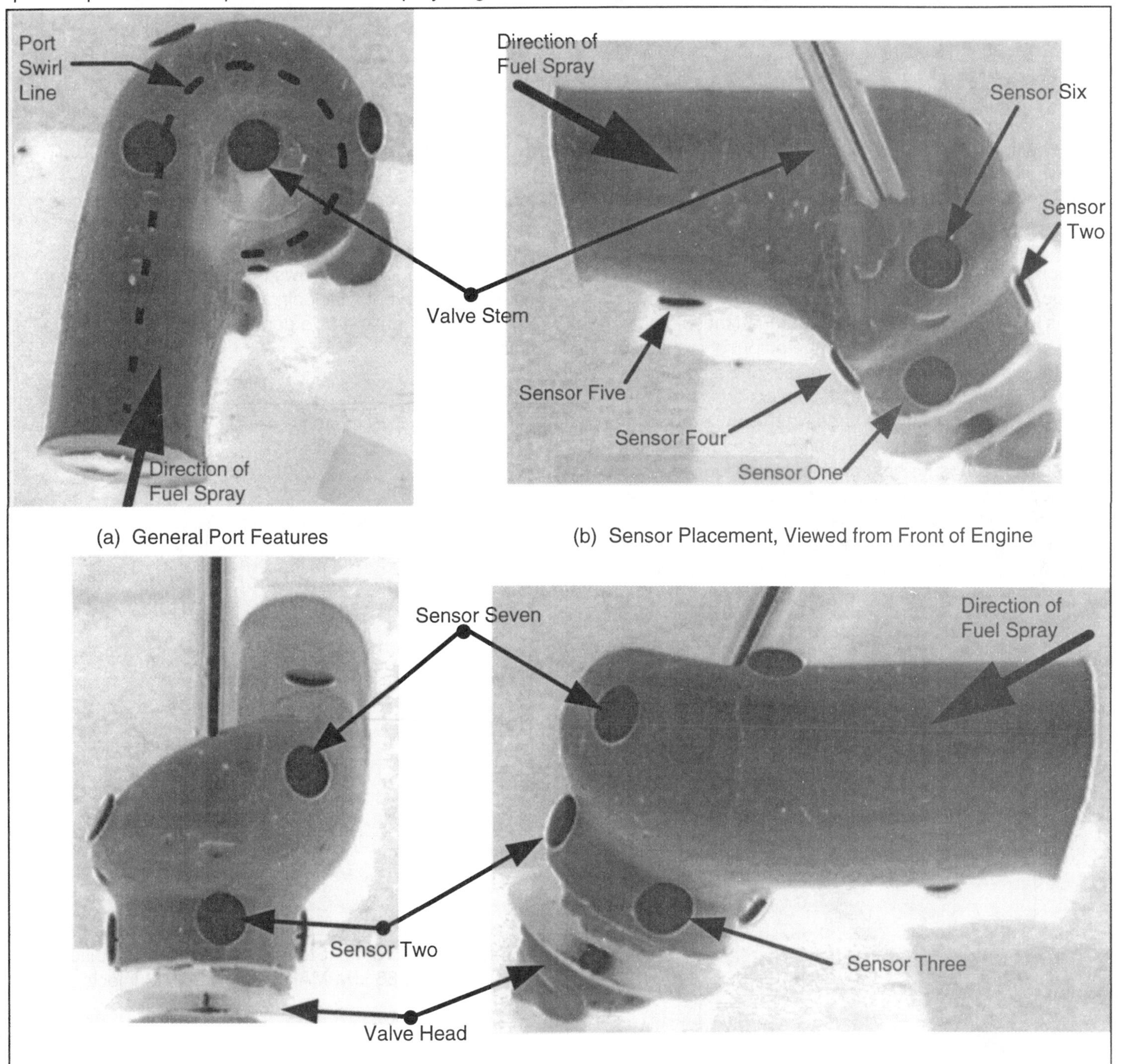

Figure 5 Sensor Placement Viewed on a Rubber Mold of the Intake Port Geometry

viewed from the rear of the engine. Sensor three, at 180° from sensor one, is visible in this view. Sensor four is 270° from sensor one, in-line with the fuel spray but facing away from the fuel injector, and is visible in Figure 5 (b). Sensor five is approximately 25 mm away from the intake valve seat along the floor of the intake port and is also visible in Figure 5 (b). The last two sensors, sensors six and seven, are aligned in the base of the swirl path of the intake port. Sensor six (visible in Figure 5 (b)) is approximately 25 mm along the swirl line from the valve seat, and sensor seven (visible in Figures 5 (c and d)) is approximately 40 mm further along the swirl line from sensor six.

SENSOR CALIBRATION

Each optical film sensor is calibrated under static film conditions, and the dynamic response of the system as a whole is determined separately. Figure 6 shows a representative calibration curve for sensor one. The curve shows voltage versus depth up to the peak depth point. In addition, the curve shows a fifth order polynomial fitted to this curve for use in converting the voltage data acquired during engine operation to film depths in mm. At film depths greater than those contained in this region, the voltage output of the sensor system falls back towards zero, presenting a possible two-valued solution to the depth conversion. However, this peak depth is relatively large when compared to the expected fuel film depths in the engine, and none of the data collected thus far has reached the voltage level corresponding to the peak depth point. Dynamically, the measurement system performs as a second order system with a natural frequency of approximately 6 kHz and a nearly constant gain out to approximately 2 kHz.

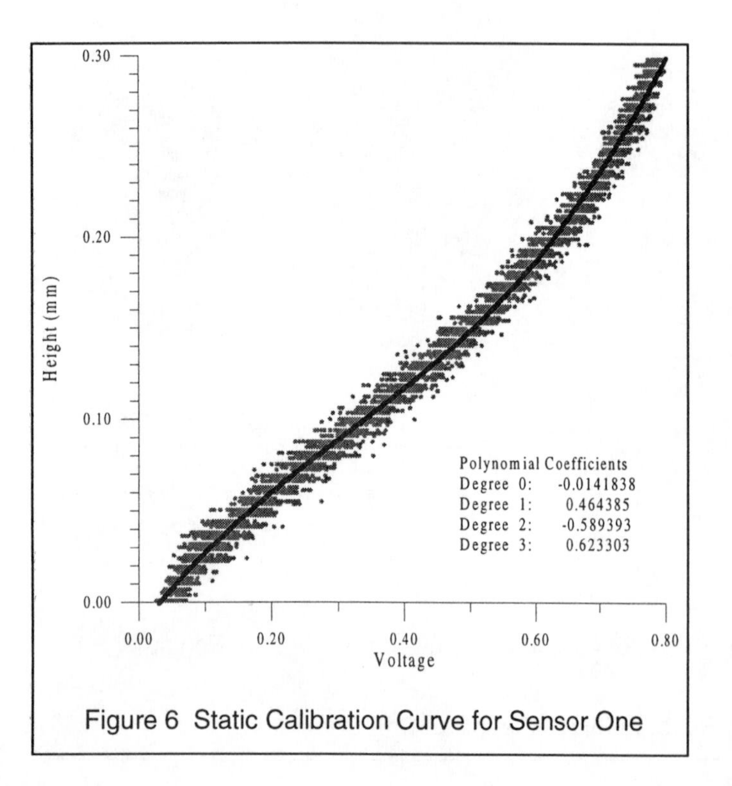

Figure 6 Static Calibration Curve for Sensor One

ENGINE TEST RESULTS

Engine testing consisted of three basic types of engine operation: steady state operation, engine start-up, and throttle transients. First, a complete series of tests were performed with a fully-warmed engine operating under steady-state conditions. This series of tests is intended to explore the boundaries of fuel film existence under warm engine conditions. The engine was run at four engine speeds (1000, 1500, 2000, and 2500 rpm) and three manifold absolute pressures (MAP) (35, 50, and 65 kPa).

CYCLIC ANALYSIS OF ONE STEADY STATE ENGINE CONDITION

Figure 7 shows the film data from film sensor two at 1500 rpm, 65 kPa MAP using the narrow CSMM injectors. This film sensor faces directly into the fuel spray and proved to be the most active sensor of the seven. The upper portion of Figure 7 shows the film sensor output in mm of film depth versus time. The lower part of Figure 7 shows the duration of the relevant intake valve events, including the valve over-lap period, the complete intake valve duration, and the time the fuel injector is open.

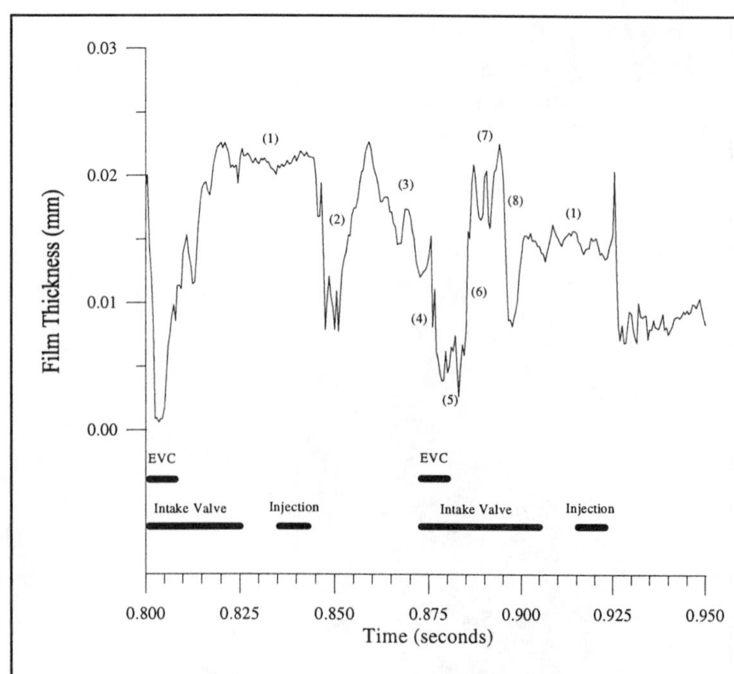

Figure 7 Film Sensor Two Showing Influence of Various Intake Port Events. Steady State Engine Conditions: 1500 rpm, 65 kPa MAP, Narrow CSMM Injectors.

During the first steady period (feature 1 of Figure 7), the intake port is undergoing a relatively quiescent period where little is occurring to induce gas or liquid motion. The valve and the injector are both closed. As the injector fires, there is no immediate influence on the film sensor signal. However, this is expected as the film sensor is approximately 80 mm downstream from the injector. Fuel spray speed has been estimated to be approximately 20 m/s. At this speed, there would be a 4 ms transport time for the fuel to reach the film sensor. Here there is approximately a 2 ms delay. The difference in delay times could also help explain the next film signal feature. In feature 2, the firing of the fuel injector appears to cause a decrease in the film depth on the sensor, followed closely by a rapid increase. Both of these phenomena could be due to air motion in the intake port induced by the fuel spray. The rapid introduction of fuel would cause air motion in front of the fuel spray as well as entraining air within the spray itself. This air motion could be blowing off the fuel that already exists on the film sensor. Then, as the main body of the fuel spray impacts the film sensor (at approximately 4 ms after injector firing in the film data), the film signal rapidly rises to a higher value.

The time during which there is a slow but steady decline (feature 3) corresponds to another period with little external influence on the air and fuel in the port. However, the port is now filled with fuel in the form of small droplets, liquid films, and vapor. This decline could be due to fuel vaporization or gravity driven liquid fuel flow toward the valve. The next period (feature 4) is the second rapid decrease in film signal magnitude. This occurs during the period in which blow back gas motion would occur. Two related possibilities exist for the rapid decrease in film signal during this period. First is the possibility that reverse gas flow causes fuel film to clear off of the film sensor through shear induced flow. Second is that there may be a rapid vaporization of any fuel film on the sensor as blow back gases are directly from the combustion chamber and therefore at high temperature.

The rapid decrease is followed by a flat, high frequency signal (feature 5). As this period corresponds to a rapidly increasing forward gas flow past the intake valve, the signal could be due to turbulence in the flow generating waves on top of a very thin fuel film, or to the occasional impact of a residual fuel droplet too large to follow the change in flow direction into the combustion chamber. The next feature (6) is rapid increase in the film signal occurring near the peak in valve lift. Two possibilities could account for this phenomenon. The most likely would seem to be liquid flow on the port walls. As the gas flow reaches its maximum speed, the shear induced flow of any fuel film on the port walls would be largest. Another possibility is a large increase in the quantity of residual droplets impacting the surface.

The next feature (7) is another flat plateau in the signal with a high frequency disturbance. Again, this high frequency characteristic is most likely due to shear induced motion in the fuel film over the film sensor. Alternatively, it could be due to further impaction of droplets, although this late after injector firing it would be assumed that small droplets would be drawn in to the cylinder already and larger droplets would have previously impacted the port walls. At the end of the intake valve event, there is a final downward spike in the film signal (feature 8). This could be due to the sudden decrease in fuel transport, either through shear flow or droplet impaction. After the valve is closed, there is again a brief quiescent period prior to the next fuel injection event, starting the cyclic process over again.

This data indicates how complex and rapidly fluctuating the fuel film may be. Due to this complexity, direct comparisons between different operating conditions and any engine performance effects are difficult to make. Therefore, a brief summary of the steady state engine testing would note that different aspects of this general signal were noticed in varying amounts in the remaining steady state data. Depending upon type of fuel injector, engine operating condition, and specific film sensor in question, none, one, or all of the noted signal characteristics could be present. A systematic examination of the relationships between engine parameters such as exhaust equivalence ratio or coefficient of variation of indicated mean effective pressure (COV_{IMEP}) and film thickness data should be performed to determine what, if any, direct impact fuel film variations have on engine performance.

TRANSIENT ENGINE TESTING

In addition to the steady state testing program undertaken, an exploration of film sensor signals under engine start-up and throttle-transient conditions was also performed. It is generally accepted that fuel transport delays, caused by large or rapidly changing fuel film, seriously affect engine operation during engine start-ups and rapid throttle transients. Therefore, understanding how fuel film behaves under these conditions is critical to meeting current and future performance goals.

COLD ENGINE START-UPS

Figure 8 shows a typical engine start up with data from film sensor two on the upper graph and from film sensor four on the lower graph. These are the two sensors in-line with the fuel spray. Sensor two faces toward the spray while sensor four faces away from the spray. This test was performed with a cold engine using production fuel injectors and the factory EEC-IV strategy.

As can be seen, film sensor two shows a film reading approximately 1/3 of a second before film sensor four. This pattern was seen in virtually all of the engine start up tests performed. Because film sensor two is in direct line with the fuel spray, its immediate reaction to the beginning of fuel injection is expected. The delay between sensor two and sensor four measuring a fuel film could be due to the fuel spray filling the port and only then impinging on the port wall where sensor four is installed, shielded from direct view of the fuel spray.

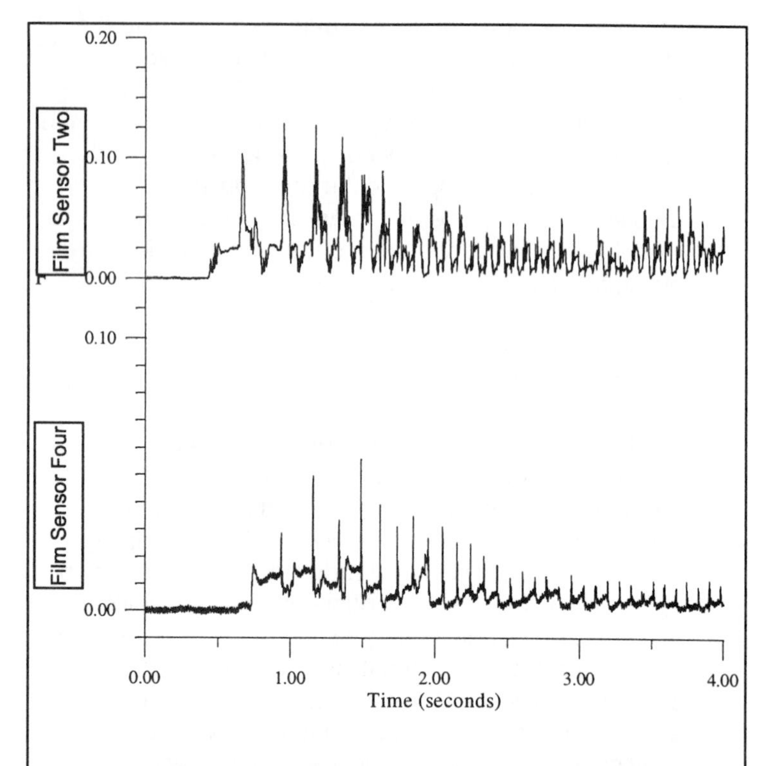

Figure 8 Cold Engine Start Film Data for Film Sensors Two and Four using the Production Fuel Injectors

Another possibility is that the film measured during a start up by sensor four is from fuel which has impinged on the port floor, upstream of sensor four. Once there is air flow into the port, this film is drawn into the cylinder, passing over sensor four.

Beyond this initial period, both sensors settle down to a typical pattern of cyclic variation. The data from both film sensors show influences from the fuel injection event and the intake valve event, with differences in the specific characteristics. In summarizing engine start-up testing, the location of the film sensor affects not only quantity and characteristic of film measurements but also the time duration between engine cranking and initial indication of film presence.

THROTTLE TRANSIENTS

Accelerating a vehicle is a typical throttle transient seen in automotive applications. In this scenario, the accelerator is depressed, possibly very rapidly, increasing engine torque output nearly as rapidly. However, due to the mass of the vehicle, vehicle speed, and hence engine speed, remain relatively constant throughout the throttle maneuver. This condition is simulated on a dynamometer by holding engine speed constant while advancing the throttle. This type of transient is known as a throttle ramp.

Figure 9 shows data from film sensors two and four, as well as the fuel injector pulse signal for reference, for a throttle ramp performed at 1500 rpm. The beginning throttle point corresponded to a MAP of 35 kPa and the ending throttle position corresponded to a MAP of 65 kPa. The ramp was performed with an engine coolant temperature of 38°C and a duration of approximately 100 ms, occurring at the approximate midpoint of the data set shown.

Film sensor two shows a slight increase in the average level and a definite increase in activity before and after the throttle ramp while film sensor four shows a distinct decay period after the throttle ramp, eventually reaching near zero film presence approximately 500 ms after the start of the transient. It is possible that increased air flow rate has shifted the distribution of fuel in the port in the direction of flow. At the beginning of throttle opening, a relatively small quantity of air is transported through the port. In this situation, it may be possible for a significant quantity of fuel to reach the port floor and flow over sensor four or perhaps reach sensor four directly. As air flow increases with the increased throttle opening, momentum imparted to the fuel spray may carry the droplets over the sensor four region of the port, striking the port walls at the rear of the port in the vicinity of sensor two.

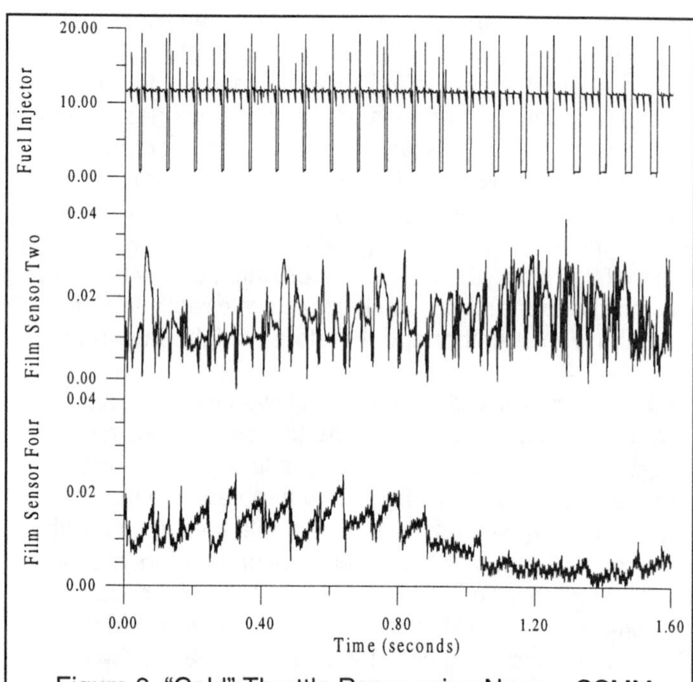

Figure 9 "Cold" Throttle Ramp using Narrow CSMM Injectors at 1500 rpm from 35 kPa MAP to 65 kPa MAP

Summarizing the results of the transient testing, changes in air flow often caused changes in location and magnitude of film presence on the walls of the intake port. These changes are likely due to changes in momentum imparted to droplets in the fuel spray and hence the location of the port on which the droplets impinge upon the port surface. The changes may also be due to changes in shear forces acting upon fuel film on the port walls or some other factor.

CONCLUSION

This work demonstrates the effectiveness of the optical film thickness sensor to provide quantitative measurements of port wall fuel film presence. We found distinct film changes with engine condition (speed and load) under steady state operation, with changes in air flow rate during throttle transients, and with changes in fuel preparation as defined by the type of fuel injector (and hence fuel spray) used.

The optical sensor presented here provides a relatively non-intrusive technique for measuring fuel films requiring only minor modifications to the engine cylinder head. It is hoped that future studies will be able to integrate information made available through use of the optical film sensor with mathematical models of fuel film dynamics and also with qualitative analysis of fuel films made with photographic and other techniques.

ACKNOWLEDGMENTS

We would like to thank Ford Motor Company for their support throughout this research. In particular, we would like to thank Charles Aquino, Debojt (Dave) Barua, and James Creehan of Ford Motor Company. In addition, we would like to thank a host of others, both with Ford Motor Company and Michigan Technological University, who played vital supporting roles in this research project.

REFERENCES

[1] U.S. Patent Number 5,396,079. "Fiber Optic Detector and Depth Sensor and Method for Doing Same."

[2] Bourke, M., and Evers, L. "Fuel Film Dynamics in the Intake Port of a Fuel Injected Engine." SAE Paper No. 940446, 1994.

DEFINITIONS, ACRONYMS, ABBREVIATIONS

MAP — intake Manifold Absolute Pressure

Engine parameter proportional to engine load.

CSMM — Compound Silicon Micro-Machined orifice fuel injector

UEGO — Universal Exhaust Gas Oxygen (λ) sensor

Indicator of engine equivalence ratio based on exhaust oxygen content.

SMD — Sauter Mean Diameter

Average droplet diameter having the same surface-area-to-volume ratio as the actual fuel spray.

950075

Model Based Air Fuel Ratio Control for Reducing Exhaust Gas Emissions

Akira Ohata, Michihiro Ohashi, Masahiro Nasu, and Toshio Inoue
Toyota Motor Corp.

ABSTRACT

In order to satisfy future demands of low exhaust emission vehicles (LEV), a new fuel injection control system has been developed for SI engines with three-way catalytic converters. An universal exhaust gas oxygen sensor (UEGO) is mounted on the exhaust manifold upstream of the catalytic converter to rapidly feedback the UEGO output signal and a heated exhaust gas oxygen sensor (HEGO) is mounted on the outlet of the converter to achieve an exact air fuel ratio control at stoichiometry. The control law is derived from mathematical models of dynamic air flow, fuel flow and exhaust oxygen sensors (HEGO and UEGO). Experimental results on FTP (Federal Test Procedure) exhaust emissions show a dramatic reduction of HC, CO and NOx emissions and a possibility of practical low emission vehicles at low cost.

INTRODUCTION

Low emission vehicles (LEVs) are social demands to improve air quality in nonattainable area. In order to develop low emission vehicles, much effort has been made, particularly toward reducing HC emission during warm up and reducing NOx emission during running. Air assisted injection systems have been developed to reduce engine-out HC emission. Precise ignition controls have also been developed to reduce unburned HC at start-up. It is well known that some methods to rapidly activate the three-way catalyst are very effective, for example, electrically heated catalytic converters (EHC) and retarded ignition timing. [1] Secondary air injection systems are often required to compensate the rich air fuel ratio during warm up. In order to achieve low NOx emission and fuel consumption, electronic controlled EGR systems have also been developed.

However, add-on hardware devices bring up costs and it will adversely impact the market for LEVs. Therefore, control software becomes one of the most important technologies to construct practical LEVs at low cost. The major reason is that the exhaust emission control system must take into account many factors. For example, EGR deteriorates accuracy of air fuel ratio control and consequently decreases conversion efficiency of the three-way catalyst, although EGR is an effective way to suppress NOx formation in combustion chambers. From experience, it is known that almost all aged catalysts have higher potentiality converting HC, CO and NOx emissions if a more accurate air fuel ratio control is adopted. This fact suggests that an accurate air fuel ratio control may be useful for low cost emission control systems.

BASIC STRATEGY FOR AIR FUEL RATIO CONTROL

It is well known that three-way catalysts can effectively convert HC, CO and NOx components to H_2O, CO_2 and N_2 only over a very narrow range of stoichiometric air fuel ratio. The strategy for this new control is to rapidly converge "the integrated air fuel ratio (IAFR)" to stoichiometry. IAFR is defined as the ratio of the integrated air mass to the integrated fuel mass. NOx is produced during lean IAFR, and CO and HC emissions are produced during rich IAFR. Considering diffusion and mixing effects during passage of the exhaust system, it is more important to control IAFR at stoichiometric ratio than to control instantaneous air fuel ratio in achieving a high conversion efficiency of HC, CO and NOx emissions.

SYSTEM CONFIGURATION

Fig. 1 shows the exhaust system of the tested engine (2.2L L4) which adopts a sequential fuel injection system. Fig.2 shows the diagram of air fuel ratio control system. The

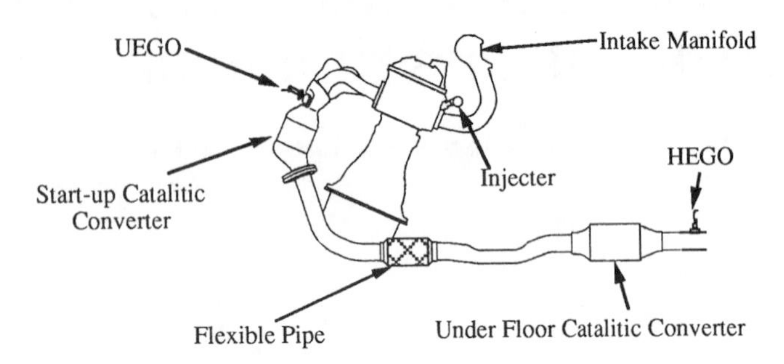

Figure 1 Exhaust System Configuration

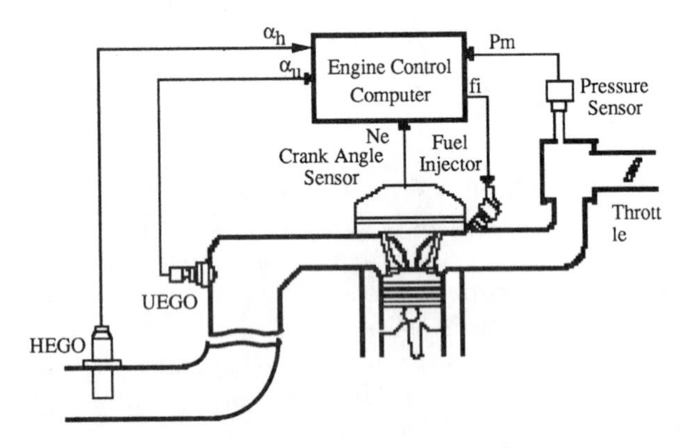

Figure 2 Air Fuel Ratio Control System

exhaust system consists of two types of three-way catalytic converters, a start-up catalyst and an under floor catalyst. The UEGO is adopted in the new air fuel ratio control system to estimate IAFR. It is mounted on the exhaust manifold to shorten measurement delay of air fuel ratio. The HEGO is mounted on the outlet of the under floor catalyst to measure the exact stoichiometry. Instead of an air flow sensor, an intake pressure sensor and a throttle position sensor are used to estimate the amount of air in the cylinders.

DYNAMIC MODEL FOR AIR FUEL RATIO CONTROL

Recent control technologies require mathematical models of controlled objects. [2][3][4] In this case, dynamic air flow and fuel behavior models and air fuel ratio detection models for the UEGO and the HEGO are required.

AIR DYNAMICS-Fuel injection must be performed in order that the ratio of the air mass to the fuel mass in the cylinder is equal to stoichiometry. It can be assumed that the cylinder pressure (Pc) is equal to intake pressure (Pm) at the bottom dead center (BDC) of intake stroke. Therefore, the amount of air induced into each cylinder (Mc) can be approximately estimated from Pm at every BDC. In fact, engine speed (Ne) and pulsation effect, which depends on Ne, affect Mc. Therefore, Mc can be estimated from the two dimensional interpolation of Mc with Ne and Pm at the intake BDC.

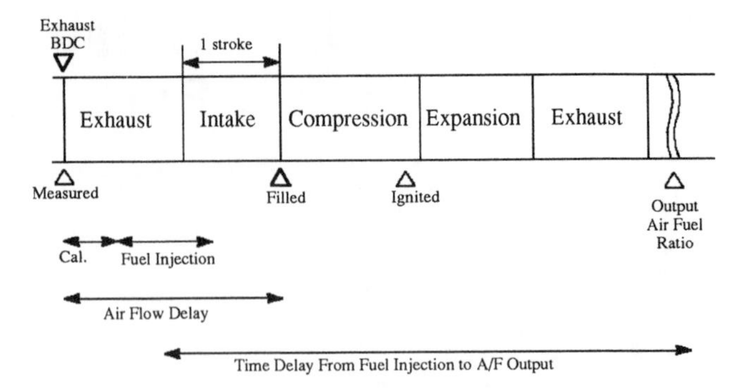

Figure 3 Timing of Measurement and Control

Fig. 3 shows the timing of measurement and control for a cylinder. The figure indicates that the amount of injected fuel mass must be calculated before the amount of air induced into a cylinder is determined at the intake BDC. Thus, the Pm at BDC must be predicted. From the conservation of intake air mass, the following equations are derived.

$$\frac{dMi}{dt} = mt - mc \qquad EQ(1)$$

$$mt = A\sqrt{2Pa\rho}\ \Phi \qquad EQ(2)$$

$$\Phi = \begin{cases} \sqrt{\dfrac{\kappa}{\kappa-1}\left\{\left(\dfrac{Pm}{Pa}\right)^{\frac{2}{\kappa}} - \left(\dfrac{Pm}{Pa}\right)^{\frac{\kappa+1}{\kappa}}\right\}}, & \dfrac{Pm}{Pa} > \left(\dfrac{2}{\kappa+1}\right)^{\frac{\kappa}{\kappa-1}} \\ \left(\dfrac{2}{\kappa+1}\right)^{\frac{1}{\kappa-1}}\sqrt{\dfrac{\kappa}{\kappa-1}}, & \dfrac{Pm}{Pa} \le \left(\dfrac{2}{\kappa+1}\right)^{\frac{\kappa}{\kappa-1}} \end{cases} \qquad EQ(3)$$

$$mc = \frac{1}{\pi} Mc\ Ne \qquad EQ(4)$$

Where

- Mi : amount of air from the throttle valve to the intake valves
- mt : mass flow rate through the throttle valve
- mc : mass flow rate induced into the cylinders
- A : effective opening area of the throttle valve
- ρ : density of air upstream of the throttle valve
- Pa : pressure upstream of the throttle valve

Mc in EQ(4) means the amount of air mass induced in an intake stroke. Mc depends on various unmeasurable factors, for example, intake and exhaust pressure pulsations, valve clearance and cylinder wall temperature. Therefore, the table of Mc with Pm and Ne obtained by measuring Mc at various Ne and throttle positions is used.

The adiabatic change supposed for the intake volume leads to

$$\frac{dMi}{dt} = \frac{V}{C^2}\frac{dPm}{dt} \quad . \qquad \text{EQ(5)}$$

V : intake volume from the throttle valve to the intake valves
C : sound velocity

From the substitution of Mi, mt and mc into EQ(1), the relationship with Pm is obtained and expressed in the following equation.

$$\frac{dPm}{dt} = -C_1 \text{ Ne Pm} + AF_1(Pm) \qquad \text{EQ(6)}$$

C_1 : constant

The function F_1(Pm) in the second term of the right side is constant when Pm is lower than the critical pressure shown in EQ(3) and F_1 depends on Pm is greater than the critical pressure. Consequently, predicting Pm is very easy when the throttle angle is small, but, difficult when it is large.

The following discrete function is derived from EQ(6). Pm at the intake BDC can be predicted using the following recursion.

$$Pm(k+1) = F_2 \text{ Ne } Pm(k) + A\ F_3(Pm(k)) \qquad \text{EQ(7)}$$

F_2 is constant and F_3 is the nonlinear function of Pm.

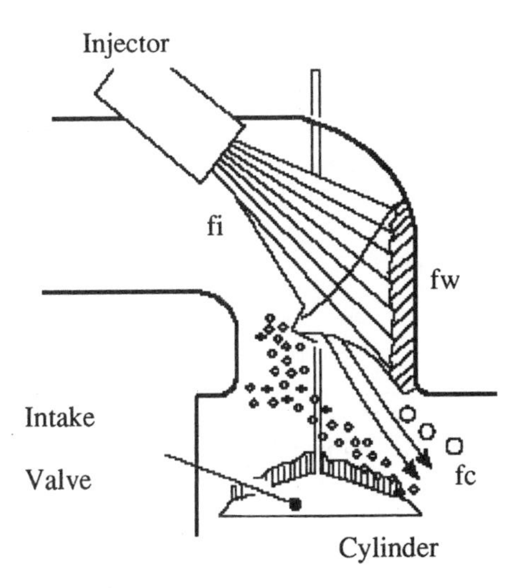

Figure 4 Fuel Behavior Model

FUEL DYNAMICS-Most of the injected fuel reaches the intake valves and is induced into the cylinder during the intake stroke. Some of it, however, adheres to the intake port walls and forms a fuel film. The film either slowly flows into the cylinder or evaporates and then is induced into the cylinder with the inflow of air at the intake stroke. Fig. 4 illustrates the phenomena with related variables. [6]-[9] A fuel behavior model is obtained as follows.

$$fw_i(k_i+1) = P_o fw_i(k_i) + R_o fi_i(k_i) \qquad \text{EQ(8)}$$

$$fc_i(k_i) = (1-P_o)fw_i(k_i) + (1-R_o)fi_i(k_i) \qquad \text{EQ(9)}$$

Where
fw : amount of fuel in the fuel film
fi : amount of fuel injected
fc : amount of fuel induced into cylinder
P_o : proportion of fuel left behind cylinder
R_o : rate of adherence
k_i : cycle number of i cylinder

Except during warm-up condition, P_o and R_o are treated as constant although in actuality they do change with engine operating condition. These errors in the models are compensated with robust UEGO feedback.

AIR FUEL RATIO DETECTION MODEL WITH THE UEGO-Air fuel ratio of exhaust gas is sensed with time delay of exhaust gas flow from exhaust ports to the UEGO. Because exhaust pressure is approximately equal to atmospheric pressure, the volume of exhaust gas from the cylinder is supposed to be proportional to the amount of air mass in the cylinder. Thus the time delay is inversely proportional to the amount of the air mass in the cylinder.

The UEGO is modeled as a first order delay system with time delay, expressed in EQ(10). The time constant changes with engine operation conditions, engine speed, intake pressure and as well as other factors.

$$\alpha_u(k+1) = a\alpha_u(k) + (1-a)\frac{Mc(k-d)}{fc(k-d)} \qquad \text{EQ(10)}$$

α_u : air fuel ratio measured by the UEGO
a : parameter corresponding to the UEGO time constant
d : cycle delay caused by exhaust gas flow

Measured UEGO signal has inevitable error caused by unequilibrium gas H_2 and HC. On the other hand, the three-way catalyst requires highly accurate control at stoichiometry. The UEGO is not sufficient to meet the demand of the three-way catalyst.

AIR FUEL RATIO DETECTION MODEL WITH THE HEGO-The HEGO mounted on the outlet of a catalytic converter can precisely detect stoichiometry of exhaust gas. However, it has been generally considered that the HEGO output indicates only whether exhaust gas is rich or lean. In this study, the voltage of the HEGO is used to measure deviation from stoichiometry quantitatively. The detection model is as follows.

$$\delta\alpha_h(k) = a_c\left(\frac{Mc(k-d_c)}{fc(k-d_c)} - \alpha_r\right) - V_r \quad EQ(11)$$

Where

da_h : voltage of the HEGO output

d_c : time delay caused by gas flow in the catalytic converter

a_c : constant

V_r : voltage of the HEGO output at stoichiometric air fuel ratio.

a_r: stoichiometric air fuel ratio

MODEL BASED CONTROL DESIGN

The basic concept is that robust control technology plays a part in compensating for various unknown dynamics and to achieve sufficient control performance even based on rough mathematical models. Robust control reduces the need to construct accurate mathematical models and assures accurate air fuel ratio control.

Considering the requirement of air fuel ratio control, we use two degree of freedom controller. Fig. 5 shows the structure of the new air fuel ratio control. [10] The amount of injected fuel mass fi is equal to $fi_m + \delta fi_u + \delta fi_h$. The control consists of four parts, the Fcr predictor, the inverse model of fc, the UEGO feedback and the HEGO feedback.

Fcr PREDICTOR-Fuel must be injected into the intake port to achieve stoichiometric air fuel ratio (α_r) in the cylinder. The Fcr predictor determines the required amount of fuel mass. The amount of air mass in the cylinder (Mc) is obtained by the two dimensional interpolation of Mc with Ne and Pm. Here, Pm is the intake pressure at BDC predicted by recursive procedure with EQ(7). The required the amount of fuel mass in the cylinder (Fcr) is calculated by Mc/α_r.

INVERSE MODEL OF fc-The inverse model calculates the amount of fuel injection mass corresponding to the fcr. For simplicity, the fuel models of all cylinders are modified to one four cylinder model, EQ(12) and EQ(13). EQ(13) gives the fc of each cylinder approximately and practically.

$$fw(k+1) = P_s fw(k) + R_s fi(k) \quad EQ(12)$$

$$fc(k) = (1-P_o)fw(k) + (1-R_o)fi(k) \quad EQ(13)$$

$$P_s = P_o^{\,0.25} \quad EQ(14)$$

$$R_s = R_o\frac{1-P_o}{1-P_s} \quad EQ(15)$$

Where k means engine stroke.

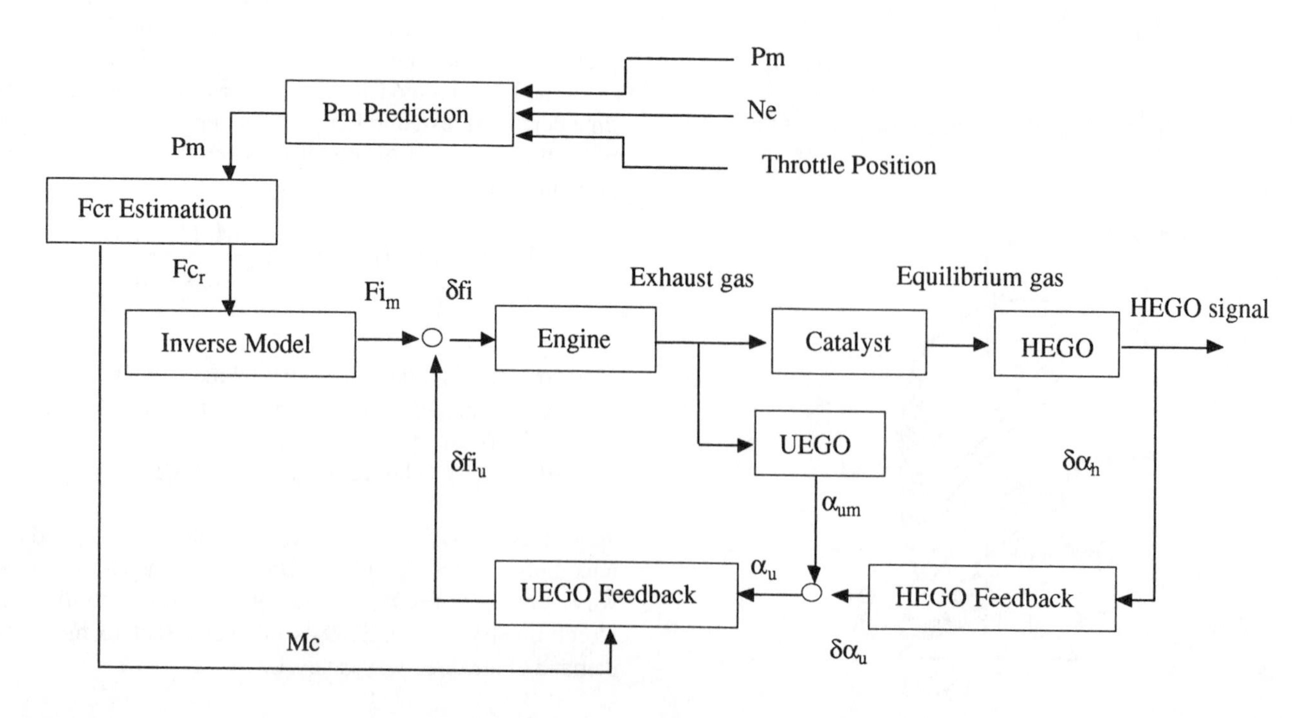

Figure 5 Block Diagram of New Air Fuel Ratio Control

From EQ(12) and (13), the inverse model is obtained as shown below.

$$fw_m(k+1) = P_s\ fw_m(k) + R_s\ fi_m(k) \qquad EQ(16)$$

$$fi_m(k) = \frac{fc_r(k) - (1-Po)\ fw_m(k)}{(1-Ro)} \qquad EQ(17)$$

UEGO FEEDBACK-According to the strategy of new air fuel ratio control, the error model is derived as follows.

$$\delta fw(k+1) = P_s\ \delta fw(k) + R_s\ \delta fi_u(k) \qquad EQ(18)$$

$$\delta fc(k) = (1-P_o)\ \delta fw(k) + (1-R_o)\ \delta fi_u(k) \qquad EQ(19)$$

Here

$$fi = fi_m + \delta fi_u$$

$$fw = fw_m + \delta fw$$

$$fc = fc_m + \delta fc$$

The purpose of the new air fuel ratio control is the convergence of y, expressed by EQ(20), to zero.

$$y(k+1) = y(k) + \delta fc(k) \qquad EQ(20)$$

Then, the following augmented system is derived by EQ(19) and EQ(20).

$$\begin{bmatrix} \delta fw(k+1) \\ y(k+1) \end{bmatrix} = \begin{bmatrix} P_s & 0 \\ 1-P_o & 1 \end{bmatrix} \begin{bmatrix} \delta fw(k) \\ y(k) \end{bmatrix} + \begin{bmatrix} R_s \\ 1-R_o \end{bmatrix} \Delta\delta fi_u(k) \qquad EQ(21)$$

The LQ (linear quadratic performance index) control design is applied to obtain the UEGO feedback, because LQ controller is one of the most reliable methods and is easy to use with various modifications. Applying LQ controller design to EQ(21), the expression for δfi_u which converges y to zero is obtained as follows.

$$\Delta\delta fi_u(k) = f_1\ \delta fw(k) + f_2\ y(k) \qquad EQ(22)$$

Where f_1 and f_2 are constant control gains. dfw in EQ(22) must be estimated by the measured value and y also includes measured error. dfcm is defined as measured δfc (= Mc/α_m) and δMc is the error of estimated Mc.

$$\delta fc_m(k) = \frac{Mc(k-d) - \delta Mc}{\alpha_m(k)} \cong \delta fc(k-d) - \frac{\delta Mc}{\alpha_r(k)} \qquad EQ(23)$$

The second term above can be regarded as constant. The estimated δfw is represented as δfw in the observer, EQ(24).

$$\overline{\delta fw}(k-d+1) = \overline{a_1}\ \overline{\delta fw}(k-d) + \overline{b_1}\ \delta fc_m(k) + \overline{b_2} \sum_{i=0}^{k} \delta fc_m(i) + \bar{j}\ \delta fi_u(k) \qquad EQ(24)$$

The δfw(k) must be estimated from δfw(k-d+1). The prediction is performed by the recursive calculation of EQ(18) with the initial condition of δfw(k-d+1). Finally, the control is the linear combination of δfc_m, δfi, the sum of δfc and the double sum of δfc. The double sum of δfc is calculated and stored in the controller memory for each session of [Pm, Ne], because calculation response is slow.

HEGO FEEDBACK-The HEGO feedback accurately compensates air fuel ratio error of the UEGO. The PI (Proportional and Integral) controller is adopted for the HEGO control strategy because the controlled object, from fuel injection to the HEGO detection, can be approximated by a first order linear system with time delay. The system can be asymptotically stabilized by a PI controller with proper feedback gains.

$$\delta\alpha_u(k) = k_i\ \delta\alpha_h(k) + k_p \sum_{i=0}^{k} \alpha_h(i) \qquad EQ(25)$$

The UEGO output is compensated as EQ(26), because $\delta\alpha_h$ corresponds to the measured error.

$$\alpha_{um}(k+1) = \alpha_{um}(k) + \delta\alpha_u(k) \qquad EQ(26)$$

EXPERIMENTAL RESULTS

We compared the new air fuel ratio control to a

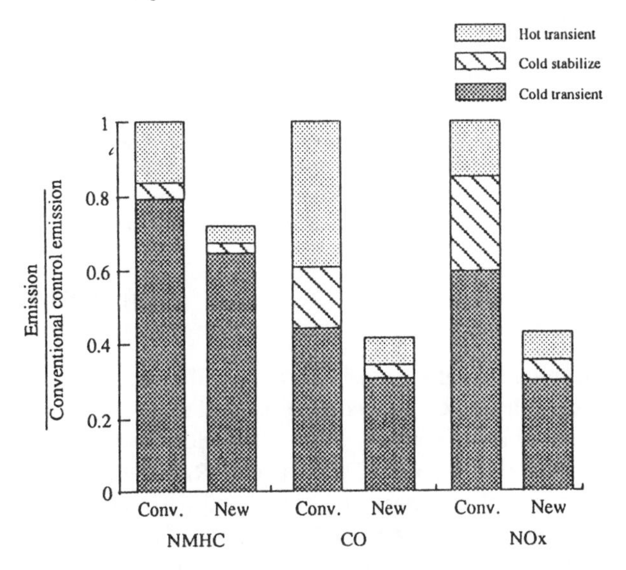

Figure 6 Comparison of FTP Emissions between New and Conventional Control

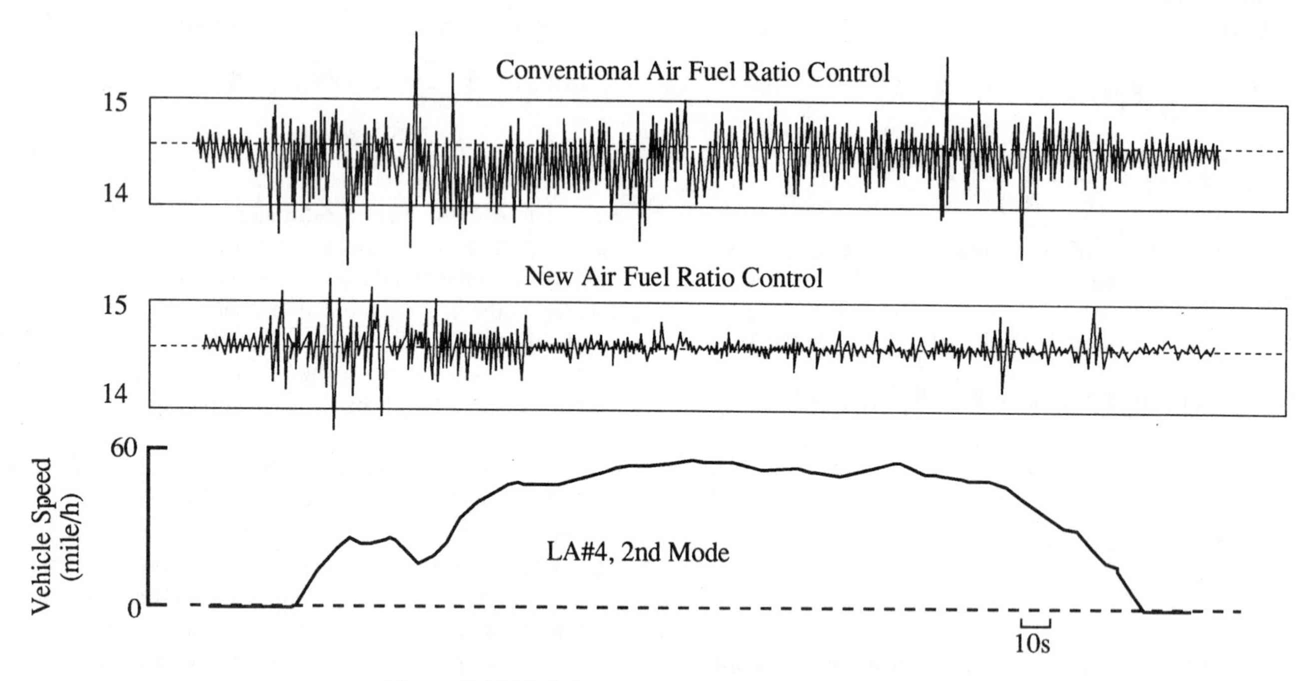

Figure 7 UEGO Output on LA#4 2nd Mode

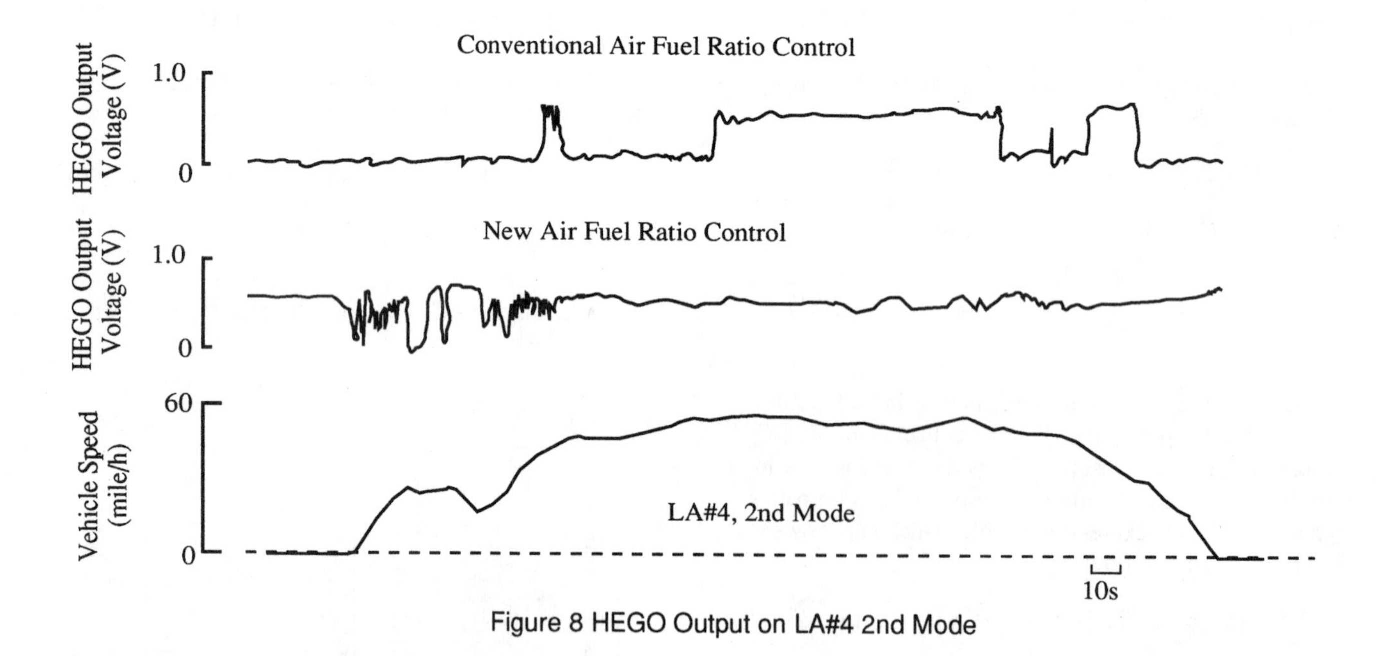

Figure 8 HEGO Output on LA#4 2nd Mode

conventional control, consisting of two HEGOs. The control logic of the tested conventional control is widely used in current emission control systems. Although the new control does not apply EGR (Exhaust Gas Recirculation), the conventional one does for optimization in the emission control system. Furthermore, the intention of the new control was to remove some devices in emission control systems while meeting LEV standard at the same time.

Fig. 6 shows a comparison of the new and the conventional control on the federal test procedure (FTP). This result indicates that the new control is very effective in reducing FTP emissions. The great effect on CO emissions indicate that the air fuel ratio is accurately controlled in the test mode.

Fig. 7 shows the UEGO signals on the new and the conventional control. Fig. 8 shows HEGO signals. In the conventional control, the UEGO is added on the exhaust manifold to measure air fuel ratio. Fig. 7 indicates that the new control suppresses the air fuel ratio perturbation more significantly than the conventional control. The HEGO output signal in Fig. 8 shows that the new control achieves precise control at stoichiometry. This is the reason why the

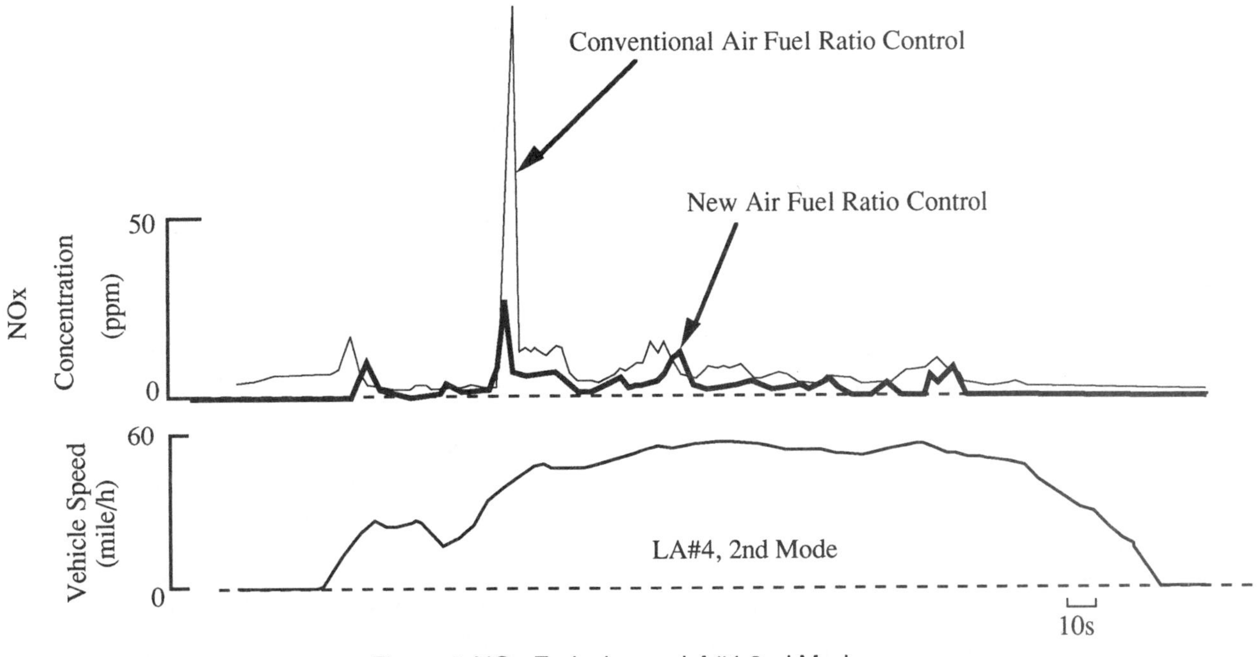

Figure 9 NOx Emission on LA#4 2nd Mode

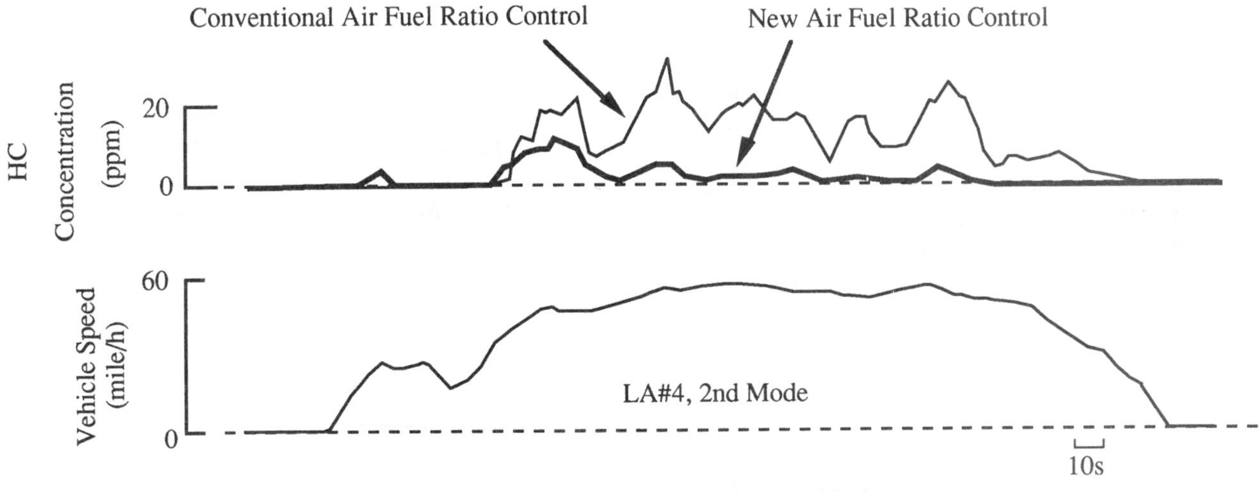

Figure 10 HC Emission on LA#4 2nd Mode

new air fuel ratio control has a great effect on FTP emissions as shown in Fig. 5, although the UEGO output signal is still perturbed.

Fig. 9 and Fig. 10 are HC and NOx concentrations of diluted gas through constant volume sampler . These results also support the effect mentioned above.

DISCUSSION

Work remains before we can introduce the new air fuel ratio control to the market. The major problem to be solved is to assure the reliability of the new air fuel ratio control. Especially, the reliability of the HEGO is an important problem because the new control depends on its accuracy .

Early activation and a wide range of the UEGO measurement are strongly required. This will reduce efforts to find optimal parameters in the new air fuel ratio control during warm-up condition and will also improve reliability.

The model based control design is useful to mechanically derive an accurate control law. However, the control is generally a high order system. Therefore, it is considered that the control must be optimized to reduce CPU time and size of memory.

CONCLUSION

The new air fuel ratio control shows significant HC, CO and NOx reduction and suggests that the more accurate the air fuel ratio control is, the lower the system cost of low emission vehicles.

(1). The new air fuel ratio control system consists two types of exhaust gas oxygen sensors, the HEGO and the UEGO. The UEGO is mounted on the exhaust manifold upstream of the start-up catalytic converter and provides rapid feedback to the fuel injection. The HEGO is mounted on the outlet of the under floor catalytic converter to accurately regulate the air fuel ratio at stoichiometric air fuel ratio.

(2). The control law is derived from the dynamic air flow, fuel flow, sensor behavior models and the consideration of the three-way catalyst. It consists of four parts, the induced fuel mass predictor, the inverse model, the UEGO feedback and the HEGO feedback. The predictor calculates the amount of fuel mass injected into the intake port before the amount of air mass in the cylinder is determined at the end of intake stroke. The inverse model roughly determines the amount of injected fuel. The UEGO feedback is devised by LQ control design combined with an observer and a time delay compensation. The HEGO feedback is designed as PI control.

(3). The basic strategy of the new air fuel ratio control is to regulate the ratio of the integrated air mass to the integrated fuel mass at exact stoichiometry by utilizing diffusion and mixing effects in exhaust gas flow.

REFERENCE

[1] T.Yaegashi, K.Yoshizaki, et al. "New Technology for Reducing the Power Consumption of Electrically Heated Catalysts", SAE Paper 940464.

[2] Y.Chujo, et al., "Development of On board Fast response Air Fuel Ratio Meter Using Lean Mixture Sensor", ISATA'89, Italy, No. 89038 459/475 Vol. 1.1.

[3] E.Hendriks, T. Vesterhold, "The Analysis of Mean Value SI Engine Models", SAE Paper 920682, 1992.

[4] T.Sekozawa, et al., "Development of a Highly Accurate Air Fuel Ratio Control Method Based on Internal State Estimation", SAE Paper 92029, 1992.

[5] Donald J. Dobner, "A Mathematical Engine Model for Development of Dynamic Engine Control"' SAE Paper 80054, 1980.

[6] C.F. Aquino, "Transient A/F Control Characteristics of the 5 liter Central Fuel Injection Engine", SAE Paper 810494, 1981.

[7] R. Nishiyama, et al., "An Analysis of Control Factors Improving Transient A/F Control Characteristics", SAE Paper 890761, 1989

[8] S.Dhires and M.T.Overrington, "Transient Mixture strength Excursions An Investigation of Their causes the Development of constant Mixture Strength Fueling Strategy", SAE Paper 810495, 1981

[9] H. Iwano, et al., "An Analysis of Induction Port Fuel Behavior"' SAE Paper 912348, 1991.

[10] H.Inagaki, A.Ohata, "An Adaptive Fuel Injection Control with Internal Model in Automotive Engines", IECON'90, 1990.

[11] R.C.Tupa and D.E.Kocheler, "Intake Valve Deposits-Effects of Engines, Fuel and Additives", SAE Paper 881645, 1988.

[12] H.Nagaisi, et al., "An Analysis of Wall Flow and Behavior of Fuel in Induction System of Gasoline Engine", SAE Paper 890837, 1989.

[13] C.F. Aquino and S.R.Fozo, "Steady-State and Transient A/F Control Requirements for Cold Operation of a 1.6 Liter Engine with Single-Point Fuel Injection", SAE Paper 850509, 1985.

[14] D.R.Hamburg, "The Measurement and Improvement of the Transient A/F Characteristics of an Electronic Fuel Injection System", SAE Paper 820766, 1982.

[15] A.J.Beaumont, A.D.Noble, "Adaptive Transient Air Fuel Ratio Control to Minimize Gasoline Engine Emissions", FISITA Congress, London, 1992.

[16] E. Hendrics, S.C. Sorenson, "Mean Value Modeling of Spark Ignition Engines", SAE Paper 900616, 1990.

[17] T. Minowa, et al. "Improvement in Torque Response during Acceleration Obtained by Using a Control System with Intake Manifold Models", JSAE Review, Vol. 13 No. 1, 1992.

FUEL LEVEL SENSORS

971072

Fuel Level Sensor Design from a System Perspective

George Nagy, Jr.
Ford Motor Co.

Copyright 1997 Society of Automotive Engineers, Inc.

ABSTRACT

Many of the current issues surrounding the achievement of accurate fuel level indication and fuel sensor design strategy revolve around systems issues, but are most often treated as component deficiencies. This paper takes a "systems" look at liquid fuel level indication as opposed to the traditional "component" view. The intent of the author is to present a framework of system considerations relative to the task of designing a robust fuel level indication system in the automobile.

INTRODUCTION

Many articles exist in the technical literature on fuel sensor technologies. Some of these articles deal with the evolution of the current technology, from wire wound to thick film resistors [1] [2] [3], while others address potential emerging technologies, such as optical, ultrasonic, capacitive, etc. [4] [5] [6] To date, the float-arm potentiometer design remains the design of choice throughout the world, primarily because of extremely low cost, high reliability, and satisfactory durability. [4] [5] Many companies are trying to penetrate this high volume component market by proposing new technologies to solve recurring customer dis-satisfaction with the performance of the system. These new technologies are often component- focused, and do not address systems issues. However, many, if not most, fuel level indication quality or performance issues are the result of system inadequacies, not component deficiencies. The challenge to achieving best-in-class system performance will depend not only upon the invention of new technology, but also upon the success in overcoming system issues. The Design Engineer must begin to look at fuel level indication as a system, not simply as a collection of components. This paper will address some of these system issues, and the associated system considerations that the design engineer may want to consider to effect a robust system design.

SYSTEM DESCRIPTION

The Fuel Indication System consists of several components. These components, working as a system, provide an indication of the remaining fuel in the tank.

FUEL SENDER- The Fuel Sender is a device mounted inside the fuel tank that measures the fluid level inside the tank, converts it into an electrical signal, and provides that signal to the indicator (gauge). Most senders typically contains a variable resistor, using a wire winding or a thick film ceramic, connected to a float through a float rod. [1] [2] [4] Changes in fuel height are followed by the float causing corresponding variations in resistance that control current flowing through the indicator. The relationship between the fuel height, and the corresponding volume is accounted for in the calibration of the variable resistor. [1] [2]

INDICATOR- The Indicator, or Gauge, is an air core (magnet-coil), bi-metallic (thermal), or electronic display device which responds to changes in electrical current controlled by the variable resistor of the fuel sender. The changing current varies the deflection of the pointer on the gauge (mechanical gauge), or the displayed value on an electronically controlled gauge.

PROCESSOR- On electronic, or electronically controlled displays, the analog sender output may be converted to a digital signal by a Processor which then updates the display to the corresponding programmed value. This processor may also account for signal fluctuations which occur under dynamic driving conditions. Some manufacturers also utilize a processor to "manage" the signal on mechanical gauges.

FUEL TANK- Contains the fuel volume being measured, and plays a major role in system considerations as will be discussed later.

OTHER- Vehicle wiring, or any other component or system which interfaces with the fuel indicating system, may influence the fuel indicating system performance.

SYSTEM FUNCTIONS

The customer's perception of fuel level indication performance consists of three essential categories: Accuracy, Linearity, and Stability. Most, if not all, customer expectations of system performance can be classified into one of these three categories.

ACCURACY- The customer expects the fuel gauge to be accurate. This sounds simple enough, but in defining accuracy the engineer may find that the customer's "perceptual definition" is different from the engineer's "analytical definition". For instance, the customer may believe that when the gauge reads "1/2" the amount of fuel necessary to fill the vehicle should equal 50% of the total fuel capacity as stated in the Owner's manual. That seems logical. However, the engineer has a much more complicated task to achieve that desired effect. First, fuel gauge, sender, tank, and other component tolerances and variations need to be taken into account. There are also uncontrollable factors such as variations in customer filling patterns (point at which customer stops trying to fill the fuel tank), and effects of environment (temperature, vapor pressure in tank) which affect the outcome. In trying to accommodate these various conditions, the engineer may need to make compromises that result in the 1/2 gauge position being targeted at something other than 50%.

Readability of the gauge may also affect customer perception of system accuracy. Is the gauge difficult to read? Are the graduation marks and fuel levels easy to interpret? Is there parallax between the customer's viewing angle of the gauge and the way the gauge was initially calibrated so as to adversely affect accuracy? Even with the best of foresight and anticipation, the engineer is challenged with overcoming many factors which lie outside his or her control.

LINEARITY- Most customers expect a fairly linear indication between F (full) and E (empty). This means that the customer expects the gauge to move at the same rate throughout its travel from "F" to "E". For example, if the gauge needle takes 200 miles of driving to go from "F" to "1/2", they expect that they can travel about 200 more miles before the gauge reads "E". If the distance traveled between the first 1/2 tank and last 1/2 tank is appreciably different, they will perceive that the system is "non-linear". Another way to think of this is in terms of refilling the fuel tank. If it takes 10 gallons of fuel to refill the tank from "1/2" to "F", the customer will expect that it will take 20 gallons to refill from "E" to "F".

Many design philosophies attempt to have a fairly linear indication between F (full) and E (empty). In addition, a large "reserve" at both empty and full insures that the gauge, regardless of tolerance stack ups, will always get to empty and full (no customer would ever run out of gas before the gauge reached empty, and after a fill-up the gauge would always read full or beyond full). See Figure #1.

This philosophy, while protecting customers who have systems that are at or near the edge of the system tolerance distributions, penalizes the majority of customers who have nominally calibrated systems by creating a perception of non-linearity. This is seen by studying Figure #1. Customers use the full reserve, so the amount of fuel in this reserve region is a part of their evaluation of fuel usage. However, the empty reserve, since it is below E, is rarely ever used by the customer, since few customers drive at that low a fuel level and therefore do not include that region in their consideration of available fuel. Look at Figure #1. The total fuel tank capacity is 76 liters (20 gallons). The amount of useable fuel from an indicated 1/2 tank to a full tank (including full reserve) is 38 liters (4+4+2=10 gallons), and the amount of fuel from "E" to 1/2 tank is 30.4 liters (4+4=8 gallons). This results in a perception of non-linearity between the upper and lower halves of the gauge.

A component viewpoint may not allow the engineer to see the system in the manner that the customer perceives it. A systems approach recognizes the customer's perceptions and wants. Customers may want a reserve when they get to empty, but they don't think of that fuel being there when they record fuel usage. The linearity of the gage, as with accuracy, must align with the customer's perceptions, not just with what the engineer knows to be "correct". The engineer may want to "adjust" the linearity in the intermediate sections of the gauge to accommodate the customer's perception and use of the full

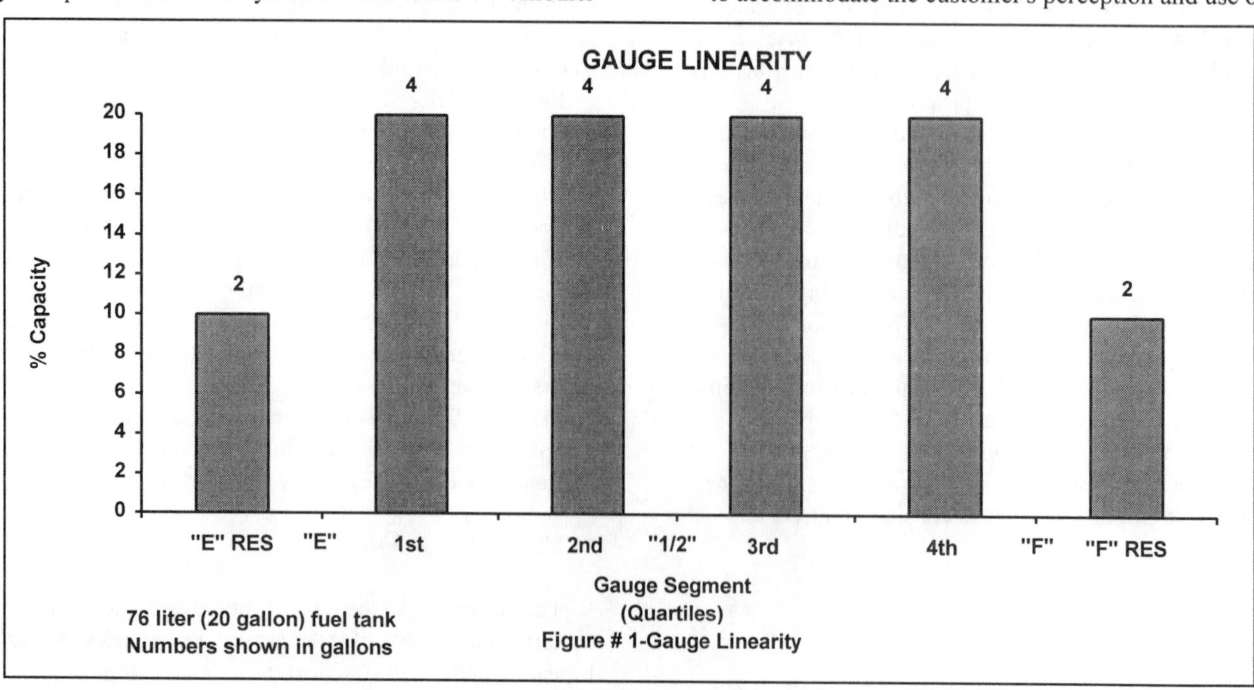

Figure # 1-Gauge Linearity

reserve (and non-use of the empty reserve).

STABILITY- This final area of functionality is possibly the most subjective. It relates to the amount of fuel gauge pointer movement that occurs under various dynamic conditions. The customer basically expects the gauge to read correctly regardless of the vehicle and environmental conditions. For example, when ascending or descending a hill, cornering, braking, etc., the customer expects the gauge to read correctly and does not like to see the pointer fluctuate. Of course, some fluctuation under these dynamic conditions is inevitable (from the engineer's viewpoint), but the driver may not share the engineer's knowledge of physics, and may actually expect a stable, non fluctuating gauge. The determination of "acceptable" pointer movement is where the subjectivity comes in. What is acceptable to one customer may be totally unacceptable to another. The engineer has the task of balancing customer satisfaction, among widely varied customer expectations, against the costs of eliminating or reducing such variations.

SYSTEM ISSUES

Typically, design engineers are tempted to focus on component parameters when evaluating performance issues when, in fact, many fuel indication issues are a result of system deficiencies. For example, excessive empty reserve may be due to the inability of the design engineer to locate the float at the desired level in the tank due to a lack of a sump, which is not packageable because of ground clearance standards. Some common system issues which can adversely affect accuracy, linearity or stability include: fuel tank distortion/ deformation due to temperature, vapor pressure or manufacturing variations; inaccessibility of the last few milliliters of fuel at the bottom of the fuel tank; variability in the fuel fill shutoff, both due to filling station variability and customer filling habit variability; tolerance stack-up errors among components; component interface or interaction effects; complex tank geometry; and differing customer perceptions of what a "best-in-class" system should look like.

SYSTEM CONSIDERATIONS

The following paragraphs address several system considerations that the engineer should consider when designing a fuel level sensor. This is not intended as an all-inclusive list, but rather as a list of thought starters to aid in the generation of system-oriented, customer driven designs.

OPTIMIZE LOW FUEL INDICATION- This is possibly the most critical aspect of fuel indication. The customer often pays little attention to the fuel gauge until the fuel level becomes low. Then the driver tends to look at the gauge much more frequently. Any abnormality in fuel indication system performance is more likely noticed if it occurs at low fuel levels. This is where accuracy, or lack of, will manifest itself.

MAXIMIZE TANK DEPTH RELATIVE TO LENGTH/WIDTH- This design goal addresses the attempt for system robustness by desensitizing the system to variations that are likely to occur. The depth vs. volume ratio is the basis of this analysis. The desired outcome is to achieve a maximum depth-to-volume ratio. In other words, the more millimeters of tank height per liter of fuel, or conversely, the less liters of fuel per millimeter height, the less sensitive the system will be to variations. The rationale is best understood by studying Figures #2 and #3. Note in Figure #2 that a 3.8 liter (1 gallon) change in fuel volume with a shallow tank

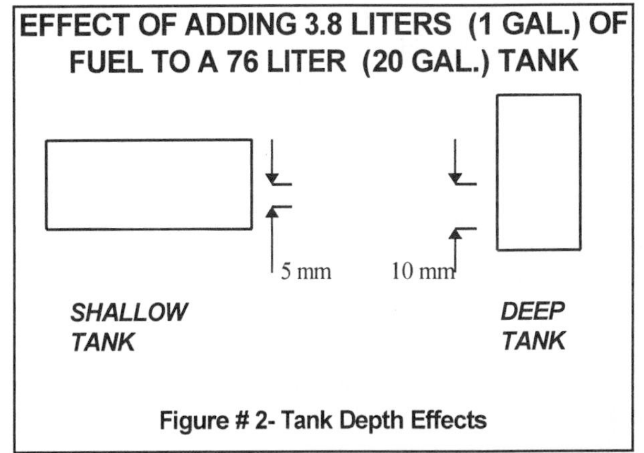

Figure # 2- Tank Depth Effects

raises or lowers the fuel height by approximately 5 mm. The same 3.8 liter change in volume in the deep tank changes the fuel level by twice that amount, about 10 mm. This is because the deep tank has about twice the depth-to-volume ratio. Note that while a 10 mm change in the deep tank accounts for only 3.8 liters of fuel, a corresponding 10 mm change in the shallow tank has twice the effect, 7.6 liters. The manifestation of this on the fuel display is shown in Figure #3 through the following example.

Figure #3 illustrates the effect on fuel gauge pointer movement due to a 10 mm variation in fuel level. This 10 mm change in level could be the effect of fuel movement due to cornering, tank distortion from pressurization, etc. The causal factors may be numerous, but the result on the gauge is the

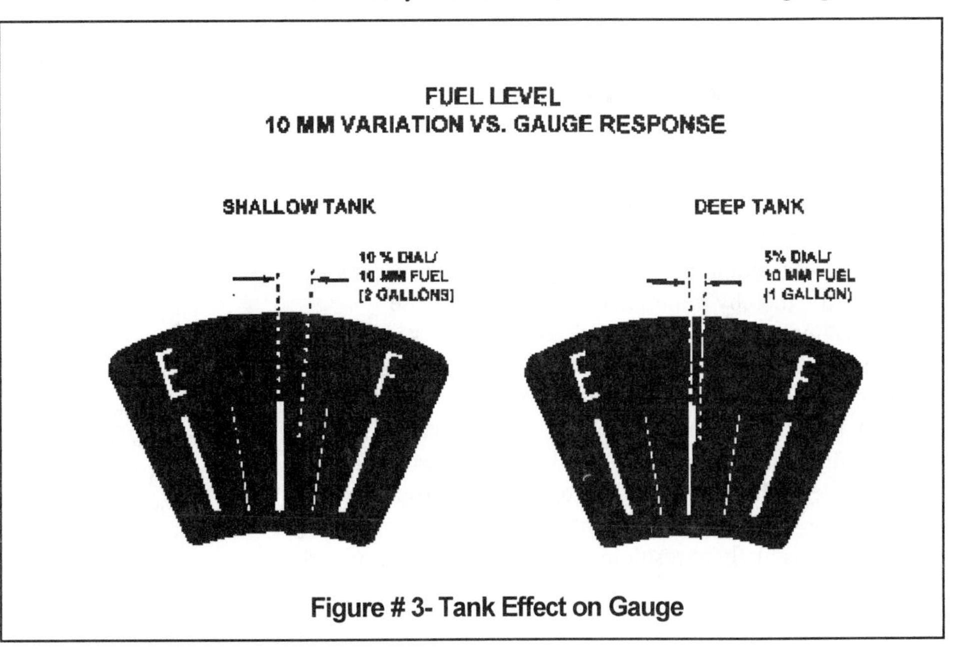

Figure # 3- Tank Effect on Gauge

same. With the shallow tank, the 10 mm variation creates a 7.6 liter (2 gallon) change which causes a pointer movement of about 10% as seen in Figure #3, based on a 76 liter (20 gallon) fuel tank. Figure #3 also shows that the deep tank results in only a 5% pointer movement for the same 10 mm variation (3.8 L. / 1 gallon variation in 76L / 20 gallon). It is more likely the customer will detect and be disturbed by the 10% fluctuation, and find the 5% variation acceptable. Thus, the deeper tank desensitizes the system to fuel gauge fluctuations.

Unfortunately, the trend toward lower, more aerodynamic vehicles is driving tank designs toward shallower configurations. This creates another engineering balancing act, this time for the vehicle engineer, to properly balance the need for a robust fuel indication system with other vehicle attributes such as styling.

STRIVE FOR UNIFORMITY IN TANK GEOMETRY- Radical, sudden changes in tank profile that suddenly change the depth/volume ratio can have an adverse impact on indication linearity. Severe changes in geometry, such as with "L"-shaped, or "saddle"- shaped tanks (Figure #4), result in dramatic changes in the depth/volume ratio. Even if this change can be accommodated by the sender calibration, any mismatch between tank and sender position on the actual parts will result in a mismatch in the transition zone, in turn resulting in poor linearity of the system.

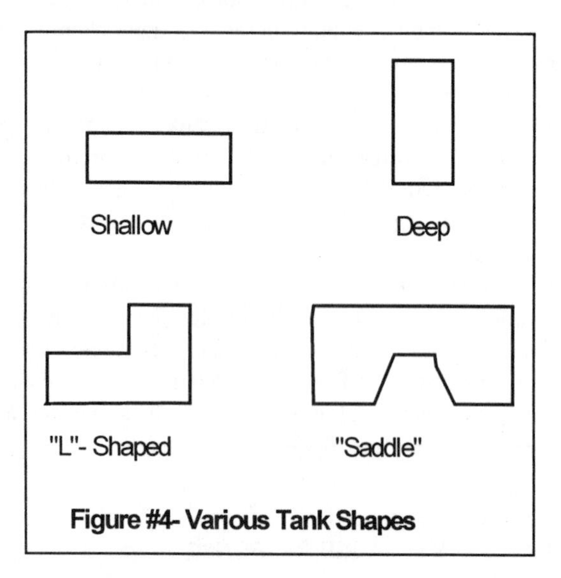

Figure #4- Various Tank Shapes

CENTRALLY LOCATE THE SENDER IN THE FUEL TANK (VOLUMETRIC CENTER)- This action will minimize the affect of fuel "stacking" at the sides of the fuel tank during accelerations, deceleration and cornering, parking on a hill, etc., thereby minimizing the effects of those conditions upon the fuel gauge indication. Figure #5 shows the effect this stacking can have on fuel gauge readings. In Figure #5, the correct fuel gauge reading is "1/2". With the centrally located fuel sender, the gauge reading has only changed slightly, by about 1/8. But the sender located at the extreme side of the tank (right side of picture) shows over a 1/4 gauge movement. This amount of movement from the correct value is likely to result in customer dis-satisfaction.

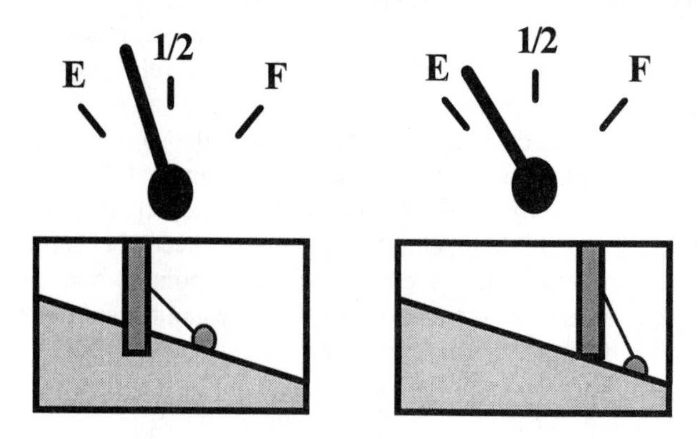

Figure #5- Effect of Location of Sender in Tank on Fuel Gauge Reading at 1/2 tank, During Vehicle Maneuver

CONTROL THE FUEL FILL LEVEL- This is the level to which the tank can be filled. Accurate, linear, fuel indication will depend on the degree to which this parameter is controlled relative to the vehicle's advertised capacity. If the fill level is not well controlled, the gauge may not go to full, or may stay on full too long, resulting in customer dis-satisfaction. While this parameter seems straightforward, it may be one of the most difficult system parameters to control. Many external factors come into play which are out of the direct control of the engineer.

For example, the filling rate of the fuel station pump will affect the point at which the pump cuts off during fill. Although the engineer can control such design parameters as filler pipe configuration, back pressure etc., they can only design to a target standard, or range, of pump fill rates. The actual variability in filling rates from one pump to another may adversely affect fuel indication from one fill to another.

Another factor affecting fill control is the customer's filling habits. Some customers may stop fueling when the pump first clicks off. This may result in the gauge not reaching "Full". Others may continue to fill at a slower rate until they reach a convenient stopping point, such as a round dollar value. And we have all waited behind the customer who will trickle-fill their tank. This will typically result in the fuel gauge remaining on "Full" for a long time. Each of these scenarios results in a different amount of fuel in the tank at fill, and thus can result in variability of the full reading. Some manufacturers attempt to explain this variability in their Owner's Manuals. But the system's approach dictates that the engineer attempt to design the system to mitigate the affects of these variables.

INSURE THAT ALL OF THE FUEL THAT IS USEABLE IS INDICATED (EXCEPT FOR ANY INTENTIONAL UN-INDICATED EMPTY OR FULL RESERVE)- Some systems evaluated by the writer have fuel tank configurations that allow fuel in areas of the tank that cannot be sensed. For example, one such design had a fuel fill cutoff level (the amount of fuel in the tank when it is fully fueled) which was 5 liters above the highest point that the sender float could reach. The sender was unable to reach the appropriate fuel level due to tank- sender geometric constraints. Another system studied had a remote cavity which contained several liters of fuel, but the sender had no way of detecting the amount of fuel in the cavity. These types of conditions result in either non-linearity, excessive full reserve, or excessive empty reserve. Any or all of these conditions may result in customer dis- satisfaction.

CONSIDER POTENTIAL EFFECTS OF FUEL VAPOR PRESSURE AND TEMPERATURE ON THE SYSTEM PERFORMANCE- Changes in ambient temperature and the fuel's vaporization characteristics may dramatically affect the resulting pressure (or vacuum) in the fuel tank. These pressure and temperature changes may cause permanent or temporary distortions in tank dimensions which may have a profound affect on fuel gauge indication. Figure #6 illustrates a fuel gauge at 1/2 tank. Note the actual gauge reading has dropped to about 1/4 tank indication due to the float position

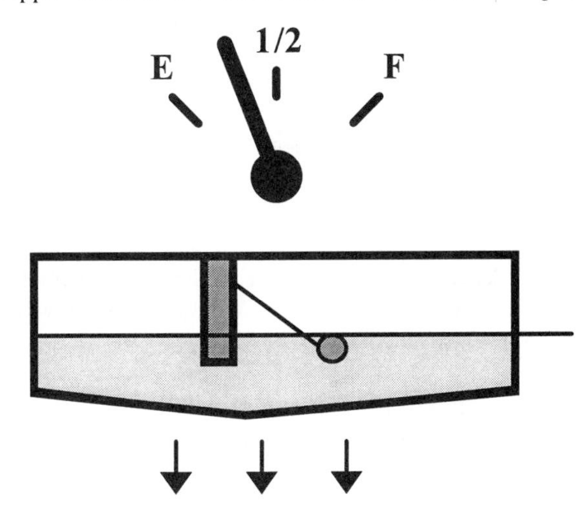

Figure #6- Effect of Tank Distortion Upon Fuel Gauge Reading

dropping lower in the tank, as the bottom of the tank distorts downward. Again, many of these variables are not directly controllable by the component engineer. The approach must be to identify the variables, and design the system around those variables to the extent possible.

OTHER SYSTEM CONSIDERATIONS- Other considerations may include: effects of electrical wiring, such as load and grounding strategies; fuel gauge graphics; vapor control strategies; tank mounting to vehicle, and component interactions.

NEW TECHNOLOGY- IS THAT THE SOLUTION?

New sender design proposals often contain claims which, under scrutiny, are unsupported. The proponent often thinks the new concept will solve all problems, when the causes of those problems are often system, not component, related. Upon such scrutiny, we find the new design suffers the same failings of the present design, because the underlying system issues remain unchanged. A review of several SAE papers on the subject supports the conclusion that new technology will not have a significant beneficial impact on fuel indication quality or cost in the near future. [4] [5]

CURRENT THICK FILM FUEL SENDER- The present, widely used, ceramic thick film (TF) rheostat with float arm design is very simple as evidenced by its very low cost. The design was developed in the early 1980's and Ford in the USA, and VDO in Germany, were among the first major companies to switch from wire wound to thick film technology. The advantages of today's thick film design is well summarized by Gaston [1], [2] and Buehler [3]. The present design is in use throughout the world primarily because of low cost, good reliability, and acceptable durability. Most major suppliers have converted from the earlier wire-wound resistor to the ceramic thick film resistor design, although at least one major USA manufacturer still uses wire wound senders. As this trend continues, virtually all wire wound should be converted to thick film by the end of the decade.

The thick film float-arm design concept offers significant packaging flexibility. For example, the sensing location of the float can be remote from the tank mounting location by varying the length of the float rod and the sender float rod pivot. Of course, one should not compromise system performance just to achieve packaging flexibility. Another example of design flexibility is in the area of sensor mounting. For example, this type of sensor can mount to the side of the tank, or to the top, or any available surface. This is not possible with vertical type sensors where the mounting of the sensor , by necessity, is directly above the level being sensed, greatly restricting the packaging flexibility.

Another design advantage is the low tooling costs associated with thick film senders. Typically only the float rod profile and the thick film card calibration require change from tank to tank. These changes are inexpensive and quick to implement.

The current potentiometer thick film design easily compensates for tank shape differences by changes in the thick film card calibration. This is done by simply changing the thick film card resistance profile through a high speed, low cost laser trimming operation. This provides a linear responding gauge from a non linear shaped tank.

ALTERNATIVE TECHNOLOGIES- Alternative emerging technologies, such as solid state devices, are shown to be accurate to within +/- 1% to +/- 2%. [3] Thick film sender technology is accurate in a range of +/- 2%. The potential for a +/-1 % improvement is not detectable by the customer. System variations usually far exceed this percentage.

The only other technology known to the author with significant worldwide volume is the vertical float/resistor design. Due to higher costs and packaging restrictions, the volumes of vertical float designs are not expected to increase, and are likely to decline, as vehicle manufacturers seek to minimize cost and complexity.

Some well-studied design approaches that continue under investigation include capacitive, differential pressure, inductive, optical, piezo-electric, TDR, and ultrasonic. Buehler summarizes several of these technologies and their current status of development in his 1993 SAE paper "Liquid Level Sender Technology". [4]

Zabler, et. al., of Robert Bosch, state in their closing paragraph comparing their wave propagation technology to the thick film design: "Initial projections of manufacturing costs would seem to indicate ... replacement for the conventional electro-mechanical sensors...is still unlikely. At the present time, it seems that the electronic sensor is unlikely to displace its conventional potentiometric counterpart on cost grounds alone in those applications where the latter represents an adequate solution". [5] Buehler further states: "At this time, the thick film ceramic sender provides the most cost effective solution for fuel level measurement." [4]

A basic failing of virtually all of these design approaches is that they all still measure liquid height to infer volume, and therefore are susceptible to the same system issues described in this paper.

New technologies will need to focus more on the system issues as opposed to component issues if they are to succeed in displacing the thick film potentiometer designs.

ROBUST DESIGN

A few general comments on this subject are valuable to point out.

Robustness in design, simply stated, is to make a product or process perform to its intended purpose, over a wide variety of expected operating conditions, for the expected life of the product (or process), and to achieve this task at a cost that represents value to the customer. The principals of robust design, also known as parameter design, were advanced by Genichi Taguchi a decade ago. Many other authors have contributed significantly to the subject since then. As Shoemaker, Tsui, and Wu state " Products and their manufacturing processes are influenced both by factors that are controlled by designers and by difficult-to-control factors such as environmental conditions, raw material properties, and aging. The idea of robust design is to select the levels of the easy-to-control factors (called control factors) to minimize the effects of the hard-to-control factors (called noise factors), thus making the product or process robust to the noise factors." [7]

Many potential performance issues with fuel indication, as seen throughout this paper, are a result of a lack of robustness of the system. The sensitivity of certain parameters on system performance, such as tank depth and fill control, were fully described. Therefore, the design or system engineer may want to consider robustness methodologies to create opportunities for system improvements.

CONCLUSION

If progress is to be made in fuel indication accuracy, linearity, and stability, design engineers must start looking at fuel indication as a system, and no longer as just a series of discrete components.

OTHER READINGS

Two recent SAE papers on the general subject of Systems Engineering that may be of interest to the reader are : SAE # 962177- "System Engineering: An Overview of Complexity's Impact", by William D. Schindel, and SAE # 962178- "Systems Engineering as a Structured Design Process", by Lee Armstrong.

ACKNOWLEDGMENTS

The author wishes to express his gratitude to Mr. Kenneth Gusfa, Mr. Phillip Pierron, and Mr. Philip Rader, all of Ford Motor Co., for their technical review of this paper and their suggestions on content, and to his wife, Marilyn, for her grammatical review.

REFERENCES

[1] SAE paper # 930458 "Fuel Sender Assembly Having Reduced Wear Including Low Wear Resistor Card", by R. Gaston, Ford Motor Co.

[2] SAE paper # 930459 "Design Evolution of the Fuel Sender Requiring No Electrical Calibration", by R. Gaston, Ford Motor Co.

[3] SAE Paper # 932402 "Thick Film Rheostat Position and Pressure Senders", by H. Wasserstrom, Stewart Warner Instrument Corp.

[4] SAE paper # 933014 "Liquid Level Sender Technology", by W. S. Buehler, Kysor/Medallion.

[5] SAE paper # 940628 "A Universal and Cost-Effective Fuel Gauge Sensor Based on Wave Propagation Effects in Solid Metal Rods", by E. Zabler et. al., Robert Bosch GmbH.

[6] SAE Paper #930359 "High Accuracy Capacitance Type Fuel Sensing System", by T. Shiratsuchi of Nippondenso Co. Ltd., M. Imaizumi, and M. Naito of Toyota Motor Corp.

[7] "Economical Experimentation Methods for Robust Design ", by Anne C. Shoemaker, Kwok- Leung Tsui, and C. F. Jeff Wu, Technometrics, Nov., 1991, Vol. 33, No. 4.

ABOUT THE AUTHOR

Mr. Nagy is a Senior Technical Specialist- Fuel Indication Systems within the Electrical and Fuel Handling Division of Ford Motor Company. Mr. Nagy has over 25 years of product design and management experience. Most of that experience is with Driver Information products, including over 13 years of involvement with Fuel Indication products/ processes.

Mr. Nagy holds B.S.- M.E. and MBA degrees from the University of Michigan.

960970

Conductive Thermoplastic Fuel Level Sensor Element

Cecil M. Williamson and Carl A. Taylor
Rochester Gauges, Inc.

Copyright 1996 Society of Automotive Engineers, Inc.

ABSTRACT

A new fuel gauge has been developed using an electrical resistance element made from a partially conductive thermoplastic material. This new thermoplastic material offers excellent resistance to most chemicals and fuels now used in the transportation industry and can easily be made into resistance elements in a variety of shapes and sizes. In this gauge, it is used to make a potentiometer type fuel level sensor with a high level voltage output to interface with a microcomputer.

INTRODUCTION

The increasing use of alternative fuels, which are often corrosive, has increased the hostility of the working environment for fuel level sensors. In addition, needs to increase fuel capacity and protect fuel tanks from damage has resulted in odd shaped tanks being designed to fit in cramped areas within the vehicle chassis. In turn, fuel level sensors must then fit in the limited space available.

Shock and vibration resistance are also important, particularly in off-highway and industrial service applications.

The proliferation of the microcomputer makes it desirable that fuel level sensors have easily scalable, high level output that interfaces directly with a microcomputer.

Last, but not least, the cost of repairs requires that the fuel level sensor be both reliable and have a long service life.

BACKGROUND

Pivot-type variable resistance fuel level sensors have long been the standard of the transportation industry. These fuel level sensors are well understood and have been refined over many years. They do, however, have several shortcomings. They require enough open space in the tank to allow movement of the floats used and thus are not well suited for some tank shapes. Also, the metals used for resistance elements are typically selected for electrical properties rather than for resistance to corrosion.

It is increasingly evident that a new type of fuel level sensor is needed to meet these challenges. The availability of a conductive thermoplastic suitable for use in making resistance elements has made possible the design of a simple, rugged fuel level sensor to fit today's needs.

CONDUCTIVE THERMOPLASTIC

The material selected for the resistance element in this new fuel level sensor is a compression molded thermoplastic. It can be molded into sheets with comolded layers of either conductive or nonconductive material in any desired pattern. As used in this fuel level sensor, the thermoplastic is conductive on one side and nonconductive on the other. The resistance elements were made by cutting strips from molded sheet.

The thermoplastic has excellent resistance to all fuels now used in the transportation industry and many other chemicals as well. Extensive immersion testing has been performed to verify this. Other properties are listed in Table 1.

Table 1. Conductive Thermoplastic Properties

Property	Value
Volume Resistivity (ohm-cm)	0.05
Tensile Strength (mpa)	62
Elongation (%)	0.6
Flex Modulus (gpa)	10.0
Density	1.82
Melt Temperature (°C)	285-315

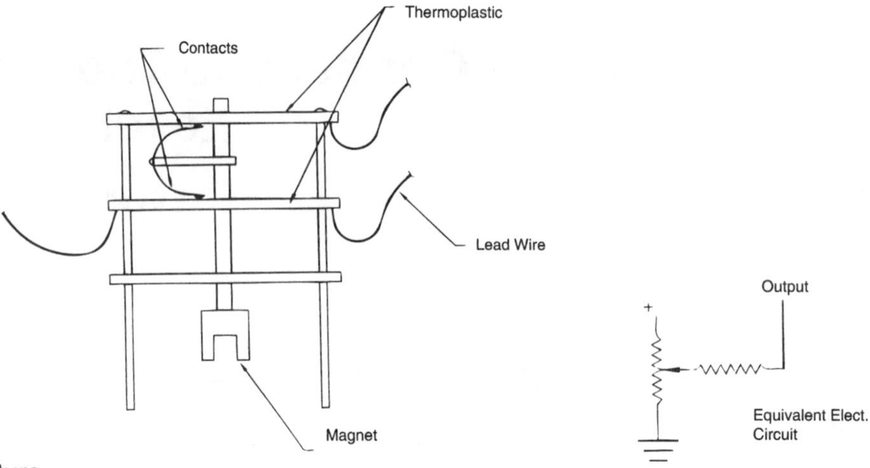

Figure 1. Contact Test fixture

Linearity testing was performed to determine uniformity of the material. Typical linearity variations were 2 to 3%. Worst-case variations from linear were 4 to 5%. This should improve as the conductive thermoplastic strip manufacturing process is fine tuned.

Various contact materials were tested using the test fixture shown in Figure 1. All contact testing was conducted with the thermoplastic material fixture immersed in SAE test fuel CMO. In operation, a magnetically coupled motor causes the shaft to rotate, which in turn rotates the contacts against the thermoplastic. The best contact material on a cost versus performance basis is the conductive thermoplastic.

FUEL LEVEL SENSOR DESIGN

The first fuel level sensor designed using conductive thermoplastic resistance elements is shown in Figure 2. This device is intended for off highway service applications and is 1 m. long. The fuel level sensor has an aluminum flange (item 1), an extruded aluminum support structure (item 2) and a float stop (item 5). These parts are held together by a stainless steel shaft (item 7) which passes through the center of the gauge. The stainless steel shaft is threaded on each end and is placed under tension by the retaining nuts. The support structure is under compression and one end is clamped tightly to the flange.

The voltage divider resistance element is a conductive thermoplastic plastic strip that is cut to the same length as the support. The strip is first inserted into an extruded nylon insulator and then into a groove in the aluminum support. Voltage is applied at one end of the resistance element and the other end is grounded. A second conductive thermoplastic strip and insulator assembly fits into a second groove on the opposite side of the extruded aluminum support.

The float carries two electrical wiper contacts. The contacts are connected to each other and the two conductive strips to form a voltage divider or potentiometer. The equivalent electrical circuit is shown as Item 8 in Figure 2. It should be noted that one of the thermoplastic strips is used as a resistor. The second strip is used as a wiper collector.

It is apparent from Figure 3 that it is possible to reduce cost and weight in many sensor applications by making most of the fuel level sensor from high performance engineering plastics. In addition, since the conductive thermoplastic can easily be molded or cut to any desired shape, a wide variety of fuel level sensors designs are possible using this material that are difficult or impossible to make in metals. Figure 3 shows one example of how plastics may be used.

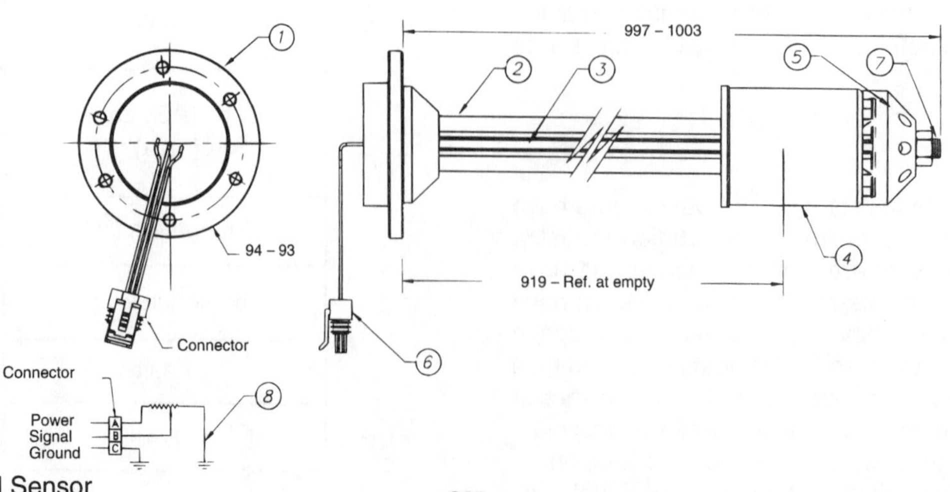

Figure 2. Fuel Level Sensor

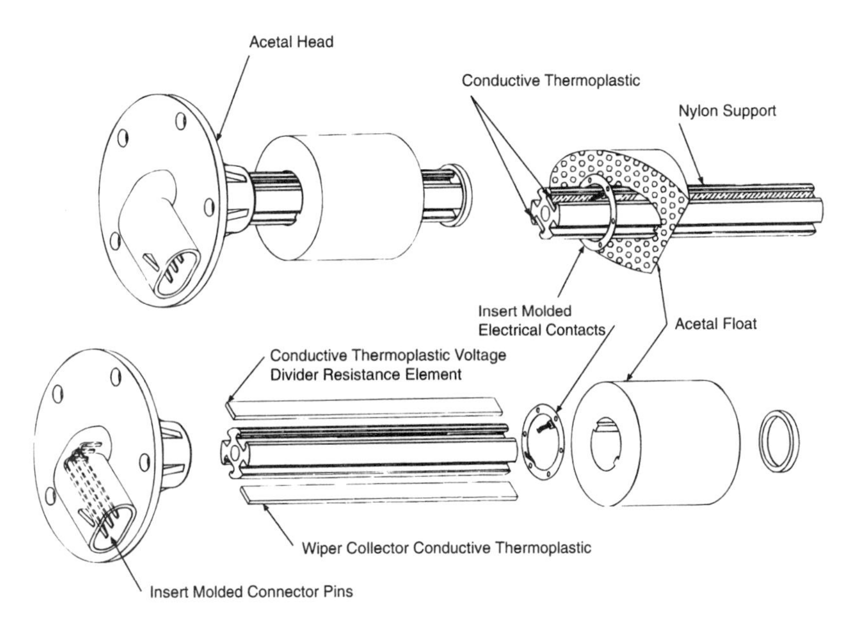

Figure 3. Thermoplastic Fuel Level Sensor

INTERFACE REQUIREMENTS

The high output impedance of this sender must be considered when interfacing it to a receiver. The output impedance is governed in large part by the contact resistance. The contact resistance may run as high as several hundred ohms.

A typical application circuit is shown in Figure 4. In this circuit, the output of the sender is being converted to a digital signal by an analog-to-digital converter that has a high impedance input. The low pass R/C filter in the circuit serves two purposes. First, it electronically dampens the signal variations caused by turbulence in the fuel tank. Second, it filters any EMI/RFI emissions that may be present.

SUMMARY

We have been able to build a simple, rugged fuel level sensor using conductive thermoplastic materials as the voltage divider and collector elements. The excellent corrosion resistance, versatility in design potential and ease of use of this plastic material make it an attractive alternative to conventional resistance element materials now in general use.

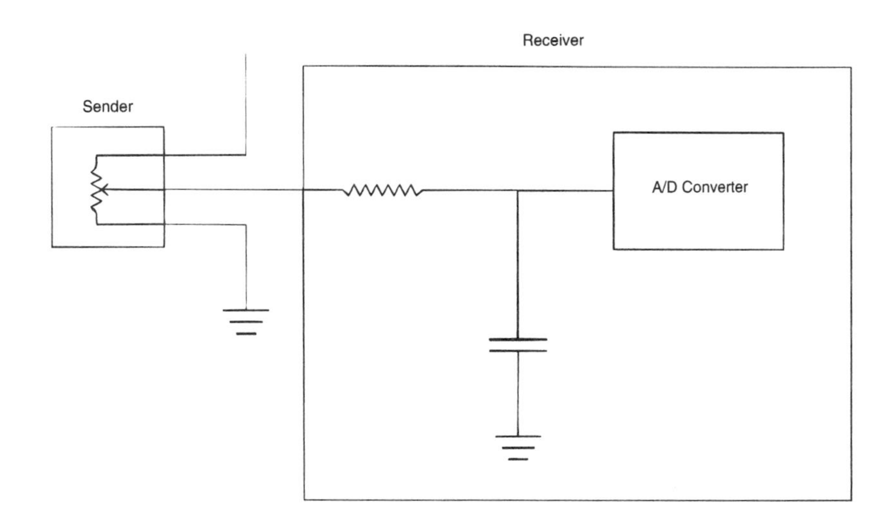

Figure 4. Typical Interface

940628

A Universal and Cost-Effective Fuel Gauge Sensor Based on Wave Propagation Effects in Solid Metal Rods

Erich Zabler, Anton Dukart, Nicolas Grein, and Frieder Heintz
Robert Bosch GmbH

ABSTRACT

In recognition of safety considerations, modern fuel tanks are frequently extremely irregular in shape. This places limits on the application of conventional potentiometric sensors. Required are more universal sensors without mechanically-moving parts. These sensors should also be characterized by especially good resolution and precision in the residual-quantity range, that is, the zero point precision should be of a high order.

One type of metal rod can be bent into any of a variety of shapes to provide an effective means of monitoring the fuel level. In this metal rod, the propagation characteristics of a certain type of sound wave, known as bending waves, display major variations according to the level of the surrounding medium: The waves spread more rapidly through the exposed section of the rod than through the area which remains submerged. Thus the rod's characteristic oscillation frequency varies as a function of immersion depth. For electric processing, electromagnetic excitation is employed to induce ultrasonic oscillations in the rod. The phase at the exposed end of the rod is regulated to a constant value in order to maintain the rod at a characteristic natural frequency corresponding to the instantaneous fuel level. It is possible to maintain temperature-induced fluctuations in both zero-point stability and sensitivity at minimal levels through careful selection of suitable rod material and an appropriate operating frequency. Suitable profiles can be employed to optimize other operating parameters and adapt the sensor for use with specific measurement ranges.

The present treatise reports on the basic principles of the process - the theory of which is simple to describe and simulate - as well as initial research results.

INTRODUCTION

Fuel-level sensors have a long tradition, having been among the earliest forms of sensor to be installed in automobiles. Initially, these devices provided an approximate display along with the certainty that a minimal amount of fuel would remain available when the reserve range was reached. Today, these sensors function together with on-board trip computers to furnish information with the maximum possible degree of precision. These data provide the basis for calculations of remaining cruising range, or the fuel consumption during the previous 100 kilometers.

At present, roughly 30 million fuel-level sensors are being installed around the world each year, with approximately 13 million of this number in Western Europe; the figure always displays a general correspondence to the number of vehicles produced. Virtually all of these devices monitor the fuel content indirectly - as a function of fluid level - and they almost always assume the form of variable-resistance wiper units employing a float to govern their position according to the fuel level in the tank. Both lever-action and immersion-tube floats are employed (Fig. 1); the operating principles will already be familiar. Despite these sensors' extremely good measuring precision, tolerances in tank shape and the relatively high degree of density fluctuation which the fuel displays in response to temperature variation combine to place limits on the use of this principle; unit precision requirements exceeding approx. 1-3 % of the terminal volume of the tank would produce no substantive benefits. Although the current sensors are contact devices and thus not immune to wear, this type of device - at a unit price of roughly $ 3.50 (tendency falling toward approx. $ 3.00 per unit) - is quite inexpensive; it would be difficult to offer alternatives at a lower price.

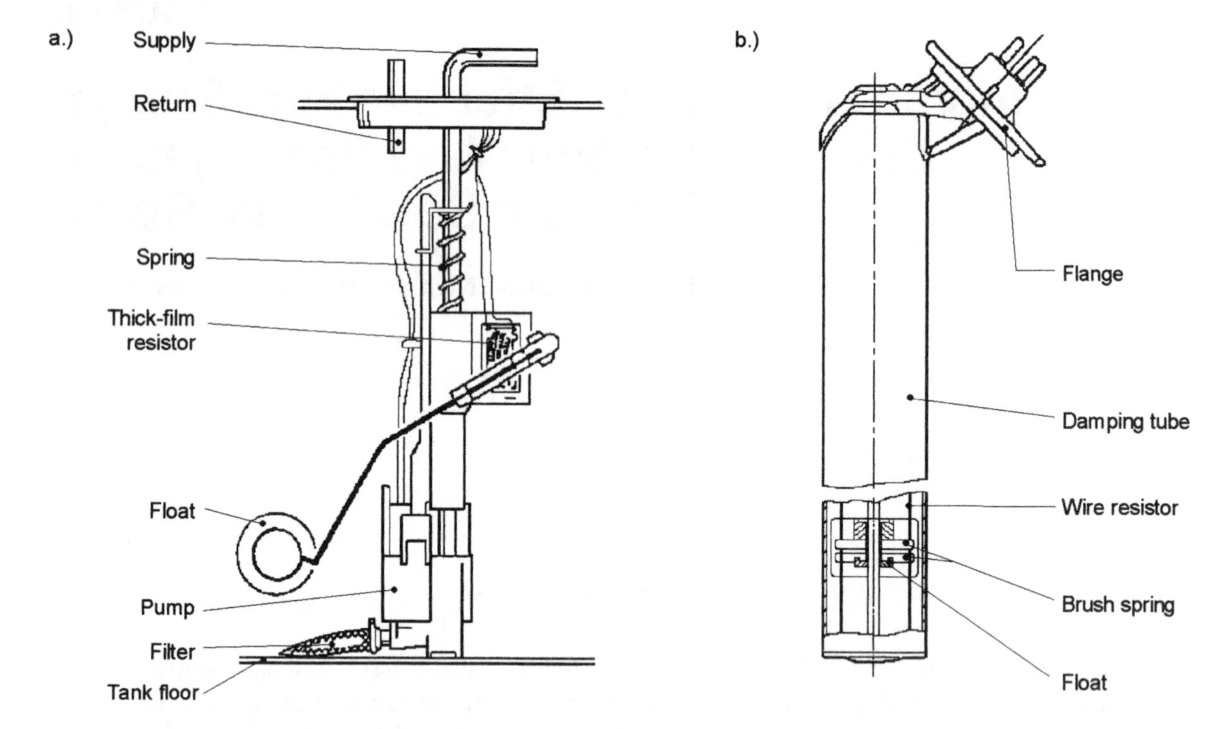

Fig. 1 - Tank-level sensor featuring a.) swing lever, b.) immersion tube

While only low absolute growth rates are anticipated for this sensor market, more stringent requirements may be expected to generate two-digit increases for new sensor designs. With their increasingly variegated range of complicated shapes, modern fuel tanks confront the conventional float-actuated potentiometer with an expanding range of difficulties; the new principles should provide the means for overcoming these disadvantages. The problem areas include:

- adequate clearance for the float
- maximum verticality for immersion-tube floats
- unsuited for extremely irregular tank configurations
- contact between float and tank floor limits precision in measuring residual levels
- wear potential (potentiometer).

Thus the new fuel-level sensors must fulfil the following requirements; this will generally entail more than merely replacing the wiper-type sensor with a contactless device:

- high precision (1-2 % of total capacity), especially in residual range,
- option of enhanced sensitivity in residual range,
- suitability for extremely irregular tank shapes,
- avoidance of moving parts to obviate wear,
- easily digitalized output signal,
- high degree of resistance to electromagnetic interference,
- design integrating sensor and electric fuel pump in a single compact assembly.

SUMMARY OF NEW SENSOR PRINCIPLES

A number of innovative sensor designs have been proposed in response to the requirements enumerated above. However, substantive difficulties have consistently stood in the way of a major breakthrough:

THERMAL - Here, a film-resistor is mounted on a flexible carrier; the level-sensitive variation in heat response serves as the basis for processing (submersed resistor heats up more slowly than exposed section). An articulated hose can be employed to endow the resistor with virtually any geometrical configuration for use in practically any tank, regardless of complexity. This sensor relies on extremely complex electronics /1/.

CAPACITIVE - Two planar condenser elements are mounted in parallel, coaxially or, occasionally, with interspersed meshing elements. The unit's capacity then varies in response to fluctuations in the amount of fuel in the space between the elements. Aside from the susceptibility to electrical interference, this design suffers from several other liabilities and sources of potential measurement error. These include sensitivity to fuel additives, the possibility of deposit formation between the capacitative planar elements, and, especially significant, fluctuations in the water content (high dielectric constant ε). All of these factors can lead to substantial display error.

HYDROSTATIC - The hydrostatic design concept is extremely simple and easy to understand. This principle can be employed with even the most unusual tank shapes, as it simply processes the pressure differential Δp between the tank floor and the atmosphere above the top of the fluid:

$$\Delta p = \rho \cdot g \cdot h \tag{1}$$

g = gravitational acceleration

The differential corresponds to the weight of a hypothetical vertical fluid column reaching from the tank floor to the fluid surface in accordance with the fuel level h. Thus this method is the only one to provide gravimetric measurements, that is, it measures the density ρ of the fuel mass remaining in the tank. However, the temperature of the fuel can then be factored in as the basis for simple calculations of fuel volume. Although this method boasts a number of attractive features, it has failed to gain acceptance up to now due to difficulties in finding suitable and inexpensive pressure-differential sensors with adequate zero-point stability. Automatic zero-adjustment when the vehicle is started or at periodic intervals during operation could represent a possible solution /2/.

ACOUSTIC (Fig. 2) - Echo range sounding with free sound waves - Methods using the travel time of a sonic pulse transmitted to the fluid surface from either the tank floor or its ceiling are simple and the results are easy to digitalize. However, the numerous possible wall echoes and problems with cavity resonance lead to severe measurement error. In addition, the shapes of today's fuel tanks can be so complicated that no direct 'line of sight' remains available between the floor and the ceiling of the tank.

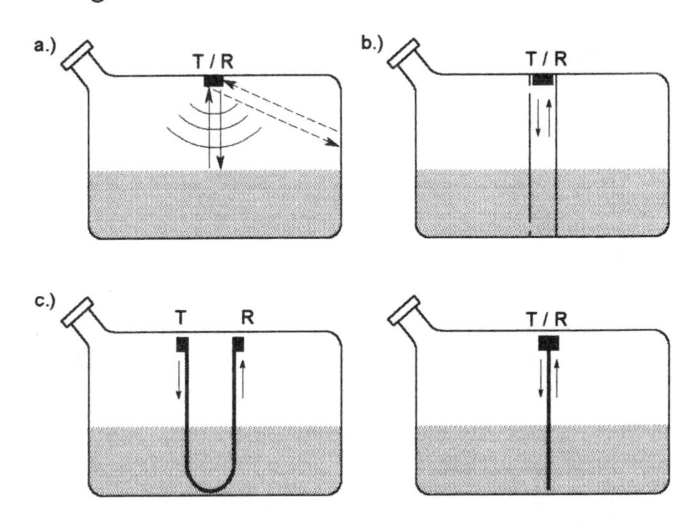

Fig. 2 - Acoustic sensor concepts: a.) Echo sounding with unguided sound, b.) with guided sound, c.) sound-wave propagation in rigid rod (U and I-shaped)

Echo range sounding with guided sound waves - A means was sought of applying the echo-sounding principle in the absence of a direct line of sight between floor and ceiling while simultaneously avoiding interference from wall echoes. This led to designs in which the sound pulses were conducted through flexible tubes which, in turn, were open to the fluid in the tank. However, the susceptibility to error stemming from cavity resonance became even greater.

Sound-wave propagation in solid rods - It was observed that specific types of acoustic waves respond to differences in the surrounding medium (for instance, tank atmosphere or liquid fuel) with substantive variations in their propagation rates within solid rods; this observation served as the basis for an extremely attractive and universally-applicable sensor configuration /3/. The geometry of these rods can also be selected to suit virtually any tank shape, allowing application as a kind of 'acoustic dipstick' /4/.

PRINCIPLES OF THE ULTRASONIC SENSOR ROD

THEORY OF SOUND-WAVE PROPAGATION IN RODS AND PLANAR ELEMENTS - The range of mechanical/acoustic waves that can be projected through solid bodies is wide and variegated. When the wave is purely longitudinal, the particles travel only in the propagation direction, while transversal waves move exclusively along the lateral axis (90° offset) of the transmission path. Bending waves (Fig. 3) display a distinctly transverse character. Because these waves induce substantial surface excursion, they differ from other types of wave by exhibiting a high degree of interaction with the surrounding medium.

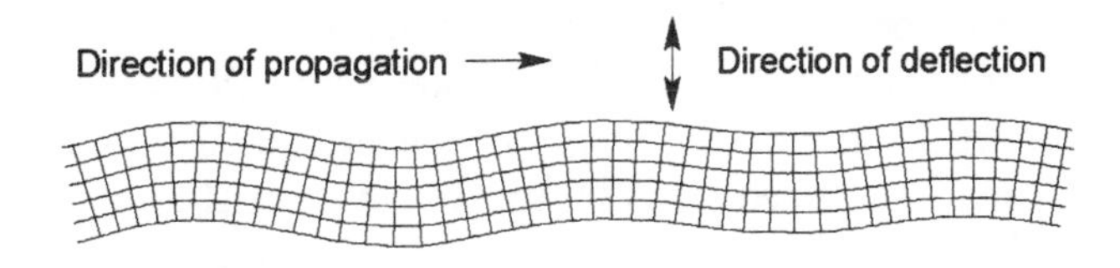

Fig. 3 - Bending wave in rigid sensor rod

In plates and rods with rectangular cross section, the propagation velocity c_b is not only a function of the modulus of elasticity E, the transversal contraction factor σ and the density ρ of the acoustic material being used. Also of determining significance is the frequency f, as illustrated in the EQ (2):

$$c_b = \sqrt{\pi \cdot d \cdot f} \cdot \sqrt[4]{\frac{E}{3 \cdot \rho \cdot (1 - \sigma^2)}} \tag{2}$$

d = thickness of the conductor

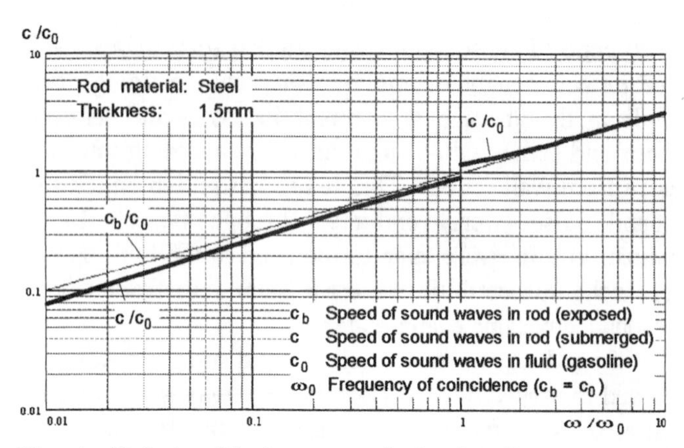

Fig. 4 - Relationship between relative bending wave velocity c/c_o and applied frequency ω/ω_o for 1.5 mm steel planar elements, with (c/c_o) and without (c_b/c_o) medium load

This relationship remains valid regardless of the density of the surrounding medium, and conforms to the curve represented by the slim line in the diagram (Fig. 4): If the bending resonator is immersed in a medium of density ρ_{Fl} and sound-wave propagation velocity c_o (e.g., c_o = 1,330 m/s for fuel at a temperature of 20° C), the result is acoustic coupling between this medium and the oscillation surface. The response pattern than varies according to whether the propagation velocity in the rod is smaller or greater than that in the surrounding medium at the selected operating frequency. The propagation velocity then corresponds to the solid line in the diagram. This point represents the state known as the frequency of coincidence ω_o. Below this frequency point, the ambient medium reduces the propagation in the rod to a substantially lower velocity c; only in the immediate vicinity of ω_o does it approach the figure c_b. The surrounding medium acts essentially as a mass load on the oscillator, while no appreciable amount of energy is radiated into the medium itself. This effect becomes progressively more pronounced as the phenomenon described as the sound-wave resistance Z (Table 1) of the rod material approaches that of the fluid.

Material	c_b / m/s	λ / cm	Z / kg/(m² · s)
Steel	580	2.23	$4.55 \cdot 10^6$
Aluminum	604	2.32	$1.63 \cdot 10^6$
Copper	510	1.96	$4.54 \cdot 10^6$
Lead	290	1.10	$3.30 \cdot 10^6$
Water	-	5.69	$1.48 \cdot 10^6$
Gasoline	-	5.11	$1.16 \cdot 10^6$

Table 1 - Wave length λ, sound-wave resistance Z and sound- wave velocity c_b in bending waves for planar elements in various materials at a thickness of 1.5 mm at a frequency of f = 26 Hz.

Above this frequency the relative influence remains diminutive, and is also inverse in character: The bending wave displays a higher propagation rate in the submersed section of the rod than in the exposed length; the resonator is merely subject to substantial damping due to wave dissemination in the ambient medium. At the frequency of coincidence ω_o, it is not possible for the sound waves to travel into the surrounding medium. Thus the frequency of coincidence ω_o is not only strongly dependent on the sound-wave propagation velocity c_o in the surrounding medium, it is also contingent on the rod parameters E, ρ, σ and d:

$$\omega_0 = 2 \cdot \pi \cdot f_0 = \frac{c_0^2}{d} \cdot \sqrt{\frac{12 \cdot \rho \cdot (1 - \sigma^2)}{E}} \qquad (3)$$

(Frequency of coincidence)

In the standard combinations of metal/fluid this figure is found mostly in the ultrasonic range, with typical values of 50...200 Hz:

Material	f_0 / Hz
Steel	163
Aluminum	146
Copper	205
Lead	610

Table 2 - Frequency of coincidence for various rod materials, 1.5 mm thick, in water (c_0 = 1,480 m/s)

POTENTIAL OF SIGNAL PROCESSING FOR MONITORING FUEL LEVEL - Two different procedures were evaluated for utilizing the bending-wave effect with rods submersed in fluid media:

- In one procedure, sinus-wave excitation at one end of a rod during gradual immersion in a fluid produces a substantial change in the phase observed at the opposite end.
- In the other process, the rod acts as a resonator with various characteristic oscillation modes. The rod's frequency varies during submersion in a fluid in response to partial changes in inner sound-wave propagation velocity /5/.

Generation of bending waves - To produce bending waves, forces acting at a cross angle to the direction of wave propagation must be transmitted into the rod. There are different, highly-variegated ways to do this. However, the preferred method is piezoelectric, or - when ferromagnetic rods are employed - electromagnetic force generation (Fig. 5). As a piezoelectric tap is already envisioned for the required phase measurement, we selected the illustrated electromagnetic excitation for the present application in order to minimize electrical crosstalk between excitation and tap.

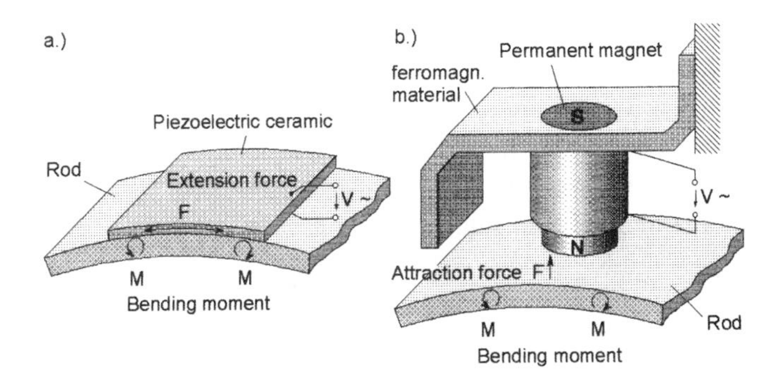

Fig. 5 - Excitation of bending waves with a.) piezoelectric, b.) magnetic force

Phase method - U-shaped rods were preferably used in order to employ the phase method without having to operate both excited and measurement ends of the level-monitoring rod below the surface of the fluid (Fig. 6).

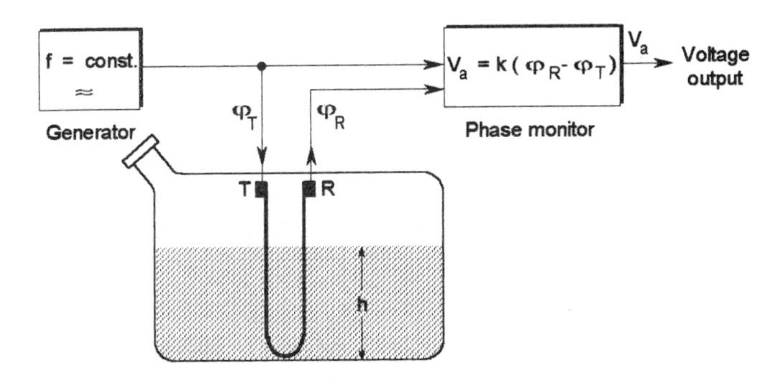

Fig. 6 - Monitoring tank level with phase method and with U-shaped rod

In this application, the rod is operated as an acoustic transmission/reception path. At one end, electromagnetic excitation induces a static sinus oscillation at a constant frequency ω (transmitter T). At the other end, a high-frequency piezoelectric bimorphic strip is bonded; this strip is used as an accelerometer for scanning the rod's oscillation phase (receiver R). The phase shift across the entire length of the exposed rod is:

$$\phi_o = \phi\ (h = 0) = \omega \cdot \frac{L}{c_b} \tag{4}$$

With rising fuel level h, the shift $\Delta\phi$ relative to this phase is:

$$\Delta\phi = \phi\ (h\ ) - \phi_0 = \omega \cdot h \cdot (\frac{1}{c} - \frac{1}{c_b}) \tag{5}$$

As Fig. 7 illustrates, the relevant curve derived from this simplified theory displays a linear progression. It will be

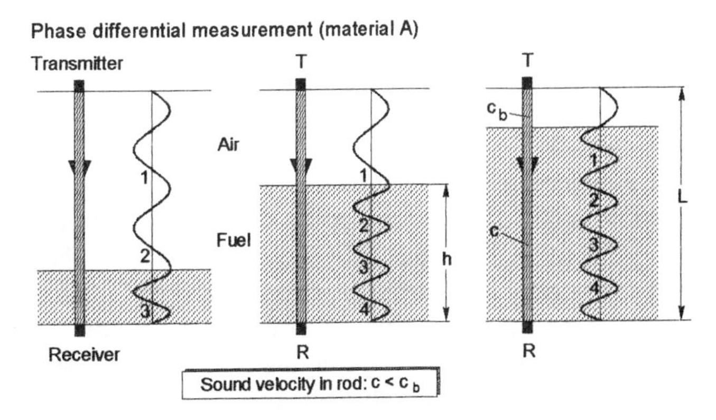

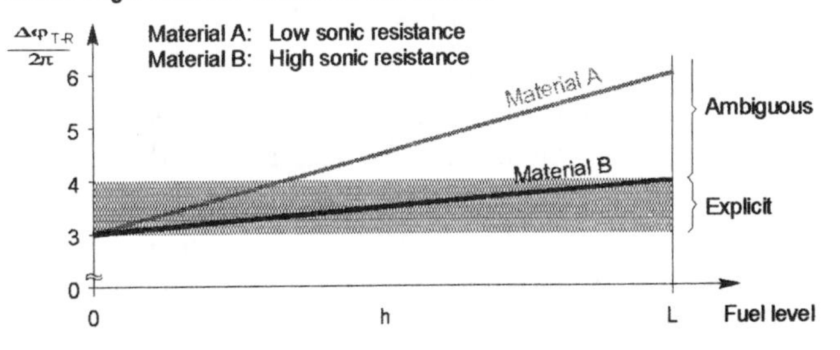

Fig. 7 - Variations in phase angle φ as function of tank level h

recognized that the angle shift $\Delta\phi$ associated with common materials, parameters and fuel levels can easily exceed the maximum acceptable figure of 2π required to avoid ambiguity:

$$\Delta\phi \leq 2 \cdot \pi \tag{6}$$

(ambiguity avoidance condition)

In designing this type of sensor, obtaining adequate sensitivity is not the main problem: Effort must be directed toward ensuring that the measuring sensitivity is not excessive. However, actual measurements provide rapid confirmation that this simplified analysis does not take into account the rod's resonance characteristics: If both rod ends are not provided with a reflection-free

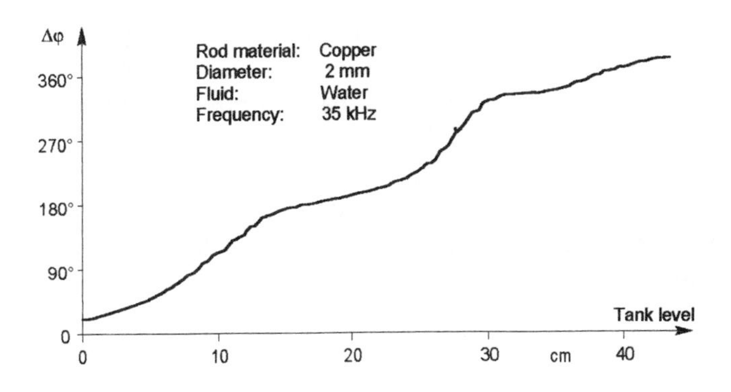

Fig. 8 - Measured phase shift $\Delta\phi$ relative to tank level for a U-shaped rod

termination, the curve defining phase shift response to increasing fuel levels will be extremely non-linear and inconsistent (Fig. 8).

Resonance method - A large degree of independence from phase fluctuation can be achieved - without sealing the rod end against reflection - by tracking the driving frequency so that the phase at the rod end maintains at a constant figure (Fig. 9). If this is, for instance, 90° relative to the excitation, then the rod will be situated directly in one of its resonance modes; there will then be a specific number of vibrational nodes and anti nodes (typical figure: n = 12) on the rod. These display a periodicity of one half a wave length λ; should no heavy masses be present at the ends, a maximum of oscillation amplitude will be found there. With this method it is important to ensure that the phase condition defined in the EQ (6) is maintained, while the measurement effect should too also intentionally be held to a lower level, as the system might otherwise spring into a different phase mode.

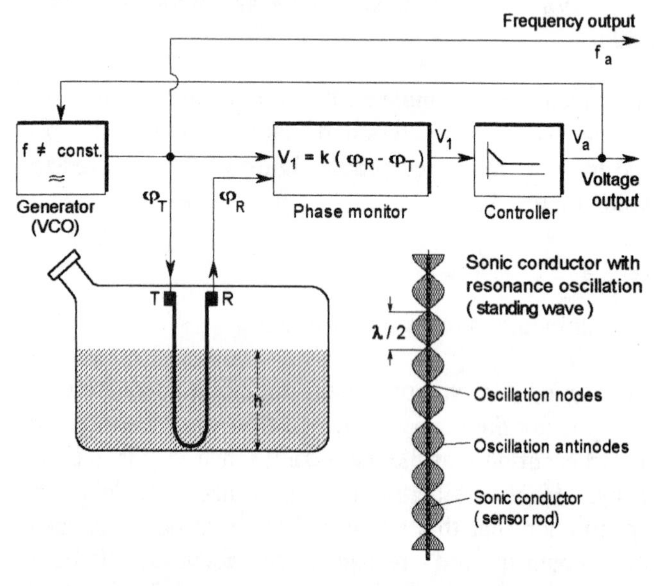

Fig. 9 - Monitoring tank level with the resonance method

This method not only provides the advantages associated with a more stable characteristic curve featuring enhanced linearity, it also makes it possible to position the excitation and the phase tap at the same end of the rod - the exposed end. The phase signal is monitored with a piezoelectric bimorphic strip as a very cost effective accelerometer. This phase signal still displays a substantial amplitude in the vicinity of the excitation point. As shown in Fig. 9, it is conducted to a phase comparator along with the excitation signal. The phase comparator's output signal modifies the frequency f via a VCO (voltage controlled oscillator) until the phase requirement at the metering rod has been met. Phase comparator and the attendant VCO are available from many sources as an inexpensive standard IC. This type of IC furnishes an extremely simple means of maintaining the oscillation within the acceptable frequency range; the control voltage of the VCO is also available for use as a voltage-analog level signal V_a.

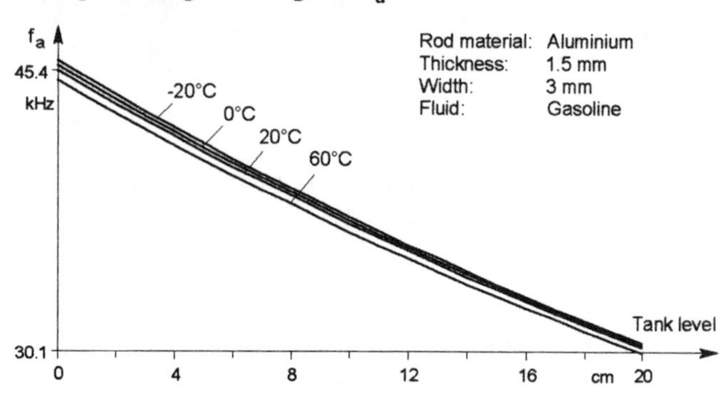

Fig. 10 - Calculated response curve for a level sensor using the resonance method

Due to the reduced propagation velocity c, the wave length λ in the submersed section of the rod would decline at a constant frequency f (λ = c/ f). The control system must thus reduce the frequency by the requisite increment as fuel level rises; both calculations (Fig. 10) and empirical measurements (Fig. 11) define the result as a declining but pretty linear relationship between initial frequency and fuel level. Despite the generally positive linearity, the details of the empirically-derived response curves indicate a small and relatively insignificant degree of ripple, the periodicity of which corresponds to half a wavelength of a bending oscillation.

This type of frequency-analog output signal f_a is also recognized as displaying additional important advantages: The signal is easy to convert to digital form, transmission remains immune against interference, and the level signal need not necessarily be processed at the sensor - no difficulties are encountered in transmitting the signal to the dashboard (dashboard controller).

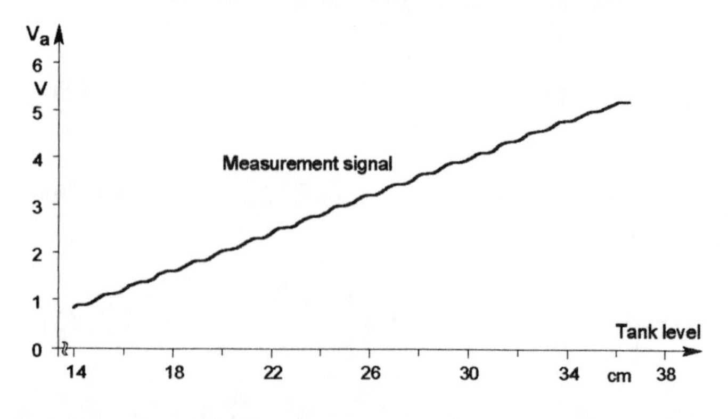

Fig. 11 - Measured response curve for a level sensor using the resonance method (voltage-analog output)

PERFECTING TEMPERATURE CHARACTERISTICS - Zero point - When any rod material is selected at random, the result can be impressive measurement effects, at least within a certain range, but these will frequently be accompanied by substantial temperature sensitivity in the zero point. The source is easily located (for instance, for the phase-measurement method) if the phase angle ϕ_0 for the fully-exposed rod (with measurement zero point h = 0) is calculated by inserting the EQ (2) in EQ (4). The result is:

$$\phi_o = \frac{L \cdot \sqrt{\pi \cdot f}}{\sqrt[4]{\dfrac{E}{3 \cdot \rho \cdot (1-\sigma)}}} \tag{7}$$

It will be noted that the parameters L, d and ρ for ϕ_0 are influenced both by the thermal expansion coefficient α in the rod material, and by the temperature coefficient TK_E of the Young's modulus E. The resulting temperature coefficient for the zero phase angle ϕ_0:

$$TK_{\phi_0} = \frac{\partial \phi_0 / \phi_0}{\partial T} = -\frac{1}{4} \cdot (\alpha + TK_E) \tag{8}$$

In metals, α is generally much lower than TK_E. If the temperature drift of this zero phase is to be maintained at a low level, then the rod material should display the lowest possible temperature coefficient TK_E for the modulus of elasticity E. However, this disappears almost completely if TK_E is deliberately selected as:

$$\alpha = - TK_E \tag{9}$$

instead of zero. Special NiFe alloys, such as Thermelast (VAC), have been specially developed to obtain minimal TK_E figures; these roughly conform to the condition (9):

Thermelast 4002: $\alpha = 8{,}5 \cdot 10^{-6}/K$; $TK_E = -5 \cdot 10^{-6}/K$

Fig. 12 illustrates the decisive effects of material selection. Thermelast can be employed to obtain the desired zero-point precision of 1...2 % of the measurement range, making it possible to dispense with a compensatory reference system or the application of other temperature-sensitive compensation elements.

Measurement sensitivity - Only the attributes of the fully-exposed rod are relevant for zero-point precision, but when the rod is fully submerged the characteristics of the medium also become important. With increasing temperature, these characteristics assume particular significance for the frequency of coincidence up to which measurement effects occur.

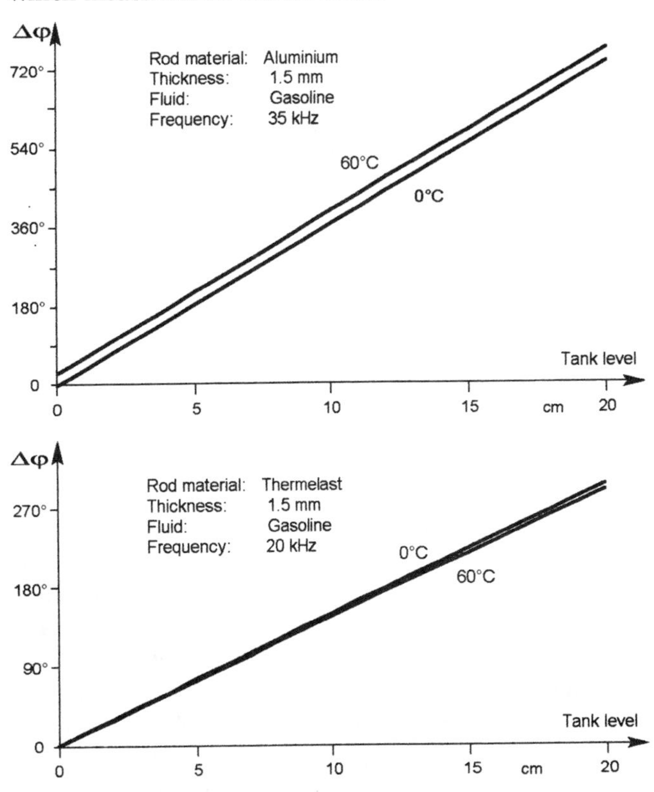

Fig. 12 - Comparison of zero-point drift (phase measurement) with various rod materials

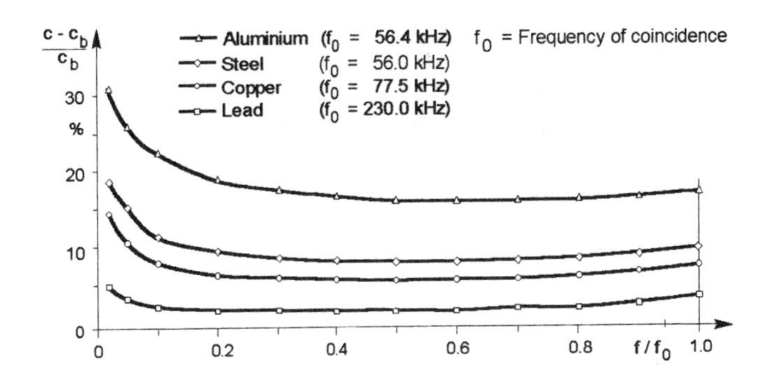

Fig. 13 - Percentual drop in bending wave velocity upon immersion in gasoline

In actual practice, the compensation frequency will virtually always be found at 15...30 % of the frequency of coincidence. As shown in Fig. 13, the measurement effect can increase or decrease in response to a lower frequency of coincidence, according to the relative location of the operating frequency. Fig. 14, produced with the aid of a simulation equation, provides a concrete example. There is always a specific operating frequency (about 35 kHz in the present case) at which the measurement effect $\Delta c/c$ does not vary along with temperature: If this compensation point is examined carefully, it

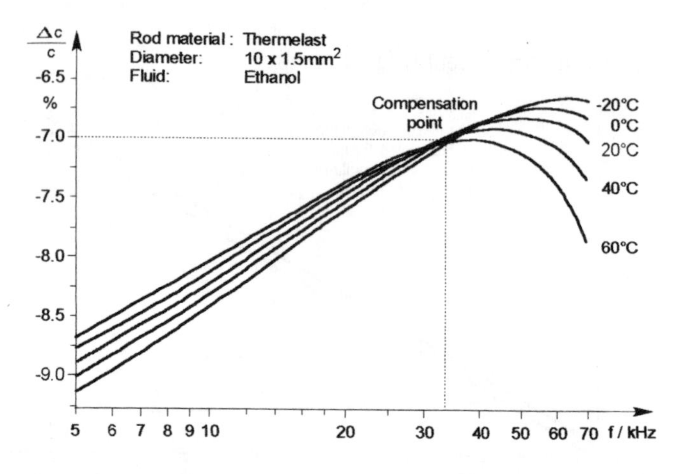

Fig. 14 - Variations in measurement effect $\Delta c/c$ relative to frequency at different temperatures

is seen that temperature-induced deviations remain below 1 % of measurement range through the entire operating-temperature range of - 40...60° C. For the resonance method described here, with which the operating frequency varies according to tank level, it is best to select the point of maximum immersion depth, and thus the lowest resonance frequency, to achieve maximum correspondence with the compensation frequency described above. For a wide Thermelast rod of 1.5 mm in thickness this point is roughly 32.5 kHz. Fig. 15 shows the results of a simulation equation under optimum operating conditions for both the phase and resonance method.

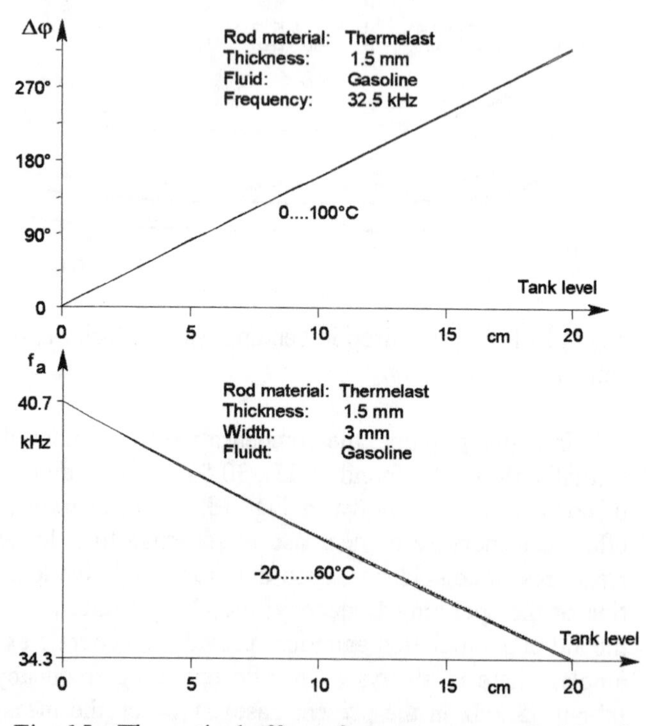

Fig. 15 - Theoretical effect of temperature on fuel-level response curve: Phase method, resonance method

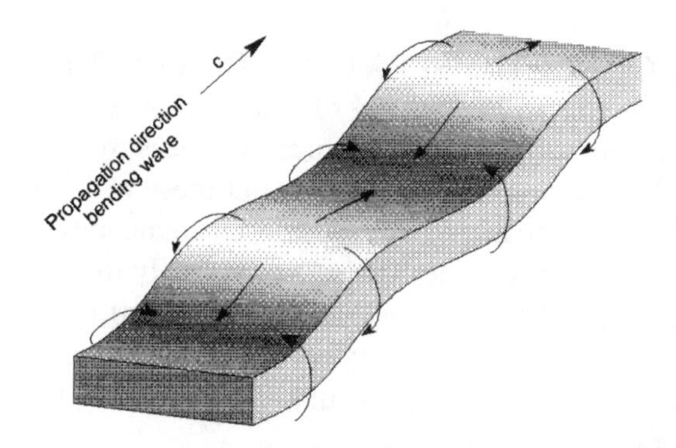

Fig. 16 - Fluid movement on narrow sensor rods ('acoustic short circuit')

OPTIONS FOR TAILORING THE SENSOR

CHARACTERISTIC - Precisely speaking, the effects described up to now are applicable only for planar surfaces of infinite extension, that is, for plates with a width that is large relative to the wavelength λ being used. When a narrower, longer rod is used, the motion which the bending wave induces in the medium may no longer be restricted to (or against) the initial direction of propagation. Instead, it can - as shown in Fig. 16 - respond to reduced width with an increasing tendency to travel the shorter distance between the front and rear sides of the rod - this phenomenon is referred to as an 'acoustic short circuit.' Of course, this will attenuate the measurement effect. Fig. 17 illustrates the empirically-determined relationship between rod width b (relative to the wavelength λ) and the reduction in measurement effect Δc related on a planar surface of infinite extension: Thus a 1.5 mm thick Thermals rod of width b = 3 mm, corresponding to 19 % of the actual wavelength at 32.5 Hz - provides only about 60 % of the measurement effect in gasoline.

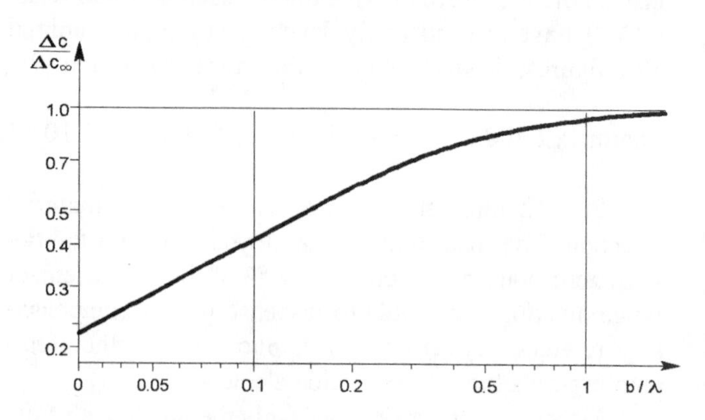

Fig. 17 - Empirically determined relationship between measurement effect $\Delta c/\Delta c_\infty$ and ratio of rod width b to wavelength λ

Thickness	Frequency of coincid.	Compens.- frequency	Width	Max. length
d / mm	f_0 / Hz	f_{comp} / Hz	b / mm	L / cm
1.5	95.0	32.5	10	13.3
			7	14.8
			5	16.9
			3	21.2
			2	25.7
			ϕ 2	44.3
2.0	71.2	24.5	10	20.5
			7	23.5
			5	27.2
			3	34.6
2.5	57.0	19.5	10	25.3
			7	29.3
			5	34.0
			3	43.4

Table 3 - Sensor-rod parameters for various rod lengths L (Thermelast, gasoline)

On the one hand, given a specific rod material and a certain operating frequency, this phenomenon makes it possible to achieve optimal measurement effects to meet the ambiguity avoidance condition described in QE (6) - provided that tank size remains within certain limits (Table 3). At the same time, a frequent priority can be achieved by expanding the lower end of the rod to extend the measurement effect in the residual range, as illustrated in Fig. 18a. In theory, it is also possible to obtain a linear relationship between tank volume and fuel depth by contouring the rod's width along its entire length; in practice, it is generally easier to achieve the same effect by storing a characteristic curve in a PROM.

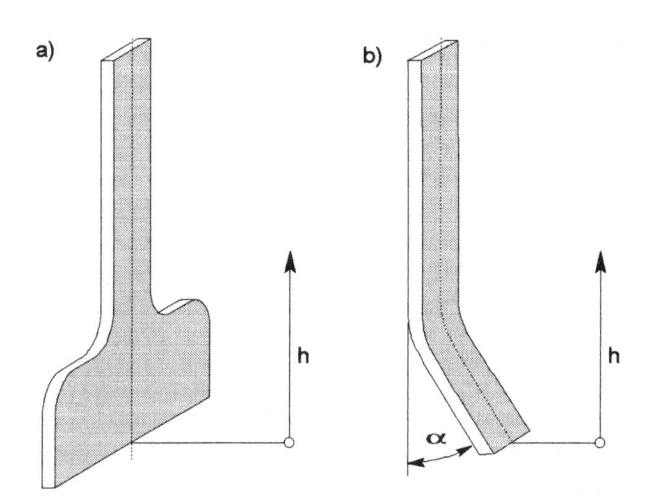

Fig. 18 - Methods for distending the response curve in the residual range, through: a.) Widening the rod, b.) Rod curvature

It is also possible to expand the residual range by using a bent or spiral-shaped rod to obtain enhanced rod length for a given fuel depth, providing a purely geometric increase in measurement effect (Fig. 18b). When the rod is curved - either to conform to tank shape or for the effects described above - the only caveat concerns the radius of the curvature, which should not drop below a certain figure. This figure generally roughly corresponds to the magnitude of the wavelength itself (typically 2.5 cm).

In a process analogous to trimming rod width, the measurement effect can also be reduced using suitable coating materials (such as Teflon, Halar, Polyethelene). This expedient also provides a side benefit in the form of enhanced corrosion protection, should such be required. Depending upon the type and thickness, this coating can be employed to obtain reductions in measurement effect of up to 50 %. The effects on temperature response generally remain minimal.

Increased rod thickness lowers the frequency of coincidence as defined in the EQ (3) and shown in Table 3. At a given frequency, enhanced material thickness reduces the relative distance to the frequency of coincidence; depending upon initial conditions, this can either raise or lower the measurement effect, as shown in Fig. 13. It must, however, be noted that this method can influence the measurement sensitivity relative to temperature.

PROTOTYPE TESTING

In order to demonstrate this measurement concept's suitability for application, the numerous laboratory prototypes were supplemented by units which could be installed in vehicles, making it possible to subject the devices to vehicle testing for specific periods under real-world conditions.

MECHANICAL DESIGN - The float units were removed from the standard tank assemblies and replaced by the appropriate sound-wave sensor rods. As an example, a straight Thermelast rod, 3 mm wide, 1.5 mm in thickness, and with length L = 20 cm, was installed in a BMW vehicle, where it was operated adjacent to the electric fuel pump installed in the same assembly. This examination was also designed to show whether any particular acoustic or electrical interference would occur when the sensor was installed in the direct vicinity of the pump.

The mechanical attachment of the sensor rod presents certain special problems, as solid contact between the rod and the adjacent housing at an unsuitable point would sharply impair - or completely cancel - the rod's ability to oscillate in a free resonance pattern. Thus a

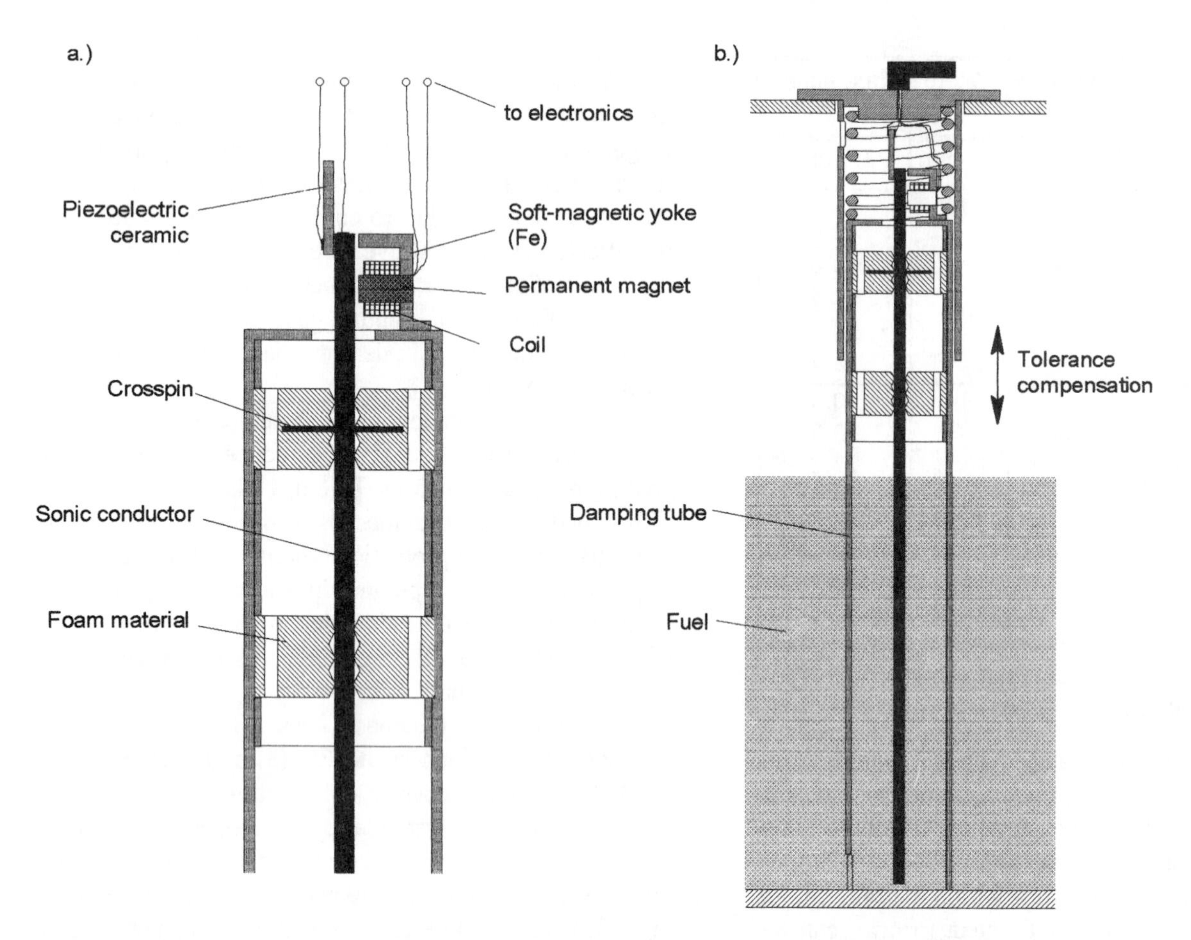

Fig. 19 - Ultrasonic sensor rod: a.) Mechanical design of prototype for vehicle installation, b.) installed in a damping tube with telescopic device for tank-floor contact

solid attachment with the tank assembly was provided at only a single point, as illustrated in Fig. 19a. A cross pin is mounted on the upper sensor rod near one of the assemblage points near the excitation area. (One design variation allows an alternative, making it possible to fix-mount the upper end of the sensor rod, producing an oscillation node at this location /5/). Supplementary support is obtained when the sensor rod is embedded in a soft material; this material must display optimal acoustic insulation properties. Interestingly, the foam materials now being used to produce floats (such as Nitrophil) have proven particularly effective in this application. Only one other attachment point was used in the vehicle prototype. This was specially shaped to allow maximum fuel penetration to the sensor rod, even at the attachment points, to prevent the rod from being 'blind' to level variations in these particular areas. Neither the components for electromagnetic excitation nor the piezoelectric tap on the upper end of the rod need be sealed against the fuel; these devices can operate in the fuel for extended periods with no damage.

In actual vehicle operation, the fuel's level fluctuates as it splashes from side to side. It is thus advisable to install the sensor rod in an damping tube to provide hydromechanical suppression of the more extreme irregularities. A sensor tube for another vehicle was mounted within such a tube featuring a telescopic section at the top (Fig. 19b) to provide spring contact on the tank floor. This compensates for geometric tolerances of the tank at installation, maintaining error in the residual range at a minimal level.

ELECTRONICS - Only those electronic components required for generating oscillations and for phase control are installed directly at the sensor. The output frequency f_a generated from this small circuit proved extremely resistant to external sources of interference. It was connected to a display mounted near the driver. The microcomputer required for the display also assumed responsibility for all other signal processing (digitalizing with counter, averaging, conversion of level to volume, corrections for limitations on measurement range, calibration, etc.). These processes are also required during testing to convert the raw signal from the sensor into a readable digital display.

An integrated unit consisting of the sensor and all inherent electronics is conceivable - among the advantages would be the option of a digital interface suitable for bus operation. However, the separation of functions used in the prototypes would seem better suited for series production, both in terms of cost and in reliability. The dashboard controller, which is already present in modern vehicles, can easily assume the more complex signal-processing functions; the zero point could also be electronically set at this location prior to the initial fill-up at the factory. For series production, it will be necessary to achieve maximum integration in those electronic components that do remain at the sensor. These should also be sealed against fuel.

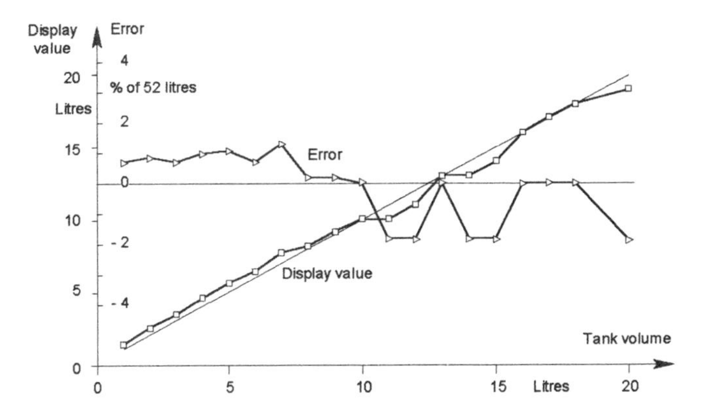

Fig. 20 - Level measurement in fuel tank, measurement error

TEST RESULTS - After laboratory calibration, the sensor was installed in a vehicle where it performed for one year (summer and winter operation) with no problems. At the end of the test period, the tank was emptied completely, then fuel was added to the tank in 1 litre increments to recheck the accuracy of the sensor: As Fig. 20 shows, the precision had not changed measurably; the deviation in the residual range (0...10 litres) did not exceed 1 %. At higher levels, it remained under 2 % (relative to total capacity of 52 litres) up to the center of the range. Within the framework defined by the exigencies of prototype testing, this examination provided an effective demonstration of the sensor's long-term durability.

EFFECT OF DIFFERING FUEL TYPES - The zero point of the ultrasonic tank-level sensor described here remains unaffected by the object medium; however, when the sensor rod is completely submersed, the differences in density, sound-propagation velocity and compressibility which characterize the various media will exercise a certain influence on the display. As an example, when the prototype under discussion is employed with a fuel tank filled with diesel fuel, the display is 5.3 % (relative to total) higher than with an equal amount of gasoline.

Half of this increment corresponds to the higher mass of the diesel fuel (approx. 12 % more than gasoline), and the variation is not considered crucial when the tank is full or nearly full. In those applications where this error is considered to lie outside the tolerance range, slight calibration adjustments would be required for the diesel vehicles. This could be taken care of at the end of the assembly process when the PROM in the display processor is programmed.

It will be remembered that a suitable operating frequency must be selected in order to optimize temperature response; variations in the type of fuel do not exercise any appreciable effect on this factor.

COST CONSIDERATIONS - As far as the basic physical design is concerned, the new sensor is much less complex than the current float/potentiometer devices; to this extent, it should also be cheaper to produce. However, initial estimates indicate that the total costs for a complete ultrasonic rod sensor would be higher than those of today's sensors due to the required local electronics. Even when the circuit is spread out to hold the number of components installed in the tank to a minimum, the remaining parts must be encapsulated for protection against the fuel and - just like any other stand-alone electronic circuit in the motor vehicle - they must be protected against electrical interference. However, there is reason to hope that an IC combining maximum monolithic integration with minimum sensitivity to voltage fluctuations will make it possible to maintain electronics costs in series production at such low levels that total sensor costs will - when the additional benefits of the new concept are included in the calculations - be competitive with conventional sensors.

CONCLUSIONS

An easily processed acoustic procedure for measuring fuel levels in the vehicle tank has been discovered, and suitability for vehicular application has been demonstrated in successful testing. The concept is based on the level-sensitive propagation of bending waves in rigid metal rods, and utilises no moving parts. The sensor rod can be shaped to adapt it for use in a wide range of irregularly-shapen modern fuel tanks, and special shapes can be employed to substantially enhance measurement sensitivity in the especially critical residual range. Selection of a suitable rod material and an appropriate operating frequency are sufficient for both offset tuning and for ensuring a large degree of resistance to temperature fluctuation. The frequency-analog output signal for the envisioned resonant operating mode is largely impervious to interference, and can be transmitted through long wires in the vehicle; counter circuits provide a convenient means of digitalizing the signals for display purposes as well as for further processing.

Initial projections of manufacturing costs would seem to indicate that broad acceptance as a replacement for the conventional electromechanical sensors - characterized by low cost, but subject to wear and friction - is still unlikely. At the present time, it seems that the electronic sensor is unlikely to displace its conventional potentiometric counterpart on cost grounds alone in those applications where the latter represents an adequate solution. In all other cases - for instance, where specific conditions make a new and more flexible sensor imperative - we believe the suggested ultrasonic sensor rod to represent the best solution.

REFERENCES

/1/ Kleinert, G.: 'Messung des Tankfüllstandes' ('Measurement of the tank level') , VI. Conference 'Elektronik im Kraftfahrzeug' June 1986, Essen

/2/ US PS 5,146,783

/3/ Kurtze, G., Bolt, R. H.: 'On the Interaction between the Plate Bending Waves an their Radiation Load', Acustica Vol. 9, 1959, S. 238-242

/4/ Grein, N.: 'Aufbau und Untersuchung von Ultraschallwellenleitern zur Positionsbestimmung von Flüssigkeitsoberflächen' ('Construction and test of ultrasonic wave guides for detecting the position of fluid surfaces'), dissertation submitted for diploma at the University of Karlsruhe, 1988, Germany

/5/ Langdon, R.M.: 'Resonator sensors - a review', J. Phys. E: Sci. instrum., Vol. 18, 1985, S.103 115, Great Britain

/6/ UK Patent GB 2 067 756 B: 'Liquid-level measurement', published 1983

/7/ EU PS 0413754 'Akustischer Tankstandgeber' ('Acoustical tank level sensor')

930359

High Accuracy Capacitance Type Fuel Sensing System

Toshiharu Shiratsuchi
Nippondenso Co., Ltd.

Motomu Imaizumi and Masaki Naito
Toyota Motor Corp.

ABSTRACT

A highly accurate fuel sensing system has been developed which provides a digital indication of the remaining fuel on a I/P cluster.

This system uses capacitance type sensors placed in the fuel tank to detect fuel level changes .

This system uses three capacitance type sensors which enables it to determine highly accurate fuel level even in sloped surfaces.

This fuel level sensor has been in production since 1991 and today is used on the TOYOTA SOARER in Japan.

INTRODUCTION

The intent of this paper is to introduce how we modeled the fuel level in a vehicle's fuel tank to digitally determine the remaining fuel , and what structure was adopted to achieve accurate measurement.

Recently, the increasing demand for high quality vehicles has led to a demand for highly accurate instrumentation. This high accuracy is also required for the remaining fuel, and especially a demand for numeral presentation in unit of liters of fuel quantity is great .

Thus highly accurate sensors are now needed such as the fuel level sensor.

Conventional fuel level sensors detect the remaining fuel using a float and a resistor plate. This float follows the fluctuations of the fuel surface as the vehicle's motion is varied. Since the signal output from the fuel sensor is influenced by the fuel surface fluctuation, accurate detection of the remaining fuel is difficult .

This new fuel sensor positively detects the slope of the fuel surface in relation to the tank even while the fuel surface is fluctuating as mentioned above.

STRUCTURE & SYSTEM

This fuel level sensor is composed of four capacitance type sensors (front , left , right and ε' sensors) (See Fig.1).

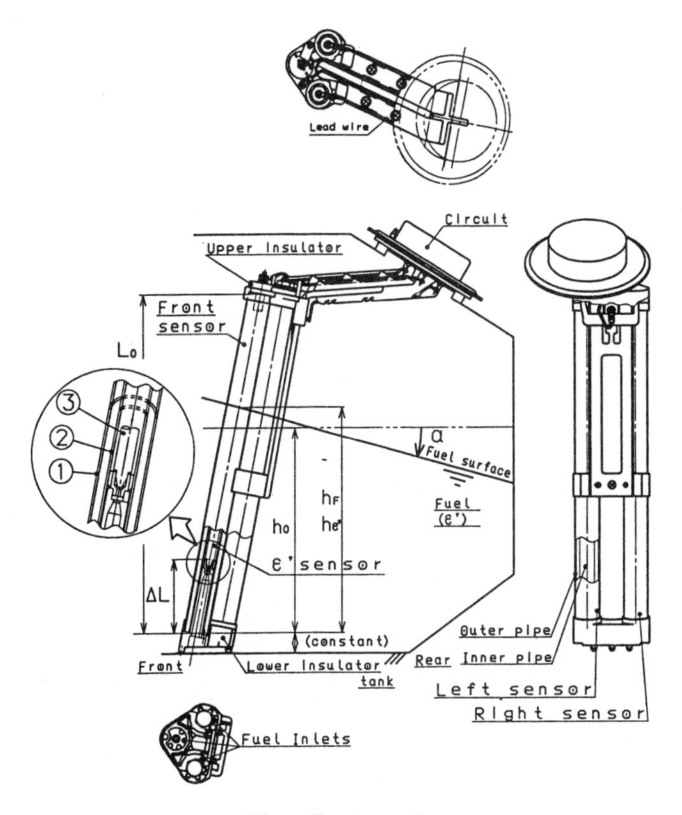

Fig1. Fuel sender

Front, left and right fuel sensors work to detect the fuel level ,while the ε' sensor detects the inductive capacity of the fuel ε'.

Each fuel sensor works with a oscillator circuit at the top of the flange to detect the fuel level. A decoder is also included in the circuit . When sensor selecting signals (clock 1, clock 2) are sent from the I/P cluster, the decoder selects a sensor and sends the signal to the I/P cluster (See Fig. 2 and 3).These signals are processed by a micro processor included in the I/P cluster.

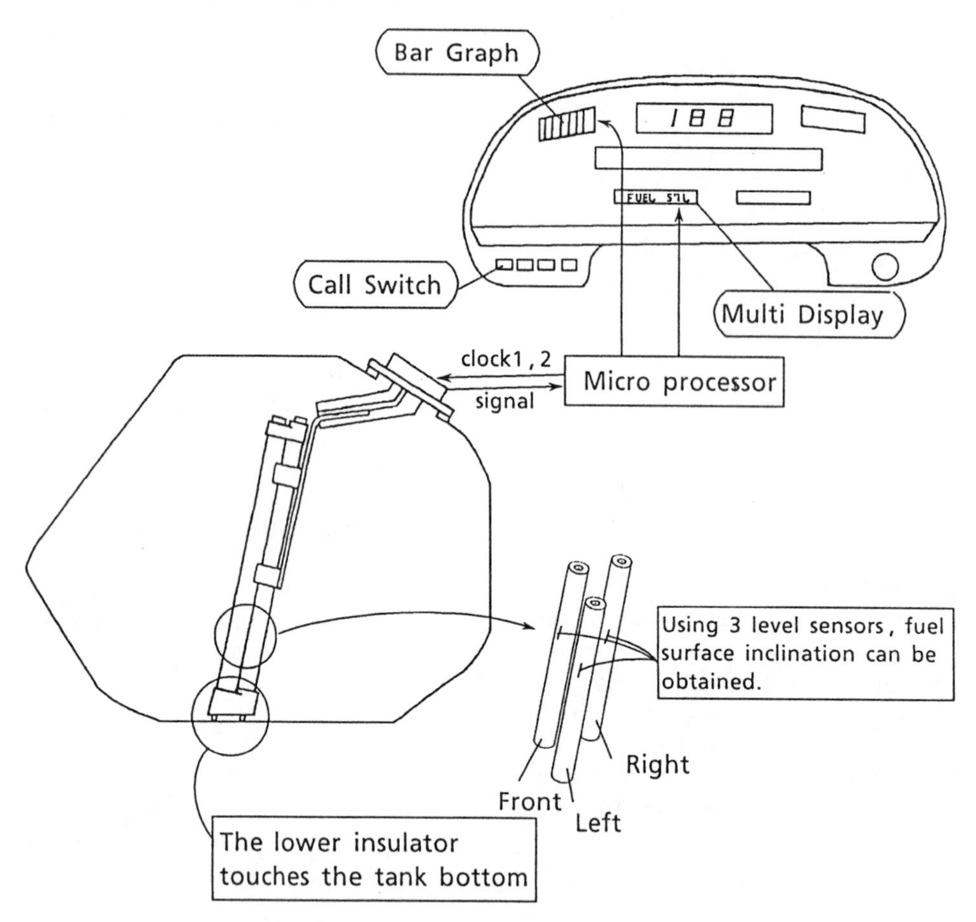

Fig2. High accuracy capacitance type fuel sensing system.

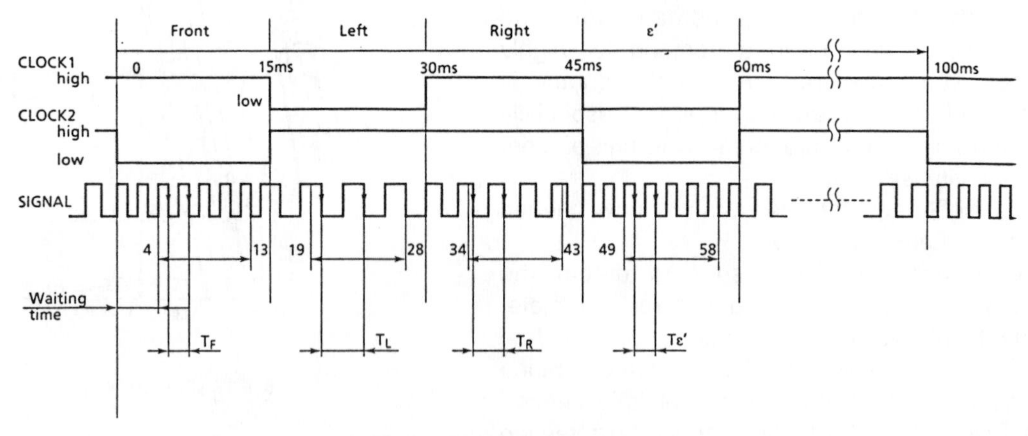

Fig3. Timing chart

The inductive capacity of fuel ε' must first be obtained based on the signal from ε' sensor. Since the inductive capacity varies depending on the fuel grade, temperature , etc . Fuel level is calculated from the inductive capacity ε', and the signals from the front, left and right sensors. Based on these measured values, the slope of the fuel level is calculated and an accurate remaining fuel quantities is calculated (See Fig. 4).

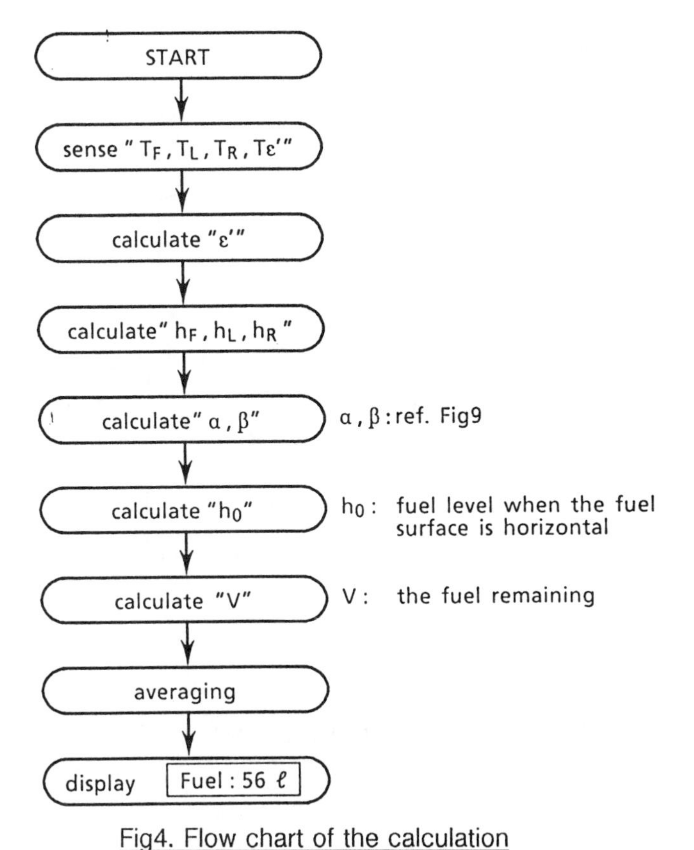

Fig4. Flow chart of the calculation

PRINCIPLE OF CALCULATING FUEL LEVEL

The fuel surface movements are considered as followed for accurate calculation of the remaining fuel quantity:

- Fuel surface is always regarded as a plane (See Fig. 5).
- Fluctuation of the fuel surface is regarded as the phenomenon that the surface changes its slope angle every moment.

When the fuel surface is regarded as a plane, the surface S can be determined by obtaining the following values:

Fuel level : h

Slope angle of the fuel surface

: α (front to rear)

β (right to left)

(See Fig. 6)

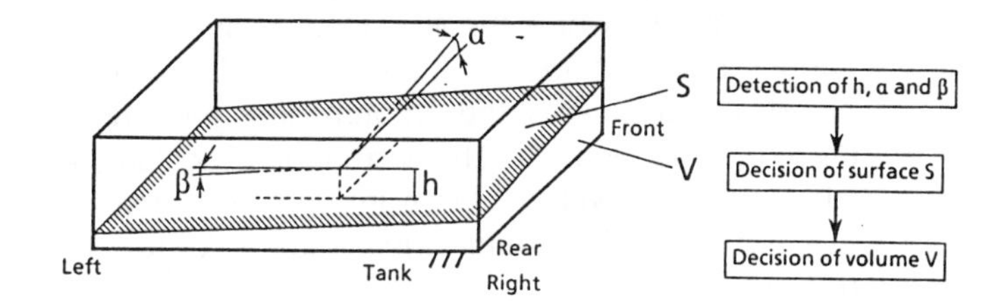

Fig6. Determination of remaining fuel

Viewpoint - Fuel surface is postulated to be always a plane

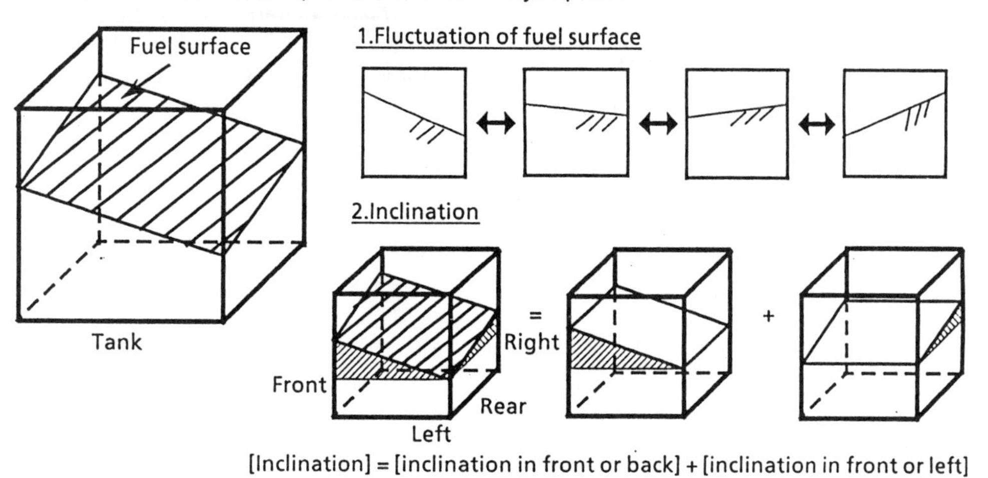

Fig5. Fluctuation of fuel surface

When the fuel surface S is determined, the volume V which is enclosed within the fuel surface and tank wall can be calculated.

The method to obtain the volume V is explained below.

When the fuel surface is sloped , the fuel level h at the position of the fuel sensor changes by Δh ($= h - h_0$) in comparison with the fuel level h_0 under leveled conditions (See Fig. 7).

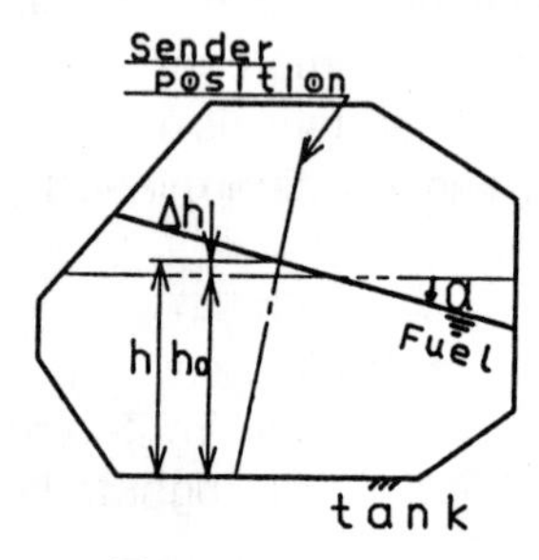

Fig7. VariationΔh

The increment Δh vs slope can be calculated in advance from the tank shape (See Fig. 8). (This data is programmed into the I/P cluster µP).

Therefore, when the fuel level h and the fuel surface slope angles α , β are obtained, the micro processor calculates the amount changed from a level fuel surface as Δh. Next the µP calculates what the fuel level would be in level conditions h_0 ($= h - \Delta h$), and obtains the volume V for the remaining fuel.

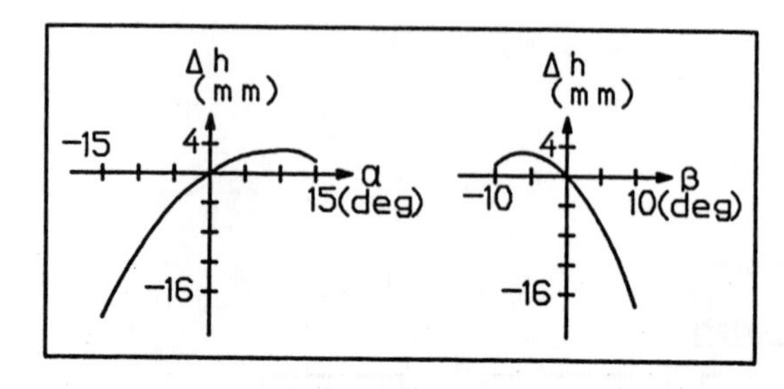

Fig8. Fuel level change caused by fuel surface inclination

DETECTION OF FUEL SURFACE SLOPE ANGLE

Three fuel level sensors are used to detect the fuel surface slope .

Slope angle in front to back direction : α

Slope angle in right to left direction : β

These values are calculated as shown in the Fig. 9.

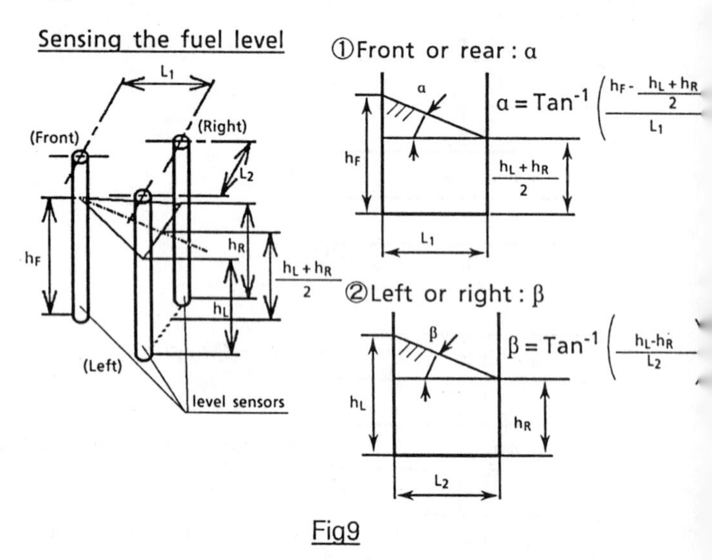

Fig9

CAPACITANCE TYPE SENSOR

Capacitance type sensors were adopted for fuel level sensing system since they are accurate and are easily mounted in the tank (three sensors are assembled as one fuel level sensor unit).

A capacitance type sensor detects fuel level change by the change of inductive capacitance between air and fuel. Details are explained in the following (See Fig. 10).

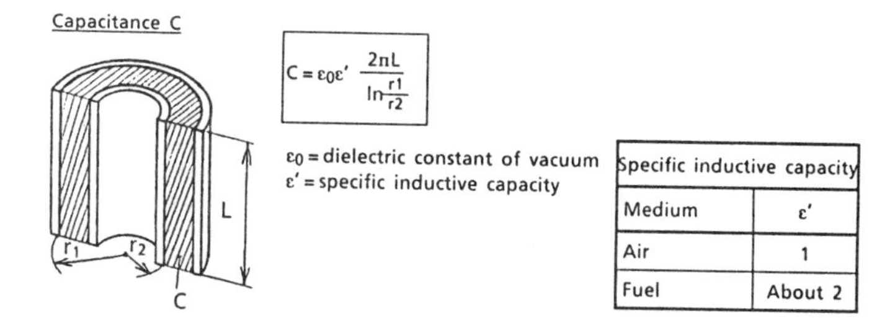

Specific inductive capacity	
Medium	ε'
Air	1
Fuel	About 2

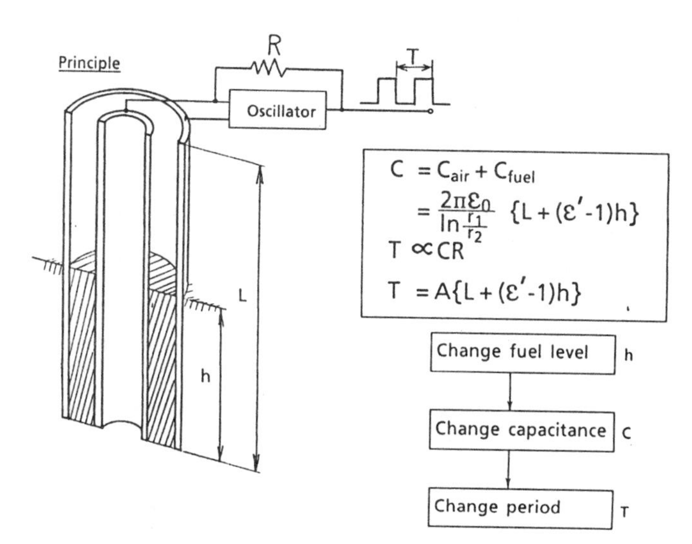

Fig10. Capacitance type fuel level sensor

A capacitance type level sensor is comprised of electrodes installed inside the tank and a oscillator circuit on the top of the flange (See Fig. 1).

Coaxial cylinder type capacitors were adopted for the following advantages when compared to a general parallel plate type:

1. Higher accuracy
2. It can be supported easily and it has no dead zone by spacers.
3. Surface fluctuation outside the cylinder is eliminated because the capacitance change depends on the fuel level inside the cylinder.
4. The capacitor for the ε' sensor which measures the inductive capacity can be included in one of the sensors .
5. Cylinder type capacitors have a higher strength than the plate type capacitors.

The capacitor pipes on all the sensors are fixed with two insulators located at the top and bottom. The lower insulator has a fuel inlet and the upper insulator has a vent hole to allow air to pass through the cylinder.

The sensor capacitance C in relation to fuel level h, when the above-mentioned capacitor structure is:

$$C \propto h$$

When this capacitor is connected to the CR oscillator circuit on the flange top, the output signal period T vs capacitance C will be as follows:

$$T \propto C$$

then,

$$T \propto h$$

So the fuel level h can be obtained by measuring output signal period T. The expression is as follows:

$$hi = \frac{Ti - A \cdot Lo - B}{(\varepsilon' - 1)\, A} \quad \ldots\ldots\ldots\ldots\ldots \quad (1)$$

(i = Front, Left, Right)

A, B : Constant

ε' : Specific inductive capacity of fuel

Determining "ε'" is explained in the following.

DETERMINING THE SPECIFIC INDUCTIVE CAPACITY OF THE FUEL

Capacitance type fuel level sensor requires the specific inductive capacity "ε'" of the fuel to accurately sense the remaining fuel in the tank where the value ε' varies with different fuel grades , temperatures , etc .

This fuel level sensing system determines the fuel's specific inductive capacity by comparing the output signals of two capacitance type sensors located in the front sensor (see Fig 1).

Electrode ③ is shorter than the other two electrodes by length ΔL and is used for determining the value ε'. The center electrode ② is commonly used for grounding between electrodes ① and ③ . Electrode ① is used for the front fuel level sensor .

The relationship between both sensors' output signal period T_F and $T\varepsilon'$, fuel level h_F and $h\varepsilon'$, and the specific inductive capacity ε' is as follows: (See Fig .11)

$$T_F = A\,[(Lo - h_F) + \varepsilon'\, h_F] + B \quad \ldots\ldots \quad (2)$$

$$T\varepsilon' = A\,[(Lo - h\varepsilon') + \varepsilon'\,(h\varepsilon' - \Delta L] + B' \quad . \quad (3)$$

$$h_F = h\varepsilon' \quad \ldots\ldots\ldots\ldots\ldots\ldots\ldots \quad (4)$$

where,

A, B, B' : Constant

Lo : Pipe length

Therefore, ε' can be obtained from the above expressions.

$$\varepsilon' = \frac{T_F - T\varepsilon' - (B - B')}{A \cdot \Delta L} \quad \text{.............} \quad (5)$$

When the value ε' obtained by EQ (5) is substituted for ε' of EQ (1), the fuel level h can be obtained.

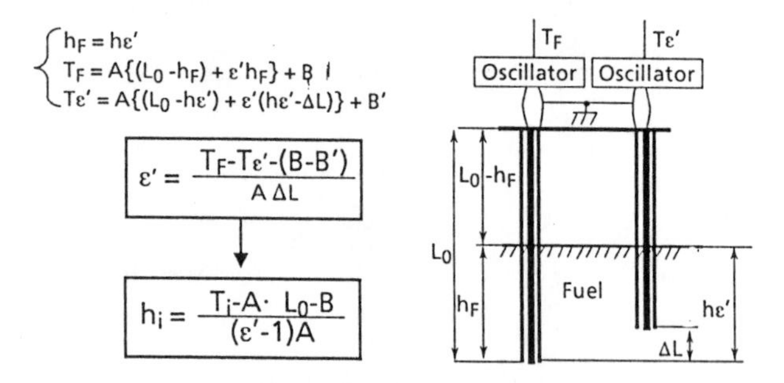

Fig11. Determination of ε'

PERFORMANCE EVALUATION

The comparison between the results obtained by this system and using an conventional fuel level sensor is shown in Fig. 12. When the vehicle moves and the fuel surface fluctuates,the conventional fuel level detection results in greatly different fuel level values from the actual . On the other hand, this capacitance system can detect the remaining fuel with high accuracy even when accelerating, decelerating, turning to right and left, driving on a slope .

Also this capacitance type sensor is not influenced by fuels with different specific inductive capacity such as other capacitance type sensors. (See Fig. 13).

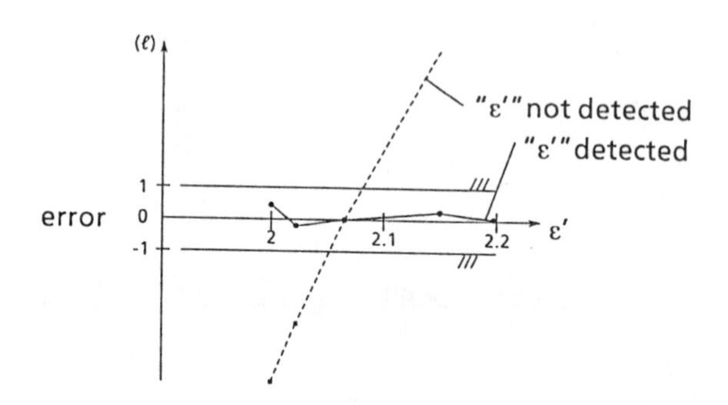

Fig13. Influence of fuels having different specific inductive capacities

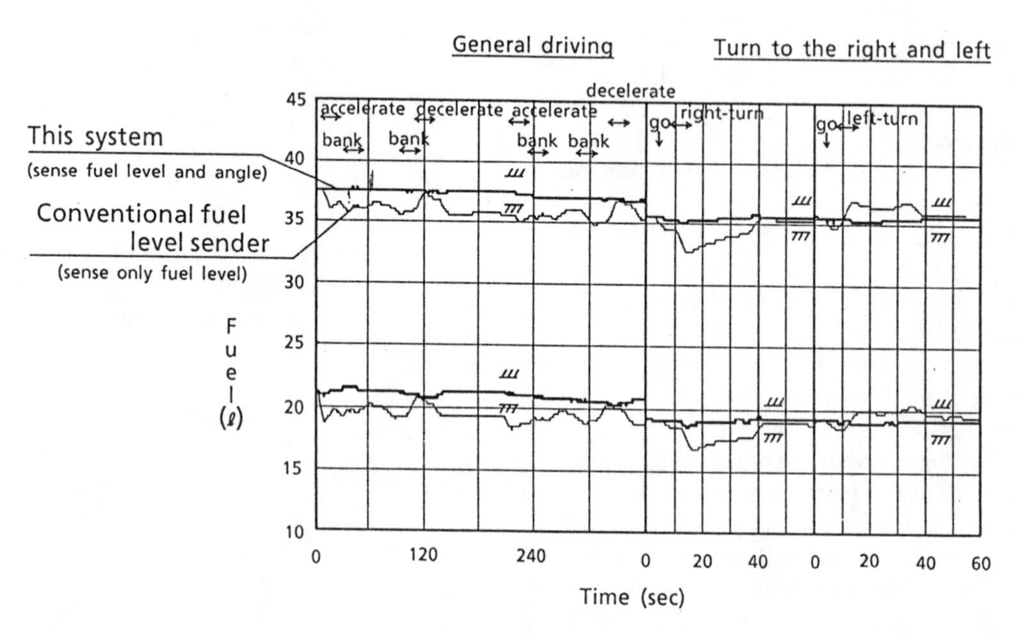

Fig12. Accuracy

OTHER INFORMATION

Remaining fuel from the tank bottom can be accurately detected by the application of capacitor structure using the bottom as the datum (See Fig. 2).

CONCLUSION

An accurate fuel level sensor has been developed using capacitance type sensors. The remaining fuel is accurately measured under any vehicle's driving condition . This is achieved through the fuel surface slope detection as well as the fuel type.

FORCE SENSORS

930258

Development of a Multi-Component Wheel Force Transducer - A Tool to Support Vehicle Design and Validation

Andreas Rupp and Vatroslav Grubisic
Fraunhofer Institut für Betriebsfestigkeit (LBF)

Juergen Neugebauer
Schenck Pegasus Corp.

ABSTRACT

The design and validation process of a new car requires detailed knowledge of the interaction of dynamic forces and moments introduced into the rotating wheel. These forces are measured under operating conditions with appropriate sensors and transducers. Due to the effects of the dynamic masses, the loads should be sensed as close to the tire/road interface as possible using a wheel load transducer. Currently, existing transducers are quite heavy, not very accurate and elaborate calibrations and computations have to be performed. With the newly developed VEhicle LOad Sensor (VELOS), these deficiencies are overcome. Examples of dynamic force and moment calibrations with the original tire are presented, as well as road load data acquisitions comparing results from the VELOS with those of the axle transducers on a passenger car under different driving maneuvers. The effect of the location of the sensor in the suspension system is demonstrated for driving over an obstacle and a rough road. Finally, the application of the VELOS during highly sophisticated road load simulation tests to prove the fatigue life of components and complete vehicles is presented and discussed.

INTRODUCTION

The interaction of dynamic forces and moments introduced into the rotating wheel plays an important role in both the design, as well as the validation process, of a new vehicle. Many aspects of suspension design, multi-body modelling of a vehicle, finite element analysis, and ride comfort evaluations are examples of methods employed during the design stages based on the dynamic road load data. The validation of components, sub-assemblies (especially tire, wheel, hub, bearing, knuckle) and of complete vehicles relies on experimental methods such as fatigue testing and road load simulation. The knowledge of the dynamic forces and moments introduced into the tire is required for the determination of load spectra for the design and testing to predict the service life, taking into account the customer usage profile, as well as the reliability requirements. Such road load data needs to be acquired on the vehicle under operational loading conditions in terms of synchronous recorded time histories of the forces and moments. Due to inertial effects, the loads should be sensed as close as possible to the points of load introduction, i.e., the road/tire interface.

At the tire patch of a wheel, the introduced forces are acting in the three coordinate directions according to Figure 1. Besides the vertical (z), the lateral (y) and the longitudinal (x) forces, three moments, drive/brake torque, (M_y), steering moment (M_z) and overturning moment (M_x) about the three coordinate axis occur resulting from the shifting of the point of load introduction due to lateral tire deformations caused by the forces. The origin of the coordinate system is located in the wheel center. The forces and moments are measured or computed with respect to this coordinate system.

ROAD LOAD TRANSDUCERS

Preparing a vehicle for road load data acquisition (RLDA) is a rather time consuming and expensive task. Methods where transducers, such as strain gages, accelerometers, and LVDTs [1, 2], are applied have been widely used. Specialized transducers such as triaxial ball joint transducers, spring and shock absorber load transducers, as well as steering knuckle transducers, are also known. The process of preparation of a vehicle with such individual sensors is complicated since extensive load calibrations of the sensors have to be done in the vehicle with appropriate loading devices in which well defined loads and load combinations are applied. The sensitivity and cross talk of individual transducers must be determined for the computation of the force time histories from the measured signals. Usually inaccuracies occur because of non-linearities of the sensors and because the point of load input moves due to the tire deformations. Furthermore, the transducers are remotely located from the point of load introduction and are scattered over the suspension components, not in one location.

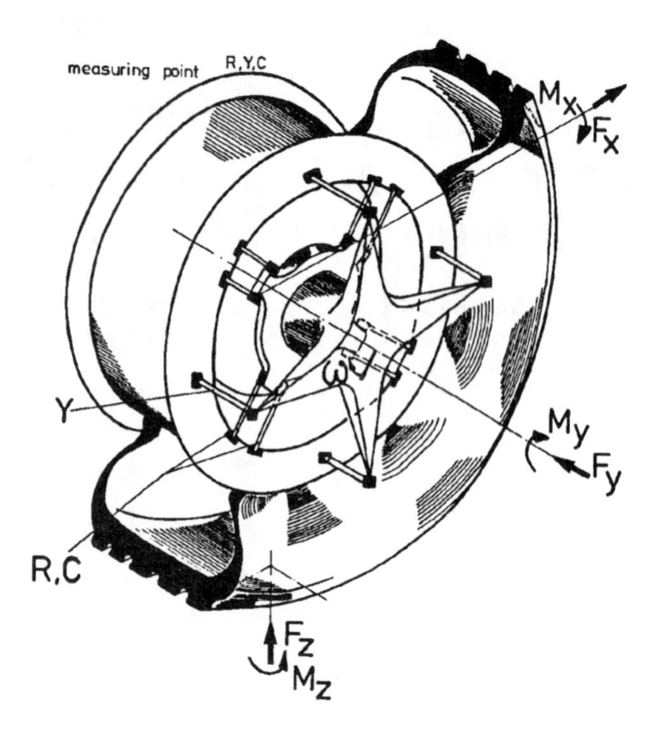

Figure 1: Forces and moments acting on the wheel of a vehicle and principle of load transfer in the VELOS

The forces measured with such transducers correspond to the road loads and the superimposed dynamic forces created by the accelerated masses between transducer and load input. In most cases, the transducers at the suspension components are not usable for another vehicle or even for changed components. In order to avoid such disadvantages, hub or wheel force transducers can be used [3-6].

Hub force transducers usually require alterations of the hub, spindle, brake and suspension components. The transducer itself also leads to an increase of the unsprung masses so that the measured forces contain the dynamic mass forces, which are not created on the original vehicle. The dynamic characteristics of the whole system is changed to a certain amount.

Most wheel force transducers require minor alterations to the axle or suspension, but they can be widely used for different vehicles. Wheel force transducers sense the forces and moments transferred from the rim to the disc of the wheel. Sensors with strain gages or RVDTs are located between rim and disc in different positions on the circumference. This location is the closest point possible to the point of load introduction which can be used to sense the loads. Nevertheless, the tire acting as a spring, and the masses of the tire and the rim still have to be taken into consideration when computing the loads at the tire/road interface.

On all wheel force transducers, the forces and moments measured in the rotating coordinate system of the wheel have to be transformed to the fixed coordinate system of the vehicle. This transformation requires an angle resolver to define the rotation of the wheel.

On the known wheel force transducers, the following problems occur. In general, RVDTs show zero shifts during longer measuring runs. Due to the design of the sensors, cross talk from one to another force component might occur. The deformation of the tire leading to a variation of the point of load introduction influences the measured force components. The computation of the forces and moments from the sensor signals requires elaborate calibrations and computations. The accuracy given with the technical data of such transducers results from calibrations without a tire, which is subjected to high deformations under the service loads. The design of the wheel disc requires extra cooling of the brake during severe braking maneuvers so not to expose the electronic components mounted on the wheel to high temperatures. The unsprung masses, i.e., tire, wheel, hub, bearing, spindle, are parts of the suspensions increased by a minimum of 30% due to the weight of the wheel force transducers. The costs for the wheel force transducers including the appropriate data processing to gain the required time histories of the three forces and three moments acting on the wheel, which are not very accurate, are quite high.

Therefore, a new VEhicle LOad Sensor (VELOS) with high accuracy, low weight, low cost and not requiring any alteration of the axle, was developed and tested during various service measurements on passenger cars and road load simulation tests. No elaborate calibrations or computations are necessary. The forces and moments can be processed with standard data acquisition equipment and software.

VEHICLE LOAD SENSOR

In Figure 1, the principle of the VEhicle LOad Sensor (VELOS) is shown. Figure 2 shows one VELOS prepared for service measurements. This force and moment transducer is capable of simultaneously measuring the vertical, lateral and longitudinal forces and the moments around these axes referring to the fixed vehicle coordinate system. Strain gaged load sensors are located between the rim and disc of the wheel using a slip ring transducer. The measured signals of the six individual sensors are transmitted to the data acquisition unit via the amplifiers. The rotation angle of the wheel is provided by the resolver built into the slip ring transducer as analog sine and/or cosine signals. Based on the angle information, the forces of the sensors in the rotating wheel are transformed to the fixed coordinate system of the vehicle.

In comparison to the existing measuring hubs and wheels, the design concept of VELOS is quite a unique one. Rather than using elaborate electronics and computational corrections to compensate the inaccuracies of the sensors due to the load path in the sensors and the significantly large tire deformations, the mechanical design provides complete separation and compensation of the different forces and moments.

Figure 2: Road load data acquisition with a VELOS

Eight specially designed flexmembers transfer the forces from the rim to the disc. Four of them, designated Y in Figure 1, are stiff in the lateral (Y) direction and flexible in the radial and circumferential direction. Through their geometric arrangement, these lateral flexmembers only transfer the lateral force F_y and the moments M_x and M_z. The four other flexmembers (R, C) are compliant in the lateral direction to ensure that the total lateral force and the moments M_x and M_z have to pass through the lateral flexmembers.

The radial flexmembers R, C are stiff in the radial and circumferential direction. Since the lateral flexmembers are weak in these directions, the radial forces (vertical and longitudinal) as well as the rotational moment M_y are transferred by the radial flexmembers. Nevertheless, the radial flexmembers show different stiffnesses in the radial and circumferential direction of the wheel coordinate system.

Thus, the most accurate transformation of the forces is possible from the rotating wheel to the fixed vehicle coordinate system. To explain this, a vertical force on the wheel is considered at a moment when the four radial flexmembers are orientated in the vertical and longitudinal vehicle direction. This force is transferred from the rim to the disc by the radial flexmembers. According to the stiffness relation of the flexmembers in radial and circumferential direction of the wheel, a certain amount of the vertical force is transferred in the circumferential wheel direction by those two flexmembers, which are orientated crosswise to the vertical force. This part of the vertical force is not sensed in the radial wheel direction on the vertically orientated flexmembers leading to inaccuracies and requiring elaborate calibrations and data processing under complex loading situations.

Additionally, to this mechanical separation of the force and moment components, the VELOS-instrumentation of the individual sensors increases the amount of decoupling between the individual load directions. On the 8 flexmembers, 12 sensors with 56 strain gages are installed. By the wiring, the strain gages are set up to 6 temperatures and speed compensated full Wheatstone bridges. 24 channels on the slip ring transducer are used for the excitation of these sensors with a voltage of 5V and the transfer of the signals to the 6 amplifiers. The slip ring transducer is bolted to the wheel disc. For the definition of the vehicle coordinate system, the outer part of the resolver is fixed to the axle or in an easy way, like in Figure 2, to the fender which leads to minor inaccuracies of the longitudinal and vertical directions under severe steering. The cables of the 6 strain gage bridges and the sine and cosine signals are lead on that frame to the amplifiers and the resolver. Via the 36 channels of the slip ring transducer, additionally strain gage bridges, e.g., the global tire deformations [1] might be transduced.

The amplifiers and the resolver provide eight analog signals which are recorded on an analog tape or are digitized and stored on a PC. The channels (Figure 1) are:

Y	for the lateral force F_y
R13, R24	for the longitudinal F_x and the vertical force F_z
C	for the braking moment M_y
Y13, Y24	for the moments M_x, M_y
SIN, COS	for the definition of the wheel revolution

The recorded data have to be processed to obtain the force and moment time histories according to the fixed vehicle coordinate system in the following way:

- A zero and calibration correction of the signals might be necessary due to inaccuracies during electrical calibration and zeroing before the RLDA.

- Possible instabilities of the measuring system (transducer, amplifier, tape) might be compensated by a zero-drift correction.
- The force and moment time histories are computed from the signals according to the formulas below. The calibration factors are determined through the VELOS-calibration in a test rig and are provided in the handbook.
  - Since the y-axis is the same in the rotating and the fixed coordinate system, the lateral F_y and the braking/acceleration moment M_y can easily be computed by:

    $F_y = K_y * Y$

    $M_y = K_c * C$

  - For the other forces and moments, the transformation from the rotating to the fixed coordinate system has to be done in the following way:

    $F_x = K_r \text{ (R13 * SIN + R24 * COS)}$

    $F_z = K_r \text{ (R13 * COS + R24 * SIN)}$

    $M_x = K_m \text{ (Y13 * SIN + Y24 * COS)}$

    $M_z = K_m \text{ (Y13 * SIN + Y24 * SIN)}$

This post processing of digitized data can be done by standard data acquisition software. Both the necessary amplifiers, as well as the data acquisition and processing equipment is already available in many cases. Thus, the transducer VELOS with the slip ring transducer can be easily connected to the standard measuring instrumentation of many users.

Measurements with 4 VELOS on one vehicle would require 24 amplifiers and 32 data acquisition channels, if all 3 forces and 3 moments were to be recorded on each wheel. Nevertheless, in many cases only the 3 forces and the braking moment are of interest so that efforts can be reduced.

The rim and the disc of the transducer require a certain stiffness relation of the rim, the flexmembers, and the disc to achieve the high accuracy. Thus, depending on the wheel size, the transducer leads to an increase of the dynamic undamped masses of about 10% to 20% compared to the original wheel.

The VELOS was designed for passenger cars with any wheel size. The disc of the wheel transducer can be adapted to accept different bolting diameters and offsets. The disc with flexmembers and sensors is bolted to the rim which consists of two halves. Other rims with different widths and diameters to accommodate a variety of different tires can easily be mounted. A VELOS for heavy trucks is in preparation. Due to the open design of the transducer by the flexmembers, the brake of the vehicle is cooled like using the original wheel. Temperatures up to about 120°C are tolerated by the instrumentation. The sensors are water protected, but still a certain risk remains when driving in heavy rain.

VELOS CALIBRATION

To prove the design concept and to investigate the accuracy of the achieved road load data measured with VELOS, a variety of calibrations and road tests were conducted. First, loads were applied statically in radial, lateral and circumferential directions in the flat base wheel roll test facility [7] to determine the sensitivities of the individual load sensors in the assembled wheel. This calibration lead to linear response characteristics of the load sensors. The calibration factors K_p, K_y, K_c and K_m for the transformation of the electrically measured signals to the forces and moments were determined in physical units. Unlike typical calibrations, which do not consider the tire with its non-linear, high deformation characteristics, a baseline calibration, as described in the following paragraph, was carried out with production tires on a flat base roll test facility.

The wheel with tire was rolled under different quasistatically applied vertical and lateral loads and load combinations. Longitudinal braking forces were applied statically. The output of VELOS was compared with the controlled forces of the test rig. With these quasistatic measurements, a good coherence was found between the measured loads of VELOS and the controlled loads of the test facility.

Additionally, braking maneuvers were measured in a biaxial wheel test machine [7,8]. The VELOS was running in a drum (Figure 3) with a speed of 100 km/h under pure vertical loading F_z = 9 kN. In this test facility, the drum is driven and the brake is mounted on the loading frame applying the braking moment on the wheel via the drive shaft. The drive shaft is equipped with strain gages to

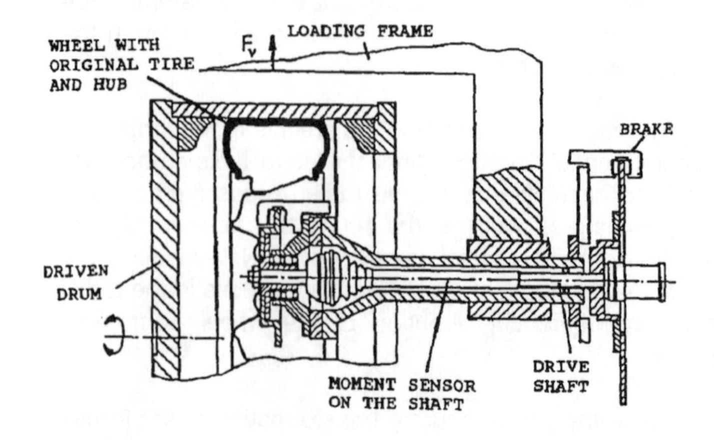

Figure 3: Wheel in a biaxial wheel/hub test machine with braking device

measure the applied moment. In Figure 4, the time histories of the vertical force F_z and the moment M_y measured with VELOS and in comparison measured with the test facility load cells are shown. In this figure, as well as in the other figures presenting time histories of forces and moments only, the important forces and moments of the individual maneuvers are given. This allows a more detailed viewing of the dynamic load interaction. Nevertheless, all three forces and moments were measured.

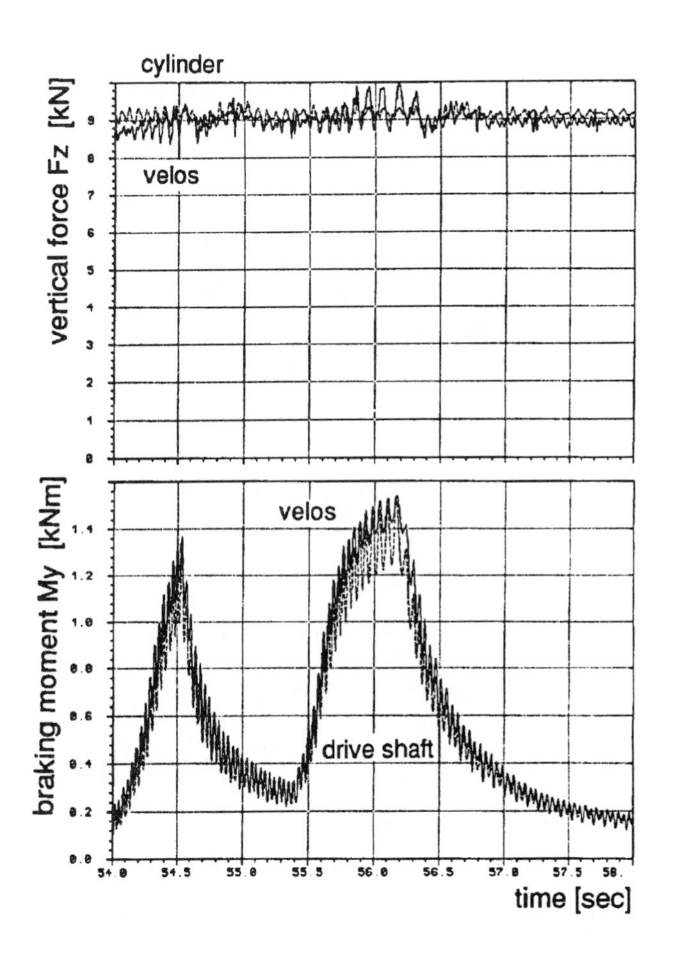

Figure 4: Time histories of the vertical force F_z and the braking moment F_y sensed with the test rig transducers and VELOS during braking

The loads measured with the test rig load cells and the VELOS correspond very well.

The content of dynamic interactions in the whole system can also be demonstrated by comparison of the braking moments acquired with VELOS and with the sensor on the drive shaft. During the two braking maneuvers, within four seconds, the braking moments sensed in the two locations are in excellent correlation. A superimposed low amplitude oscillation (frequency of about 25 Hz) due to the dynamic effects coming from the tire and the test facility were registered by all sensors.

SERVICE MEASUREMENTS WITH VELOS

Road load data was acquired on a small medium class passenger car under severe driving maneuvers with a VELOS mounted on the right rear position of the vehicle. As an example in Figure 5, the load time histories (approximately one second) are presented when driving over a sequence of five obstacles, consisting of beams with a height of 3.5 cm and a length of 10 cm. Here, the vertical and the lateral force and the moment M_y are shown as relevant loading data. High dynamic forces occurred in the vertical direction superimposed to on the rated wheel load which is based on the GVW of the vehicle. During rebound, the wheel was nearly unloaded. The absolute values differ from obstacle to obstacle according to the vehicle response to these load inductions with a frequency content of approximately the axle frequency of 15 Hz. The lateral forces are low and originate from the jounce and rebound of the wheel. The low moments M_y occurring during this maneuver are dynamic mass moments of the wheel, hub, and brake disc in combination with the rotational spring stiffness of the tire.

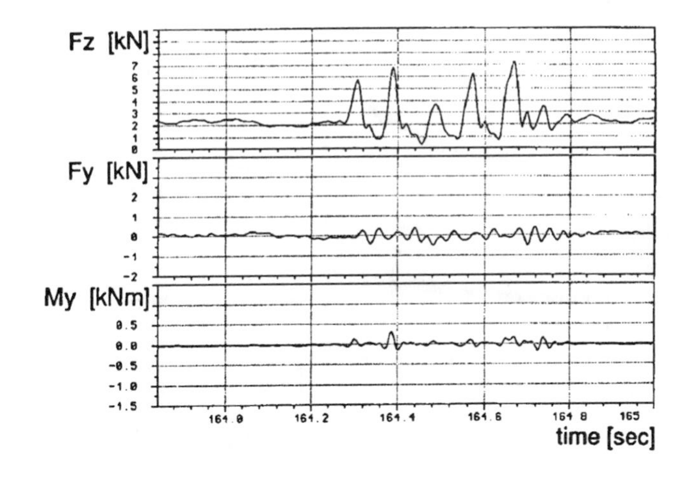

Figure 5: Relevant force and moment time histories measured when driving over obstacles

These are usually the road load data to be acquired with VELOS. Subsequently, such data needs to be analyzed and validated.

COMPARATIVE RLDA — To investigate the quality and accuracy of such road load data, service measurements were performed with VELOS. A large medium class passenger car was equipped with additional force transducers on different suspension components. On the front axle, the lower area of the MacPherson Strut was strain gaged to measure the vertical and lateral forces. Longitudinal forces were sensed at the ball joint of the lateral arm. The sensitivities and cross talks of these load transducers were determined during extensive calibrations of the total vehicle on load platforms under the different forces and their combinations. Vertical, lateral and longitudinal forces were independently applied covering the total range of these loads occurring in service. The calibration characteristics were linearized and sensitivity matrices were derived. With the inversion of these matrices, the individual force components were computed from the signals of these transducers.

In Figures 6 through 8, examples of those service measurements are presented. The time histories of the forces and moments obtained with the transducers on the axle and with the VELOS in the right position on the front axle are compared for different maneuvers. Not all measured forces and moments are plotted, but only those which are relevant for the individual maneuvers.

The data during 70 seconds left cornering on a skid pad with increasing speed up to 60 km/h are plotted in Figure 6. In addition to the vertical and lateral forces, the speed and the vertical acceleration of the hub are given. The two transducer systems provide lateral forces which correspond very well to each other. There are differences in the quasistatic vertical force during this left hand turn with the transducers on the outer position of the vehicle. The vertical forces obtained with the axle transducer are higher than with the VELOS.

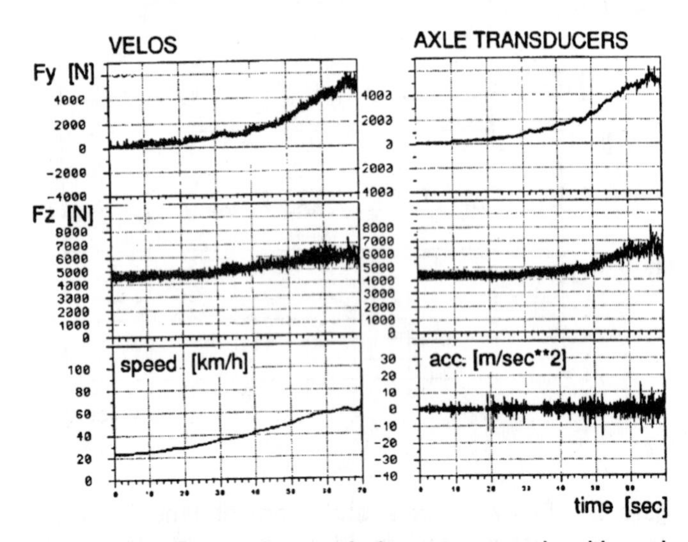

Figure 6: Comparison of forces sensed with axle transducers and VELOS — cornering on the skid pad with increasing speed

In Figure 7, the measured forces during so-called step in tests to the right (a) and to the left (b) are plotted together with the speed and the vertical hub acceleration. When driving straight with a speed of 100 km/h, the steering wheel was suddenly turned by 40°, once to the right resulting in an unloading of the right front wheel, and once to the left. Also, here the lateral forces obtained with the two transducer systems are in excellent agreement. Especially the highly dynamic lateral force measured while driving over a small obstacle during the right bend is reproduced very well by the two sensors. Also during these maneuvers, the vertical forces measured with the two systems differ. In the right bend, when the transducer wheel position was on the inner cornering side of the vehicle, the change of the vertical force obtained with the axle transducers was smaller than the one obtained with VELOS. But driving through the left bend, the vertical force measured with the axle transducers was higher than the one delivered by VELOS, as it also was seen during cornering on the skid pad. These results cannot be explained with any insufficiency of the wheel transducer, because the behavior is different for right and left curves. It was concluded that there was a problem with the cross talk on the axle transducers in combination with the activity of the stabilizer. The force calibration of the axle transducers was done by symmetrically loading the left and right wheel of the car. In a bend, mainly the outer wheel is loaded, whereas the inner wheel is almost unloaded.

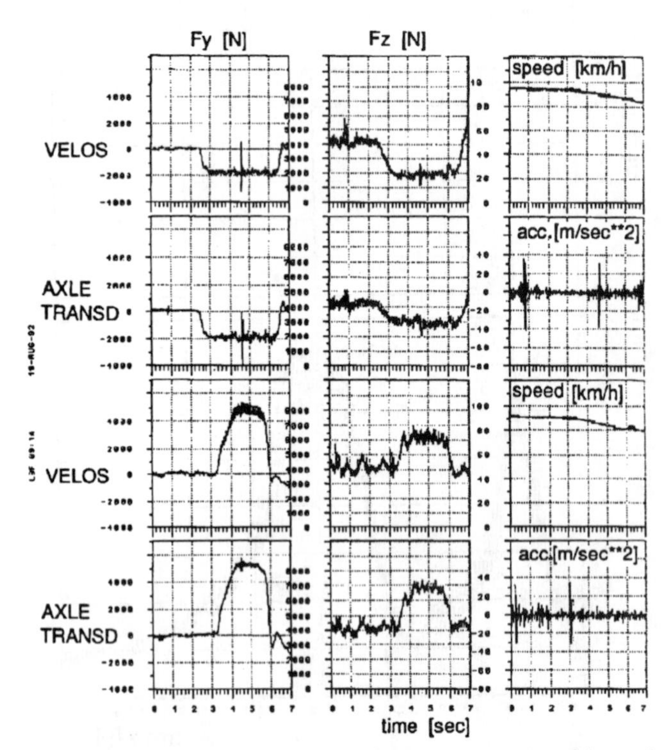

Figure 7: Comparison of forces sensed with axle transducers and VELOS — step-in test to the right and left with 100 km/h

Severe acceleration and subsequent braking data are presented in Figure 8. The most appropriate output of the axle transducers is given in terms of the longitudinal force F_x. The moment M_y cannot be measured with only one sensor on the axle since there are two different physical sources, braking and acceleration by the drive shaft. The relevant output of VELOS for this maneuver is the braking/acceleration moment M_y, which could easily be compared with F_x via the rolling radius of the tire of 0.32 mm. The signs of the signals are opposite due to the coordinate system. Additionally, the time history of the speed is plotted. The car was accelerated from first to fourth gear to a speed of 100 km/h. Subsequently, severe braking was induced during which the wheel was blocked. Excellent coherence is proved for the two transducer systems during the quasistatic loading conditions of acceleration and braking, as well as during the dynamic events of gear change and block of the wheel.

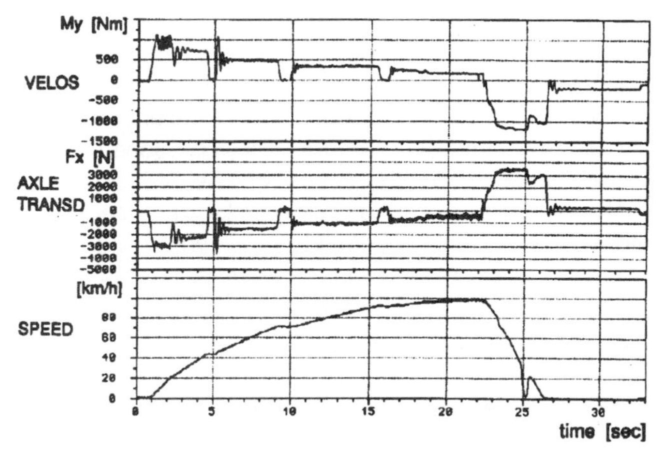

Figure 8: Comparison of forces sensed with axle transducers and VELOS — maximum acceleration to 100 km/h and subsequent braking with block of the wheel

INVESTIGATION OF DYNAMIC MASS EFFECTS — In Figure 9, data acquisition was conducted when driving over an obstacle 3.5 cm in height and 10 cm wide. The time histories of one second are plotted for F_z obtained with VELOS (9a) for F_z obtained with the axle transducers (9c) and for the vertical hub acceleration (9d). The vertical forces achieved with the two transducer systems during this short event with a frequency of about 15 Hz are quite different. Even the obtained force time histories are not in phase. For such dynamic interactions, the dynamic mass forces are not negligible any more. The location of the

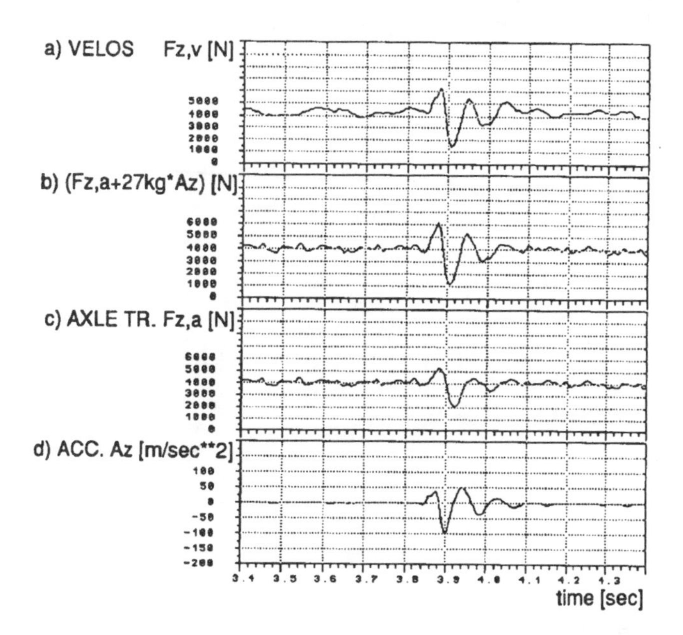

Figure 9: Comparison of vertical forces sensed with axle transducers and VELOS and the influence of dynamic mass forces when driving over an obstacle

transducers in the system of the axle is decisive for the measured forces. With a wheel force transducer, the loads are sensed as close to the point of load input as possible. Between these sensors in the wheel and the sensors on the axle, there are undamped, dynamic masses — the wheel disc with the bolting, the hub, bearings, joints, brake and the spindle (Figure 10). On the investigated vehicle, the dynamic mass between the two transducer systems was 27 kg. In a first approach, the vertical forces on the wheel could be computed by a simple model regarding the different components as a point mass. Then the force at the wheel corresponds to the force sensed on the axle and the superimposed dynamic mass force originating from the vertical acceleration of these components:

$$F_{z,\, wheel} = F_{z,\, axle} + A_z * 27 \text{ kg}$$

The time history of the vertical hub acceleration is plotted in Figure 9d. The sum of the vertical force sensed at the axle $F_{z,\, axle}$ and the dynamic force according to the vertical hub acceleration of the 27 kg point mass results in the time history shown in Figure 9b. This transformed vertical force time history corresponds very well to the vertical force measured on the VELOS. Thus, the very simple model describes the dynamic interactions between wheel and axle rather accurately.

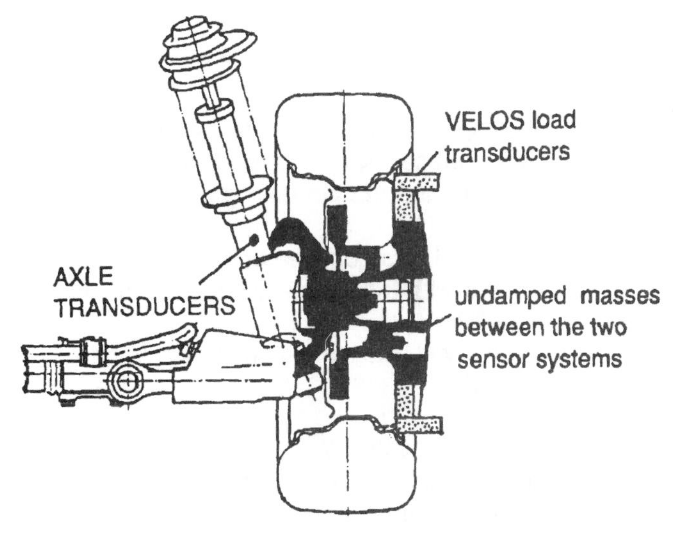

Figure 10: Simple point mass model of the undamped masses between the transducers on the axle and in the VELOS

To extend the investigation of these effects on the acquired road loads obtained with the two transducer systems, measurements were carried out on a road with defined roughness. The load time histories measured with the VELOS and the axle transducers were recorded and then reduced to the frequency distributions by level crossing and range pair counting. In Figure 11a, the frequency distributions of the vertical forces derived from the two transducer systems show large differences, although the maximum forces are rather the same. Especially in the mean and low range, the vertical forces acting on the axle are much smaller than on the wheel. By adding the dynamic mass forces, the vertical hub acceleration A_z * 27 kg, to the axle force and counting this computed time history, the frequency distributions in Figure 11b were achieved. Very good correlation was obtained by this simple model.

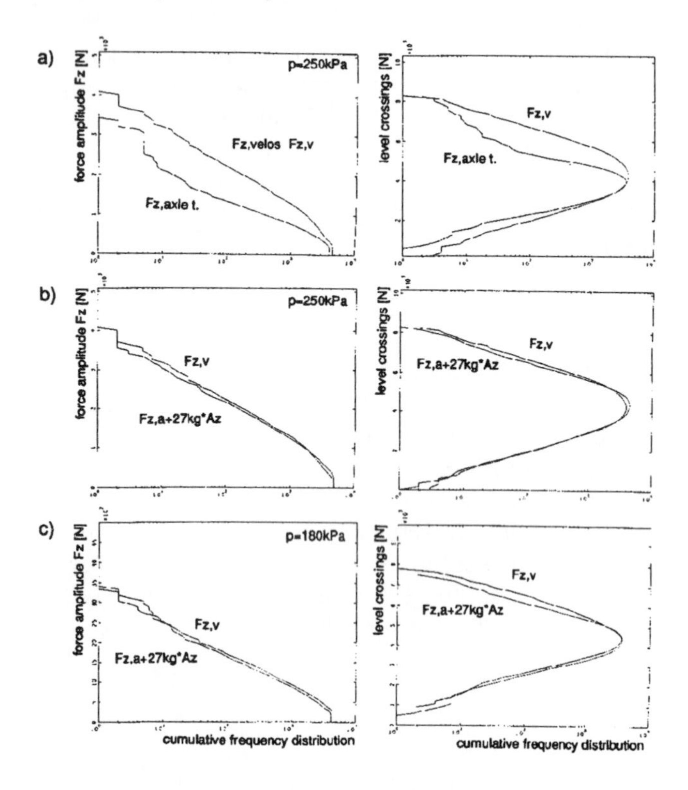

Figure 11: Comparison of the frequency distributions of the vertical forces sensed with axle transducers and VELOS driving on a rough road

a) uncorrected force derived from axle transducer
b) corrected with the simple point mass model - p=250 kPa
c) corrected with the simple point mass model - p=180 kPa

The measurements and analysis were conducted for two different tire pressures, the standard pressure of p = 250 kPa and a reduced pressure of p = 180 kPa. For both tire pressures, the vertical forces obtained from VELOS and the axle transducers, in combination with the simple point-mass-model, correspond to each other very well (Figure 11c). The low tire pressure resulted in lower vertical forces according to the smaller stiffness of the tire.

By this investigation, the effect of the location of the sensors in combination with dynamic mass forces could be demonstrated clearly.

VELOS USAGE AT MULTIAXIAL ROAD LOAD SIMULATION

To reduce time, cost and efforts during the validation process of cars, automotive components, and subassemblies, increasingly computer-controlled multiaxial servohydraulic test facilities are used instead of driving on the proving ground. Such tests can be performed 24 hours a day and by proper editing techniques [8], the testing time can be reduced from one test to another. The loading conditions are exactly reproducible. The tests can already be carried out on components and subassemblies before the actual vehicle is available for proof out tests. During such tests, the loads and deformations must correspond to those under operational conditions. The load program of the tests must include all service conditions decisive for the fatigue behavior of the investigated components.

Such test facilities could be divided in two groups:

A. Test facilities in which components and subassemblies are tested, for which only the introduced forces and their correlation is important, e.g., biaxial wheel/hub test machines or two and three axis component test rigs.

B. Multiaxial test facilities in which components or structures are tested, which are excited by load interactions.

A. BIAXIAL WHEEL TEST MACHINE — Road load simulation tests were done with the VELOS in a biaxial wheel test machine to investigate the actual loads applied to a wheel in the test rig (Figure 12). With the biaxial wheel test facility, it is possible to simulate realistically the service like deformations of a wheel under a load program that corresponds to the mission in service. This test facility is used for reliable validation tests on the total subassembly consisting of the wheel, the hub, the brake drum or disc, as well as the bearings and the threaded connectors [9].

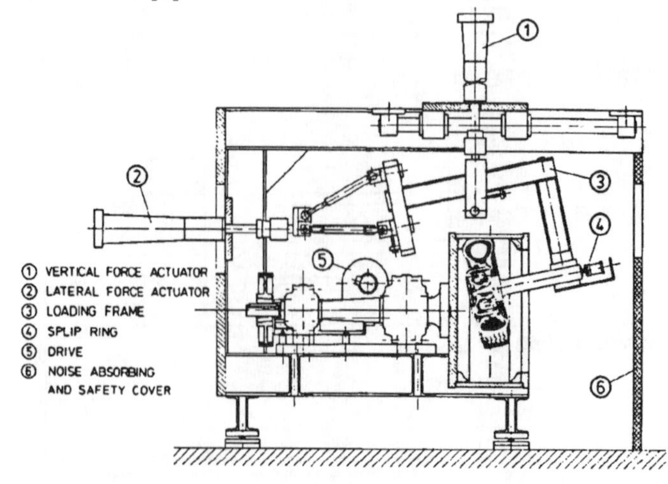

Figure 12: Biaxial wheel test machine

The wheel with tire is mounted to the original hub with brake drum (or disc) and the whole assembly then runs on the inside of a drum with an inside diameter that is slightly larger than the outside diameter of the tire. The vertical as well as the horizontal wheel forces are created by two independently acting servohydraulic cylinders, and the drum is driven by an infinitely variable DC motor. The forces of the servohydraulic cylinders are transmitted to the wheel by means of loading beams whereby the lateral forces are transmitted to the wheel in an inclined position, by means of form fitting side rings. In order to assure that the lateral forces are introduced into the wheel in a service like manner, namely via the tread of the tire and not via its sidewall, it becomes necessary to tilt the wheel while simulating turns. This "machine camber" differs from the camber known from vehicle mechanics. For this reason, the applied cylinder loads during cornering simulation do not correspond to the vertical and lateral loads in the wheel coordinate system.

Using VELOS, the wheel loads applied during the tests were controlled and compared to the cylinder loads.

In Figure 13, the recorded time histories for the vertical and lateral forces sensed with the VELOS and the forces of the vertical and lateral cylinder of the test rig are plotted for one round of the "Eurocycle" (standardized load program developed to test vehicle wheels under customer usage conditions described in [9]).

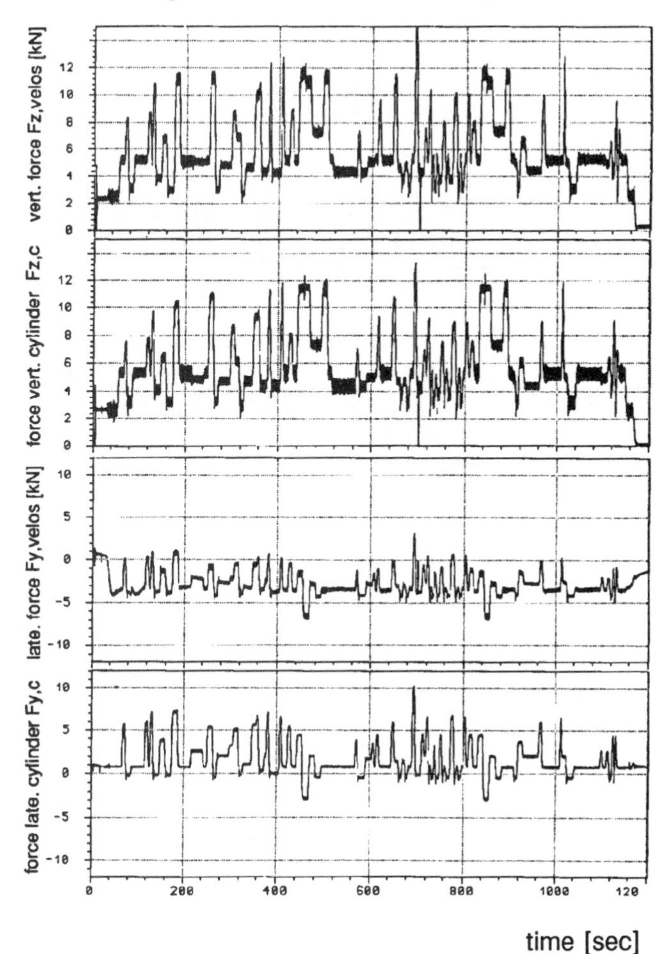

Figure 13: Comparison of the wheel forces and the cylinder forces in the biaxial wheel test machine for one round of the test program EUROCYCLE

This data acquisition was carried out during parameter studies about the influence of the machine camber.

Such measurements have shown that the established load programs, in terms of demanded time histories of the cylinder forces and the machine camber, actually produce the required wheel loads and deformations.

B. MULTIAXIAL TEST FACILITY FOR CAR AXLES — Usually the methodology for the road load simulation on an axle or a car requires the following steps:

- Road load data acquisition on test tracks, during which sensors and transducers with different sensitivities and cross talks to the individual force components are installed. The recorded time histories of these sensors will be used as desired time histories during the simulation.

- The entire car or axle is mounted on the simulator with the same sensors. As an example, an 8-channel multiaxial test facility is shown in Figure 14.

Figure 14: 8-channel axle simulator

- Following an identification phase, whereby the dynamic characteristics of the entire system (electronics, servovalve, actuator and test specimen) are determined, drive signals are computed using iterative procedures.

The drive signals obtained are then used to dynamically load the structure such that the transducer response signals coincide well with the desired time histories. Cross coupling effects between all actuators and all feedback transducers are thereby properly considered.

These computations are typically carried out on the frequency domain using iterative procedures to minimize the simulation error. These iterations are necessary because the entire system contains non-linearities due to axle kinematics and non-linear components such as shock absorbers, elastomer bushings, etc. A more recent development in simulation technology is represented by RTC®[10], a real-time transfer function compensation algorithm, which operates in the time domain. This algorithm has a number of very distinct advantages such as significantly speeding up the test preparation (identification/iteration) phase and providing on-line monitoring capabilities allowing the operator to view the desired and actual feedback signals on one screen.

Such simulation tests can be significantly enhanced by the use of a measuring wheel since it:

- shortens the time for vehicle preparation to about two to three days vs. weeks,
- improves the quality and accuracy of the simulation,
- provides real-time load data which is important for all dynamic modelling and optimization efforts.

Already during the road load data acquisition, the correct load data are achieved. The location where these forces are sensed is close to the point of load input, where no dynamic masses are included in the measured forces. Mounted in the test facility, a VELOS could directly be used as actuator feedback and only the axle kinematics would have to be taken into account. The acquired service loads can be used for different parameter variations or vehicles. All original axle components can be used and do not have to undergo modifications. The loads can easily be applied by substituting the tire and rim of a VELOS by a loading ring onto which the loading bars of the different cylinders are connected.

ACKNOWLEDGEMENT

This publication includes the data of measurements under operational conditions with a passenger car, which was instrumented with axle sensors for a research project "Verification of Procedures for Fatigue Evaluation," sponsored by Centro Richerche Fiat (CRF) S.C.p.A., Torino, Italy. The authors would like to thank the responsible engineers of CRF for carrying out this road load data acquisition and for the permission to present the data in this publication. A publication from CRF about the research project itself is planned.

REFERENCES

[1] Neugebauer, J.; Grubisic, V.; Stalnaker, D.O.; Fleischmann, T.S.; "Analysis of Tire Loads and Deformations under Operational Conditions," SAE Paper 880 578 (1988)

[2] Neugebauer, J.; Grubisic, V.; Fischer, G.; "Procedure for Design Optimization and Durability Life Approval of Truck Axles and Axle Assemblies," SAE Paper 892 535 (1989)

[3] Shoberg, R.S.; Wallace, B.; "A Triaxial Automotive Wheel Force and Moment Transducer," SAE Paper 750 049 (1975)

[4] Martini, K.H.; "Rotierende und stehende Mehrkomponenten - Rafkraftdynamometer für die Automobil- und Reifenindustrie," Kistler Instrumente AG., INTERCAMA 1983

[5] Kötzle, H.; "Mehrkomponenten-Messung am PKW-RAD im Betrieb" Berichtsband 10, Sitzung des AK Betriebsfestigkeit des DVM, DVM Berlin (1984) S. 77-98

[6] Loh, R.; Nohl, F.W.; "Mehrkomponenten Radmeβnabe - Einsatzmöglichkeiten und Ergebnisse" Automobiltechnische Zeitschrift 94 (1992), S. 44-53

[7] Grubisic, V.; Fischer, G.; "Automotive Wheel, Method and Procedure for Optimal Design and Testing," SAE Paper 830135 (1983)

[8] Neugebauer, J.; Bloxsom, K.; "Fatigue-sensitive editing reduces simulation time for automotive testing," - Test Engineering & Management, October/November 1991

[9] Fischer, G.; Grubisic, V.; Klock, J.; "Computer Controlled Proof Tests of Wheels and Wheel-Hub-Assemblies — Principle and Procedure," LBF Report No. TB-187 (1989)

[10] "RTC™ Brings the Road into the Lab," Schenck Pegasus brochure SPG-15.392

960759

A Magnetostrictive Pedal Force Sensor for Automotive Applications

Eric J. Hoekstra
SSI Technologies, Inc.

Copyright 1996 Society of Automotive Engineers, Inc.

ABSTRACT:

A pedal force sensor which offers substantial improvements in cost effectiveness and ease of application is described. The sensor uses magnetostrictive techniques to measure the stress within the pedal arm and provide an output signal proportional to pedal force. The sensor is applicable to virtually any application using a ferromagnetic pedal arm and will accommodate a broad range of pedal dimensions and forces. The sensor will work well for brake pedal force measurement in both manual and power assisted braking systems.

INTRODUCTION:

Many potential applications exist for a low-cost automotive pedal force sensor. Drive-by-wire systems are an obvious application, requiring an electronic signal output in proportion to the force applied to pedals. Major performance and safety advantages may be realized by including applied brake pedal force in addition to wheel speed as inputs to ABS system controllers. Additional anticipated applications of the pedal force sensor include use as an input source in:

- active ride control systems.
- engine control systems (both brake and throttle pedal sensing)
- towed vehicle braking systems

Current technologies suitable for sensing of pedal force include strain gauge techniques, measuring stress through use of ultrasonic devices and similar mechanisms. These techniques typically result in expensive devices suitable only for laboratory use; few are applicable to volume production at prices which can be supported by the automotive market. Most applications which need an applied brake pressure input infer this signal from pedal position, time duration of pedal application, or vehicle accelerometers and suffer from limited accuracy.

The sensor described uses the magnetostrictive properties of steel under stress, which is proportional to pedal force, to obtain the required signal. The sensor can be easily mass-produced using techniques common to automotive sensors currently in production. Costs of the sensor and associated electronics should be attractive to many potential users of the product. The sensor can also be used with most existing designs of automotive pedals with little or no modification . Brake pedals fitted with the sensor retain the overstress capacity commonly built into the pedal design.

DESCRIPTION:

The sensor uses the decrease of magnetic permeability of ferrous alloys under compression (with a corresponding increase in permeability when under tension) to obtain a DC output voltage proportional to this change. For the mechanical system of the brake pedal, applied forces can be resolved into compressive and tensile forces at 45° angles to the pedal arm. The sensor produces a DC signal proportional to the pedal force based on the permeability changes along these lines.

The sensor package and configuration will vary by application. Mechanical options include differing mounting features, varying the proportions of the sensor, and connecting via an attached cable or through a connector. Electronic options consist of including the signal conditioning electronics within the sensor or placing them within the module which uses the sensor signal.

The fundamental sensor, similar to sensors used for torque sensing, is of the "cross" configuration as commonly described in literature [1]. This configuration is easily miniaturized and has good response to both compressive and tensile stresses. Figure 1 depicts a typical brake pedal application.

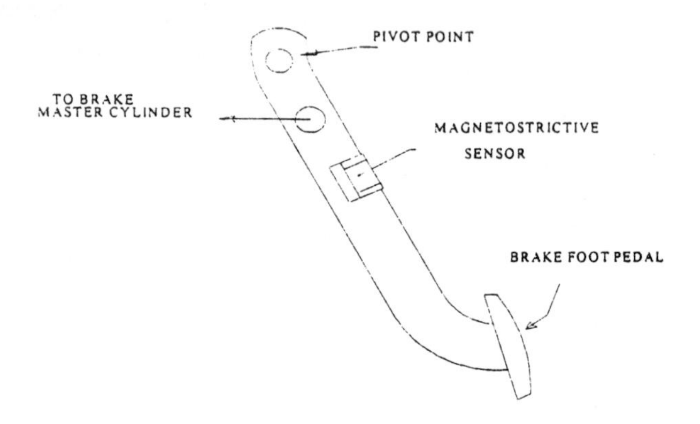

Figure 1. Pedal Force Sensor.

The sensor consists of a pickup coil on a "C"shaped ferrite core mounted within and at right angles to a similar but larger drive coil. The faces of all coil pole pieces are ground parallel to the surface of the sensor. The sensor assembly is overmolded with mounting features as required for the specific application. The encapsulated sensor is mounted in contact with the flat side of the brake pedal arm (see Figures 1 and 2).

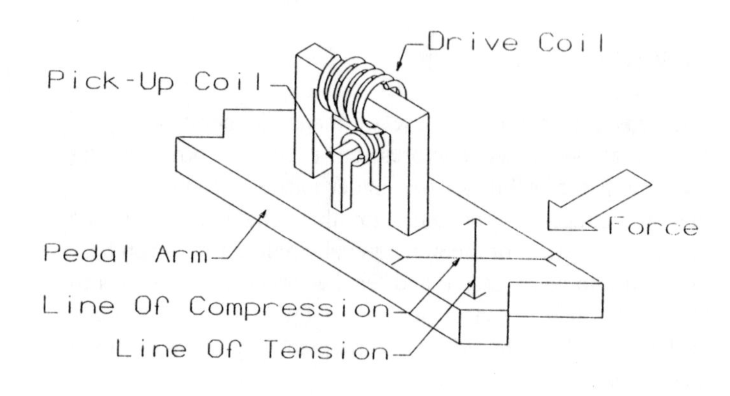

Figure 2. Major Sensor Components

Sensor Operation:

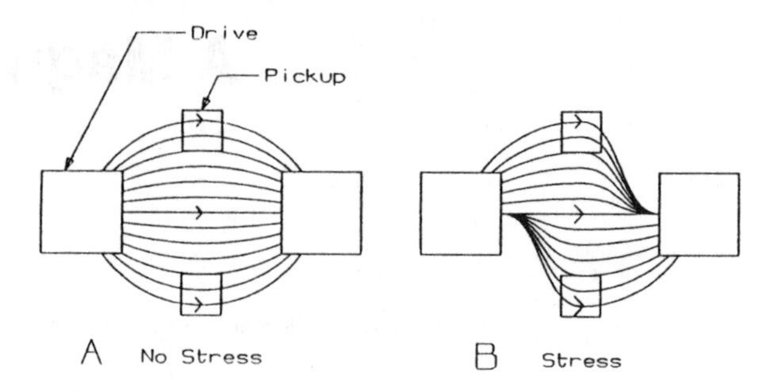

Figure 3. After Dahle [2].

Sensor operation can be most easily understood by visualizing the magnetic flux lines of the magnetic circuit. In operation, a sinusoidal voltage of about 20 KHz is applied to the drive coil which provides the magnetic bias required for sensor operation. Referring to Figure 3a it can be seen that for the unstressed system the magnetic flux lines are symmetrical about the pick-up pole pieces. There is no net change in the magnetic field across the pick-up coil, therefore no signal is produced. Figure 3b depicts the flux lines that are skewed by the permeability changes about the compressive and tensile stress lines. It should be noted that the field is no longer symmetrical about the pick-up coil. Since there is a flux difference across the pick-up coil core, an EMF is developed across the coil which is proportional to the magnetic imbalance. Through signal conditioning this EMF is converted into the final sensor output signal.

The signal from the pick-up coil is nominally a sine wave with an amplitude proportional to the stress applied to the magnetic member. This signal is amplified as necessary and rectified to provide the final DC output signal. A phase sensitive rectifier is necessary to distinguish between positive and negative stresses as both will produce an EMF within the pick-up coil; the resulting sine wave will be 180 degrees out of phase with the negative stress condition. Since "negative" stresses do not normally occur within the pedal application

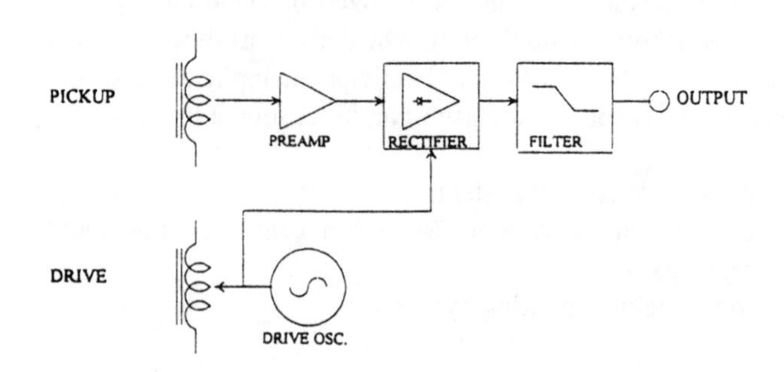

Figure 4. Sensor Electronics

this rectifier stage can be simplified in many cases. A block diagram showing the required signal processing appears in Figure 4.

An important note is that the above discussion is based on ideal magnetic materials with no "baseline" stresses. This is not true of production brake pedal arms which may have substantial stresses, surface blemishes and magnetic irregularities built in from the fabrication process. These non-idealities result in an offset to the sensor output as the "unstressed" (no pedal force) magnetic flux map is not symmetrical about the pickup pole pieces. This offset can be compensated for by various electronic or magnetic means.

Physical Implementation:

The sensor assembly is overmolded into a block form, with mounting features as required by the specific application. Connections are made with three leads for power, ground and signal output for the "smart" sensor configuration or with a cable containing two shielded pairs for the sensor with separate electronics. Figure 5 depicts a conceptual package for a sensor suitable for aftermarket installation. Application-specific mounting features will be provided to the overmolded case to optimize assembly in OEM settings.

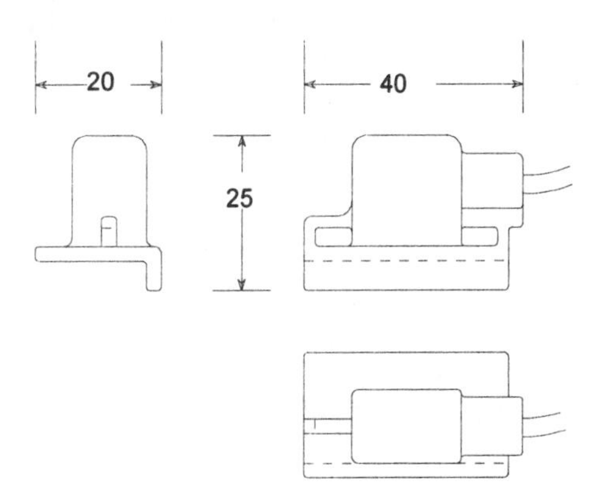

Figure 5. Packaged Sensor.

Results/Performance:

The sensor shows good linearity over the entire operating range of braking pressures as can be seen in Figure 6. Current drain for the sensor with discrete electronics unit is on the order of 100mA at operating voltages of 9 to 16Vdc, most of which is current to the drive amplifier. Work is underway to reduce operating current below this level and maintain desired sensitivity through the development of a custom ASIC.

Most pedals tested have shown good performance with the pedal force sensor although some designs have shown high sensitivity to the direction with which the force is applied (i.e. left foot Vs right foot pedal actuation). This is to be expected as the torsional forces induced by off-center actuation of the pedal will introduce stresses along the same

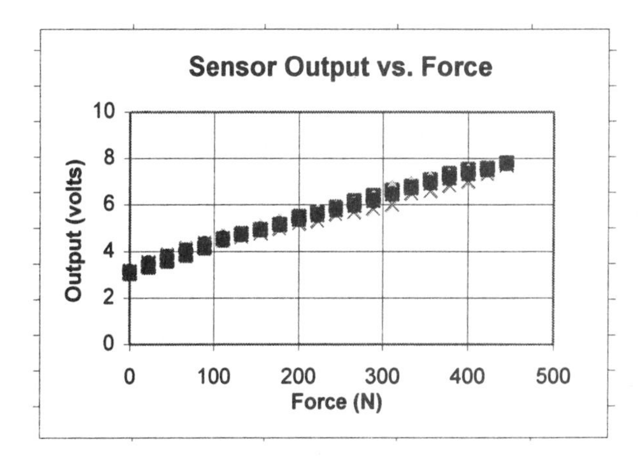

Figure 6. Sensor Performance

axis as will the "normal actuation force. Proper placement of the sensor on the brake pedal arm will limit errors introduced by this effect although for some brake pedal designs it will always be a factor.

For applications which must eliminate this torque effect or pedal designs which exaggerate this effect a second sensor can be added. By mounting the sensor to the opposite side of the brake pedal arm it will have an equal and opposite response to the torsional stresses and an equal response to the applied pedal force. These signals can simply be added to obtain a response that excludes the effects of torque on the pedal shaft.

Alternate sensor configurations may give better sensitivity and/or selectivity to stress along the axis of interest. Work is underway to evaluate these alternatives.

Example Application:

As an example the pedal force sensor will be applied to the control of the electrically actuated brakes of a towed trailer. Since this is an aftermarket application the sensor must be field installable on a wide variety of existing brake pedal designs and must interface with an existing electric brake controller with a minimal amount of modification. Accuracy and resolution requirements may be slightly relaxed, but reliability requirements are high. Figure 7 depicts the overall system; a proposed sensor package for the aftermarket installation is shown in Figure 5.

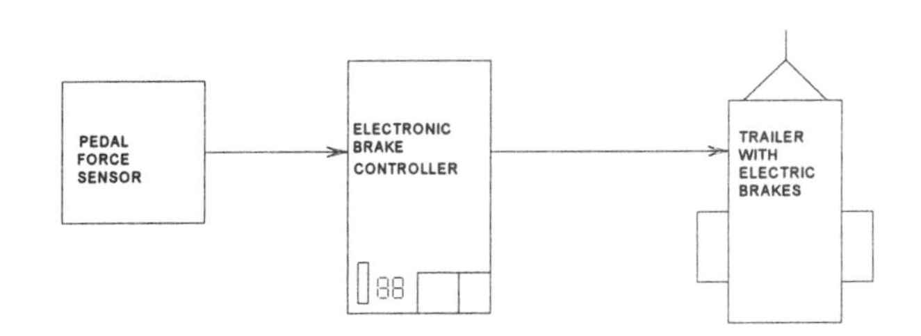

Figure 7. Trailer Braking System

The aftermarket sensor package includes a molded rib to hold the sensor in position with the edge of the pedal arm, molded in loops to retain fasteners used to hold the sensor to the pedal arm, and a molded in strain-relief to protect the lead-wire attachment.

A commercially available electronic trailer brake controller was modified to demonstrate the system. The pedal force sensor was used as the primary input when the controller operated in the "AUTOMATIC" mode, the "MANUAL" mode remained unchanged. In this way the trailer's electrically operated brakes were applied in proportion to the applied brake pedal force and hence in proportion to the towing vehicle brake force.

The system was installed in a 1994 Chevrolet Astro van and was used to tow a 2000 pound test trailer. The improvement to the drivability of the system was dramatic. As the trailer brakes now responded in proportion to the vehicle brakes the trailer no longer pushed or pulled the towing vehicle; driving the van/trailer combination was nearly as smooth as driving the van alone. Additionally, since the driver had immediate control of both the application and release of the trailer brakes the maximum braking level of the trailer could now be set 50% higher, allowing shorter stopping distances.

CONCLUSIONS:

A three-terminal sensor is described to sense the applied force on an automotive brake pedal. The sensor uses magnetostrictive properties of the ferromagnetic pedal arm to produce a DC output signal proportional to the stress within the arm. Major advantages offered by the sensor include:

- **Reliability.** The sensor is packaged in a rugged plastic case and is not dependent on small precision etchings, flexing members, or optics. Overstress capacity of the existing pedal design is retained.
- **Simplicity.** In most applications the sensor is a one-piece unit with three connecting wires. Output is high level, low impedance offering good resistance to automotive EMC concerns.
- **Cost effectiveness.** The sensor does not depend on precision strain sensing components or precision optics. Sensor construction is compatible with low cost high volume production methods. The sensor uses the magnetic properties of materials commonly used for pedal arms - no exotic materials are necessary and little or no redesign of the pedal arm is necessary.
- **Ease of installation.** Aftermarket versions of the sensor can be field installed to a wide variety of vehicles with a minimum of training. In OEM applications the sensor can be mounted at the time of pedal manufacture.

References:

The sensor described is very similar to and uses techniques common to sensors used to measure torque within shafts. Many authors have written on the subject. For additional information the following papers are suggested:

1. W.J. Fleming, "Basic Understanding Of Magnetostrictive Torque Sensors," Sensors Expo International 1989 109A

2. O. Dahle, "Method and Device for Indicating and Measuring Mechanical Stresses Within Ferro-Magnetic Material", U.S. Patent #2,912,642. November 10, 1959.

3. W.J. Fleming; "Magnetostrictive Torque Sensors - Derivation of Transducer Model," SAE paper 890482. 1989.

4. T.H. Barton and R.J. Ionides, "A Quantitative Theory of Magnetic Anistropy Torque Transducers," IEEE Trans. on Instrumentation and Measurement, Vol. IM-14, pp. 247-254, December 1965.

5. T.H.Barton and R.J. Ionides, "A Precision Torquemeter Based on Magnetic Stress Anisotropy," IEEE Trans. on Power Apparatus and Systems, Vol. PAS-85, No.2, pp.152-159, February, 1966.

6. B.D. Cullity, "Introduction to Magnetic Materials," Chapter 8, Magnetostriction and the Effects of Stress. Addison-Wesley Publishing Co., Reading, Massachusetts. 1972

VARIOUS PARAMETER SENSORS

950471

Simulation and Comparison of Infra-Red Sensors for Automotive Collision Avoidence

Sridhar Lakshmanan
University of Michigan-Dearborn

Thomas Meitzler, Euijung Sohn, and Grant Gerhart
U. S. Army TARDEC

ABSTRACT

This paper presents a simulation and comparison of two different infra-red imaging systems in terms of their use in automotive collision avoidence applications. The first half of this study concerns the simulations of an "cooled" focal plane array infra-red imaging system, and an "uncooled" focal plane array infra-red imaging system. This is done using the United States Army's Tank-Automotive Command Thermal Image Model - (TTIM). Visual images of automobiles - as seen through a forward looking infra-red sensor - are generated, by using TTIM, under a variety of viewing range, and rain conditions. The second half of the study focuses on a comparison between the two simulated sensors. This comparison is undertaken from the standpoint of the ability of a human observer to detect potential (collision) targets, when seeing through the two different sensors. A measure of the target's detectability is derived for each sensor by using the United States Army's Tank-Automotive Research Development and Engineering Center Visual Model (TVM).

1. INTRODUCTION

Collision avoidence systems are seen as an integral part of the next generation of active automotive safety devices [1, 2]. Automotive manufacturers are evaluating a variety of imaging sensors for their usefulness in such systems [1]. Sensors that operate at wavelengths close to the human vision (such as video cameras) provide images that have good spatial resolution. However, the quality of the images (in terms of relative contrast and spatial resolution) acquired by such a sensor degrades drastically under conditions of poor light, rain, fog, smoke, etc.. One way to overcome such poor conditions, is to choose an imaging sensor that operates at longer (than visual) wavelengths. The relative contrast in images acquired from such sensors do not degrade as drastically under poor visibility conditions. However, this characteristic comes at a cost; the spatial resolution of the image provided by such sensors is less than that provided by a video camera.

Passive infra-red sensors operate at a wavelength slightly longer than the visual spectrum.[1] Hence they perform better than a video camera (in terms of relative contrast) when the visibility conditions are poor. Also, since their wavelength of operation is only slightly longer, the quality of the image provided by an infra-red sensor is comparable to that of a video camera (in terms of spatial resolution). As a result, infra-red sensors have much potential for use in automotive collision avoidence systems [1, 3].

There are two state-of-the-art infra-red detectors, and they offer two alternatives when it comes to infra-red sensor system for automotive collision avoidence applications. The first alternative is based on a cooled focal plane array of infra-red detectors that operate in the $3-5\mu m$ wavelength. The second alternative is based on an uncooled focal plane array of infra-red sensors that operate in the $8-12\mu m$ wavelength. The first one provides images with better spatial resolution than the second. However, the cost of manufacturing and operating such a sensor system is more than the second one.

The TACOM Thermal Image Model (TTIM) is a

† THIS WORK WAS SUPPORTED IN PART BY THE UNITED STATES ARMY RESEARCH OFFICE UNDER CONTRACT DAAL03-91-C-0034

[1]The visual spectrum is between $0.4-0.7\mu m$, where as the infra-red spectrum is between $0.7-12\mu m$

computer model that simulates the appearance of a thermal scene as seen through infra-red imaging system [6]. TTIM can simulate the sampling effects of the older single detector scanning systems, as well as more modern systems that use focal plane staring arrays. TTIM can also model image intensifiers. A typical TTIM simulation incorporates the effects of atmospheric conditions on the image, and it is accomplished by using LOWTRAN - a computer model of the effects of atmosphere conditions on thermal radiation that was developed at the United States Air Force's Geophysics Laboratory. A particularly attractive feature of TTIM is that it produces a simulated image for the viewer, not a set of numbers as some of the other simulations do. We refer the reader to Fig. 1 for schematic representation of TTIM.

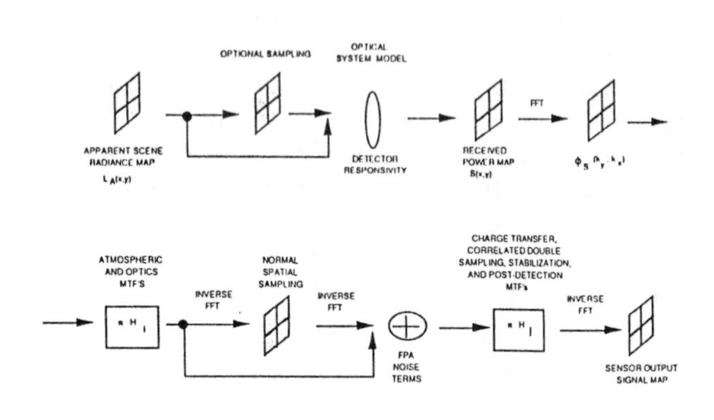

Figure 1: Schematic reprsentation of TTIM

In the first half of this paper we use TTIM to simulate the cooled and uncooled infra-red imaging systems, and compare their performance from the standpoint of automotive collision avoidence applications. Analogous comparisons exist in current literature (see [4, 5] for example). However, it is our opinion that such studies are not applicable for the situation at hand. TTIM allows us to compare the performance of the two infra-red systems in terms of how good is the quality of their images for subsequent human perception/interpretation. The existing studies do not allow such comparisons.

The comparison of system performance leads us to the second half of this paper. Given that we have two images of the same scene, captured by using the two different infra-red systems, we use TVM to assess which of the two is "better". TVM is a computational model of the human visual system [7]. The model consists of two parts: the first part is a color separation module, and the second part is a spatial frequency decomposition module. The color separation module is akin to the human visual system. The spatial frequency decomposition system is based on a Gaussian-Laplacian pyramid framework. Such pyramids are special cases of wavelet pyramids, and they represent a reasonable model of spatio-frequency channels in early human vision [8]. We refer the reader to Fig. 2 for a schematic representation. of TVM.

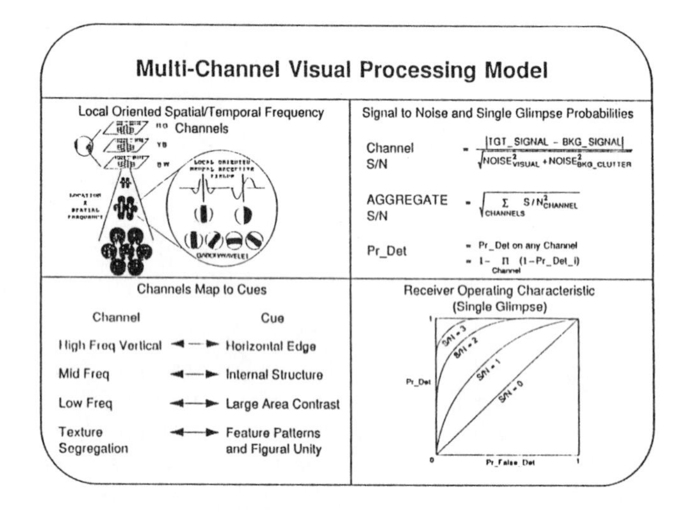

Figure 2: Schematic reprsentation of TVM

Given two images of the same scene, an object of interest, and the background, we use TVM to produce a measure of detectability for the object in each of the images. The image in which the object has measure of detection is the "better" one among the two.

2. SIMULATION OF INFRA-RED SENSORS

This section presents the simulation of cooled and uncooled infra-red imaging systems using TTIM. Specifically, we generate (simulated) images of commercial vehicles in a typical road scene as seen through such infra-red systems using TTIM. We present examples of how viewing range and rain affects the quality of the acquired image.

We see this simulation as a substantial first step, and as providing a means to comprehensively evaluate and compare the two sensor systems in the near future. Our ability to simulate the sensors provides a means for exactly repeating imaging experiments and measurements, something that is difficult to achieve in field trials. Also based on our experience, the ability simulate the sensors provides us with the ability to exercise precise control over the imaging conditions. In the cooled infra-red systems, for example, it is important to provide proper temperature shielding (and control) during field trials. Otherwise, the quality of the images acquired from the infra-red system is badly affected, and it negatively impacts the validity of subsequent comparisons between sensor systems. By simulating cooled infra-red systems we can overcome such difficulties.

In Figs. 6 and 7 we present (simulated) images of typical commercial vehicles when the viewing distance (the distance between the vehicle and the sensor) increases. This done for both the cooled and uncooled cases, by inputing into TTIM the thermal image in Fig. 3.

Figure 3: Input image to TTIM

In Figs. 8 and 9 we present (simulated) images of the same set of vehicle when the viewing distance is fixed, but when the amount of rain fall under which the image is acquired increases.

3. SENSOR COMPARISON

In this section we use TVM to compare the quality of images acquired from the cooled and the uncooled infra-red imaging systems. Specifically, we input into TVM two (simulated) images, corresponding to the two infra-red systems. Then, using TVM we obtain a measure of detectability in each of the images for a vehicle of interest.

The detectability measure obtained from TVM is the signal-to-noise ratio (SNR) between the vehicle of interest and the background (as explained in Fig. 2). In Fig. 4 we plot this SNR for both cooled and uncooled systems as a function of spatial frequency.

Next, in Fig. 5 we plot the SNR for both systems (actually the maximum SNR among the different frequency channels) when the amount rain fall under which the image is acquired increases.

4. CONCLUSIONS

In this paper we provided a simulation of and a comparison between cooled and uncooled infra-red imaging systems. This was done with a view towards using such systems for automotive collision avoidence applications. Using TTIM, we successfully simulated both the infra-red imaging systems. We provided (simulated) images as seen through these sensors, when the viewing distance changes and when the amount of rainfall under which the images are acquired increases. Next, by using the TVM we compare the two sensors. In each of the spatial frequency channels found in early vision among humans, we obtain a measure of detectability (in terms of SNRs) for an object and background interest. We plot the SNR versus spatial frequency for both the sensors, and obtain the variation in the SNR as the amount of rain fall under which the images are acquired increases. Sensor comparisons, are just but one aspect of collision avoidence. There are a number of other human factors and social issues as well associated with the "science of collision avoidence" as pointed out in [2].

5. REFERENCES

[1] *IEEE Micro: Special Issue on Automotive Electronics,* Vol. 13, 1993.

[2] R. K. Deering and D. C. Viano, "Critical success factors for crash avoidence countermeasure implementation," *Proc. Intl. Congress on Transp. Electronics,* pp. 209–214, 1994.

[3] S. Klapper, B. Stearns, and C. Wilson, "Low Cost Infrared Technologies Make Night Vision Affordable for Law Enforecment Agencies and Consumers," *Proc. Intl. Congress on Transp. Electronics,* pp. 341–345, 1994.

[4] G. A. Findlay, and D. R. Cutten, ""Comparison of performance of 3-5 and 8-12 micron infrared system," *Applied Optics,* Vol. 28, pp. 5029–5037, 1989.

[5] T. W. Tuer, ""Thermal Imaging Systems Relative Performance: 3-5 vs 8-12 Microns," *Technical Report,* AFL-TR-76-217.

[6] C. S. Hall, E. T. Buxton, and T. J. Rogne, *TACOM Thermal Image Model Version 3.1: Technical Reference and Users Guide,* Optimetrics Incorporated Technical Report OMI–405, 1993.

[7] G. H. Lindquist, G. Witus, T. H. Cook, J. R. Freeling, and G. Gerhart, "Target discrimination using computational vision human perception models," *Proceedings of the SPIE Conference on Infrared Imaging Systems: Design, Analysis, Modeling, and Testing V,* Vol. 2224, pp. 30–40, Orlando, 1994.

[8] J. Malik and P. Perona, "Preattentive texture discrimination with early vision mechanisms," *J. Op. Soc. Am.,* Vol. 7, pp. 923–932, 1990.

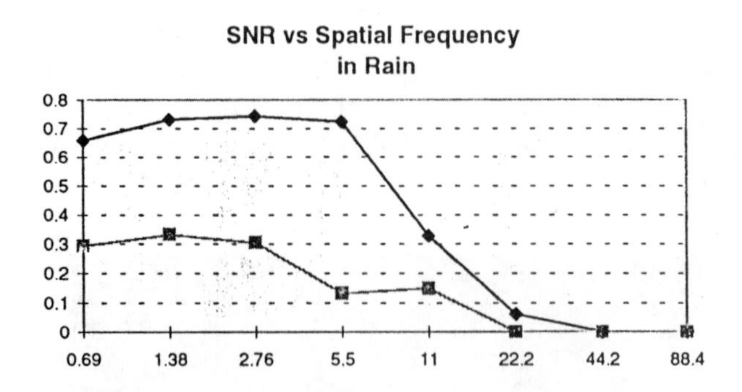

Figure 4: SNR versus spatial frequency

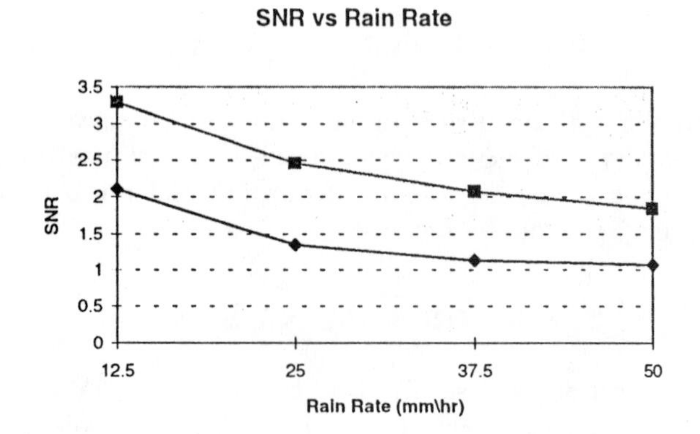

Figure 5: SNR versus rain fall amount

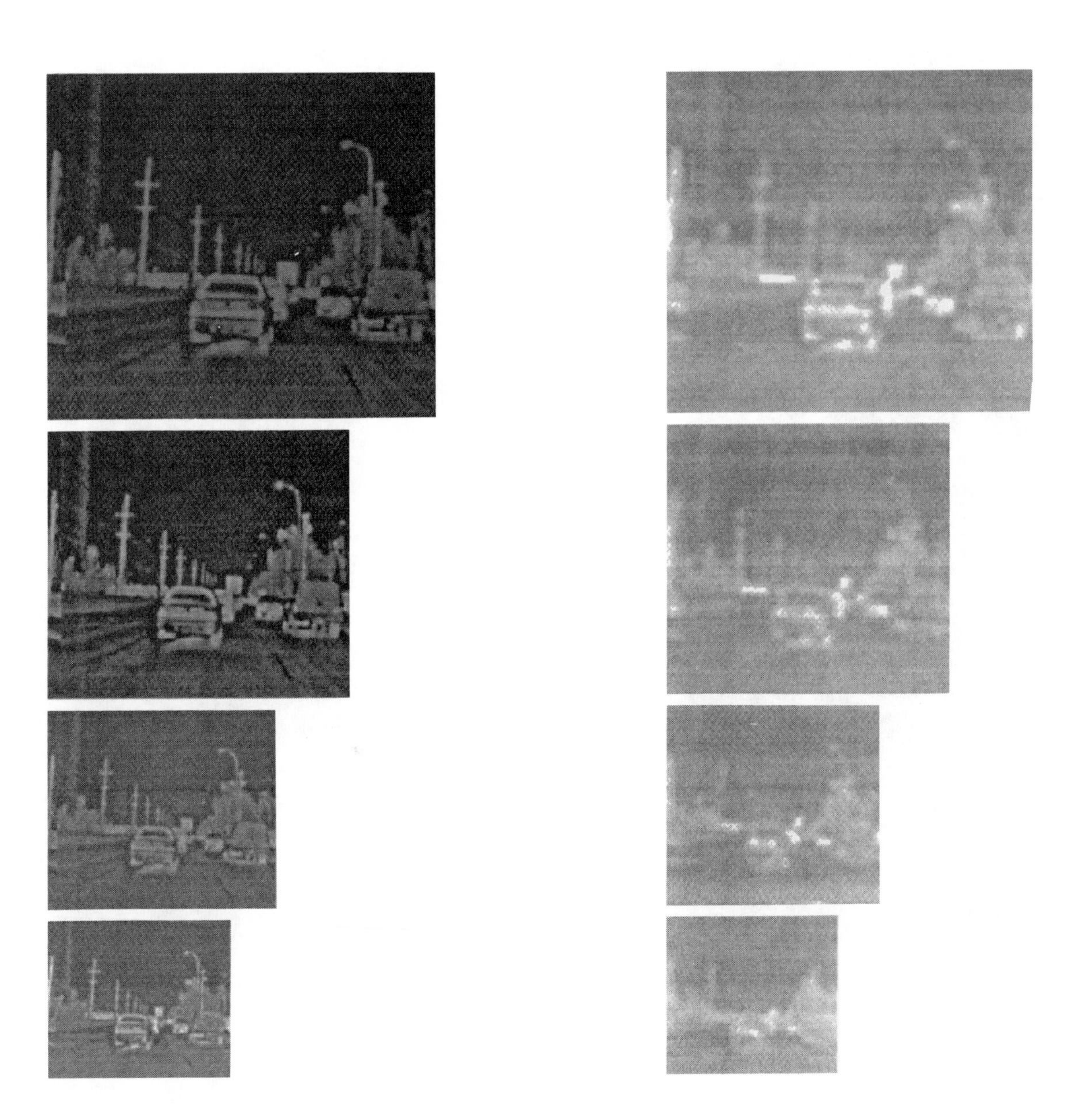

Figure 6: Simulation of the effects of viewing distance on spatial resolution in images acquired via cooled infra-red imaging systems - 70m, 90m, 120m, 150m (Top to Bottom)

Figure 7: Simulation of the effects of viewing distance on spatial resolution in images acquired via uncooled infra-red imaging systems - 70m, 90m, 120m, 150m (Top to Bottom)

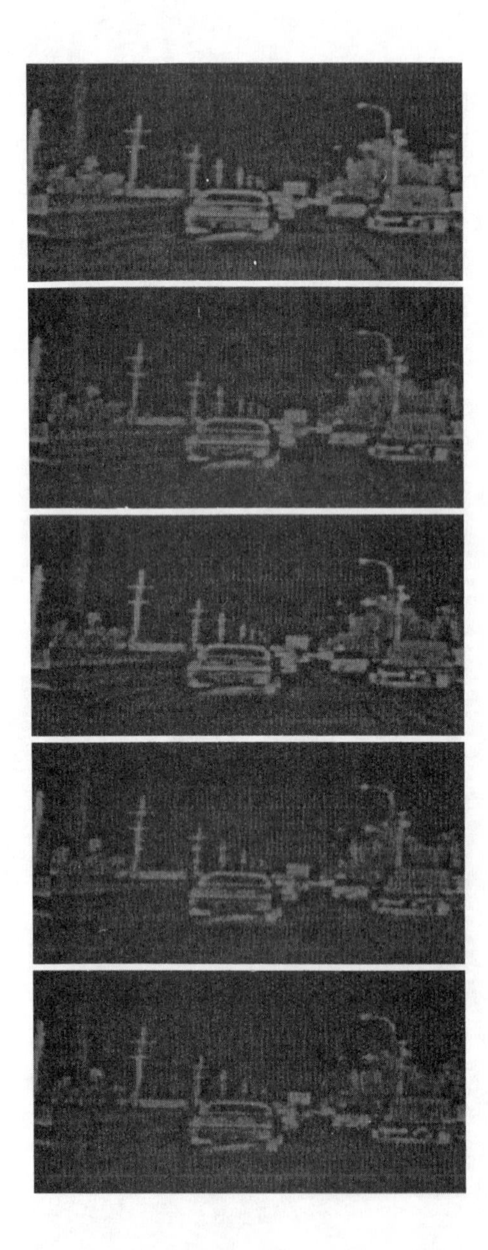

Figure 8: Simulation of the effects of rain fall on relative contrast in images acquired via cooled infra-red imaging systems - 0mm/hr, 12.5mm/hr, 25mm/hr, 37.5 mm/hr, 50mm/hr (Top to Bottom)

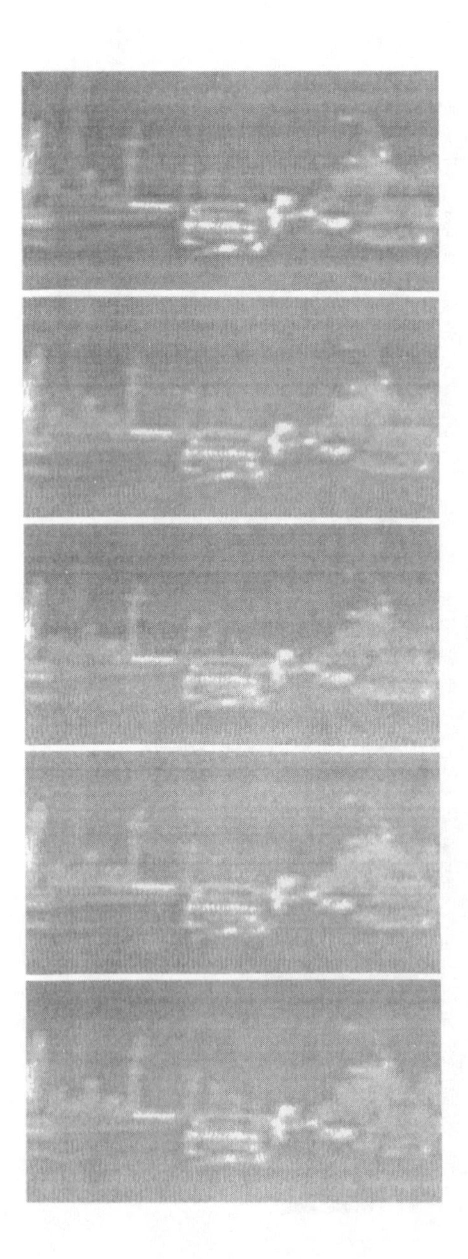

Figure 9: Simulation of the effects of rain fall on relative contrast in images acquired via uncooled infra-red imaging systems - 0mm/hr, 12.5mm/hr, 25mm/hr, 37.5 mm/hr, 50mm/hr (Top to Bottom

961199

Development of a Time Resolved Spectroscopic Detection System and Its Application to Automobile Engines

P. Moeser and W. Hentschel
Volkswagen AG Research and Development

ABSTRACT

A novel in-cylinder diagnostic technique for time-resolved investigation of intermediate combustion products during the combustion of one engine cycle is discussed. UV/VIS emission spectra are recorded from inside the combustion chamber with temporal resolution in the μs-range. By means of a spark light-conducting sensor the investigations are applied to production engines. The special arrangement and the setup of the high speed detecting system, a modified CCD-camera with a streak-mode operation, are discribed. The general design concept is outlined and first experimental results are presented.

Experimental results were obtained both on a SI engine and a diesel engine. The results are plotted as 3-d-images with time and wavelength and the intensity as colored 3rd dimension. They are time-resolved for the complete or a chosen part of the combustion cycle. The highly resolved spectral and temporal images are used to chart the progress of the combustion by observing the emissions of intermediate combustion products, molecule formation processes or particle emissions. This high speed investigation may yield helpful information for the improvements of engine design and control.

INTRODUCTION

The combustion process taking place in engines is a complex phenomenon, which is not yet fully understood. The technical limits for improving fuel efficiency and reducing pollutants are not well known. Thus, the investigation of the flame may yield essential information for developing combustion models for calculations, or helpful information for the improvement of engine design and control.

The principle of this optical in-cylinder diagnostic technique is based on emission spectroscopy. Several intermediate combustion products at a time and molecule formation processes can be measured by emission spectroscopy simultaneously. Thus emission spectra contain information on the state of the flame.

The light emissions of reaction species during combustion are presumed to exist in ultraviolet and visibles ranges [1,2]. When recording these spectra by means of a high-speed detection system, the recorded data show the progress of the flame. High-resolution temporal measurements are required to investigate the progress of the internal combustion of an automobile engine. At 2000 rpm, one complete engine cycle requires 60 ms and the fuel is totally consumed in a period of only 6 ms.

Spectroscopic measurement techniques are quite common as applied to flames [1,2,3] but have scarcely been applied to engines because of optical and mechanical problems. The application to an engine involves more complex considerations [4,5]. Internal combustion engines, in particular high pressure engines, usually do not provide apertures for observing the flame. Optical access is always a question of constructional expense. Severe restrictions are imposed on transparent materials used in engines. First, they have to be adaptable to the design environment of the engine. Then they also have to withstand high pressure and temperature gradients. The technical solution will be described in the next chapter.

The use of streak cameras with a deflection plate are known even for time-resolved investigations of the strong emission of a diesel engine [6,7], but the weak intensity of light during the combustion stage of a SI engine is difficult to detect. An accurate handling of the light transmission and separation, a powerful light intensifier and a sensitive detection system is required. A description of the detection system will be given before the obtained results of SI and diesel engines will be presented.

EXPERIMENTAL

EXPERIMENTAL APPLICATION TO AN ENGINE - Optical access to the combustion chamber is required to analyse the combustion light signal of engine combustion. Most optical, in particular the laser-based methods are carried out on a transparent engine [4,5]. For this, quartz glass windows are fitted in the cylinder block and in the piston crown. Although emission spectroscopy does not require a laser, it does require optical access for detection. Especially for the study of the combustion process it must be ensured that the combustion chamber, which has a decisive effect on combustion, particularly the flow conditions, is not changed. Any modification should be avoided. Optical access for the studies was achieved using a special design of spark plug which did not result in any change to the combustion chamber. Light-conducting plugs for optical in-cylinder diagnostics are known mostly as modified glow plugs for diesel engines [8]. For this investigation a rigid optical fibre rod with an effective diameter of 1.5 mm is integrated into the central electrode of the spark plug. When designing the light-conducting spark plug it was ensured that

- the spark plug with the light-conducting material is resistant to temperature and pressure
- the light-conducting material used for the VIS and UV range is optically transparent
- the heat range affected by the modification is corrected by selecting the appropriate design of spark plug.

A significant advantage of the observation from above through the spark plug is that the flame front is located in the observation volume right from the start. Other optical measuring techniques, working with a laser light section which is emitted from the side, do not always have the flame front in sight. It is thus impossible to observe the complete combustion cycle.

INFLUENCE OF THE SPARK PLUG ON LIGHT EMISSION - Single and triple electrode spark plugs as well as a surface gap spark plug (not covering the centre electrode) were used for the study. In many cases, it was possible to detect light emissions during the time-resolved measurements which are caused directly by the ignition spark. However, the surface gap spark plug was an exception. The spark only propagates laterally due to the long spark gap and is therefore not visible for geometric reasons.

A measurement was also carried out with a spark plug with an offset centre electrode. The bore for the light pipe is made next to the centre electrode. However, in this case the disadvantage is that it is not possible to detect any light in the interesting time range of ignition and the first ° CA of combustion.

INFLUENCE OF THE LOCATION OF THE FLAME FRONT - The observation cone (angle of approximately 20°) of the light pipe at two different piston positions can be seen in Fig 1. The propagating flame front is also drawn in at the respective section through the combustion chamber, spherical propagation being assumed for the sake of simplicity.

The spherical surface increases by r^2 with the propagation of the flame front, i.e. the observation area increases by this factor. However, the solid angle of the radiating molecules decreases by $1/r^2$ at the same time. Both processes compensate each other so that a correction factor depending on the radius is not required.

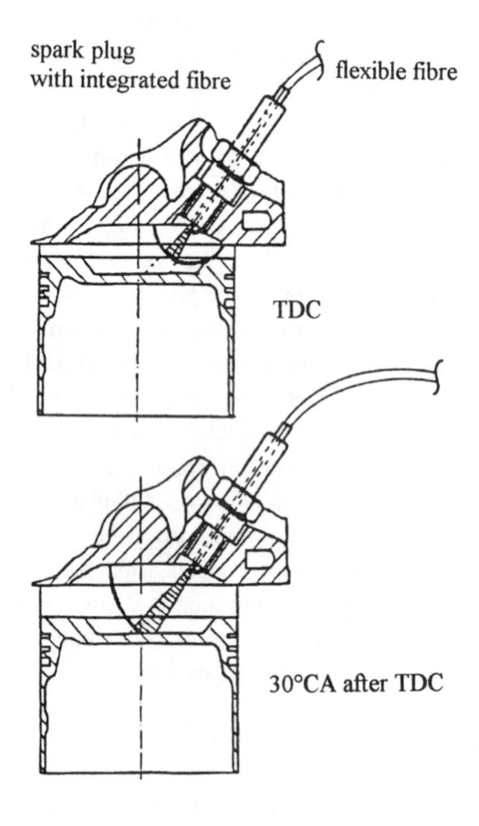

Figure 1: Observation cone of the light pipe

With the propagation of the flame front, the spectra emitted by radicals of the flame front are expected to be affected by absorption from burnt gas located in the optical path to the light-conducting probe. However, emission spectroscopic measurements using the transparent piston crown of a transparent engine (observation from beneath) from [9] do not show any significant changes, even when the flame front has reached the piston crown.

Additional interference is caused by the running engine or test bench operation. Fluorescences excited in the light pipe and contamination of the optical coupling due to oil vapours and soot, if the usual burning free of the light pipe surface is prevented, are additional inevitable difficulties. However, the measurement time is not limited by the contamination of the optical surface.

DETECTING ARRANGEMENT - The light emission from inside the combustion chamber is guided by an UV-transmission fibre with an effective diameter of 1.5 mm and 2 m in length to the entrance slit of a 125 mm - Czerny-Turner spectrograph (F-number of 3.7) with a high light performance (Fig. 2). The optical fibre is mounted without any additional aperture. The slit width chosen has to be a compromise of spectral resolution and the amount of light intensity and can be varied from 50 μm and wider. Different gratings for light separation are used depending on the spectral resolution and the wavelength section. The spectral resolution of the detection system is 0.6 nm using a slit width of 50 μm and a grating of 2400 lines per mm [10].

The gratings used are optimized for near UV region sensitive with a blaze wavelength of about 200 nm. The spectrum obtained is projected onto the camera head. The camera head of the detecting system is composed of a single stage image intensifier and a CCD sensor. The image intensifier consists of an UV-sensitive photocathode, a multichannel plate (MCP) and a phosphor screen [10]. A fast type of phosphor (P46) was selected. This has a very short decay time of 100 to 300 ns. Thus, a chronological "blurring" of the signal due to phosphor afterglow does not occur.

The image intensifier is connected to the CCD sensor via a tapered image-conducting bundle of glass fibres, reducing the effective cross-section from 25 to 11 mm. This results in a higer information density on the chip. The light signal is chronologically integrated on the CCD chip and transformed into the charge signal which can subsequently be read out. The chip has a pixel field of 576 lines and 384 columns.

The required time resolution is realized using an electronically controllable streak mode of the camera. The special timing electronics of the camera stage provides a stepwise transfer of pixel charges line by line along the CCD chip (Fig.3) with a time resolution in the μs-range. The cycle frequency of the line charge transfer is adjustable between 28 and 3160 kHz.

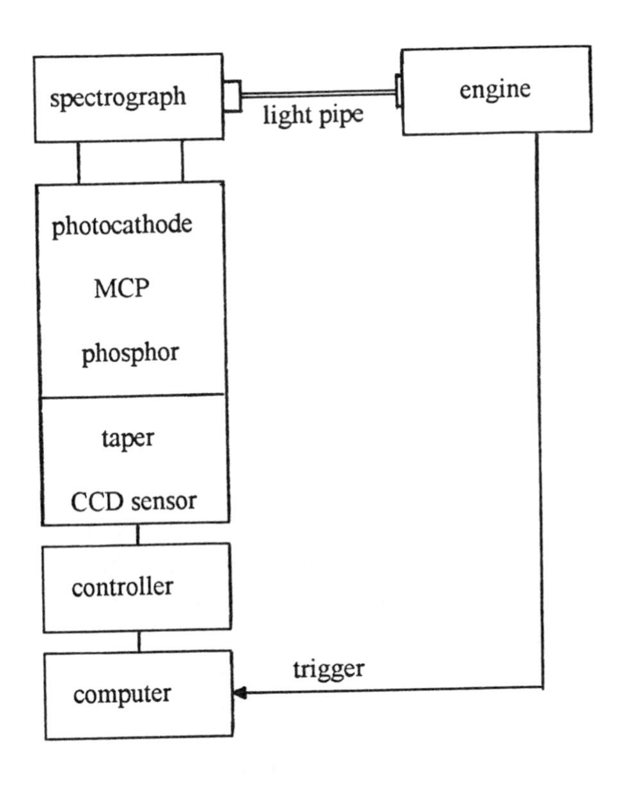

Figure 2: Configuration of the equipment

In practice the light signal is focused on several adjacent lines. The resulting image height depends on the height of the coupled light beam. As the pixels are exposed during charge transfer, there is a time-blur. When recording signals of very low light intensity, this has a very positive effect on the signal-to-noise ratio.

The camera device is controlled by the main trigger pulse. The triggering pulse is specified externally by a marking on the engine flywheel. The camera exposure time is syncronised on the engine cycle and the combustion stage of investigation.

The read-out of the CCD sensor is carried out by the slow scan mode, achieving a 12-bit dynamic range, which is sent to the A/D-converter. The digitized intensities are stored and processed in a computer. The final image contains the records of spectrally resolved light emission vs. time. The images are presented on a monitor in false colors at real time.

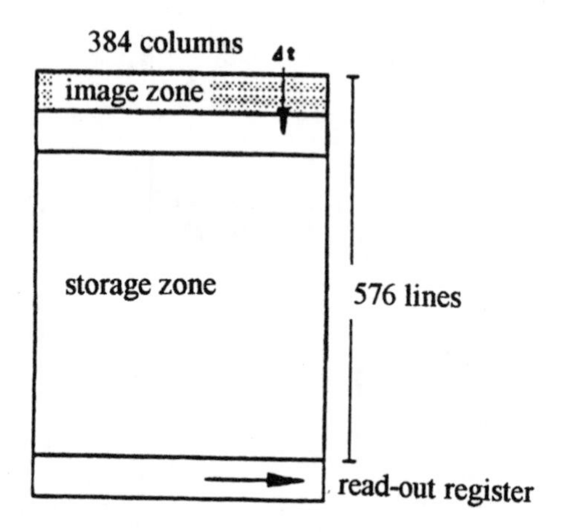

Figure 3: Streakmode operation

CALIBRATION AND CORRECTION FUNCTIONS - The line spectrum of a mercury vapour lamp, which covers the full sensivity range of the comera system, is used for wavelength calibration.

The calibration curve of spectral sensitivity E(λ) of the measuring system in total is determined for the various gratings employed by the use of a xenon lamp, as it emits a continuous spectrum. The spectral sensitivity of the system results from the measurement spectrum of the xenon lamp and the known calibrated spectral emission intensity of the xenon lamp $S_{xenon}(\lambda)$:

$$E_{system}(\lambda) = S_{measuring\ spectrum}(\lambda)/S_{xenon}(\lambda)$$

EXPERIMENTAL RESULTS

IN-CYLINDER INVESTIGATION OF A SI ENGINE

Experimental results were obtained on a SI engine with multipoint injection. The engine operation point was at 2000 rpm and part load. The fuel used was gasoline and the fuel-to-air ratio was stoichiometric.

The time-resolved emission spectroscopic measurements of the combustion are shown as 3-d spectra, the light signal intensity recorded as a function of light wavelength and time. A prerequisite for the interpretation of the combustion spectra is the identification of the molecule bands and the continuum using the relevant literature [1, 2, 11]. Identification is also made more difficult by the transient nature of engine combustion, since in addition to the occurrence of pronounced changes in pressure and temperature, mixture ratios and the condition of the fuel may also vary locally.

Engine combustion shows highly turbulent flow. The flow process dominated by the turbulence and laminar flame propagation dominated by diffusive processes combine. The flame front is highly fissured, which would make a spatially resolved measurement of the flame front difficult. In these spectroscopic measurements the flame front is not resolved spatially, but considered integrally in the direction of propagation. However, a reaction sequence can be recognized by means of the time-resolved observation of the progressing flame front.

The separation of the reaction kinetics and the consideration of three-dimensional propagation of the flame front in the turbulent flow zone is considered in [12, 13].

Fig. 4 shows the 3-d-spectrum as an example of the capability of the detection device. Due to the spectral emissions of species of interest, the wavelength range chosen in this experiment covers 400 to 450 nm with a high dispersion. Since combustion takes place around top dead centre (TDC) we limited the camera shutter opening time to a 40°CA window at TDC. The whole combustion course of one working cycle takes place in 6 ms, whereas in this investigation the important period at the beginning will be observed with a resolution of 56 kHz. All images are a pixel by pixel integration of 10 consecutive cycles to improve the signal-to-noise ratio for demonstration.

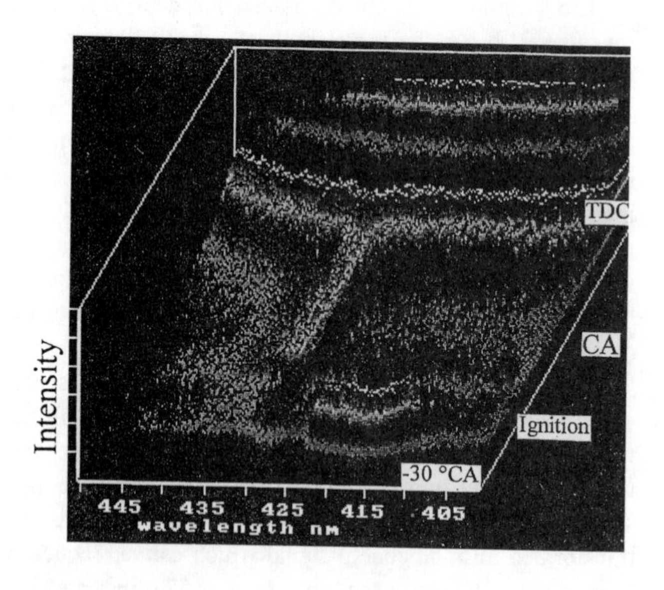

Figure 4: SI engine combustion process

At Fig. 4 the first signal is observed at the time of ignition. The shortness and intensity is characteristical for the igniton event. The spectra show CN-bands. Herden et al. [14] were also able to observe the occurrence of CN at the ignition point. They attributed the formation of CN to the ignition plasma. During the disruptive discharge phase of the ignition spark, in which a thin plasma channel of less than 50 μm develops between the electrodes, CN already forms at the edge of the channel [14].

Shortly after the ignition a discrete band spectrum is observed - identified as the emission of the CH-radical - at the first stage of combustion. Its appearance indicates the primary combustion of hydrocarbon or related components [2].

At this operation point a continuum starts shortly before TDC. It dominates to such an extent that most of the existing discrete band spectra are superposed. This continuum is due to the formation process of CO to CO_2 [2].

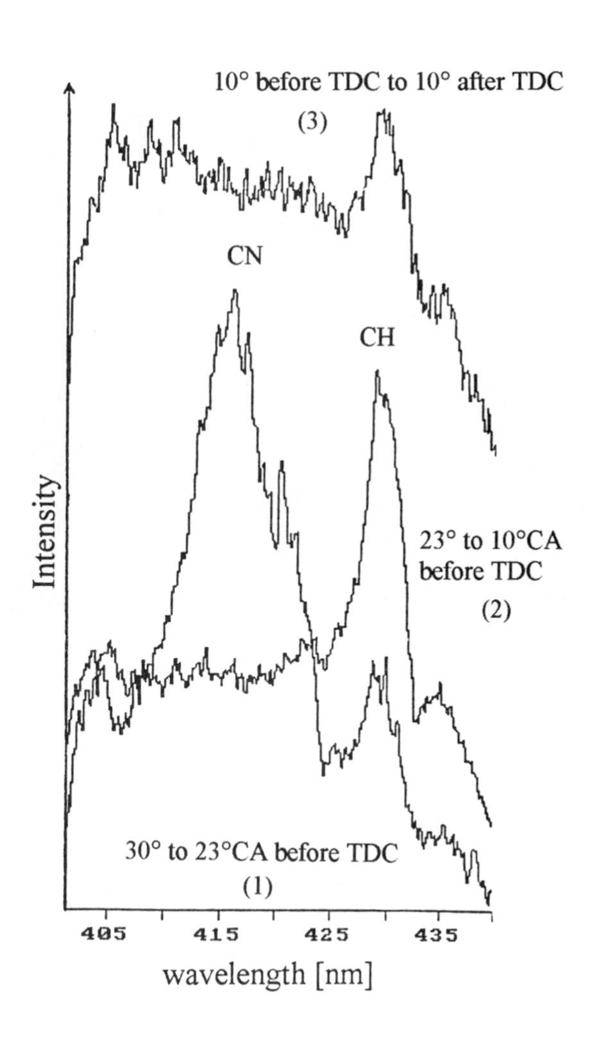

Figure 5: Single spectra representing the different stages out of Fig. 4

Fig. 5 shows single spectra, each representing the different combustion stages out of Fig. 4 for the identification of the molecule emissions. For identification a high spectral resolution of the spectrum is advantageous. The spectrum is compared to data from flame diagnostic [15] or from literature [1,2,11]. The spectrum 1 represents the light emission during the ignition event. The emitting molecule of the band spectra is identified as CN. At spectrum 2 a strong line intensity appears at 431 nm which is due to the emission of CH during the first stage of combustion. At spectrum 3 the emission of the formation process of CO_2 occurs as continuum at this wavelength range. Radiation is at a maximum in the wavelength range between 350 to 450 nm. This continuum is largely identical to that of the CO flame. Gaydon [2] assumes that it is caused by the excited CO_2 molecules which are formed by recombination of CO with atomic oxygen:

$$CO + O = CO_2 + \text{radiation}$$

Concluded, at this wavelengh range and at this time sequence the different stages of the combustion of the SI engine are obtained. The appearances of the molecule emissions and the CO-continuum indicate the different stages. Looking at Fig. 4 the progress of the combustion can be observed, e.g. the determination of the time of ignition is possible, primary combustion lasts for 20°CA at this circumstances until the combustion stage of the CO-oxidation starts.

Time resolved measurements are possible on a single cycle. In that respect cycle-to-cycle variations can be detected by means of this method. Investigations due to time events can be done with different operation points, different components like e.g. injection valves or different fuels. One example of an investigation will give a comparison of different fuels.

COMPARISON OF FUELS AT THE SI ENGINE - In addition to the standard production unleaded Euro-Super gasoline, M85 methanol fuel mixture standardized for alcohol fuels was used for the studies. M85 consists of 85% by volume of methanol and 15% by volume of super gasoline. The physical data of gasoline and M85 differ considerably and a change in engine combustion characteristics can therefore be expected. Gasoline fuel consists of a mixture of more than 200 different hydrocarbon compounds, in particular higher hydrocarbons and polycyclic compounds. Methanol (CH_3OH) consists of an OH group and a methyl radical. The analytic data of both fuels are listed in [15, 16].

Fig. 6 shows the 3-d-spectrum of the gasoline combustion. The investigation was done on a SI engine at 2000 rpm at low part load. The fuel-to-air ratio was stochiometric. With this investigation the UV region is included. The wavelength range observed of 240 to 500 nm provides an overview of the molecular emissions occuring. The camera exposure time is limited to a ±30°CA window at TDC.

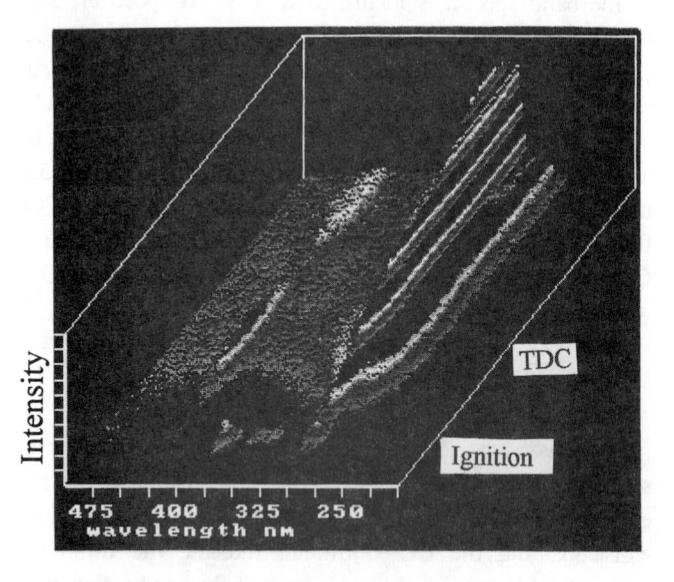

Figure 6: SI engine combustion process of gasoline

At Fig. 6 the brief spectra are characteristic at ignition point. Apart from the CN-bands the spectra show O_2-bands. O_2-bands do occur in very hot environments when excess O_2 is available [11]. At the time of ignition consumption of O_2 has not yet occurred, thus excess air is present [15].

In the UV region OH bands appear at 306 nm and exhibit a strong rise in intensity. The increase takes place to a maximum at 20° CA after TDC. The OH-radical strongly dominates the combustion process. In the visible range, the CH bands show a different progress rate than the emissions attributed to OH. In the first instance the rise is similar to the OH signal at 306 nm. The CH signal already reaches its maximum very early at 5° CA before TDC and then drops. The discrete light emission at 431 nm does not increase anymore after this. At the instant of the maximum CH signal the OH signal increase at 306 nm flattens.

By means of the measurements it can also be seen that the hydrocarbons react strongly with the OH radicals. Fig. 6 shows an increase in the intensity of the OH signal only once the CH radicals are consumed. This relationship is confirmed by [17-19].

The subsequent oxidation of CO to CO_2 is usually delayed until the original hydrocarbons or the fragments of intermediates are consumed [17-19]. In Fig. 6 it can clearly be seen that the CO flame continuum does not occur, while there is CH emission at the wavelength of 431 nm. At this operating point which has a low load, the CO flame continuum only starts once the CH radicals are consumed, except for isolated local zones, in which uncleaved fuel is still present.

Fig. 7 shows a 3-d spectrum, recorded under the same conditions as the previous 3-d gasoline spectra: 2000 rpm and low part load. The fuel-to-air ratio was stochiometric. However, this time the engine was operated with the M85 methanol fuel mixture, which is in most cases described simply as methanol in the following, as the physical properties of methanol predominate. The time window of the camera system for the time-resolved spectrum recording is shifted slightly to later times as an ignition signal is not visible and is also not expected earlier. The wavelength range under consideration covers 240 to 500 nm.

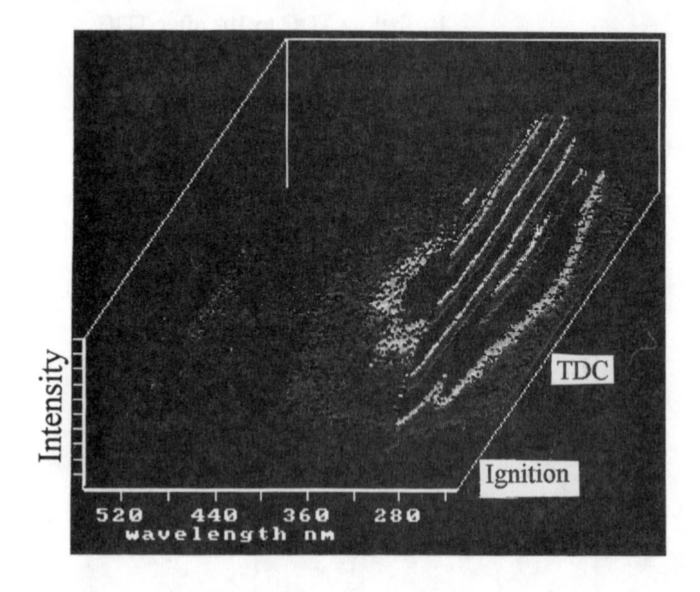

Figure 7: SI combustion process of methanol fuel (M85)

In Fig. 7 the OH band at 306 nm is visible very early on, immediately after the ignition point. One reason for the immediate occurrence of the OH emission may be that the first OH radical result from the direct cleavage of methanol [19]. After this, as in the case of gasoline combustion, the OH radiation extends over the entire combustion period of the operating cycle, but occurs at a significantly higher intensity. OH is an active intermediate combustion radical and higher concentration could lead to an increased flame velocity [17-19]. Thus, the whole combustion sequence is shorter compared to the one with gasoline.

Hardly any CH occurs during methanol combustion unlike during gasoline combustion, which can be related to the different reaction kinetics at the primary combustion stage. However, the CO flame continuum also appears, both for gasoline and for alcohol fuels, in addition to OH.

Summarized, the appearance of distinct radicals depending on the fuel leads to different combustion characteristics. During the combustion process, the methanol molecules are initially broken down to a large extent into reactive radicals, like OH radicals, which results in a more rapid burning through or in a higher flame velocity of the methanol/air mixture. Such information one can get by using this investigation method.

A detailed discription of the investigation of the fuel comparison and the interpretation are in [15].

IN-CYLINDER INVESTIGATION OF A DIESEL ENGINE - In comparison to the emission spectroscopic results of the SI engine, the emission spectra out of the diesel engine show a basically different character, which is discribed in the following.

The experimental results were obtained on a diesel engine with direct injection. The operation point was at 2000 rpm and low part load. The start of injection was set at 2 °CA before TDC. Apart from the modified glow plug instead of the spark plug, the same arrangement of the measurement was used as for the SI engine.

Fig. 8 shows the result of the diesel engine. The image spans a wavelengh ranging from 300 to 700 nm.The time scale ranges from TDC to 33° CA after TDC.

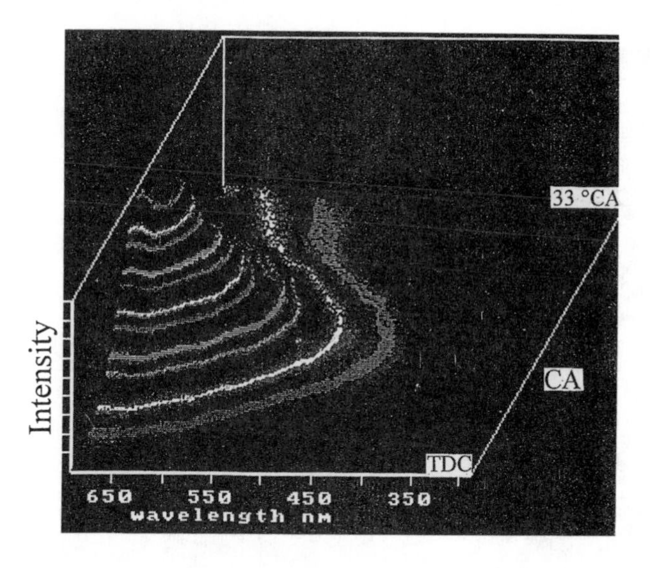

Figure 8: Combustion process of a diesel engine

As distinguished from the SI engine the spectra of the diesel engine show mostly a spectrally continuous course. The reason is the thermic emission due to particles [20,21]. There are no line structures obvious necessary for the identification of molecules.

The continuous spectra is commonly interpreted as the thermic emission of soot particles. If the soot emission behaves like a thermic blackbody emission, then the temperature of the emitting particles is direct related by Planck's law to the shape of the light intensity. This approach is used to calculate the temperature already in [20,21]. One has to consider, this approach regards the emission of the hottest particles in the observation volume. Emissions, like broad-banded molecule bands or process continua can influence the spectral curve, which are required for the temperature calculation.

Nevertheless the spectral and time-resolved images of the diesel engine show the progress of combustion in different stages. They are of course not as informative as the line structured spectra of the SI engine. What information can be obtained will be discribed in the following. Regarding the time lapse in Fig. 8 of the light emission, at the beginning there is a low light intensity at longer wavelength indicating low temperature. This light emission is due to the ignition stage of the diesel combustion and holds on during the ignition delay. Comparative measurements of the in-cylinder pressure confirm this. The main combustion follows, the spectra start in the UV region and their intensity is higher. It is known, that particles emitting in the UV region can be related to a higher temperature. After the main combustion the afterglow takes place.

One example for a broad-banded molecule emission is shown in Fig. 9. The emission is due to polycyclic aromatic hydrocarbons [3,22]. Further emission attributed to molecules could not be observed. NO exhaust emission is, especially in the combustion of the diesel engine a pollutant, which should be reduced. Therefore information of NO processes during combustion would be helpful and recording NO by means of time resolved emission spectroscopy would be desirable. The wavelength region from 200 to 250 nm was investigated. In this UV region bands of NO are expected, but no line structure was recognised. Even in SI engines no NO bands could be observed, but NO was detected in a propane-air-flame at atmospheric pressure [15]. Reason for the missing band spectrum of NO could be absorption processes, which are present at the pressure and temperature existing in engines.

With investigation prior self-ignition, no bandspectra were observed yet. The radical formaldehyde takes part during those processes. The occurrence of the band spectrum of formaldehyde would be in the visible wavelength range and therefore detectable by the measurement system. A reason for the lack of emission spectra could be that the temperature for the exitation process is not high enough.

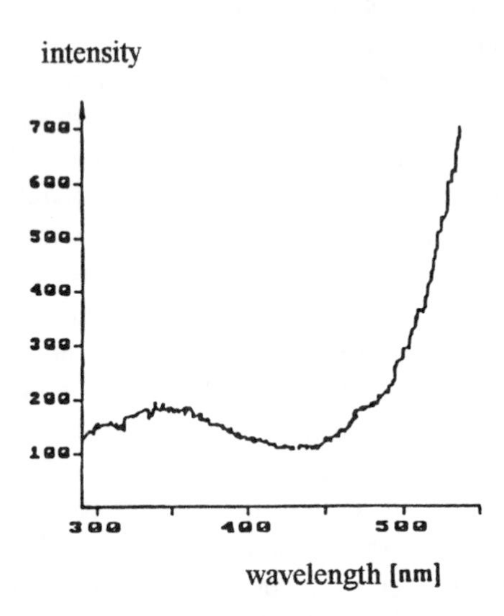

Figure 9: Broad-banded molecule emission of the diesel engine

The formation of NO can be related to high combustion temperatures. By applying exhaust gas recirculation (EGR) the temperature is reduced and subsequent the NO exhaust emission. Fig. 10 shows as an example the investigation of the diesel combustion with different EGR rates. The detected bandwith was 300 to 700 nm. The images demonstrate the spectral and time-resolved light emission of 3 stages of EGR rates: a) 0% (without EGR), b) 35% and c) 53%. Comparing the 3 images, the light emission decreases with the increasing rate of the EGR. Due to the increased exhaust concentration, the inert gas proportion of the combustion gas increases and the temperature of the combustion decreases. The light intensity and the spectral location of the emission are influenced by the temperature of the particles, the light intensity decreases constantly with the increasing EGR and the location of the emission shifts away from the UV region. The drop off of the light intensity with increasing EGR can be interpretated with the decreasing temperature.

In addition the start of the light emission is delayed, an increasing ignition delay with higher EGR rates can be observed. The main combustion starts later and the stage is shorter. But the afterglow does not decrease at higher EGR rates. Reason is that at higer EGR rates the main combustion is incomplete, a longer afterglow takes place.

Despite there is almost no identification of molecules possible in the diesel combustion, some temporal information about the combustion process can be obtained.

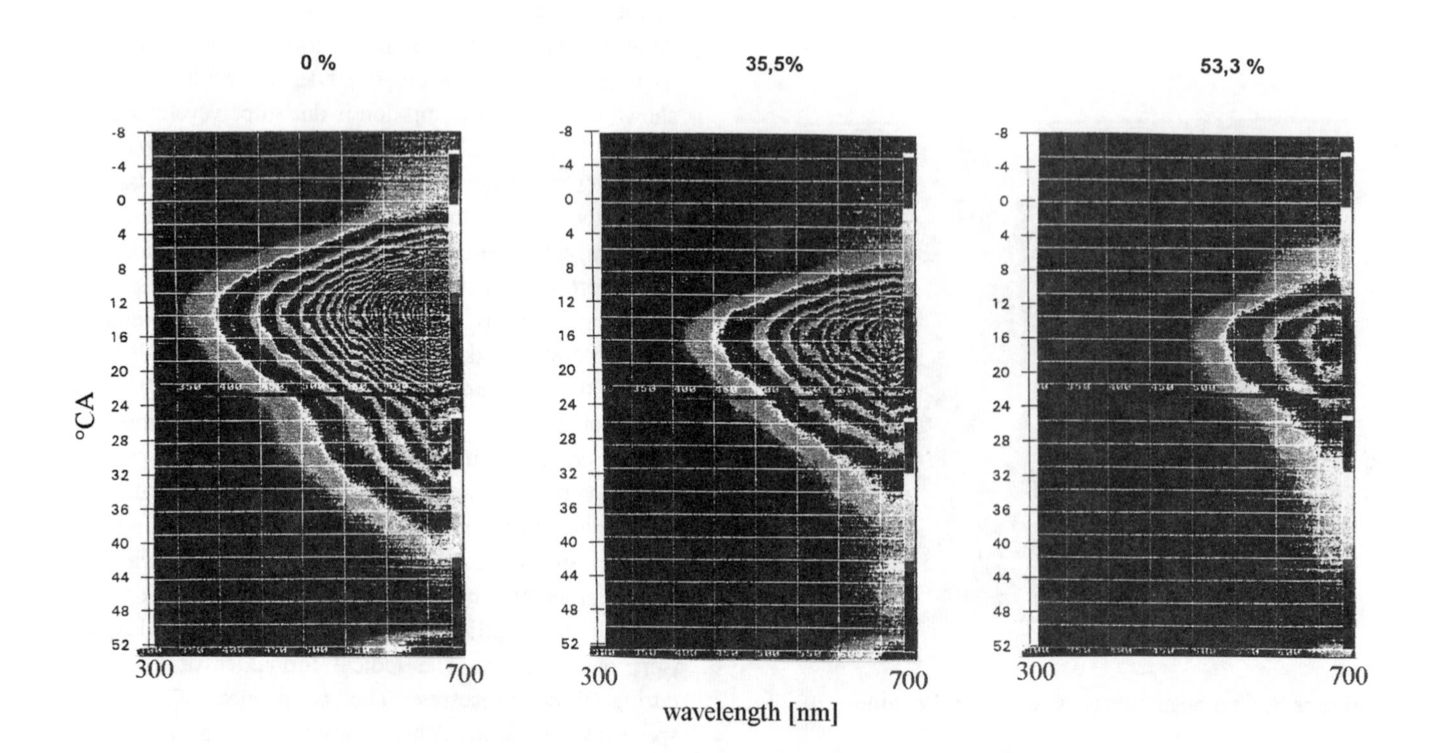

Figure 10: Diesel combustion process with different EGR rates

CONCLUSIONS

The use of CCD readout systems on streak cameras permits immediate data collection and storage. Thus it is a very useful tool for investigation of instantaneous processes. In combination with optical fibres it can also be applied to an automotive engine as an in-cylinder investigation instrument.

The first result of a SI engine presents highly time-resolved molecule emissions. The appearance of the molecule emissions indicates the different stages of the combustion process, e.g. time of ignition, the active combustion phase and the CO-oxidation phase. It was shown, with the help of spectral and time resolved emission spectroscopy the combustion process of SI engines can be investigated. Investigations can be done like e.g. cycle-to-cycle variations, different operation points or different fuels.

The combustion of a SI engine with different fuels, gasoline and a methanol fuel, was investigated. The appearance of distinct radicals depending on the fuel leads to different combustion characteristics, which can be observed with this investigation method.

In comparison to the line structured spectra out of the SI engine, the spectra out of the diesel engine show a continuous character. Despite there is no identification of the molecules possible, the different stages of the combustion process can be observed. The influence of the EGR on the combustion process was investigated.

We could prove the applicability of the high speed detecting system to automobile production engines. Further investigations of the combustion behaviour at varied engine parameters are intended to yield a detailed understanding of the in-cylinder combustion process at different circumstances.

ACKNOWLEDGEMENTS

We would like to thank Professor Knoche (RWTH Aachen) for the valuable contributions and his co-workers, especially M. Haug and J. Meyer, for the stimulating discussions.

REFERENCES

[1] Gaydon, A.G.: The Spectroscopy of Flames, Chapman and Hall, London, 1960

[2] Gaydon, A.G., Wolfhard, H.G.: Flames, Their Structure, Radiation and Temperature, Chapman and Hall, London, 1960

[3] Cignoli, F., Benecchi, S., Zizak, G.: Simultaneous one-dimensional visualization of OH, polycyclic aromatic hydrocarbons, and soot in a laminar diffusion flame, Optics Letters, Vol. 17, No. 4, p. 229-231, 1992

[4] Andresen, P., Meijer, G., Schlüter, H., Voges, H., Koch, A., Hentschel, W., Oppermann, W., Rothe, E.,: Fluorescence imaging inside an internal combustion Engine using tunable Excimer Lasers, Applied Optics, Vol. 29, No. 16, p. 2392-2404, 1990

[5] Koch, A., Voges, H., Andresen, P., Schlüter, H., Wolff, D., Hentschel, W., Oppermann, W., Rothe, E.: Planar Imaging of a Laboratory Flame and of Internal Combustion in an Automobile Engine using UV Rayleigh and Fluorescence Light, Applied Physics B, 56, p. 177-184, 1993

[6] Nagase, K., Funatsu, K.: Spectroscopic Analysis of Diesel Combustion Flame by Means of Streak Camera, SAE Technical Paper Series 881226

[7] Nagase, K., Funatsu, K.: A Study of NO_X Generation Mechanism in Diesel Exhaust Gas, SAE Technical Paper Series 901615

[8] Schindler, K.-P., Hentschel, W., Hötger, H., Leipertz, A., Münch, K.U., Voges, H.: Untersuchung der Verbrennung im Dieselmotor: Vergleich der spektralen Emissionslichtverlaufsanalyse mit 2-dimensionalen Lasertechniken, VDI-Berichte Nr. 922, S. 435-451, 1991

[9] Winkelhofer, E.: private communication, Wolfsburg, 1993

[10] Fa. LaVision, Manual, Streak Star Camera-System, Göttingen, 1992

[11] Pearse, R.W.B., Gaydon, A.G.: The Identifikation of Molecular Spectra, Chapman & Hall, London, 3. Auflage 1965

[12] Peters, N.: Laminar Flamelet Concepts in Turbulent Combustion, 21. Symposium (Int.) on Combustion, The Combustion Institute, p.1231-1250, 1986

[13] Wirth, M.: Die turbulente Flammenausbreitung im Ottomotor und ihre charakteristischen Längen, Dissertation RWTH Aachen, 1993

[14] Herden, W., Maly, R., Saggau, B., Wagner, E.: Neue Erkenntnisse über elektrische Zündfunken und ihre Eignung zu Entflammung brennbarer Gemsiche, 2. Teil, Automobilindustrie 2, S. 15-21, 1978

[15] Möser, P.: Zeitlich hochaufgelöste emissionsspektroskopische Untersuchung des Verbrennungsvorgangs im Otto-Motor, Verlag Augustinus Buchhandlung, Aachen, 1995

[16] Menrad, H., König, A.: Alkoholkraftstoffe, Springer-Verlag, Wien, 1982

[17] Warnatz, J., Bockhorn, H., Möser, A., Wenz, H.W.: Experimental Investigations an computational Simulation of Acetylene-Oxygen Flames from near Stoichiometric to Sooting Conditions, Nineteenth Symposium (International) on Combustion, The Combustion Institute, p. 197-209, 1982

[18] Warnatz, J.: Resolution of the Gas Phase and Surface Combustion Chemistry into Elementary Reactions, Twenty-Fourth Symposium (International) on Combustion, The Combustion Institute, p. 553-579, 1992

[19] Westbrook, C.K., Dryer, F.L.: Chemical Kinetic Modeling of Hydrocarbon Combustion, Prog. Energy Combust. Sci., 1984, Vol 10, p. 1-57

[20] Matsui, Y., Kamimoto, T., Matsuoky, S.: A Study on the Time and Space Resolved Measurement of Flame Temperature and Soot Concentration in a D.I. Diesel Engine by the Two-Color Method, SAE Technical Paper Series 790491

[21] Matsui, Y., Kamimoto, T., Matsuoky, S.: A Study on the Application of the Two-Color Method to the Measurement of Flame Temperature and Soot Concentration in Diesel Engines, SAE Technical Paper Series 800970

[22] Jander, H., Wagner, H.G.: Soot Formation in Combustion, An International Round Table Discussion, Nachrichten der Akademie der Wissenschaften in Göttingen, Vandenhoeck and Ruprecht, Göttingen, 1990

971122

Application of Neural Networks in the Estimation of Tire/Road Friction Using the Tire as Sensor

W. R. Pasterkamp and H. B. Pacejka
Delft University of Technology

Copyright 1997 Society of Automotive Engineers, Inc.

ABSTRACT

The importance of friction between tire and road for the dynamic behavior of road vehicles has been emphasized in many publications. Continuously updated knowledge of the friction potential and the friction demand can help to improve maneuverability and thereby safety of vehicles under slippery road conditions. An on line estimation method, based on combination of side force and self aligning torque, generated by the tire, is theoretically founded on a simple brush type tire model.

The system is implemented in the front wheel suspension of a passenger car. To cope with the highly non-linear behavior of the wheel suspension and the actual tire, various static neural networks have been applied in the estimation procedure. Experiments have been carried out both in simulation using a full vehicle multi-body model and with an actual vehicle. Conclusions are drawn regarding the estimation principle, the application of neural networks and the implementation in a test vehicle.

INTRODUCTION

Tires form the link between the vehicle and the road; their ability to transmit forces between tire and road determines the limits of the vehicle's dynamic operating range. Since virtually all tire forces in the road plane are frictional forces, the friction coefficient is a key parameter in this process. The friction coefficient is not measurable directly, However, introduction of advanced vehicle motion control systems such as Vehicle Dynamic Control and Intelligent Cruise Control has increased the need for knowledge about the current state of friction potential and friction demand.

For these reasons, many research efforts have been made to estimate the friction coefficient between tire and road, either based on effects of the frictional process itself (e.g. [1, 3, 15]), or based on parameters affecting the frictional process (e.g. [3, 4, 8]). This article discusses the continuing development of a system to estimate the tire/road friction coefficient using the tire as a sensor and artificial neural networks for identification [10, 11].

First, the principle of the proposed method of friction estimation is based on the simple brush tire model. Then, the application of neural networks in the actual estimation procedure is regarded, followed by a discussion of an experimental implementation in a vehicle and a presentation of some results.

MODEL BASED FRICTION ESTIMATION

The proposed friction estimation method can be based upon the simple steady-state brush tire model. Since this model is well known in literature (e.g. [6, 7, 11]), only the case of pure side slip will be recapitulated. Then the friction estimation principle will be explained.

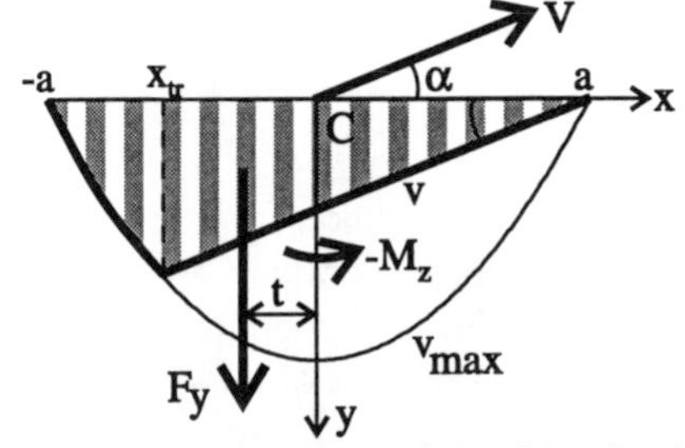

Fig. 1. Contact line of steady state side slipping brush type tyre model

BRUSH TIRE MODEL - The brush tire model assumes elastic bristles or tread elements connected to the rim at one side and touching the road surface at the other end, thus assuming infinite carcass stiffness.

The effects of tire width are neglected; the contact patch is modelled as a contact line (Fig. 1).

If the direction of the wheel speed vector V encloses a slip angle α with respect to the wheel rim, the tread elements deflect as they travel through the contact area (from right to left in the figure).

Adhesion only - As long as the friction force between a tread element and the road surface is large enough to deflect the tread element, it is in the adhesion zone: the tread element remains in contact with the road surface and consequently the contact line is parallel to the wheel speed vector. The tread element is assumed to generate side force proportionally to its deflection (Eq. 1 and 2).

$$\left.\begin{aligned} v &= (a-x)\tan\alpha \\ q_{y_{ad}} &= c_p \cdot v \end{aligned}\right\} \text{for } c_p \cdot v < \mu q_z \qquad (1, 2)$$

Herein a is half the contact length, q_y the side force distribution, μ the friction coefficient, q_z the load distribution and c_p the lateral tread element stiffness.

For $\alpha \to 0$ and/or $\mu \to \infty$, these expressions hold for the entire contact region; there is complete adhesion. The asymmetry of the side force distribution with respect to the y-axis causes the side force F_y to act at a point shifted over a distance t, the so-called pneumatic trail, from the center C of the contact line. The side force acting on an arm t creates the self aligning torque M_z. For complete adhesion, side force and self aligning torque do not depend on friction μ, while both are linearly related to the side slip angle α and the pneumatic trail has a constant value (Eq. 3-5).

$$F_y = \int_{-a}^{a} q_y dx = 2c_p a^2 \alpha \qquad (3)$$

$$-M_z = -\int_{-a}^{a} q_y x dx = \frac{2}{3} c_p a^3 \alpha \qquad (4)$$

$$t = -\frac{M_z}{F_y} = \frac{1}{3} a \qquad (5)$$

Adhesion and Sliding - For a finite friction value μ and a load distribution q_z gradually dropping to zero at both edges of the contact area, both adhesion and sliding may exist in the contact area. The tread element is under the regime of adhesion (Eq. 1 and 2) up to the transition point x_{tr}, after which the friction force is not able to deflect the tread element any further. From that point up to the trailing edge of the contact area, the tread element is sliding over the road surface. In this sliding zone, the generation of side force by the tread elements is governed by the friction coefficient μ and the load distribution:

$$q_{y_{sl}} = \mu q_z \quad \text{for } c_p \cdot v \geq \mu q_z \qquad (6)$$

For simplicity, a parabolic load distribution over the contact line is assumed:

$$q_z = \frac{3F_z}{4a}\left\{1 - \left(\frac{x}{a}\right)^2\right\} \qquad (7)$$

The transition point x_{tr} is obtained from Eq. 2 and 6:

$$q_{y_{ad}} = q_{y_{sl}} \qquad (8)$$

Introducing the parameter φ

$$\varphi = \frac{2c_p a^2 |\tan\alpha|}{3F_z} \qquad (9)$$

the transistion point is given by

$$x_{tr} = a\left(\frac{2\varphi}{\mu} - 1\right) \qquad (10)$$

By integrating the contributions of adhesion over $(x_{tr}, a]$ and sliding over $[-a, x_{tr}]$, expressions for side force, self aligning torque and the pneumatic trail as functions of friction and side slip for the partially sliding tire are obtained:

$$F_y = \frac{1}{\mu^2} F_z \varphi\left(3\mu^2 - 3\mu\varphi + \varphi^2\right)\operatorname{sgn}(\alpha) \qquad (11)$$

$$-M_z = \frac{1}{\mu^3} F_z a \varphi (\mu - \varphi)^3 \operatorname{sgn}(\alpha) \qquad (12)$$

$$t = -\frac{M_z}{F_y} = \frac{a(\mu-\varphi)^3}{\mu\left(3\mu^2 - 3\mu\varphi + \varphi^2\right)} \qquad (13)$$

For a partially sliding tire, side force and self aligning torque are linearly independent functions of side slip angle α (through φ) and friction coefficient μ.

Sliding only - Total sliding of the tire starts if x_{tr} equals a, which results in the condition

$$\varphi = \mu \qquad (14)$$

In this case, the side force distribution is symmetrical with respect to the y-axis. Thus, t and consequently M_z equal zero and the side force is proportional to the friction coefficient while only its direction is determined by the slip angle:

$$F_y = \mu F_z \operatorname{sgn}(\alpha) \qquad (15)$$

$$M_z = 0 \qquad (16)$$

$$t = -\frac{M_z}{F_y} = 0 \qquad (17)$$

BASICS OF THE FRICTION ESTIMATION METHOD - The previous section shows that for the brush tire model, given the tire parameters (a and c_p) and the load, the friction μ can be determined through combination of side force and self aligning torque by (numerically or graphically) solving Equations (11) and (12) in the case of partial sliding, or just from the side force by solving Equation (15) in the case of total sliding. Only in the case of complete adhesion, μ cannot be determined. At the other hand, the side slip angle α can be determined either from side force or aligning torque in the case of complete adhesion using Equations (3) or (4), from side force and self aligning torque again by solving Equations (11) and (12) in the case of partial sliding, and not at all in the case of total sliding.

This is illustrated by Figure 2, a so-called Gough-plot, depicting F_y/F_z versus $-M_z/F_z$ for various slip angles and friction levels using the brush tire model at a given load. Lines of constant friction (solid), constant side slip angle (dotted) and constant pneumatic trail (dash-dotted) are shown in this diagram.

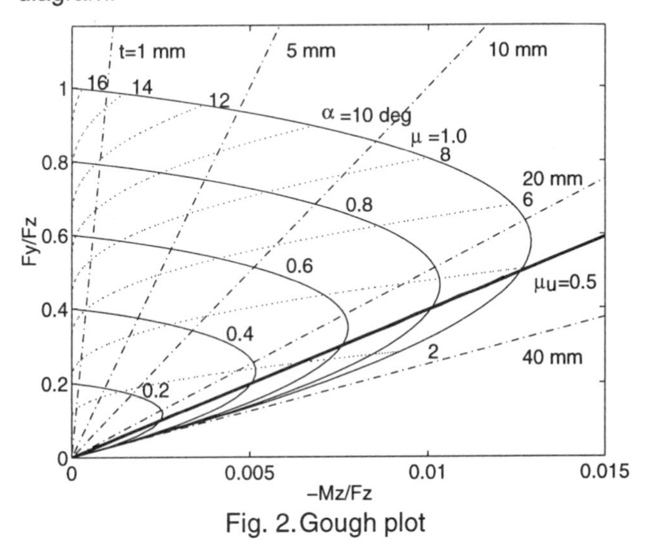

Fig. 2. Gough plot

Over a wide range of variables, the values of α and μ can be derived from F_y and M_z: a pair of values (F_y, M_z) results in a unique pair of values (α,μ), corresponding to a value of t. This corresponds to the case of partial sliding. Complete adhesion is found where the lines of constant friction (almost) coincide. In theory complete adhesion only exists at vanishing slip angle or infinite friction, however, from the figure it becomes clear that even at finite friction levels the distinction between various friction levels becomes harder where their respective curves almost coincide. Along the y-axis, the lines of constant side slip angle coincide, this is the case of total sliding. In this case, the various side slip angles cannot be distinguished, but the friction coefficient can be distinguished.

To put the above in a more realistic perspective, Figure 2 also shows a (fat) line of constant utilized friction potential μ_u, given by

$$\mu_u = \frac{F_y}{\mu F_z} = 0.5 \tag{18}$$

In every day driving behaviour, especially at higher friction levels, this value of 50% utilization of friction potential is hardly exceeded. In that case, only the small part of the graph below/right of this line remains available for friction estimation.

NEURAL FRICTION ESTIMATION

So far, a theoretcial situation has been discussed, where the tire behaves according to a highly simplified tire model and all required signals are available and undisturbed. This section first discusses the actual tire behavior, then explains the use of neural networks for identification. Thereafter, the neural networks are applied to the friction estimation problem and the sensitivity of the friction estimate for signal disturbances is studied.

COMPLEX TIRE BEHAVIOR - The actual tire behavior is considerably different from the results obtained by the brush type tire model. The main reasons for this difference are the non-constant friction coefficient, the lateral compliance of carcass and belt, and the varying pressure distribution. In addition, there is variations in load, speed and camber angle, combined longitudinal and lateral slip, residual forces and torques, and dynamic tire behavior. Their various influences have been discussed earlier [9]; we only mention here that the dynamic behavior of the tire, as induced by the relaxation length, is of minor importance to this research. However, at relatively fast steering manoeuvres, it may become more important.

The complexity of the actual tire behavior disqualifies the simple brush tire model to be used in actual friction estimation. In [9], a table lookup method is regarded based on the (empirical) Magic Formula (MF-) tire model. This method proved to be valid, albeit laborious. This article discusses Artificial Neural Networks as identification tools.

NEURAL NETWORKS - Due to their learning and non-linear input-output mapping capability, Artificial Neural Networks (NN's for short) are becoming increasingly popular in automotive research (e.g. [8, 9]). In the vast variety of neural networks (see e.g. [5, 13] for a survey), we focus on so-called mapping NN's. This mapping is called static if there is no time-dependency involved, otherwise it is called dynamic. Since the dynamics of the tire are of less importance to this research, we narrow our focus to static networks. The most commonly applied networks of this type are the Multi Layer Perceptrons (MLP's) and the Radial Basis Function (RBF) networks. They share a common architecture of an input layer, a number of hidden layers (usually just one is sufficient) and an output layer (Figure 3).

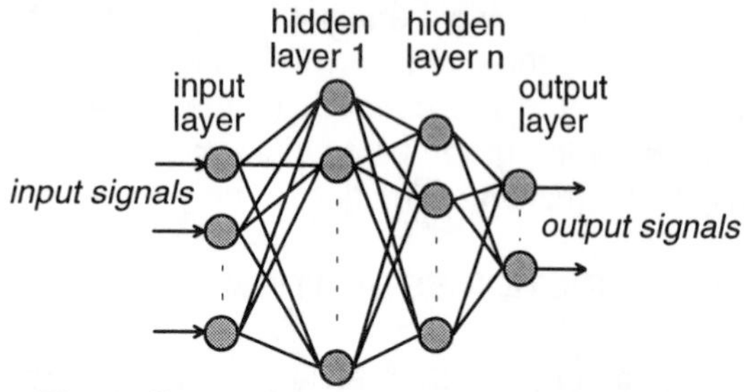

Fig. 3. General structure of a static network

To have a nonlinear mapping, at least some of the hidden or output nodes need to contain a nonlinear activation function. For mapping NN's, the output layer is usually chosen to have linear nodes. In the case of MLP's, the hidden layer nodes usually inhibit sigmoid-alike functions, of the kind

$$f(x) = \frac{1}{1+e^{-x}} \qquad (19)$$

which squashes the output of the node between some boundaries, in this case (0,1]. In the case of RBF networks, the hidden nodes contain radial basis functions, usually of Gaussian type:

$$f(x) = e^{\frac{-(x-c)^2}{r^2}} \qquad (20)$$

The main difference between MLP's and RBF networks is due to the nature of their respective activation functions [5]. The argument of the activation function of each hidden unit in an RBF network computes the distance between the input vector x and the center c of that unit. Thus, RBF networks construct local approximations to nonlinear input-output mappings. This results in fast learning networks if the number of input nodes and/or the number of training samples is small. However, to span a large input space, usually a large number of hidden units is needed. The argument of the activation function of each hidden unit in an MLP, is the inner product of the input vector and the weight vector of that unit. The result is that MLP's construct global approximations to nonlinear input-output mappings. This implies that generalization in regions of the input space where little data is available (extrapolation) is possible to some extent. In the next section, both MLP's and RBF networks are applied and compared.

NEURAL NETWORKS FOR FRICTION ESTIMATION - A simulation experiment shows the strength of NN's for friction estimation. A training and a test data set are generated by calculating F_x, F_y and M_z for different values of α, μ and F_z using the MF-tire model. The training and test sets contain 343 and 64 data points respectively. The results on the test set after training are shown in Figure 4 for a one-hidden-layer MLP network containing 20 hidden sigmoid nodes and for a RBF network containing 80 hidden Gaussian nodes.

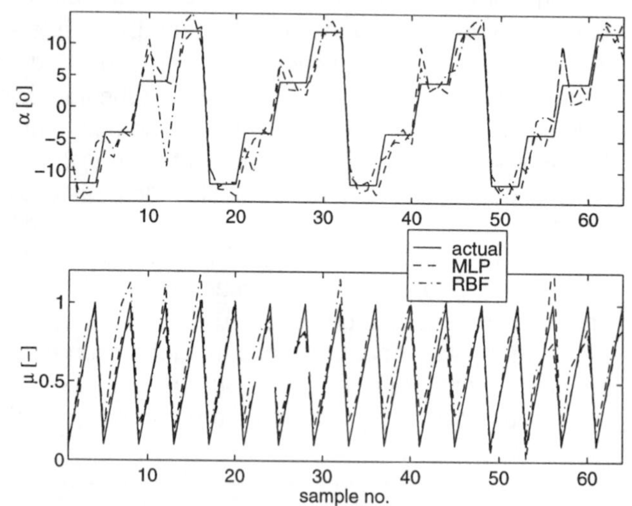

Fig. 4. Friction estimation using MLP and RBF networks

Clearly, both MLP and RBF networks are able to make the required mapping from tire forces and torque to slip angle and friction coefficient. In agreement with the model based friction estimation, the friction estimate becomes less accurate at vanishing slip angle, while the slip angle estimate degrades at low friction. Actually, the mapping can be regarded as an inversion of the tire model; this inversion becomes ill conditioned for these cases. Finally, it appears that although the MLP contains less hidden nodes, it makes a slightly more accurate mapping than the RBF in this case.

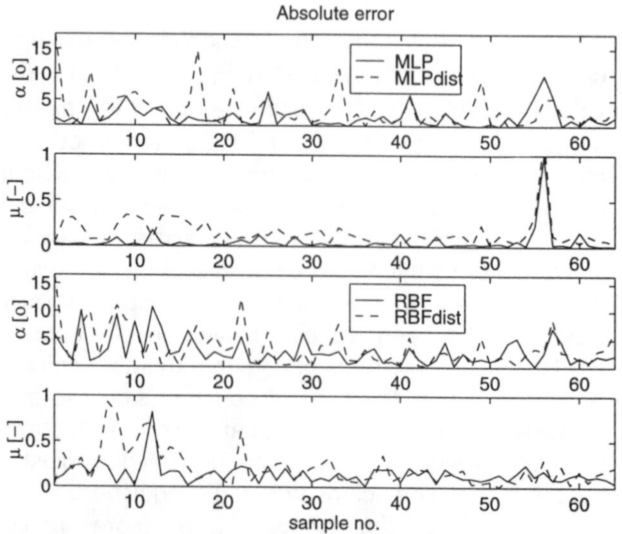

Fig. 5. Robustness with respect to bias on F_y

Next, the robustness of the estimation is investigated by disturbing the input signals in the test set. To avoid a lengthy discussion, only the effect of a bias of 500 N on the side force is shown in Figure 5 for both the MLP and the RBF network. Other robustness tests may be performed in similar manner. Figure 5 shows the absolute error of the side slip and friction estimates without and with bias on F_y for both the MLP and the RBF network. Clearly, the bias causes an increased error, that manifests itself primarily at small side forces, which is in accordance to the brush tire model theory. Both kinds of networks appear to be fairly robust to this kind of disturbance; the MLP appears to perform just slightly better.

The better accuracy, slightly better robustness and smaller size make MLP's preferable over RBF networks for this mapping problem. In the sequel, we will therefore only consider MLP's.

IMPLEMENTATION IN A TEST VEHICLE

EXPERIMENTAL SETUP - The estimation method proposed so far assumes the availability of the signals F_X, F_y, F_Z and M_Z of the wheel. Unfortunately, these signals cannot be measured directly on a regular vehicle, unless equipped with expensive rotating wheel dynamometers. Instead, we choose to measure the required signals indirectly by a set of relatively inexpensive sensors in the front wheel suspension.

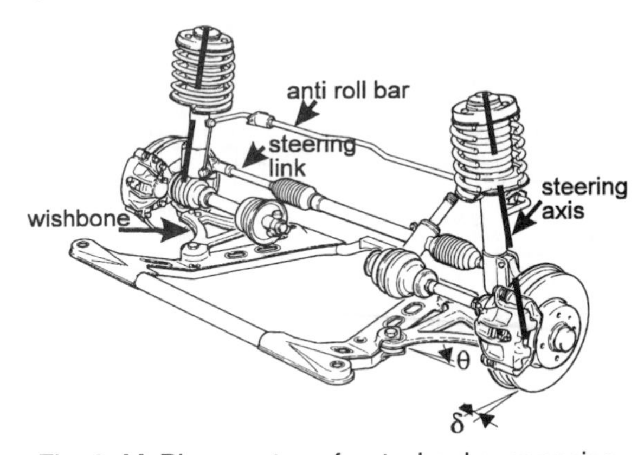

Fig. 6. McPherson type front wheel suspension

The test vehicle is a mid-size front wheel driven car with a McPherson type front wheel suspension, depicted by Figure 6. It is instrumented to measure:

- The force F_S in the steering link by a load cell in the link.
- The angle θ of the wishbone relative to the car body, using a potentiometer, which gives a measure for the suspension strut force.
- The (horizontal) forces acting on the king pin by four very small strain gauges, forming two half Wheatstone-bridges.
- The steering wheel angle using a potentiometer, related to the steering angle δ at the wheel by the steering ratio.

The measured signals are processed by a PC on board the test vehicle.

IDENTIFYING F_X, F_y, F_Z, AND M_Z - The complex kinematics of the suspension mechanism complicate in particular the determination of the torque M_z from the measured signals. This torque, generated by the tire, causes a torque M_a around the (virtual) steering axis (Fig. 6), and will thus result in a force F_s in the steering link that is measured by the load cell in the link. However, the longitudinal, lateral and vertical tire forces also contribute to M_a, due to caster trail, king pin inclination, tire width, etc. [12]. Moreover, a torque in the anti roll bar contributes to M_a through the small linkages connecting it to the suspension struts. Finally, all contributions to M_a and F_s depend on the steering angle δ and inclination θ of this mechanism.

Some of these contributions are illustrated by Figure 7, assuming stiff links and joints and a tire force acting point in the center of the tire/road contact patch. In this figure, the vertical force is assumed to be a result of the compression of the suspension spring. The longitudinal and lateral forces are assumed to be equal and the modulus of their vectorial sum is assumed to be equal to the vertical force (μ = 1). (Note that the maximas in Figure 7d are not commonly encountered, they represent situations of counter steering.)

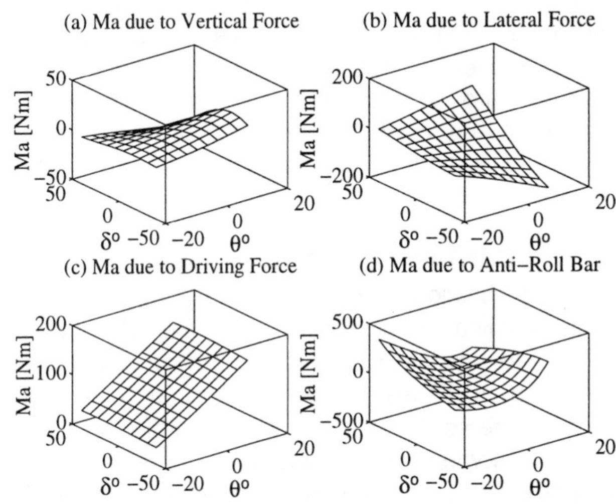

Fig. 7. Contributions to torque around the steering axis

Considering the computational burden of an (on line) model that accurately describes the suspension kinematics, including flexible joints connecting the suspension parts to the vehicle body, and considering the inaccurate knowledge of the acting point of the resultant tire force in lateral direction, which has a major influence on the contribution of longitudinal and vertical forces on M_a, it has been chosen not to use such a model, but instead to describe the suspension's behaviour by a non linear function, formed by a neural network.

A full vehicle multi body model, initiated by [14], with accurate suspension and tire description has been used to generate reliable data of forces and torques in the vehicle suspension in relation to the forces and torques generated by the tire. These data have been used to train a 3-layer (one hidden layer) MLP that outputs F_x, F_y, F_z, and M_z, when fed by the steering angle δ, the suspension inclination angle θ, the forces on the king pin in longitudinal and lateral directions and the force in the steering link.

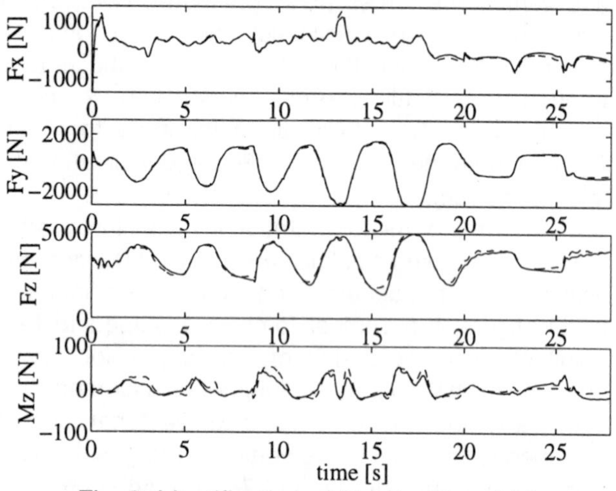

Fig. 8. Identification of F_x, F_y, F_z and M_z

A training and a test set have been filled with simulation results of slalom manoeuvres on 4 different friction levels and 3 different forward speeds using the vehicle model. The results on the test set (not used for training) are shown in Figure 8. It shows that identification of F_x, F_y, F_z, and M_z from the measured entities using an MLP is indeed possible with sufficient accuracy.

IDENTIFYING α AND μ - Having shown that both identification of α and μ from F_x, F_y, F_z, and M_z and identification of F_x, F_y, F_z, and M_z from the measured entities are possible, we could put these networks in series to identify α and μ from the measurements. However, it is more efficient, both in terms of computational burden and in terms of error propagation, to integrate these networks into one MLP that identifies α and μ directly from the measurements.

To show this approach, a number of experiments have been conducted with the test vehicle on a test track at three different friction levels. All outputs of the sensors as described in a previous section are recorded. The speed and steering data from these experiments have been used to 'replay' the manoeuvres using the full vehicle simulation model. The latter has the advantage of a noise free environment and, most importantly, a precisely defined actual road friction. Since we do not have such accurate information on the actual friction of the test track, we have to settle with plausible assumptions of the actual friction for the experiments with the actual vehicle.

The data sets from simulation and actual experiments have been split in training and test sets, and have been used to train a MLP to identify α and μ. Figure 9 shows the results of these experiments using the simulation results, both in training (upper two graphs) and in testing (lower two graphs). This figure shows that the neural network is capable of estimating α and μ. The spikes in the friction estimate correspond to cases of (almost) zero slip angle. The estimation is best when the magnitude of the quotient α over μ is large, which is in accordance to the brush tire theory. This quotient may serve as a weighting factor to form a weighted mean over a number of previous samples, thus trying to eliminate poor estimates, however, at the expense of introducing phase lag. This has been done in the last graph of the figure, using 15 previous samples.

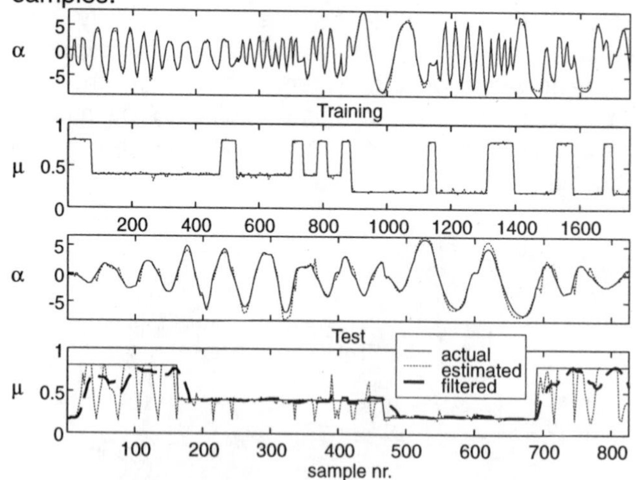

Fig. 9. Estimation α and μ from BAMMS simulations

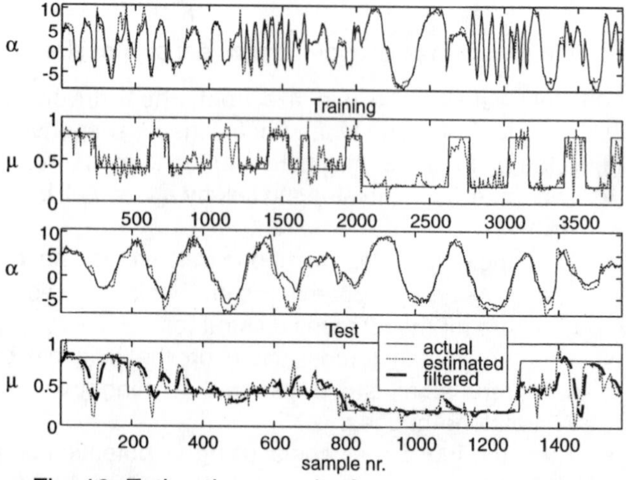

Fig. 10. Estimation α and μ from test vehicle data

Figure 10 shows the corresponding results on the data from the actual vehicle. Even though these results are not as well as the simulation results, for the reasons mentioned above, it is clear that the estimation method is working in actual conditions.

Finally, we remark that since the estimation only involves a static MLP, it is suitable for on line implementation in a vehicle.

CONCLUSION

This article presents a method to estimate friction between tire and road based on measurements of side force, self aligning torque and vertical load of the tire. Theory built on the simple brush tire model shows that an estimate of the friction can be made as soon as the tire is partially sliding. For actual implementation, it has been shown that Artificial Neural Networks can perform this estimation adequately.

The behaviors of two kinds of NN's, namely MLP's and RBF networks, have been compared. Although both kinds can make the required mappings, the MLP's are favored because of their smaller size and slightly better robustness properties.

The set up of a low cost measurement system in the front suspension of a passenger car requires computation of the complex suspension kinematics. Instead of having a complex on-line model of the suspension kinematics, MLP's are trained to describe the suspension's behavior.

Simulation experiments and experiments with a test vehicle have shown the possibility to estimate side slip angle and friction coefficient directly from the measured entities using MLP's, thus integrating the above mentioned tasks. Extension of the estimation method to include dynamic behavior of tire and suspension is being considered for further research.

REFERENCES

[1] Dieckmann Th. (1992). Der Reifenschlupf als Indikator für das Kraftshlußpotential. Dissertation IKH Universität Hannover.

[2] Demuth H. and Beal M. (1994), *Neural Network Toolbox User's Guide.* The Mathworks, Inc., U.S.A.

[3] Eichhorn U. and Roth J. (1992). *Prediction and monitoring of tyre/road friction.* XXIV.FISITA-Congress, London, Great Britain.

[4] Görich H.J., Jacobi S. and Reuter U (1993). *Ermittlung des aktuellen Kraftschlußpotentials eines Pkws im Fahrbetrieb.* VDI Berichte Nr. 1088, 1993

[5] Haykin S. *Neural Networks, A Comprehensive Foundation.* Macmillan College Publishing Company, Inc., 1994

[6] Pacejka H.B. and Sharp R.S. (1991). *Shear force development by pneumatic tyres in steady state conditions: A review of modelling aspects.* Vehicle System Dynamics Vol. 20, pp. 121-176

[7] Pacejka H.B. (1996). *The Tyre as a Vehicle Component.* XXVI FISITA Congress, Prague, 1996

[8] Pal C., Hagiwara I., Morishita S. and Inoue H. (1994). *Application of Neural Networks in Real Time Identification of Dynamic Structural Response and Prediction of Road-Friction Coefficient μ from Steady State Automobile Response.* Proceedings of AVEC'94 Symposium, Tsukuba, Japan

[9] Palkovic L., El-Gindy M. and Pacejka H.B. (1994). *Modelling of the cornering characteristics of tyres on an uneven road surface: a dynamic version of the 'Neuro-Tyre',* Int. Journal of Vehicle Design, Vol. 15, Nos. 1/2, pp. 189-215

[10] Pasterkamp W.R. and Pacejka H.B. (1994). *On line estimation of tire characteristics for vehicle control.* Proceedings of AVEC'94 Symposium, Tsukuba, Japan

[11] Pasterkamp W.R. and Pacejka H.B. (1996). *The Tyre As A Sensor To Estimate Friction.* Proceedings of AVEC'96 Symposium, Aachen, Germany.

[12] Reimpel J. (1984). *Fahrwerktechnik: Lenkung.* Vogel-Buchverlag, Würzburg, Germany, ISBN 3-8023-0739-2

[13] Pham D.T. and Xing L. (1995). *Neural networks for identification, prediction, and control.* Springer-Verlag London, ISBN 3-540-19959-4.

[14] Venhovens P.J.Th. (1993). *Optimal Control Of Vehicle Suspensions.* Dissertation, Delft University of Technology, The Netherlands

[15] Witte B., Zuurbier J. (1995). *Detection of the Friction Coefficient in a running Vehicle and Measurement of Tire Parameters on different Road Surfaces.* VDI Berichte Nr. 1224, pp. 61-79

962112

An On-Board Method of Measuring Motor Oil Degradation

Paul J. Voelker and Joe D. Hedges
Voelker Sensors, Inc.

Copyright 1996 Society of Automotive Engineers, Inc.

ABSTRACT

The performance of an on-board oil quality sensor is evaluated and shown to reliably measure motor oil degradation in real-time over the temperature range found in a vehicle. This sensor operates by utilizing a conductive polymeric matrix to track the change in the solvent properties of an oil. Tracking oil degradation by a change in its solvent properties provides a means of directly measuring 1) oil oxidation and 2) water and antifreeze contamination. The sensor shows a relatively constant signal between fresh oils but offers a clear separation between degraded oils. Results are shown to compare favorably with complementary methodologies including infrared spectroscopy, differential scanning calorimetry, thermogravimetic analysis, dielectric analysis, and metal analysis.

INTRODUCTION

A means of reliably measuring motor oil degradation and determining the end of an oil's useful life provides economic and environmental benefits to both automotive manufacturers and consumers. These benefits include: helping ensure minimum exhaust emission requirements, reducing costs related to engine warranty, and optimizing oil change intervals. An on-board sensor would provide an additional advantage to oil quality measurement by enabling oil conditioning to be monitored in real-time.

Generally oil deterioration results when one or more of the following conditions occur:

Contamination
- water / fuel / coolant

Change in the Physical / Chemical Properties
- oxidation /sludge
- viscosity
- acids

Wear Debris
- metals
- soot

Several instrument methods that are currently employed to evaluate oil quality include: metal analysis, viscometry measurements, acid & base number determinations, infrared spectroscopy and differential scanning calorimetry (these last used to estimate oxidation). Unfortunately, the price of these instruments prohibit their everyday use and their size and design constraints make them impractical to include in a vehicle. Consequently, these techniques remain limited to bench or laboratory environments.

A number of on-board oil quality sensors have been proposed over the years. [1-4] But because of production costs or reliability issues they have remained of theoretical interest only. Some of the requirements for a successful on-board oil quality sensor include:[5,6]

- reliably monitor the quality of an oil
- operate over the temperature range of a vehicle
- detect water, fuel, sludge and varnish
- survive severe vibration conditions
- remain operational over the life of the vehicle
- remain functional with different oil grades and additives
- inexpensive to manufacture

A unique approach to on-board oil quality measurement has been developed that appears to satisfy these requirements.[7] The methodology utilizes a sensor to monitor oil quality by directly measuring 1) oil oxidation and 2) water and coolant contamination. Tracking both oxidation and aqueous contamination provides an effective means of monitoring oil quality since oxidation is considered the leading indicator of oil quality and aqueous contamination is correlated with a decline in lubricity. [6,8,9] Oxidation has also been correlated with the formation of acids and hydroperoxides which can ultimately lead to bearing corrosion.[10]

DESCRIPTION

The oil quality device consists of a sensor; a mechanical interface and; a signal conditioning unit (Figure 1).

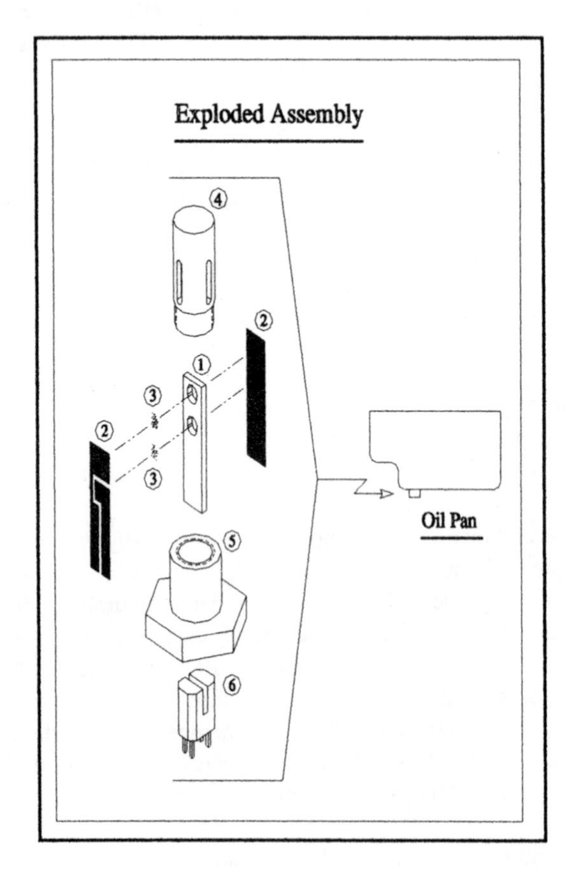

Figure 1.

The sensor, is composed of (1) a nonconductive plastic strip having two holes covered by (2) stainless steel screens that each contain (3) an insoluble polymeric matrix. While one hole allows the degrading oil to flow freely through the matrix, the other hole is permanently sealed with clean oil in order to serve as an internal standard or constant reference. A cover (4) protects the sensor during installation.

The mechanical interface (5) secures the sensor in the motor oil. A drain plug currently serves as the interface. The sensor's electrodes (6) protrude through the oil drain plug and connect to the signal conditioning unit.

The conditioning unit (not pictured) consists of a simple circuit board that includes a microprocessor that performs the signal conditioning. The assessment of the oils current quality is transmitted to either an LCD or LED which can display oil quality, or a simple alarm indicating that it is time to change the oil.

The sensor operates by measuring the electrical properties of a polymeric matrix held between the two electrode surfaces i.e., the screens. (Figure 2.) The polymeric matrix consists of resin beads holding charged groups which serve as a a conducting medium. A change in the electrical properties results when the interactive behavior between the charged groups adjust to the changing solvent properties of an oil. In a nonpolar environment, such as clean oil, the charged groups form a series of tightly bridged clusters that facilitates electrical transfer. This condition can be measured as resistance. As the polarity of the oil increases, indicating oxidation/degradation, the charged groups no longer form as tight a series of bridged clusters, which results in less efficient electrical transfer and a relative increase in resistance. As a result, a change in the electrical properties of the charged beads can be correlated with a change in the solvent properties of an oil.

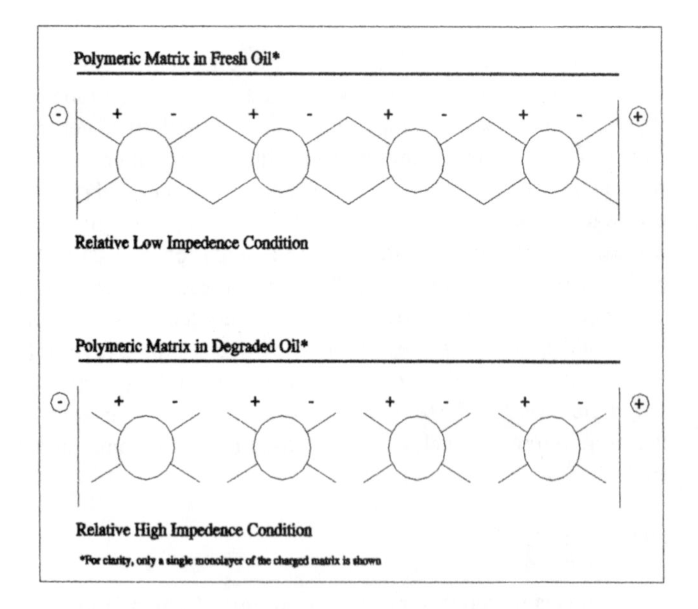

Figure 2.

TESTING METHODOLOGY

Testing was designed to establish the reliability and sensitivity of the method with respect to changes in oxidation and water contamination. The device was tested on the bench using a combination of fresh / dirty oils and oils treated with water. On-board testing was conducted in select vehicles driven over the course of an oil change.

Bench tests were set-up to simulate conditions found in an engine using different brands and grades of oil under normal operating temperatures. Flow conditions were simulated using a beaker equipped with a magnetic stir bar. The tests included temperatures encountered in stop-and-go city driving (20 - 80°C) as well as sustained highway driving (80 - 100°C). The reliability of the method was evaluated by measuring signal repeatability at elevated temperatures for extended periods, through cycling between extremes in temperature, and by repeated cycling between a fresh and degraded oil. The sensitivity of the method was tested using samples of closely degraded oils. It was also of particular interest to confirm the insensitivity of the method to different additive packages. Different additive packages have been responsible for certain brand fresh oils to behave as dirty oils in other oil quality sensors.

The sensitivity of the method to aqueous contamination was evaluated by adding known quantities of water to the oil and measuring a change in the signal.

On-board tests were conducted under a mixture of city and highway driving conditions. Vehicles included older models requiring oil every 2,000 miles as well as newer models that did not require any oil.

Materials

* The fresh oils used for testing included:
  - Oil A 10W-40
  - Oil B 20W-50
  - Oil C synthetic 10W-30
  - Oil D 5W-30
  - Oil E 30W
  - Oil F 5W-30
  - Oil G diesel 15W-40
  - Oil H base stock
* Dirty oil used for testing was collected from the crankcase of the following vehicles:
  - -Oil A from a 1991 V-6 minivan after two separate oil changes: 3,200 & 11,200 kilometers of use.
  - -Oil A from a 1989 4 cylinder turbo after two separate oil changes: 9,600 & 11,200 kilometers of use.
  - -Oil E from a 1968 V-8 following 14,500 kilometers of use. The oil in this engine had to be topped off every 3,200 kilometers. This oil was diluted 1:1 with Oil E in subsequent tests.
* Aqueous tests used deionized water with Oil C.
* The sensor consisted of:
  - approximately 8 mg. of ionic polystyrene resin beads
  - a single strip of 3 cm. x 1.5 cm.of polysulfone as the nonconductive housing
  - two strips of 500 mesh stainless steel cloth as the electrodes
* Output from the sensor was measured in megaohms using a Fluke 87 multimeter in both DC and AC modes.
* Flow conditions were simulated using a beaker equipped with a magnetic stir bar running at a medium rate.
* Temperature conditions were attained using a stirrer hot plate.
* Temperature was measured using either a 150 °C thermometer or a YSI thermistor (the sensor operates over the temperature range of -40 to 150 °C, however for clarity only readings at 80°C are presented).

Methodology

Sensor measurements were made in approximately the same manner for both bench and vehicle tests. To ensure complete oil saturation of the resin beads, the sensor was allowed to set overnight in the oil before taking readings. Readings were taken over a temperature range of 20- 100°C (in increments of 5 degrees). For bench tests approximately 10 mL of oil were used in the tests. Vehicle tests measured the entire volume in the oil pan.

Analysis

Data from the sensor are represented graphically and are reported as megaohms. Each data point is based on an average of three to four successive readings at temperature. The error associated with each value is expressed as two standard deviations about the mean and are depicted as error bars.

DISCUSSION

As a way of validating the overall ability of the sensor to track oxidation, conditions were created to promote oxidation and test the device. Four identical devices were each treated on the bench with four different fresh oils and run at 80°C for 12 hours (Figure 3).

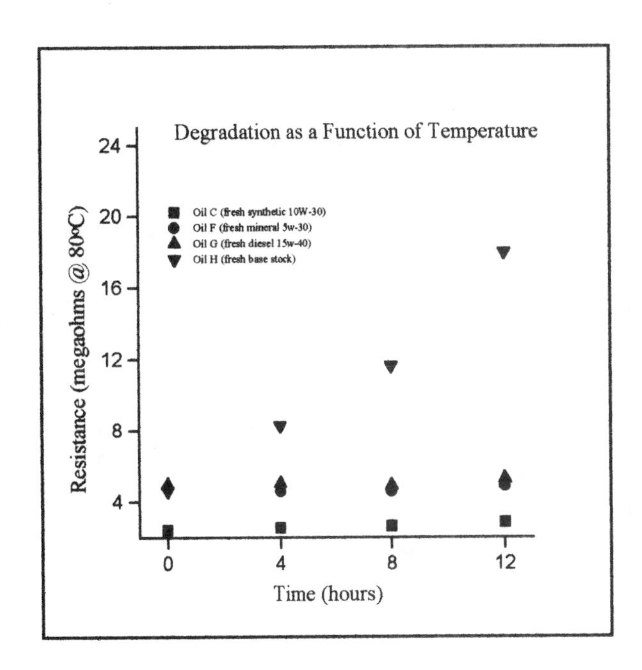

Figure 3.

The four oils selected were; three commercial grade oils (Oil C, Oil F, Oil G) and a base stock oil (Oil H). Under these conditions, the sensors measuring the three commercial oils showed approximately the same signal for the entire 12 hours. This result was consistent with expectations since a motor oil typically operates at 80°C without any appreciable degradation. The sensor measuring the base stock however, began to show a steady increase in resistance (indicating oxidation) in as little as four hours. After twelve hours the oil had developed an odor and was slightly discolored, suggesting an onset in oxidation.[8]

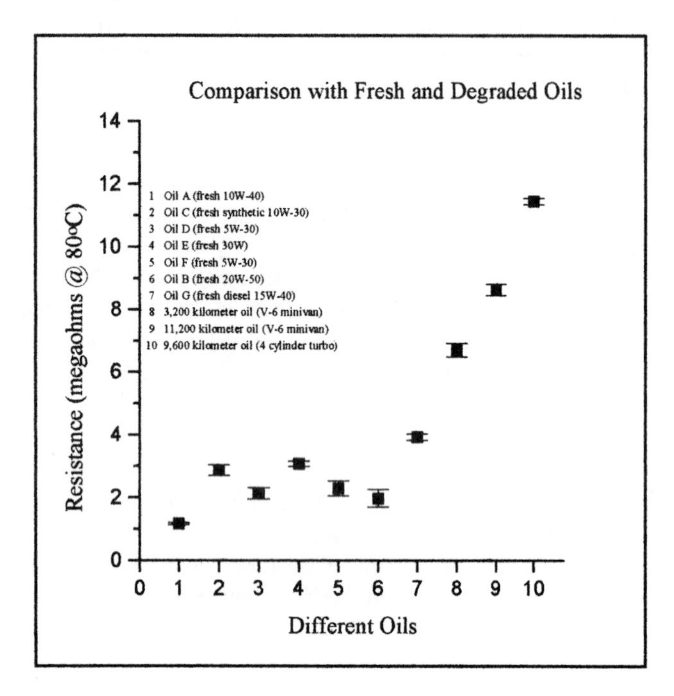

Figure 4.

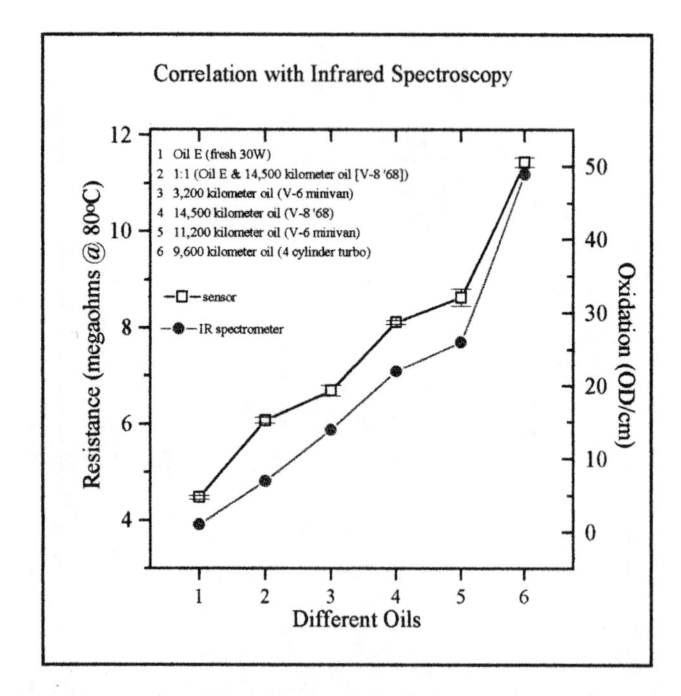

Figure 5.

Since motor oils can vary depending on viscosity, composition (mineral vs. synthetic) and additives (commercial oils vs. base stock), the sensor was evaluated for a consistency in signal response using a variety of different fresh oils (Figure 4). Runs were made using a single sensor. Results showed the signal was approximately the same (ranging from 1 to 4 megaohms) for the seven fresh oils tested. A series of three dirty oils, included as part of the tests, provided a clear separation in signal and a trend consistent with complementary methodologies (Table 1.) including oxidation readings collected using infrared (IR) spectroscopy. The trend observed with dirty oils is particularly noteworthy since it shows oil from the 9,600 kilometer 4 cylinder turbo is further oxidized than the oil from the 11,200 kilometer V-6 minivan which is the reverse of the expected results based on mileage, illustrating that mileage is not the only factor to consider in establishing the extent of oil degradation.

Sensor readings from a variety of oils were tested for sensitivity and correlated to oxidation, measured as optical density (OD) by IR spectroscopy (Figure 5). Oils that were evaluated ranged in quality from clean to heavily oxidized (4 cylinder turbo with 11,200 kilometers of use). A first order test of sensitivity provided clear separation in signal using a serial

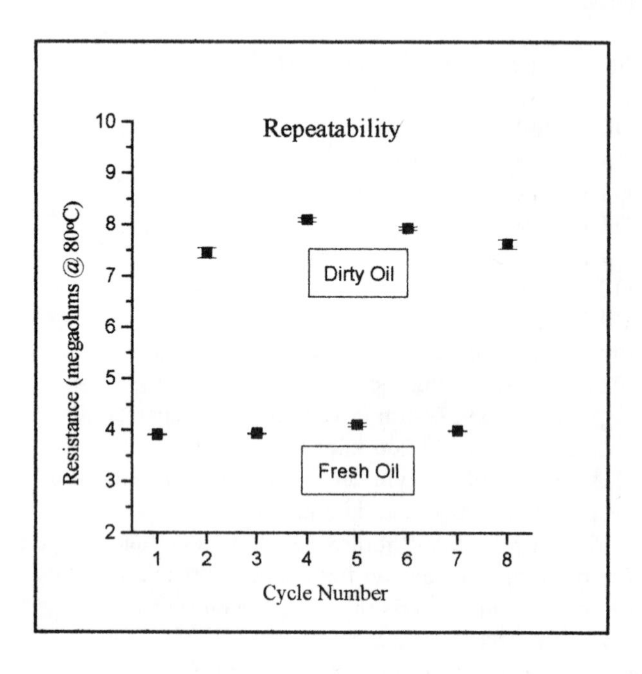

Figure 6.

dilution of a clean oil, a dirty oil and a 1:1 mixture of each. More closely degraded oils were also shown to provide signal resolution and oils of similar OD gave approximately the same sensor output (i.e., 11,200 and 14,500 kilometer oils). A difference in signal between the fresh oil and oil collected after 14,500 kilometers of use was ~2:1 at 80°C and ~3:1 at 20°C. The 14,500 kilometer oil gave an oxidation and nitration reading of 22 OD/cm and 24 OD/cm, respectively. A nitration value of 20-25 OD/cm is the recommended oil change interval, according to Mercedes/Shell.[11] An independent testing lab (Herguth Labs, Vallejo CA) offers this oxidation scale (in OD/cm) based on a statistical database: normal is 15, high is 27, and severe is 39. The 14,500 kilometer oil is therefore close to the oil change interval, according to both sources.

Table 1. Complementary Methodologies Used in Correlating the Oxidation Level of a Mixture of Fresh and Dirty (30W) Motor Oil

	Fresh Oil	1:1 Mixture of Fresh & Dirty Oil	Dirty Oil (14,500 kilometers of use)
Infrared Spectroscopy [1] (optical density/cm)	1	7	22
Thermogravimetric Analysis[2] (% wt loss)	98.89	98.14	97.48
Differential Scanning Calorimetry[3] (oxidative induction time, in seconds)	>7200	1643	585
Dielectric Analysis[4] (pmho/cm)	281.3	145.3	120
Oil Quality Sensor[5] (megohms)	19	36	50
Trace Metal Analysis (ppm)			
Iron	2	34	85
Copper	1	7	18
Lead	1	49	124

1) The following oxidation scale is based on a statistical database (at room temperature): normal is 15, high is 27, and severe is 39. Repeatability is within 21%.
2) Samples were heated under nitrogen from room temperature to 600 °C at a ramp rate of 5 °C /min. At 600 °C the reaction gas was switched to air for ignition. Repeatability is within 0.1%.
3) Oxidative induction time was measured from when the samples reached 195 °C. Repeatability is within 5%.
4) Samples were run at room temperature and at a frequency of 1000 Hz. Repeatability is within 5%.
5) Measurements were taken at 20 °C. Repeatability is within 5%.

Sensor readings are based on an average of three to four successive readings at temperature and afford an error of < 0.2 standard deviations. The inherent precision of the sensor therefore provides a clear means of distinguishing between even closely degraded oils.

The repeatability of the sensor was tested by cycling the sensor between a fresh oil and a degraded oil (Figure 6). Results show a relatively flat response for each of the two oils after 4 separate cycles. (A slight difference in readings between certain oils shown in Figures 4 , 5 and 6 result from sensor-to-sensor variation present in early prototypes.)

Additional sensor correlations were made using other methodologies (Table 1.) These other methods include, differential scanning calorimetry (DSC), thermogravimetric analysis, dielectric analysis, and metal analysis. Differential scannining calorimetry is commonly used in estimating oxidation by measuring the amount of antioxidants present in an oil. Thermogravimetric analysis measures the amount of residue remaining after igniting an oil, and appears to correlate the amount of residue remaining to a decrease in oil quality. Dielectric analysis measures the dielectric constants of an oil and equates an increase in signal with a change in polarity, indicating oxidation. These last two methods are not widely used, although dielectric analysis is gaining acceptance as a way of estimating oxidation. It should be noted that although a warning limit of <5min. is suggested for DSC, differences in experimental conditions from different labs make direct comparisons difficult.[6] Metal analysis is used to measure the level of metal contamination present in an oil and provides information concerning the extent of wear in an engine. Results showed successively higher levels of metal contamination present (iron, lead and copper) in the more degraded oils.

On-board tests were collected over short commute distances on a V-6 minivan for 5,000 kilometers without

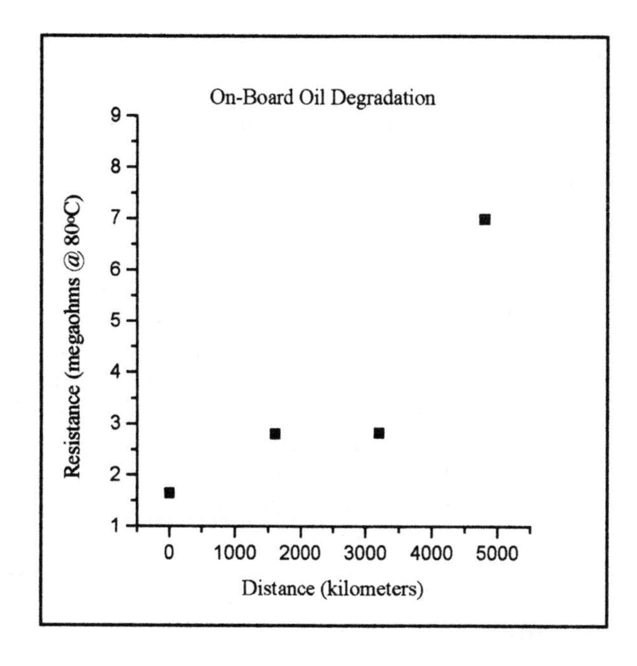

Figure 7.

topping off. A distance of 5,000 kilometer (3,000 miles) is

often quoted as the recommended oil change interval for severe driving conditions i.e., frequent trips of less than 4 to 5 miles.[12] Figure 7 shows sensor output being relatively flat for 3,200 kilometers. At 5,000 kilometers the resistance has markedly increased indicating an onset in degradation. Comparison with resistance values of oil standards presented in Figure 5 show the 5,000 kilometer oil is still providing adequate protection

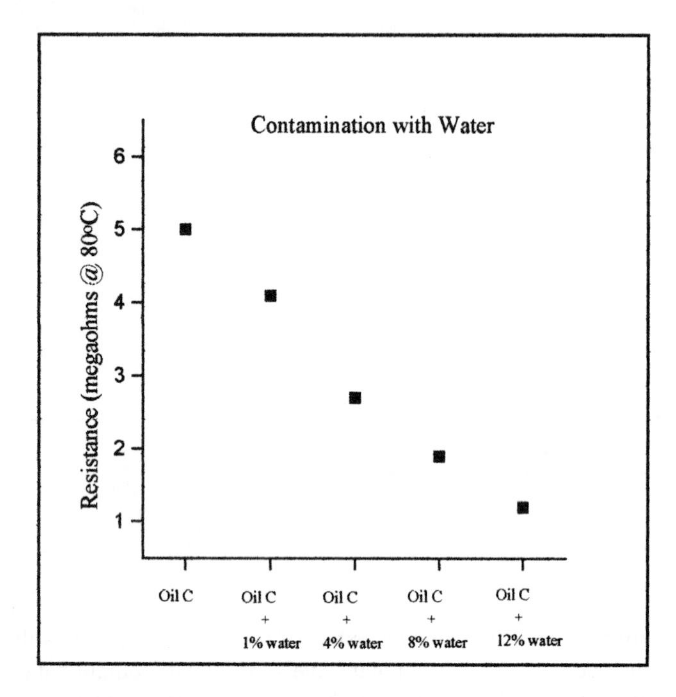

Figure 8.

and does not need to be changed. Analysis of the 5,000 kilometer oil by IR spectroscopy confirms a "normal" reading of 15 OD/cm.

A separate sensor was used to detect water. This sensor relies on a different formulation of resin and is designed to be used in combination with the degradation sensor. The sensitivity of the sensor to aqueous contamination was evaluated by treating a fresh oil with increasing amounts of water (Figure 8).

As shown, the method can detect the presence of as little as 1% water. The warning limits for water have been reported as > 0.1 - 0.5% for long-trip service, although >4% has been observed in short-trip service without engine failure.[6] What is significant about the operation of the sensor in detecting water is that the signal trend is effectively reversed, i.e., the signal decreases in resistance with increasing contamination. This behavior provides a unique approach to distinguishing water contamination from oxidation.

CONCLUSIONS

An oil quality sensor was tested and shown to reliably measure oil degradation by a change in the solvent properties of the oil resulting from oxidation or aqueous contamination. The sensor was evaluated using both bench and on-board vehicle testing. Results from these tests suggest the following conclusions concerning the efficacy of this on-board oil quality sensor:

* The device tracks oil oxidation in real-time
* Aqueous contamination can be detected independently of oxidation.
* The device remains operational over the temperature range normally found in a vehicle.
* The device remains relatively insensitive to fresh oils varying in viscosity, composition (mineral vs. synthetic) and additive formulations (commercial oils vs. base stock).
* A clear difference in signal is obtained between clean and degraded oils.
* IR spectroscopy was found to correlate closely with the device and provides a means of calibrating the sensor.
* Sensor output was shown to correlate closely with other established methods of oil analysis.

REFERENCES

1. Nattrass, S. R., Thompson, D. M., McCann, H., "First In-Situ Measurement of Lubricant Degradation in the Ring Pack of a Running Engine" SAE Technical Paper 942026 (1994)
2. Saloka, G. S. and Meitzler, A. H. " A Capacitive Oil Deterioration Sensor" SAE Technical Paper 910497 (1991)
3. Sato, A. and Oshika, T., "Electrical Conductivity Method for Evaluation of Oxidative Degradation of Oil Lubricants" Lubrication Engineering, 48, 539-544 (1992)
4. Kato, T. and Kawamura, M. "Device to Detect the Degradation Level of Oils" Lubrication Engineering 42, 694-699 (1986)
5. Johnson, M. D., Korcek, S., Schriewer, K., "In-Service Engine Oil Condition Monitoring Opportunities and Challenges" SAE Technical Paper 942028 (1994)
6. Smolenski, D. J. and Schwartz, S. E., "Automotive Engine-Oil Condition Monitoring" CRC Handbook of Lubrication and Tribology, Vol. III 17-32 (1994)
7. Voelker, P. J. and Hedges, J. D "Method and Apparatus for Fluid Quality Sensing" U.S. Patent 5,435,170 (1995)
8. Williams, J. A., "Engineering Tribology", Oxford University Press p 15, (1994).
9. Hiroshi, M., et. al, "Investigation on Oxidation Stability of Engine Oils Using Laboratory Scale Simulator", SAE Technical Paper 952528, 1995
10. Sato, A. and Oshika, T., "Electrical Conductivity Method for Evaluation of Oxidative Degradation of Oil Lubricants" Lubrication Engineering, 48, 539-544 (1992)
11. Thom, R., Kollmann, K, Warnecke, W., and Frend, M. "Extended Oil Drain Intervals: Conservation of Resources or Reduction of Engine Life", SAE Technical Paper 950135 (1995)
12. "The Surprising Truth About Motor Oils" Consumer Reports, 10-15 July 1996

970774

The Use of Magnetostrictive Sensors for Vehicle Safety Applications

Tony Gioutsos
Artistic Analytical Methods, Inc.

Hegeon Kwun
Southwest Research Institute

Copyright 1997 Society of Automotive Engineers, Inc.

ABSTRACT

New sensor approaches termed magnetostrictive sensor (MsS) and nonlinear harmonics (NLH) for vehicle safety applications such as crash detection and occupant seat weight sensing are described. Both sensor approaches utilize the changes in the magnetic properties of ferromagnetic materials that occur when a stress (or a strain) is applied to the material. The MsS is a passive device suitable for vehicle crash sensing. The NLH is an active device suitable for occupant seat weight sensing. Technical features of these sensors are also discussed together with preliminary results of ongoing testing.

1.0 INTRODUCTION

Vehicle safety sensing applications have increased dramatically with the increased use of airbag technology. Of the various vehicle safety sensing applications, there are three areas which have generally produced inadequate performance, they are: 1) single point crash detection for frame/body vehicles; 2) side impact detection and; 3) occupant seat weight sensing.

In this paper, we describe two sensor approaches termed "magnetostrictive sensor (MsS)" and "nonlinear harmonics" (NLH)" for the above mentioned application areas. Both sensor approaches utilize the changes in the magnetic properties of ferromagnetic materials that occur when a stress (or a strain) is applied to the material. The MsS [1,2] is a passive device suitable for vehicle crash sensing. The NLH [3] is an active device suitable for occupant seat weight sensing

In the following, an overview of the currently proposed sensor solutions and their limitations for the three application areas mentioned above is first given in Section 2. Then a description of the MsS and NLH is given in Section 3 including their physical principles, sensor design configurations, and their general properties and technical advantages. Specific applications of these sensor approaches to crash detection and occupant seat weight sensing are then given in Section 4 followed by conclusions in Section 5.

2.0 BACKGROUND ON VEHICLE SAFETY APPLICATIONS

The use of electronic accelerometer based single point sensing for vehicle airbag deployment has increased substantially in the last few years. Their use has been slower than anticipated but still climbing. An overview of this technology is given in [4] and [5]. A natural extension of their use is for other applications including side impact detection and frontal impact detection for frame/body vehicles.

2.1 FRAME/BODY VEHICLE CRASH SENSING - For many frame/body vehicles, acceleration values near zero can be encountered until the required Time-To-Fire (TTF) for a single point module located in the passenger compartment. This leaves even the best algorithm designers with no chance of achieving the desired requirements. A robust sensing concept capable of producing information at a faster rate from the passenger compartment is desired.

2.2 SIDE IMPACT DETECTION - Similarly, side impact sensing with accelerometer based approaches has produced at best marginal results. The inherent problem with an accelerometer based approach is the variation encountered for ON/OFF crashes across the side of the door. For example, a pole crash located at the B-pillar (assume an accelerometer is mounted at the B-pillar), will produce a substantially different sensor output than a similar crash at the A-pillar. Yet, the crashes to the occupant are similar in terms of potential injury. A detailed discussion of the disadvantages of crush zone sensing is given in [5].

There are other approaches to side airbag deployment as well (Autoliv [6] and Siemens [7]). In [6], a mechanical firing pin approach is described. This sensor approach suffers from the same inherent weakness as an accelerometer based approach, namely, the large variation in sensor output based on location of the crash on the side of the vehicle. The idea behind both the accelerometer based approach and the mechanical based approach is to reduce this effect by creating an "array" sensor from a "point" sensor.

In both [5] and [8], the benefits of array sensors are described . In [8] a crush zone sensor is described. Unfortunately, this array sensor is truly an array causing other problems. In [5], frontal crash detection using a single point module approach is described as the best of both worlds (assuming an appropriate algorithm): array sensing from a point sensor. The front part of the vehicle acts as the array, the sensor simply produces similar waveforms for similar crashes (e.g. an offset pole crash on either side of the vehicle).

Therefore, it is beneficial to create an array sensor for side impact detection. By stiffening the vehicle or adding a crossbeam, the accelerometer/firing pin approaches begin to look more like a vehicle side array sensor. However, the inherent weaknesses still exists.

In [7], an air pressure approach is described that again expands on the notion of array sensing from a point. However, disadvantages of this approach include: variation over life, mounting in the door and the limited capability of the approach in sensing only those crashes impacting the door. Without dwelling on this concept, it does not provide the robustness that a magnetostrictive sensor approach can provide.

2.3 OCCUPANT WEIGHT SENSING - With the recent attention on fatalities due to airbag deployment, increased emphasis has been placed on passenger sensing. In particular, weight has been deemed the appropriate parameter for disabling the airbag. Several approaches have been proposed and are being proposed. Without getting too specific, several problems exist with current and proposed approaches including:

- Cost
- Implementation without disrupting the seat
- Temperature compensation
- Robustness to variability in seat occupancy scenarios
- Area coverage

These issues and potential solutions will be addressed further in section 4.

3.0 TECHNICAL BACKGROUND ON MsS AND NLH APPROACHES

3.1 MAGNETOSTRICTIVE SENSOR (MsS) - The MsS is a passive sensing approach which relies on a specific physical phenomenon, that exists in ferromagnetic materials, called "inverse magnetostrictive (or Villari)" effect [1,2]. The Villari effect refers to a change in magnetic induction (B) of material with application of stress, in comparison to the magnetostrictive (or Joule) effect which refers to changes in physical dimension of the material with magnetization. Being a passive device, the MsS is limited to detection of only time-varying or transient stresses (or strains) in the material such as those produced by mechanical impacts or those acoustic emission signals produced by cracking [1,2]. The MsS has a very broad frequency response, ranging from a few Hz to a few 100 kHz. The sensor can also be applied to nonferrous material such as plastics, if a thin layer of ferromagnetic material is plated or bonded to the material surface in a local area where the sensor is to be placed. In addition, the MsS requires neither a direct physical contact to the material nor a couplant for sensing.

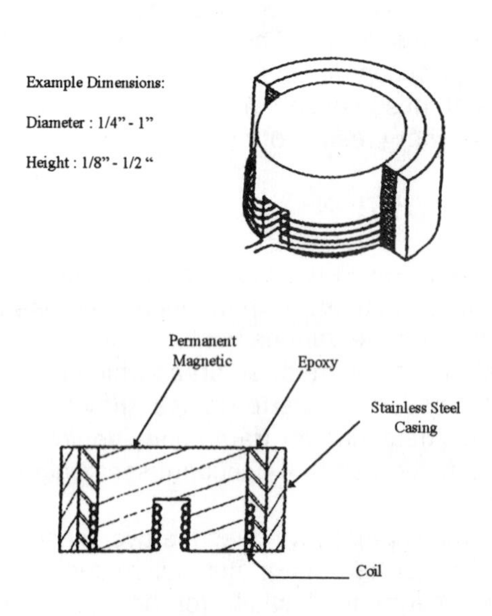

Figure 1 : Typical MsS Sensor

A typical MsS is depicted in Figure 1. It consists of an inductive coil and a permanent magnet. When stress in the material changes with time, the resulting changes in B due to the Villari effect induce an electrical voltage in the coil. The permanent magnet provides a static bias magnetic field to the material which will enhance the sensor sensitivity. The transient stresses produced by a crash or impact will propagate through the material. The attenuation of such impact stress waves in vehicles is negligibly small. Therefore, the MsS can sense the crash or impact event at a location far away from the exact impact site as long as the path for the wave to propagate is good. The MsS can survey a large area and, thus, functions like an "array" sensor from a point and is ideally suited for vehicle crash detection.

3.2 NON-LINEAR HARMONIC SENSOR (NLH) - The NLH sensor operates in a similar fashion to the MsS, but it is an active device [3]. An example NLH sensor which consists of a U-shaped ferrite core with an excitation coil and a detection coil wound on each leg is depicted in figure 2. When a ferromagnetic material such as steel is excited by a sinusoidal magnetic field (H), the corresponding magnetic induction (B) of the material is no longer sinusoidal but shows a distorted waveform. This distortion is due to nonlinear magnetic permeability and the magnetic hysteresis of the material. The distorted B-waveform contains harmonic frequencies of the applied H. In addition, the magnetic hysteresis curves of a ferromagnetic material change significantly when the material is subjected to mechanical stress or strain. The NLH sensor transmits a sinusoidal H to a ferromagnetic material and then detects the resulting B waveform.

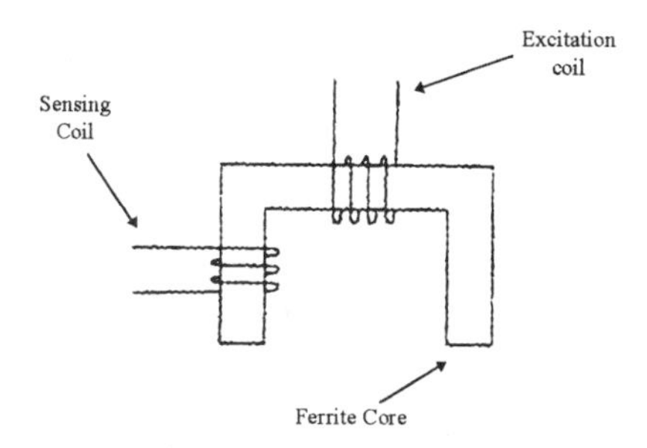

Figure 2 : Typical NLH Sensor

Typically, a coil is used for applying H and detecting B. The H is achieved by sending a sinusoidal current of a given frequency through an excitation coil (e.g. 1kHz). The resulting time-varying B induces an electrical voltage in the detection coil. The stress on the material of interest is determined by harmonic analysis of waveform. In general, the magnitude of the third harmonic of the resultant waveform is related to the stress on the material. In steel, up to about 50 percent of the yield strength, the relationship is linear (i.e. the greater the magnitude of the third harmonic the greater the stress).

The chief differences between the NLH sensor and the MsS are:

- The MsS is a passive device, whereas the NLH requires power
- The NLH sensor can sense constant stress

The NLH approach is similar to those approaches described as magnetostrictive sensors used for torque measurements [9] except that the NLH utilizes nonlinear harmonic components, whereas, the other utilizes the fundamental component.

3.3 TECHNICAL FEATURES - The two sensing approaches have the following features in common:

- Temperature insensitivity
- Non-contacting
- No moving parts
- Small and inexpensive

The MsS also has the following features:

- Passive (produces a signal without power)
- No DC offset (i.e. detects transient stresses)
- An "array" sensor from a point

The NLH also has the following features:

- Active
- Detects DC or constant stress

4.0 VEHICLE SAFETY APPLICATIONS

4.1 FRAME/BODY VEHICLE CRASH SENSING - The MsS can be used in the same fashion as an accelerometer for vehicle crash sensing [10]. Since the frontal crash sensing field is more mature, we will focus on the application of using a single point module for frame/body vehicles (e.g. trucks).

Because of the structure of frame/body vehicles, accelerometer signal values are near zero until the required TTF for passenger compartment mounting locations. This is due to the fact that the frame encounters the crash object well before the body has. By cutting a ferromagnetic hole (e.g. rubber or plastic) in the body above the frame and then facing the MsS at the frame, we are able to detect stress waveforms in the frame from the body. This will allow faster response from a single point sensor module located in the passenger compartment.

4.2 VEHICLE SIDE IMPACT DETECTION - The MsS is ideally suited for side impact detection. The device can be mounted on or near the door without necessarily being mounted on a surface. For example, one could mount the sensor on the crossbeam and face it at the door skin. Therefore, stress on the skin can be measured. Again, the "array" nature of the sensor will allow the sensor to produce waveforms even if the sensor is not contacted directly or is nearby the "hit" area. A good propagation path is all that is needed. There are other locations on or near the door that the sensor can be placed to provide a good signal with ease of mounting. For example, placing the sensor on the sill facing the base of the door will allow the door to be sensed but the sensor to be mounted within the passenger compartment. Various isolated side impacts can be detected with this arrangement. If a pole hits the A or B pillar or is on-line with the sill mounting location, the signal strength should be similar for the same impact speed. This response should be differentiated from that of an accelerometer, where depending on the mounting location (e.g. A/B pillar location or the middle of the door) substantially different waveforms will be produced.

These two features (i.e., non-contacting and array sensing from a point) make the MsS ideally suited for this application.

4.3 OCCUPANT WEIGHT SEAT SENSING

The occupant seat weight sensing problem has become a very important issue in the automotive safety community. The key to the problem is to determine the occupant's or object's weight with a very high resolution. The goal is to change the airbag deployment characteristics (e.g. do not deploy) if the weight is less than a given threshold(s). There are many issues involving this approach but basically there are three main concerns: performance, cost, and ease of implementation.

The NLH sensor is suited to this application. By placing the sensor under the seat on either the pan, springs or mounting points; stress upon the seat can be measured. For example, for seat designs which feature an array of wires holding the seat cushion in place, stress on the wires can be measured as follows. As an object is placed in the seat the foam causes stress on the wires. The stress applied to the wire by the occupant is detected by using a NLH sensor placed on the wire. The NLH output can then be converted to weight. The sensor location, design and placement will be related to the given seat, but should not affect the seat's deign. This will allow a seat weight sensor that is inexpensive, robust, and easy to implement.

5.0 CONCLUSION

We have shown that the MsS and NLH sensor have good potential for automotive safety applications. There are several Tier One suppliers testing these devices with so far excellent results. In addition, the MsS has also shown good potential for condition monitoring of combustion engines such as knock and misfire.

REFERENCES

[1] Kwun, H., "Back in Style: Magnetostrictive Sensors," *Technology Today* (Southwest Research Institute), March 1995, pp. 2-7

[2] Kwun, H. and Teller, C.M., Patent #5456113, "Nondestructive Evaluation of Steel Cables and Ropes Using Magnetostrictive Induced Ultrasonic Waves and Magnetostrictively Detected Acoustic Emissions"; also Patent #5457994 and other pending patents. Assignee: Southwest Research Institute

[3] Kwun H. and Burkhardt G. L., "Nondestructive Measurement of Stress in Ferromagnetic Steels Using Harmonic Analysis of Induced Voltage", *Non-Destructive Testing International*, June 1987, pp. 167-171

[4] Gioutsos, T. and Gillis, Ed, "Testing Techniques for Electronic Single Point Sensing Systems," *SAE International Congress and Exposition*, Paper # 940803, 1994

[5] Gioutsos, T. and Gillis, Ed " Tradeoffs and Testing for Frontal Crash Sensing Systems," *SAE Worldwide Pass. Car Conf. and Expo.*, Paper # 932911,1993

[6] Dahlen, M. "Side Airbag Systems: Seat-mounted vs. Door-mounted", *SAE Side Impact Protection Toptec*, 1994

[7] Hartl, A. Mader, G. Pfau, L. and Wolfram, B., "Physically Different Sensor Concepts for Reliable Detection of Side-Impact Collisions", ," *SAE International Congress and Exposition*, 1995

[8] Breed, D., et al. ,"Performance of a Crush Sensor For Use With Automobile Air Bag Systems," *SAE International Congress and Expo.*, Paper # 920122, 1992

[9] Klauber, R., et. al., "Miniature Magnetostrictive Misfire Sensor," Paper # 920236

[10] Gioutsos, T., Patent # 5,580,084, "System and Method for Controlling Vehicle Safety Device", Assignee : Artistic Analytical Methods, Inc.

MULTIPARAMETER SENSORS

961801

Improving Mobile Equipment Performance Using New Flow and Pressure Feedback Technology

George Kadlicko and Greville Hampson
Microhydraulics Inc.

Copyright 1996 Society of Automotive Engineers, Inc.

Abstract

A recently patented transducer introduces new options for controlling and monitoring mobile equipment by measuring flow, pressure, and temperature. Traditional flow transducers lack the robustness, accuracy, or responsiveness to be effectively used in actual hydraulic circuits outside the laboratory environment. Pressure transducers are used, but present packaging and cost obstacles to machine designers. Temperature sensors are also available in various subsystems in mobile equipment.

Accurate monitoring or control of a fluid system depends upon dynamic measurement of the flow and pressure in a system, but at the same time it is important that the monitoring apparatus is capable of measuring small variations in relatively large flow rates.

A new device has been developed in a cartridge format which solves previous limitations. The cartridge mounts easily into a manifold block, and can be produced economically. Output signals are analogue voltages suitable for digital display or processing by a control unit. Robust design as well as efficient low power operation round out the packaging criteria.

The cartridge is capable of measuring flow, pressure and temperature in a single unit with response fast enough to be used for feedback in circuit applications as well as for diagnostic information. This transducer is an example of how available technologies and well understood principles can be applied and packaged into a format easily adaptable to the next generation of electrohydraulic control systems and to bring the long promised benefits of electrohydraulics to users.

Introduction

Microhydraulics Inc., an engineering company, has developed a flow, pressure, and temperature transducer suitable for use in hydraulic circuits for feedback or diagnostic purposes.

Hydraulic circuits are no different from other control circuits in that they require a continuous and iterative feedback from the controlled output to achieve a desired result. The state of the output is verified and a signal generated for corrective action.

The feedback loop needs to be considerably more responsive than the desired response of the output. If not, the control functions will be impaired by the feedback loop.

In many cases, static measurements are sufficient to establish performance characteristics of a circuit. However, in closed loop applications, the dynamic conditions are the relevant factor for circuit control.

Transducers develop dynamic (AC) and static (DC) signals. If the AC component of the waveform cannot be detected by the transducer, then the circuit will be controlled from the DC part of the signal. This averaging will mask significant information from the controller. See Figure 1.

The need for rapid response to control dynamic systems is even more pronounced in today's high performance circuits. In addition to the need for response are the concurrent requirements for low cost, effective packaging, and robust design. Expensive and cumbersome instrumentation can often be practical in a laboratory but not in the field.

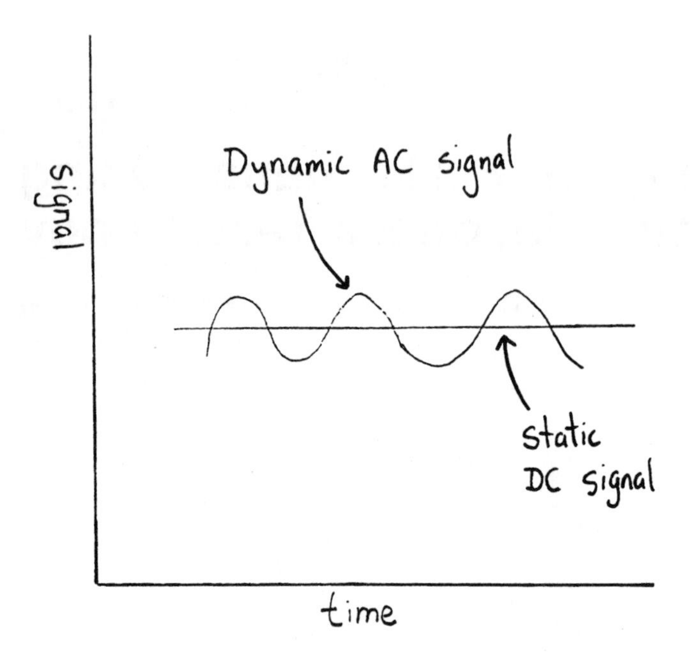

Figure 1 *Dynamic picture can be lost with low response*

All pressure, flow, and temperature transducers are comparative instruments. The AC capabilities are established through design and the DC capabilities are established through calibration to a known static condition.

In hydraulics the need for responsive measuring devices for flow, temperature and pressure has been long established and is well known. As closed loop electrohydraulic circuits become more common, appropriate electrohydraulic feedback devices are needed.

Evolution of Electrohydraulic Feedback Transducers

Early devices satisfied basic static conditions. In pressure sensors the earliest devices were pressure gauges based on Borden tube feedback elements (relying on deflection of a metal tube, geared to drive a pointer). Stability and accuracy were always questionable, particularly for large deflections. Vibration, temperature changes, and material variations all affected its performance. Initial attempts to turn the Borden tube into an electrohydraulic device using potentiometers and subsequently LVDTs were tried but met with the same underlying problems. Evolution continued through to current times and is based on measuring microstrain in a metal diaphragm.

Temperature sensors have also gone through a similar evolution by using the differential expansion bimetallic strips and producing an interpreted electrical signal. More recently, solid state devices have evolved. These metal-doped silicon-based devices have a very small mass with limited energy storage so are highly suitable for dynamic applications.

Flow measuring devices have not kept up with the evolution of the dynamic capabilities of temperature and pressure sensors. Displacement of a cylinder of known cross-sectional area in a given time yields an accurate volume. Problems arose when the cylinder reached the end of its stroke and was ready to reverse. Also, accuracy of measuring displacement was difficult for long strokes. To overcome these problems various types of hydraulic motors were used, including gear, vane, and piston motors. Difficulties arose due to leakage paths, and due to the fact that the rotating mass of the motor causes reluctance to respond to change. Finally the turbine flow meter became widely used, especially after considerable effort went into reducing inertial mass and bearing friction. The basic principle is placing a runner on its own bearings inside a metal case in the flow path. Inductive transducers (pulse pick-up) detect the change in the magnetic field caused by the rotating runner and produce a waveform. The density of the wave is dependent on the speed of the runner. This magnetic interference with the rotation provides the energy for the inducted current, but also acts as a brake in the system. Output is often adequate for intended uses, but linearity is questionable, ripple of the transducer or of the flow is hard to distinguish, and great efforts in the geometry of the impeller are required to achieve any kind of precision. Fairly complex electronic conditioning is required, but responsiveness remains low.

Transducer Design Objectives

This brief history shows the independent evolution of flow, pressure and temperature instruments to meet the dynamic needs of today's closed loop systems.

The Microhydraulics development objectives were as follows:

- single package for pressure, temperature and flow
- compact physical dimensions
- suitable for widely varied hydraulic circuits
- minimum mass in moving components
- clean electrical output
- minimum pressure drop in flow path
- insensitive to temperature change
- highly responsive
- robust
- cost effective

A cartridge style was selected based on its flexibility, consistency with other hydraulic products, and suitability to be a building block in a circuit.

System Description

The measuring section of the transducer consists of a sleeve separated by an O-ring between input and output, as shown in Figure 2. A movable piston is placed in the sleeve. This piston is biased to its null position by a calibrated spring.

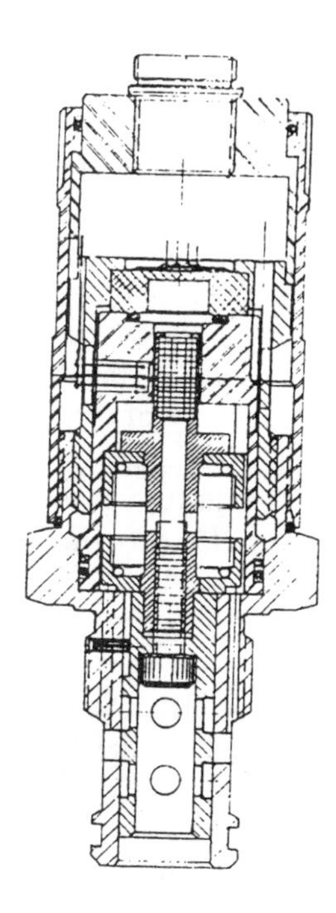

Figure 2 *Assembly drawing of -12 size triple transducer*

The piston position is decoded by a cylindrical, samarium cobalt iron magnet located in a carrier attached to the piston. A non-magnetic stationary sleeve encapsulating the magnetic area provides the barrier for the fluid, yet a magnetically transparent communication to a Hall Effect sensor. The movement of the magnet relative to the Hall Effect sensor causes the location of the magnetic field at the Hall Effect sensor to change, causing the voltage output of the sensor to change in proportion to the piston position.

As a practical matter, in permanent magnetic bars the flux pattern emitted will not be uniformly concentric. The effect of this magnetic eccentricity would cause an unpredictable error at the Hall Effect sensor. The potential problem is addressed by preventing the magnetic carrier from rotation, thereby eliminating the false readings.

If there is flow, the pressure differential from input to output will move the piston. The piston will cover all passages in its null (no flow) position and will move up or down dependent on the flow direction. For a specific flow it will open until a precise pressure drop is achieved across the transducer. This will place the piston on its sleeve in a relative position indicative of flow.

In a flow condition, the piston is trapped in the flow stream and responds as part of the fluid. The bias spring is not a reference and serves only as a "homing" element. The fluid dampens the piston motion. The geometry and material of the piston serves to minimize reluctance. The force in the fluid stream provides ample energy to move the piston assembly.

The flow transducer is effectively at the same pressure at the input and output. This allows a pressure diaphragm to be installed (with a strain gauge pressure transducer element) into the cartridge unit. Since the fluid and the transducer body are at the same temperature (particularly in the flow affected areas) a solid state temperature sensor was installed in the cartridge.

The transducer with the triple sensors effectively provides all the relevant feedback for a hydraulic circuit in one independently functioning cartridge.

Packaging

The three electrical sensors (Hall Effect, pressure, and temperature) all have their electrical connections facing the same way, sealed from the fluid and providing a convenient method of connecting the sensors to the electronics built in to the top of the cartridge. The electronics in the upper half of the transducer are hermetically sealed from the pressurized fluid inside as well as from contamination outside. The processed signal is carried out through a multipin connector ready to be used in control applications.

All exposed parts can be produced from high tensile corrosion resistant materials such as stainless steel, although for cost reasons a plated steel is normally selected. All seals are compatible with regular cartridge techniques dealing with various oil and alternative substances such as water soluble oil, water glycol, vegetable oil, and fuels.

The electronic package is also used to linearize the actual flow and pressure curves against a calibrated standard. This calibrated reference is maintained in a reference circuit within the transducer. The temperature sensor is used in the circuit to compensate for temperature drifts, viscosity changes or other temperature related characteristics.

Utilizing electronics in this way has made it possible to use a considerably simpler mechanical device in place of complex geometry and hole patterns.

For different flow rates different size cartridges are available in standard SAE sizes.

The transducer is formatted in a screw-in cartridge for lower flow rates and a slip-in cartridge for the higher flows, although alternative packages could be made available for tailored systems.

Screw-in cartridges are available in the -10, -12, -16, and -20 sizes. Larger than -20 is not practical in the screw-in format due to material strength, thread, wrench size, and other practical considerations. The slip-in cartridge format would be used for flow rates exceeding 300 litres per minute.

A -12 size transducer is shown in Figure 3.

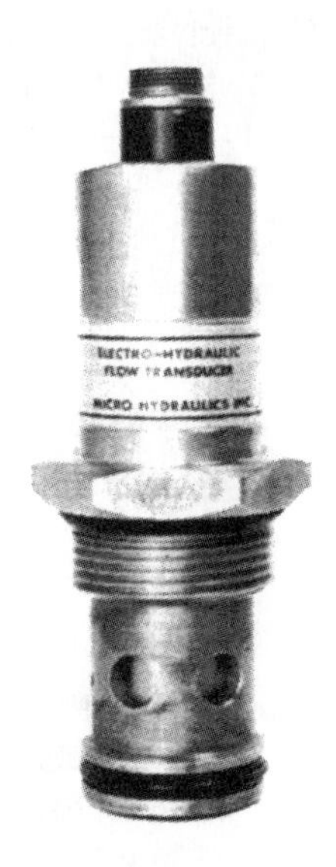

FLOW TRANSDUCER "C" SIZE

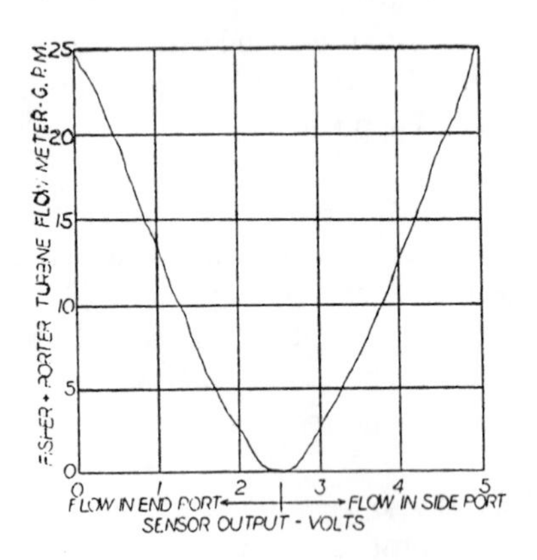

TYPICAL PERFORMANCE

Figure 3 *Photograph and typical performance of flow transducer*

To minimize pressure drops in a circuit due to this device, careful consideration is given to sizing. For example, a -10 size directional poppet valve may be matched with a - 12 size flow transducer to manage pressure drops in the system.

Dynamic Capabilities

In a testing environment, multipoint testing is often used to characterize the behavior of control elements, pumping elements, and circuits. Using this kind of multipoint sensor guesswork will be supported by solid data and art will start to turn into science.

In the Microhydraulics' development of the flow transducer, there were difficulties in evaluating the dynamic capabilities of the device - problem that all development engineers can appreciate. Using a high response servovalve was the logical choice. However, it was difficult to create a reference at those frequencies with available transducers. Variations in the signal could be the result of the flow transducer or the result of the actual flow in the servovalve.

In place of this approach, a multibarrel diesel fuel injection system was used as a reference, initially using only one barrel. In the injection system, a cam causes fluid to be expelled from the pump barrel. Driving the cam at a constant speed such that the only variable in the flow was the cam profile, the electrical output from the transducer matched the mathematically correct geometry of the cam. Subsequently, 5 barrels were tied into a common output and the combined flow was measured. The signature of the 5 cams at 3600 rpm was visible, as was the difference between barrels due to unequal leakage of the pistons.

With 5 cams rotated at 3600 rpm, this translates into a frequency of the output of 300Hz. Subsequent tests on a high performance servovalve displayed the asymmetric behavior on the servovalve spool at 200Hz.

Some of the empirical results obtained are outlined in Figures 4, 5, and 6. Figure 4 shows the approximate accuracy of a flow transducer compared to a traditional turbine meter. Figure 5 tracks the output of the flow transducer compared to the spool position of an in-line proportional directional valve. The step response of the transducer is shown against the 10 msec step response of a high speed proportional directional valve in Figure 6. No attempt was made to eliminate the effects of hoses or fittings in these tests, and the transducer did not have any of the digital linearization features enabled.

Applications

Applications are diverse, but a few of the most apparent are:

- laboratory and test
- diagnostic - during machine production and subsequent servicing
- feedback element for control of circuits
- electrohydraulic dynamometer

An OEM may use the transducer cartridge for evaluating circuit performance in new equipment. The transducer could be removed after on-site delivery and reinstalled later for regular servicing. Alternatively the transducer could be designed into the system and form an integral part of the hydraulics.

In an electronically controlled directional valve, a flow transducer provides an effective flow output using it in place of a pressure compensator element by using the electrical

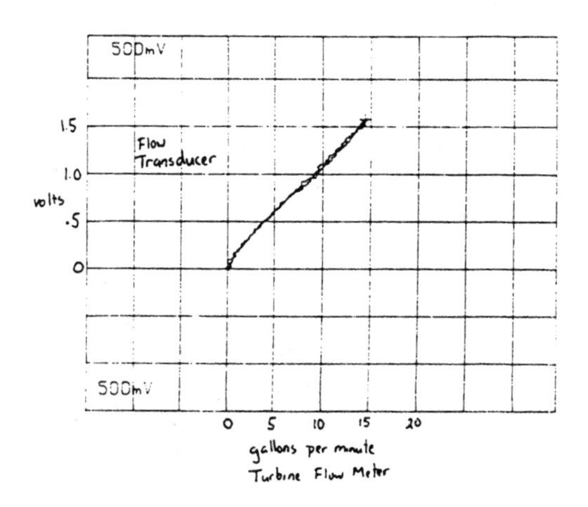

Figure 4 *High response flow transducer plotted against a traditional turbine meter.*

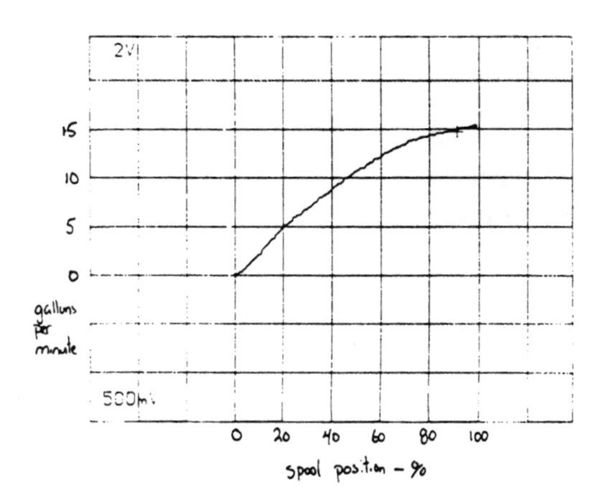

Figure 5 *The flow transducer responds to a change in the spool position of a high speed proportional directional valve.*

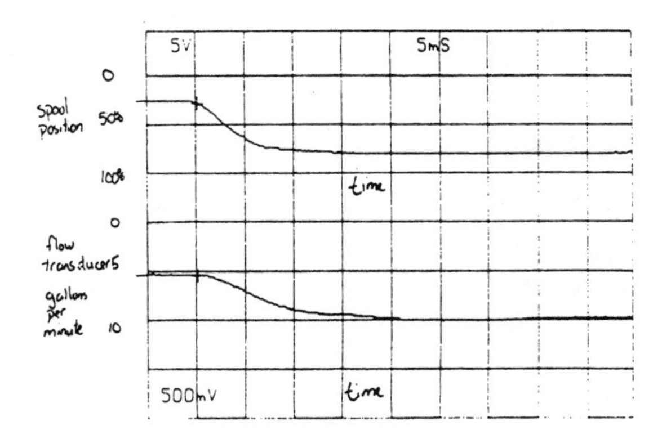

Figure 6 *An indication of step response is shown by stepping the spool of a high speed proportional directional valve and monitoring the change of the flow transducer output.*

signal given by the sensor in an electronic loop. The control loop modifies the spool movement in accordance with pressure changes in the circuit. By matching an actual input signal to flow output this valve and transducer combination becomes an effective electrohydraulic pressure compensator. This eliminates the sluggishness of the traditional pressure compensator function which otherwise inhibits the performance of these systems.

In a pump scenario, the position sensing of the swashplate would be an accurate reference to flow if there were no leakage and the rpm were monitored at all times. Theoretically it is feasible to relate a pump input signal to flow using mathematics and knowledge of the swashplate position and shaft speed. If a flow sensor was placed on the output flow path from the pump, the swashplate and shaft speed information would not be relevant. The transducer will measure actual flow from the pump. An electronically closed loop input versus flow would be the result. Since temperature and pressure is included in this transducer the temperature information can be used to correct for viscosity and other temperature effects, and the pressure information can be used to tie the loop for pressure conditions (assuming the pump is fast enough to respond). With the triple transducer, many pump control scenarios are now possible:

- input versus flow
- input versus pressure
- load sensing
- horsepower limiting
- torque control
- multiflow capability

A dynamometer is naturally formed in every pumping circuit. This provides a continuous monitoring between the pump, load, and prime mover and functions on demand or continually for circuit control. In a multipump installation maintaining the sum of all loads not to exceed a fixed maximum output from the prime mover is desirable yet difficult. This kind of installation is one of the most justifiable places to install transducers and appropriate control systems since alternatives are more costly.

A flow measuring instrument cannot provide positional information about a cylinder or actuator. There are, however, many applications where velocity or rate of change is the critical performance measurement. The transducer is well suited for use in a closed loop with a high response valve/actuator arrangement.

The overall package is such that no exotic materials are required. Elegant for manufacturing, this naturally provides the opportunity for the cost to be kept low. The sensing elements are in mass production by third parties, are extensively used elsewhere, and are widely available.

The proposed electronic module is based on well proven techniques in other industries which have received acceptance in demanding applications including aerospace, military, and automotive.

Conclusion

To recognize the benefits of electrohydraulics, considerable improvement in feedback sensors is required. These feedback sensors should not be the limiting element in a control circuit. High response with accurate results is essential. Packaging into a modular, robust, and cost effective form is also needed so that elements can be practically and economically applied.

The triple transducer is a significant step in this direction.

950536

A Single Transducer for Non-Contact Measurement of the Power, Torque and Speed of a Rotating Shaft

Ivan J. Garshelis, Christopher R. Conto, and Wade S. Fiegel
Magnetoelastic Devices, Inc.

ABSTRACT

Simple constructional modifications extend the capability of polarized ring torque transducers, when applied to rotating shafts, to include the measurement of power and speed. Gear-like teeth on thin, high permeability rings affixed to the polar regions of the magnetoelastically active ring spatially modulate the magnetic field arising with the torque. Speed is found from the frequency of the ac component of the field sensor signal while torque is found from its amplitude. Drift in quiescent outputs of low cost Hall effect sensors have no effect on these ac signal features. Signals proportional to transmitted power are generated in sensing coils by shaft rotation.

INTRODUCTION

Basic to the operation of modern machinery is the transmission of mechanical energy from source locations to points of utilization by means of rotating shafts. In a typical machine the energy is first imparted to a rotating shaft after conversion from chemical, thermal, electrical or kinetic sources within some prime mover such as an engine, turbine or motor. Machines often contain systems of shafts whose rotational motions are interconnected by couplings, belts, gears, or related devices in order to better match the prime mover to the load or to distribute the energy to a multiplicity of loads. Clutches between shafts allow for purposeful decoupling of their rotational motions. The mechanical energy imparted to the output shaft of the prime mover eventually is used to perform useful work in forms and at locations that characterize the function of the specific machine, e.g., propulsion of a vehicle, compression of a fluid, forming or machining of a manufactured part, electrical generation, etc. The ubiquity of utilization of rotating shafts to transmit and distribute mechanical energy is readily illustrated even with the very abbreviated listing in TABLE I.

The *rate* at which work is performed is termed *Power* [1]. *Power* is also defined as "the time rate of transferring or transforming energy" [2]. When the mechanical energy performing the work is transmitted by a rotating shaft, *power* therefore describes the rate of energy flow along the shaft. Transmitted power is thus clearly a measure of the functionality of a rotating shaft. From this perspective it is clear why "output power" is the primary quantitative factor used to rate both mechanical [3] and electrical [4] prime movers. It is also understandable why so many shaft driven machines such as pumps and compressors, spindles on lathes, mills and grinders and other machine tools, and even some appliances such as vacuum cleaners and garbage disposal units are often sized and compared by their power capacities [5]. On-line measurement of the power actually being transmitted along key shafts in a machine can, by quantifying the machine's performance, enable its more precise control and adjustment and also help to ensure its safe and efficient operation. Noted departures from normally generated or utilized power can even provide an early indication of a developing fault.

TABLE I. Machines transmitting power with rotating shafts.

INDUSTRY	TYPICAL MACHINE
Agriculture	Tractor, Combine, Harvestor
Construction	Concrete mixer, Crane, Excavator
Food	Mixing, Bottling, Canning
Lumber/Paper	Sawing, Planing, Pulping
Mining/Oil	Boring, Loading, Pumping
Manufacturing	Machine tools, Conveying
Metals, Plastics	Rolling, Slitting, Extruding
Recreation	Ski lift, Amusement park rides
Textile	Weaving, Knitting, Sewing
Transportation	Land, Sea, Air vehicles

The importance of on-line power measurement on rotating shafts in working machinery has long been recognized with the resulting development of more or less standardized measurement methods and apparatus [6]. Since the power transmitted through a cross section of any shaft is the product of its instantaneous angular velocity and the torque transferred across the section, the measurement of this power generally reduces to the separate measurement of these two, more basic quantities. Whatever technologies and specific types of rotational speed and torque measuring devices are actually employed, the determination of power still requires the multiplication of these separately measured quantities. Conventional power measuring instruments therefore include in the overall apparatus, besides means for measuring both speed and torque, some computational circuitry for on-line multiplication of these two, separate signals.

MEASUREMENT METHODS

SPEED - A great variety of methods and apparatus exist for measuring angular velocity of rotating shafts. Many of these are classical, having been in use for more than half a century [7]. Speed measuring devices may be classified into two general types: those based on counting and timing discrete rotational events and those which develop a measurable physical quantity, e.g., a force or a voltage, proportional to velocity. Modern devices employing the counting and timing principles typically use non-contacting magnetic or optical means to sense the passage of salient features on an "encoder" wheel that rotates with the shaft of interest. Magnetic encoders are now routinely incorporated directly into the ball bearings used to support the shaft [8, 9]. Determining rotational speed with such devices requires either some form of clock (nowadays generally incorporated into the digital computer concerned with the processing and utilization of the speed information) or some electronic "frequency to voltage" conversion circuitry (e.g., a diode pump). A variety of technologies are utilized to directly develop analog indications of rotational speed. Most common among these are "drag cup" devices based on the forces associated with velocity dependent eddy currents. Use of this type of device requires the further conversion of the developed force into a proportional electrical signal. Small, permanent magnet, electric "tachometer" generators are often used to develop dc (or ac) voltages in direct proportion to rotational speed. Devices of this type are usually arranged to be driven by (or in synchronism with) the rotating shaft of interest rather than being mounted directly thereon. A different type of non-contacting tachometric device [10] utilizes only a stationary combination of a permanent magnet and magnetic field sensor(s) to develop a signal proportional to the velocity of a conductive target, e.g., an aluminum disc mounted on the shaft of interest. There are clearly a variety of technologies and a wide choice of devices suitable for developing the rotational speed signals needed for on-line determination of the power being transmitted by a rotating shaft.

TORQUE - The net (i.e., resultant) torque from all rotating shafts entering or leaving a machine having an identifiable non-rotating frame can be simply determined by measuring the reaction forces which prevent the frame from rotating. The torque on the output shaft of engines, motors, etc. is often measured in this manner, either by measuring the forces directly at the mounts [11] or on a torque balancing arm of a cradled, absorption type of dynamometer [6] within which the transmitted energy is converted to heat (either directly or via its first conversion to electricity). Sometimes torque is inferred from measurements of acceleration [12] or, with specific applicability to reciprocating engines, from dynamic measurements of cylinder pressure [13]. It should be clear that such methods, by their very nature, are limited in applicability.

Measurement of the torque actually being transmitted on a rotating shaft generally relies on either the elastic or magnetoelastic properties of the shaft itself or of locally attached ad hoc materials. Whatever the underlying principle at work in a particular torque measuring device, the on-line determination of power is facilitated if the measured value of torque is presented in the form of an electrical signal. Measuring devices of this type are classified as *torque transducers*.

Torque transducers relying on shaft elasticity actually measure torsional strain, either in terms of the angular deflection over a dedicated section of the shaft having a length typically 10 or more times its diameter, or of the unit surface strain at a gauging "point", typically only a fraction of the diameter in length. Angular deflections of rotating shafts are often measured by utilizing magnetic [14] or optical [15] methods to sense the time difference between the passage of two, initially (circumferentially) aligned, axially separated features on the shaft (e.g., gear teeth, notches, reflective marks). Many of the recently developed torque transducers (especially those aimed at power steering applications) utilize the variable permeance associated with the twist dependent tooth/notch alignment of a pair of gear-like ferromagnetic discs effectively attached near opposite ends of a high compliance "torsion bar" [16]. Work on differential capacitive methods for sensing the angular twist in rotating shafts for automotive applications has also been reported [17]. Operation of most commercially available torque transducers depends on sensing the surface shear strain by the change in electrical resistance of "strain gauges" adhesively bonded to the rotating shaft. Various methods are employed to bring electrical power to the gauges and to extract the electrical signal representing the torque. The reliability benefits of replacing the conventional method, using brushes and slip-rings, with non-contacting methods has encouraged the development of several alternative designs, e.g., using rotary transformers [18], infra-red light energy [19] or radio frequency telemetry [20]. By their nature, elastic types of torque transducers require either a long gauging length, high compliance, or a considerable number of shaft mounted electrical components. These requirements discourage their application wherever size, reliability in rugged environments and/or cost are important considerations.

Magnetoelastic torque transducers rely on the fundamental interaction, found in most ferromagnetic materials, between certain measurable physical quantities generally classified separately as either an elastic or a magnetic property. The conventional constructions of these types of transducers depend on torque induced variations in the permeability of a dedicated portion either of the shaft surface itself or of a localized area of attached material (chosen specifically for its superior magnetoelastic properties). These constructions have been described, analyzed and compared in detail by Fleming [21]. Magnetoelastic torque transducers sense torsional stress rather than strain and therefore are generally torsionally stiffer than similarly rated elastic types. This leads to smaller envelope dimensions and mechanical robustness. Moreover, since these devices operate by an inherently non-contacting, magnetic mode of sensing, as a class they appear better suited for the measurement of torque on rotating shafts. However, the performance of these conventional types of magnetoelastic torque transducers is highly dependent on magnetic excitation parameters and material variations, factors which complicate temperature compensation and calibration stability [22]. This makes it difficult to design low cost, mass producible units.

A different type of magnetoelastic torque transducer technology has recently been described, different in that a circumferentially polarized ring of magnetoelastically active material actually creates a magnetic field that, in polarity and intensity, is a near perfect linear analog of the torque carried by a shaft on which the ring is mounted [23, 24, 25, 26]. These transducers use no excitation power and require only a magnetic field sensor in addition to the ring to construct a complete device. Second generation transducers, using two oppositely magnetized rings [27] or a single ring having contiguous regions of opposite circular polarizations [28, 29], together with symmetrically located field sensors, also offer effective immunity to ambiguous effects of ambient magnetic fields. Still, in many applications, realization of the full benefits from this basically simple, low cost and readily adaptable construction is hindered by a limitation of economically compatible field sensors. Thus while integrated circuit silicon Hall effect sensors are available in small, standardized packages, are electrically simple and low in cost, even the best commercially available devices [30] have output drifts with temperature that are commensurate with the torque respondent output signals. Compensation techniques, e.g., using matched pairs, can add sufficient cost to mitigate the economic attractiveness of this technology. The non-availability of "off the shelf" components for other field sensing technologies, e.g., "flux gate" methods, necessitates a specific field sensor design for each newly contemplated torque transducer.

POWER - Most shaft power measuring instruments use clearly distinct devices for developing the speed and torque signals [11, 13, 31]. Sometimes speed and torque are determined from the measurement of two related time intervals [12, 14, 15, 32] and some devices, while employing different sensing technologies for the two measurements, combine both sensors within the same housing [33, 34]. The desirability of providing both torque and speed information in one signal was recognized more than 50 years ago [35] although in that early device speed information was derived from the (necessary) presence of a fixed pattern of torque variation. Nevertheless, with all such devices, multiplication of the two measured values is required to determine the power transmitted by the rotating shaft.

NEW METHODS - In this paper we first describe a simple variation of the polarized ring torque transducer, exclusively for use with *rotating* shafts, that avoids the undesirable consequences of drift in the quiescent output signal of low cost field sensors. This new construction also provides easily separable rotational speed and torque information in one output signal, thus enabling power to be determined from a single transducer. We also describe a further variation in which an output signal proportional to the transmitted power is *generated* by the rotation of the shaft [36]. presence of a fixed pattern of torque variation.

DESCRIPTION

Evolution of this new power/torque/speed (PTS) transducer from the basic polarized ring transducer is schematically illustrated in Fig. 1.

As indicated in (b) of this figure (and as explained in detail in [23-29]), a magnetic field is created in regions of space surrounding the magnetoelastic ring whenever a torque is applied to the shaft on which the ring is mounted. Narrow "homogenizer" rings, of Permalloy or similarly magnetically

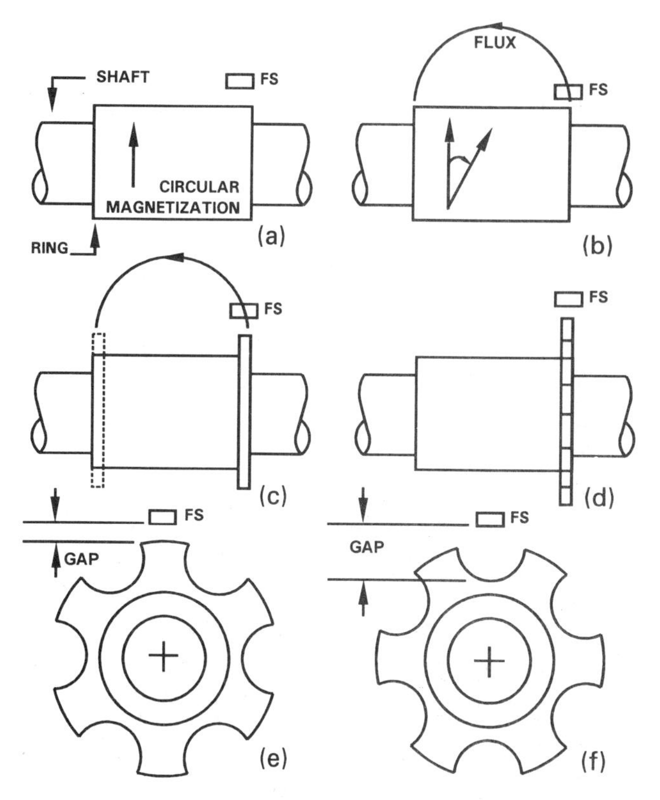

Fig. 1. Evolution of the PTS transducer.
(a) Basic construction. A circumferentially polarized ring of magnetoelastically active material is affixed to the shaft. FS is a magnetic field sensor.
(b) Stresses in the ring associated with torque on the shaft tilt the ring magnetization into a helical path. The resulting magnetic flux is sensed by FS.
(c) Soft magnetic "homogenizer" rings eliminate localized flux variations.
(d)Teeth on the periphery of the homogenizer ring modulate the permeance of the flux paths through FS as the shaft rotates.
(e) Small effective air gap when FS is opposite a tooth.
(f) Larger effective air gap when FS is opposite a space.

soft material, affixed to one or both (or all) polar regions of the field generating ring, as indicated in Fig. 1(c), assure the circumferential uniformity of this field. If the periphery of a homogenizer ring is provided with some uniformly spaced salient features, such as the gear-like teeth illustrated in Fig. 1 (d,e,f), the field intensity in space near this ring will vary in intensity with a circumferential pattern representative of the peripheral features. The effects of a smooth homogenizer ring, and of one having a toothed periphery, on the variation of field intensity at a fixed radius in a polar region of the magnetoelastic ring, is shown in Fig. 2. The actual field intensity at any particular point along the circumference of any one such circle, will vary linearly with the torque. The polarity of the field, being (for any one ring) dependent only on the CW or CCW directionality of the torque, will therefore be constant around any such circle.

In the presence of torque T on the shaft, the field intensity H_t at FS when FS is directly opposite a tooth, as in Fig. 1(e), may be found from

$$H_t = k_t T. \quad (1)$$

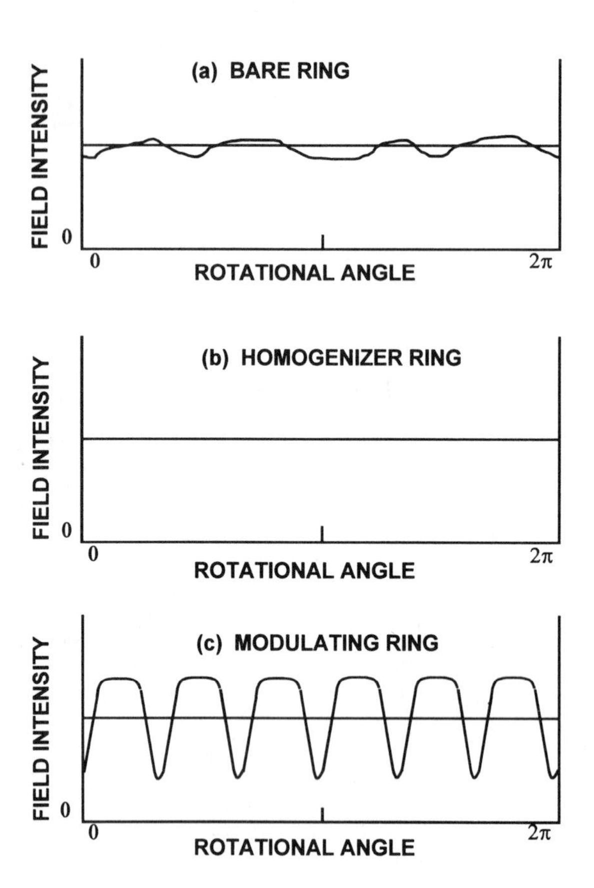

Fig. 2. Circumferential variation in field intensity (in arbitrary units) at constant radius with fixed torque.

where the value of k_t depends on the ring material and its dimensions as well as on the actual location of FS relative to the ring. Thus k_t is a constant associated with the specific transducer construction. Similarly, when the shaft rotates into a position such that FS is centered over a space, as in Fig. 1(f), the field intensity H_s becomes

$$H_s = k_s T. \tag{2}$$

where k_s is clearly smaller than k_t and also reflects the specifics of the transducer construction. Between these two locations the field intensity will vary continuously following some function of the rotational angle determined by both the precise tooth/space shapes and the point in space at which it is measured. Nevertheless, at any rotational angle and for any air gap, the field intensity will follow the transfer function of the ring, i.e., it will be a linear function of the torque. If the shaft rotates continuously, the field intensity at FS thus varies between the two extreme values, H_t and H_s, each a linear function of the torque. The output signal V_o from FS in any field H will in general be found from

$$V_o = QV_o + SH \tag{3}$$

where QV_o is the quiescent output signal (i.e., the offset voltage in zero field) and S is its sensitivity factor. Thus the output signal will vary between

$$V_{ot} = QV_o + SH_t \tag{4}$$

when FS is opposite a tooth, and

$$V_{os} = QV_o + SH_s \tag{5}$$

when FS is opposite a space. As the shaft rotates the output signal will periodically vary between these two extreme values. The peak variation in output signal, ΔV_{op}, is found from the difference between V_{ot} and V_{os} as

$$\begin{aligned}\Delta V_{op} &= (QV_o + SH_t) - (QV_o + Sh_s)\\ &= S(H_t - H_s)\end{aligned} \tag{6}$$

which is both a linear function of the torque and independent of QV_o. ΔV_{op} is thus free from the thermal or other drifts in QV_o peculiar to any one field sensing device.

Recognizing that the *shape* of the periodic waveform resulting from the variation in V_o accompanying shaft rotation is constant for any specific transducer construction, it should be clear that the measurement of any characteristic *amplitude* of the variational (ac) component of the V_o waveform, e.g., peak to peak, rms, average of absolute value, etc. will provide a value that is linearly dependent on ΔV_{op} and hence will be a valid measure of the torque. This is illustrated in Fig. 3.

It should also be clear that the *frequency* of this waveform is proportional to the rotational speed, being simply the number of teeth in the ring (a constant for any one transducer) multiplied by the speed. Thus by spatially modulating the sensed portion of the torque responsive field from a magnetically polarized ring, a single signal containing both torque and speed information can be obtained with neither signal component being affected by the drift characteristics of the field sensing device.

If the ac component of V_o (i.e., Voac) vs time waveform is assumed for the moment to be represented by the simple sinusoidal function:

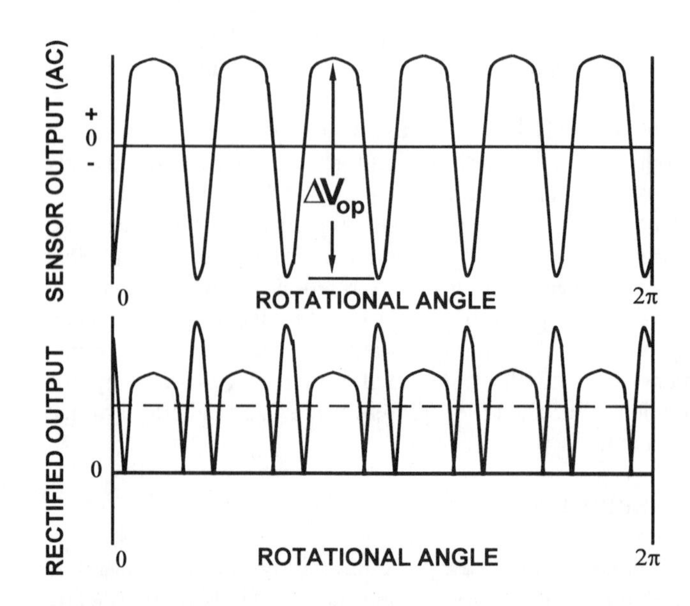

Fig. 3. (a) Variational (ac) component of signal from field intensity sensor having raw output signal shown in Fig. 2(c). Peak to peak amplitude is a linear function of the torque.

(b) Rectified waveform of (a). Dotted line represents average of absolute values over full rotation. For any fixed waveshape, this average varies with ΔV_{op}.

$$V_{oac} = kT\sin(n\omega t) \quad (7)$$

where n is the number of teeth on the field intensity modulating ring and ω is the angular velocity of the shaft, its time derivative becomes

$$dV_{oac}/dt = knT\omega\cos(n\omega t). \quad (8)$$

Other than the constructional constants k and n, the amplitude of the periodic function defined by equation (8) is seen to be proportional to the product Tω and this product is the *power* being transmitted by the shaft! Since more complex waveforms can be represented by the sum of fundamental and harmonically related sinewaves, the sum of the derivatives of these components will similarly have an amplitude that is proportional to power.

It is not necessary to compute the time derivative of the output voltage from a field intensity sensor in order to develop a signal that is proportional to power. As the shaft rotates, the time varying magnetic field in the vicinity of the modulator ring will induce a periodic voltage, e, in a nearby coil, having N turns, that is at all times proportional to the instantaneous rate of change of flux linking the coil, i.e.,

$$e = N d\varphi/dt \quad \text{(Faraday's Law)}. \quad (9)$$

While changing periodically during rotation of the shaft, the flux change dϕ during any small angular rotation of the shaft dθ is always a linear function of the torque (since the flux density is proportional to the field intensity). Thus

$$d\phi/d\theta = cT \quad (10)$$

where c varies both with constructional details and the rotation angle. Since the angular velocity ω = dθ/dt, the induced voltage can be expressed as

$$e = N(d\phi/d\theta)(d\theta/dt) = (NT\omega)c. \quad (11)$$

Thus while the voltage induced in the coil changes periodically with rotation angle, it's *amplitude* is proportional to Tω, i.e., the power transmitted along the shaft. Rotational speed may, if desired, be obtained from the *frequency* of this signal. It is to be noted that with this simple construction, no electrical power source is required to obtain a signal proportional to the power being transmitted by the shaft since this signal is actually *generated* by the rotation of the shaft.

EXPERIMENTAL RESULTS

An experimental PTS transducer was constructed and tested under various conditions of torque and rotational speed using both Hall effect and coil type of field sensors. Details of construction of the shaft and magnetoelastic ring are shown in for Fig. 4. Matching tapers were provided in the bore of the ring and the on the mating surface of the shaft to simplify attainment of an assembly having a controlled interference fit; in this case the ring was axially pressed 5 mm (from the hand tight position) on to the stainless steel shaft. The resulting contact pressure at their interface provided sufficient friction

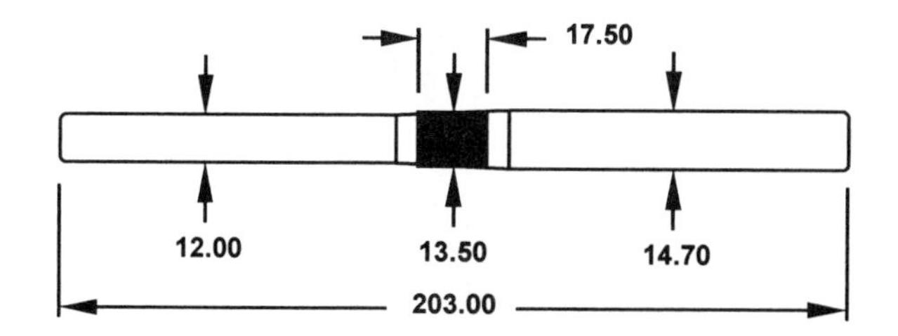

Fig. 4. Dimensions of shaft and field generating ring used for experimental PTS transducer. Ring of 18% Ni maraging steel [23] is axially centered on type 303 stainless steel shaft. The ring ID and shaft OD having matching 1:48 tapers.

for the ring and shaft to act as a mechanical unit while the tensile hoop stress in the ring established the circumferential direction as the magnetic "easy axis" [23]. Though the ring was physically a single piece it was effectively divided into two magnetically distinct regions, A and B, by polarizing each axial half in respectively opposite circumferential directions. Simultaneous polarization of both regions was accomplished by rotation past an assembled pair of oppositely poled electromagnets [28]. This "dual region" design was chosen to avoid possibly troublesome effects from the relatively strong (> 10 Oe at the highest levels of torque) magnetic fields created by the dynamometer [27, 28].

Three modulator rings, detailed as shown in Fig. 5, were affixed, by light press fits, one at each end and one in the center of the magnetically active ring. The circumferentially aligned modulator rings were accurately located and held square to the shaft axis by means of stainless steel spacer rings fitting closely over the active ring. The assembled field generating portion of the transducer is shown in Fig. 6.

Each region A and B in this figure is seen to effectively comprise a separate field generating region of the type described in Fig. 1. Since the active ring is oppositely polarized in the two regions, under an applied torque, oppositely directed fields arise in the two axial spaces between an end and the center modulator rings. Two field sensors mounted to the frame of the experimental set-up were inserted in these spaces as indicated in Fig. 6. Connecting these sensors to sense the differential field in these two spaces eliminated effects of fields from other sources (e.g., those from the dynamometer) having the same direction in each space.

The shaft was driven by a nominally 1.5 HP - 2300 RPM DC shunt motor with the transmitted power absorbed in a water cooled eddy current dynamometer (Borghi & Saveri

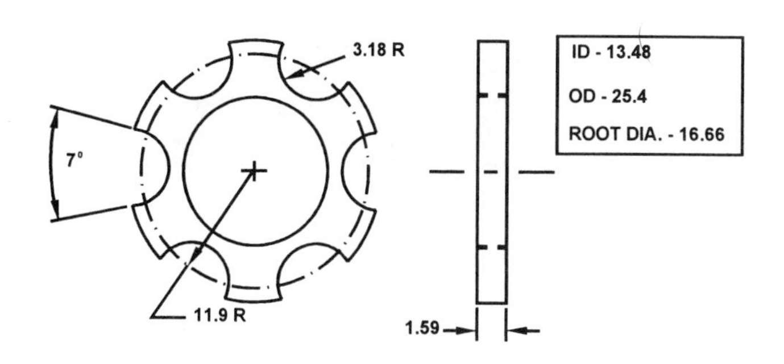

Fig. 5. Modulator ring. Material: 78 Permalloy, annealed in Hydrogen for 1 hour at 1100° C.

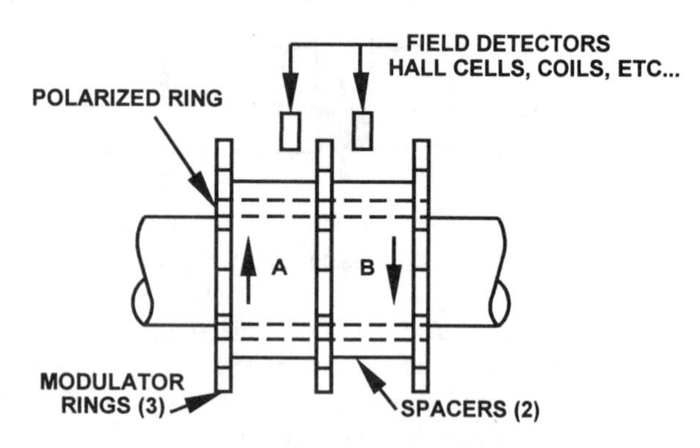

Fig. 6. Field generating and sensing portion of the experimental PTS transducer.

Model FA. 100/30 SL). The reaction torque developed on the dynamometer casing was measured with a load cell (Interface Model SSM 500) calibrated with weights on a measured lever arm. Rotational speed was indicated directly in RPM using a digital frequency meter displaying the count of electrical pulses generated in one second (i.e., hertz) by the teeth on a 60 tooth steel gear passing a magnetic proximity sensor. Two Hall effect integrated circuit field sensors (Allegro type 3506 UA [30]) or two coils (to be described), mounted on small circuit boards, were used to sense the generated field. See Fig. 7.

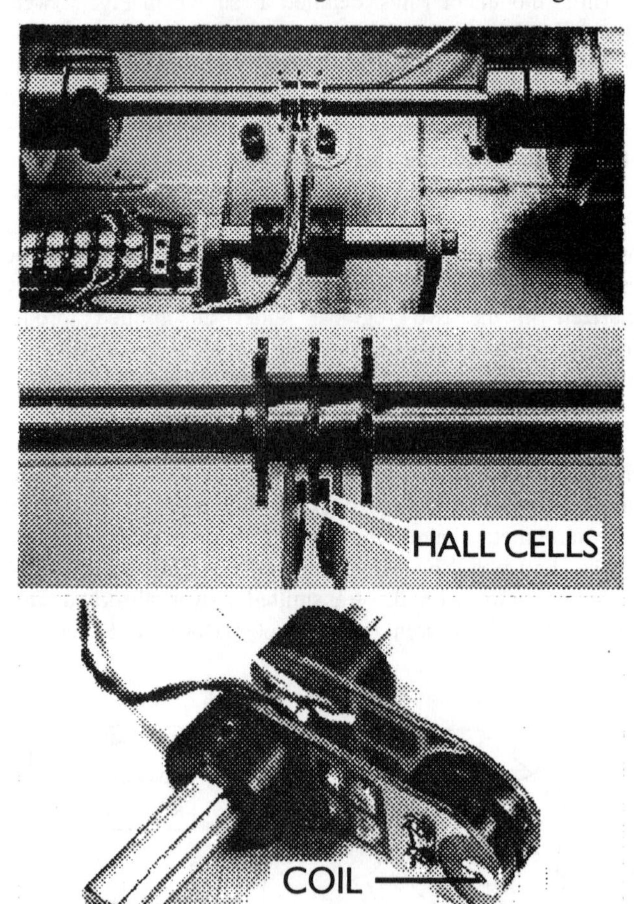

Fig. 7. Photographs of the experimental set-up. (Top) General view. (Center) Close up of sensor section with Hall cell field sensors. (Bottom) Close up of coil sensors and mounting assembly. This entire assembly is interchangeable with the Hall cell sensor assembly seen in top photo.

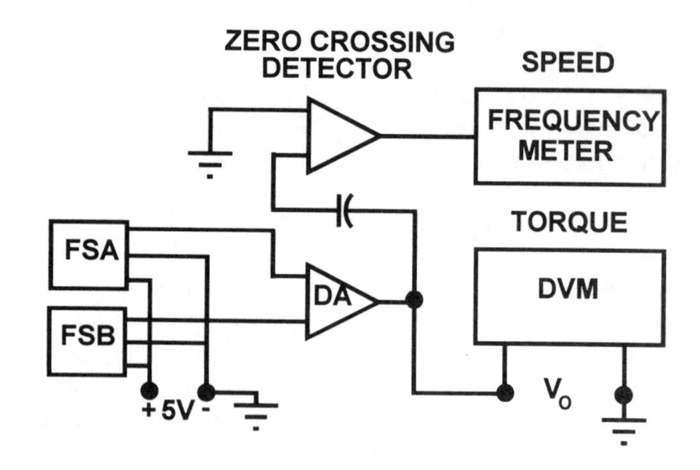

Fig. 8. Circuit for sensing torque and speed using two Hall effect field sensors FSA and FSB. DA is a differential amplifier; DVM is an ac digital (RMS) voltmeter.

The electrical circuit used for extracting and separately displaying the torque and speed information from the Hall sensor output signals is schematically shown in Fig. 8. An (ac coupled) oscillogram of a typical signal appearing across the output terminals of the differential amplifier (V_O in Fig. 8) is shown in Fig. 9. The effectiveness of the modulator rings in creating a periodic variation in the magnitude of the sensed field is clearly evident. The relatively flat and prolonged peaks representing the sensed field opposite a tooth suggests that a modulator ring having narrower teeth and wider intertooth spaces would provide a signal having a greater peak to peak amplitude for the same torque.

The chart recording in Fig. 10 shows, for conditions of fixed torque and rotational speed, the transient variations in V_O when first one Hall sensor and then the other was rapidly cooled by a momentary spray of a liquid refrigerant (chlorodifluoromethane). The dramatic changes in V_O seen in this recording resulted from the "thermal drift" in QV_O of the sprayed sensor. While it is not likely that such steep thermal gradients would exist in any practical application, the allowable variation in QV_O of any one Hall sensor over the rated temperature range [30] is itself comparable to the intensity of the field created by the ring under rated torque. On the other hand, since the nominal variation of sensitivity (S in equations (3)-(6)) was designed [30] to be -0.02%/°C, individual variations in S will expectedly have a far smaller

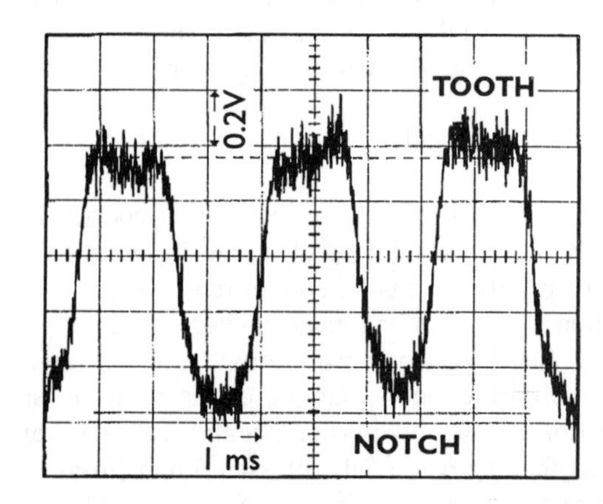

Fig. 9. Oscillogram of V_{oac} (see fig. 8). DA gain was 66.

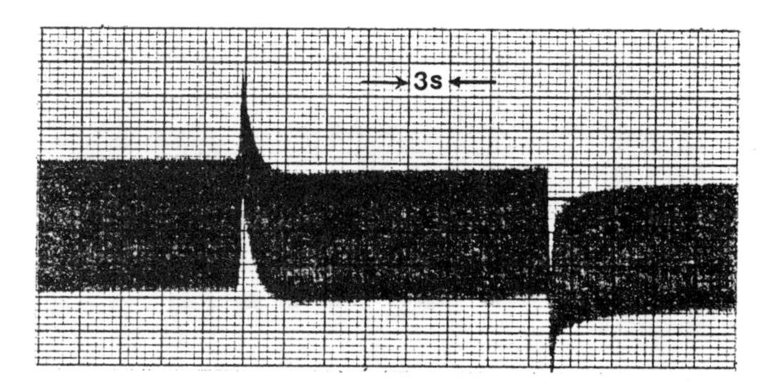

Fig. 10. Chart recording showing immunity of ac component to the stark changes in V_o induced by thermal transients,

(and if required, a more easily compensated) effect on V_o. The lack of visible change in the peak to peak signal throughout the entire recording period shown in Fig. 10 attests to the usability of Hall sensors, in spite of their propensity for thermal drifts, to provide a reliable measure of torque by simply separating out the amplitude of the field modulation.

Measurements of the rms value of the output signal (Voac) at three fixed values of torque are plotted against rotational speed in Fig. 11. While relatively small there are nevertheless some notable deviations of the indicated signals from perfect constancy over the test range of speeds. Three sources of speed dependence in the indicated signals are readily identified: 1) Meter error. Vo was measured with a Triplett Model - 4800 digital multimeter having no rated capability for voltage measurements at frequencies below 20 Hz, i.e., below 200 rpm with 6 tooth modulator rings. 2) Torque error. Torque on the test transducer exceeded the torque measured by the dynamometer by both the frictional torque in the input bearing and the drag torque (windage) originating primarily at the coupling to the dynamometer. 3) Dynamic error. Motor torque is never free of ripple components. Moreover the transducer was mounted on a relatively long slender shaft test attached at its ends to rotating masses having substantial moments of inertia, exactly the conditions required to stimulate and maintain tor-

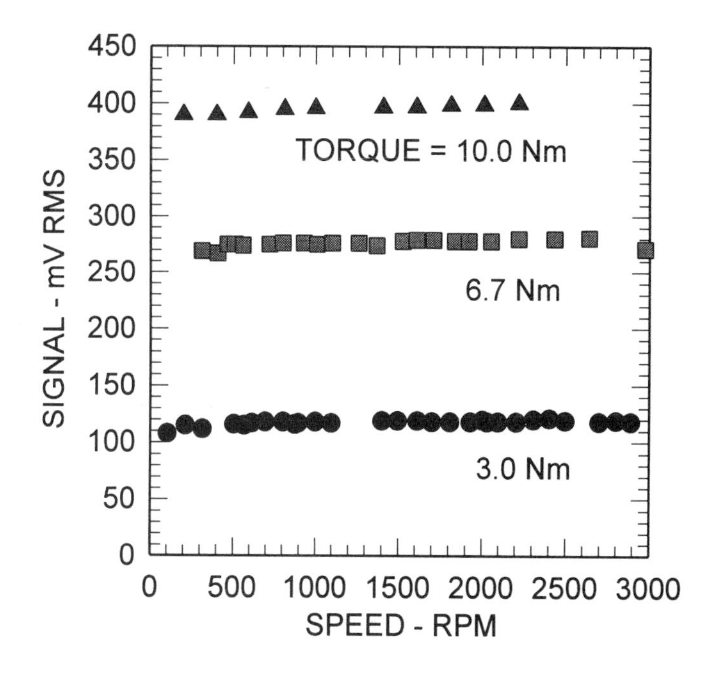

Fig. 11. Variation in V_{oac} with speed at torques indicated.

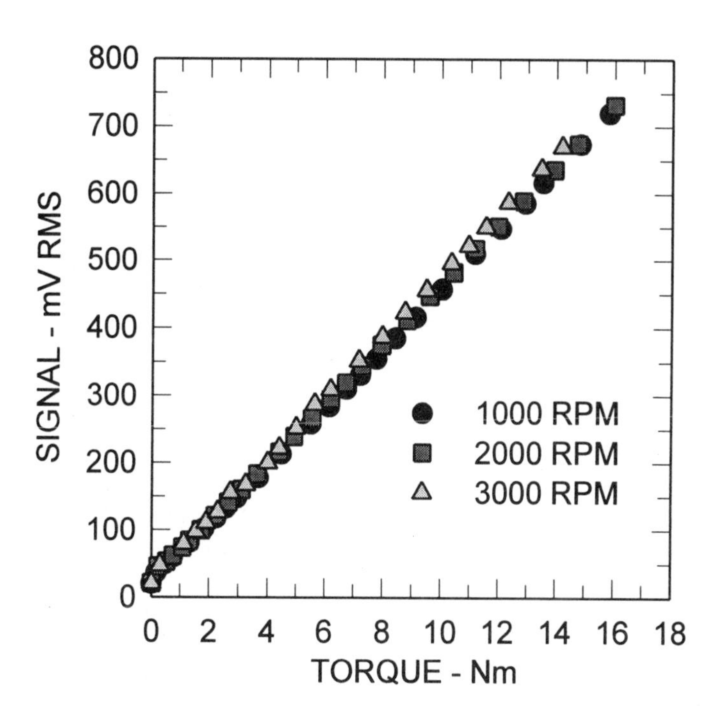

Fig. 12. Variation in V_{oac} with torque at speeds indicated.

sional oscillations at several frequencies within the measurement speed range. These torsional resonances were manifested by unequal peak signals for each of the 6 teeth on the modulator rings, a condition clearly affecting the ac component and hence the rms value of the output signal.

The data plotted in Fig. 12 indicates the expected linear variation in output signal with torque. The small signals seen to be present at zero torque represent both circuit noise and drag torque. There appears to be both the expected increase in drag torque with increasing speed and a further slight increase in signal with increasing torque. This latter is thought to be of magnetic origin, associated with the increasing stray field from the dynamometer. (Such fields can affect the measurements via an additional drag torque associated with the interaction of eddy currents induced in external rotating parts with magnetized parts of the adjoining frame, or via unequally altering the fields at the sensor locations; appearing thereby as an additional signal component. This latter effect seems likely since the stray field falls off with distance from the dynamometer and the modulator rings, being axially separated from each other, are therefore situated in unequal ambient fields. Shielding (as by encasing the transducer) would eliminate this effect.)

Power measurements were made using two small coils, instead of the Hall effect devices, as field sensors. The coils were mounted on the physically interchangeable assembly shown in the bottom photo of Fig. 7. Details of the sensing coils and the associated measuring circuit are shown in Fig. 13. (The "home brewed" cores increased the signal voltages to nearly 3 times the air core values.) Since the signals were directly generated in the coils, no electronics were needed other than were contained within the digital voltmeter.

Fig. 14 is an oscillogram of typical output signals. The waveshape should be recognized as the time derivative of the (inverted) waveform shown in Fig. 9, as expected from consideration of equation (9). Since the period of the wave seen in Fig. 14 is the time between passages of successive teeth on the modulator rings, six such periods represent the

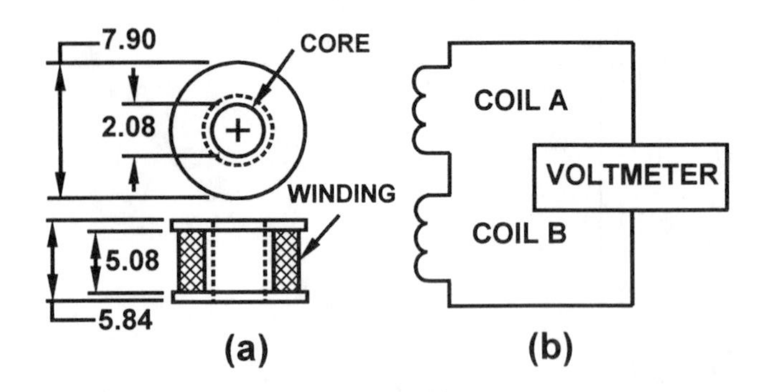

Fig. 13. (a) Details of field sensing coils. The core was made from a mixture of permalloy powder and epoxy. The winding consisted of 1000 turns of polynylon insulated #40 AWG wire. (b) The measuring circuit, comprised of series connected coils.

time for a single rotation of the shaft. Hence,Fig. 14 indicates that the shaft rotates once each 18 ms, i.e., at 3300 RPM.

Fig. 15 shows the relative constancy of the output signals over wide speed ranges at three fixed levels of transmitted power. The imperfect flatness of these data sets reflects all of the previously identified sources of measurement error, further exacerbated by the experimental difficulty in maintaining a precisely constant product of torque and speed. Tests at the higher power levels needed to be performed quickly, as can be appreciated by considering that the nominal rating of the driving motor was only 1.12 kW (1.5 Hp).

The data plotted in Fig. 16 shows that the linear variation in output signal with transmitted power at constant speed, predicted by equation (11), is closely realizable even with this unshielded experimental transducer.

CONCLUSIONS

The PTS transducer has been shown to be a physically predictable device of simple construction that is suitable for simultaneous measurements of the power, torque and speed of a rotating shaft. Proper application also requires recognition of its limitations. For example it should be recognized that (at least in the basic constructions described here) there will be no output signal if the shaft is either not rotating or not transmitting torque. This should not be a serious limitation since most shafts are functional only when they are both rotating and transmitting torque. Also, to obtain a meaningful

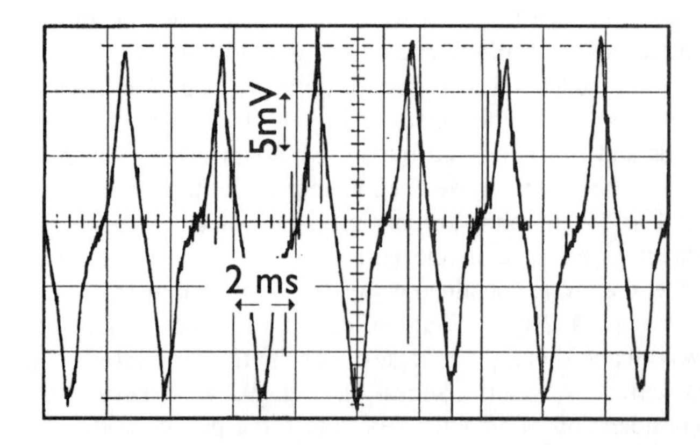

Fig. 14. Oscillogram showing typical output signal across series connected coils.

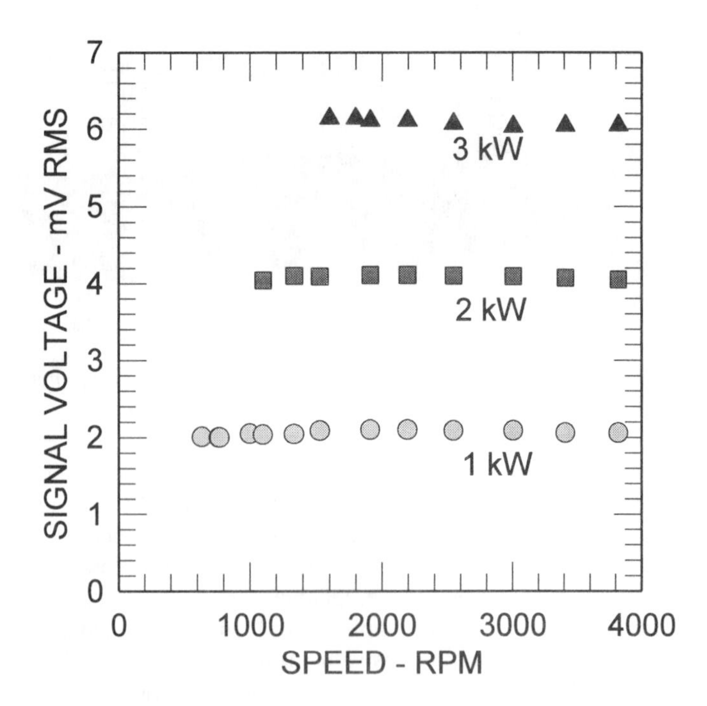

Fig. 15. Variation in output signal with speed at transmitted power levels indicated.

signal, the torque must remain essentially constant while the shaft rotates through an angle that includes at least one whole tooth and one whole space. While these conditions exclude its use for detailed measurement of dynamic torque, in many applications, torque, speed and power change little over one or more complete shaft revolutions.

The PTS transducer should find application in all of the industries and quite possibly in many of the machine types listed in TABLE I. Directly and indirectly, therefore, the PTS transducer promises significant benefit to the mobility industry.

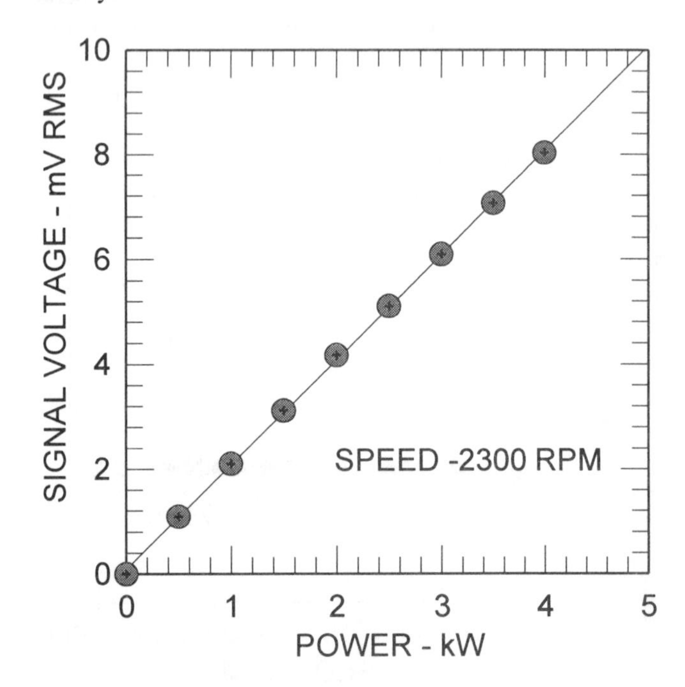

Fig. 16. Variation in output signal with transmitted power at constant speed indicated. The line is the linear regression calculated from the experimental data points.

REFERENCES

1. Alfred Del Vecchio, Dictionary of Mechanical Engineering, Peter Owen Ltd., London, 1961, p 222.
2. Kents Mechanical Engineers Handbook, 12th Ed., 1962, POWER, J. Kenneth Salisbury, Wiley, p 16-99.
3. Mechanical Prime Movers, Edited by P.C. Bell, The Macmillan Press Limited, Hampshire UK, 1971.
4. Standard No. MG1-1993, National Electrical Manufacturers Association (NEMA), 1993.
5. See manufacturer's catalogs, e.g.: Ingersoll-Rand Co.-Air Compressors; Goulds Pumps, Inc.-Submersible Pumps; Bridgeport Machines, Inc.-Series Standard Mill; Craftsman-Industrial Vacuum Cleaners.
6. Supplement to ASME Performance Test Codes, Measurement of Shaft Power, ANSI/ASME PTC 19.7-1980 (Reaffirmed 1988).
7. Supplement to ASME Performance Test Codes, Measurement of Rotary Speed, ASME PTC 19.13-1961. See also C. F. Shoop and G. L. Tuve, Mechanical Engineering Practice, McGraw-Hill, NY, originally published in 1930, final edition in 1956.
8. S. F. Brown and C. Rigaux, Speed sensor integration in wheel bearing hub units, SAE Paper No. 910899, 1991.
9. Bayer Oswald, A new generation: ABS capable wheel bearings, SAE Paper No. 910699, 1991.
10. Ivan J. Garshelis, A new type of magnetic motion sensor and its application, SAE Paper No. 930235, 1993.
11. U.S. Pat. #3,978,718, R. W. Schorsch, ELECTRONIC DYNAMOMETER, 1976.
12. U.S. Pat. #3,729,989, D. R. Little, HORSEPOWER AND TORQUE MEASURING INSTRUMENT, 1973.
13. U.S. Pat #4,064,748, Leshner et al., POWER INDICATING MEANS AND METHOD FOR AN INTERNAL COMBUSTION ENGINE, 1977, Assigned to Fuel Injection Development Corp..
14. U.S. Pat. #3,273,386, H. Sipler TORQUE TRANSDUCER UTILIZING A MAGNETIZED SHAFT HAVING A SURFACE DISCONTINUITY REFERENCE MARK, 1966, Assigned to U. S. Gov't. (Army).
15. Noncontact measurement of shaft speed, torque and power, NASA Tech Briefs, January 1993, p 30.
16. U. S. Pat. #4,876,899, D. B. Strott and K. W. Kawate, TORQUE SENSING DEVICE, 1989, Assigned to Texas Instruments Inc..
17. J. D. Turner, Development of a rotating-shaft torque sensor for automotive applications, IEE Proceedings, Vol. 135, Pt. D, No. 5, 1988, p 334-8.
18. E. Zabler, F. Heintz, A. Dukart and P. Krott, A non contact strain gage torque sensor for automotive servo driven steering systems, SAE Paper No. 940629, 1994.
19. TorXimitor™ , Bently Nevada Corporation, Minden, Nevada, 1991.
20. Susan J. Woodward, Catching shaft torque data on the fly, Power Trans. Des., Vol. 35, No. 9, 1993, p 24.
21. William J. Fleming, Magnetostrictive torque sensors-comparison of branch, cross and solenoidal designs, SAE Paper No. 900264, 1990.
22. U.S. Pat. #4,920,809, S. Yoshimura; R. Ishino; S. Takada; H. Kimura, MAGNETICALLY ANISOTROPIC TORQUE MEASURING DEVICE WITH ERROR CORRECTION, 1990, and U.S. Pat. #5,307,690, A. Hanazawa, TEMPERATURE COMPENSATING DEVICE FOR TORQUE MEASURING APPARATUS, 1994. Both assigned to Kubota Corporation.
23. I. J. Garshelis, K. Whitney and L. May, Development of a non-contact torque transducer for electric power steering systems, SAE Paper No. 920707, 1992.
24. I. J. Garshelis, A torque transducer utilizing a circularly polarized ring, IEEE Trans. Magn. Vol. 28, No. 5, 1992, p 2202-4.
25. I. J. Garshelis, Investigations of parameters affecting the performance of polarized ring torque transducers, IEEE Trans. Magn, Vol. 29, No. 6, p 3201-3, 1993.
26. U. S. Pat .# 5,351,555, I. J. Garshelis, CIRCULARLY MAGNETIZED NON-CONTACT TORQUE SENSOR AND METHOD FOR MEASURING TORQUE USING SAME, 1994, Assigned to Magnetoelastic Devices, Inc.,.
27. I. J. Garshelis and C. R. Conto, A torque transducer utilizing two oppositely polarized rings, IEEE Trans. Magn. Vol. 30, No. 6, 1994, p 4629-31.
28. I. J. Garshelis and C. R. Conto, A torque transducer utilizing a ring containing oppositely polarized regions, submitted for presentation at 1995 Intermag Conference.
29. Patents applied for.
30. Allegro Microsystems Inc., Worcester, Mass., Data sheet type 3506UA.
31. U.S. Pat. #4,100,794, Meixner, 1978; #4,106,334, Studtmann, 1978; #4,306,462, Meixner, 1981; # 4,406,168 Meixner, 1983; #4,479,390 Meixner, 1984, Assigned to Borg-Warner Corporation..
32. U.S. Pat. #5,195,382, F. Peilloud, DEVICE FOR MEASURING SPEED AND TORQUE ON A SHAFT, 1993, Assigned to The Torrington Company.
33. U.S. Pat. #5,323,659, Wakamiya et al., MULTI-FUNCTIONAL TORQUE SENSOR, 1994, Assigned to Matsushita Electric Industrial Co. Ltd.
34. Lebow Load Cell and Torque Handbook, Eaton Corporation, Troy, Mich., 1991.
35. U. S. Pat. #2,365,073, N. L. Haight, MEANS FOR INDICATING HORSEPOWER AND HORSEPOWER FACTORS, 1944.
36. Patents applied for.

SMART SENSORS

920471

Automotive Silicon Sensor Integration

William Dunn and Randy Frank
Motorola Semiconductor Products Sector

Abstract

Silicon by virtue of its electrical and mechanical properties is eminently suited for use in mechanical and other types of sensors. The integration of these sensors with signal conditioning circuits is being used to develop a wide variety of cost effective devices for use in automotive applications.

A number of micromachining techniques and methods of converting mechanical force to electrical signals are available. Each of these must be evaluated for ease of integration with respect to the types of signal conditioning that are required to obtain the most cost effective system solution. Additional factors that have to be considered are analog versus digital outputs, temperature operating range, linearity, self test features, reliability, packaging, testing and assembly problems. This paper will explore and evaluate these features for silicon pressure, position and acceleration (crash) sensors developed for automotive applications.

INTRODUCTION

The increasing need for sensors in automotive systems has provided several new opportunities for sensor manufacturers. These sensors must have (relatively) high accuracy with high reliability and low cost. The new systems are based on MCU (microcontroller unit) or DSP (digital signal processing) technology and the sensor's ability to easily interface to these control devices can either be a cost advantage or disadvantage depending on the technology that is used by the sensor manufacturer.

The ultimate in cost effective micromachined sensors will be achieved through integration. The approach to this integration is occurring from two different directions: the makers of discrete sensors are attempting to add control electronics for integration and semiconductor manufacturers are approaching integration from the chip processing stand point - developing sensors with technology that is compatible with bipolar and/or CMOS processing. The dilemma on one hand is that discrete micromachined sensors were developed, and indeed have proven themselves over the years, without integration due to the low volume requirements, and the ability of the market to support a high price. Consequently the sensor processing is not compatible with the processing of the control electronics. Processing development and modifications are required for cost effective integration to meet the demands of a high volume low cost market.

SENSORS AND INTERFACE CIRCUITS

The type of sensing technology and the appropriate interface circuit can vary considerably depending on the parameter to be sensed and the system requirements such as accuracy, resolution and noise immunity. The ability to integrate the interface circuit on the same substrate as the sensing element is an inherent advantage of silicon sensors. The usage of semiconductor processing techniques for the sensing element also means that process revisions can also allow interface circuits to be designed that can be specifically tailored for the sensing application. Three examples of different sensing, processing and interface designs will be discussed to demonstrate the range of variations that are possible.

PRESSURE SENSORS Silicon pressure sensors are an excellent example of a technology that has gained almost universal acceptance for the sensing of manifold absolute pressure. Both capacitive and piezoresistive sensors have been in high volume production for several years. The piezoresistive approach also has the capability to be applied to several differential or gauge pressure measurements that are, or will be, required in vehicles including: oil pressure, tire pressure, fuel vapor pressure and EGR pressure. Pressure sensors can also be used to measure liquid and air flow and liquid (i.e., fuel) level.

The basic piezoresistive silicon pressure sensor combines semiconductor processes of photolithography, ion implanting, diffusion, deposition of thin film resistors, on-chip laser trimming and chemical etching. The piezoresistive sensing element(s) are either diffused or ion implanted into the silicon to provide a stress sensitive resistor. Silicon's mechanical properties of high modulus of elasticity, excellent elastic properties with little or no hysteresis and high tensile strength are a significant improvement over a traditional strain gage sensor. A much higher gauge factor is also obtained using silicon. A single stress sensitive resistor can be placed at a location in the diaphragm that maximizes sensitivity and linearity. The silicon diaphragm is obtained by chemically etching (bulk micromachining) silicon material from silicon substrate in a batch assembly process. Several hundred units can be obtained simultaneously from a single wafer, and several thousands from a single processing lot, to provide a very low cost, yet consistent performance, sensing element.

In addition to the basic sensing element it is also possible to integrate the control circuitry necessary to provide the amplification, temperature compensation, zero and full scale calibration in the same monolithic structure. The circuit in Figure 1 shows the sensor and interface circuit of a fully integrated, MCU compatible manifold absolute pressure sensor. Bipolar IC technology was combined with semiconductor sensor technology to produce a sensor that utilizes the same voltage reference as the MCU and provides an input directly to the A/D converter. Laser trimming of thin film resistors that are on the sensor means that an extremely small package can be used for the sensor. The reference vacuum is sealed inside the etched cavity by sealing a second wafer to the back of the top wafer by a glass frit. The full functionality of a complex sensing module is achieved in a single silicon chip. With 20 to 105 kPa applied, the output is capable of swinging from 0.2 to 4.9V with a 5.1V supply voltage. A photomicrograph of the integrated pressure sensor is shown in Figure 2. The chip carrier package in which the sensor die is assembled is shown inside the diaphragm area and the single sensing element is on the right of the package.

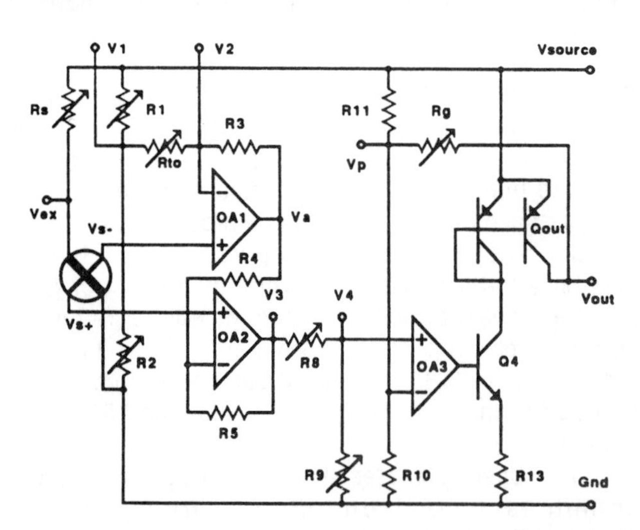

Figure 1 - Single Piezoresistive Silicon Pressure Sensor with Integrated Signal Conditioning Circuit.

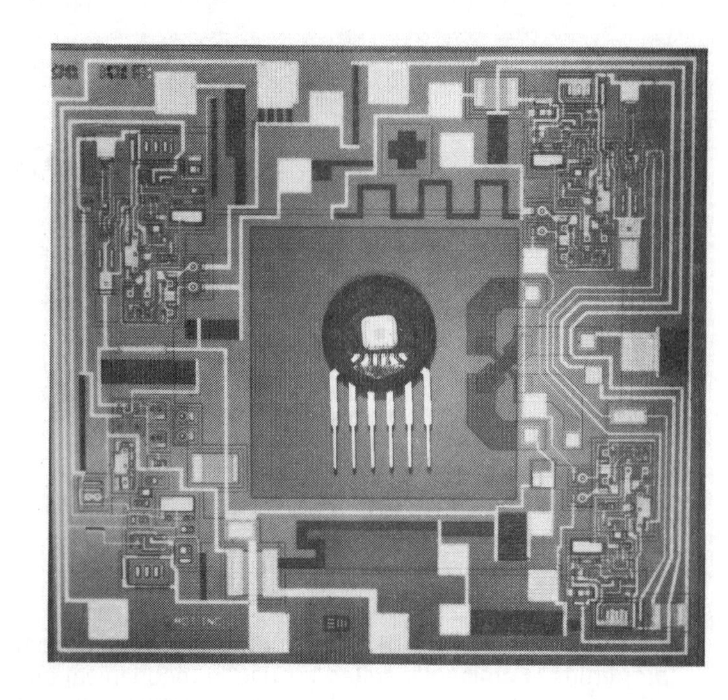

Figure 2 - Photomicrograph of Integrated Silicon Pressure Sensor.

POSITION & SPEED SENSING Automotive systems require position/speed sensing inputs for several parameters including: crank angle, steering wheel angle, vehicle height and wheel speed. Opto sensors are among the techniques that are used to provide these inputs. A typical optical sensor uses infrared GaAs emitters and silicon detectors with an interrupter to provide a countable, time-based input to the MCU. The detector has a Schmitt trigger integrated

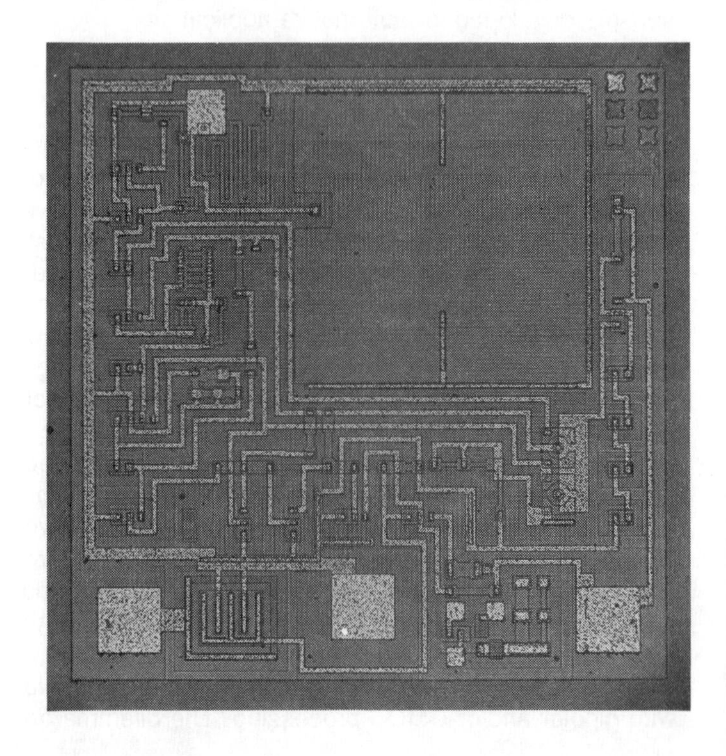

Figure 3 - Opto Detector Die Photomicrograph.

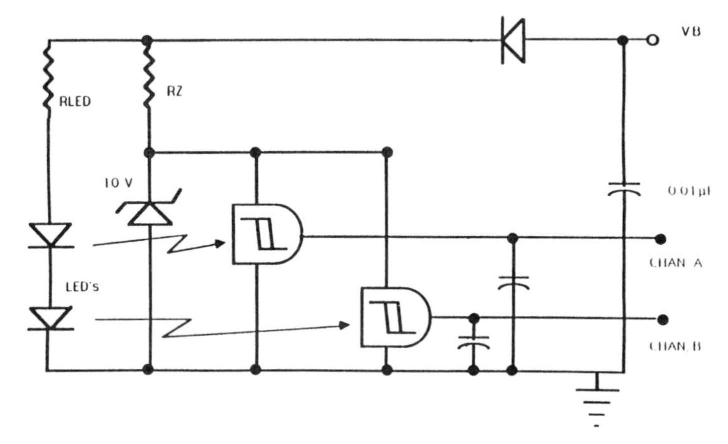

Figure 4 - Interface Circuit for Schmitt Trigger Opto Sensor.

on the chip to minimize the number of components in the sensor interface circuit. The sensor (photodetector) die shown in Figure 3 utilizes bipolar IC processing to achieve both the detector and interface circuit. The opto sensor interface circuit used to achieve a quadrature output signal to provide both position and direction is shown in Figure 4.

ACCELERATION APPLICATIONS Automotive applications for accelerometers are in passive restraint systems, suspension, anti-lock braking systems and traction control systems. The operating range of these devices varies widely from 1 to 100 g's. At present automotive restraint systems use multiple mechanical trigger devices. The goal is to have one cost effective centrally located sensor and control system. In order to meet the reliability requirements, and to have a linear output device for signature monitoring, with self test features, automotive designers are turning to solid state sensors. These sensors have a proven track record in the area of pressure monitoring and are now being developed for monitoring acceleration.

ACCELERATION SENSORS Solid state sensors normally use one of three methods to convert mechanical force into electrical parameters, these are piezoresistive, capacitance, or piezoelectric effects. The characteristics of these three alternatives are shown below.

Because of offset and drift with temperature (pyroelectric effect) in piezoelectric devices, they are unsuitable for use below 5 Hz. Also, overloading of the amplifier can occur due to transient forces. With capacitive micromachined devices there is no shift due to shock, they have a wide temperature operating range, and the ability to provide self test features make this type of sensing preferred.

Capacitive sensing[1] has two modes of implementation: open and closed loop operation. Sensing in either case, is accomplished by differential capacitor measurements, using a charge redistribution technique[2]. The input circuit and the cross section of a capacitive sensor are shown in Figure 5. The capacitors are alternately switched between a reference voltage (Vref) and ground, using switch capacitor techniques. The switching frequency is not critical (typically 250 kHz) and the integrator circuit is used in a virtual ground configuration so that the output is independent of stray capacitances. This circuit converts the differential capacitance to a voltage. With this type of sensing, differential capacitive changes as low as 0.5 fF can be detected, which translates into sensing capacitors of 250 fF (500 fF if not using differential sensing). This gives a capacitor size of 160 microns x 160 microns with 2 microns spacing. This assumes requirements of 1% accuracy, and for good linearity the deflection for full scale output is limited to 10% of the spacing. In closed loop operation, electrostatic forces are used to counter balance the accelerating forces, and hold the seismic mass in its central or neutral position. This type of sensing has a number of advantages over open loop operation. These advantages are:

Reduction of fatigue in the suspension members.

Wide dynamic operating range.

Improved linearity.

No latch up problems.

Elimination of material parameters.

Bandwidth set by control circuits.

Air damping gives high resistance to shock.

Characteristic	Capacitive	Piezoelectric	Piezoresistive
Loading effects	high	high	low
Size	small	small	medium
Temperature range	wide	wide	medium
DC response	yes	no	yes
Sensitivity	high	medium	medium
Shift due to shock	no	yes	no
Self test features	yes	no	no
Damping available	yes	no	yes

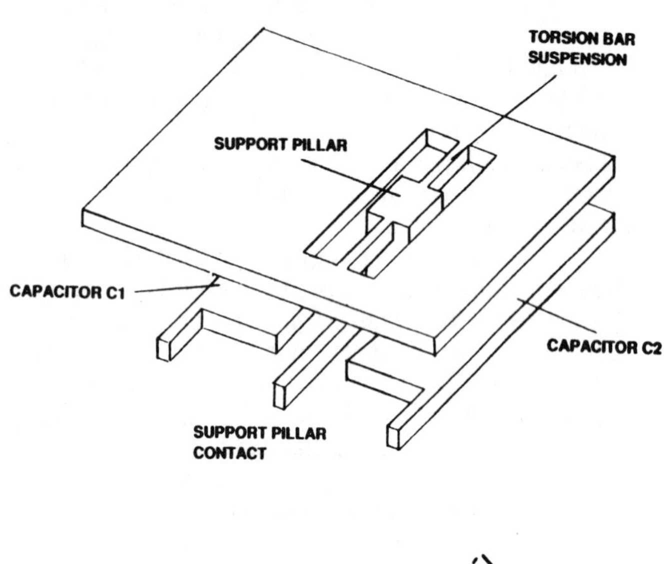

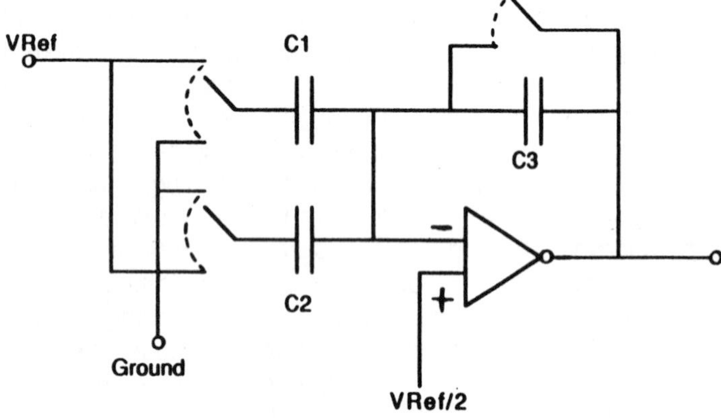

$$V_{out} = \frac{VRef\,(C1 - C2)}{C3}$$

Figure 5 - Capacitive sensing. (a) Capacitive layout; (b) Sensing circuit.

These factors go towards making a device that is reliable and has longevity. The block diagram and waveforms used for closed loop PWM operation are shown in Figure 6. Capacitive sensing can be used at high temperatures (≥ 150°C) as the capacitors are dielectrically isolated. Temperature correction is also not required.

There are two methods of micromachining available: bulk and surface. In bulk micromachining such as the pressure sensor described previously, shapes defined by masking are etched into the silicon wafer, using isotropic and anisotropic etch techniques. The lower shapes are defined by diffusions that are also used as etch stops, and/or by using wafer bonding techniques[4]. With surface micromachining[3], layers of structural material (polysilicon), and sacrificial material (silicon dioxide) are deposited and patterned, and the sacrificial material is then etched away to give the operational structures shown in Figure 7. Either micromachining

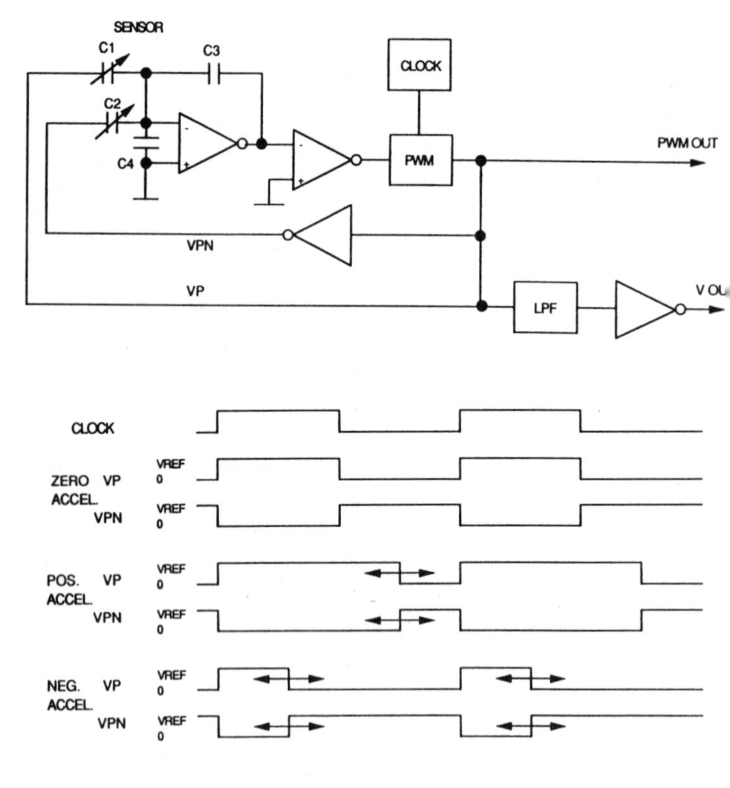

Figure 6 - Closed loop operation.

methodology is suitable for sensors. Surface micromachining technology provides a smaller, more cost effective structure, however, stresses in the layers have to be annealed out, to prevent warping. Bulk material is typically stress free.

The switched capacitor techniques used for sensing small capacitance changes, make CMOS the preferred technology for the signal conditioning circuits with capacitive types of sensors. Using PWM or other modulation techniques, the signals are converted into the time domain, which can be measured very accurately. This also gives a digital format which opens the door for cost effective integration with MCU's and the development of a smart sensor.

Using surface micromachining techniques the structures are dielectrically isolated so that leakage is not a problem at elevated temperatures. The maximum operating temperature, is therefore, limited by the control electronics. With bulk micromachining, the same temperature constraints apply to both the sensor and control electronics.

SYSTEM COMPATIBILITY A number of questions on mounting and trimming have to be addressed - there is a big difference between mounting a 1 and 100 g accelerometer. In a ride control application the 1 g device has a bandwidth of only 100 Hz, so that direct printed circuit mounting is feasible. However, in the case of the 100 g crash (air bag) sensor with a required bandwidth of 1 kHz, rigid chassis mounting is imperative for transmission of the shock wave signature to the sensor.

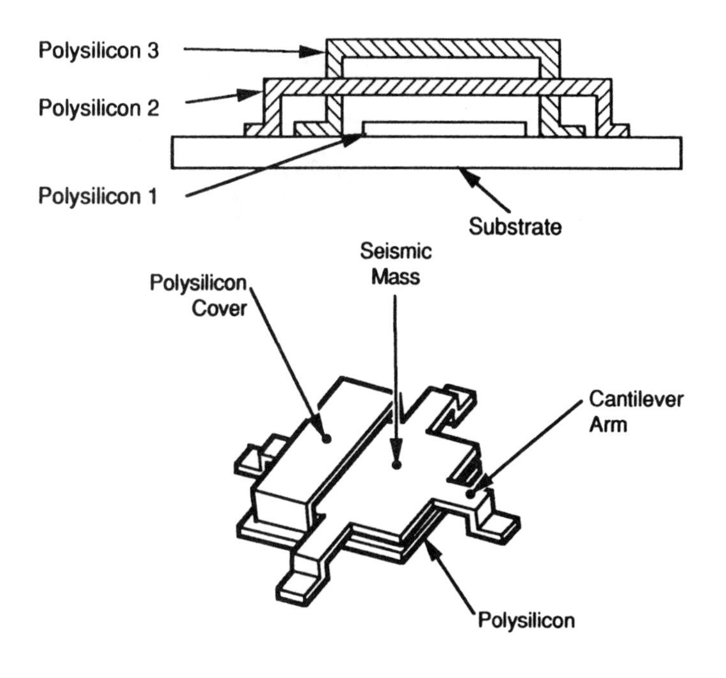

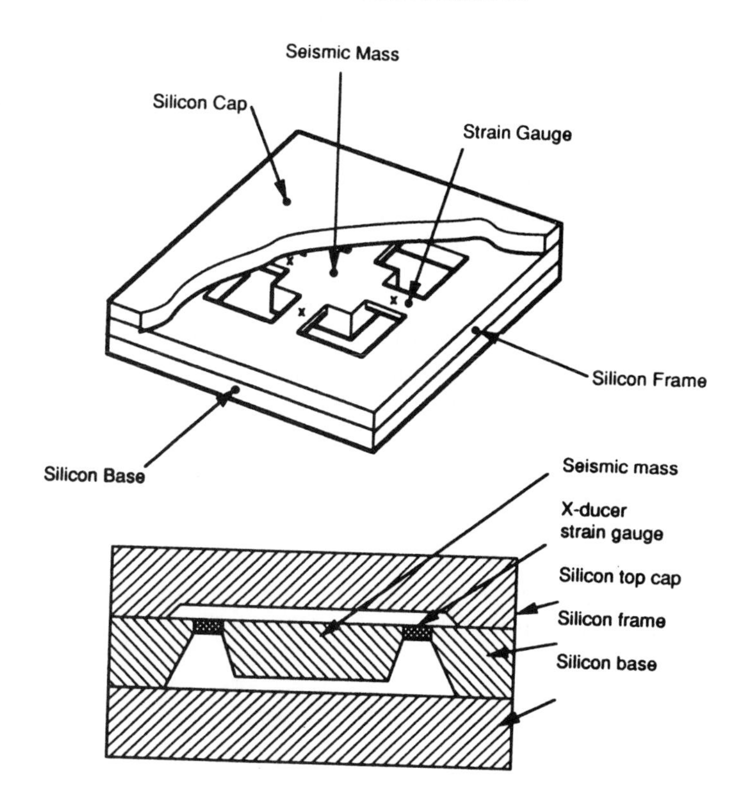

Figure 7 - Micromachined structures. (a) Surface; (b) Bulk.

This is particularly true when the sensor is mounted in the cabin well removed from the point of impact. The signature of a crash will also vary from chassis type to chassis type so that the sensor characteristics may be have to be adjusted on a chassis basis by the end user making integration with an MCU advantageous. In the case of ride control where remote sensing is used, the output from the sensor may need to be bus compatible and also have its own load dump protection. In this case integration with a high voltage technology may be preferred.

OTHER TYPES OF SENSORS

For position sensing without direct contact, the Hall effect sensor or magnetic resistance element (MRE) are often used. Both of these devices detect changes in magnetic flux. In the Hall effect device an output voltage is obtained that is proportional to the magnetic field across the device. In the MRE the resistivity or resistance changes, and the change is proportional to the angle of the magnetic field across the device. Both of these devices can be integrated onto silicon with the necessary interface circuitry. These sensors are used for throttle positioning, vehicle speed sensing, ignition timing and crank shaft positioning. They have an advantage over optical sensing in that they can be used in higher temperature operation.

SYSTEM DESIGN CONSIDERATIONS

Additional design considerations must be taken into account when determining the choice of sensor and optimum approach for the interface circuit. These include the maximum operating temperature, operating environment, reliability and test/calibration of the sensor. For example, the operating temperature of a wheel speed sensor makes the choice of Hall effect or MRE preferable to opto. In other applications, such as steering wheel sensor, opto techniques are acceptable, easy to implement and cost effective.

MEDIA COMPATIBILITY Although silicon has excellent mechanical properties for sensing applications, active circuitry built into the silicon must be protected from the routine and harsh aspects of the automotive environment. Other semiconductor components have made the transition from hermetically sealed, expensive, ceramic and metal can packages to low cost injection molded plastic packages. Potting compounds and encapsulating materials usually provide an additional barrier for moisture and corrosive materials in the modules that utilize these components. However, the need for sensors to operate in the environment requires different solutions.

The pressure sensor package shown in Figure 3 has improved reliability in oil pressure applications by using high temperature, oil resistance materials for both the package and die attachment.

A more visual approach to coping with the

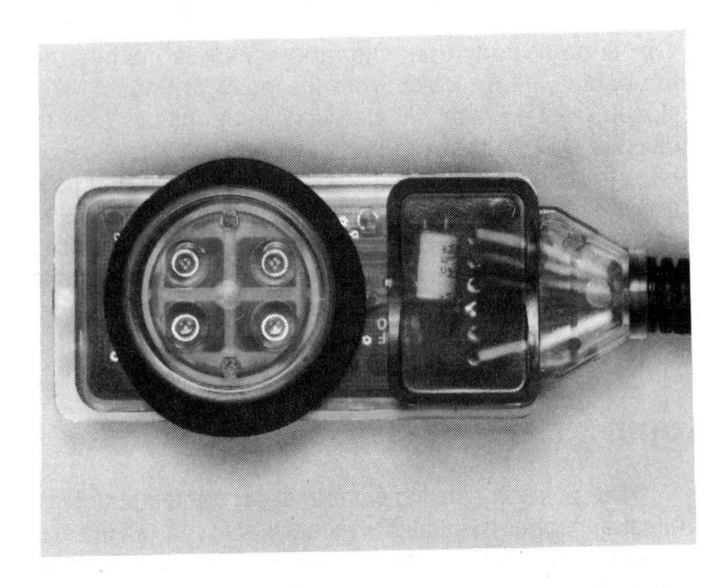

Figure 8 - Shock Absorber Mounted Opto Sensor.

automotive environment is shown in the opto sensor in Figure 8. In this case, opto techniques are used for sensing vehicle height by using the reflective surface of the shock absorber strut. The opto devices are sealed inside the shock absorber in a clean environment by the rubber gasket that is around the two sets of emitter and detector pairs.

PACKAGING INTEGRATION In addition to electrical integration, mechanical components can also be integrated. Figure 9 shows an opto steering wheel position sensor that integrates the interrupter with the opto emitters, detectors and sensing circuitry in one package. This results in a small form factor, self-aligned assembly which is easily installed in the steering column.

FUTURE SENSOR POSSIBILITIES

In addition to the units that have been described, many new solid state sensors are being developed such as chemical and gas sensors for use in automotive systems for such applications as multi-fuel monitoring for efficient engine operation. There are a number of reasons for the demand for more sensors: more monitoring for better performance and improved engine efficiency; sensors are becoming cost effective, and smarts are being added to give many new features, for the replacement of older less cost effective, less reliable mechanical types; and because of legislation requiring better fuel economy, more diagnostics and improved

Figure 9 - Opto Steering Wheel Position Sensor with Integrated Interrupter.

emission control. As systems become more customized, there is a need for close cooperation between supplier and end user to ensure that the right sensors are being developed with the correct specification to give complete system enhancement.

REFERENCES

1.] Watanabe, K.; Chung, W-S.; "A Switched Capacitor Interface for Intelligent Capacitive Transducers," IEEE Trans. on Inst. and Meas., Vol. IM35, No. 4, December 1986.

2.] Kung, J.T.; Lee, H-S.; Howe, R.T.; "A Digital Readout Technique for Capacitive Sensor Applications," IEEE Journal of Solid-State Circuits, Vol. 23, No. 4, August 1988, pp. 972-977

3.] Tang, W.C.; Nguyen, T.C.H.; Howe, R.T.; "Laterally Driven Polysilicon Resonant Microstructures," IEEE Micro Electro Mechanical Systems Workshop, February 1989.

4.] Peterson, K.; Barth, P.; "Silicon Fusion Bonding: Revolutionary New Tool for Silicon Bonding and Microstructures," Wescon, (November 1989), pp. 220-224.

960757

A One Chip, Polysilicon, Surface Micromachined Pressure Sensor with Integrated CMOS Signal Conditioning Electronics

Michael F. Mattes and James D. Seefeldt
SSI Technologies, Inc.

Copyright 1996 Society of Automotive Engineers, Inc.

ABSTRACT

In this paper the integration of a polysilicon, surface micromachined, absolute pressure sensor with analog CMOS (Complementary Metal Oxide Semiconductor) signal conditioning circuitry is a cavity filled with sacrificial oxide. The evacuated cavity under the polysilicon plate is created by removing the sacrificial oxide and sealed using reactive sealing. The pressure is sensed by a Wheatstone bridge formed from dielectrically isolated polysilicon piezoresistors deposited on top of the polysilicon diaphragm [4].

The integrated sensor chip is formed by interweaving a standard CMOS process with the sensor fabrication processes. All of the sensor process steps are compatible with standard CMOS processing. A total of 17 mask steps is required for both the sensor and the signal conditioning electronics, including a passivation layer over the CMOS electronics. The sensor and the signal conditioning electronics have been fabricated on a prototype die measuring 2.54 mm on a side. A production die 1.78 mm on a side is in wafer fabrication.

The signal conditioning electronics amplifies and temperature compensates the sensor bridge output. The output of the sensor is ratiometric and can drive a 1K Ohm load to within 200 mV of either power supply rail. The chip requires a 5V regulated supply. The temperature compensation reduces the temperature error to less than 2.5% of Vref over a temperature range of -40 to 150 °C. Polysilicon thin film resistors are laser trimmed to adjust offset and gain of the signal conditioning electronics.

Absolute pressure sensors have been fabricated in pressure ranges from 0-to-450 KPa up to 0-to-17 MPa. Performance data over temperature for various pressure ranges is presented, and results of overpressure and pressure cycling tests are shown.

INTRODUCTION

Silicon based pressure sensors have been fabricated since the 1960's. The most common form is the bulk micromachined piezoresistive transducer (bulk PRT). Typically, these sensors include a thin diaphragm region, formed by etching from the back side of the wafer, and diffusing or implanting piezoresistors configured in a Wheatstone bridge. The chip may also include resistors used for temperature compensation and sensor offset voltage adjustment. The electronics to amplify and temperature compensate the sensor output are external to the sensing die. Several companies have successfully integrated the signal conditioning electronics on the same die as the bulk PRT sensor [1-3]. However, the ability to scale the size of the die is limited by the size of the bulk PRT sensor.

A solution to the size problem of the bulk PRT sensor is to form the sensor diaphragm from a thin layer of polysilicon on the surface of the die (surface PRT). Because the thickness of this polsilicon film can be made approximately 10X thinner than the bulk PRT diaphragm the sensor element area for surface PRT is approximately 100X smaller than for bulk [5-6]. Furthermore the process used to fabricate the sensor element is compatible with CMOS processing. This results in a sensor which is easily integrated with the sensor signal conditioning electronics. Using surface PRT technology pressure sensor die 1.78mm X 1.78mm square are feasible.

DESIGN GOALS

The automotive market offers opportunity for manufacturers to provide components in large volumes provided that certain criteria are met. In addition to meeting a very specific customer application, the component must be low cost, have a very low defect rate, high reliability, and be able to operate in a very harsh environments. Low power consumption is also becoming a valuable feature as the electronics content of the automobile continues to grow.

In order to meet specific customer applications and at the same time be low cost, two decisions were made at the beginning of the project. Plastic would be used whenever possible since this material can be reconfigured to meet the customer's shape and space requirements and has a low material cost. Second, the sensing device would have to take advantage of the batch processing nature of the semiconductor industry in order to be low cost. These two decisions are reasonable as evidenced in present day automotive sensors such as the Manifold Absolute Pressure (MAP) sensor.

It is not sufficient, however, to simply re-create present day technology in our competitive market. It is necessary to supply the customer with some added benefits in order to justify their decision to choose a new product. Lower cost is always an incentive and it was clear that this would have to come from technology improvement. A pressure sensor with integrated signal conditioning electronics seemed to be the best way to achieve the low cost goal since the electronics typically consumes the largest share of a sensor's material cost budget. A CMOS compatible, surface micromachined, diaphragm structure with isolated piezoresistive strain sensors was chosen as the core technology. This was merged with a proven CMOS process in order to create the support circuitry.

The chosen surface micromachined structure offered two distinct advantages: 1) All of the processes were CMOS compatible. 2) A thinner diaphragm offered a huge reduction in the silicon area required for the sensing device, leaving enough room to integrate the electronics and still create a small die. See Figure 1.

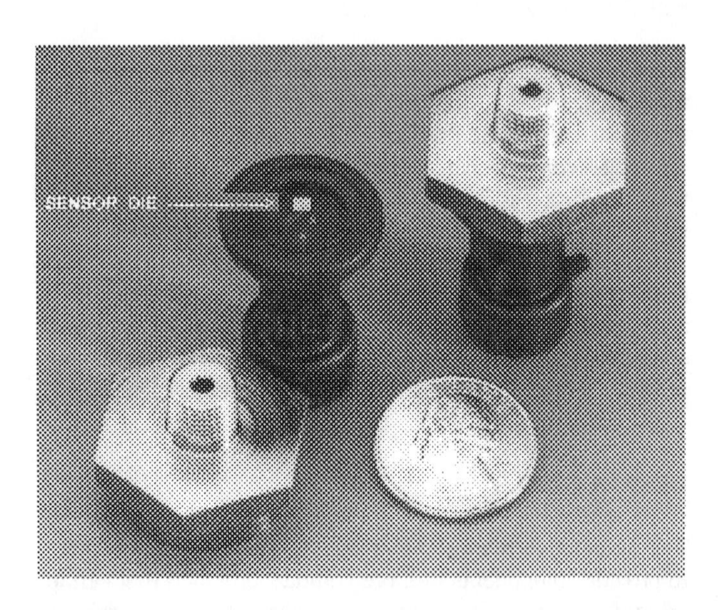

Figure 1a.
Pressure sensor with signal conditioning exposed within a package.

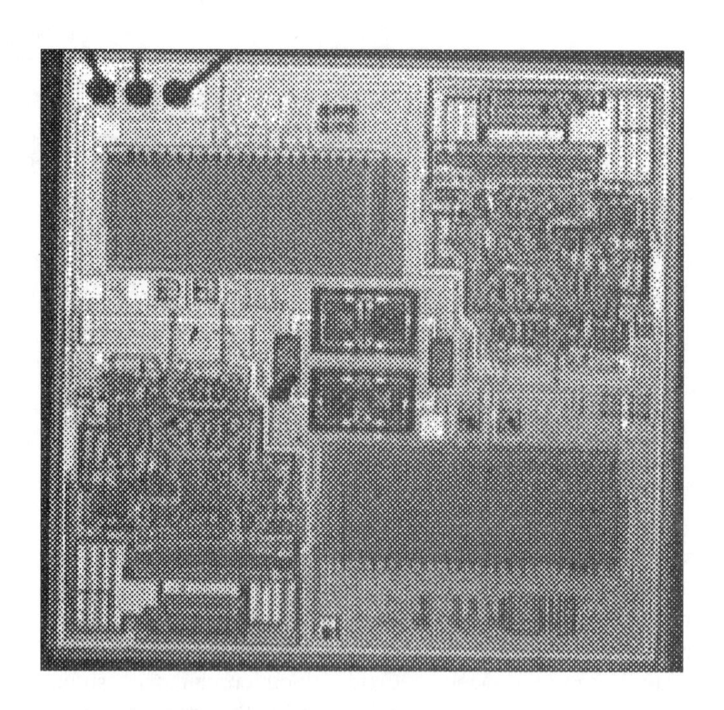

Figure 1b.
Pressure sensor die with CMOS signal conditioning circuitry.

MICROSTRUCTURE BACKGROUND

The sensor microstructure is a square plate with clamped edges supported over a cavity. It is created by oxidizing silicon in the shape of a square. This silicon dioxide is removed and the process is repeated. The oxide is allowed to grow until it reaches the height of the original silicon surface. (We now have a bathtub in silicon which is filled to the top with silicon dioxide). This time the oxide is left in place and covered with silicon nitride. The silicon nitride is patterned to create small etch channels, which will be used later, and clear areas for the polysilicon diaphragm to attach to the silicon wafer. Then the polysilicon diaphragm is deposited using Low Pressure Chemical Vapor Deposition (LPCVD). The unneeded polysilicon is etched away leaving the square diaphragm attached to the silicon with the silicon nitride etch channels exposed. Finally, hydrofluoric acid is used to remove the etch channel material as well as the silicon dioxide within the cavity. (We now have a diaphragm suspended over a cavity). The last step requires the process known as reactive sealing. The wafers are placed in an oxidation furnace which is pumped down to 300 mTorr. This causes the pressure inside and outside of the cavity to be at this very low pressure. While held at this pressure, oxygen is pumped through the furnace, causing the small etch channels to oxidize and slowly seal shut. Once the channels are sealed, oxygen within the sensor cavity will continue to react with the silicon cavity and polysilicon diaphragm until there are no free oxygen molecules left. (We now have an evacuated, sealed cavity under the polysilicon diaphragm).

TECHNOLOGY ADVANTAGES

The cost savings in electronics is obvious when compared to a bulk micromachined diaphragm with support circuitry implemented in thick film, however there are a number of additional benefits to the customer. The small die has the potential to fit in places which were once considered totally inaccessible to a sensor without integrated signal conditioning. The device also has a built-in overpressure stop to prevent the diaphragm from being damaged during abnormal, high pressure conditions. See Figure 2. Diaphragms have been taken to 36 times their rated pressure and have survived with little or no change in characteristics. Bulk micromachined devices do not have an inherent overpressure stop which makes them prone to shattering under similar conditions. At first glance it might be assumed that a thinner diaphragm would be more prone to fracturing. However, if we examine the equation for stress at the edge of a square diaphragm it can be seen that ;

$$Stress = k_1 * pressure * (L/t)^2$$

where,

k_1 is a constant dependent on material properties
pressure is the pressure applied to the diaphragm
L is the length of one side of the square diaphragm
t is the thickness of the diaphragm

So, it can be seen that if we make the diaphragm 2 μm thick and 100 μm long the stress will be no higher than if we make a diaphragm 200 μm thick and 10,000 μm long.

If we are not careful we can still exceed the fracture stress of the diaphragm. The present 1 MPa diaphragm has been designed to reach critical stress (827 MPa) at greater than 3 Mpa pressure. Before this is reached the diaphragm will have begun to make contact with the overpressure stop, thus preventing its destruction.

Surface micromachining also affords a number of advantages in the back end packaging process. Since all of the semiconductor processing is performed on the top side of the wafer, alignment errors between the piezoresistors and the cavity are minimized. Piezoresistor geometry is also symmetrical. This results in typical mismatch of only 1% to 2% between the bridge outputs before calibration. Thus, offset calibration and temperature drift are minimized at the back end. Bulk micromachining is prone to alignment errors since it is difficult to align patterns on opposite sides of a wafer to the same accuracy as surface alignments. Finally, the piezoresistors are dielectrically isolated, thus preventing leakage problems associated with diffused piezoresistors.

Another benefit of the CMOS processing is that using LPCVD provides control of diaphragm thickness to within 1000 Angstroms, or 5 % for a 2 μm thick diaphragm. This is important since the amount of diaphragm deflection at a given pressure is inversely proportional to the thickness cubed;

$$Deflection = \frac{k_2 * pressure}{t^3}$$

where,

k_2 is a constant dependent on material properties
pressure is the pressure applied to the diaphragm
t is the thickness of the diaphragm

As a result of small diaphragm thickness variation, the die-to-die and wafer-to-wafer variation in bridge sensitivity is kept to a minimum. This minimises the trim range requirements which helps minimize the die size.

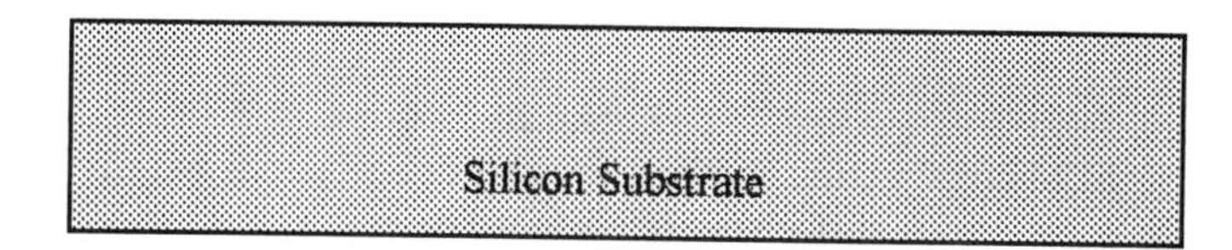

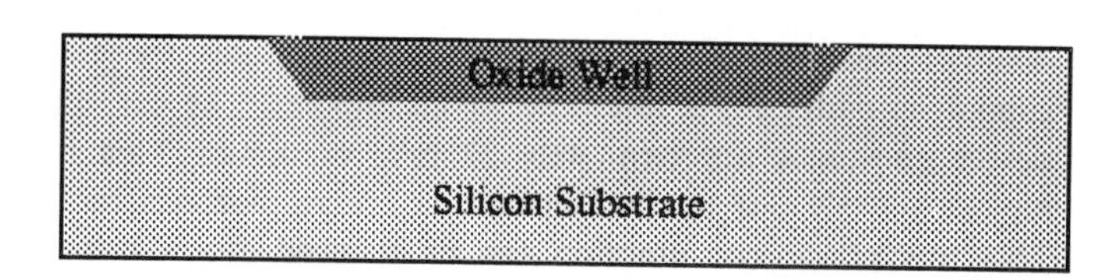

Polysilicon Diaphragm

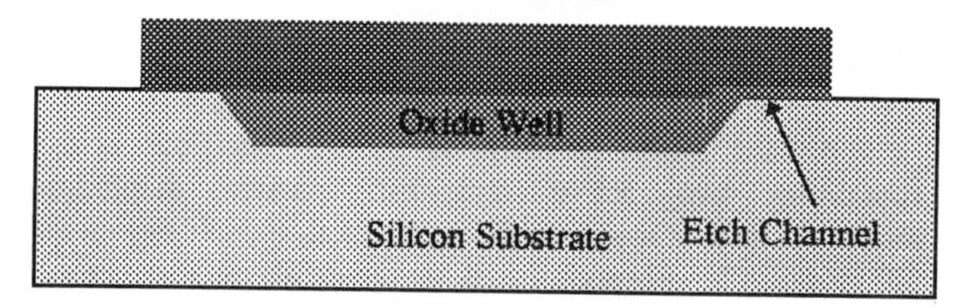

Polysilicon Diaphragm

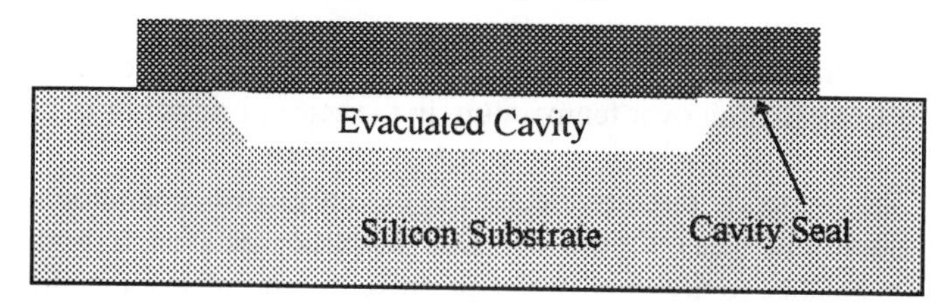

Figure 2a.
Cross section of procees flow for polysilicon surface micromached pressure sensor.

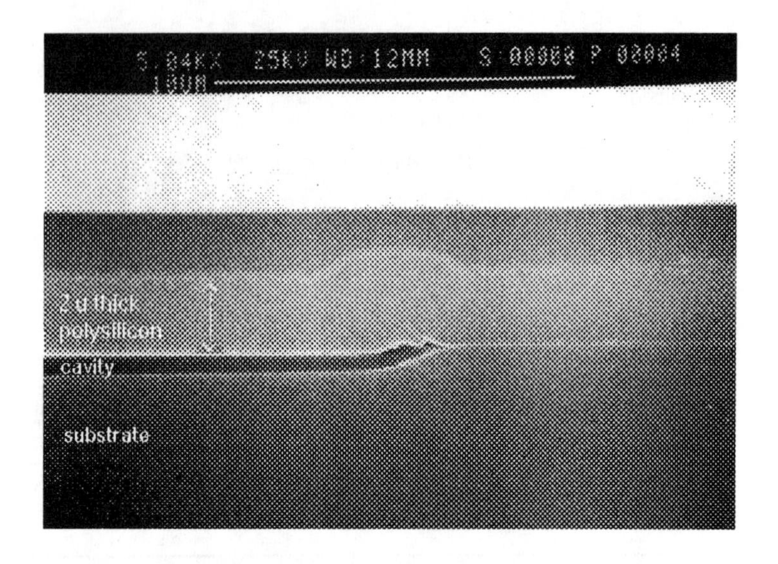

Figure 2b.
SEM cross section showing overpressure stop inherent in surface micromachined device.

Another benefit of choosing CMOS technology for the support circuitry is its low power consumption. The current device consumes approximately 25 mW of power. This is almost entirely due to the Wheatstone bridge piezoresistors. This low power consumption can be reduced even further with a modest effort in optimization of piezoresistor and diaphragm geometries.

SUPPORT CIRCUITRY

Presently the sensor structure is a full Wheatstone bridge containing two active and two passive diaphragm microstructures. One structure is devoted to each leg of the full bridge. The process for making the active microstructures has already been described. The passive diaphragm microstructures, however, still have the sacrificial oxide in place. This oxide prevents the diaphragms from responding to pressure. These two cavities serve as references to prevent offset variation over temperature in the sensor bridge.

The support circuitry consists of only two operational amplifiers. (See Figure 3). One amplifier acts as a high impedance, unity gain buffer in order to prevent loading the negative going half of the Wheatstone bridge. The buffer amplifier feeds the gain amplifier along with the positive going half of the Wheatstone bridge. Both signals are amplified using the same gain stage which depend on only two passive devices for the gain factor. Offset is adjusted with a pullup resistor.

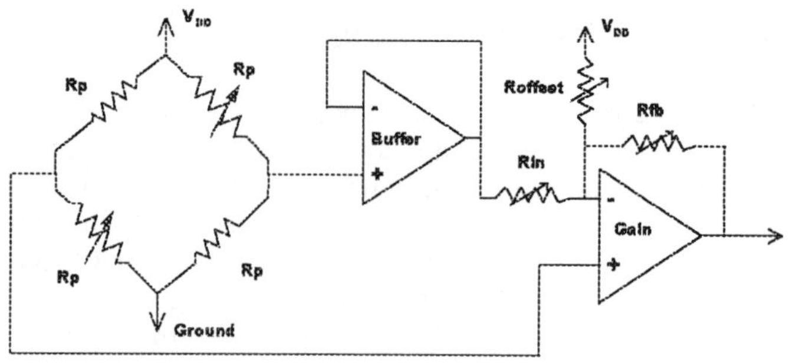

Figure 3.
Schematic of integrated pressure sensor circuit including sensor Wheatstone bridge and support circuitry.

TEST RESULTS

The graph in Figure 4 illustrates the linearity obtained from one of these devices. Linearity is typically better than 0.1% of full scale. This is within the range of our current measurement error. Linearity is a function of piezoresistor geometry and not a function of temperature. At present, manual calibration is performed using standard laser trimming methods which allow for accuracy better than 0.2 % of full scale at 25 °C. Various trim geometries are being investigated which should allow us to obtain better resolution. In production the trimming will be automated which should also improve our resolution.

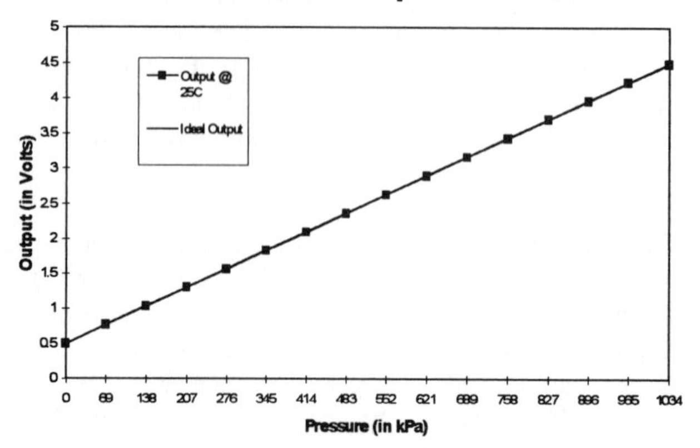

Figure 4.
Output voltage vs. pressure at 25 °C for 1000 kPa device.

Figure 5. demonstrates the temperature compensation capability of one of the devices. As stated previously, temperature does not affect the linearity, however it does have some effect on offset and sensitivity. These effects are reflected in the graph below. Devices are specified to be within 2.5% of the reference voltage over the entire automotive temperature range (-40 to +150 °C).

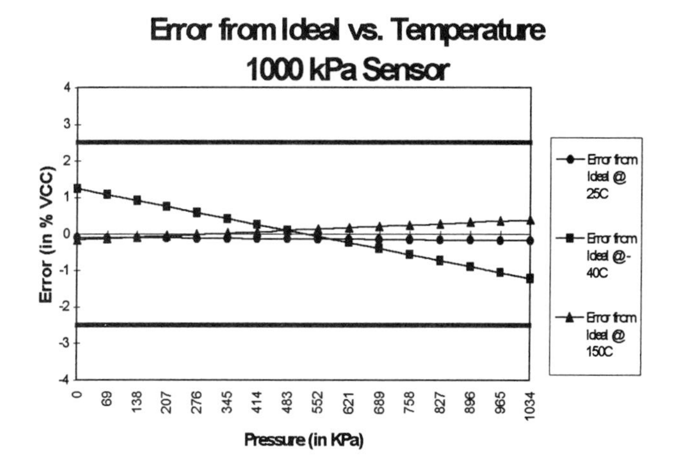

Figure 5.
Error in sensor output at -40, +25, and +150 Degrees Centigrade.

It would be nice to know just how much extra protection the overpressure stop provides. A 1000 kPa device was subjected to approximately 20.7 MPa for a period of 1 minute. This was the highest pressure that our equipment could achieve. Figure 6. illustrates the robust nature of the microstructure. It not only survived, there was no measurable change in its output characteristics.

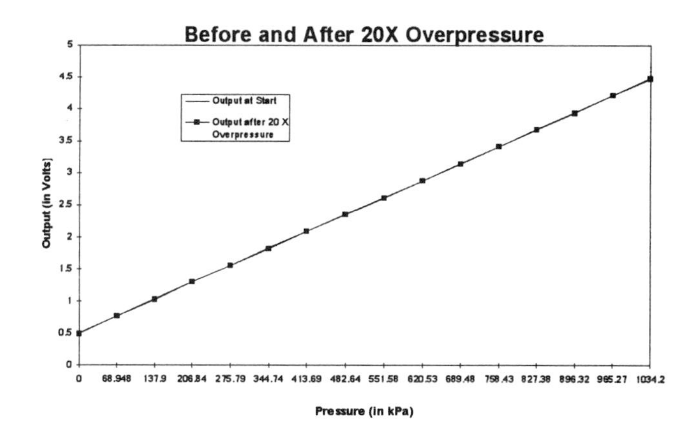

Figure 6.
Initial sensor output compared to output after being subjected to 20.7 MPa.

Durability is of paramount interest to the customer. Earlier mechanical type pressure sensors were based on technologies such as potentiometers and LVDTs (Linear Variable Differential Transformers). These devices had moving parts which were subject to wear. The polysilicon diaphragm, as stated earlier, has a maximum deflection well below its plastic deformation point. Life test equipment cycled pressure on the device represented in Figure 7. from room pressure to full scale pressure at 25 °C for 2.0 million cycles. (The overpressure stop does not come into play in this region of operation). There was approximately a 1.1 % change in the output at full scale pressure pressure.

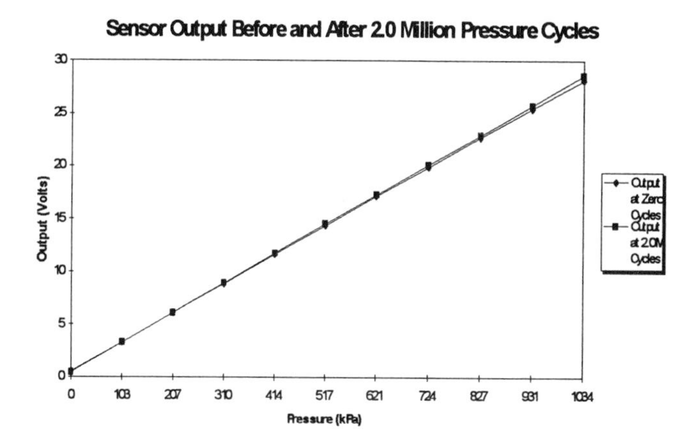

Figure 7.
Sensor output before and after 2.0 million pressure cycles at full scale pressure.

CONCLUSION

The sensor described in this paper has exceeded our design goals in all respects. Its small size allows it to be used in applications which were inaccessible with previous devices. Its low power consumption will also open up new applications where power is at a premium. Additional effort is being directed towards increasing bridge sensitivity, tightening up process parameters that affect sensitivity and improving the signal-to-noise ratio of the fully integrated device. This device has also paved the way to many sophisticated variations due to the ability to integrate CMOS circuitry with the microstructure. For example, future devices could incorporate network interface electronics for various protocols such as Bosch's CAN or SAE Standard J-1850. Opportunities also exist to combine multiple sensing elements such as pressure and temperature on the same device. The integration of surface micromachining and CMOS circuitry offer huge opportunities for technological and financial growth.

Acknowledgments:

We would like to thank Robert Cooney, Paul Rozgo, Barbara Cleasby, and Lloyd Fager for all of their efforts in assembly, testing and characterization of the devices. We would also like to thank Tom Pumo and Karen Berry of SMC for their successful fabrication of these challenging devices.

REFERENCES

[1] K. Yamada, et. al. "A piezoresistive integrated pressure sensor", *Sensors and Actuators*, vol. 4 (1983), pp. 63-69

[2] E. Obermeier, et. al. "A smart pressure sensor with on chip calibration and compensation capability", Sensors, March 1995, pp. 20-22, 52-54

[3] "Integrated pressure sensors with electronic trimming" Automotive Engineering, April 1995, pp. 65-68

[4] R. Frank "Pressure sensors merge micromachining and microelectronics", *Sensors and Actuators* A, vol. 28 (1991) pp. 93-103.

[4] H. Guckel and D. W. Burns, "Planar-processed polysilicon sealed cavities for pressure transducer arrays", IEDM Tech. Digest, 1984, pp. 223-225.

[5] E. Obermeier, "Polysilicon Layers Lead to a New Generation of Pressure Sensors", IEDM Tech. Digest, 1985, pp. 430-433.

930354

Methanol Concentration Smart Sensor

John J. C. Kopera and Mark E. McMackin
Chrysler Corp.

Richard K. Rader
Borg-Warner Automotive

Abstract

A Methanol Concentration Smart Sensor has been developed to support the demand for alternately fueled vehicles operating on blends of methanol and gasoline in any mixture up to 85% methanol.

The sensor measures concentration by exploiting the difference in dielectric properties between methanol and gasoline. The measurement is made based on the distributed capacitance of a coil of wire, contained in a reservoir through which the fuel passes. This signal, along with temperature compensation inputs, is then fed to an integral microprocessor, which provides a voltage output proportional to the methanol concentration of the fuel. The Powertrain Controller uses this information to modify injector pulse width and provide proper spark advance. This paper will explain the sensor's development methodology and function.

1. Introduction

In the search for independence from finite oil sources and to improve vehicle environmental performance, methanol has become a prime fuel alternative for internal combustion engines. Methanol may be produced from a number of readily available or renewable resources - biomass, wood, coal and natural gas. Its transport and handling requirements are similar to gasoline, allowing the existing fuel transport infrastructure to deliver methanol blends in much the same method as today's gasoline transports. Environmentally, the reductions in photochemically reactive exhaust components provide for a decrease in ozone breakdown and smog.

Given the above advantages, FFV's (Flexible Fueled Vehicles) have been developed to capitalize on the use of methanol and retain the flexibility of operating on gasoline. Flex Fueled Vehicles may use methanol in concentrations from M85 (85 percent methanol, 15 percent gasoline) to M0 (100 percent gasoline) or any blend in between.

The minimum 15 percent gasoline in M85 is required for vehicle starting at cold temperatures and flame luminosity.

Unfortunately, methanol fuel has approximately half the caloric energy content of gasoline. This requires considerably more fuel be consumed when running on methanol. To address this variation in fuel flow quantity and mixture, a sensor has been developed to detect these changes in fuel composition and provide a feed forward signal to the Powertrain Controller so that fuel delivery and spark timing may be adjusted for proper engine operation.

The Methanol Concentration Smart Sensor (MCS) was developed to provide an analog signal in relationship to the percentage of methanol, allowing the engine to adjust during refueling, cold starts and wide open throttle operation. After initial warm-up conditions, the standard feedback strategies utilizing the oxygen sensor are still employed, in addition to the MCS's output, to maintain stoichiometric A/F ratios.

2. Methods of Measurement

During vehicle development, a number of technologies were emerging - capacitive, infrared, reflective/refractive, conductive, and ultrasonic/conductive. After intense study, the refractive and capacitive approach were determined to be viable options and a parallel path to create a sensor was taken.

Refractive Index - The principal of the refractive or optical sensor centers on the difference in the refractive index between gasoline (1.40 - 1.46) and methanol (1.33). Sensor construction uses a glass rod with a light emitting diode placed on one end and a photo transistor on the other. The fuel flows around the rod. The reflection from fuel/ glass interface then determines the amount of light transmitted to the photo transistor, based on the refractive indices of the glass/ fuel interface. In working with optical sensors it became apparent that the glass tube used to transmit the light from the LED to the photodiode would fog over with a varnish film from the fuel as mileage was accumulated, disturbing the critical angle and shifting the output curve.[1] Output shifts would also take place when the composition of the hydrocarbon compounds varied - as with the differences in aromatic content between premium and regular gasolines. In the case of fogging, the sensor was disrupted when small amounts of water would mix with methanol, producing a milky fluid, which resulted in abnormally high and unpredictable readings of methanol concentration.[2] Resolving these physical and systemic problems proved difficult to overcome and the optical approach was abandoned.

Capacitive or Dielectric - The capacitive approach provides a much larger measurable difference between gasoline and methanol in that their respective dielectric constant values are over an order of magnitude apart (gasoline = 2 and methanol = 32) offering excellent resolution. Unlike the refractive sensor, the aromatic content of gasoline could now be ignored, as its effect on dielectric constant is negligible. The dielectric constant of methanol is heavily dependant on temperature, which necessitates use of temperature compensation in the determination of fuel composition. Fortunately, gasoline has a

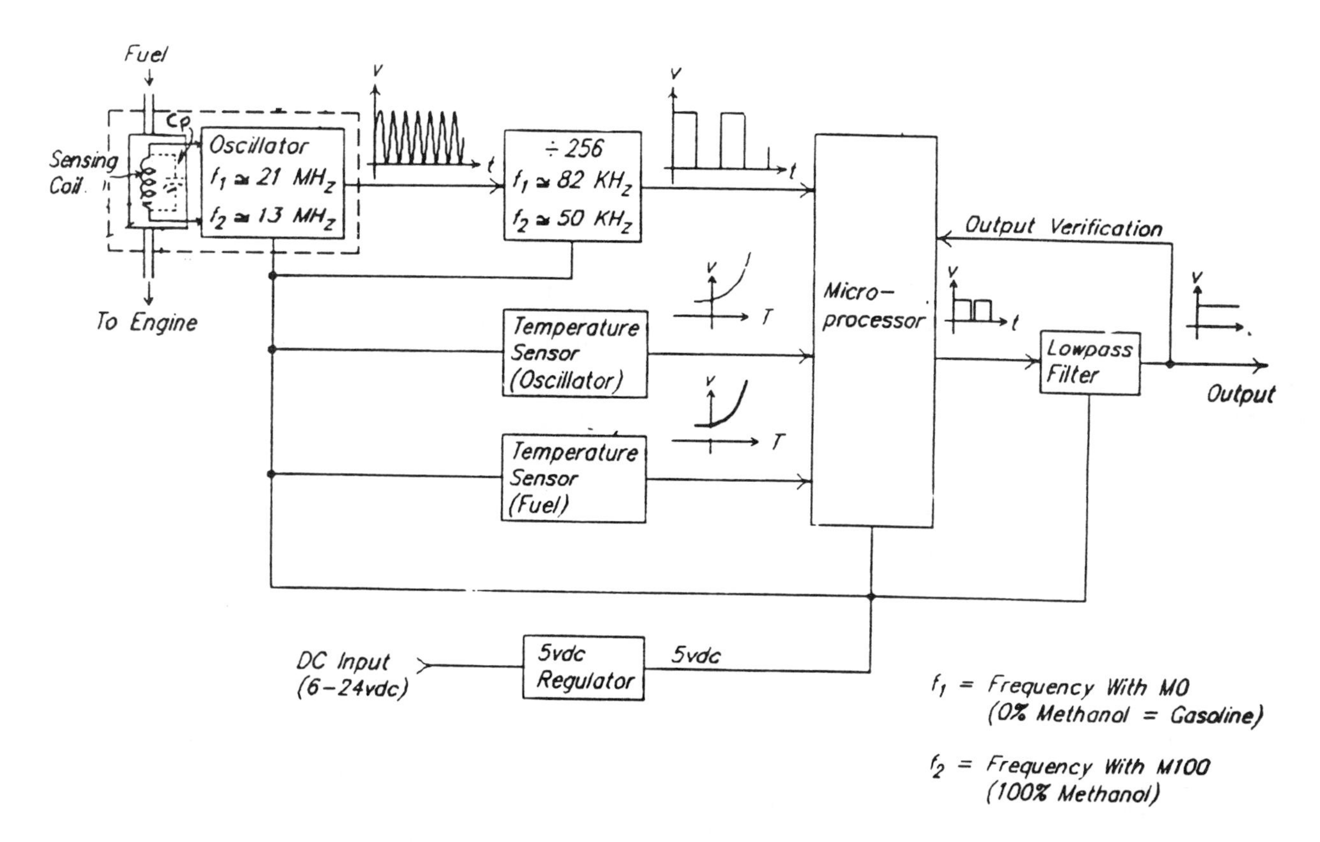

Figure 1

dielectric constant relatively independent of temperature, thereby simplifying the calculation of fuel composition.[3]

3. Operational Theory

The Methanol Concentration Smart Sensor consists of a wound coil which is immersed in the fuel mixture. A distributed capacitance exists between the windings of the coil upon energization of the coil. The turns of the coil act as equivalent electrodes of a capacitor and the fuel mixture acts as a dielectric medium. This distributed capacitance of the coil is directly proportional to the dielectric constant of the mixture which is proportional to the percentage of methanol in the mixture.[4] The equivalent circuit is designated as the dashed line capacitor symbol and labeled C_p as shown in figure 1.

The leads from the sensing coil to the circuitry are kept short. Without this type of arrangement, long interconnection cables would be needed which would add undesirable stray or parasitic impedance to the circuit, thus rendering it a less accurate indicator of the fuel mixture.

The dielectric constant of the mixture is determined by using the coil as the resonant element in the oscillator, thereby generating an oscillating frequency which is inversely proportional to the square root of the dielectric constant of the fuel mixture. The frequency is converted to a voltage output and sent to a Powertrain controller for processing the fuel and spark implications.

The capacitance between the coil's two electrodes is defined as the charge stored per unit potential difference between them. Capacitance is dependant on the area, spacing, and the character of the dielectric material between the electrodes as shown in the formula:

$$\text{CAPACITANCE} = (\epsilon)\ [A / d]$$

where "ϵ" is the dielectric constant, "A" is the effective area of the intrinsic capacitor formed by the loops of wire, and "d" is the distance between the electrodes. We are concerned here with the dielectric constant of the mixture since the area and the spacing will remain constant.

Dielectric materials are electrical insulators in which an electrical field can be sustained with a minimum dissipation of power. The fluid with the higher dipole moment will have the higher dielectric constant. In a homogenous mixture of two known fluids a proportional relationship exists between the percentage of each fluid in the mixture and its cumulative dielectric constant. This also means that capacitance will increase with an increase in percentage of the fluid possessing the higher dielectric constant. It is this relationship which the Methanol Concentration Smart Sensor exploits.

The frequency of operation of the oscillator circuit is determined by the resonant frequency of the coil since the inductance does not vary. A change in dielectric constant causes a change in resonant frequency due to the change in intrinsic capacitance of the coil and its housing. In this manner, resonant frequency decreases as capacitance increases according to the formula:

$$f = \frac{1}{2\pi\ (LC)^{1/2}}$$

$$\text{and } C = C_0(k + \epsilon_{fl})$$

where "f" is frequency in Hz, "L" is the inductance of the sensing coil, "C" is the capacitance, ϵ_{fl} is the relative dielectric constant of the fluid being measured, C_0 is the base distributed capacitance of the sensing coil, and k is a constant which accounts for the effects of the coil insulation, housing oscillator circuitry, and all stray or parasitic impedances reflected across the inductor terminals.

From the above relationships, it can be seen that as the percentage of the fuel with the higher dielectric constant increases, the resonant frequency of the coil decreases. The sensor can be set to oscillate at any frequency, however for our purposes, the range between 13 Mhz and 21 Mhz was chosen. This frequency range is a function of the size of the sensing coil and of the fuel to be sensed.

The output from the oscillator stage is then fed to a divide by 256 circuit as shown in the block diagram [figure 2]. This circuit converts the sensor output signal into a more manageable frequency ranging from ≈50 Khz to ≈82 Khz which is then forwarded to the microprocessor for further analysis.

To compensate for board and oscillator temperature fluctuations, a thermistor is mounted on the circuit board. Its output is fed to the microprocessor for processing with the divided sensor output signal. Also provided is a fuel temperature thermistor to sense

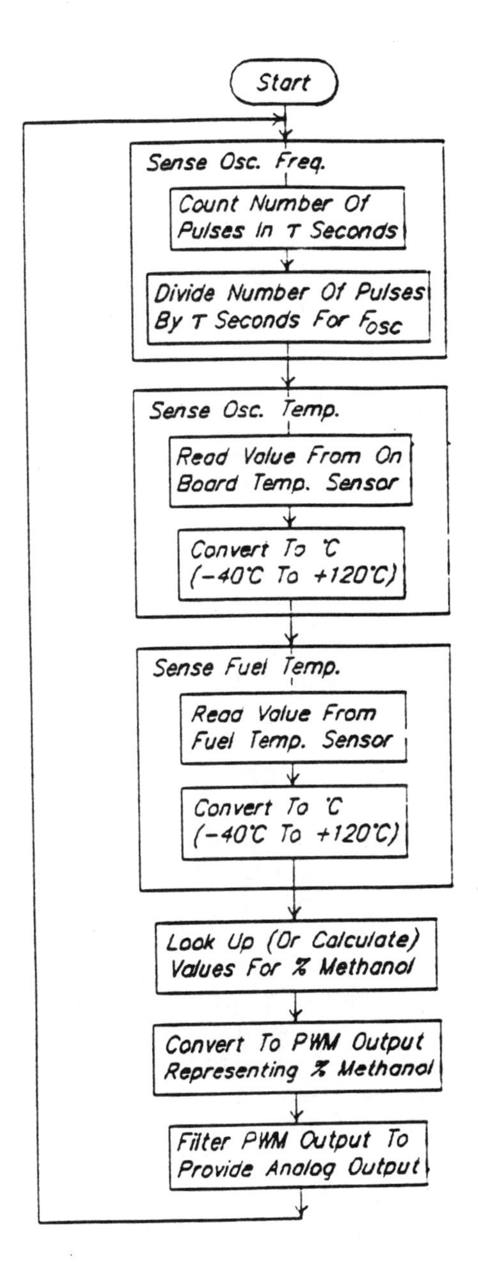

Figure 2

variations in fuel temperature. This is shown as the fuel temperature output signal and is processed by the microprocessor.

The microprocessor compares the divided sensor signal along with the oscillator temperature sensor signal and the fuel temperature sensor signal to predetermined calibration tables to determine the methanol concentration of the sensed fuel. A low pass filter converts the pulse width modulated output signal from the microprocessor into an analog voltage output.

Referring to figure 3 shown is a graph of the analog output signal from the sensor vs the percentage methanol sensed in the fuel. Figure 4 shows a graph of the percentage methanol sensed in the fuel vs the frequency output of the Methanol Concentration Smart Sensor. These two curves are representative of the relationship between the voltage and frequency outputs as related to the percentage methanol contained in the fuel. The shapes of the curves will change slightly depending on the scale used and the choice of the hardware used, the amount of fuel present to sense, the temperatures involved, etc.

4. Diagnostics

The microprocessor has been given additional functions beyond processing the two temperature compensation inputs and the frequency signal to provide an analog output. These are related to its interface with the Powertrain control module and its requirements during vehicle start up and run modes. Because of the sensors rear mounting location and the length of wire routing from the Powertrain controller, software was added to provide the controller with a calibration signal. This calibration signal (a fixed voltage within the analog output range) is enabled during the first seconds of engine start up and allows the Powertrain controller to perform ratiometric calculations to make up for wiring losses and component tolerances between the sensor and controller.

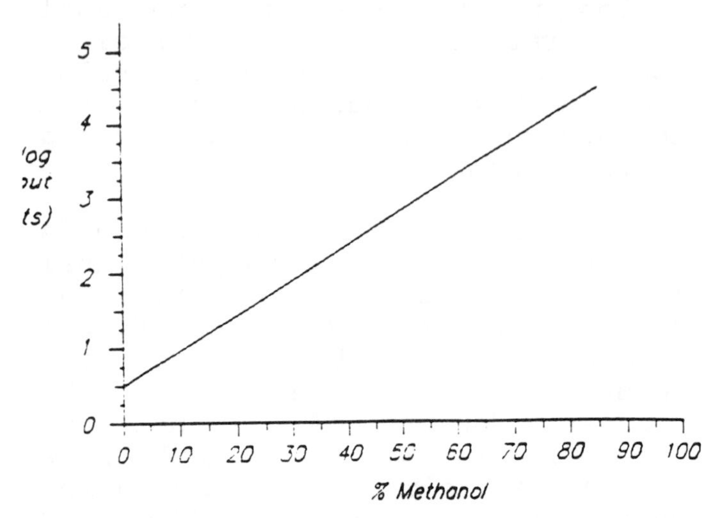

Figure 3

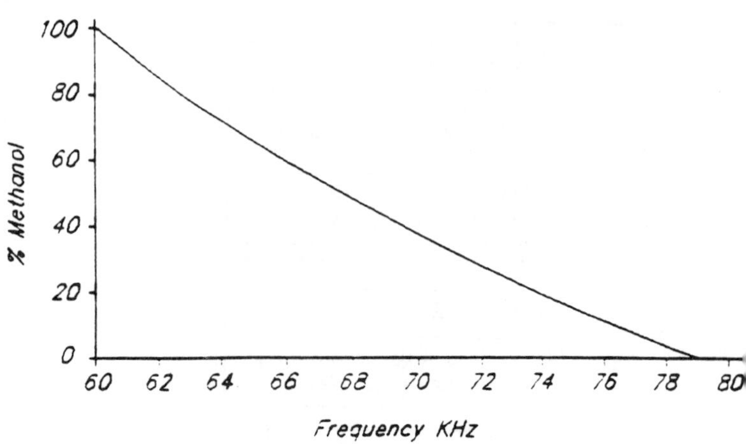

Figure 4

After the calibration period, the sensor resumes its normal output to the controller.

The microprocessor's second function is to perform self-diagnostics, including but not limited to, temperature sensor out of range, sensing coil not responding, and output verification feedback incompatible with microprocessor output. When a fault is detected, the microprocessor will drive the output below the normal minimum output of 0.5V (M0), alerting the Powertrain controller to a potential problem.

Given these features - the uniqueness of the sensing coil approach, and that the sensor is microprocessor based, we have added the term "smart" to the sensor's designation.

5. Mechanical Design

The mechanical design of the housing required a number of factors be taken into account - structural integrity, environmental/chemical compatibility and EMI/RF performance. To meet our structural requirements , Polyarylsufone, one of the newer engineering plastics, was chosen for the following advantages. This material, with a 40% fill of mineral and glass, provided the rigidity for the mounting ears which attach the sensor to the vehicle; and support the fuel system connections. It would also withstand fuel line pressures associated the vehicle's supply pressure of a maximum 690 Kpa. And, most importantly, allowed the encapsulation of the sensing coil within the chamber by overmolding the coil/bobbin assembly. This overmolding process seals the sensing coil leads which enter the circuit area from the methanol/gasoline coil chamber.

The material tested acceptable for the methanol application after experimenting with the M0, M85, and blends considered the most corrosive over time and temperature cycling.

Another reason for using polyarylsufone is that this particular formulation could be plated providing the means to meet EMI requirements while maintaining light weight and avoiding expensive machining operations. This process involves dipping the completed housing in an etch bath to prepare the plastic, then electroless plating with copper and following with nickel overplate application. In the process, the polyamidimide on the coil wire and plastic bobbin resist the etching and plating, thereby retaining the functionality requirement for the sensor circuit, i.e. that the coil be in direct contact with the fluid to be measured.

The sensing coil is fabricated from magnetic wire which is insulated with a polyamidimide coating. The polyamidimide coating provides a long life resistance to gasoline and alcohol fuels while providing the thinnest coating possible to preserve capacitive properties.

Additionally, the coating is needed to eliminate shunt resistance (being generated between the loops of wire that form the coil) and fluid ion mobility effects (i.e., the conductivity of the fluid itself), yet it was still thin enough so as not to mask the dielectric constant of the mixture being measured.

Figure 5 shows the physical locations of the various components which make up the sensor's construction - 1. coil chamber with fuel connections attached, 2. Fuel temperature sensor, 3. circuit board (microprocessor is mounted below the

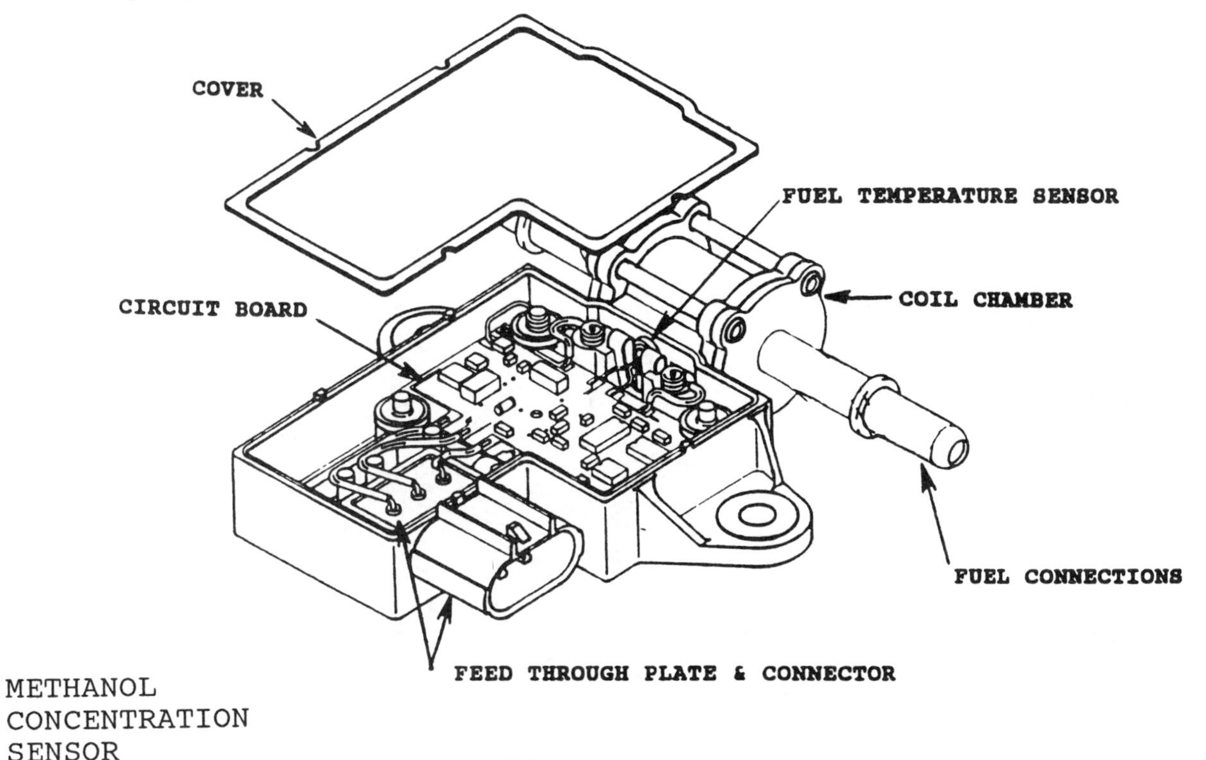

Figure 5

board to conserve space), and 4. feed through plate to the output connector.

The sensor can be mounted in the fuel supply line anywhere between the vehicle's fuel tank and the engine. In our application the sensor was mounted at the rear of the vehicle near the fuel tank as shown in figure 6. Mounting the sensor in the fuel supply line provides the Powertrain control module with predictive information as to the nature of the mixture being delivered to the injectors.

Methanol Concentration Sensor - Production Mounting Location

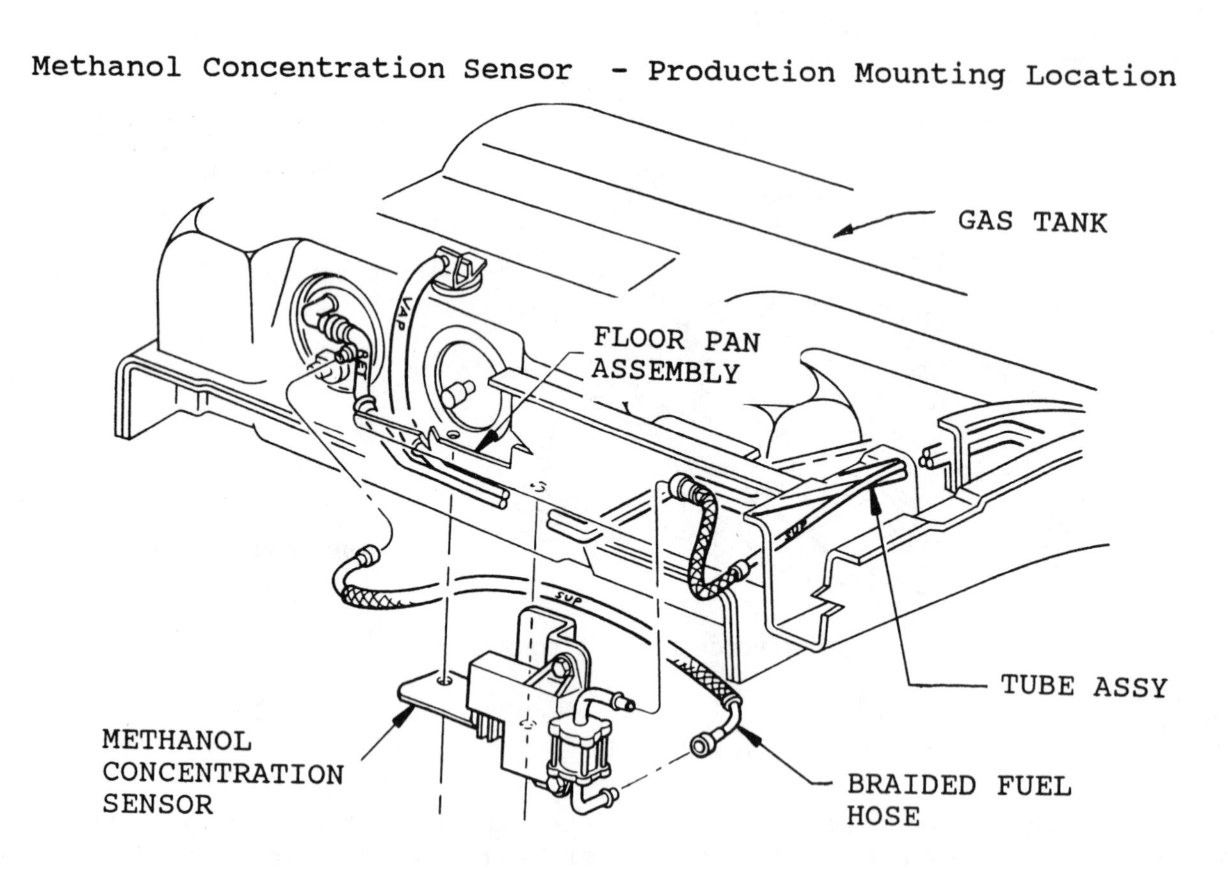

Figure 6

6. Vehicle Testing

In the spring of 1990 an in-house fleet of 30 vehicles was launched to gain more real world information and feedback on the acceptability of methanol powered vehicles. Participants were allowed ready access to methanol and were required to report any abnormalities in driveability, starting, handling, etc. A preproduction version of the Methanol Concentration Smart Sensor was installed on each vehicle along with other modifications standard with flexibly fueled operations. To date this fleet has logged approximately 1,560,000 KM. The average mileage accumulated was 51,500 KM, with a maximum achieved mileage of 86,900 KM and a minimum of 22,200 KM. The sensors performed through a year and a half of

Michigan weather without malfunctions. Upon inspection of the fuel/coil chamber a slight discoloration in the nickel coating could be observed, most likely due to varnishing from the fuel, otherwise the sensor functioned within the +/- 5% accuracy specification.

concentration returned to the previous reading made the day before. A second test was performed at a concentration of M45. Again the concentration had risen, but only to M53 and then returned to the previous day's reading. From this data, it can be inferred that as the percentage of methanol in the

Cold Start and Idle Data Showing Methanol Separation at Low Temperatures

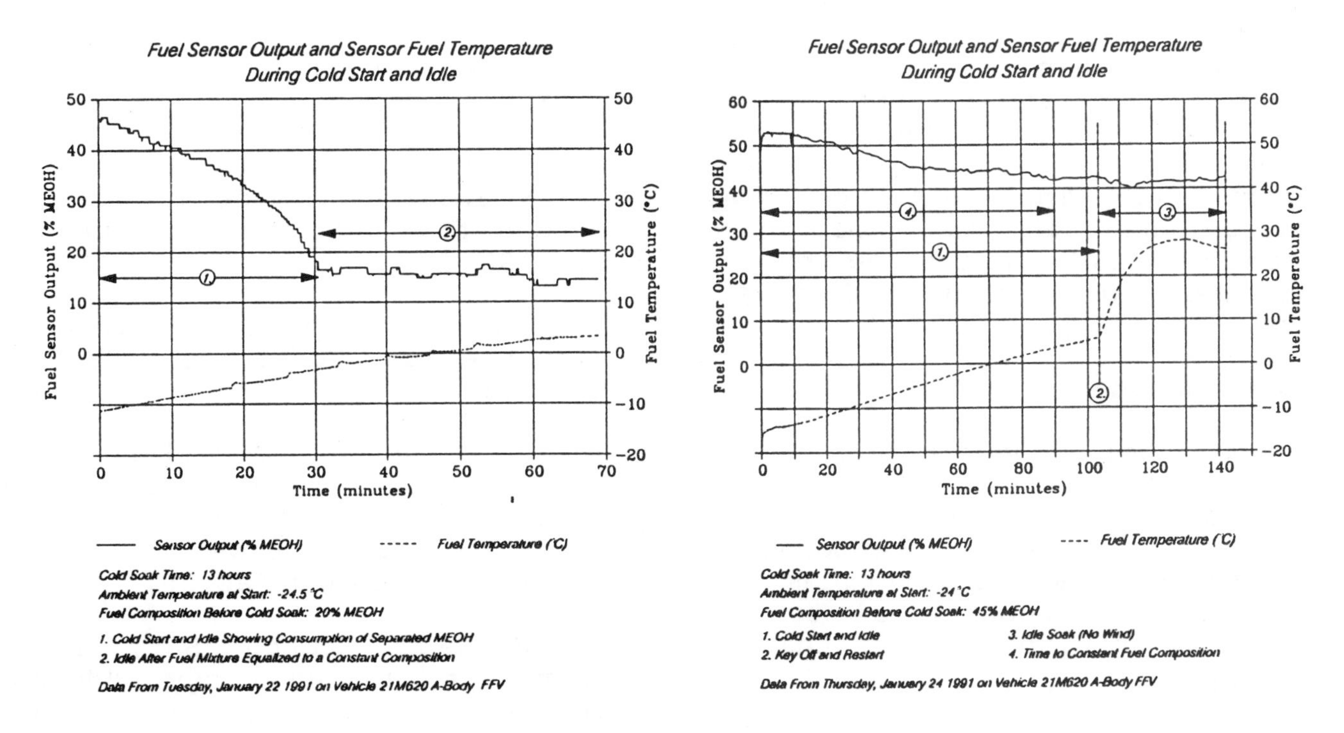

Figure 7

More interesting were the results of cold testing performed during January of 1991, as shown in figure 7, after an overnight soak at -24.5 C a successful start was made, but the concentration of methanol had changed dramatically. The night before the concentration was noted as M20. Upon starting the concentration rose to M45. As the engine and fuel system warmed (stationary at idle), the "heavy" methanol was consumed and the mixture is decreased, the more concentrated (phase separated) methanol component becomes from the gasoline at a given sub-zero (0 C) temperature.[5] However, this did not become a driveability problem as the MCS sensor correctly indicated the increased percentage and the injector pulse width/ spark advance were modified accordingly.

This phenomena had also been observed during bench testing at sub-zero temperatures, wherein the methanol would precipitate at the bottom of the test reservoir forming a more and more concentrated mixture at low circulating pump flows. It should be noted that, as figure 7 indicates, the methanol concentration will fall somewhat below the earlier reading over time because, as we hypothesized, a more concentrated mixture of methanol has been consumed during the period before the fuel warms sufficiently to blend the remaining methanol back into the gasoline present in the fuel tank. This is of concern, because many of the FFV vehicles to be sold have methanol concentration readouts in the passenger compartment. The operator may construe, and rightly so, that there is some malfunction with the vehicle when operating during cold conditions.

7. Conclusions

- A Methanol Concentration Smart Sensor has been developed with microprocessing capabilities, self-diagnostics, and temperature compensation, which provides an analog signal proportional to the concentration with a better than 5% accuracy.

- The capacitive / dielectric constant approach to sensor design with a small sensing coil provides a small, robust, and capable package for the requirements of feedforward methanol sensing to a Powertrain Controller.

- Until a fueling infrastructure is in place, vehicles designed to operate on alternative fuels will be of necessity FFV's equipped with a methanol concentration sensor allowing operation on either gasoline or methanol.

References

1. H.Suzuki and K. Ogawa, "Development of an Optical Fuel Composition Sensor", SAE Paper 910498, also in 1991 SAE-SP-242

2. Allen H. Meitzler and George S. Saloka, "Two Alternative, Dielectric-Effect, Flexible Fuel Sensors", SAE Paper 920699, Also in 1992 SAE-SP-903

3. G. Schmitz and M. Siedentop, "Intelligent Alcohol Fuel Sensor", SAE Paper 900231, Also in 1990 SAE-SP-819

4. H.Depa, J. J. C. Kopera, "Flexible Fuel Sensor System", U.S. Patent No. 5,150,683, Issued Sept 29,1992

J. J. C. Kopera, "Method For Flexible Fuel Control", U.S. Patent No. 5,119,671, Issued June 9, 1992

J. J. C. Kopera, "Oscillator Having Resonator Coil Immersed In A Liquid Mixture To Determine Relative Amounts Of Two Liquids", U.S. Patent 5,091,704, Issued Feb 25, 1992.

5. Student Branch SAE, Florida Institute of Technology, "Methanol Innovation: Technical Details of the 1989 Methanol Marathon", SAE-SP-804

930357

An Efficient Error Correction Method for Smart Sensor Applications in the Motor Vehicle

Rainer Dietz, Erich Zabler, and Frieder Heintz
Robert Bosch GmbH

ABSTRACT

In conventional sensor systems, mechanical and electronic components are generally operating at separated locations. Smart sensors integrate mechanical and electronic elements to a single system, thus offering new facilities for a common error compensation.

In this concept, a unit-specific temperature dependence and a non-linear characteristic curve of the mechanical sensor element can be tolerated, thus saving a lot of costs in the manufacturing process of the mechanical components. The behaviour of the mechanical sensor element is described by a two-dimensional sensor correction function: Given the output of the mechanical sensor element and a measured value for the temperature, the true measurement value can be calculated by an error correction unit.

In this paper, different error correction methods are examined and evaluated which can be used for a wide range of sensor types. They are applied to the example of a short-circuit ring displacement sensor. It is shown that the approximation of the error correction function by a two-dimensional characteristic offers advantages in both accuracy and simplicity of the correction procedure. The correction algorithm can be carried out by a simple dedicated processor which requires small chip area if the smart evaluation circuit is integrated into a single chip.

INTRODUCTION

The measuring signal generated by a real sensor element is always disturbed by different influencing variables. In most applications, the temperature T turns out to be the most important of them. Great efforts are made to minimize its influence on the sensor output signal. Another error source is the non-linear characteristic curve of the sensor element. In general, the form of this curve and its offset vary from specimen to specimen. These errors are called systematic errors.

Fig. 1 shows the principle of a conventional sensor system. The mechanical sensor element transforms the analog process quantitiy X into an electrical quantity $x(X,T_1)$ with T_1 denoting the temperature of the sensor element. In conventional systems, the evaluation circuit is located at a separate location (usually in the respective control unit) with a different temperature T_2. The sensor signal $z(X,T_1,T_2)$ contains both the errors of the mechanical and electronic system. In the worst case, these errors must be added because these systems must be optimized independently from one another.

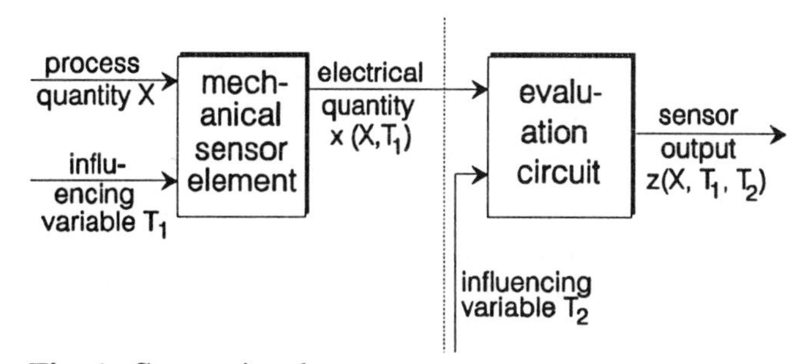

Fig. 1: Conventional sensor concept

The idea behind the smart sensor concept is a common error compensation of the mechanical and electronic sensor components. This is done by integrating them into a single system as shown in Fig. 2.This concept is also designated by the expression "electronics on site".

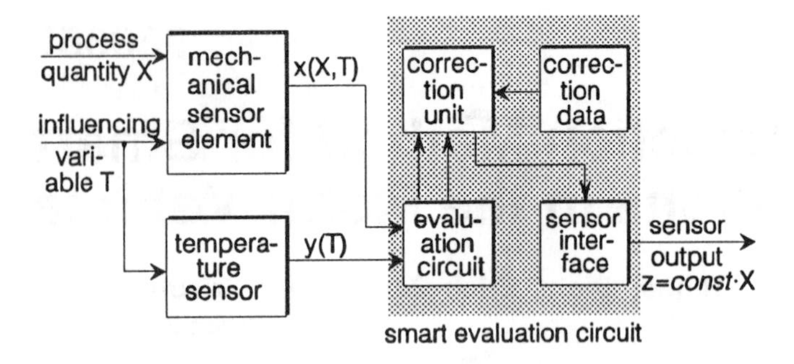

Fig. 2: Smart sensor concept

Compared to conventional systems, a second sensor element is necessary in order to get an electrical quantity y depending on the common temperature T. This can be a simple temperature dependent resistor. The smart evaluation circuit consists of 4 parts. The evaluation circuit measures the quantities x and y and converts them into digital signals. These are processed in the correction unit. It uses the correction data describing the unit-specific characteristics x(X,T) and y(T), thus being able to calculate a corrected sensor signal which is fault-free in the ideal case, i. e. linear in X and not depending on T. This is sent to the sensor output via the sensor interface. A digital sensor interface offers the advantage of interference-free signal transmission. Furthermore, with a bus capable interface, new sensor concepts with multiple sensor usage as well as sensor self-diagnosis are possible.

AN EXAMPLE: SHORT-CIRCUIT RING DISPLACEMENT SENSOR

Displacement and angle measurement sensors are the most widely used sensors in motor vehicles. As the well known potentiometer sensors have serious problems in reliability, they are replaced by advantageous contactless and wear-free sensors in ever more applications.

Sensors constructed according to the short-circuit ring (SCR) principle are an important alternative for displacement and angle sensors. The principle which is described in detail in /1,2/ is illustrated in Fig. 3. The magnetic flux of a soft iron core, such as a laminated plated E-core, is limited according to the position of a moving short-circuit ring made of copper or aluminium, thus varying the inductance at the excitation coil terminals. This sensor is used, for instance, in an electronically controlled diesel pump.

Fig. 4 shows the concept of the conventional system. The shape of the iron core has turned from the simple E-form to a complicated shape with non-constant air

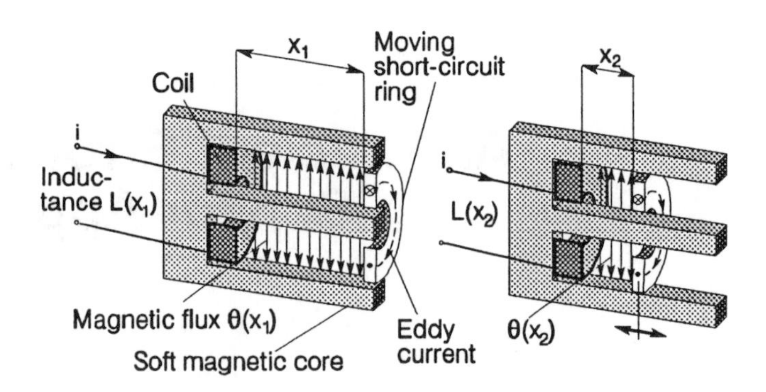

Fig. 3: Principle of SCR-sensors

gap which is done to achieve a characteristic with better linearity. A second reference system is added to minimize temperature errors by evaluating the quotient of two inductances.

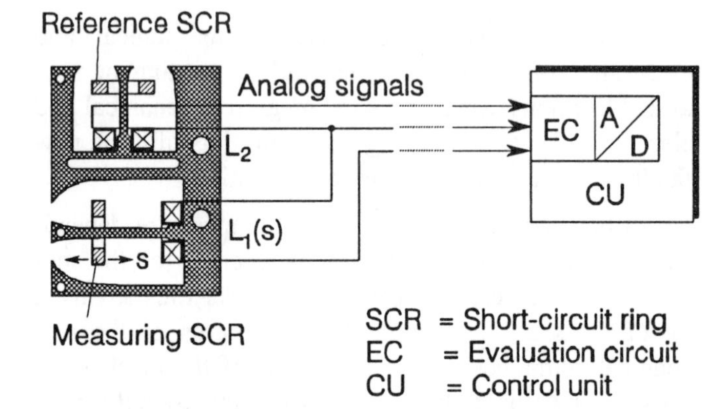

Fig. 4: Conventional SCR-sensor

The smart sensor concept as shown in Fig. 5 uses the simple E-shaped core.The reference system is no longer necessary. The temperature is measured by a simple resistor and both systematic errors and temperature influences are compensated together in the correction unit.

In a first approach, the correction unit was realized by a microcontroller, using large storage for the unit specific correction data /3/. The accuracy could be improved by at least a factor of 4 over the whole measurement range.

The next step is to optimize the correction process in order to find a smart evaluation circuit with minimum hardware complexity. For a one chip implementation, the minimization of chip area is essential due to the cost factor. Both the processor hardware for the correction unit and the memory size for the correction data must be taken into consideration. Their sizes depend on the

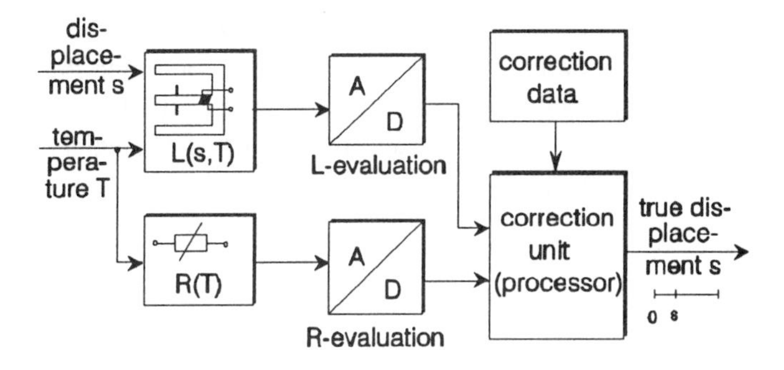

Fig. 5: Smart SCR-sensor

applied error correction method. In the following, different methods are discussed and compared to one another.

PROBLEM FORMULATION

After the calibration process, a specific correction function $f_{k,i}$ is known for each sensor unit i. In order to calculate $f_{k,i}$, measurements without correction must be carried out at some operating points in the specified measurement and temperature range. The true value for X is determined with the aid of a reference sensor, but no reference is required for the influencing variable T. It must merely be ensured that the characteristic curve y(T) remains reproducible in calibration and in operation.

The two-dimensional correction function $f_{k,i}$ describes the true measurement value X as a function of the non-ideal (i.e non-linear and temperature-dependent) measurement x(X,T) and the measured temperature value y(T): $X = f_{k,i}(x,y)$.

The correction function must now be approximated by way of a sensor model f_m. The model function f_m is common to all units of a given sensor type, but it depends on a set of specific model parameters $p_{1,i} \ldots p_{n,i}$. Thus the following approximation applies:

$$f_{k,i}(x,y) \cong f_m(p_{1,i}, \ldots, p_{n,i}, x, y). \qquad (1)$$

For clarity, the unit-specific index i is omitted below. The sensor model f_m is implemented in the processor of the smart evaluation circuit, and the parameter set $p_1 \ldots p_n$ is stored in its correction data memory.

For an approximation according to EQ (1), an approximation error e can be defined as follows:

$$e = \underset{\forall(x,y)}{\mathrm{MAX}} \{ \mid f_k(x,y) - f_m(p_1, \ldots p_n, x, y) \mid \}. \qquad (2)$$

The approximation error e is thus defined by the greatest deviation between the correction function and its approximation. EQ (2) assumes that the measurements x and y are discrete and that therefore exists a finite number of pairs (x,y).

In addition to the error e, there are two further important evaluation criteria in the formation of a model. The first of these is the number of model parameters $p_1 \ldots p_n$, which largely determines the memory requirement of the smart evaluation circuit. The second criterion is the complexity of the sensor model f_m, which defines the stages of calculation to be carried out during error correction. The complexity of the sensor model ultimately determines both the hardware required for the correction processor and the processing time.

ANALYTICAL METHODS

The analytical approach uses mathematical equations that describe the model function in its whole definition area. The equations contain coefficients which form the model parameters. If the main equation contains only one variable x or y, the coefficients depend on the other variable and are defined by further equations. Only one equation is needed if a two-dimensional analytical function is used, for instance a polynominal which considers the powers of x as far as the cube and the powers of y as far as the square:

$$\begin{aligned} f_m(x,y) = {} & a_{00} + a_{10}\cdot x + a_{20}\cdot x^2 + a_{30}\cdot x^3 \\ & + a_{01}\cdot y + a_{11}\cdot x\cdot y + a_{21}\cdot x^2\cdot y + a_{31}\cdot x^3\cdot y \\ & + a_{02}\cdot y^2 + a_{12}\cdot x\cdot y^2 + a_{21}\cdot x^2\cdot y^2 + a_{32}\cdot x^3\cdot y^2. \end{aligned} \qquad (3)$$

In this case, the 12 coefficients a_{ij} (i = 0...3, j = 0...2) would form the model parameters. In the analytical methods, the coefficients, i.e. the model parameters, are resolved from the correction function by an averaging process according to the least error squares method.

Although the analytical methods are notable for their extremely low memory requirement, the calculations are long-winded. And since the approximation error e is intolerably large in the sensors examined, these methods have proven to be impracticable and are not discussed further in this paper.

CHARACTERISTIC METHODS

In the characteristic methods, the values of the correction function $f_k(x,y)$ for given values (x,y,), i.e. vertices, are stored as model parameters. The characteristic thus provides a number of sample values of the correction function. The vertex interval selected is conventionally equidistant in both directions and is denoted below with Δx and Δy. The smaller Δx and Δy, the greater the number of model parameters and thus the greater the memory requirement.

The model function $f_m(x,y)$ is defined in this characteristic by an interpolation rule. In contrast to the analytical methods, only a few model parameters are relevant for a given measurement (x,y). This section describes two interpolation methods which utilize 4 and 3 vertices respectively.

The principle of 4-point interpolation is illustrated in Fig. 6. For a given measurement pair (x,y), 4 adjacent vertices (x_0, y_0), $(x_0 + \Delta x, y_0)$, $(x_0, y_0 + \Delta y)$, $(x_0 + \Delta x, y_0 + \Delta y)$ are initially determined. The corresponding function values $f_k(x,y)$ stored in the characteristic are referred to as z_{00}, z_{10}, z_{01}, z_{11}. This gives 4 points through which a bilinear surface is plotted which defines the model function in the range under consideration. This is the same as calculating the corrected measurement value according to the following equation:

$$f_m(x,y) = z_{00} + (z_{10} - z_{00})\cdot(x-x_0) / \Delta x + (z_{01} - z_{00})\cdot(y-y_0) / \Delta y + (z_{11} + z_{00} - z_{10} - z_{01})\cdot(x-x_0)\cdot(y-y_0) / \Delta x\cdot\Delta y. \quad (4)$$

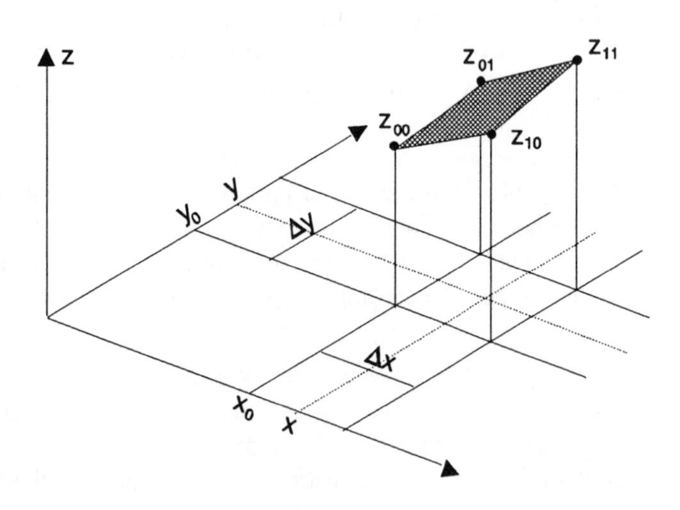

Fig. 6: Principle of the 4-point-interpolation

With 3-point interpolation, the 4th vertex $(x_0 + \Delta x, y_0 + \Delta y)$ is omitted. A plane is plotted through the three remaining points, which again defines the model function. Thus the last summand drops out of the calculation, and the equation is as follows:

$$f_m(x,y) = z_{00} + (z_{10} - z_{00})\cdot(x-x_0) / \Delta x + (z_{01} - z_{00})\cdot(y-y_0) / \Delta y. \quad (5)$$

EQ (4) and EQ (5) clearly show the simplicity of the sensor model: only one equation needs to be processed and this contains no powers. The division is also straightforward because there is no reason not to select a square power for the vertex intervals Δx and Δy.

The interpolation according to EQ (4) is referred to in section 7 as the 4-point approach, that according to EQ (5) as the 3-point approach.

COMBINED METHODS

It is also possible to combine the two methods. Conventionally, the vertex method is used for the actual measured value x (because it provides greater accuracy), and the analytical description for the influence of the variable y (to save memory).

One approach is to approximate the temperature behaviour by means of a linear characteristic curve. For each measured value x, two characteristic values a(x) and b(x) must be available which describe the offset and the gradient of the temperature curve. These coefficients are stored for given vertices x and thus form the model parameters. The a- and b-values for the intermediate x-values are obtained by means of linear interpolation. The two x-vertices adjacent to the value x of the measurement pair (x,y) are again denoted as x_0 and $x_0+\Delta x$. With $a_0 = a(x_0)$, $a_1 = a(x_0+\Delta x)$ and with $b_0 = b(x_0)$, $b_1 = b(x_0+\Delta x)$, the following equation for the sensor model results:

$$f_m(x,y) = \{ a_0 + (a_1-a_0)\cdot(x-x_0) / \Delta x \} + \{ b_0 + (b_1-b_0)\cdot(x-x_0) / \Delta x \} \cdot y. \quad (6)$$

A second approach is to approximate the temperature behaviour with the aid of a quadratic characteristic curve. This correspondingly requires three one-dimensional curves a(x), b(x) and c(x). If $c(x_0) = c_0$ and $c(x_0+\Delta x) = c_1$, the following model equation results:

$$f_m(x,y) = \{ a_0 + (a_1-a_0)\cdot(x-x_0) / \Delta x \} + \{ b_0 + (b_1-b_0)\cdot(x-x_0) / \Delta x \} \cdot y + \{ c_0 + (c_1-c_0)\cdot(x-x_0) / \Delta x \} \cdot y^2. \quad (7)$$

As in the analytical methods, the coefficients a, b and c are determined by means of averaging according to the least error squares method.

The sensor model according to EQ (6) is referred to below as the linear approach, that according to EQ (7) as the quadratic approach.

RESULTS

Fig. 7 shows the correction function for one SCR-sensor unit. The value x is here a period duration P into which the variable inductance is converted by the analog part of the evaluation circuit. The variable y is the temperature T itself, since the principles of temperature evaluation have not yet been determined.The correction function gives the true displacement value s, thus x=P, y=T and X=s must be set to find the correspondence with the generalized formulation used in the equations above.

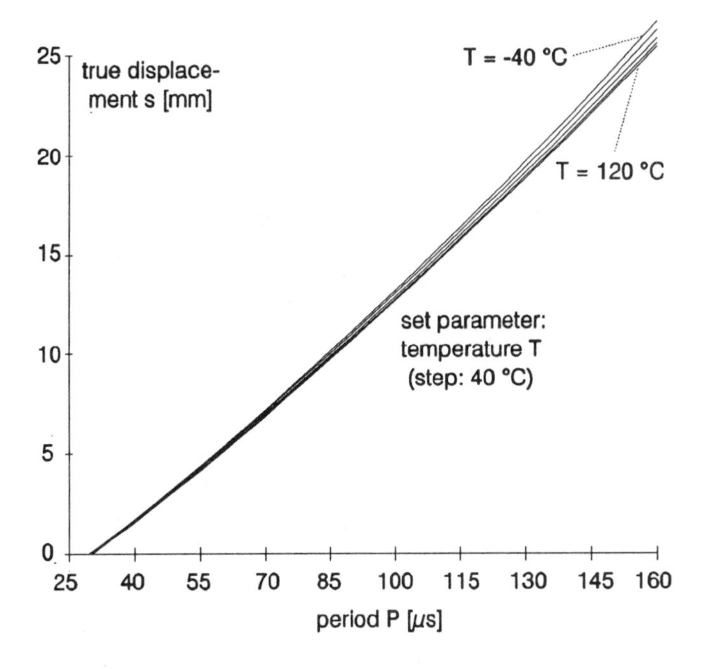

Fig. 7: Correction function for a typical SCR-sensor

The correction function is given as a set of curves, whereby T is used as the set parameter. The curves clearly show the non-linear relationschip between the period duration P and the displacement s. It also can be seen that the same value for P corresponds to a shorter measured displacement at a higher temperature.

The four approaches described in section 5 and 6 were applied to the correction data of 4 SCR-sensor units. The accuracy of the models was calculated by simulating the correction unit on a computer, whereby the whole possible range for the number of vertices was examined. In assessing the results, a tolerable approximation error of 5 µm over a measuring range of 25 mm is assumed.

For the 4-point model, the approximation accuracy for one typical sensor is given in Fig. 8. The number of vertices for the period duration is entered on the horizontal axis and the number of temperature vertices is the set parameter. In order to meet the necessary accuracy of 5 µm, a memory capacity of 32 * 16 = 512 model parameters is required. This is also valid for the other sensor units. In the given example, an approximation accuracy of 3.60 µm is obtained at this point.

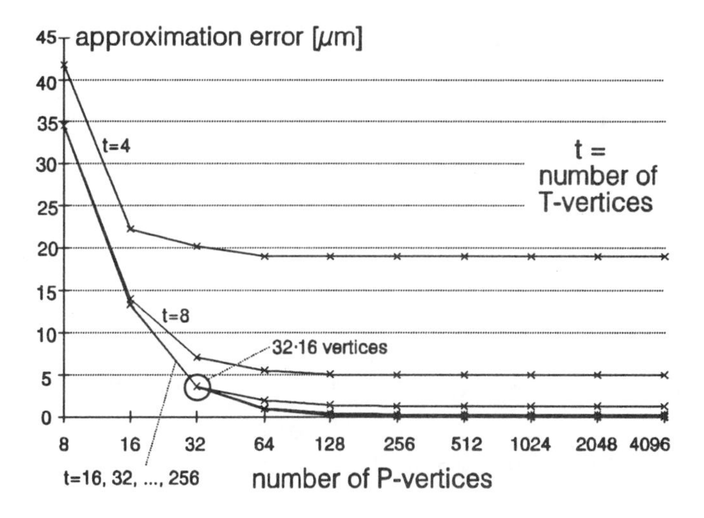

Fig. 8: Accuracy of the 4-point approach

Fig. 9 illustrates the results of the 3-point-approach for the same sensor unit. In this case, a memory capacity of 32 * 32 = 1024 is required for an approximation error less than 5 µm (4.27 µm for the given example).

For the linear approach, the resulting approximation error of all four sensor units is given in Fig. 10. In the best case, an accuracy of only 112.65 µm was obtained with sensor 2. This value cannot be improved by an ever increasing number of vertices. With the desired level of accuracy, the linear apporach falls far below the requirements.

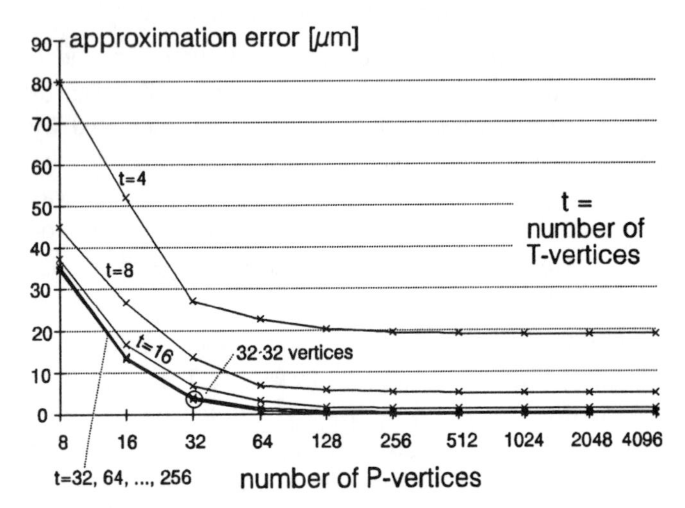

Fig. 9: Accuracy of the 3-point-approach

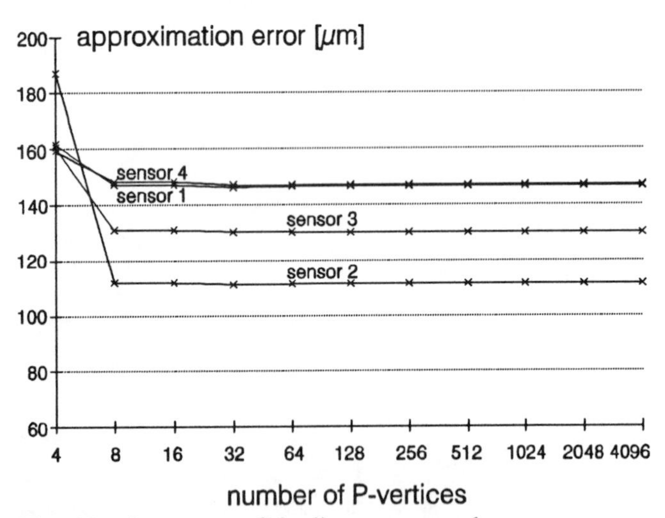

Fig. 10: Accuracy of the linear approach

Fig. 11 shows the simulation results for the quadratic approach. Although the accuracies obtained are far better than those of the linear approach, they still cannot meet the requirements.

CONCLUDING REMARKS

The characteristic methods for the approximation of the correction function has been proven to be suitable for implementation in a smart evaluation circuit. In this paper, this has been shown for the example of a SCR displacement sensor, but the same results have already been obtained for other sensor types (e. g. a pedal position sensor).

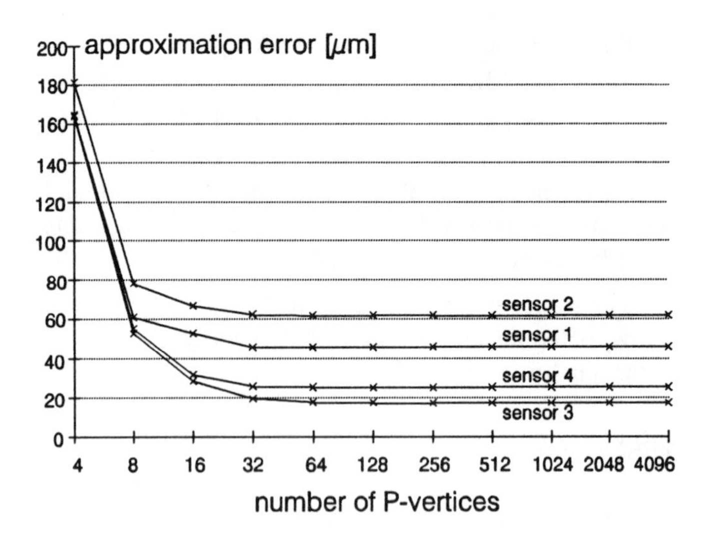

Fig. 11: Accuracy of the quadratic approach

The main advantage is a sensor model function of low complexity, thus the correction algorithm can be carried out by a simple dedicated processor as shown in /5/. The desired approximation accuracy determines the number of model parameters. As the better accuracy is obtained with the 4-point-approach, this method is preferred in most cases. Only if this advantage is considered insignificant, then preference should be given to the 3-point approach on account of its somewhat lower model complexity.

REFERENCES

/1/ A. Weckenmann; "Ein neuer Weggeber zum elektrischen Messen großer Verschiebungen"; Archiv für technisches Messen 14 (1973), pp. 1121-31.

/2/ E. Zabler, F. Heintz; "Shading-Ring Sensors as Versatile Position and Angle Sensors in Motor Vehicles"; Sensors and Actuators 3 (1982/83), pp. 315-326.

/3/ H. Friedrich; "Entwicklung und Programmierung eines intelligenten Kurzschlußring-Wegsensors", Diplomarbeit, Fachhochschule Karlsruhe, 1990.

/4/ E. Zabler, F. Heintz, R. Dietz, G. Gerlach; "Mechatronic Sensors in Integrated Vehicle Architecture", Sensors and Actuators 31 (1992), pp. 54-59.

/5/ R.Dietz, E. Zabler, F. Heintz "Prozessor für die Fehlerkorrektur intelligenter Sensoren in einem zweidimensionalen Kennfeld", Technisches Messen 59 (1992), pp. 353-360.

950538

Surface Micromachined Angular Rate Sensor

Jack D. Johnson, Seyed R. Zarabadi, and Douglas R. Sparks
Delco Electronics Corp.

ABSTRACT

A surface micromachined angular rate sensor utilizing a vibrating metal ring structure on a silicon IC has been developed. Substantial signal conditioning circuitry is included on the IC with the vibrating structure. Tests of the sensor demonstrate that its performance is equivalent to that required for implementation of a yaw control system. Vehicle handling and safety are substantially improved using the sensor to implement yaw control.

INTRODUCTION

Consumer demand for increased automotive safety features has sparked interest in vehicle chassis control systems. These systems, which perform active control of braking, steering or suspension, require real time chassis dynamics data. Yaw, pitch, and roll rates as well as linear accelerations in three axes are required for complete knowledge of chassis dynamics. Several automotive manufacturers are using or have announced intentions to use yaw rate sensors to implement vehicle yaw control. Yaw control appears to be the first step toward totally integrated chassis control systems.

At this time, the cost of angular rate sensors restricts the implementation of yaw control to the luxury vehicle market. Advances in micromachining technology may provide the breakthrough necessary to make yaw control cost effective for a broad automotive market.

AUTOMOTIVE APPLICATIONS FOR YAW RATE

Yaw rate, or angular velocity about a vehicle's vertical axis, is a critical piece of information to a chassis control system. The control system might also use steering wheel angle, wheel speeds and tire normal force information as part of the control algorithm. The steering wheel angle information provides an indication of the driver's intentions, the desired yaw rate. The yaw rate sensor output represents the actual yaw rate. The difference between the desired and actual yaw rates is the error signal in the yaw control loop [1].

The chassis control system reacts to undesired yaw rates by applying a counter-torque about the vehicle's vertical axis. The most cost effective method of creating this torque is likely to be independent control of wheel brake pressure through the vehicle's existing anti-lock brake system (ABS) and traction control system (TCS). A particular example of the value of yaw control is a vehicle operating under poor traction conditions. Unequal coefficients of friction (μ) on the right side and left side tire surfaces may create difficult problems for the driver. This condition will cause an undesired yaw torque to result when the brakes are applied due to unequal braking forces at the tires (Figure 1).

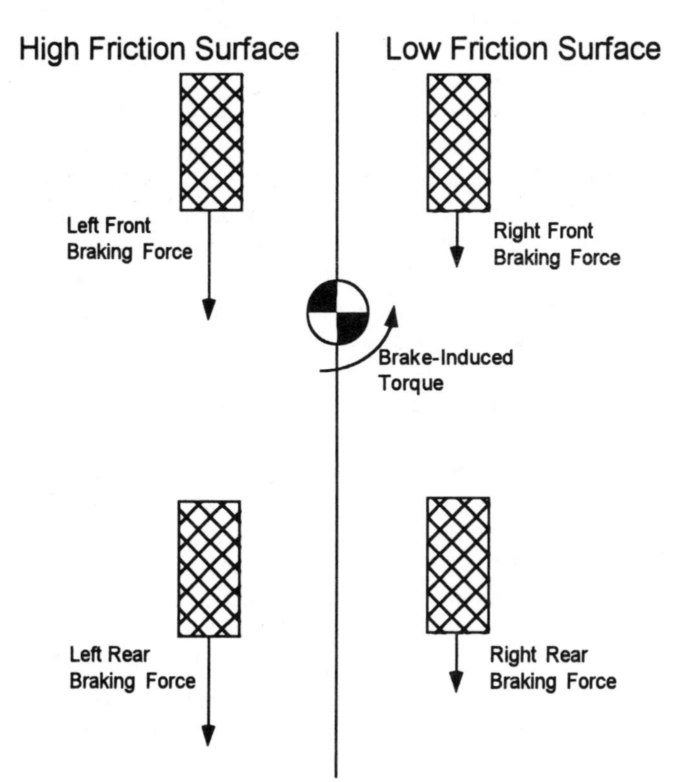

Figure 1. Undesired yaw torque caused by braking. Tires are on surfaces with different coefficients of friction.

In vehicles without active yaw control, the driver uses the steering wheel to apply the counter-torque required. In many cases the steering wheel alone provides insufficient control over the heading of the vehicle [1].

Yaw control is also of value in creating desired yaw rates. In this case, the system assists in achieving the desired yaw rate indicated by the steering wheel angle. The control system may apply the brakes selectively to initiate a turn or evasive maneuver. The brakes may be used by the control system even when the driver is not applying the brakes [1].

SENSING ELEMENT DESCRIPTION

The sensing element developed is a vibrating ring based on the principle of a ringing wine glass as described by G.H. Bryan in 1890. Bryan listened to the ringing wine glass as he rotated it and noted that the vibration pattern in the wine glass predictably lagged the angle of rotation [2]. Delco Systems Operations successfully applied this concept to a navigation-quality sensor almost one hundred years later [3]. The new micromachined vibrating ring sensor is an application of the same concept to create a low cost automotive angular rate sensor.

The ring sensor is similar in function to G.H. Bryan's wine glass. A horizontal slice taken out of the glass would form a ring shape and retain the rotation sensing capabilities of the glass. The micromachined ring is driven into resonance in the plane of the chip. The vibration of the ring forms an elliptical-shaped pattern. The four points on the ring which have no radial deflection are the nodes. Lines drawn through the nodes form the nodal lines as shown in Figure 2. The four points with maximum radial deflection are located 45° away from the nodes. These points are called the anti-nodes.

The sensor may measure either angle of rotation or angular rate of rotation depending on the control electronics used. These two modes of operation are referred to as the "whole angle" mode and the "force to rebalance" mode [6].

Operating in the "whole angle" mode, Coriolis force causes the angular position of the nodes in the elliptical vibration pattern to lag the angular position to which the sensor is rotated. The locations of the nodal points may be used to measure the angle of rotation [2,6]

Used in "force to rebalance" mode, angular rate output signals are obtained. As the device is rotated, the nodal points begin to rotate causing the amplitude of vibration at the defined nodal points to increase. This vibration signal is fed back to reduce the amplitude, thereby maintaining a stationary pattern with nodes at the defined locations. The signal that is fed back to maintain the stationary pattern is proportional to the angular rate and is used as the output [6].

Scanning electron photomicrographs of the sensor element are shown in Figures 3a and 3b. Semi-circular springs support the ring and store the vibrational energy. Sacrificial micromachine fabrication techniques are used to allow the ring and springs to vibrate over the surface of the IC. The springs are attached to the substrate with a symmetric post.

Packaging stress effects on the sensor are small due to the single point connection between the IC and the metal micromachine. Since the center of the ring is the only connection to the IC, stresses imposed by packaging the IC have little effect on the sensing element. Package-induced stress has been a long-standing problem for many other sensor programs [4,5,6].

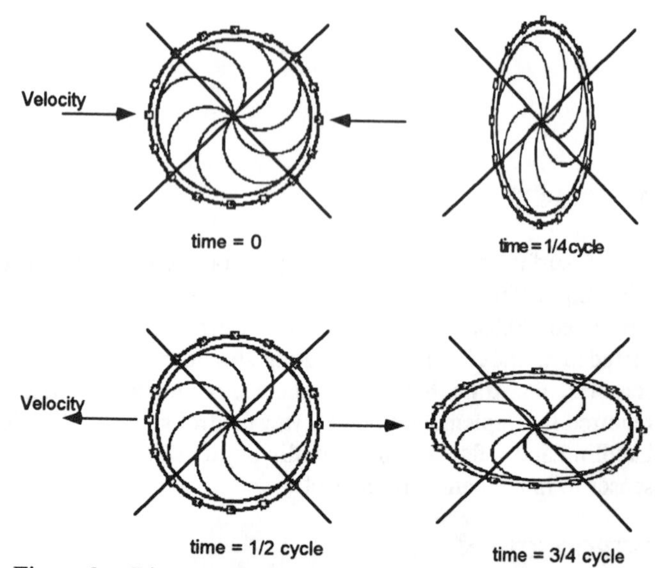

Figure 2. Ring oscillation and nodal lines.

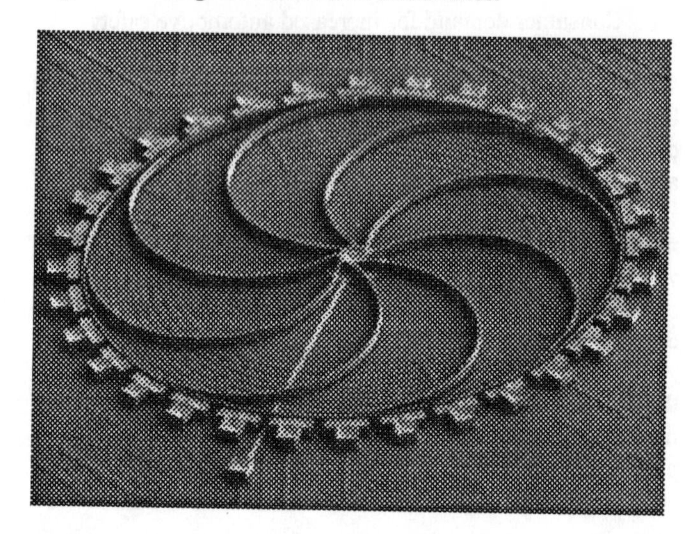

Figure 3a. Photomicrograph of rate sensor.

Figure 3b. Close-up view of ring, electrode gap and electrodes.

SENSING ELEMENT FABRICATION

Using recently developed micromachining technologies such as LIGA and more inexpensive LIGA-like photoresist methods [7,8], ring-shaped micromachined rate sensors can be fabricated on the surface of integrated circuit (IC) wafers. This implementation allows hundreds of devices to be fabricated on each silicon wafer, which are processed together in batches of 10-25 wafers at a time. This high volume manufacturing technique along with the circuit up-integration allows for the low cost production of micromachined rate sensors. It is this low cost manufacturing method that is so appealing for automotive applications.

The fabrication of the rate sensor begins with a mature, high-volume 1.5 micron, dual level poly-silicon, dual level metal CMOS process. The use of a 0.8 micron CMOS process could further reduce the size of the IC. After the final traditional CMOS processing step, pad etch, the micromachining begins. The micromachining process is a modular add-on to an existing production CMOS process. This is a big advantage if the infrastructure needed to produce, trouble-shoot and shrink the circuit portion of the rate sensor is already in place.

The first step in the LIGA-like micromachining process is the deposition of barrier metal and plating seed metals onto the open CMOS bond pads (Figure 4a). Barrier metal prevents the undesired diffusion of adjoining metals. A sacrificial layer is then applied (Figure 4b). The sacrificial layer allows for the eventual motion of the structure by providing a space between the ring structure and the IC. The thick (15-50 micron) photoresist mold layer is then put down and patterned (Figure 4c). Vertical photoresist side walls as thin as 2-3 microns are used to form the plating mold for the metal micromachine. A thick metal layer is selectively electroplated into the open portions of the mold to create the ring structure (Figure 4d). The mold is then removed, followed by the etching away of the sacrificial layer leaving the free-standing micromachine (Figure 4e).

Following the micromachining, vacuum packaging is performed. A vacuum is required to generate a high Q (quality factor) since air molecules would dampen the resonant vibration. The prototype sensors are placed in ceramic DIP packages and sealed using a solder reflow process in a vacuum. Vacuum packaging process improvements will permit SOIC packaging of production parts. After vacuum packaging the devices may be assembled into modules suitable for automotive applications.

SENSE AND CONTROL ELECTRONICS

External control electronics are necessary to maintain the ring oscillation, to sense and process the angular rate signal and to compensate for nonuniformities of the ring. All drive, sense and balance functions are performed electrostatically on the ring structure through the surrounding electrodes. The 32 electrodes are separated from the ring by a vacuum dielectric gap.

A DC polarization voltage is maintained on the ring. It creates an attractive force between the ring and the surrounding electrodes.

An AC signal is applied to the drive electrode. This drive signal at the ring's resonant frequency acts to drive the ring capacitively into resonant oscillation.

Electrodes dedicated to sensing ring vibration are connected to low input capacitance CMOS buffer amplifiers. The capacitance change due to the ring vibration creates the buffer input signal. Low input capacitance buffers are essential because of the small change in capacitance that must be observed. Sensing electrodes around the circumference are located at 45° intervals.

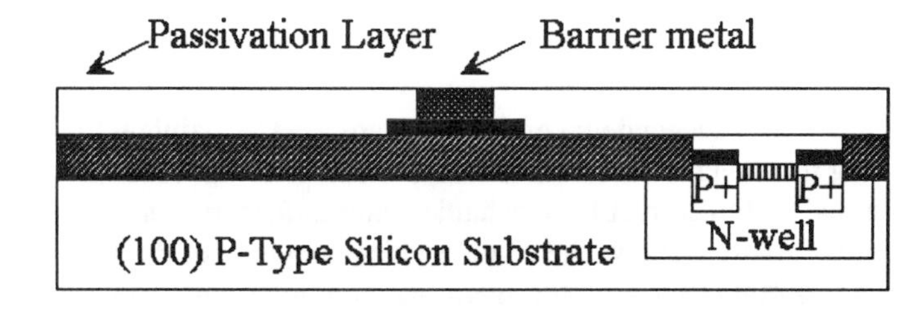

Figure 4a. CMOS circuitry and barrier metal deposition.

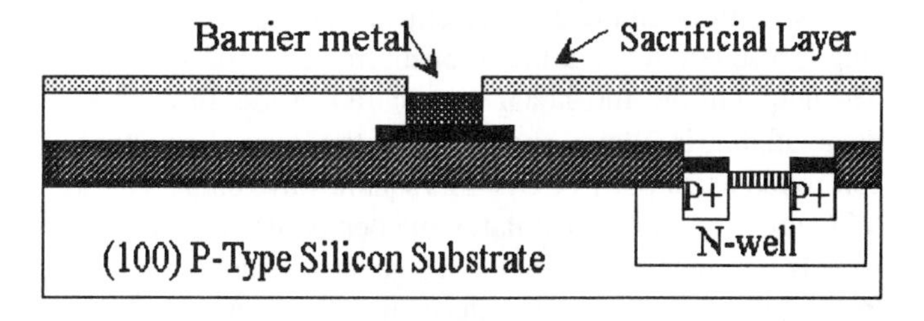

Figure 4b. Sacrificial layer applied.

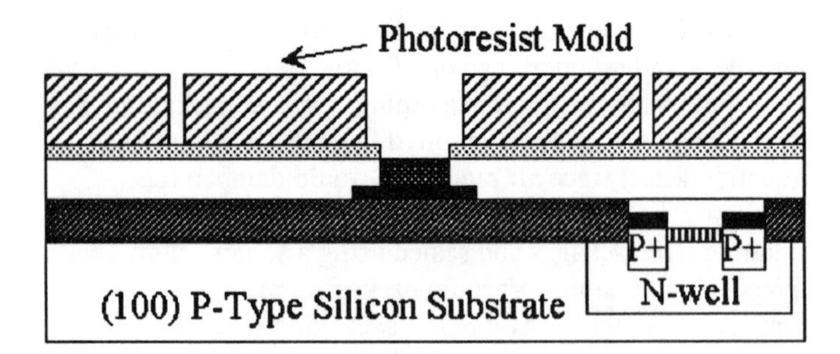

Figure 4c. Photoresist mold on sacrificial layer.

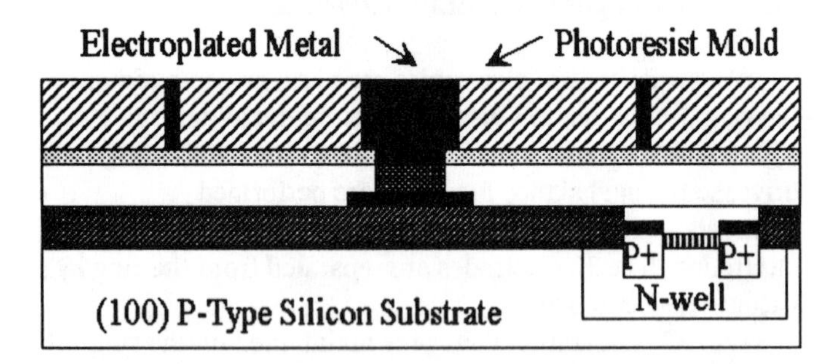

Figure 4d. Metal ring electroplated into photoresist mold.

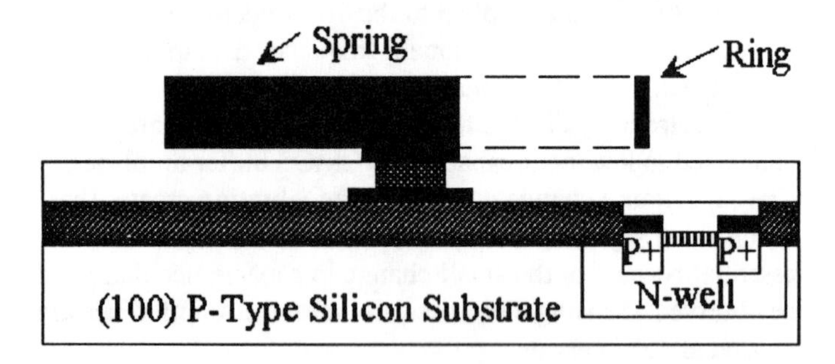

Figure 4e. Free standing ring, mold and sacrificial layer removed.

Four independent control loops are used to maintain the ring in resonance at a constant amplitude, to obtain the rate signal and to correct for mechanical nonuniformities in the ring. Figure 5 shows the complete block diagram of the sense and control electronics for operating the rate sensor. The function of each control loop is described here in detail.

A phase-locked loop is used to drive the ring into resonance through the drive electrode. The drive signal to the ring is a square wave at the resonant frequency of the ring. The output of the anti-nodal sense buffer passes through a demodulator, is filtered and passed to the input of a voltage controlled oscillator (VCO). The square wave outputs of the VCO are used as the modulator and demodulator control signals throughout the circuit. To guarantee that the PLL locks, an error signal is introduced into the loop during power-up. This signal causes the VCO to sweep its frequency range until the resonant frequency is generated and the loop locks.

An amplitude control loop is included in the drive circuitry to maintain a constant amplitude of oscillation. Amplitude control is essential to maintain a constant scale factor of the rate output. The amplitude of the antinodal sense signal is compared to a reference level. The loop adjusts the magnitude of the drive signal until the antinodal sense signal is equal in magnitude to the reference level. In this way, changes in device parameters such as Q (quality factor) will not affect the device scale factor [6].

The force-to-rebalance or rate feedback loop maintains the angular location of the vibratory pattern. The location is controlled through the feedback of the rate signal from the nodal sense buffer to the rate feedback electrode. As described previously, the signal that is fed back is a measure of the angular rate and provides the rate output signal. The loop gain and compensation selected for the loop set the bandwidth of the rate output signal. If the loop was opened and none of the rate signal was fed back, the rate output bandwidth would be

$$BW = \frac{\omega}{2Q} \qquad EQ(1)$$

where BW is the bandwidth, ω is the resonant frequency of the ring and Q is the quality factor. Rate feedback is used to increase the bandwidth of the sensor to 50Hz. The ratio of the rate feedback to the antinodal drive is expressed by

$$\frac{V_{RF}}{V_{DRIVE}} = 4AgQ\frac{\Omega}{\omega} \qquad EQ(2)$$

where V_{RF} is the rate feedback signal, V_{DRIVE} is the anti-nodal drive voltage, Ag is the angular gain constant of the ring, and Ω is the rotation rate [6].

The balance control loop is necessary to compensate for mass and stiffness nonuniformity of the ring. These imperfections unintentionally introduced during fabrication cause the ring to have two unequal resonant frequencies. Rate sensitivity is reduced if the two frequencies are not equal. The equations for the two resonant frequencies are

$$\omega_1 = \sqrt{\frac{k_1}{m_1}} \qquad EQ(3)$$

$$\omega_2 = \sqrt{\frac{k_2}{m_2}} \qquad EQ(4)$$

where ω is the resonant frequency, k is the spring constant and m is the mass. The electrostatic attractive force between the balancing electrodes and the ring may be used to adjust the effective ring stiffness. In this way, the spring constants may be independently adjusted to equalize the resonant frequencies. The nodal sense signal may be used as a

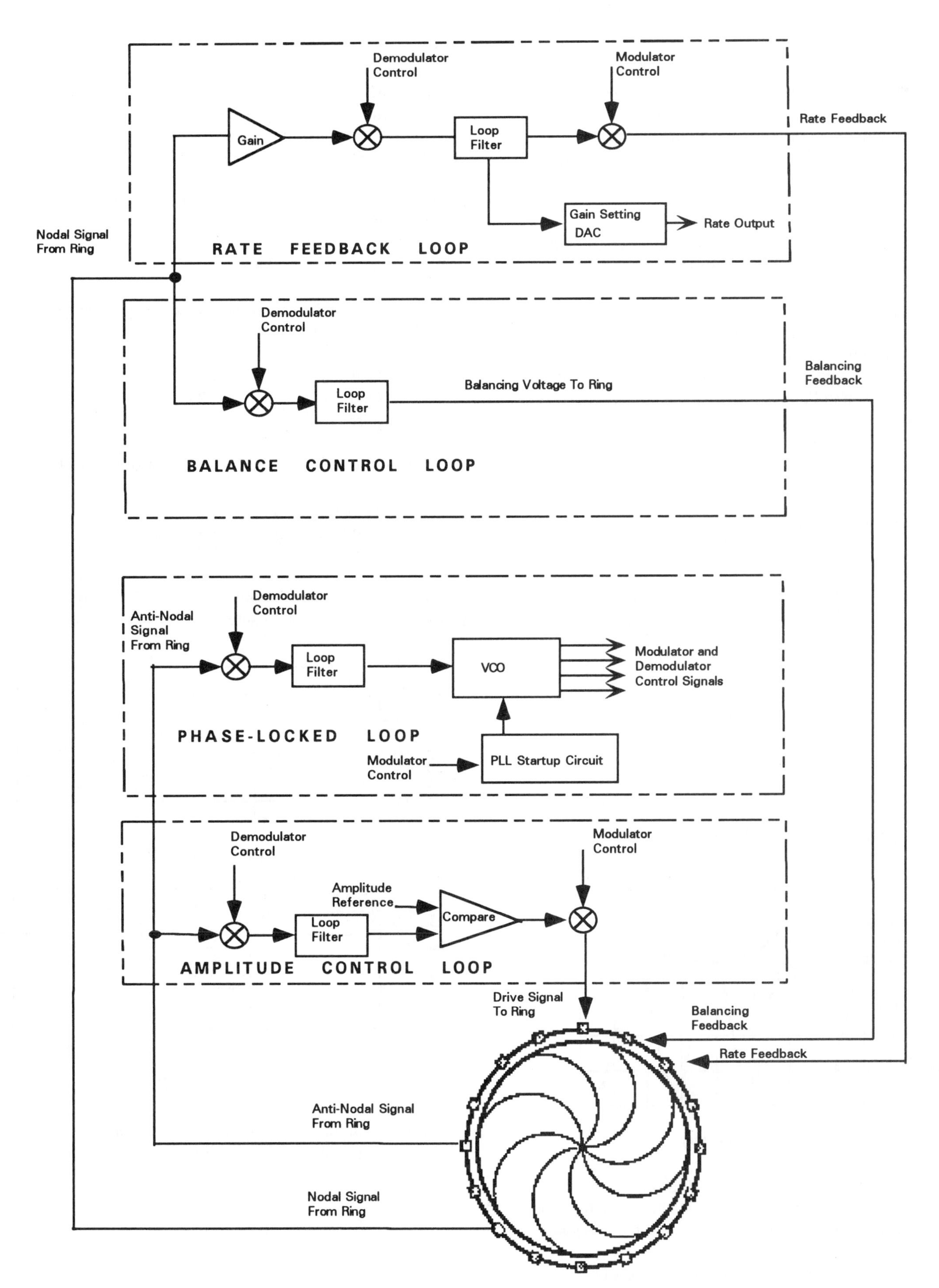

Figure 5. Block diagram of the control IC functions

measure of the nonuniformity. This signal is demodulated and used as the error signal in the balance control loop. The control loop minimizes the error signal through the application of DC voltages at the balancing electrodes [6].

IC IMPLEMENTATION OF THE SYSTEM

The system is partitioned into two ICs, the sensor IC and the control IC. The sensor IC has been implemented in 1.5μm CMOS process. It includes the sensor, low-input capacitance buffer, and the amplifier. The buffer is a self-biased PMOS source follower whose terminal capacitances are canceled using bootstrapping and shielding. The simplified circuit diagram and a sample input/output measurement of the buffer amplifier is shown in Figures 6 and 7 respectively. The buffers exhibit 15fF input capacitance and 0.5μVrms/(Hz)$^{1/2}$ noise.

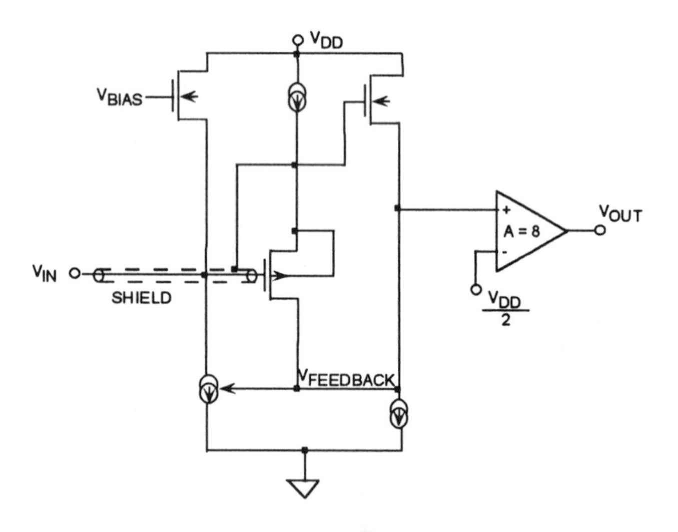

Figure 6. Simplified circuit schematic of the buffer.

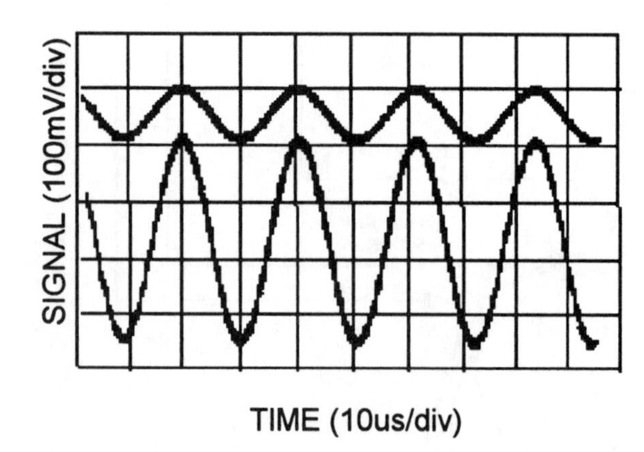

Figure 7. The measured input/output of the buffer; input to buffer is fed through a 10fF on-chip capacitor. Top trace is buffer input, bottom trace is buffer output.

The control IC uses a high voltage 1.2μm CMOS process. In addition to the four aforementioned loops, the IC incorporates serially programmable digital-to-analog converters which are used to cancel any offset or gain errors. The ring polarization voltage is internally boosted from the 5V supply. The polarization voltage is regulated using an on-chip bandgap reference voltage. To eliminate the DC offset caused by the electronics in the rate loop, all amplification is achieved prior to the demodulation. Any DC offset is rejected by the demodulator.

PERFORMANCE SUMMARY

Figures 8,9 and 10 show that the sensor has achieved the following performance:

Bandwidth:	50Hz
Nonlinearity:	< 0.2%
Noise Level:	< 1°/sec

Targeted performance parameters of the sensor also include:

Temperature Range:	-40C to +125C
Total Error Band:	< +/- 5°/sec
Power:	5V, 35mA

This level of performance seems to be typical of the type of performance required for an automotive yaw rate sensor.

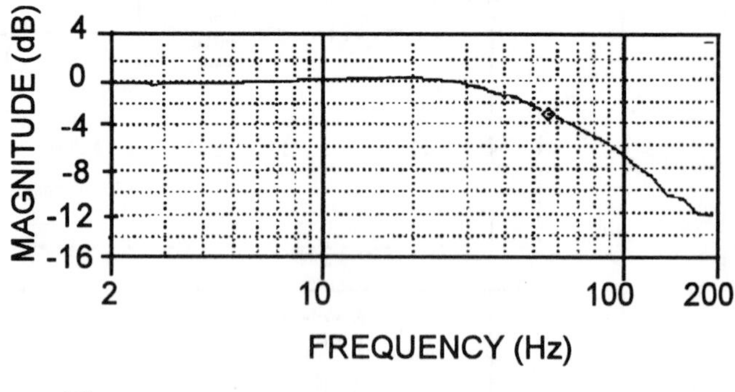

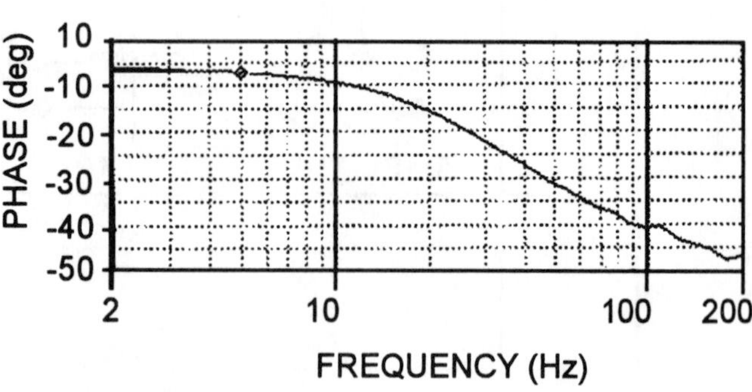

Figure 8. Magnitude and phase plots vs. frequency. Magnitude at 50Hz is -3dB. Phase shift at 50Hz is 32 degrees.

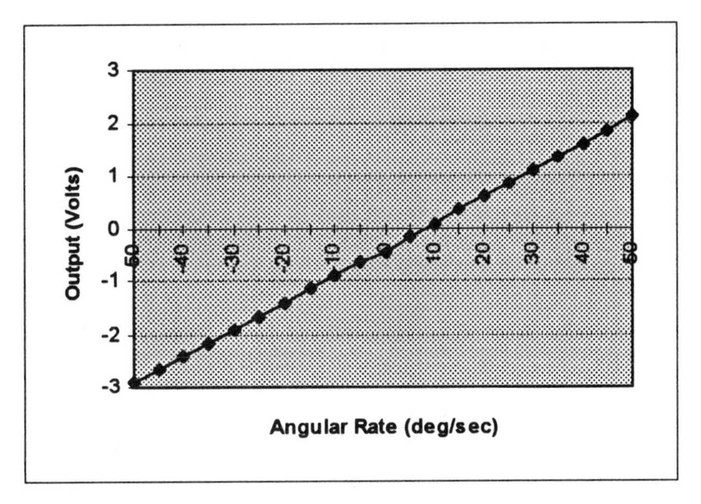

Figure 9. Non-linearity less than 0.2%.

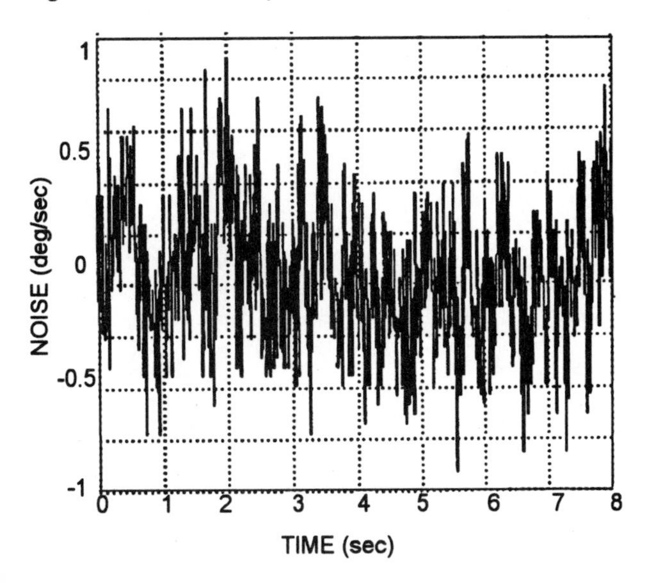

Figure 10. Peak noise level less than 1°/sec.

CONCLUSIONS

The surface micromachined vibrating ring rate sensor has demonstrated performance equivalent to that required for implementation of yaw rate control in vehicles. The added safety and handling provided by yaw rate control and the inexpensive nature of high volume micromachined sensors suggests that the use of yaw control may be implemented broadly throughout the automotive market as micromachined sensors become more readily available. Additionally, the market for low cost angular rate sensors may extend into roll and pitch control systems and complete vehicle dynamics systems as the technology matures.

ACKNOWLEDGEMENTS

The authors would like to acknowledge the contributions of Mike Putty, Larry Oberdier, Scott Chang, Jenny Shi, Dave Hicks and others at the General Motors Research and Development Center and Delco Electronics personnel including Pedro Castillo-Borelly, Mike Chia, Fie-An Liem, Oya Larsen, Bill Higdon, Jean Tolin and Larry Jordan for their support and dedication to this effort.

REFERENCES

1. T. Mathues, “Extending the Scope of ABS”,Automotive Engineering, July 1994, Vol. 102, no. 7, pp.15-17.

2. G. H. Bryan,”On the Beats in the Vibrations of a Revolving Cylinder or Bell”, 1890, Proceedings of the Cambridge Philosophical Society, Volume VII, Cambridge: The University Press, 1892.

3. E. Loper and D. Lynch, "Projected System Performance Based Recent HRG Test Results," Proceedings of the 5th Digital Avionics System Conference, IEEE, Nov., p.18.1.1-18.1.6, 1983.

4. J. Soderkvist, “Micromachined Gyroscopes”, Sensors and Actuators, Vol. A43, pp. 65-71, (1994).

5. J. Burdess, T. Wren, “The Theory of a Piezoelectric Disc Gyroscope”, IEEE Transactions on Aerospace and Electronic Systems, Vol. AES-22, pp. 410-418, (1986).

6. M. Putty and K. Najafi, "A Micromachined Vibrating Ring Gyroscope," Solid-State Sensors and Actuators Workshop, June 13-16, p.213-220, (1994).

7. J. Bond, "The Incredible Shrinking Disk Drive," Solid St. Tech., 36, p.39-45 (1993).

8. A. Frazier and M. Allen, "High Aspect Ratio Electroplated Microstructures Using A Photosensitive Polyimide Process," Micro Electro Mech Systems '92, p.87-91 (1992).

Figure [illegible]

Figure [illegible]

[illegible]CLUSIONS

[illegible] rate sensor has [illegible] demonstrated [illegible] [illegible] feature [illegible] ground [illegible] vehicles. [illegible] and [illegible] and the [illegible] of high [illegible] sensor [illegible] become [illegible]. Additionally [illegible] rate sensors may extend into [illegible] and pitch [illegible] systems [illegible] dynamics systems [illegible].

ACKNOWLEDGEMENTS

The authors would like to acknowledge the contributions of [illegible] Mike Petty, Larry [illegible], Scott [illegible], Jenny [illegible], Dave [illegible] and others [illegible]

[illegible] Development Center [illegible] including [illegible] Lisa Larsen, [illegible] for their support [illegible]

REFERENCES

1. [illegible]

2. [illegible] Proceedings [illegible] Cambridge [illegible] Press, 1992.

3. [illegible] Proceedings [illegible]

4. [illegible] Sensors and Actuators [illegible]

5. [illegible] Sensors and Actuators [illegible]

6. [illegible]

7. [illegible]

8. [illegible]

871589

New Merged Technology Combines Hall Effect Sensors with Intelligence and Power on a Single Chip

Mark F. Heisig
Sprague Electric Co.
Semiconductor Div.

ABSTRACT

An innovative IC development "merges" Hall effect sensing with the control circuitry, protective functions, and the high-current outputs necessary to power a new series of brushless dc (fan) motors. This proprietary and single-chip design replaces a separate Hall sensor IC and the various other components (many discretes) and provides the position and rotational sensing, the commutation circuitry, and switching functions required in a 2-phase, unipolar motor. Additionally, this custom IC includes several advantages and features not (usually) included in "discrete" designs: internal zener diode for limiting "flyback" voltage, output current limiting (with an option for further reduction), output enable (PWM speed control), and thermal shutdown. This paper will highlight the functions, limits, merits, and potential for the "merging" of magnetic sensing and power.

For years, as Brushless DC motors evolved it became apparent that drive electronics represented the majority of the cost. In cooling fan applications, where cost was the driving force, simple single phase discrete designs emerged as the cheapest way to implement the drive function. Both fan manufacturers and semiconductor vendors realized the potential benefits of merging the hall magnetic sensor and the power drives into a single package or chip but, either vendor inexperience in both technologies (semiconductor processing, testing, packaging) or fears relating to thermal instability or packaging stress caused many semiconductor manufacturers to shy away from this task.
Now, a new chip has been designed to drive either 12 or 24V DC brushless fans up to 1 amp of output current. This device contains a hall effect device, power driver and protection circuitry all on one chip. The ultimate design goal of this concept was to eliminate all external components to drive the fan (except for the chip) thus reducing drive electronic cost, adding inherent protection features and possibly eliminating the need for a printed circuit board. The concept, in its most advanced stage would make the single chip fan driver a reality.

The functional block diagram for this device (UGN-3625M) is shown in figure 1.A highly sensitive, thermally stable hall effect element is included along with commutating logic, output drive transistors, zener flyback protection diodes, thermal shutdown and user adjustable short circuit current control. In addition, a buffered hall cell output (open collector NPN) is provided for use as a tachometer pulse train or simply for monitoring motor movement.

In its simplest form only the hall cell/power IC is required. In some applications an external resistor is added to adjust output current, and a series polarity protection diode is added for applications without polarized connection schemes (Figure 2).

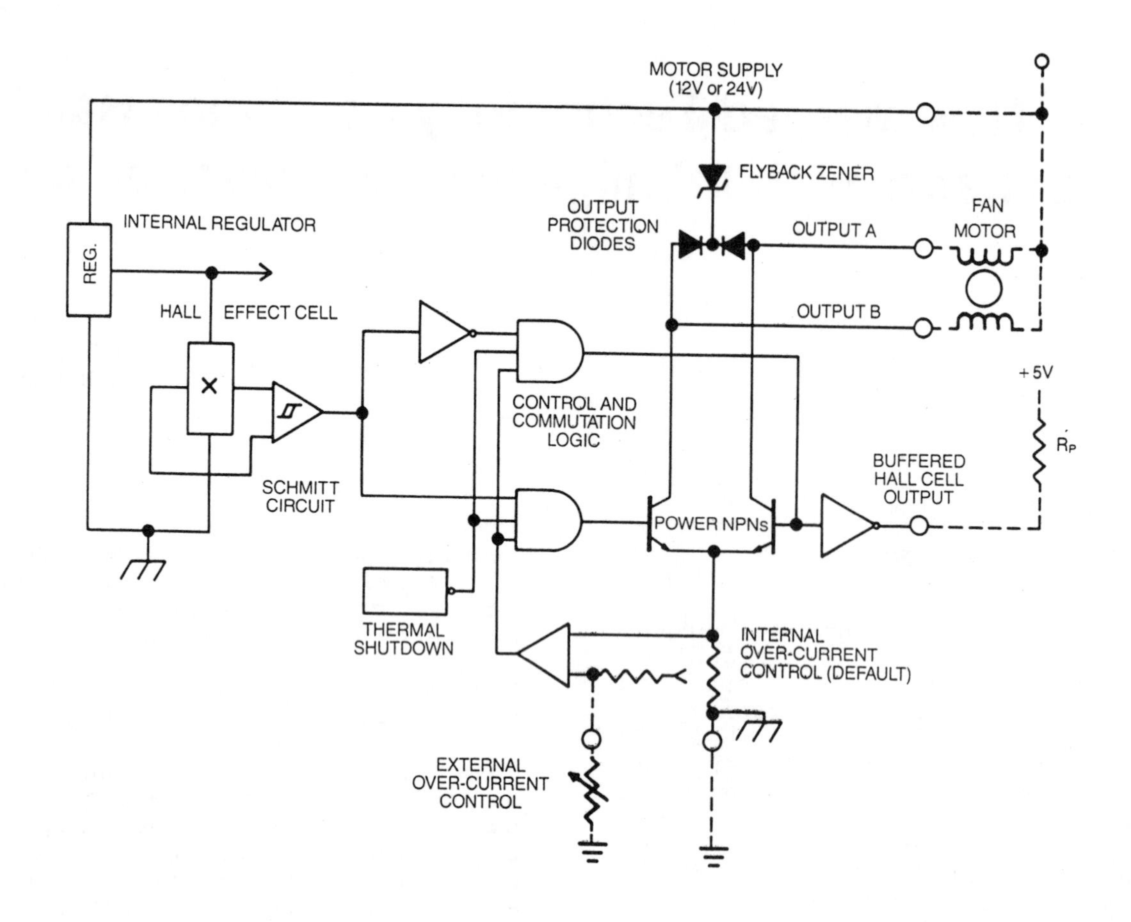

Figure 1

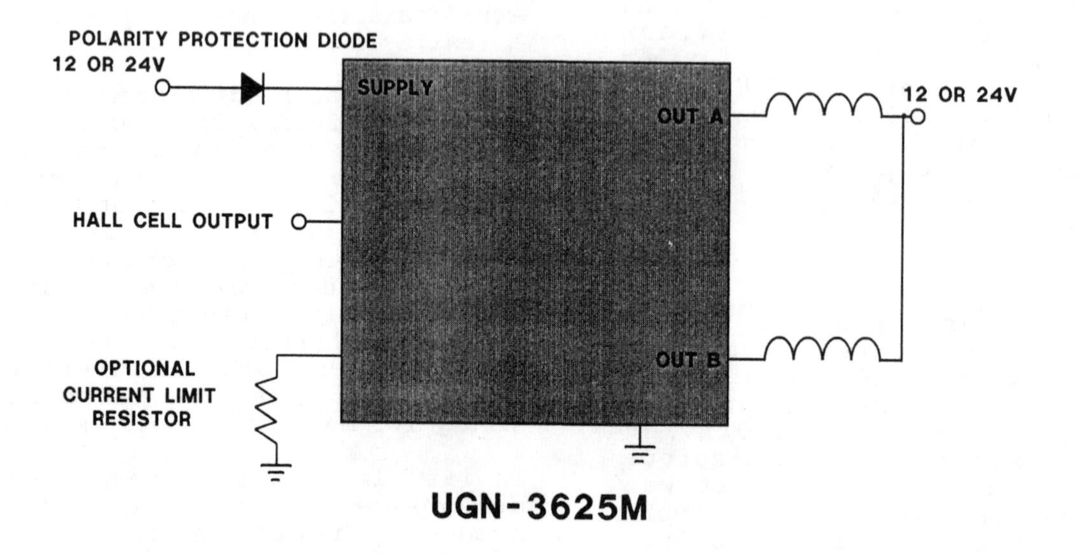

Figure 2

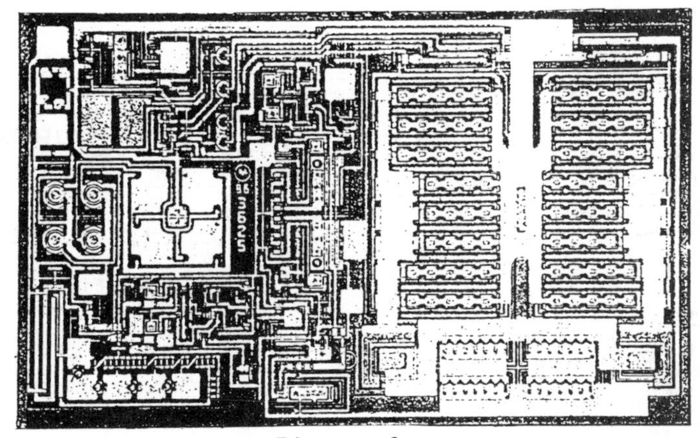

Figure 3

Figure 3 shows a chip photograph of the device. The hall effect device is located in the left half of the circuit (center) and all the amplifier schmitt and protective circuitry surrounds the sensor. The power drive function and clamping networks are on the right hand side of the chip.

The UGN-3625M consists of the following circuit elements:

- Hall Effect Sensor
- Amplifier Circuitry
- Schmitt Trigger (Magnetic Hysteresis)
- Control and Commutating Logic
- Power NPN Outputs
- Regulator
- Output (Transient) Protection Diodes
- Zener (Flyback) Diode
- Overcurrent (Start-up) Protection
- Thermal (Shutdown) Protection

In operation, the circuit function is quite simple. As the rotor cup turns, a ring magnet attached to the cup alternates the magnetic field polarity applied to the hall element. With +100 gauss applied to the chip the commutation logic turns output A on (low state). With -100 gauss applied output A is turned off (high state) and output B is turned on. The trip points are symetrical with 200 gauss of hysteresis. Figure 4 shows that unlike typical motor designs, the hall element is not vertical and inside the rotor cup but horizontal and flat on the PC board. Granted, the magnitude of magnetic field strength in this direction is smaller than in previous designs however, hall effect sensitivity is enough to allow for reliable operation of this configuration Device truth Table and Transfer Characteristics are shown in Figures 5 and 6.

OUTPUT DRIVERS AND FLYBACK CHARACTERISTICS

Electrical characteristics for this device are shown in Table I. As previously mentioned, a 12V and 24 version are available so only one single supply is required to the chip. Devices can be operated as low as 6.5V for varying supply, speed, and performance requirements.

The 12V device permits inrush circuitry up to 1.5A peak with default current limiting set at ~1.3A. The higher voltage device allows 0.75A peak inrush with internal current limit set at ~0.65A.

Outputs are saturated NPN devices for low collector-emitter saturation voltage drops (0.8V max at 500mA) (i.e. more efficient, lower power dissipation). In addition, outputs can sustain up to 50V at 1 amp for the provision of allowing inductive flyback voltages to exceed supply voltage by ~ 2X. Internal zener diodes allow energy to be rapidly discharged out of the load winding for improved motor performance. Allowing the inductive flyback to rise to ~2X supply during turn-off prevents current to continue to recirculate in an "OFF" winding in opposition to the "ON" winding. Also transformer action between windings can be reduced if energy discharge during turn-off is rapid. Figure 7 shows the output configuration and output waveform for zener clamping during turn-off.

OVERCURRENT (STARTUP OR "LOCKED ROTOR" SURGE) PROTECTION

Without any external resistor to select a "surge" limit the 12V

PREVIOUS POSITIONING OF HALL CELL

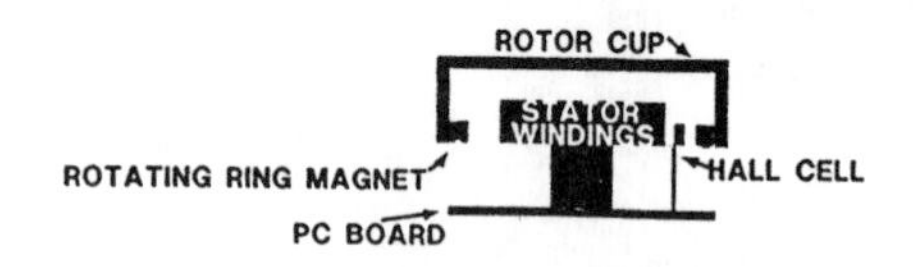

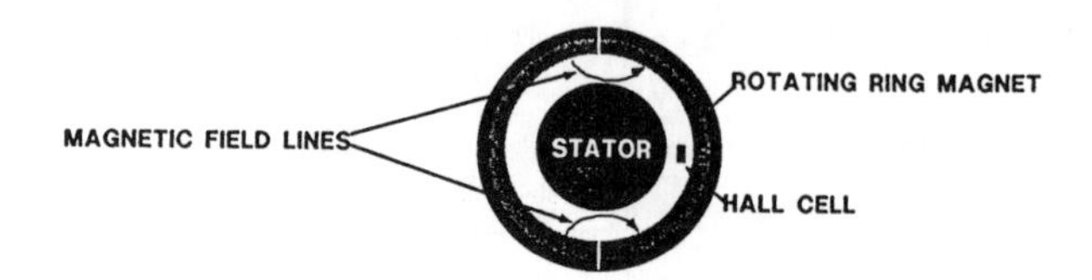

PRESENT POSITIONING OF HALL CELL

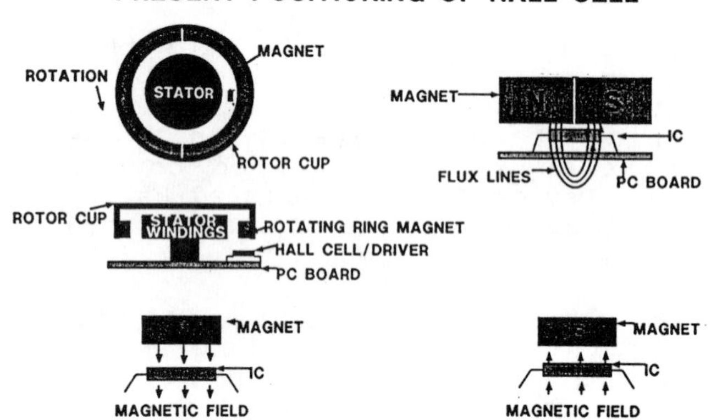

Figure 4

TRUTH TABLE

B	Out_A	Out_B
$>B_{tr+}$	Low	High
$<B_{tr-}$	High	Low

Figure 5

TRANSFER CHARACTERISTICS

Figure 6

TABLE I

Characteristic	Ratings at Nom. Motor Supply						Unit
	Nom. V_s = 12 V			Nom. V_S = 24 V			
	Min.	Typ.	Max.	Min.	Typ.	Max.	
Motor Supply Range, V_s	6.5	—	15	6.5	—	25	V
Output Current, I_{OUT}							
(Continuous)	—	—	1.0	—	—	0.5	A
(Peak)	—	—	1.5	—	—	0.75	A
(Over-Current)	—	1.3	—	—	0.65	—	A
Output Voltage, V_{OUT}	—	—	50	—	—	50	V
Zener Flyback Voltage, V_z	—	—	12	—	—	24	V
Magnetic Trip Points, B_{tr}	—	±100	—	—	±100	—	G
Thermal Shutdown Temp.	—	165	—	—	165	—	°C

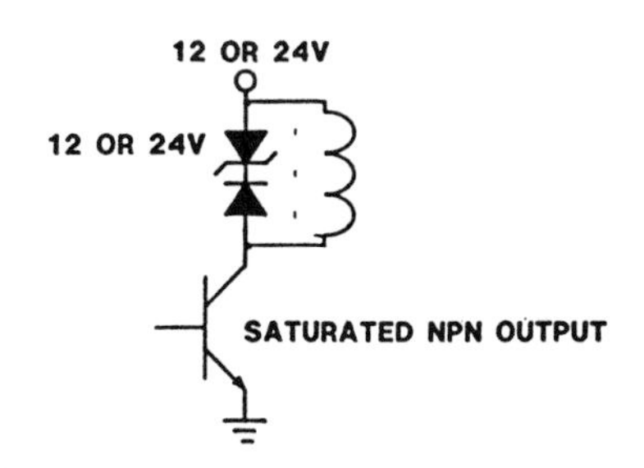

OUTPUT DRIVER AND CLAMPING NETWORK

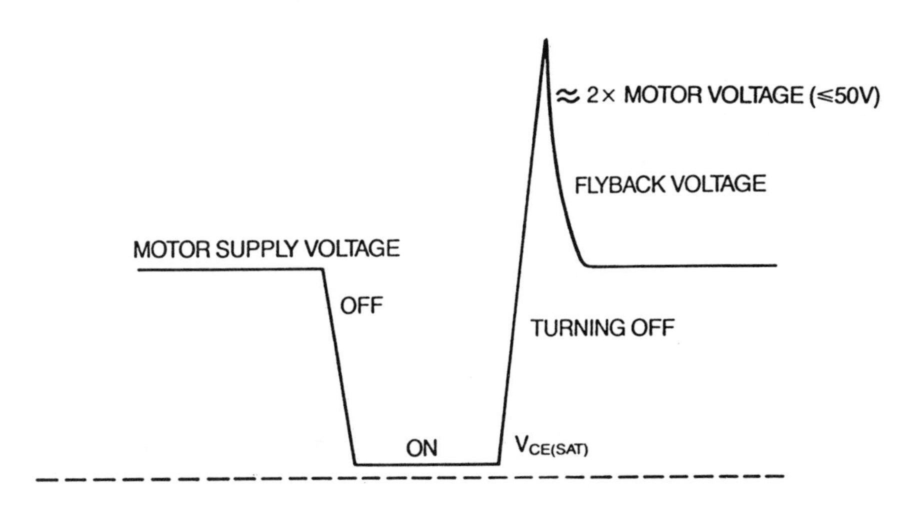

Figure 7

version has a typical limit of 1.3A, and the 24V device has a nominal value of 0.65A. Using a "current mirror" circuit, and a "dominant pole" (state) to avoid instability, the circuit senses a voltage drop across an internal (very low value) resistor. Motor startup is the greatest concern, but a "locked rotor" condition can also trip the device into current limiting. During startup this high-current surge is brief, and will quickly decay as the motor accelerates to rated speed and produces its (normal) "back emf".

The current limit may be reduced by an external resistor, but is an option that will only be used in certain motors and/or applications. Additionally, this "ADJUST" function may be used to "inhibit" the chip; grounding this device control function shuts OFF the IC (option: pulse-width speed control realized via "monitoring" the hall sensor output and use of system logic). The chief, although not extremely obvious, advantage of this overcurrent function stems from its relationship to the motor supply. Reducing this startup surge translates into a power supply with; lower current rating, reduced size and weight, and lower cost. Thus, a function readily incorporated in an IC (output current limiting) becomes quite an advantage and benefit.

THERMAL (SHUTDOWN) PROTECTION

Thermal protection has become, increasingly, a part of power ICs; and is a function that can only be effectively implemented in monolithic circuits. It is customary to rate ICs for operation and allowable chip temperatures to +150°C. Any condition that produces excessive power dissipation (resulting in extremely high chip temperature) may severely affect IC reliability (including very swift failure) and/or the motor windings as well. Should the chip temperature exceed +165°C (typical) the circuit senses this condition and disables both outputs. Effectively, this is accomplished by grounding (via internal circuitry) the ADJUST function.

PACKAGING CONCERNS

The UGN-3625 device will be offered in three separate package options. The initial package option will be an 8-pin dual-in-line plastic mini-dip for standard through applications. This package offers ~ 1.56W of allowable power dissipation at 25°C ambient temperature. Future surface mount options include a 14 pin small outline 150mil wide plastic package which dissipates ~1.14W at 25°C. Also, a modified 20 pin SO wide body (300 mil) batwing style package (gull lead style) has been designed to alleviate some of the power limitations associated with surface mount devices. This device has a metal tab connected to 3 pins on either side of the package. The die is mounted to this metal bar to provide improved thermal and electrical performance. This package can dissipate up to 2.2 W at 25°C and improvements in this capability can be seen with appropriate heatsinking to the tab. Figure 8 shows the batwing configuration for the 20 pin device and Table II shows the thermal characteristics of all three packages.

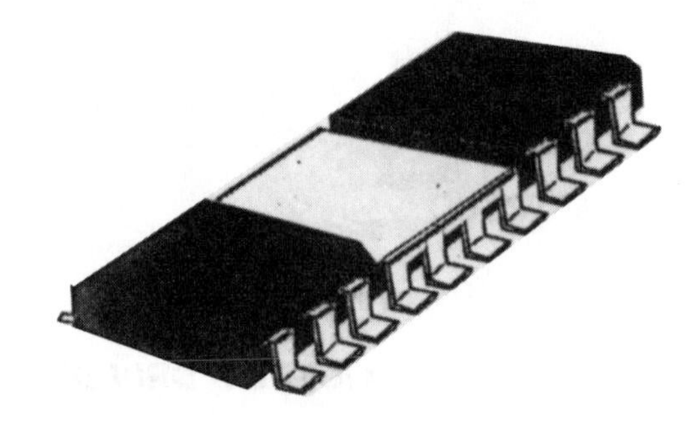

SOL-20B

Figure 8

PACKAGE THERMAL CHARACTERISTICS

	THETA Ja	THETA Jc	Pd(25 Degrees C)
	Degrees C/W	Degrees C/W	Watts
8-PIN DIP	80	55	1.56
14-PIN SOIC(150mil)	110	35	1.14
20-PIN SOLB(300mil)	56	6	2.23

Table II

SUMMARY

Figure 9 shows the actual evolution of current board complexity from the design (discrete components) to the hall/power combination (1 IC plus 1 diode). The diode is required for reverse polarity protection on some fans. Eventually, the hope is that this function can be incorporated into the chip, the PC board can be eliminated and a lower cost, single chip fan, emerges.

Figure 9

A LOOK TO THE FUTURE

Future Sensing in Vehicle Applications

Randy Frank
Motorola Semiconductor Products Sector
Transportation Systems Group

SENSING IN AUTOMOTIVE SYSTEMS

The sensors that have been discussed in this book represent those that have been developed for and/or implemented in production vehicles. These sensors are used for driver interface and/or system inputs. Today's typical production vehicle has about 20 sensors of various types which use several different technologies. A high-end luxury vehicle has over 60 sensors for almost everything—for measuring the intake manifold air pressure to detecting the presence of rain on the windshield.

Sensing represents an area that can improve significantly for future vehicles. Today's sensors are frequently identified as the weak link in any vehicle system. Their position on the front line, interfacing to the vehicle environment without the benefits of traditional isolation and protection techniques, makes their ability to survive, while maintaining expected performance, one of the more difficult system challenges. As a result, R&D efforts are occurring in automotive, supplier, national, and university laboratories to find improvements and achieve advances for sensors.

This paper will reveal what is coming out of R&D laboratories and into production vehicles in the near future, investigate technologies that will change significantly over the coming years, assess the progress in technology that can be expected, and examine the trends that will impact sensing and the systems that use it.

DRIVERS FOR NEW SENSORS

Historically, three driving forces have significantly impacted sensors: requirements for driver information, legislation, and driver demands. In some cases, auto manufacturers have designed electronic controls systems to offer improved capability or a new feature, anticipating driver demands. In addition, the transition from electromechanical to electronic systems has been integral to these systems as well as the need to reduce warranty and complexity.

INFORMING THE DRIVER

The earliest applications of sensors were driven by the driver's need for vehicle information. Sensors measured vehicle speed, amount of fuel, and critical engine parameters, including water, temperature, and oil pressure. Today's electronics, combined with the latest sensing technologies, can extend driver information to include other critical vehicle measurements that previously were performed only periodically using manual techniques and sometimes required an expert to interpret the results. These measurements include oil level and quality and tire pressure. Today these measurements can be made using sensing technology developed over the past few years. In fact, the number of measurements that can and will be made on future vehicles will cause information overload for the driver—or at least some drivers. Electronic technology can be used to handle the information and customize it to fit the driver's needs. Depending on driver's preferences, the sensor's information could be displayed continuously (i.e., road speed, oil pressure, etc.) or only when driver attention is required (low fuel, low tire pressure, excessive coolant temperature).

LEGISLATION

For over 20 years, legislation for improvements such as reduced vehicle emissions, on-board diagnostics (OBD-2), and safety has been the dominant driver for new technology in vehicles. New control systems have usually meant new parameters to be sensed and frequently new technologies to sense these parameters. The powertrain control system is the first electronic control system to use a microcontroller and a number of sensors. Fig. 1 shows a typical engine and transmission or powertrain control system. The oxygen sensors use chemical sensing ceramic technology. The manifold pressure sensor can be silicon piezoresistive or ceramic or silicon capacitive, and the knock sensor is piezoelectric.

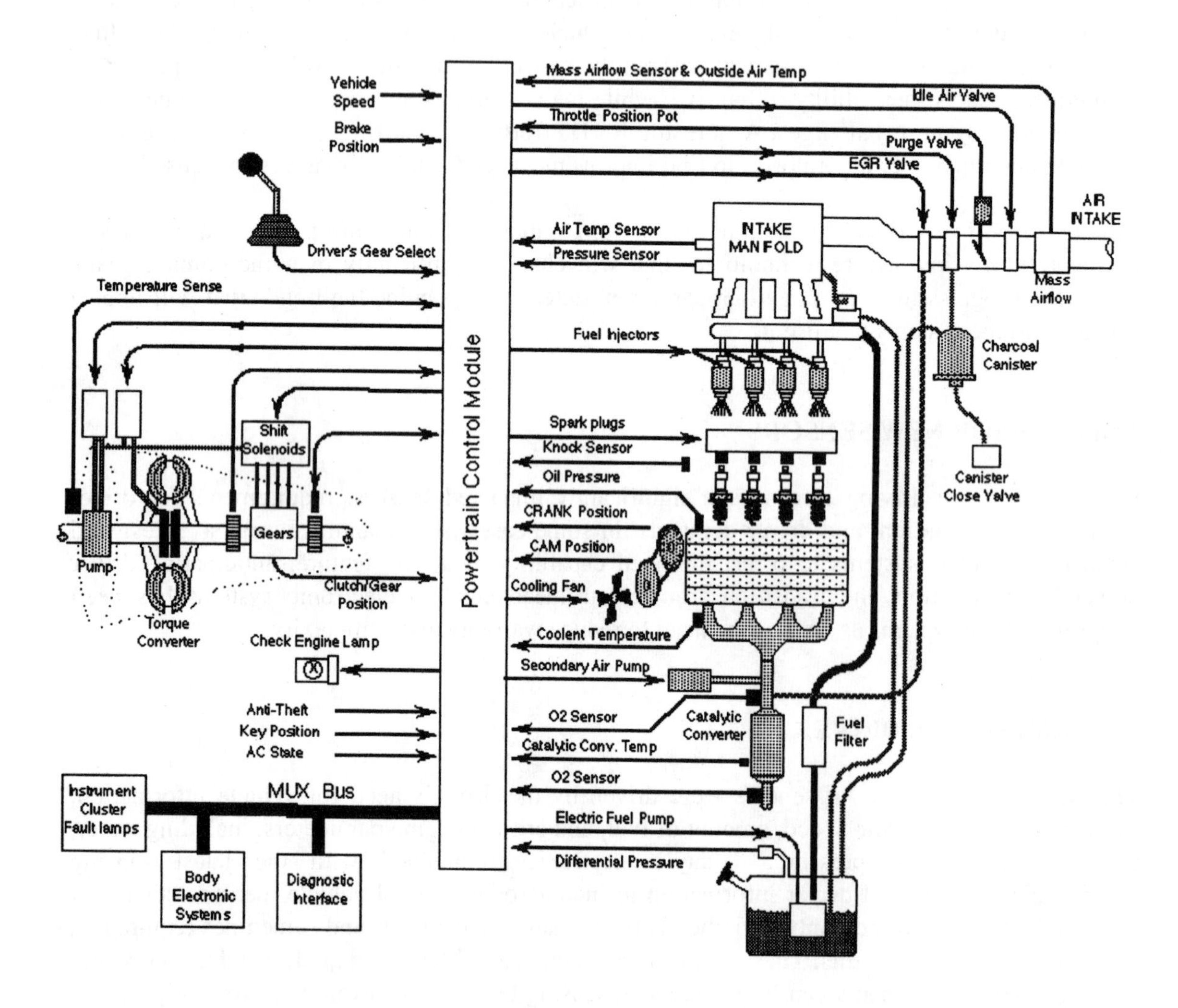

Fig. 1. Sensors in a Powertrain Control System

CONSUMER DEMANDS

Today, meeting driver demands includes: added levels of security against vehicle theft, identifying where the vehicle is in real time, recognizing a specific driver for adjusting vehicle settings (i.e., entertainment center, heating and air conditioning, and adjustable seat location), recognizing the size and position of the driver for appropriate air bag deployment, and a vast array of diagnostic measurements in all the existing and proposed vehicle control systems.

The major measurements that are made in today's vehicles and a projection of future growth is shown in Fig. 2.[1] The total automotive sensor market is expected to be over \$6 billion. While the compound annual growth is expected to be just over 11% from 1996 to 2000 for the total sensor market, the growth rate for automotive solid-state sensors is projected at almost 52%. Micromachined sensors are projected to grow over 85% in this same period, displacing some traditional technologies.

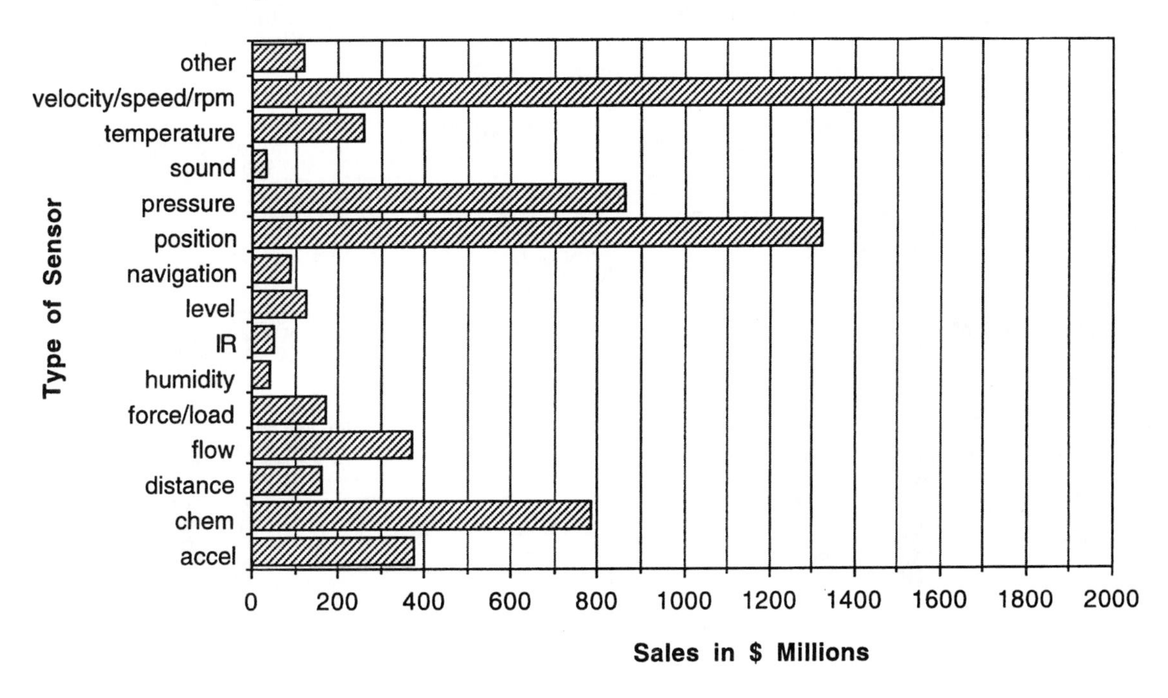

Fig. 2. Types of Sensors and Projected Sales in 2000

Projections include the use of several new sensors in vehicles in 2000. Table 1 shows potential sensor applications based on measurements both during development or during system operation that will be required in the future[2]. Well over 100 measurements are identified. (Note: Sensor used in electronic motor controls and entertainment systems are not included.) Some of these measurements are used for more than one system indicating a need for multiplexing (MUX). Several "sensors" are also simple switches that indicate system status.

Table 1 - SENSING REQUIREMENTS VERSUS VEHICLE SYSTEMS [AFTER 2]

SYSTEM	PARAMETER	SYSTEM	PARAMETER
ENGINE CONTROL		NODS	
	MANIFOLD ABSOLUTE PRESSURE	(NEAR OBSTACLE DETECTION)	RELATIVE SPEED VEHICLE #2
	BAROMETRIC PRESSURE (ALTITUDE)	COLLISION AVOIDANCE	DISTANCE VEHICLE #2
OBD-2	EXHAUST GAS OXYGEN (2)		
	NOx	INTELLIGENT HIGHWAYS	
	FUEL COMPOSITION		[SEE NODS]
	MASS AIR FLOW		[SEE NAVIGATION]
	ENGINE KNOCK		TRAFFIC FLOW
	EGR VALVE POSITION		
	EGR PRESSURE	HVAC	
	THROTTLE POSITION	(CLIMATE CONTROL)	HUMIDITY (PASSENGER COMPARTMENT)
	COMBUSTION PRESSURE		TEMPERATURE (PC)
	EXHAUST BACK PRESSURE		CHEMICAL COMPOSITION OF AIR (PC)
	FUEL COMPOSITION/FLEX FUEL		AIR FLOW / FILTER STATUS (PC)
	FUEL RAIL PRESSURE		OUTSIDE AIR TEMPERATURE
	VEHICLE SPEED		CHEMICAL COMPOSITION (OUTSIDE AIR)
	AIR CHARGE TEMPERATURE		
	ENGINE SPEED	CRUISE CONTROL	
	CRANK ANGLE/POSITION		VEHICLE SPEED
	ENGINE COOLANT TEMPERATURE		BRAKE SWITCH
	HUMIDITY		SELECTOR SWITCH
	TORQUE		
	OIL QUALITY	MUX / DIAGNOSTICS	
	FLUID CONDITION		MULTIPLE USAGE OF SENSORS
	ENGINE NOISE		
	AMBIENT TEMPERATURE	IDLE SPEED CONTROL	
OBD-2	MISFIRE DETECTION (NON-INTRUSIVE)		AC CLUTCH SENSOR/SWITCH
			POWER STEERING PRESSURE
ANTI-SKID BRAKES			SHIFT LEVER (PRNDL) SWITCH
	WHEEL SPEED		ENGINE SPEED
	BRAKE SWITCH		
	BRAKE PRESSURE	ELECT TRANSMISSION	
	DECELERATION	(CONTINUOUSLY	THROTTLE POSITION
	FLUID LEVEL	VARIABLE	SHIFT LEVER POSITION (PRNDL)
	FISH TAILING	TRANSMISSION)	TRANS OIL PRESSURE
			SHAFT SPEED
TRACTION CONTROL/ABS			TRANSMISSION OIL TEMPERATURE
	ENGINE SPEED		VACUUM MODULATION
	STEERING WHEEL ANGLE		
	STEERING WHEEL RATE OF CHANGE	DRIVER INFORMATION	
			VEHICLE SPEED
AIR BAGS			ENGINE SPEED
	CRASH DETECTION (FRONT)		OIL PRESSURE
	CRASH DETECTION (SIDE)		FUEL LEVEL
	SAFING SENSOR		OIL LEVEL
SMART SENSING	PRESENCE OF PASSENGER		OIL QUALITY/CONDITION
SMART SENSING	SEATING LOCATION		COOLANT PRESSURE
SMART SENSING	WEIGHT		COOLANT CONDITION
	BAG PRESSURE		COOLANT TEMPERATURE
	VEHICLE SPEED		AMBIENT AIR TEMP
			COOLANT LEVEL
SUSPENSION			WINDSHIELD WASHER LEVEL
	VEHICLE HEIGHT (WH-BODY DISPL)		TRANSMISSION OIL LEVEL
	STEERING WHEEL ANGLE		TIRE PRESSURE
	STEERING WHEEL DIRECTION		TIRE SURFACE TEMPERATURE
	ACCELERATION		BATTERY FLUID LEVEL
	VEHICLE WEIGHT		RAIN SENSOR
	ROAD SURFACE CONDITIONS		SUN SENSOR
	SIDE SLIP ANGLE		
	SPEED OF YAW ANGLE	MEMORY SEAT	
	VERTICAL ACCELERATION		DRIVER SELECTOR SWITCH
	LATERAL ACCELERATION		LUMBAR PRESSURE
	LATERAL DISPLACEMENT		SEAT POSITION
	VEHICLE SPEED		
	PNEUMATIC SPRING PRESSURE	NAVIGATION	
	RIDE SELECTOR SWITCH		GYROSCOPE
			STEERING WHEEL TURNS
ELECT POWER STEERING			GEOMAGNETIC (MAGNETIC FIELD) SENSORS
(ALSO ELECT ASSISTED)	STEERING WHEEL TORQUE		DISTANCE TRAVELED
	HYDRAULIC PRESSURE		GPS
	VEHICLE SPEED		YAW RATE
	STEERING WHEEL POSITION		
		DRIVER	
4-WHEEL STEERING			FATIGUE / DROWSINESS
	VEHICLE SPEED		PERSONAL INJURY
	STEERING WHEEL ANGLE		BLOOD PRESSURE
	FRONT-REAR WHEEL ANGLE		RESPIRATION RATE
			PULSE RATE
SECURITY &			GLUCOSE LEVEL
KEYLESS ENTRY	ILLEGAL ACCESS INDICATION		BLOOD ALCOHOL CONTENT
	REMOTE ACCESS INDICATION		DRIVER IDENTIFICATION
			VOICE RECOGNITION
			FINGERPRINT RECOGNITION

ELECTRONIFICATION

The transition from electromechanical to electronic techniques is one of the trends that has been occurring since the first electronic control systems were implemented in vehicles. The aneroid barometer and LVDT (linear variable differential transformer) pressure sensor in early engine control systems were quickly replaced by a semiconductor pressure sensor once the new technology had demonstrated that it could survive in the automotive environment. In fact, both pressure measurements can be accomplished by a single sensor and a measurement technique that makes the barometric reading when the engine is not running. More recently, micromachined accelerometers have replaced mechanical sensors in air bag systems and even changed the system design from a multiple number of remote inputs to central detection with an algorithm providing the crash discrimination. New technologies, some originally developed for the military and/or aerospace applications are being applied to vehicles. These include: night vision, radar, LIDAR (laser intensity direction and ranging), GPS (global positioning system), etc.

ISSUES FOR AUTOMOTIVE SENSORS

The automotive environment, accuracy, response time, reliability (packaging, EMC/EMI), and cost (including ease of interface) pose significant challenges for new sensors. Major improvements can be expected in the future based on new approaches.

THE AUTOMOTIVE ENVIRONMENT

The automotive environment is recognized to be one of the more difficult applications for electronic systems and for sensors, especially microelectronic sensors. The environment includes a wide temperature range, high humidity conditions, the need to withstand several chemicals, the ability to operate under high electromagnetic interference (EMI) conditions and yet cause relatively low levels of radio frequency interference (RFI), and also the capability to accommodate wide voltage swings. At the same time, the acceptance of electronics is extremely customer driven and demands low cost and high reliability. Projected sales volumes can decrease (or increase) quickly depending on customer acceptance of the system, the vehicle in which the system is offered, and external economic factors which affect purchasing decision for high-cost items such as automobiles[2].

One of the major differentiating factors for automotive usage is whether the location of sensors is in the passenger compartment or under the hood. Table 2 (after SAE J1211) shows the variation that can occur to several variables depending on the location of the electronic component. High temperature and humidity can significantly reduce the useful life, especially in underhood applications of electronics. The SAE (Society of Automotive Engineers) has developed SAE J1211 "Recommended Environmental Practices for Electronic Equipment Design" to address the unique problems of automotive electronics[2].

Table 2 The Automotive Environment

Variable	Underhood	Passenger Compartment
Storage Temperature	-40°C to +150°C	-40°C to +85°C
Operating Temperature	-40°C to +125°C	-40°C to +85°C
Vibration	15g, 10 to 200 Hz	2g, 20 Hz
Humidity	up to 100%	up to 100%
Chemicals	Salt spray, water, fuel, oil, coolant and solvent immersion	----
Thermal Cycling	> 1000 Cycles from -40°C to +125°C	> 1000 Cycles from -40°C to +85°C
EMI Protection	up to 200 V/m	up to 50 V/m

The system voltage in passenger vehicles is nominally 12 V. However, several normal and abnormal voltage variations occur which must be taken into account at either the component or system level. Normal operating charging systems regulate the output of the alternator to attempt to provide sufficient voltage to keep the battery charged under various temperature and load conditions. This voltage can range from 16 V when very cold (–40°C) to about 12 V when maximum underhood temperatures are reached. Electronic assemblies on the car must be able to withstand reverse battery conditions (–12 V), jump start from tow trucks (24 V), short duration voltage transients (that can easily exceed ±100 V), and long duration (>400 ms) alternator dump transients which can be as high as 120 V. The load dump is clamped at some point or points in the system to restrict its maximum to a level of 60 V or less. To prevent damage due to excessive voltage, microelectronic sensors must have a regulated and protected power bus[2]. The impact of higher system voltage in future vehicles will need to be evaluated for sensors.

The increasing complexity of modern vehicles has increased the possibility of EMI causing problems between the various vehicle systems. In addition, electronics can be susceptible to and radiate RFI causing poor radio reception and malfunctioning vehicle systems with a resulting no-fault condition found during service diagnostic procedures. The potential problems in this area have been addressed by SAE with several information reports and a recommended practice as shown in Table 3[2].

Table 3 SAE Documents Relative to Sensor Applications in Automobiles

Document	Number
Recommended Environmental Practices for Electronic Equipment Design	SAE J1211
Performance levels and Methods of Measurement of Electromagnetic Radiation from Vehicles and Devices	SAE J551
Performance Levels and Methods of Measurement of EMR from Vehicles and Devices (Narrowband RF)	SAE J1816
Electromagnetic Susceptibility Procedures for Vehicle Components (Except Aircraft)	SAE J1113
Vehicle Electromagnetic Radiated Susceptibility Testing Using a Large TEM Cell	SAE J1407
Open Field Whole-Vehicle Radiated Susceptibility 10 kHz - 18 GHz, Electric Field	SAE J1338
Class B Data Communication Network Interface	SAE J1850
Diagnostic Acronyms, Terms and Definitions for Electrical/Electronic Components	SAE J1930
Failure Mode Severity Classification	SAE J1812
Guide to Manifold Absolute Pressure Transducer Representative Test Method	SAE J1346
Guide to Manifold Absolute Pressure Transducer Representative Specification	SAE J1347

SENSOR PACKAGING

The sensor packaging problems specific to the application must be thoroughly addressed to ensure acceptable performance over the sensor's expected lifetime. One approach is described in a recent paper[3]. Identifying minimum expectations and critical application factors that limit device lifetime are part of a methodology known as the *physics of failure* approach to reliability testing. It involves analyzing the potential failure mechanisms and modes. Once these are understood, application-specific packaging development can proceed, the proper test conditions can be established, and critical parameters can be measured and monitored[3].

One key failure mechanism in harsh media applications is corrosion, specifically galvanic corrosion caused by dissimilar metals that are in electrical contact in aqueous solutions. Using a physics of failure approach, one can start to ask the appropriate questions: Which environmental factors contribute to the failure mechanism? What accelerates it? What should be done to the sensor package to prevent corrosion or minimize its impact over the sensor's lifetime? What is the expected sensor lifetime? The answers to these questions allow simulations to be performed and costs to be analyzed before lengthy and potentially expensive testing is initiated. The testing will ultimately prove the acceptability of a particular proposal. Unless the proper mechanisms are known, over-designing can add unnecessary cost to the packaging, such as a passivation layer, a hard die attachment, or any nonstandard process. Furthermore, the design tradeoffs that result may not prove beneficial. For example, adding a hard die attach for a pressure sensor to improve the strength of the die bond could limit device performance in other areas such as temperature coefficient of offset[3].

IMPROVING SENSOR PERFORMANCE

The low-level output of most sensors and the requirement to meet specifications over a wide operating range are among the problems that must be solved to use a sensor in a particular application. Table 4 summarizes some of the common shortcomings of sensors and how electronics is used to overcome these limitations[4]. Improved sensor performance can result when the design of the sensor and capability of subsequent components are taken into account in the overall design of the sensor. Digital logic provided by either an MCU (microcontrol unit) or DSP (digital signal processor) plays a vital role in smart sensing and in improving the sensor's performance.

Table 4 Common Undesirable Sensor Characteristics and Improvement Techniques

Characteristic	Sensor Design	Sensor Interface	MCU/DSP
Nonlinearity	consistent		reduce
Drift	minimize		compensate
Offset		calibrate	calibrate/reduce
Time dependence of offset	minimize		auto-zero
Time dependence of sensitivity			auto-range
Nonrepeatability	reduce		
Cross-sensitivity to temperature and strain		calibrate	store value & correct
Hysteresis	predictable		
Low resolution	increase	amplify	
Low sensitivity	increase	amplify	
Unsuitable output impedance		buffer	
Self-heating	increase Z		PWM technique
Unsuitable frequency response	modify	filter	
Temperature dependence of offset			store value & correct
Temperature dependence of sensitivity			store value & correct

Error budgeting in the system can allow or restrict a specific sensor in a particular application. As shown in Fig. 3, a maximum error band of ±1.5% has contributions from many sources in a pressure sensor. The eight parameters identified for this example contribute as little as ±6.90 mV. The applications requirements must be specified accurately to use available or next-generation sensors. Electronic sensors with a long production history and high volumes, such as pressure sensors, have achieved improved accuracy over time.

Circuit Error

Description	Value	Units
Pressure Range	0 to 6	kPa
	0 to 611.85	mmH_2O
Span	4.6	V
Sensitivity	0.7667	V/kPa
	0.00752	V/mmH_2O
Gain		
Offset	0.2 Typical	V
Error	±1.5%	% of 4.6V FSS
	±9.18	mmH_2O
	±69	mV
	±0.09	kPa
Vcc	5.0	V

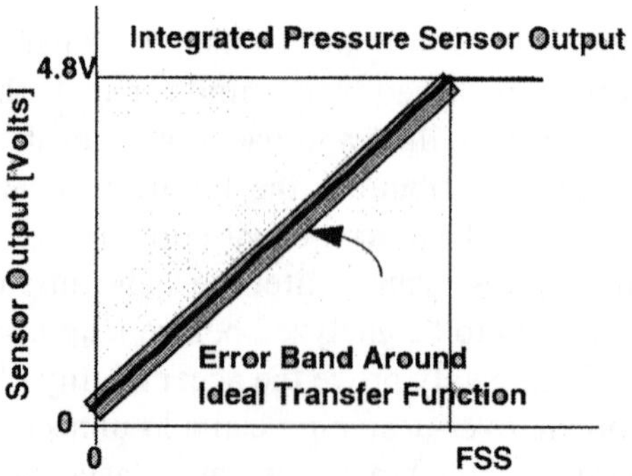

Integrated Pressure Sensor Output

Sensor Output [Volts]

4.8V

0

0

FSS

Error Band Around Ideal Transfer Function

Preliminary Error Budget Partitions

Parameter	Value [mV]	% of Error Budget	Error [%FSS]
Creep (Stability)	±6.90	10%	0.15%
Zero Shift (Offset)	±6.90	10%	0.15%
Sensitivity Shift (Span)	±6.90	10%	0.15%
Nonlinearity	±13.80	20%	0.30%
Test Guard Band	±6.90	10%	0.15%
Thermal Hysteresis	±6.90	10%	0.15%
TC Offset	±13.80	20%	0.30%
TC Span	±6.90	10%	0.15%
Total	±69.0	100%	1.50%

Fig. 3. Sensor Error Budget Specification

MEETING SYSTEM OBJECTIVES AND CUSTOMERS' NEEDS

The sensors that progress from the laboratories or on-vehicle development testing to production will be the major winners over the next decade. These will be the ones that produce the highest volume, meet system cost objectives, and provide a distinct advantage to the systems that use them. The sensors themselves will have addressed the issues described previously. In some instances the sensor's success will be determined by the system's success. A new sensor technology that enables a new system/feature may never make it if the system/feature is not accepted. Customers may desire a system with antilock brakes but they do not care if the system uses a Hall effect sensor or variable reluctance wheel speed sensor. However, customers may expect performance from the system that dictates a specific sensor technology or drives technology development to meet customer expectations.

ALTERNATIVES TO SENSING

The MCU and digital control have been significant driving forces for developing microelectronic sensors. At the same time, an MCU can also limit the number of sensors that are used in a particular system. The ability of the MCU to take information provided by existing sensors and calculate and obtain an inferred or indirect measurement can avoid the cost of adding a sensor—even though the sensor could provide more accurate data. The cost-to-benefit ratio must be evaluated for every new sensor that could be added to a a system. One example of a sensing alternative was briefly mentioned previously. Early requirements to measure manifold absolute pressure and barometric pressure utilized separate sensors. By reading the barometric pressure when the input to the MAP sensor is not affected by manifold pressure (during cranking and at wide open throttle conditions which other sensors or switches indicate to the MCU), the additional barometric sensor has been eliminated. It should be noted that the manifold absolute pressure measurement is also an indirect measurement compared to measuring mass air flow[2].

TECHNOLOGIES

The critical design aspects for automotive sensors can be divided into three different areas: the transducer element, interface electronics, and packaging. The basic transducer element is expected to meet the performance requirements of the system and be able to tolerate the normal over-range occurrences that could be anticipated in certain failure modes of other components. The output should be stable over the operating temperature range and the lifetime of the vehicle. As mentioned earlier, electronics is frequently required to enhance the output signal; calibrate the zero and sensitivity; compensate for the effects of temperature and other variables; tolerate electromagnetic interference; provide protection for short circuits, reverse battery and other voltage conditions; and increasingly to provide diagnostic and self-test features. The ease of interface for digital processing also requires the output of an analog sensor to be converted through an (integral) analog-to-digital interface or serial communication interface. The sensor package required for the automotive environment must be small; easy to handle and mount in the application; extremely robust and capable of withstanding thermal cycling, thermal shock, vibration, and various chemicals and high levels of humidity; and provide the EMI protection for the sensor and interface electronics[2].

Progress is being made in several areas for sensors in future vehicles. These improvements are targeting the basic sensing technique, manufacturing processes, and sensor interface (both electronic and mechanical).

resonant beam, interdigitated structures should be used for both sensing and actuating. More advanced deep photolithography techniques, such as the LIGA process, will add new sensor structures and performance capability for future sensors.

Most sensors address only one parameter, although temperature information is frequently derived simultaneously. In the future, multiple sensors created on a single sensing chip will be more common. These will include sensors such as tri-axial accelerometers to sense three directions on a single chip or sensors for different functions such as combined pressure, temperature, humidity, and vibration sensors for HVAC (heating, ventilation, and air conditioning) applications. The cost of designing an additional separate sensor will be greater than simply including the sensor with other devices on the same chip. From a manufacturing perspective, proper design will mean that no additional process complexity occurs from the addition of the new sensor. The additional material for the sensor will be insignificant compared to existing sensors and especially to the electronics already required for interfacing, calibrating, and adding intelligence to the other sensor(s). Sophisticated testing will determine which sensors are ultimately used for a particular system. This is the point where cost increases will occur. However, it is also the point where value will be added to the sensor.

An example of how micromachining is extending the types of measurements and improving the performance compared to previous approaches is a recently introduced chemical sensor. Thin-film metal oxide technology is combined with an embedded microheater on a thin micromachined silicon diaphragm. The cross-section and top view of the die is shown in Fig. 4. The small sample area of a micromachined sensor can more readily be raised to the higher operating temperature that is required for detecting the presence and actual value of specific chemicals. The polysilicon heaters in the chemical sensor operate at temperatures as high as 450°C. The ability to change the temperature in a short time period is used to minimize the effects of humidity and also to reduce the power consumption[5]. Sensitivity is 10:1 at 100 ppm and the response time of the sensor is less than two minutes. Initially these products have been aimed at industrial and consumer applications, but on-going development should provide future units for automotive applications.

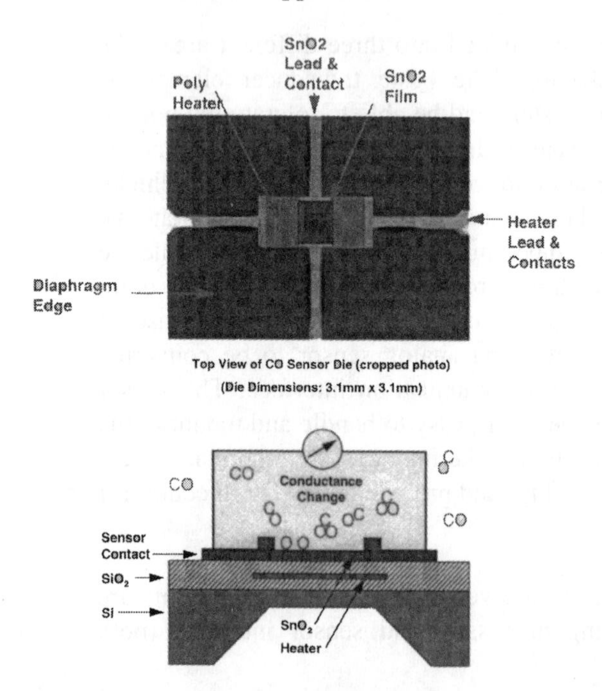

Fig. 4. Chemical Sensor Using Micromachining Technology

MICROMACHINING PLUS MICROELECTRONICS

Increased electronics will be used with future sensors either by simultaneously fabricating electronic circuits with the mechanical structure, in the case of some semiconductor sensors, or by packaging techniques such as multichip modules. Fig. 5 shows a sensor technology migration path that ultimately has the semiconductor sensor(s) fabricated directly on CMOS MCUs[4]. While the cost of this final form may be several years away from reaching a level low enough for automotive usage, the technology capability has already been demonstrated[6]. Intermediate forms of integration with CMOS memory components used for calibration and localized digital logic providing decisions are extremely likely in the near future (before 2000). The electronics system design will change as inherently digital signals, instead of analog, can be input directly to the MCU. These signals could be at the level that allows usage by several vehicle systems using a multiplex bus.

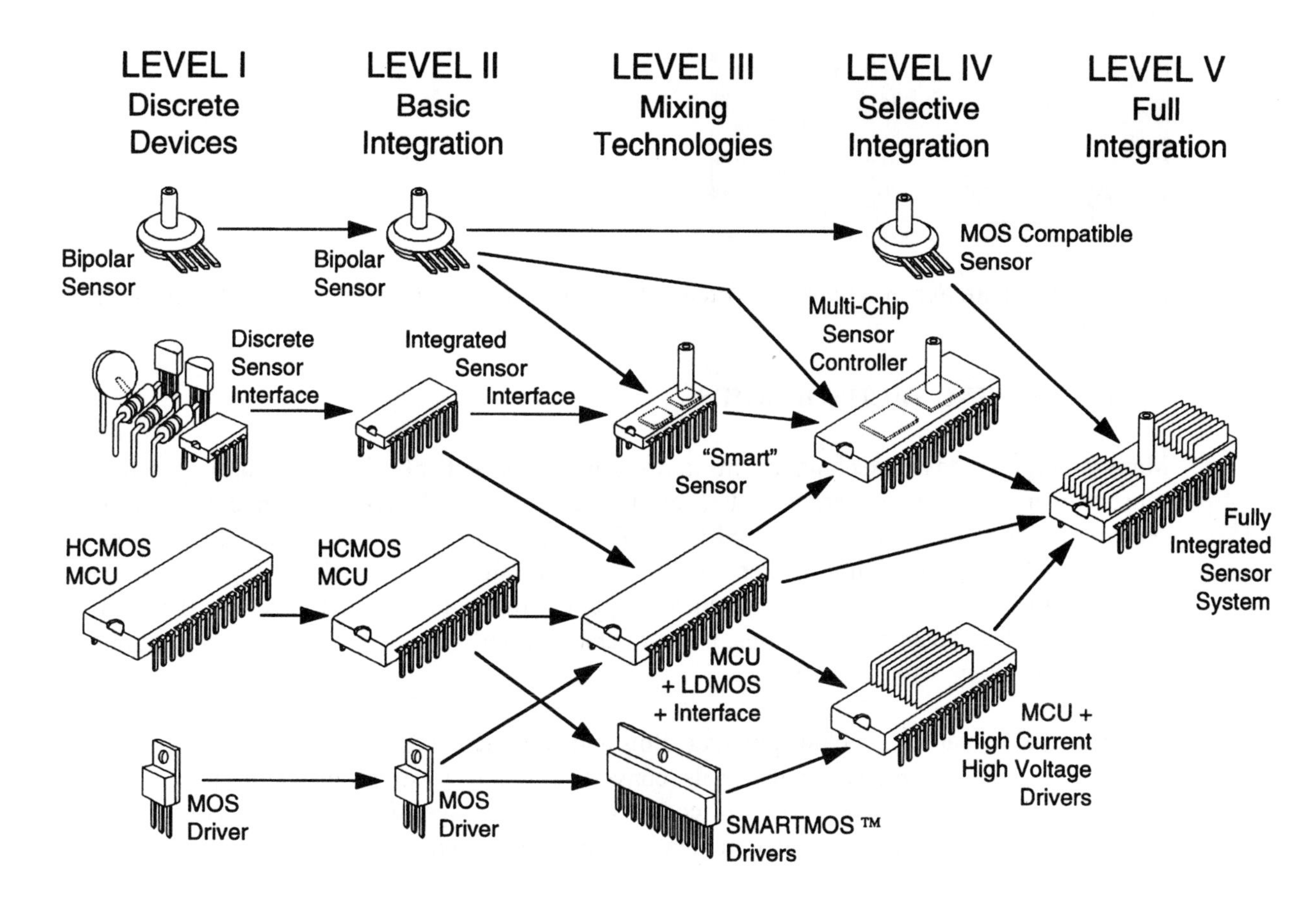

Fig. 5. Sensor Technology Migration Path

An example of a merged micromachining and microelectronics is a microcontroller with an integral micromachined sensor. Fig. 6 shows a possible die layout of such a device[4]. The sensor is treated as a building block in the design and processing of the microcontroller. Today's techniques use about 10% of the die area for the sensor but add several masking steps to the processing. Future processing techniques that reduce masking steps will make this technology viable for automotive applications. However, testing and packaging are also difficult problems that will not be as easy to solve. Testing, for example, is complicated by requiring both electrical measurements for the MCU and mechanical measurements for pressure.

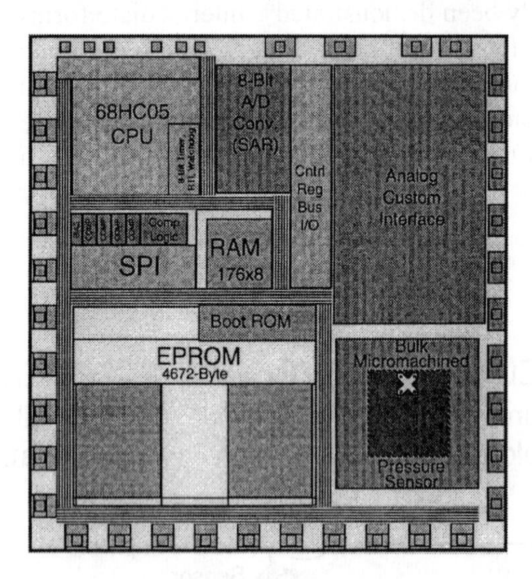

Fig. 6. A Pressure Sensor Diaphragm Etched in a Microcontroller Using a Bulk Micromachining Process

OTHER SENSING TECHNOLOGIES

Many emerging sensing techniques are in various stages of R&D. These technologies include sensors using advanced materials, radar, infrared, chemical, and even next-generation magnetic sensors. Silicon has been the advanced material of choice for electronic components in vehicle systems including computing, power control, transient suppression, and sensing. But alternate materials that operate at higher frequencies and/or temperatures continue to be explored for sensing and other vehicle applications. Gallium arsenide and silicon carbide are two examples of advanced semiconductor materials.

Other advanced materials include polymer films and ceramic materials in which the piezoelectric effect is present. Piezoelectric materials convert mechanical stress or strain into electrical voltage. These materials also expand or contract when a voltage of opposite polarity is applied. As a result, a sensing and actuating combination can be achieved.

Radar sensing techniques are used in automatic cruise control, collision avoidance, and near obstacle detection systems (NODS) or blind-spot monitoring. NODS with shorter distance measurements can operate at frequencies in the 10.5 to 24 GHz range. Several companies have incorporated radar sensing in vehicles for longer range detection in collision avoidance systems using 77 GHz technology. Developmental vehicles that incorporate radar sensing may use different types of radar, both millimeter and microwave, and can have as many as 16 radar sensors. Fig. 7 shows the sensors and other components in a collision-avoidance system[7]. These systems demonstrate the viability of radar sensing for sophisticated vehicle applications. At present, the cost of these systems, specifically the sensors, limits their vehicle applications. Efforts to reduce costs could increase usage in future vehicles.

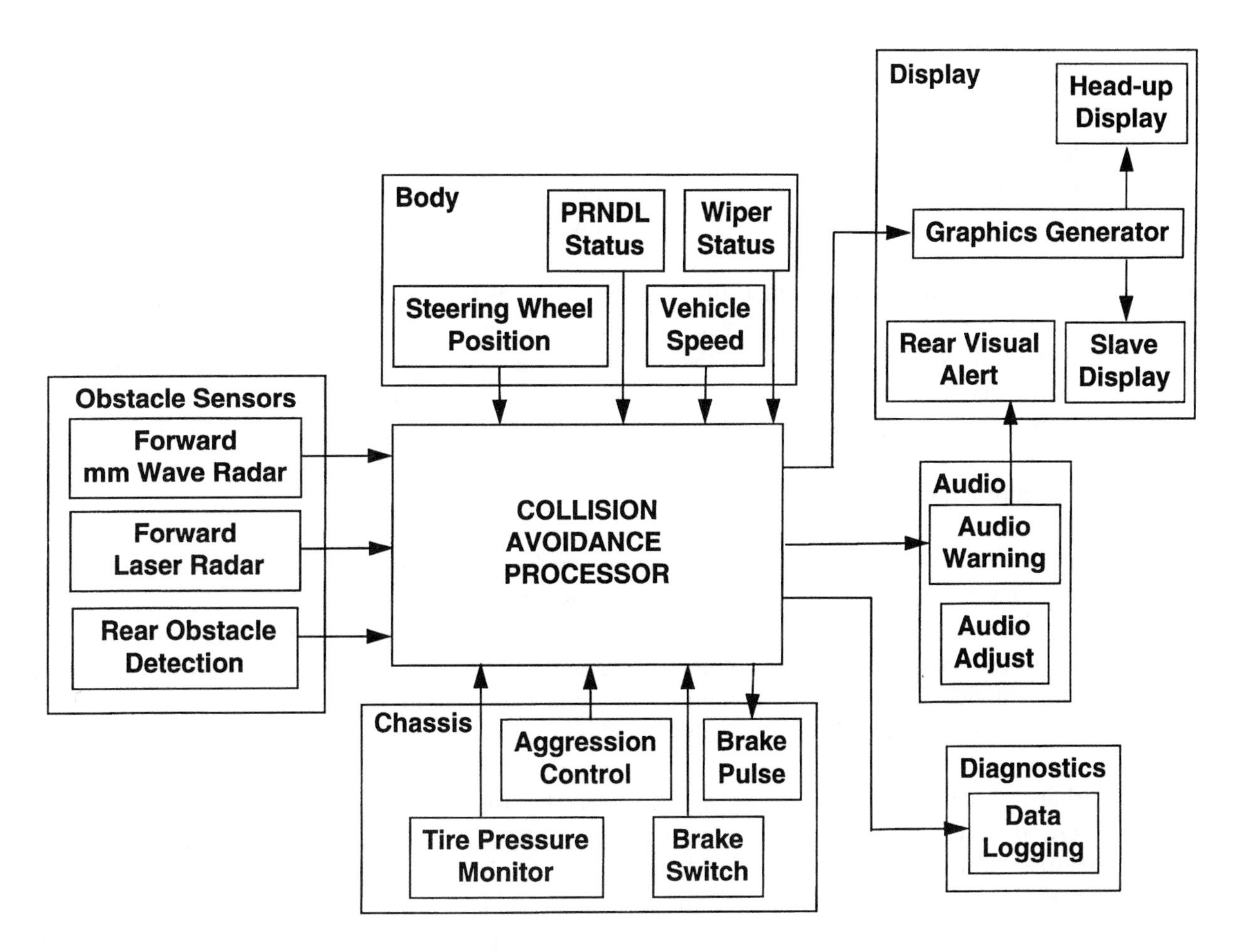

Fig. 7. Radar and Other Sensors in a Collision-Avoidance System[7]

A micromachined calorimeter has been created in a research laboratory environment that can detect extremely subtle chemical reactions on surfaces. The monolithic combination of several of these sensors to serve as a "nose" for detecting multiple chemicals is among future possibilities.

Infrared sensors are the heart of night vision systems. Near infrared sensors and head-up displays provide vision enhancement for night-time driving. Information from the sensors requires real-time processing to display images to the driver. The improved vision is one more approach to reducing accidents, especially those involving pedestrians.

Magnetic sensing has evolved from reluctance to Hall effect sensors. Magnetic resistive elements using materials such as permalloy, a ferromagnetic alloy composed of 20% iron and 80% nickel, are being developed and are already in use in vehicles. Continued R&D efforts to find a low-cost solution for speed sensing, especially at very low speeds, should provide some interesting sensor technology for future vehicles.

NEW APPLICATIONS

Emerging applications such as torque sensing, NO_x, and other chemical sensing, misfire detection, oil quality, night vision, obstacle detection/collision avoidance, unauthorized vehicle entry, and passenger position/size detection in smart air bag systems will provide new applications for sensors and drive R&D for new sensor technology.

Safety applications such as smart tires and run flat tires that incorporate sensors demonstrate both potential growth and technology evolution. A recent survey indicated that 89% of drivers surveyed chose run flat tires as their number one choice in safety innovations. Projections indicate that tire monitoring systems could be installed in as many as 10% to 40% of new passenger cars by 2000[8]. While the sensors could be similar to units installed in vehicles today, lower power consumption and improved RF communication techniques are among the areas that could improve in future systems.

Detecting illegal access in security systems can be accomplished by several different sensing techniques including switches at doors and windows, compartment pressure, vibration, and infrared sensors. Anti-theft systems are increasingly being purchased by consumers to protect their investments. In the future, global positioning systems could be used extensively to locate vehicles under normal driving conditions in navigation systems or, if a vehicle is stolen, quickly locate the vehicle prior to disassembly by a chop shop.

Perhaps some of the most exotic applications of future sensors could be aimed at the driver. These include: speech recognition, measuring fatigure or blood alcohol content, fingerprint identification, and measurement of blood pressure, pulse and repetition rates, and glucose level. Determining the state of the driver and even identifying the driving could be features of future systems. Many of these approaches are already being developed, although not necessarily for vehicle applications. The ability to bring these systems into the vehicle could depend on the success of other ongoing efforts such as the development of an intelligent transportation system data bus. The vision of the group working on this standard is an open architecture that allows consumer electronics to be used in the vehicle[9]. Interaction with most vehicle control systems would be restricted by vehicle manufacturers but interaction with some systems on the vehicle would be allowed, especially for communications purposes.

IMPROVING VEHICLE SYSTEMS THROUGH SENSING TECHNOLOGY

Future vehicles will use more sensors to provide more information directly to the driver, to control systems in the vehicle, and to provide remote monitoring stations for locating the vehicle and analyzing the vehicle's and the driver's status. As long as drivers see value in the systems that employ these sensors and recognize improvements in the driving experience from these systems, manufacturers will develop and offer them. Technology evolution to reduce cost and improve functionality and reliability will continue in those applications where sensors are already an integral part of the vehicle.

REFERENCES

1. Selantek Inc., Report, San Francisco, Calif. 1997
2. Ljubisa Ristic, Sensor Technology and Devices, Artech House, Boston, Mass., 1994.
3. Theresa Maudie, David J. Monk, and Randy Frank, "Packaging considerations for predictable lifetime sensors," *Proceedings Sensors Expo Boston*, 1997.
4. Randy Frank, Understanding Smart Sensors, Artech House, Boston, Mass., 1996.